10th International Conference on Topics in Astroparticle and Underground Physics 2007

(TAUP 2007)

Journal of Physics: Conference Series Volume 120

Sendai, Japan
11 - 15 September 2007

ISBN: 978-1-61738-424-0
ISSN: 1742-6588

Printed from e-media with permission by:

Curran Associates, Inc.
57 Morehouse Lane
Red Hook, NY 12571

Some format issues inherent in the e-media version may also appear in this print version.

Printed by Curran Associates, Inc. (2010)

For permission requests, please contact the Institute of Physics
at the address below.

Institute of Physics
Dirac House, Temple Back
Bristol BS1 6BE UK

Phone: 44 1 17 929 7481
Fax: 44 1 17 920 0979

techtracking@iop.org

Additional copies of this publication are available from:

Curran Associates, Inc.
57 Morehouse Lane
Red Hook, NY 12571 USA
Phone: 845-758-0400
Fax: 845-758-2634
Email: curran@proceedings.com
Web: www.proceedings.com

TABLE OF CONTENTS

PART 4: DARK MATTER IN UNIVERSE, ANTIMATTER IN SPACE

PART 5: NEUTRINOS

PART 6: COSMIC RAYS AND NEUTRINOS, GAMMA RAYS

PART 7: 20th ANNIVERSARY

PART 8: UNDERGROUND LABORATORIES

The tenth meeting of the TAUP Workshop Series, TAUP 2007, was organized by the Research Center for Neutrino Science, Tohoku University.

In TAUP 2007 all the various aspects of Astroparticle Physics have been covered, from Cosmology and Dark Constituents, to Gravitational Waves, to Neutrino Physics and Astrophysics, to High Energy Astrophysics, to Cosmic Rays and Gamma-Rays Astronomy. New and important scientific results were presented and debated in the plenary review talks and in a very large number of contributions in topical parallel sessions.

As editors of these proceedings, we hope that this volume, which contains most of the talks and contributions presented at TAUP 2007, will provide a detailed state-of-the-art account of the various facets of Astroparticle Physics. We thank all the invited speakers, conveners, and contributors who made this possible. Full coverage of the transparencies presented at the conference can be found on the website http://www.awa.tohoku.ac.jp/taup2007.

The TAUP 2007 Organizing Committee thanks IUPAP/PaNAGIC, Sendai Tourism and Convention Bureau, COE program: Exploring New Science by Bridging Particle-Matter Hierarchy, SEIKO EG&G, and REPIC corporation for sponsoring the Conference, and Sendai Civic Auditorium, where the meeting was held, for their hospitality.

We wish to thank Alessandro Bottino, Junpei Shirai, Fumihiko Suekane, David Sinclair, Takaaki Kajita, Takeo Moroi, Masaki Mori, Masahiro Kawasaki, Yoshihito Gando, Sei Yoshida, Kyoko Tamae, Sanshiro Enomoto, Alexandre Kozlov, Yasuhiro Kishimoto, Itaru Shimizu, Kengo Nakamura, Haruo Ikeda, and Kyo Nakajima for their invaluable contribution in the scientific shaping of the conference and in the preparation of the present volume.

The Organizing Committee is grateful to the members of the International Advisory Committee and of the TAUP Steering Committee for assistance and advice on the scientific program.

Very special thanks are due to Ms Rika Bizen, Mr Fujio Miura, Ms Akemi Otsuka, and Ms Yuri Endo, our workshop secretaries, for their continuous and excellent work in the organization of the conference, and to Ms Chiyo Itoh, and Ms Machiko Mizutani, for their invaluable assistance during the conference. We also gratefully thank the technical staff: Tomoaki Takayama, Hiromitsu Hanada, Takashi Nakajima, for their invaluable help.

As announced at the end of the conference, TAUP 2009 will be held in Gran Sasso, Italy, hosted by the Istituto Nazionale di Fisica Nucleare (INFN) with the chair of Professor Eugenio Coccia.

Kunio Inoue, Atsuto Suzuki and Tadao Mitsui

COMMITTEES

TAUP Steering Committee

TAUP 2007 International Advisory Committee

TAUP 2007 Organizing Committee

A Bottino *U Torino/INFN*

T Kajita *ICRR, U Tokyo*

K Inoue (co-chair) *RCNS, Tohoku U*

F Suekane *RCNS, Tohoku U*

T Moroi *Tohoku U*

M Kawasaki *ICRR, U Tokyo*

D Sinclair *U Carleton*

A Suzuki (co-chair) *KEK/Tohoku U*

J Shirai *RCNS, Tohoku U*

T Mitsui (scientific secretary) *RCNS, Tohoku U*

M Mori *ICRR, U Tokyo*

Dates

Issue 00 (2008)

Astroparticle Physics: Puzzles and Discoveries

V. Berezinsky
INFN, Laboratori Nazionali del Gran Sasso, Assergi (AQ) 67010 Italy
and Institute for Nuclear Research, Moscow

Abstract. Puzzles often give birth to the great discoveries, the false discoveries sometimes stimulate the exiting ideas in theoretical physics. The historical examples of both are described in Introduction and in section "Cosmological Puzzles". From existing puzzles most attention is given to Ultra High Energy Cosmic Ray (UHECR) puzzle and to cosmological constant problem. The 40-years old UHECR problem consisted in absence of the sharp steepening in spectrum of extragalactic cosmic rays caused by interaction with CMB radiation. This steepening is known as Greisen-Zatsepin-Kuzmin (GZK) cutoff. It is demonstrated here that the features of interaction of cosmic ray protons with CMB are seen now in the spectrum in the form of the dip and beginning of the GZK cutoff. The most serious cosmological problem is caused by large vacuum energy of the known elementary-particle fields which exceeds at least by 45 orders of magnitude the cosmological vacuum energy. The various ideas put forward to solve this problem during last 40 years, have weaknesses and cannot be accepted as the final solution of this puzzle. The anthropic approach is discussed.

1. Introduction

ALL GREAT DISCOVERIES IN ASTROPHYSICS
APPEARED UNPREDICTABLY AS PUZZLES.
WHAT WAS PREDICTED WAS NOT FOUND.

Not many good things fall down on us from the sky, but discoveries do. They arrive as the puzzles, and usually disappear as the errors. But sometimes they become real, leaving behind the great discoveries. I will give below a short list of astrophysical discoveries of the last four decades, separating intuitively astrophysics from cosmology.

Quasars were discovered in early 1960s as compact radio sources. G. Mathews and A. Sandage in 1960 identified radio source 3C48 with a stellar-like object. M. Schmidt in 1963 deciphered the optical spectrum of quasar 3C273 assuming its redshift, $z = 0.158$. Surmounting resistance of sceptics, this explanation moved the source to the distance of 630 Mpc and made its luminosity uncomfortably large, $L \sim 10^{46}$ erg/s. It was a puzzle, and many respectable astrophysicists spent years trying to squeeze the source back into Galaxy. All of them failed, and in the end the puzzling energy release resulted in the discovery of a *black hole*, an object of general relativity.

Pulsars were discovered first in 1967 by a student of A. Hewish, Jocelyn Bell. She observed a puzzling periodicity of radio-pulses from an unknown source. After short but intense discussion of different possible sources, including extraterrestrial civilizations and "little green men", the magnetized rotating neutron stars, the pulsars, were found to be responsible. It opened a new field of cosmic physics: *relativistic electrodynamics.*

The atmospheric neutrino anomaly and the solar neutrino problem went along most difficult road to the status of discovery. The puzzling phenomenon in both cases was a neutrino deficit as compared with calculations. The solar neutrino experiments have been started by Ray Davis in 1960 in Brookhaven. With time the solar-neutrino deficit raised, but scepticism of the community raised too. Pushed first by Ray Davis and John Bahcall, the solar neutrino problem moved like a slow coach along a road of three decades long. Fortunately, physics differs from democracy: opinion of majority means usually less than that of ONE. In the end the two obscure puzzles (solar and atmospheric neutrino deficits) have turned into discovery of the most fascinating phenomenon, *neutrino oscillations*, theoretically predicted by Bruno Pontecorvo.

Supernova SN 1987a became an elementary-particle laboratory in the sky for a study of properties of neutrinos, axions, majorons *etc.* Detection of neutrinos became a triumph of the theory: the number of detected neutrinos, duration of the neutrino pulse and estimated neutrino luminosity have appeared in agreement with theoretical prediction. Gravitational collapse as a phenomenon providing the SN explosion has been established ... and a puzzle appeared. It was the triumph of the incomplete theory, without rotation of collapsing star. The asymmetric ring around SN 1987a implies that the presupernova was a rotating star (it would be a surprise if not!). Rotation should diminish the neutrino luminosity or even change the collapsing scenario. Why then there is a beautiful agreement between theory and observation? This problem still expects its solution.

1.1. Greatness of false discoveries

WHEN FIRST APPEARED THE PUZZLES LOOK WEAK.
SAVE YOUR TIME SAYING: IT'S RUBBISH.
IN 90% OF CASES YOU WILL BE RIGHT,
BUT YOU MAY MISS THE GREAT PHYSICS.

False discoveries often have great impact on physics, and *Cyg X-3* is a famous example.

Cyg X-3 is a galactic binary system well studied in all types of radiations, most notably in X-rays. In 80s many EAS arrays detected from it the 4.8 hour periodic "gamma-ray" signal in VHE (Very High Energy, $E \geq 1$ TeV) and UHE (Ultra High Energy, $E \geq 0.1 - 1$ PeV) ranges. The list of these arrays included Kiel, Haverah Park, Fly's Eye, Akeno, Carpet-Baksan, Tien-Shan, Plateau Rosa, Durham, Ooty, Ohya, Gulmarg, Crimea, Dugway, Whipple and others. Probably it is easy to say that there was no single EAS array which claimed no-signal observation. Additionally, some underground detectors (NUSEX, Soudan, MUTRON) marginally observed high energy muon signal from the direction of this source. Apart from the Kiel array, which claimed 6σ signal, the confidence level of detection was not high: $3 - 4\sigma$.

In 1990 - 1991 two new-generation detectors, CASA-MIA and CYGNUS, put the stringent upper limit to the signal from Cyg X-3, which excluded early observations.

Apart from two lessons:

(i) good detectors are better than bad ones,

(ii) "3σ" discoveries should not be trusted, even if many detectors confirm them,

experience of Cyg X-3 has taught us how to evaluate statistical significance searching for periodic signals.

The false discovery of high energy radiation from Cyg X-3 had great impact on theoretical high energy astrophysics, stimulating study of acceleration in binary systems, production of high energy gamma and neutrino radiation and creation of high energy astrophysics with new particles, such as light neutralinos, gluinos *etc.*

2. Ultra High Energy Cosmic Ray (UHECR) Puzzle

This is the oldest puzzle in physics which exists for 40 years, and most probably now is close to being resolved. It appeared in 1966 together with prediction of the Greisen-Zatsepin-Kuzmin (GZK) cutoff [1]. As physical phenomenon the GZK cutoff is explained by production of pions by UHE extragalactic protons interacting with CMB photons. The energy of CMB photon in the Lorentz system, where proton is at rest is Γ times larger than in the laboratory system (where Γ is Lorentz factor of a proton) and when this energy exceeds the mass of pion, the proton energy loss becomes large, and the spectrum of UHE protons obtains steepening which is (incorrectly but impressively!) called the "GZK cutoff". For the diffuse spectrum the GZK feature (steepening) starts at energy $(3-5) \times 10^{19}$ eV. The UHECR puzzle was born simultaneously with the

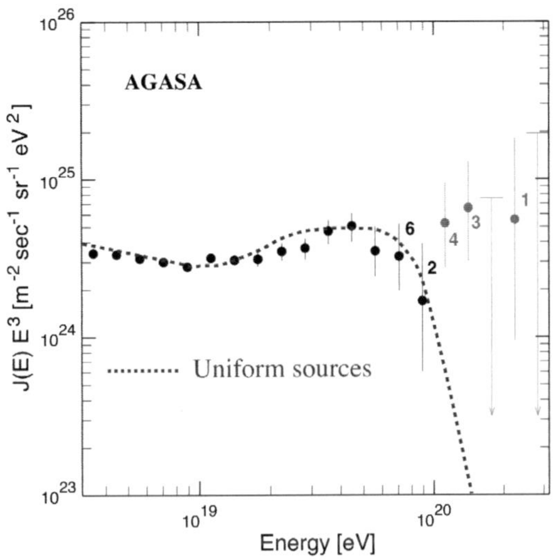

Figure 1. UHECR spectrum measured by AGASA detector in 1990 -1996 in comparison with calculated diffuse spectrum. The excess of events over predicted GZK cutoff is clearly seen. At present "AGASA excess" includes 11 events at energy $E \geq 1 \times 10^{20}$ eV.

prediction of the GZK feature: already in 1966 there were known at least three events with energies above the GZK cutoff. With time the number of these particles increased and in 1994 the famous golden Fly's Eye event [2] with energy $\sim 3 \times 10^{20}$ eV was detected. The energy of this event was determined very reliably and with the high precision. It questioned the presence of the GZK cutoff. However, most serious arguments against presence of the GZK feature were given by AGASA data [3]. In Fig. 1 the excess of AGASA events over predicted GZK cutoff is shown for data of 1996 - 1998 with 8 detected particles at energy $E \gtrsim 1 \times 10^{20}$ eV, in 2004 this number reaches reached 11 (for a review of observational data see [4]). The GZK cutoff is predicted for extragalactic protons, but already in 70s it was known that other conservative UHE signal carriers, UHE nuclei, suffer the steepening of the spectrum approximately at the same energy as protons [5, 6]. The other reasonable carriers, UHE photons, have too small absorption length interacting with radio-photons (see [7]).

2.1. New-physics solutions

No viable astrophysical solution to the "AGASA excess" was found, and the solutions with "new physics" have been proposed. I will give here the short list of the new solutions without references which can be found in [8]

- *Superheavy Dark Matter.* Long-lived Superheavy Dark Matter Particles are accumulated in galactic halos. These particles are naturally produced gravitationally at post-inflationary epoch and at mass $m_x \sim 10^{13}$ GeV they provide CDM density as observed by WMAP. These particles can be long-lived, with lifetime exceeding the age of the Universe. Decays of these particles produce UHECR without GZK cutoff (most of UHE particles come from Galactic halo).
- *Topological Defects* (TD). There are various mechanisms of production of UHE particles by TD. In some cases TD become unstable and decompose to constituent fields (superheavy Higgs and gauge bosons), which then decay to ordinary particles. This mechanism works for cusps and superconducting cosmic strings. In case of monopoles and antimonopoles connected by strings, high energy particles are produced at annihilation of monopole-antimonopole pairs. The most promising candidates are necklaces and monopole-antimonopole pairs connected by string. UHECR from TD have a spectrum with a soft GZK cutoff which does not contradict observations.
- *Resonant neutrinos.* Very high energy neutrinos are resonantly absorbed by target neutrinos comprising Hot Dark Matter (HDM): $\nu + \bar{\nu}_{HDM} \rightarrow Z^0 \rightarrow hadrons$. Very large flux of primary neutrinos with superhigh energies is needed for this hypothesis.
- *Light gluino.* Light gluinos can be effectively produced by TD or in pp-collisions in astrophysical sources. They weakly degrade in energy interacting with microwave radiation. The interaction of UHE light gluino with nucleons is similar to that of UHE proton. Light gluino is disfavored by accelerator experiments.
- *Strongly interacting neutrino.* In extra-dimension theories, for example, neutrino can have large cross-section of scattering off the nucleon. In this case neutrino can be a carrier of UHE signal from remote astrophysical sources.
- *Lorentz invariance breaking.* In this case for protons with energies 10^{20} eV and higher, the c.m. energy could be not enough for production of pions in collisions with microwave photons.

It is interesting that these new ideas, in particular, Superheavy Dark Matter and Lorentz-invariance violation, first proposed in 1972 [9] to solve UHECR puzzle, are developed now independently of UHECR.

2.2. Towards astrophysical solution

GZK cutoff is nothing but a signature of UHE proton interaction with CMB.

Are there some other signatures of the same interaction, and what they tell us about UHECR problem?

Such signature is known since long time [10]: this is the *dip*; the tiny feature in the proton spectrum left behind by production of electron-positron pairs in collisions with CMB photons ($p+\gamma_{\rm CMB} \rightarrow p+e^+ + e^-$). This feature is seen better when analyzed in terms of the modification factor

$$\eta(E) = J_p(E)/J_p^{\rm unm}(E), \quad (1)$$

where $J_p(E)$ is proton spectrum calculated with all energy losses included and $J_p^{\rm unm}(E)$ is one calculated with adiabatic energy losses (red-shift) only. The advantage of modification factor is its model-independence [11, 12, 13], in contrast to GZK cutoff which at $E \geq 1 \times 10^{20}$ eV is strongly model dependent.

The calculated modification factor is shown in comparison with experimental data in Fig. 2. *As Fig. 2 shows the pair production dip and beginning of GZK cutoff up to energy 1×10^{20} eV is reliably confirmed by all experimental data including AGASA.* As to AGASA excess at

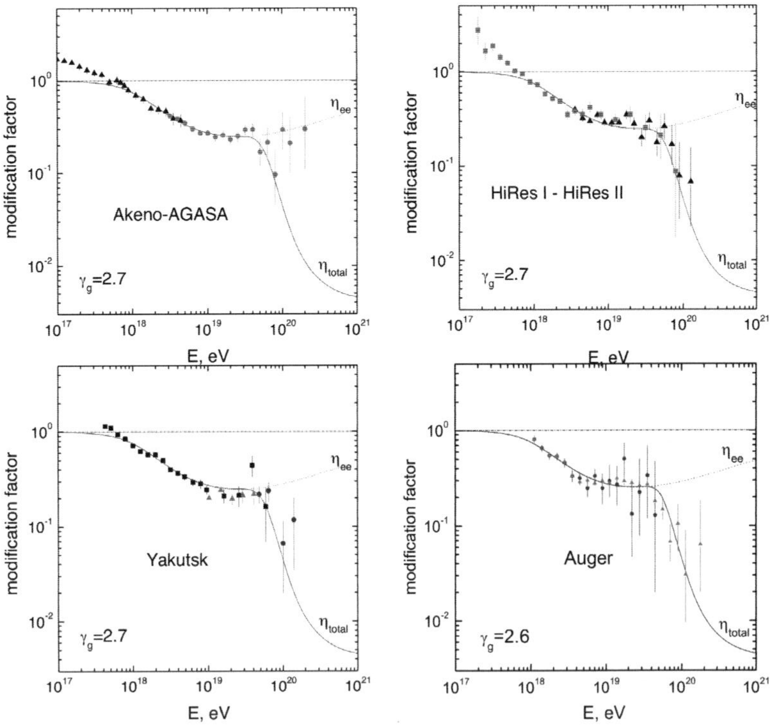

Figure 2. The predicted pair-production dip in comparison with Akeno-AGASA, HiRes, Yakutsk and Auger data. The first three experiments confirm dip with good χ^2/d.o.f. $\approx 1.0-1.2$, while the Auger data are characterized by larger χ^2/d.o.f. (see the text). The data of Fly's Eye (not presented here) confirm the dip as well as AGASA, HiRes and Yakutsk.

$E > 1 \times 10^{20}$ eV it can be explained by some other reasons, e.g. [14] by systematic energy errors combined with statistical fluctuations. The large χ^2 in comparison of the dip with Auger data is explained by energy errors not included in the analysis.

2.3. Evidence for GZK cutoff

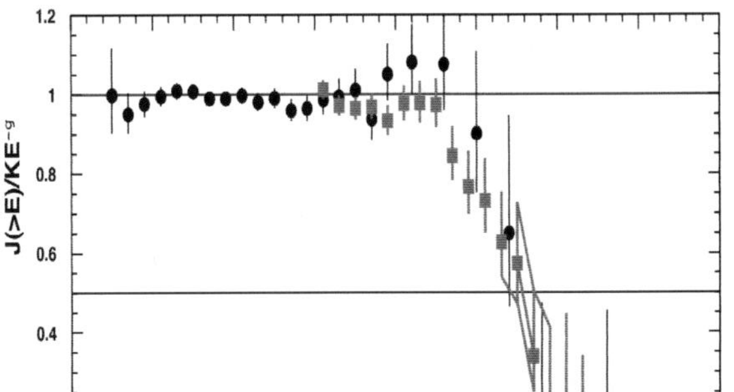

Figure 3. $E_{1/2}$ as numerical characteristic of GZK cutoff in the integral spectrum (see text).

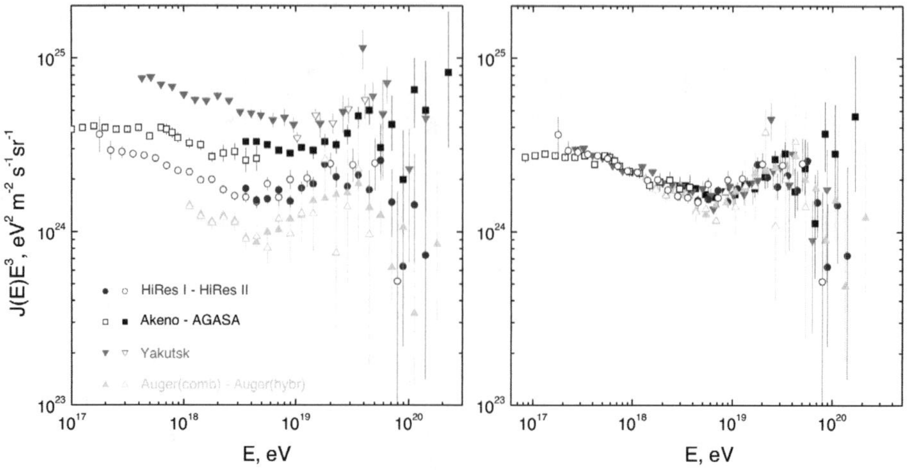

Figure 4. The spectra and fluxes measured by Yakutsk, AGASA, HiRes and Auger before (left panel) and after (right panel) energy calibration.

The Fig. 2 shows that data of HiRes, Auger and Yakutsk are consistent with GZK cutoff. The presence of GZK cutoff is seen most clearly in the HiRes data. However, low statistics and a possibility of imitation of the observed steepening by other reasons, e.g. by "acceleration cutoff", precludes one from making the final conclusion. Recently, HiRes collaboration obtained the numerical confirmation that this steepening is really the GZK cutoff [15]. In the integral spectrum the GZK cutoff is characterized by energy $E_{1/2}$, where calculated spectrum $J(>E)$ becomes half of power-law extrapolation spectrum $KE^{-\gamma}$ from low energies. As calculations [10] show this energy is $E_{1/2} = 10^{19.72}$ eV for a wide range of generation indices from 2.1 to 2.8. HiRes collaboration found $E_{1/2} = 10^{19.73 \pm 0.97}$ eV in a good agreement with the theoretical prediction. In Fig. 3 we reproduce the HiRes graph from which $E_{1/2}$ was determined. The plotted value is given by ratio of measured flux $J(>E)$ and its power-law approximation $KE^{-\gamma}$. Extrapolation of this ratio to the higher energies is given by 1, while intersection of measured ratio with horizontal line 1/2 gives $E_{1/2}$.

2.4. Calibration of detectors with help of dip and GZK cutoff

The fluxes measured by Yakutsk, AGASA, HiRes and Auger are displayed in Fig. 4 (left panel). To the large extent the great discrepancy seen there is caused by comparison of the values $E^3J(E)$ and thus systematic energy errors affect strongly this contradiction. As was demonstrated in [16, 12, 13] the energy calibration with help of the dip results in good agreement between the absolute fluxes of all detectors. Here we use the different approach for calibration of the detectors, based on both features, dip and GZK cutoff. Since energies as measured by HiRes fit well the both features and especially GZK numerical characteristic $E_{1/2}$, we assume that HiRes energies are correct and the energies of all other detectors must be shifted by factor λ to reach the best agreement in fluxes (see also [17]). This procedure gives values of λ equal to 1.2, 0.75, 0.83, and 0.625 for Auger, AGASA, Akeno and Yakutsk, respectively. This calibration does not give minimum χ^2 for the dip shape, but describes better the dip and beginning of GZK cutoff together. The fluxes after this energy calibration are shown in Fig. 4 (right panel).

2.5. Conclusions

The dip and beginning of the GZK cutoff at $E \leq 1 \times 10^{20}$ eV are signatures of proton interaction with CMB. They are confirmed with good accuracy by all experimental data, including AGASA. The GZK cutoff is seen most clearly in HiRes data and does not contradict Auger and Yakutsk data. HiRes experiment has numerical proof in the form of value $E_{1/2}$ that observed steepening is really GZK cutoff. AGASA excess can be explained by systematic energy errors combined with insufficient statistics. The UHECR fluxes and spectra measured by all four detectors coincide well after energy calibration of detectors with help of dip and GZK cutoff. As recent Auger data [18] indicate, the sources of UHECR can be AGN (see [19] for analysis).

3. Cosmological Puzzles in the Past and Present

The theoretical basis of cosmology is given by Friedmann solution of the Einstein equation. After discovery of highly isotropic CMB radiation and observational indications to the flat universe, it was understood that Friedmann solution leads to two puzzles: *horizon* and *flatness* problems.

The *horizon problem* can be explained in following way.

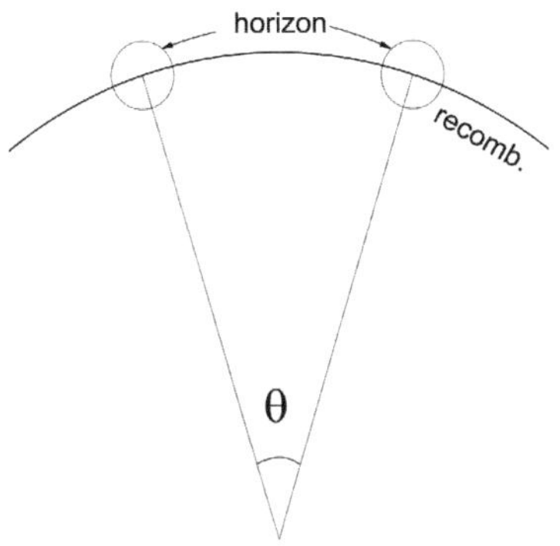

Figure 5. Two regions on recombination sphere ($z = z_{\rm rec}$ separated by the horizon distance $ct_{\rm rec}$.)

CMB radiation decouples from matter after recombination time at red-shift $z \approx 1100$ or $t_{\rm rec} \approx 1.2 \times 10^{13}$ s. Two regions separated by horizon length $ct_{\rm rec}$ (see Fig. 5) are seen at angle $\theta \approx (1 + z_{\rm rec})ct_{\rm rec}/ct_0$, where $t_0 \approx 13.7$ Gyr is the age of the universe. These regions are causally disconnected and therefore the regions in the sky separated by angle $\theta \gtrsim 2^\circ$ cannot have equal CMB temperature. This conclusion is in contradiction with observed isotropy of CMB radiation.

Why the universe is *flat* now?

Within Friedmann solution it can be only due to initial conditions at $t_{\rm Pl} \sim 1/m_{\rm Pl}$. To have $\Omega - 1 \sim O(1)$ now, it is necessary to have at $t \sim t_{\rm Pl}$ $\;\Omega - 1 \sim \xi$ with $\xi \sim 10^{-30}$.

If $\Omega - 1 >> |\xi|$ the universe collapses during time much shorter than the age of universe now. If $\Omega - 1 << -|\xi|$ galaxies and stars do not have enough time to be formed. Such fine-tuning is highly unnatural.

The third cosmological puzzle was observation of rotational curves in galaxies and cluster of galaxies, which showed that virial (gravitational) masses of the objects are much bigger than visible (luminous) mass. It was interpreted as dominance of weakly interacting Dark Matter (DM). At present there are many particle candidates for DM with detailed description how these particles can be dominant in universe and produce Large Scale Structures (LSS).

As it is well known the first two puzzles were solved assuming an early stage in expanding universe different from Friedmann regime: *inflation.* At the different level of physical realization, this stage was proposed in works by E. Gliner (1965) [20], A. Starobinsky (1979) [21], K. Sato (1981) [22], A. Guth (1981) [23], A. Linde (1983) [24], A.Albrecht and P. Steinhardt (1982) [25], and spectrum of density perturbations ("galaxy formation") imposed by this stage, was first studied by S. Mukhanov and G. Chibisov [26].

Inflation is also expanding solution of the Einstein equation, but driven by potential of the scalar field ϕ, inflaton. Expansion proceeds exponentially, much faster than in the Friedmann regime.

The Einstein equation for flat universe and energy conservation result in the following equations [27]:

$$\begin{aligned} \dot{a}^2(t) &= \frac{8\pi}{3}Ga^2(t)\rho \\ \ddot{a}(t) &= -\frac{4\pi}{3}G(\rho+3p)a(t) \\ \dot{\rho} &= -3H(p+\rho), \end{aligned} \tag{2}$$

where G is gravitation constant, $a(t)$ is the expansion factor of the metric, $H(t) = \dot{a}/a$ is the Hubble parameter, p and ρ are pressure and density, connected by equation of state $p = p(\rho)$. Let us consider the matter with equation of state $p = -\rho$ and $\rho(t) = \rho_0$, which follows from the third equation above. Such equation of state in particular is realized by scalar field ϕ rolling down in quasi-flat potential. Eqs. (2) are easily solved in this case:

$$a(t) = a_0 e^{H_0 t} \text{ with } H_0^2 = \frac{8\pi}{3}G\rho_0, \tag{3}$$

The solution (3) describes exponentially expanding bubble with constant Hubble parameter H_0. This regime is called inflation.

Inflation obviously solves the horizon problem, because the size of bubble $a(t)$ is much larger at any time t than the size of the observed universe and thus all observed regions are causally connected.

The problem of flatness is also solved. To demonstrate it let us generalize the first equation (2) to the cases of closed ($k = 1$), open ($k = -1$) and flat ($k = 0$) universes.

$$\dot{a}^2(t) - \frac{8\pi}{3}Ga^2(t)\rho = -k. \tag{4}$$

Dividing this equation by a^2, and using $H = \dot{a}/a$ and $\Omega = \rho/\rho_c$ we obtain

$$\Omega - 1 = \frac{k}{a^2H^2}. \tag{5}$$

For inflationary solution ($p = -\rho$), $H = H_0$ and $a(t) \sim \exp(H_0 t)$ we see that at the end of inflation, e.g. $t \sim 100H_0^{-1}$ the r.h.s. is negligibly small, providing the flatness at inflation and at all later universe ages up to $t \sim t_0$.

3.1. Where is Dark Matter?

Observational cosmology is characterized now by very precisely determined parameters, and the main contribution to accuracy of their determination is given by WMAP data [28]. WMAP confirms well ΛCDM model with 6 parameters and the main parameters have the following values: $H_0 = 73.2$ km/sMpc, $\Omega_{tot} = 1 + \Omega_k$, $\Omega_k = -0.011 \pm 0.012$, $\Omega_b = 0.0416$, $\Omega_m = 0.238$, $\Omega_\Lambda = 0.716$. However, the precise determination of cosmological parameters are given by combination of WMAP data with other observations such as SNI, lensing, Large Scale Structures (LSS) and others. The most important results are determination of fraction of Dark Matter (DM) and Dark Energy (DE). We summarize here the status of DM according to these and other observations.

Direct search for DM gives positive result only in observation of modulation signal by DAMA [29]. It does not imply the direct contradiction between DAMA and other experiments: the comparison is always model dependent and difference in target affects the comparison (see [29]).

The strongest evidence for DM is given by *indirect* methods. They include:

- According to WMAP data density of matter Ω_m exceeds much the baryon density Ω_b (see the data above). The presence of DM is seen directly from WMAP data: without DM the height of the third acoustic peak would be much lower than observed.
- Virial (gravitational) mass of galaxies and clusters of galaxies is much larger than baryonic gas: $M_{\rm vir} \gg M_b$.
- Theory of LSS formation (hierarchical clustering model) successfully explains the LSS formation using dominance of DM.

As was mentioned above only in DAMA the positive direct signal from DM (modulation) is observed. Inspired by negative results in other experiments, some authors develop the alternative theories based on modified gravity. The pioneering works in this direction, Modified Newtonian Dynamics (MOND), have been performed by Milgrom [30]. MOND is characterized by critical acceleration $a_0 \sim 1 \times 10^{-8}$ cm s^{-2}. At $a < a_0$ the Newtonian mechanics is modified, and the effects of DM are described by modified gravity. The essential step forward in construction of a theoretical model for modified General Relativity (GR) was made in the work by Bekenstein [31]. He has build theoretically correctly the relativistic model of gravitation, TeVeS, introducing three gravitational fields $g_{\mu\nu}$ (tensor, like in GR), U_μ (vector) and ϕ (scalar). Apart from it the model includes a non-dynamical scalar σ, dimensional constants G and l, two dimensionless parameters k and K and one arbitrary function $F(\sigma)$. The model tends to the standard GR in the proper limit and describes effects of MOND at low acceleration. Thus, the model is much more complicated than the standard GR and the only physical motivation of it is given by possibility to avoid DM. This model successfully describes (with baryonic matter only) flat rotation curves in galaxies, high velocities in clusters and lensing. Recent work [32] indicates that TeVeS may provide the perturbation spectrum needed for LSS formation.

There are, however, two contradictions of MOND with observations:

(i) In the absence of DM the height of the third acoustic peak must be much lower than observed by WMAP. In the paper [28] one can find more general statement: "Models without dark matter are very poor fits to the data".

(ii) In the observed "bullet cluster" 1E0657-558 [33] (two colliding clusters) the gravitational potential is not centered by X-ray emitting gas, which is the dominant baryon component in this cluster (the ratio of the masses for gas and galactic components is $M_{\rm gas}/M_{\rm gal} \sim 5-7$).

In conclusion, I think that there is no DM puzzle. DM particles are not seen in the directly-search experiments, either because sensitivity is still low or because DM particles interact superweakly with ordinary matter (e.g. gravitinos or SHDM particles). MOND and TeVeS should be considered just as interesting alternatives.

3.2. Accelerated expansion of the universe

As follows from the second equation (2) for matter with equation of state $p = \omega\rho$ and with $\omega < -1/3$, acceleration $\ddot{a} > 0$, i.e. we have accelerating expansion of metric. This is generalization of the inflation case $\omega = -1$ which we considered above: $a(t) \propto \exp(H_0 t)$ and $\rho = \rho_0$ (vacuum energy). In the general case there can exist both the matter described by energy-momentum tensor $T_{\mu\nu}$ in the Einstein equation and vacuum energy described by term $\rho_{\rm vac} g_{\mu\nu}$. The latter is Lambda term, where $\Lambda = 8\pi G \rho_{\rm vac}$. In the general covariant form the Einstein equation reads

$$R_{\mu\nu} - \frac{1}{2} g_{\mu\nu} R = -8\pi G (T_{\mu\nu} - \rho_{\rm vac} g_{\mu\nu}). \qquad (6)$$

Presence of Λ term results in acceleration of expansion. While density of ordinary matter in $T_{\mu\nu}$ diminishes with time, $\rho_{\rm vac}$ remains constant, and finally it dominates. Starting from this moment universe enters again the phase of exponential expansion.

The presence of acceleration in the universe is established experimentally in terms of non-zero Λ-term, but more generally the data can be explained also by other terms which cause accelerated expansion. For example, the SN data are explained by non-standard dependence $H(z)$, where H is the Hubble parameter. Such dependence might appear due to matter with equation of state $p = \omega\rho$ and $-1 < \omega < -1/3$. This matter is usually referred to as Dark Energy (DE) fluid.

In principle there are three sources which cause cosmic acceleration:

1. Vacuum energy (Λ-term) in Eq.(6).
2. DE fluid in $T_{\mu\nu}$ with equation of state $p = \omega\rho$ and $\omega < -1/3$. It can be realized as *quintessence*, ultra-light scalar field rolling down the potential; *phantom*, ghost field with equation of state given by $\omega < -1$; *k-essence*; *Chaplygin gas* etc. (for review see [34, 35]).
3. Modified gravity. In this case the modified part of Eq. (6) is its l.h.s. No Λ-term or DE is needed for acceleration and explanation of observational data. The most interesting example is given by DGP [36] extra dimension model, where gravity can leak into the bulk only at large distances. This is one-parameter model which explains the observed acceleration.

Is it possible to distinguish between these possibilities by observations?

Some options differ by spectrum of fluctuations and by equations of state. WMAP data efficiently distinguish the models with different ω. The combined data of WMAP, LSS and SN gives $\omega = -1.08 \pm 0.12$. With a prior assumption of flat universe the above data give $\omega = -0.967 \pm 0.073$. It coincides well with prediction of theory with Λ-term ($\omega = -1$) and is allowed by some other models. Thus, the model with Λ-term (ρ_{vac} =const) is allowed and preferable by its simplicity.

3.3. Vacuum energy problem

Associated with Λ-term, the vacuum energy density is given numerically as

$$\rho_\Lambda = \Lambda/8\pi G = \Omega_\Lambda \rho_c = 4 \times 10^{-47}\ \text{GeV}^4, \tag{7}$$

for $\Omega_\Lambda = 0.73$ [28]. The vacuum energy ρ_Λ can be presented as sum of energy density of some cosmological field(s) σ and zero-modes of all known fields (particles) i. Taking them as quantum oscillators with ground-state energy $\omega_k/2$ one obtains vacuum energy of field i

$$\rho^i_{\text{vac}} = \int_0^{k^i_{\max}} \frac{d^3k}{(2\pi)^3} \frac{\omega_k}{2}. \tag{8}$$

Then the total vacuum energy is given by

$$\rho_\Lambda = \rho_\sigma + \sum_i \rho^i_{\text{vac}}. \tag{9}$$

The problem is that the second term in r.h.s. of Eq. (9) is too large. For example, the reliably known and existing now in the universe quark-gluon condensate has vacuum energy between -3×10^{-3} GeV^4 as estimated in pioneering work [37] and -0.01 GeV^4 in modern lattice QCD calculations. Its absolute value exceeds the cosmological vacuum energy density (7) by 45 orders of magnitude (for further discussion see [38]).

The problem of too large zero-mode energy of ordinary fields in comparison with cosmological constant is very old. It was already discussed in late sixties (see [39]). It is important to emphasize that this problem consists in cosmologically too large vacuum energy of known fields and thus it is present for example in all DE models. It may be a problem of elementary-particle physics. For example with unbroken supersymmetry there is exact compensation between negative vacuum energies of bosons and positive of fermions. However, realistic breaking of this symmetry leaves the vacuum energy too large. The 40 years of existing the problem were 40 years of intensive brain attack ... and the problem still exists. Among the proposed solutions there were the beautiful ones. Among the authors there were the stars of theoretical physics such as S. Coleman, S. Hawking, S. Weinberg, E. Witten, Ya. Zeldovich. It makes me think that this puzzle will have some unexpected solution.

S. Weinberg in his review "The cosmological constant problem" [40] describes five roads which could lead to the solution:

1. supersymmetry and supergravity,
2. anthropic approach,
3. adjustment mechanisms,
4. alternative gravity,
5. quantum cosmology.

All approaches are qualified as interesting and promising (quantum gravity especially) but for all of them are indicated the weak points, which do not allow to consider them as the correct, final solution.

As the new direction we can add to the Weinberg list the extra-dimension approach studied recently (for a review see [34]).

3.4. Acceleration and Anthropic approach

- Why Λ-term is zero or very small?
- Why does acceleration start now?
- Why physical parameters are tuned to produce life?

These questions might have answers not in terms of physical principles, but because in a universe with the "wrong" parameters there is nobody to measure them. This is an anthropic solution. For many interesting details I direct a reader to the exiting book of A. Vilenkin "Many Worlds in One" [41].

The answer in the Anthropic approach to the above questions consists in assumption that there is the infinite number of universes where Λ (or $\rho_{\rm vac}$) takes huge variety of different values. In fact, these quantities do not have the fundamental character to be the universal ones. The matter density ρ_m diminishes with time faster than $\rho_{\rm vac}$. Then if Λ is small (negative) a universe collapses early and galaxies (observers) are not produced. If Λ is too large the exponential expansion begins early and galaxies are not produced, as well. The galaxy formation occurs roughly when $\rho_{\rm vac} \sim \rho_m$. This is the basics of answering to the first two questions and relevant calculations one can find in [42], where the value of Λ was predicted.

The third question has a name "coincidences". To provide the life many physical constants cannot be changed even by very small amount. The famous example is given by F. Hoyle [43]: the resonance in reaction $3He^4 \to C^{12}$ is the most important channel of carbon production in stars, without it we would not have life (observer). There are many other coincidences, described e.g. in [44].

3.5. From Inflation to Anthropic approach

Anthropic approach very naturally appears in chaotic-inflation scenario [45]. In this scenario the distribution of the inflaton field ϕ in early universe is assumed to be chaotic and respectively the potential $V(\phi)$ is characterized by different values from very low ones up to $M_{\rm Pl}^4$. In fact, this postulate of chaotic inflation is very reasonable because different initial domains, separated by distance larger than horizon, must have the different ϕ and V. Inflation is characterized by

$H(\phi) = const$ for each domain, where ϕ is quasi-homogeneous and it proceeds independently in each horizon domain of size $H^{-1}(\phi)$, which are called mini-universes. At present epoch $t = t_0$ the size of each mini-universe is many orders of magnitude larger than the horizon size $\sim 10^{28}$ cm.

Since value of Λ ($\rho_{\rm vac}$) is determined by ϕ the different mini-universes are characterized by different values of Λ. It provides a possibility of anthropic selection of Λ.

The chaotic inflation automatically results in the "eternal inflation" [46], a process which has no beginning and end. It occurs because the chaotic inflation includes two competing processes of regeneration of inflationary regions and their decay. There is some critical value ϕ_c below which ϕ oscillate, potential energy is transferred to particles, and the Friedmann expansion begins. On the other hand the region with relatively small ϕ can through quantum tunneling turn into region of much larger size. After tunneling the matter density ρ_m becomes smaller, while $\rho_{\rm vac}$ remains the same. As a result $\rho_{\rm vac} > \rho_m$ provides the new round of inflation. In Ref.[46] the process of increasing ϕ and $\rho_{\rm vac}$ is described as diffusion of ϕ to higher values as quantum fluctuation process. Thus the Universe always consists of exponentially large number of mini-universities, part of which inflate and part experiences Friedmann expansion. They have the different values of Λ, and therefore some of them might have an observer.

All mini-universes have connected space-time. However, there can be also the large or infinite number of universes with disconnected space-time. This scenario is provided by "creation of a universe from nothing"[47]. A closed universe with zero total energy can be created by quantum tunneling from initial object which has a vanishing size and mass. This object can be considered as "nothing". If after-tunneling object has the Planck size or less, it immediately collapses, if more - it inflates, producing a disconnected space-time and a universe. All these universes have different mini-universes with different Λ.

3.6. Status of anthropic approach

Anthropic approach to the problem of vacuum energy is different from all other physics solutions and its status will depend on whether or not the irreproachable ordinary-physics solution (e.g. the compensation solution) will be found or not. I will remind again that cosmological constant problem exists for more than 40 years, attracting much attention of physics community. The negative attitude of large part of physics community to the anthropic approach could be conservatism of *present generation*, and some indication of it is given by discussion of anthropic solution by such deep thinkers of our generation as S. Weinberg, S. Hawking, M. Rees, A. Vilenkin, A. Linde and others. It could be that next generation of physicists will consider the anthropic approach as an ordinary physics. In fact, the anthropic approach reminds that of quantum mechanics. The quantum state of a universe with a fixed value of Λ can be described by the wave function [48], which gives the distribution of universes over different values of Λ. In another version of this formalism [49] each universe is characterised not by single value of Λ, but by their superposition, with the wave function giving again a probability to find a universe with fixed value of Λ. The presence of observer in a universe (anthropic selection) plays the role similar to the the macro-detector in quantum mechanics.

Acknowledgements

It is my great pleasure to thank Alex Vilenkin for helpful discussions.

[1] Greisen K *Phys. Rev. Lett.* 1966 **16** 748;
Zatsepin G T and Kuzmin V A 1966 *Pisma Zh. Experim. Theor. Phys.* **4** 114

[2] Bird D J *et al* [Fly's Eye collaboration] 1994 *Ap. J.* **424**, 491

[3] Shinozaki K *et al* [AGASA collaboration] 2004 *Nucl. Phys.* B (Proc. Suppl.) **136** 18

[4] Nagano M and Watson A A 2000 *Rev. Mod. Phys.* **72** 689

[5] Berezinsky V S, Grigorieva S I and Zatsepin G T 1975 *Proc. 14th Int. Cosm. Ray Conf. (Munich)* **2** 711 1976 *Izv. Acad. Nauk USSR (ser. phys.)* **40** 524

[6] Puget J L Stecker F W and Bredekamp J J 1976 *Astroph. J.* **205** 638
[7] Berezinsky V S 1970 *Soviet Journ. Nucl. Phys.* **11** 399;
Prothroe R J and Biermann P 1966 *Astrop. Phys.* **6** 45
[8] Berezinsky V 2000 *Nucl. Phys.* B (Proc. Suppl) **87** 387
[9] Kirzhnitz D A and Chechin V A 1972 *Pisma ZhETP* **14** 261;
1972 *Sov. Journ. of Nucl. Phys.* **15** 1051
[10] Berezinsky V S and Grigorieva S I 1988 *Astron. Astroph.* **199** 1
[11] Berezinsky V, Gazizov A and Grigorieva S 2005 *Phys. Lett.* B **612** 147
[12] Berezinsky V, Gazizov A Z and Grigorieva S I 2006 *Phys. Rev.* D **74** 043005; hep-ph/0204357
[13] Aloisio R *et al* 2007 *Astrop. Phys.* **27** 76
[14] De Marco D, Blasi P and Olinto A V 2006 *J. Cosm. Astrop. Phys.* **01** 002
[15] Hires collaboration, arXiv:astro-ph/0703099
[16] Berezinsky V, astro-ph/0509069
[17] Kampert K-H These Proceedings, astro-ph/0801.1986
[18] The Pierre Auger collaboration, astro-ph/0711.2256
[19] Berezinsky V, Gazizov A Z and Grigorieva S I, astro-ph/0210095
[20] Gliner E B 1965 *Sov. Phys. JETP* **22** 378
[21] Starobinsky A A 1979 *JETP Lett.* **30** 682
[22] Sato H 1981 *MNRAS* **195** 467
[23] Guth A H 1981 *Phys. Rev.* D **23** 347
[24] Linde A D 1982 *Phys. Lett.* B **108** 389
[25] Albrecht A and Steinhardt P J 1982 *Phys. Rev. Lett.* **48** 1220
[26] Mukhanov V F and Chibisov G V 1981 *JETP Lett.* **33** 549
[27] Weinberg S 1972 *Gravitation and Cosmology, John Wiley and Sons*
[28] Spergel D N *et al* for WMAP collaboration 2007 *Ap. J. Suppl.* **170** 377
[29] Bernabei R *et al* 2003 *Riv. N. Cim.* **26** 1 ; 2004 *Int. J. Mod. Phys.* D **13** 2127; 2006 *Int. J. Mod. Phys.* A **21** 1445; 2007 *Int. J. Mod. Phys.* A **22** 3155; 2008 *Phys. Rev.* D **77** 023506
[30] Milgrom M 1983 *Ap. J.* **270** 365-84
[31] Bekenstein J D 2004 *Phys. Rev.* D **70** 083509
[32] Dodelson S and Liguori M 2006 *Phys. Rev. Lett.* **97** 231301
[33] Clowe D *et al* 2006 *Ap. J. Lett.* **648** L109
[34] Copeland E C, Sami M and Tsujikawa S 2006 *Int. J. Mod. Phys.* D **15** 1753
[35] Wetterich C, these Proceedings.
[36] Dvali G G, Gabadadze G and Porrati M 2000 *Phys. Lett.* B **495** 208
[37] Shifman M, Vainstein A and Zakharov V 1979 *Nucl. Phys.* B **147** 448
[38] Dolgov A D 2005 *Proc. of LIX Yamada Conf. (eds. Suzuki H, Yokoyama J, Suto Y and Sato K), Tokyo, Japan* 105
[39] Zeldovich Ya. B 1967 *JETP Lett.* **6** 316
[40] Weinberg S 1989 *Rev. Mod. Phys.* **61** 1
[41] Vilenkin A 2006 *Many worlds in one, Hill and Wang, New York*
[42] Weinberg S 1987 *Phys. Rev. Lett.* **59** 2607;
Vilenkin A 1995 *Phys. Rev. Lett.* **74** 846, (gr-qc/9406010v2);
Carriga J and Vilenkin A, hep-th/0508005.
[43] Hoyle F 1953 *Ap. J.* **118** 513
[44] Carr B J and Rees M J 1979 *Nature* **278** 605;
Davies P C W 1982 *The accidental universe (Cambridge University Press, Cambridge)*
[45] Linde A D 1983 *Phys. Lett.* B **129** 177
[46] Vilenkin A 1983 *Phys. Rev.* D **27** 2848;
Linde A D 1986 *Phys. Lett.* B **175** 395
[47] Zeldovich Ya B 1981 *Sov. Astron. Lett.* **7** 322;
Vilenkin A 1982 *Phys. Lett.* **117** 25
[48] Hartle J B and Hawking S W 1983 *Phys. Rev.* D **29** 2960
[49] Hawking S W 1987 *Phys. Scr.* T **25** 202

The Cosmological Model: an overview and an outlook

Alan Heavens

Institute for Astronomy, University of Edinburgh, Blackford Hill, Edinburgh EH9 3HJ, UK

E-mail: `afh@roe.ac.uk`

Abstract. The Standard Model of Cosmology has emerged from a number of independent lines of evidence, and accounts very well for most large-scale observations of the Universe. I present a brief review of the model, focussing on the underpinning observational data, and commenting on the strengths and weaknesses of each method. I review some of the current implications for inflation, Dark Matter and Dark Energy, and take a look forward to what may be learned from ambitious future cosmological surveys. The future prospects include the possibility of determining definitively whether the Dark Energy is Einstein's cosmology constant or alternatively an evolving scalar field, or even testing whether gravity is not General Relativity, but a manifestation of a higher-dimensional theory based on strings.

1. Introduction

Over the last ten years, a number of lines of evidence have pointed towards a consistent picture for the state of the Universe. The observational status is considered by many to be sufficiently strong that the term 'Standard Cosmological Model' is justified as a description of our understanding of the Universe and its contents. The picture which emerges is, however, a curious one, in which the Universe is unexpectedly accelerating, with the vast bulk of its energy density in a form which has not been detected directly on Earth. Indeed, the picture is even more strange, in that the data appear to demand two unknown ingredients, Dark Matter and Dark Energy. However, with the inclusion of these ingredients, a wide variety of quite detailed cosmological observations can be explained, to the extent that the parameters in the theory can be consistently determined with an accuracy typically of a few percent. Confidence in the model is such that the data are increasingly being use for rather detailed examination of the model ingredients. Subtle effects of a number of features, such as the masses and types of neutrinos, the types of interactions of Dark Matter, and the equation of state of the Dark Energy, have been calculated and confronted with theory.

The security of the standard model, and its implications, are such that ambitious plans are being made to probe its ingredients in more detail. With about 75% of the energy budget in Dark Energy, it is perhaps little surprise that this forms the focus for a number of planned observational programmes over the next decade. The main immediate question here is whether the Dark Energy is Einstein's cosmological constant, or a previously undiscovered field with an unusual equation of state which is driving the acceleration of the Universe. More radically, the acceleration may be a manifestation of a breakdown of General Relativity, as may happen if the Universe has extra dimensions. Some string-inspired braneworld models have been investigated theoretically to the point where confrontation with observation is possible, and we have the intriguing possibility of being able to test such models with future cosmological surveys.

In this short review, I outline the Standard Cosmological Model, and summarise the principal observations underpinning the theory. I discuss the points where the standard model comes under some pressure, and try to give some indications of where the comparison between theory and observation is relatively secure, and where our capability is limited by uncertain physics. I look at the upcoming and planned cosmological surveys, and their capability to measure Dark Energy properties, and to detect beyond-Einstein gravity.

2. The Standard Cosmological Model

The notion that there might be Dark Matter in the Universe is very old, dating back to the 1930s when Fritz Zwicky noted that the combined gravity of galaxies in clusters was not sufficient to keep them bound to the cluster, given their large velocity dispersion. The matter cannot be entirely baryonic, since there are stringent limits on the density parameter of baryonic matter, $\Omega_b = 0.04 \pm 0.004$ (e.g. [1]), arising from the primordial nucleosynthesis of the light elements. This is short of the mass requirements of clusters ($\Omega_m \simeq 0.2$), thus establishing the need for non-baryonic Dark Matter.

The idea that there might also be a cosmological constant (or Dark Energy) component was hinted at by a number of observations, such as the power spectrum of fluctuations in the APM survey of galaxies [2]. However, it is probably fair to say that the idea that the Universe was dominated by such a component only gained wide acceptance when observations of distant supernovae in the late 1990s [3] implied that the Universe was not only expanding, but accelerating. This is quite unexpected, as if it is driven by a source field (rather than a cosmological constant), it requires the field to have a very unusual equation of state, with negative pressure. Specifically, the equation of state parameter $w \equiv p/(\rho c^2) < -1/3$. Einstein's cosmological constant is equivalent as far as the expansion of the Universe is concerned with a source field with $w = -1$.

The Standard Cosmological Model therefore consists of the following: the Universe began with the Big Bang, went through a period of near-exponential inflation at early times ($\sim 10^{-35}s$), and has expanded ever since, under gravitational influence described by General Relativity. The expansion rate is controlled by its contents, which are baryonic matter (and electrons) with density parameter $\Omega_b \simeq 0.04$, photons at 2.73K and neutrinos (with negligible energy density), cold Dark Matter with $\Omega_m \simeq 0.26$ (0.22 in CDM), and Dark Energy, with $\Omega_{DE} \simeq 0.74$). The current expansion rate is described by a present-day Hubble parameter $H_0 \equiv 100h \simeq 72$ $\mathrm{km\,s^{-1}\,Mpc^{-1}}$. Errors on these quantities are a few percent. It is spatially flat, or very nearly so, and structure grew from adiabatic fluctuations with a near-scale-invariant spectrum, with a primordial power spectrum, $P(k) \propto k^n$ with $n \simeq 1$, generated during inflation. The present-day amplitude of fluctuations is characterised by a parameter $\sigma_8 \simeq 0.75$ and the optical depth by $\tau \simeq 0.08$. A flat model with 6 free parameters ($\Omega_m h^2$, $\Omega_B h^2$, h, n, τ, σ_8) provides a good fit to most observations, but extensions to this model, where n varies with scale, or neutrinos have a significant non-zero mass. This relatively simple model accounts for nearly all cosmological observables, including geometrical measurements of the Universe, the power spectrum of fluctuations, and the light element abundances arising from primordial nucleosynthesis. In the simplest model, the Dark Energy is simply Einstein's cosmological constant Λ.

2.1. Probes of geometry: supernovae

The Cosmological Model can be tested by probing the distance-redshift relation $r(z)$. Type 1a supernova have roughly the same intrinsic brightness and can therefore be used as standard candles to measure the luminosity distance, $D_L \equiv (1+z)r(z) = c(1+z)\int_0^z dz'/H(z')$.

Geometrical measures like this essentially constrain the Hubble parameter vs redshift, which depends on several of the cosmological parameters, including Ω_m, Ω_{DE} and w.

Use of the luminosity distance requires objects of known brightness, and Type 1a supernovae are good standardiseable candles for this purpose. They are thought to arise from the collapse of a stellar core which for reasons of mass transfer reaches the Chandrasekhar limit. In principle one might expect these to be identical, but in practice there is some variation in the peak luminosity of the supernovae, leading to a scatter which is half a magnitude or more. The variations are, however, correlated with the colour and the decay lifetime of the lightcurve, and empirical corrections can be made which reduce the scatter to around 0.13 magnitudes (about 13%; fig. 2). The supernova Hubble diagram (Fig. 3) can then be used to confront theory with observations.

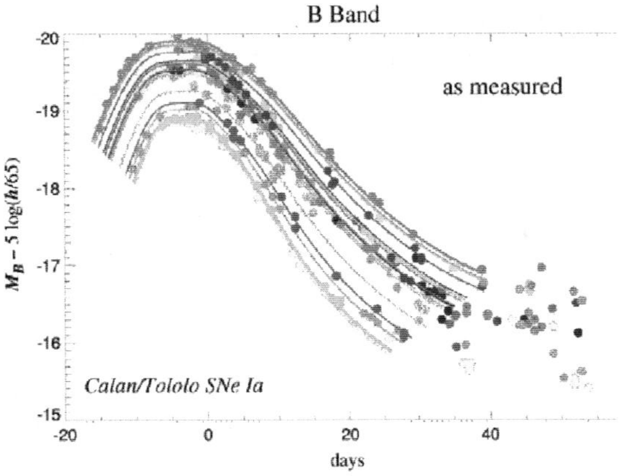

Figure 1. Light curves of Type 1a supernovae before empirical corrections [4].

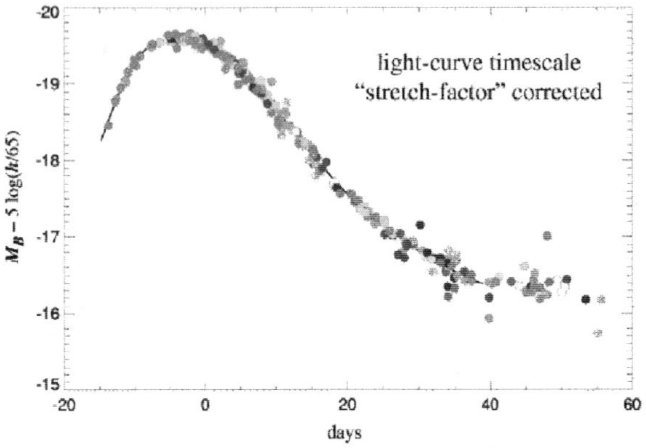

Figure 2. Light curves of Type 1a supernovae after empirical corrections [4].

The most remarkable implication of the diagram is the evidence that the Universe is accelerating, which is not expected in simple models of Big Bang evolution, and is indicative of an effectively repulsive gravity component, which may be provided by the cosmological constant, or Dark Energy with a sufficiently negative pressure. Of particular interest is that the supernovae are now being detected at sufficiently high redshift that they probe an earlier era before the Universe started its current acceleration. Interpreting the data in terms of the standard model with a cosmological constant (or Dark Energy as a genuine vacuum energy with $w = -1$) indicates a positive Dark Energy density. One caveat to bear in mind is that the analysis assumes that type 1a supernovae are standard candles (after correction) at all redshifts, and there is increasing evidence that there are at least two types of progenitors for these supernovae, associated with a young population [7] (perhaps as young as 75 Myr [8]) and an old (probably several Gyr) population. The proportions of the two types of progenitor must change with cosmic time, and it is not known yet whether the corrections apply equally to both types, at least sufficiently accurately to use these as detailed cosmological probes of the Dark Energy equation of state parameter.

2.2. Cosmic Microwave Background observations

The cosmic microwave background (CMB) gives us the most robust cosmological constraints, due to a combination of a very firm theoretical footing, and high-quality data. Theoretical understanding is very sound because the principal processes are interactions between photons and matter which formed an almost uniform density fluid close to thermal equilibrium. Linear perturbation theory describes the Universe at the time of emission to high accuracy, and there

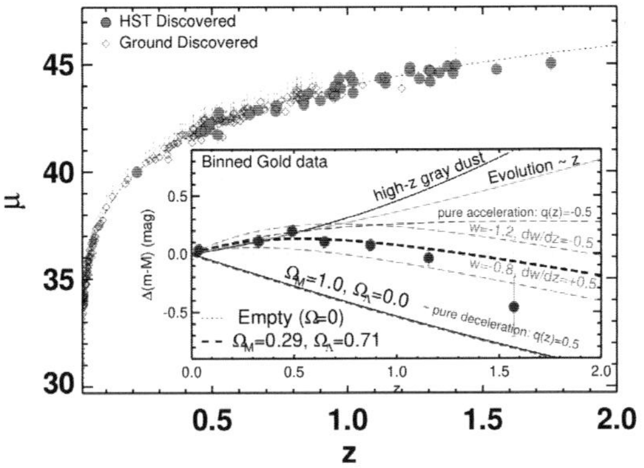

Figure 3. The supernova Hubble diagram [5].

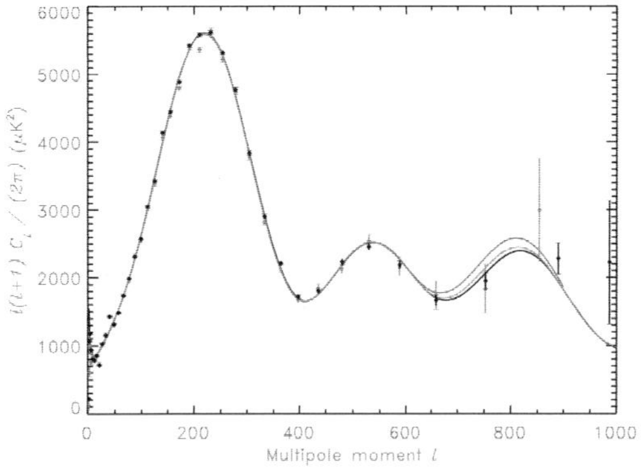

Figure 4. The WMAP angular power spectrum [6].

were no astrophysical objects to confuse the issue. The main uncertainties are foreground effects, such as dust and synchrotron radiation from the Milky Way galaxy and other sources. With observations at many frequencies these foregrounds seem able to be subtracted to high accuracy over much of the sky. The WMAP satellite currently provides by far the most powerful dataset.

Although the analysis is not quite done this way, the confrontation with theory is most easily illustrated by comparing the angular power spectrum of CMB fluctuations with theory, as shown in Fig. 4. The agreement is quite remarkable. Perturbations which enter the Hubble radius before recombination will soon oscillate as the sound speed is roughly $c/\sqrt{3}$ which means the Jeans length is comparable to the Hubble radius. The first peak corresponds to the smallest waves which don't oscillate; the second peak to those waves which enter before recombination, compress and then reach maximum expansion at recombination. Subsequent peaks have similar interpretations. The features in the power spectrum are dependent on a number of factors, but the first peak corresponds roughly to a pre-recombination Jeans' length fluctuation, of a more-or-less fixed physical size. It acts like a standard ruler, so its angular size (or wavenumber ℓ) measures the geometry of the Universe - flat or curved. For a flat Universe, the peak should be at $\ell \simeq 200$, as observed.

2.3. Large-scale structure

The power spectrum of matter fluctuations emerging after recombination can also be computed with high confidence. Following the growth of these fluctuations, the power spectrum can be compared with structures today. This involves two complications. One is that the fluctuations do not remain small, so numerical simulations are required to evolve the matter power spectrum. This is relatively well understood. Potentially more difficult is that we do not observe matter directly in large-scale structure, but luminous baryonic matter in the form of galaxies. The connection between the galaxy distribution and the matter distribution is often referred to as *bias* and represents a significant source of systematic uncertainty, especially on small scales. The power spectrum of galaxies from the 2dF galaxy redshift survey is shown in Fig. 5, along with theoretical curves with cosmological parameters which fit the WMAP CMB Data. The agreement is reassuring. Details in the power spectrum, such as baryon acoustic oscillations

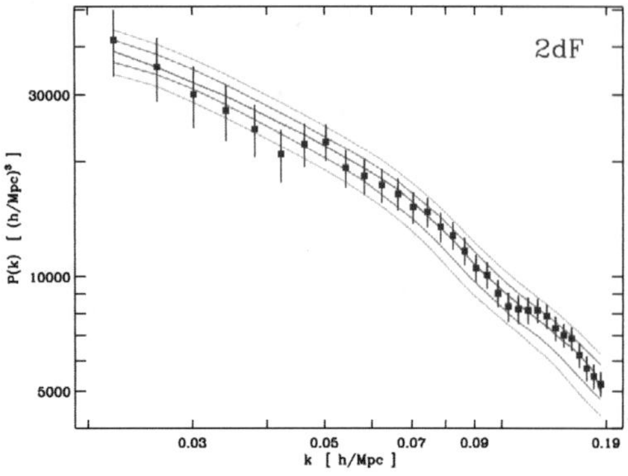

Figure 5. Power spectrum of 2dF galaxies from [9], along with expected power spectrum with best fit parameters from WMAP, with errors [6].

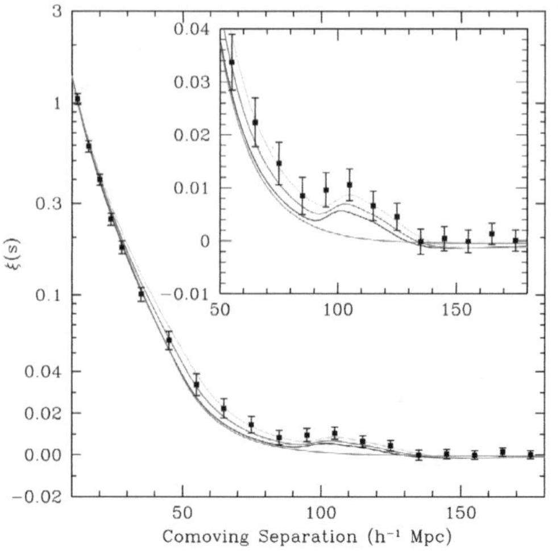

Figure 6. BAOs in SDSS [10].

(BAOs), remnants of the oscillations present in the CMB, may be used as probes of geometry, as they act as more-or-less standard rulers. BAOs have been detected in both the 2dFGRS and the SDSS (Fig. 6). The main uncertainty is bias, and how the positions of the peaks may shift when the field becomes nonlinear.

2.4. Weak gravitational lensing

Weak gravitational lensing avoids the problems with bias by probing the matter density directly. Fluctuations in the matter distribution perturb the paths of light from background galaxies, whose images get distorted. The very appealing aspect of lensing is that the physics is simple and well-understood. The distortions are, however, rather small ($\sim 1\%$) and control of systematics in optics is challenging. Results of cosmological weak lensing measurements have been reported since 2000, and provide complementary constraints on cosmological parameters. Latest results [11] have reduced the previous tension with the CMB, and are now in agreement with WMAP. The main change is a better estimate of the redshift distribution of the lensed sources. The clean physics involved makes this a very promising probe for future surveys, with the main physical source of systematic error being a possible correlation between orientations of foreground galaxies and distorted background images [12].

2.5. Lyman α forest and neutrino masses

Absorption of light from bright distant quasars by systems containing neutral hydrogen gas can be used to probe the clustering of matter on smaller scales than can be probed by galaxy surveys. It is also possible (indeed necessary) to do this at high redshift, which has the advantage of investigating the evolution of clustering, and it also probes the Universe at a time when it was less clustered, which reduces uncertainties. On the downside, the relationship between the matter density and the characteristics of the absorption rely on rather complex physics, and it is fair to say that the connection between observables and theory is harder for this probe than the others considered here. In principle, the Lyα forest can put strong constraints on the sum of the neutrino masses, as non-zero masses alter the shape of the power spectrum at large wavenumbers. Currently the strongest limit claimed is $\sum m_\nu < 0.17$ eV [13], using Lyα forest clustering statistics. The most conservative limit restricts analysis to the well-understood

CMB, which puts a much weaker limit of 1.8 eV [6]. This limit is secure; a middle course is to include galaxy clustering statistics, which are independent of the complex Lyα physics, but which could be affected by scale-dependent bias. Combining WMAP and large-scale structure from 2dF/SDSS gives an upper limit of $\sum m_\nu = 0.66eV$ [6].

3. Problems with the Standard Cosmological Model

The main challenges to the standard model come from small scales. Numerical simulations show that many small-mass haloes are expected, but far fewer satellite galaxies of the Milky Way have been found. SDSS has recently found more low-mass galaxies [14], but there is still a deficit. One possibility is that the comparison has not been done fairly - the full halo circular velocity may not be achieved until beyond the visible outer radii of the dwarf galaxies [15]. However, in the end it is likely that feedback processes from star formation hold the key - the potential wells are very shallow, and observed dwarf galaxies have very low baryon fractions, so probably there is a population of even fainter dwarfs with few stars. The other point of tension is the inner profile of the Dark-Matter dominated dwarfs, which are shallower than theory predicts. This is to a certain extent an open issue, but there are possible resolutions with bars or triaxial haloes. Self-interacting Dark Matter can remove the theoretical cusps [16], but the cross-section required may already be almost ruled out (see next section).

4. Constraints on inflation, Dark Matter, and modified Gravity

There are consistency relations which inflation imposes on the tensor-to-scalar ratio and the slope of the power spectrum. Space constraints here preclude a full discussion, but the CMB and large-scale structure data are beginning to constrain inflationary potentials. Constraints on the number of coupled neutrinos can be obtained from their influence on the growth of perturbations, with current data favouring none [17].

A very interesting object which puts a number of interesting constraints is the bullet cluster, which shows two clusters post-collision [18]. X-ray emission shows that the hot gas is left between the two concentrations of galaxies, which have passed through each other with little direct interaction. Most interestingly, gravitational lensing analysis provides a 'convergence' map with peaks at the positions of the galaxy centres, not the (dominant baryonic) x-ray emitting gas. In General Relativity, the convergence is proportional to the surface density, so the conventional interpretation is that there is a dominant Dark Matter component which is more-or-less collisionless. This puts severe pressure on gravity models such as MOND and TeVeS, which explore whether one can do away with Dark Matter. One caveat here is that in such models the convergence is not simply proportional to the surface density, but it will be a challenge for these models to account for the convergence peaks at the galaxy concentrations. Within the standard model, the separation of the Dark Matter concentrations puts constraints on the Dark Matter cross-section of $\sigma/m < 0.12\,\mathrm{m}^2\,\mathrm{kg}^{-1}$ [19].

5. Prospects for the Future

We can expect major improvements in data on several fronts. In the CMB, Planck promises measurements of the power spectrum to smaller scales than WMAP, as well as wider frequency coverage which should aid foreground subtraction. CMB polarisation experiments, of which many are underway or planned, can in principle detect gravitational waves from inflation, which would be a significant validation of the theory. In weak lensing, the Pan-STARRS 1 survey of up to 3/4 of the sky promises a huge leap in statistical accuracy over the $\sim$ 100 square degrees of the largest optical surveys to date. Further ahead, more powerful photometric surveys are envisaged from space or the ground (DUNE, SNAP, LSST, Pan-STARRS 4). Many galaxy surveys motivated by BAOs are planned, both with spectroscopic redshifts (WFMOS, MegaMOS, ADEPT etc) and with photometric redshifts, some using a very large number of filters (PAU). Many of the multi-band photometry surveys would also obtain supernova samples in large numbers out to high redshifts ($z \sim 1.7$).

What can we expect to know in 2020? Dark Energy constraints should be very tight, with weak lensing in combination with Planck potentially achieving sub-1% accuracy on w, and useful constraints on its evolution. Currently w is known to $w = -1.04 \pm 0.06$, assuming that it does not vary [13]. Future errors will be dominated by systematics, and the requirements are demanding [20]. BAOs and supernovae on their own are somewhat less powerful, but the systematics in each technique are sufficiently complex that it is vitally important to pursue all viable methods and seek independent verification.

Neutrino masses and properties could potentially be measured with high accuracy, with claimed achievable limits on the sum of 0.06eV from Lyα forest [21], 0.05eV from weak lensing [22], and 0.025eV from high-z clustering [23]. Here, the physics is complex in all methods (including weak lensing, because of baryonic physics on small scales), so one would demand consistent measurement from multiple probes.

One intriguing possibility is to test General Relativity itself. Methods which probe both the geometry and the growth rate, which both depend on $H(z)$ in General Relativity can look for an inconsistency. Large weak lensing surveys from space, of the size and depth proposed for DUNE and SNAP, could theoretically find strong evidence for gravity theories arising from extra dimensions, such as proposed by string-based braneworlds [24].

6. References

[1] Yao W-M et al 2006 *J. Phys.* G **33** 1

[2] Efstathiou G P, Sutherland W and Maddox S 1990 *Nature* **348** 705

[3] Riess A G et al 1998 *AJ* **116** 1009; Perlmutter S et al 1999 *ApJ* **517** 565

[4] Garcia-Bellido J 2004 *Proceedings of XXXII International Meeting on Fundamental Physics, Alicante, 2004 hep-ph* 0407111

[5] Riess A G et al 2007 *ApJ* **659** 98

[6] Spergel D N et al 2007 *ApJL* **170** 377

[7] Sullivan M et al 2006 *ApJ* **648** 868

[8] Aubourg E, Tojeiro R, Jimenez R, Heavens A F, Strauss M A and Spergel D N 2007 *astro-ph* 0707.1328

[9] Cole S et al 2005 *MNRAS* **362** 505

[10] Eisenstein D J et al 2005 *ApJ* **633** 560; Percival W J, Cole S, Eisenstein D J, Nichol R C, Peacock J A, Pope A C and Szalay A S 2007 *MNRAS* **381** 1053

[11] Benjamin J, Heymans C, Semboloni E, van Waerbeke L, Hoekstra H, Erben T, Gladders M D, Hetterscheidt M, Mellier Y and Yee H K C 2007 *MNRAS* **381** 702

[12] Hirata C and Seljak U 2004 *Phys. Rev.* D**70** 063526

[13] Seljak U, Slosar A and McDonald P 2006 *JCAP* **10** 014

[14] Belokurov et al 2007 *ApJ* **654** 897

[15] Stoehr F, White S D M, Tormen G and Springel V 2002 *MNRAS* **335** L84; Kazantzidis S, Mayer L, Mastropietro C, Diemand J, Stadel J and Moore B 2004 *ApJ* **608** 663

[16] Spergel D and Steinhardt P 2000 *Phys. Rev. L* **84** 3760

[17] Friedland A, Zurek K M and Bashinsky S 2007 *astro-ph* 0704.3271

[18] Markevitch M, Gonzalez A H, Vikhlinin A, Murray S, Forman W, Jones C and Tucker W 2002 *ApJ* **567** 27; Clowe D, Gonzalez A and Markevitch M 2004 *ApJ* **604** 596

[19] Randall S W, Markevitch M, Clowe D, Gonzalez A H and Bradac M 2007 *astro-ph* 0704.0261

[20] Amara A and Refregier A 2007 *astroph* 0710.5171; Kitching T D, Taylor A N and Heavens A F, 2007 in preparation

[21] Gratton S, Lewis A and Efstathiou G P 2007 *astro-ph* 0705.3100

[22] Hannestad S 2006 *Prog. Part. Nucl. Phys.* **57** 309

[23] Takada M, Komatsu E and Futamase T 2007 *Phys. Rev.* D**73** 3520

[24] Ishak M, Upadhye A and Spergel D 2006 *Phys. Rev.* D**74** 043513; Heavens A F, Kitching T D and Verde L 2007 *MNRAS* **380** 1029; Amendola L, Kunz M and Sapone D 2007 *astro-ph* 0704.2421

The impact of non-linear evolution of cosmological matter power spectrum on the measurement of neutrino masses

Shun Saito[1], Masahiro Takada[2] and Atsushi Taruya[3]

[1] Department of Physics, The University of Tokyo, Tokyo 113-0033, Japan
[2] Astronomical Institute, Tohoku University, Sendai 980-8578, Japan
[3] Research Center for the Early Universe, School of Science, The University of Tokyo, Tokyo 113-0033, Japan

E-mail: ssaito@utap.phys.s.u-tokyo.ac.jp

Abstract. Next-generation galaxy redshift surveys will open up an exciting opportunity for precision determinations of neutrino masses. Here, we discuss the prospects for measuring the neutrino masses, including properly the non-linear gravitational evolution of matter power spectrum for a mixed dark matter model (neutrinos plus cold dark matter). Based on the perturbation theory, the effect of non-linearity is quantified. Moreover, using Fisher matrix analysis, we present how the neutrino masses will be determined for future galaxy redshift survey.

1. Introduction

One of the most important recent discoveries in physics is that neutrinos have non-zero masses, since neutrinos are massless in the Standard Model. Massive neutrinos necessarily involve new physics, and we need to have improved knowledge of neutrino masses. Intriguingly, the total mass of neutrinos can be determined by the cosmological data through two effects. One is the effect on background expansion of the universe, and another is suppression of matter perturbation. The second effect comes from the fact that the neutrino perturbations affect the structure formation by suppressing the growth of matter density fluctuations at smaller scales below the free-streaming scale owing to their large velocity dispersion. In fact, the most stringent upper bound on the total neutrino mass is obtained from cosmology as $\sum m_\nu \lesssim 0.6\mathrm{eV}(95\%\mathrm{CL})$ [1], compared to the results of the terrestrial experiment limit $\sum m_\nu \lesssim 2\mathrm{eV}$. It is interesting to note that the neutrino's free-streaming scale is comparable to the scale of baryon acoustic oscillation (BAO), $\approx 100\mathrm{Mpc}$, which is aimed by the future wide-field galaxy redshift surveys such as Wide-Field Fiber-fed Multi-Object Spectrograph (WFMOS) in order to reveal the nature of dark energy [2]. The observation around BAOs or neutrino's free-steaming scale requires accurate theoretical template of the matter power spectrum including the non-linear gravitational evolution.

Here, we investigate the non-linear evolution of matter power spectrum including the effect of massive neutrinos for the first time [3]. Based on the fluid approximation of mixed dark matter model (baryon, cold dark matter, and neutrinos) and the perturbation theory, we derive the next-to-leading order correction of gravitational non-linear evolution. Moreover, we demonstrate how

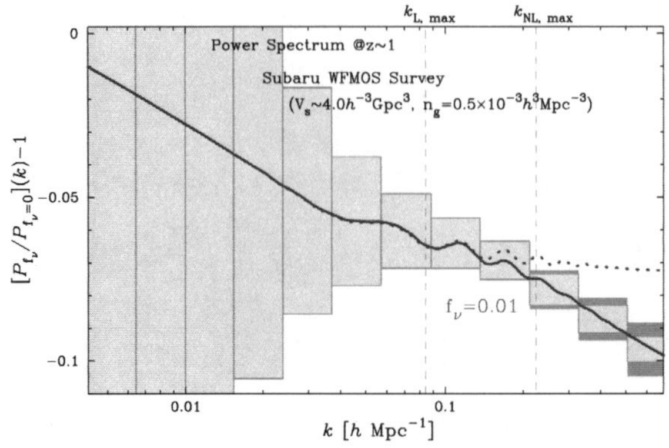

Figure 1. The ratio of matter power spectrum to one without massive neutrinos for the fiducial model. The cases for linear theory, $P^{\rm L}_{f_\nu\neq 0}/P^{\rm L}_{f_\nu=0}$, ( ) and for the non-linear theory, $P^{\rm NL}_{f_\nu\neq 0}/P^{\rm NL}_{f_\nu=0}$, ( ) are plotted. The lines represent the validity of linear theory, $k_{\rm Lmax} = 0.084h{\rm Mpc}^{-1}$ and our non-linear theory, $k_{\rm NLmax} = 0.213h{\rm Mpc}^{-1}$, respectively.

neutrino masses are constrained by future galaxy redshift survey based on Fisher information matrix formalism.

2. Perturbative Approach

Here we treat the evolution of mass distribution for a mixed dark matter model which consists of the cold dark matter (CDM), the baryon, and the neutrinos. One may write the fluctuation of mass density as $\delta_{\rm m} \equiv (\delta\rho_{\rm cdm} + \delta\rho_{\rm b} + \delta\rho_\nu)/(\rho_{\rm m}) = f_{\rm cb}\delta_{\rm cb} + f_\nu\delta_\nu$, where the coefficients, $f_{\rm cb}$ and f_ν, are the fraction of each matter component: $f_{\rm cb} = (\Omega_{\rm cdm} + \Omega_{\rm b})/\Omega_{\rm m}$ and $f_\nu = \Omega_\nu/\Omega_{\rm m}$. The matter power spectrum is evaluated as $P_{\rm m}(k) = f_{\rm cb}^2 P_{\rm cb}(k) + 2 f_{\rm cb} f_\nu P_{{\rm cb},\nu}(k) + f_\nu^2 P_\nu(k)$,. Since the tiny fraction of massive neutrinos gives a negligible contribution to the non-linear growth and the non-linear gravitational potential contributed from CDM plus baryon does not affect the neutrino's density contrast, it would be sufficient to include the non-linear correction to the term of CDM plus baryon, $P_{\rm cb}(k)$, only.

For our interest on large-scales, the CDM plus baryon component can be regarded as a pressure-less fluid. From standard perturbation theory (PT), next-to-leading order corrections for the solutions of fluid equations called one-loop corrections can be calculated in a similar way to the cases without massive neutrinos [4]. Note that the effect of neutrino's free-streaming is encoded entirely in $P^L_{\rm cb}(k;z) = D_{\rm cb}(k;z) P^L_{\rm cb}(k;z_{\rm d})$ through the scale-dependent linear growth factor, $D_{\rm cb}(k;z)$.

In Fig.1, our PT results for matter power spectra for $f_\nu = 0.01$ with fiducial parameters. We plot the ratio of power spectrum with neutrinos. $f_\nu = 0.01$, to one without neutrinos, $f_\nu = 0$, for linear ( ) and PT ( ) at $z = 1$. Here, we adopt the following cosmological parameters as fiducial model parameters: $\Omega_{\rm b}h^2 = 0.0223$, $\Omega_m h^2 = 0.1277$, $h = 0.73$, $w = -1$, $\Delta^2_{\mathcal{R}} = 2.35 \times 10^{-9}$, $n_s = 1$, no running spectral index and $N_\nu = 3$. As one can see from Fig.1, the suppression of non-linear power spectra becomes significant due to the non-linear gravitational evolution. Notice that PT may invalid at some large-k. We define the upper limit of validity of our prediction based on the N-body results without neutrinos [5]; $\Delta^2(k_{\rm NLmax}; z) \equiv k^3 P_{\rm m}^{f_\nu=0}(k_{\rm NLmax}; z)/2\pi^2 = 0.4$. Further, applicability of linear theory is restricted to the range within the 1% relative difference between linear and PT results. The dashed vertical lines in Fig.1 indicate these limits, $k_{\rm Lmax}$ and $k_{\rm NLmax}$. The suppression of non-linear matter power spectrum is noticeable compared to the linear theory even at $k \lesssim k_{\rm NLmax}$.

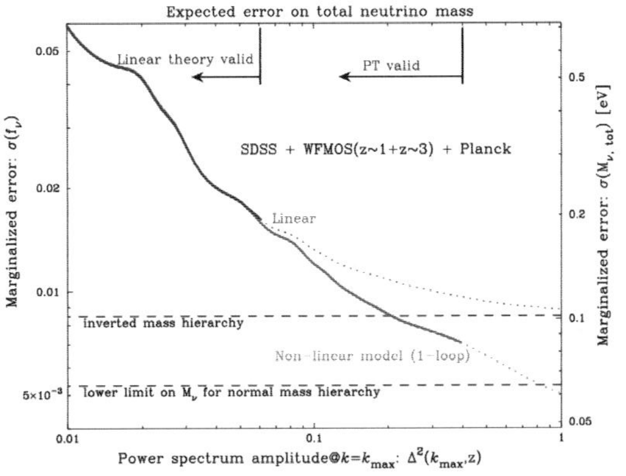

Figure 2. Expected 1σ error on total neutrino mass obtained from linear and non-linear theory for future galaxy redshift survey.

In Fig.1 we also plot the statistical error, $\Delta P(k)/P(k)$, expected from WFMOS survey, for linear (box) and PT results (box). For PT prediction, the statistical error becomes relatively smaller than linear theory due to the reduced shot-noise term. The relative enhancement of the suppression and reduction of error shown in Fig.1 lead to the improvement of the neutrino mass constraint as discussed below.

3. Parameter forecast with Fisher matrix analysis

Finally, we investigate how the difference of suppression between linear and non-linear theory affect the determination of neutrino masses for future galaxy redshift survey with Fisher matrix analysis developed in [6]. Here we assume the linear biasing and linear redshift distortion, but include non-linearity of gravitational clustering empirically by setting the parameter β as free parameter. The survey parameters can be found in [6]. Then, we consider a set of parameters, $p_\alpha = \{\Omega_{\mathrm{m}}, \Omega_{\mathrm{m}}h^2, \Omega_{\mathrm{b}}h^2, \Delta^2_{\mathcal{R}}, n_s, \alpha, w_0, f_\nu, N_\nu, b_1(z_i), \beta\}$. Here, we adopt the fiducial model discussed above with $f_\nu = 0.01$ and fix the parameters, $N_\nu = 3$ and $b_1(z_i)$ for each redshift slice. As for the maximum wavenumber, we adopt the value of k_{max} so that $\Delta^2(k_{\mathrm{max}}, z_i) = x$, and obtain the result against the value of x. Moreover, we assume Planck prior for CMB data.

In Fig.2, we show the 1σ error on total neutrino mass, f_ν, or equivalently $\sum m_\nu$ for reference. The   and   lines show the expected 1σ errors of neutrino masses. We adopt the validity range for both theories conservatively; for non-linear theory, $\Delta^2(k_{\mathrm{NLmax}}, z_i) \leq 0.4$, and for linear theory, 3% accuracy between linear and non-linear theory. The specific values on constraint on neutrino masses are 0.195eV for linear theory and 0.085eV for non-linear theory, which is about factor of two improvement. This result is remarkable, since it suggest that improvement from non-linear theory may lead to potential to distinguish the difference of mass hierarchy. Note that our result may be optimistic because we ignore several non-linear effects. Nevertheless, we stress that, since the effect of neutrino on matter power spectrum gives characteristic suppression, constraint derived here would be achievable.

References

[1] D. N. Spergel *et al.*, Astrophys. J. Suppl. **170**, 377 (2007),
[2] J. A. Peacock, *et al.*, arXiv:astro-ph/0610906. A. Albrecht *et al.*, arXiv:astro-ph/0609591.
[3] S. Saito, M. Takada and A. Taruya, in prep (2007).
[4] N. Makino, M. Sasaki, and Y. Suto, Phys. Rev. D **46**, 585 (1992).
[5] D. Jeong and E. Komatsu, Astrophys. J. **651**, 619 (2006).
[6] M. Takada, E. Komatsu and T. Futamase, Phys. Rev. D **73**, 083520 (2006).

A Subaru galaxy redshift survey: WFMOS Survey

Masahiro Takada

Astronomical Institute, Graduate School of Science, Tohoku University, Sendai, 980-8578, Japan

E-mail: takada@astr.tohoku.ac.jp

Abstract. A planned galaxy redshift survey with the Subaru 8.2m telescope, the WFMOS survey, offers a unique opportunity for probing detailed properties of large-scale structure formation in the expanding universe by measuring clustering strength of galaxy distribution as a function of distance scale and redshift. In particular, the precise measurement of the galaxy power spectrum, combined with the cosmic microwave background experiments, allows us to obtain stringent constraints on or even *determine* absolute mass scales of the Big-Bang relic neutrinos as the neutrinos imprint characteristic scale- and redshift-dependent modifications onto the galaxy power spectrum shape. Here we describe the basic concept of how the galaxy clustering measurement can be used to explore the neutrino masses, with particular emphasis on advantages of the WFMOS survey over the existing low-redshift surveys such as SDSS.

The neutrino oscillation experiments such as Kamiokande, SNO and KamLAND have given strong evidence that the standard three-flavor neutrinos finite masses. These experiments are, however, only sensitive to mass square differences between neutrino mass eigenstates, and the most fundamental quantity of neutrinos, the absolute mass scale, has not been determined yet.

On the other hand, exploring large-scale structure formation in the Universe can provide a powerful, albeit indirect, means to constraining the neutrino masses. The current concordance scenario for structure formation is given by the cold dark matter (CDM) dominated model, where the hierarchical structures seen today have arises as a result of the gravitational instability and amplification of the primordial seed fluctuations. The Big-Bang relic neutrinos having finite masses lead to intriguing influence on structure formation processes. The neutrinos behaved like radiation until the temperature of the Universe cooled down to the neutrino mass scale, afterward became non-relativistic and have given contribution to the total matter density by a fraction of $f_\nu \equiv \Omega_{\nu 0}/\Omega_{\rm m0} \gtrsim 0.5\%$ [1]. However, the neutrino mass is very likely to be much lighter than the CDM mass, and therefore have much larger thermal velocity dispersion than that of CDM particles. The consequence for structure formation is that the massive neutrinos can not be trapped by the gravitational potential well mainly due to the CDM density fluctuations, if the length scale of the potential well is smaller than the neutrino free-streaming scale, the distance that the neutrino can travel within a cosmic expansion rate. The neutrino free-streaming scale is of the order of $\sim$ 100Mpc at redshifts relevant for a galaxy survey or more explicitly the corresponding wavenumber $k_{\nu,\rm fs}(z)$ is given in [1] as

$$k_{\nu,\rm fs}(z) \simeq 0.017 \left(\frac{1+z}{2}\right)^{-1/2} \left(\frac{m_\nu}{0.1\ \rm eV}\right) \left(\frac{\Omega_{\rm m0}h^2}{0.127}\right)^{1/2} \rm Mpc^{-1}. \qquad (1)$$

[1] Note that f_ν is defined by the total neutrino mass, rather than the mass of each mass eigenstate, as $f_\nu \propto \sum_i m_{\nu(i)}/\Omega_{\rm m0}$.

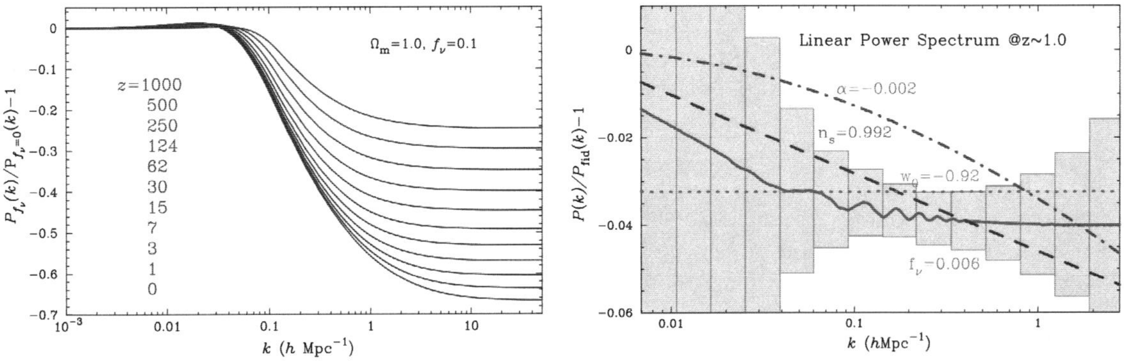

Figure 1. : Shown is how an inclusion of the massive neutrino contribution affects the power spectrum shape, compared to that for a pure CDM model, as a function of wavenumber and redshift. : The shaded boxes show the expected 1-σ errors in measuring the band powers of galaxy power spectrum at each wavenumber bin for the WFMOS survey probing the redshift range $0.5 \lesssim z \lesssim 1.3$. The solid curve show the neutrino effect for a more realistic case of the neutrino mass (~ 0.07 eV). For comparison, the dashed, dot-dashed and dotted curves show how changing each of the tilt (n_s) and running (α_s) of the primordial power spectrum and the dark energy equation of state parameter (w_0) from the fiducial values, with other parameters being fixed, affects the power spectrum shape, respectively.

As a result, the presence of the massive neutrinos suppresses the growth of structure formation on small scales, compared to a pure CDM model with the same matter density $\Omega_{\rm m0}$. More precisely, the left panel of Fig. 1 shows how an inclusion of the massive neutrino contribution ($f_\nu = 0.1$) modifies the shape of linear power spectrum $P(k)$, relative to the pure CDM model. The massive neutrinos give characteristic influences on $P(k)$ as a function of wavenumber and redshift. First, the fractional suppression in the power spectrum amplitudes is more significant at lower redshifts as the neutrino suppression effect has accumulated since the epoch when the neutrinos became non-relativistic. Second, the growth of power spectrum at smaller scales is more suppressed as the smaller scale fluctuations have been, from earlier on, below the neutrino free-streaming scale (see Eq.(1)). The amount of the suppression at sufficiently small scales and at low redshifts relevant for a galaxy survey is roughly expressed as $\Delta P(k)/P(k) \sim 10 f_\nu$, a factor 10 boost compared to the contribution to the matter density $\Omega_{\rm m0}$. These results demonstrate that measuring the power spectrum from a galaxy survey provides a powerful way of constraining the neutrino mass parameter f_ν. In fact, the SDSS galaxy power spectrum, combined with WMAP, has been used to obtain an upper bound on the total neutrino mass, $m_{\nu,\rm tot} \lesssim 0.5$ eV at 95%C.L.[3], more stringent than that from the direct experiment, ~ 1 eV.

Thus a large-area galaxy redshift survey probing the Universe at redshifts allows an even more powerful opportunity for constraining the neutrino mass, by measuring the clustering strength of galaxy distribution in the -dimensional space (redshift plus two-dimensional angular positions for each galaxy). For a galaxy survey, only the clustering information at large scales greater than a few 10Mpc is useful to extract clean cosmological information, because the galaxy clustering at the smaller scales is polluted by poorly understood baryonic physics, referred as to the galaxy biasing uncertainty. An ambitious galaxy redshift survey has been proposed with the next-generation instrument for the 8.2m Subaru telescope, a new wide-field ($\sim 1^\circ\!.5$-diameter field-of-view), highly-multiplexed (~ 3000 fibers) multi-object spectrograph (WFMOS) [2]. Fully utilizing the unique capability of the Subaru telescope and WFMOS, the baseline surveys probing two slices of the Universe at $0.5 \lesssim z \lesssim 1.5$ and $2.5 \lesssim z \lesssim 3.5$ with areas of 2000 and 300 deg^2, respectively, has been proposed.

Such high-redshift WFMOS surveys are particularly useful for the precise measurements of galaxy power spectrum. First, given a fixed solid angle, the comoving volume in which we can observe galaxies is larger at higher redshifts than in the local universe. Second, since large-scale structures are not yet strongly evolving in higher redshifts, more Fourier modes of the density fluctuations staying in the linear or weakly non-linear regimes can be used in extracting cosmological information from the measured galaxy power spectrum, avoiding uncertainties due to the poorly understood non-linear effects (non-linear clustering, baryon physics, redshift distortion etc.). Assuming that the density fluctuations considered obey Gaussian statistics, the statistical precision of measuring the band powers of galaxy power spectrum at a given wavenumber bin k_i with width $\Delta(\ln k)$ is given by

$$\left[\frac{\sigma(P_g)}{P_g(k_i)}\right]^2 \simeq \frac{2\cdot(2\pi)^3}{4\pi k_i^3\cdot\Delta(\ln k)\cdot V_s}\left[1+\frac{1}{\bar{n}_g P_g(k_i)}\right]^2, \tag{2}$$

where V_s is the surveyed comoving volume and $\bar{n}_g$ is the average number density of galaxies. As stated above, a high-redshift survey allows the use of galaxy power spectrum down to a larger maximum wavenumber, $k_{\rm max}$, to extract the cosmological information due to the weaker non-linearities. The equation above implies that a factor 2 gain in $k_{\rm max}$, for fixed V_s and $\Delta\ln k$, is equivalent to a factor 8 increase in the surveyed volume. In fact, the planned WFMOS survey is designed to achieve much greater precisions of $\sigma(P)/P$ than SDSS by a factor of 10.

In the right panel of Fig. 1, the shaded boxes (around the solid curve) represent the expected 1-σ measurement error on $P(k)$ at each k-bin for the WFMOS $z\sim 1$ survey, showing that the power spectrum amplitude around $k\sim 0.1{\rm Mpc}^{-1}$ can be measured with $\sim 1\%$ accuracy. The error bars can be compared with the solid curve that demonstrates the fractional suppression in the linear power spectrum as in the left panel, for a more realistic case of the total neutrino mass, $f_\nu = 0.006$ or equivalently $m_{\nu,{\rm tot}}\simeq 0.07{\rm eV}$, similar to the lower bound implied from the normal mass hierarchy. Clearly the small-scale suppression seems to be detected with high significance. However, this is an optimistic guess, because a combination of other cosmological parameters could mimic the neutrino mass dependence of $P(k)$, within the framework of CDM dominated structure formation model. The dashed, dot-dashed and dotted curves show how changing each of the tilt (n_s) and running (α_s) of the primordial power spectrum and the dark energy equation of state (w_0) from the fiducial values, with other parameters being fixed, affects the galaxy power spectrum shape, where the parameters considered may be considered to be most degenerate with the neutrino mass effect on $P(k)$. Apparently the different parameters affect the power spectrum shape in modestly different ways. In particular, the galaxy power spectrum measurement combined with the CMB constraints turns out to be a very powerful, efficient way of lifting the parameter degeneracies [1]. However, to realize the genuine power of the WFMOS survey for constraining the neutrino mass, a more careful study is needed. In particular, it is of great interest to address whether the precision of $\sigma(m_{\nu,{\rm tot}})\sim 0.05$ eV, needed to discriminate between the normal and inverted mass hierarchies, can be attained with WFMOS survey, carefully taking into account various systematic issues. These issues will be presented elsewhere [4].

References

[1] Takada, M., Komatsu, E., and Futamase, T., 2006 *Phys. Rev.* D **73** 083520
[2] Grazebrook, K. et al., 2005, astro-ph/0507457
[3] Tegmark, M. et al. The SDSS collaboration, 2006 *Phys. Rev.* D **74** 123507
[4] Saito, S., Takada, M. and Taruya, A., 2007 in preparation

Neutrino mass constraint from CMB and its degeneracy with other cosmological parameters

Kazuhide Ichikawa

Institute for Cosmic Ray Research, University of Tokyo, Kashiwa 277-8582, Japan

E-mail: kazuhide@icrr.u-tokyo.ac.jp

Abstract. We show that the cosmic microwave background (CMB) data of WMAP can give subelectronvolt limit on the neutrino mass: $m_\nu < 0.63\,\mathrm{eV}$ (95% CL). We also investigate its degeneracy with other cosmological parameters. In particular, we show the Hubble constant derived from the WMAP data decreases considerably when the neutrino mass is a few times 0.1 eV.

1. Introduction

It is known that cosmological considerations give more stringent constraints on the neutrino mass than the present tritium β-decay experiments. For example, [1] obtained $m_\nu < 0.21\,\mathrm{eV}$ and [2] obtained $m_\nu < 0.58\,\mathrm{eV}$, using the WMAP 1st year data (WMAP1) combined with the galaxy power spectrum. Although these limits are tighter than those of ground experiments, they also show some dependence on how we use the galaxy data. The former used the data from the 2dFGRS (and the bias information) and the latter from the SDSS main sample. Meanwhile, we found $m_\nu < 0.66\,\mathrm{eV}$ from WMAP1 alone as reported in [3]. We note that this is the first to point out CMB data (precisely speaking, WMAP1) alone can give a sub-eV upper bound on the neutrino mass, which is comparable to the limits obtained from the CMB and galaxy clustering data combined.

Before [3] (and some time after that too), there seems to be a lack of consensus about whether the CMB experiment with the WMAP-level precision can derive a sub-eV neutrino mass limit and, in fact, the WMAP1 alone limit reported in [2], $m_\nu < 3.8\,\mathrm{eV}$, which allows 100% HDM was seemingly accepted (incidentally, the WMAP group did not report the WMAP1 alone limit on the neutrino mass). We, on the contrary, have derived the upper limit 0.66 eV as quoted above from the same data by the χ minimization method which is independent from the MCMC method adopted by [2]. Our conclusion is later confirmed by [4–6] ([4] does not report the WMAP1 alone limit in a number but judging from their likelihood figure, it looks less than 1 eV. [5] and [6] gives $m_\nu < 0.70\,\mathrm{eV}$ and $m_\nu < 0.63\,\mathrm{eV}$ respectively).

Below, we first discuss the WMAP alone limit on the neutrino mass comparing results from WMAP1 and the WMAP 3rd year data (WMAP3). Then, we will investigate its degeneracy with other cosmological parameters, especially the Hubble constant.

We here summarize our notations. We derive neutrino mass constraint in the flat ΛCDM model with the power-law adiabatic perturbations. Namely, cosmological parameters we consider are: baryon density ω_b, matter density ω_m, hubble parameter h, reionization optical depth τ, spectral index of primordial spectrum n_s, its amplitude A and massive neutrino density ω_ν. Here,

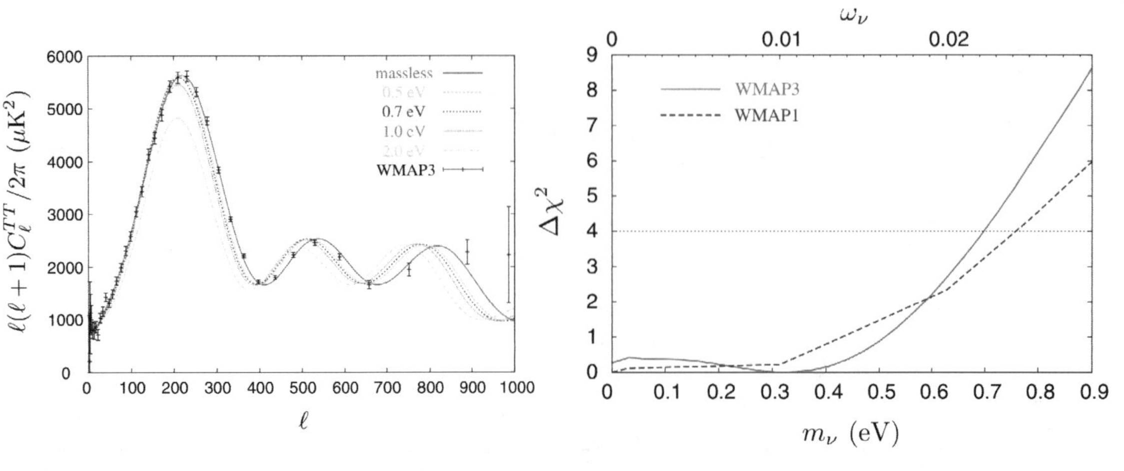

Figure 1. (a) Effects of massive neutrinos on the CMB TT power spectrum. (b) $\Delta\chi$ of WMAP data as functions of the neutrino mass m_ν. $\Delta\chi$ = 4 roughly corresponds to 95% C.L. limit. The blue dashed line uses the WMAP1 (TT+TE) [3] and the red solid line uses the full WMAP3 including temperature and polarization.

$\omega \equiv \Omega h$ where Ω is the energy density normalized to the critical density and $\omega_m \equiv \omega_b + \omega_c$ where ω_c is CDM density (Caution that many literatures define $\omega_m \equiv \omega_b + \omega_c + \omega_\nu$ to include the massive neutrino in the matter density). ω_ν is related to neutrino masses by $\omega_\nu = \sum m_\nu/(94.1\,\mathrm{eV})$ and we assume three mass degenerate generations of neutrinos so that $\omega_\nu = m_\nu/(31.4\,\mathrm{eV})$.

2. WMAP alone limit

We begin with showing how CMB power spectrum is modified by increasing neutrino mass in Fig. 1 (a). The other cosmological parameters are fixed here. In this figure, we see horizontal shift and suppression around the first peak. The horizontal shift comes from the fact that the larger m_ν (more non-relativistic particles at present epoch) implies that the distance to the last scattering surface is shorter and the peaks move to smaller ℓ. However, this shift is easily cancelled by the shift in h. Therefore this does not produce a neutrino mass signal. The suppression of the 1st peak takes place only when $m_\nu \gtrsim 0.6\,\mathrm{eV}$. This corresponds to 0.3 eV in terms of photon temperature T_γ. Meanwhile, the recombination takes place at $z \approx 1088$ or $T_\gamma \approx 0.3\,\mathrm{eV}$. In other words, massive neutrinos become non-relativistic before the epoch of recombination if they are heavier than 0.6 eV. Therefore, only in this case, the neutrino mass can significantly imprint a characteristic signal in acoustic peaks (specifically, the matter-radiation equality occurs earlier due to less relativistic degrees of freedom and the enhancement of the 1st peak by the early-integrated Sachs-Wolfe effect is smaller).

This signal, however, could be accidentally mimicked by some combination of other cosmological parameters. So we searched a large cosmological parameter space in order to find the degree of degeneracy between m_ν and the other cosmological parameters. For each value of m_ν, we varied 6 other ΛCDM cosmological parameters to find minimum χ . The results are shown in Fig. 1 (b) [7]. The WMAP1 result [3] is also shown. We obtained the upper bound of 0.63 eV from WMAP3 alone and notice that the WMAP3 constraint is not improved much from the WMAP1 limit. This is reasonable because the neutrino mass (larger than 0.6 eV) characteristically modifies the acoustic peaks around 1st and 2nd peaks in the temperature

[1] There is small difference from the limit reported in [7] (0.68 eV) because we here report the one obtained with the updated likelihood code (ver. Nov. 2006) by the WMAP team. Our new limit here is consistent with that of published version of the WMAP 3-year paper [8], $m_\nu < 0.60\,\mathrm{eV}$.

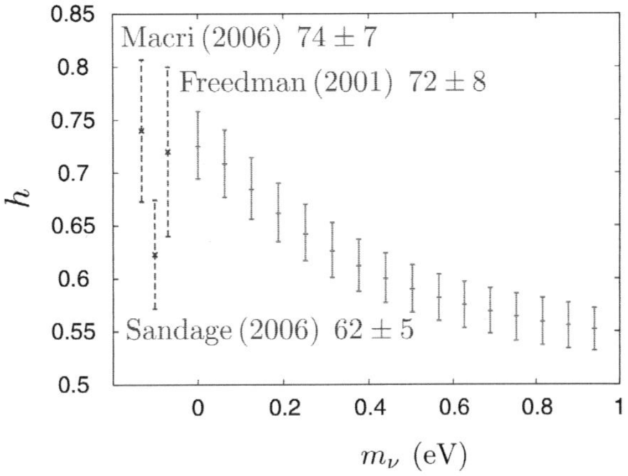

Figure 2. The constraints on h for several fixed values of neutrino mass (red solid bars). The constraints from distance ladder measurements [9–11] are also shown (blue dashed bars).

power spectrum and these regions are already well measured by the WMAP1.

We stress that this bound is robust in a sense that it is obtained from CMB data of the WMAP which is considered to be the cleanest cosmological data and that it is obtained from a single experiment. Also, CMB can be dealt with the linear perturbation theory so it does not suffer from non-linearity or biasing which appear in galaxy clustering data.

3. The degeneracy between the neutrino mass and other cosmological parameters

So far, we have argued that the Hubble constant is degenerate with the neutrino mass in the direction to keep the acoustic peak position constant. Namely, when m_ν is larger, h should be smaller in order to fit the CMB data. We show this explicitly in Fig. 2 for the case with WMAP3. Note that if $m_\nu \gtrsim 0.3\,\mathrm{eV}$, the best fit h is significantly lower than $h = 0.7$, which is usually considered as the WMAP value assuming massless neutrinos. More interestingly, among the measurements of h via the cosmic distance ladder, there is a group [10] deriving somewhat lower h than the famous value of [9]. We can expect both "direct" measurement of h and m_ν will improve in near future: [11] notes that 1% accuracy could be obtained through measuring maser distance to a large number of galaxies in the Hubble flow, and [12] states that $m_\nu = 0.35\,\mathrm{eV}$ could be determined with 5σ by KATRIN experiment. These observations and experiments may find the true cosmological parameters to be different from what we usually consider.

We have here shown the effect on h but other parameters are affected too (see Fig. 2 in [3]). In particular, the negative correlation between n_s and m_ν is notable. I would say that uncertainty of m_ν is one of the largest systematic errors for estimating cosmological parameters from CMB.

4. Conclusion

We have reported the CMB alone constraint on neutrino mass. In addition, we have discussed its degeneracy between other cosmological parameters especially the Hubble constant. One may consider that the Hubble constant is well determined to be around $h = 0.7$ from WMAP but this assumes massless neutrinos. If the neutrino mass is a few times $0.1\,\mathrm{eV}$, the estimated value of h from WMAP would be much lower than $h = 0.7$. Therefore, it is very important to measure the neutrino mass and the Hubble constant independently from CMB. We hope to obtain interesting results from the β-decay experiment and cosmic distance ladder measurement of the Hubble constant in future.

References

[1] D. N. Spergel *et al.* [WMAP Collaboration], Astrophys. J. Suppl. **148**, 175 (2003).
[2] M. Tegmark *et al.* [SDSS Collaboration], Phys. Rev. D **69**, 103501 (2004).
[3] K. Ichikawa, M. Fukugita and M. Kawasaki, Phys. Rev. D **71**, 043001 (2005).
[4] C. J. MacTavish *et al.*, Astrophys. J. **647**, 799 (2006) [arXiv:astro-ph/0507503].
[5] S. Hannestad, Ann. Rev. Nucl. Part. Sci. **56**, 137 (2006).
[6] J. Lesgourgues and S. Pastor, Phys. Rept. **429**, 307 (2006).
[7] M. Fukugita, K. Ichikawa, M. Kawasaki and O. Lahav, Phys. Rev. D **74**, 027302 (2006).
[8] D. N. Spergel *et al.* [WMAP Collaboration], Astrophys. J. Suppl. **170**, 377 (2007).
[9] W. L. Freedman *et al.*, Astrophys. J. **553**, 47 (2001).
[10] A. Sandage *et al.*, Astrophys. J. **653**, 843 (2006).
[11] L. M. Macri, K. Z. Stanek, D. Bersier, L. Greenhill and M. Reid, Astrophys. J. **652**, 1133 (2006).
[12] J. Angrik *et al.* [KATRIN Collaboration], "KATRIN design report 2004"

Capturing Relic Neutrinos with β–decaying Nuclei

Alfredo G. Cocco[1], **Gianpiero Mangano**[1,2], **Marcello Messina**[3]

[1] INFN, Sezione di Napoli, Monte S.Angelo I-80126, Naples, Italy

[2] LAPTH, Université de Savoie et CNRS, F-64941, Annecy-le-vieux Cedex, France

[3] Laboratorium für Hochenergiephysik - Universität Bern, CH-3012, Bern, Switzerland

E-mail: cocco@na.infn.it, mangano@na.infn.it, marcello.messina@cern.ch

Abstract. We summarize a novel approach which has been recently proposed for direct detection of low energy neutrino backgrounds such as the cosmological relic neutrinos, exploiting neutrino/antineutrino capture on nuclei that spontaneously undergo β decay.

1. Introduction

The electron **N**eutrino **C**apture from a nucleus A which spontaneously decays via **B**eta decay to a daughter nucleus B (in the following NCB)

$$\overset{(-)}{\nu}_{\rm e} + A \rightarrow B + e^{\pm}\,, \tag{1}$$

shows the remarkable property that it has no energy threshold on the value of the incoming neutrino energy. In the limit of vanishing value for neutrino mass m_ν and energy the neutrino contributes to (1) uniquely via its lepton flavor quantum number and in this case the electron in the final state has exactly the β decay endpoint energy Q_β. However, for finite m_ν the electron kinetic energy is $Q_\beta + E_\nu \geq Q_\beta + m_\nu$, while electrons emerging from the corresponding β decay has at most an energy $Q_\beta - m_\nu$, neglecting nucleus recoil energy.

The idea of using NCB to measure the cosmological relic neutrino background predicted in the framework of the Hot Big Bang model was already suggested many years ago in [2]. The original idea was that if relic neutrinos have a large chemical potential μ, the electron (positron) energy spectrum for β decays and NCB would get quite a typical signature in a interval of order μ around the zero neutrino mass endpoint Q_β. However, Big Bang Nucleosynthesis constrains relic neutrino–antineutrino asymmetry to the small value $\mu/T_\nu \leq 0.1$, see e.g. [3, 4]. This implies that, unless more exotic scenarios are considered, as a larger amount of relativistic degrees of freedom in the Early Universe, the effect of neutrino degeneracy in β decays and NCB is too small to be detected experimentally. However, for massive neutrinos a gap around Q_β is expected of the order of twice the neutrino mass, which for $m_\nu \sim 1$ eV is several orders of magnitude larger than the corresponding effect due to neutrino-antineutrino asymmetry. At least in principle, this allows to distinguish between β decay and NCB interaction. In this paper we briefly summarize the results of [1] where it is argued that if m_ν is in the eV range, future NCB experiments could represent an almost unique way to detect cosmological neutrinos.

2. The Neutrino Capture Rate

Assuming an isotropic neutrino flux corresponding to a distribution function in phase space $f(p_\nu)$, the NCB integrated rate can be expressed as an integral over the electron (positron) energy

$$\lambda_\nu = \int \sigma_{\rm NCB} v_\nu \, f(p_\nu) \, \frac{d^3 p_\nu}{(2\pi)^3} = \frac{G_\beta^2}{2\pi^3} \int_{W_o+2m_\nu}^{\infty} p_e E_e F(Z,E_e) C(E_e,p_\nu)_\nu E_\nu p_\nu \, f(p_\nu) \, dE_e \, , \tag{2}$$

where $F(Z, E_e)$ is the Fermi function, with E_e and p_e the energy and momentum of the outgoing electron and W_o the corresponding β decay endpoint. The rate contains the nuclear shape factor $C(E_e, p_\nu)_\nu$, an angular momentum weighted average of nuclear state transition amplitudes, which depends upon the nuclear properties of the parent and daughter nuclei and represents the main source of uncertainty in $\sigma_{\rm NCB} v_\nu$, the product of the cross section times neutrino velocity.

On the other hand, NCB rate is strongly related to the corresponding β decay process rate

$$\lambda_\beta = \frac{G_\beta^2}{2\pi^3} \int_{m_e}^{W_o} p_e E_e F(Z,E_e) C(E_e,p_\nu)_\beta E_\nu p_\nu \, dE_e \, , \tag{3}$$

where a simple relation between the two shape factors holds

$$C(E_e, p_\nu)_\nu = C(E_e, -p_\nu)_\beta \, , \tag{4}$$

though both variables have different kinematical domains in the two processes.

Therefore, the β decay rate can be used to provide a relation giving the mean shape factor, defined as

$$\overline{C}_\beta = \frac{1}{f} \int_{m_e}^{W_o} p_e E_e F(Z,E_e) C(E_e,p_\nu)_\beta E_\nu p_\nu dE_e \, , \tag{5}$$

in terms of observable quantities, W_o and the product $f t_{1/2}$

$$f t_{1/2} = \frac{2\pi^3 \ln 2}{G_\beta^2 \, \overline{C}_\beta} \, , \tag{6}$$

Defining

$$\mathcal{A} = \int_{m_e}^{W_o} \frac{C(E'_e,p'_\nu)_\beta}{C(E_e,p_\nu)_\nu} \frac{p'_e}{p_e} \frac{E'_e}{E_e} \frac{F(E'_e,Z)}{F(E_e,Z)} E'_\nu p'_\nu dE'_e \, , \tag{7}$$

where a prime denotes all variables depending on E'_e which should be integrated over, the NCB cross section times neutrino velocity can then be conveniently written as

$$\sigma_{\rm NCB} v_\nu = \frac{2\pi^2 \ln 2}{\mathcal{A} \cdot t_{1/2}} \, , \tag{8}$$

Notice that $\mathcal{A}$ contains the ratio of NCB and β decay shape factors. As discussed in details in [1], in several relevant cases (super-allowed transitions, unique k-th forbidden transitions) the evaluation of $\mathcal{A}$ is particularly simple so that Eq. (8) can be computed in an exact way. Furthermore, in all cases where this is not possible, systematic uncertainties affecting the nuclear matrix element evaluation largely cancel in the shape factor ratio appearing in $\mathcal{A}$, thus providing a reliable estimate of the NCB cross sections. Results for both allowed and unique forbidden decay cross sections having branching ratios greater than 5%, namely 1272 β^- decays and 799 β^+ decays can be found in [1]. Indeed, there are several nuclei spanning a wide range in Q_β for which interesting high values are reached in the range $\sigma_{\rm NCB}(v_\nu/c) = 10^{-41} - 10^{-43}$ cm^2. As an example, in the interesting case of ^{3}H one gets $\sigma_{\rm NCB}(v_\nu/c) = 7.84 \times 10^{-45}$ cm^2, which for the standard homogeneous flux of cosmological neutrinos corresponds to 7.5 events per year of data taking for a mass of 100 g. In general, this estimate represents a lower bound, as massive neutrino density is expected to be locally larger because of gravitational clustering. This effect in a Cold Dark Matter Halo is quite relevant for order eV neutrino masses, see Table 1.

Of course, the finite energy resolution of any experimental apparatus and the extremely low cross section make relic neutrino detection via NCB a real challenge due to the large background events produced by standard β decay. In particular, the ratio of the event rate $\lambda_\beta(\Delta)$ for the last β decay electron energy bin $W_o - \Delta < E_e < W_o$, compared with the total NCB event rate is typically very large, since $\Delta >> T_\nu$

$$\frac{\lambda_\beta(\Delta)}{\lambda_\nu} = \frac{2}{9\zeta(3)}\left(\frac{\Delta}{T_\nu}\right)^3\left(1+\frac{2m_\nu}{\Delta}\right)^{3/2} >> 1\,, \tag{9}$$

It is therefore a crucial issue to reach an energy resolution, the electron energy bin dimension of the apparatus Δ, smaller than m_ν. For example, the expected background electron events which are produced by β decay, yet having an energy which corresponds to the relic neutrino capture energy bin centered at $E_e = W_o + 2m_\nu$ are smaller than a factor three with respect to NCB processes if $\Delta = 0.2$ eV for $m_\nu = 0.7$ eV, while a smaller neutrino mass of 0.3 eV requires $\Delta = 0.1$ eV. Presently, to obtain such an energy resolution seems very demanding. Nevertheless, if a large neutrino mass will be found by the ongoing β decay experiment KATRIN [5], it is conceivable that more efforts could be devoted to future generation of experiments with an improved energy resolution as good as 0.1 eV.

Table 1. Relic neutrino capture rate for 100 g of ^{3}H, for a standard Fermi-Dirac distribution with $T_\nu = 1.7 \cdot 10^{-4}$ eV (FD). Results are also shown for a Navarro Frenk and White profile (NFW) and for present day mass distribution of the Milky Way (MW) for two values of m_ν.

m_ν (eV)	FD (events yr^{-1})	NFW (events yr^{-1})	MW (events yr^{-1})
0.3	7.5	23	33
0.15	7.5	10	12

3. Conclusion

In this paper we have summarized the analysis of NCB performed in [1]. These processes have the remarkable property of having no energy threshold on the incoming neutrino energy and thus they might represent a good and numerous class of interactions suitable for low energy neutrino flux measurements. The possibility to pursue the ultimate goal of cosmological relic neutrino background detection via a future experimental implementation of this approach depends upon two crucial issues, a high value for the expected order of magnitude of NCB event rate, as well a very good energy resolution of the outgoing electron or positron, of the order of neutrino mass. Both these aspects should be optimized by a careful choice of the β decaying nuclei.

References

[1] Cocco A, Mangano G and Messina M 2007 *J. Cosmol. Astropart. Phys.* JCAP0706(2007)015
[2] Weinberg S 1962 *Phys. Rev.* **128** 1457
[3] Cuoco A *et al.* 2004 *Int. J. Mod Phys.* A **19** 4431
[4] Hansen S H *et al.* 2002 *Phys. Rev.* D **65** 023511
[5] Osipowitcz A *et al.* [KATRIN Collaboration] 2001 arXiv:hep-ex/0109033
[6] Ringwald A and Wong Y Y Y 2004 *J. Cosmol. Astropart. Phys.* JCAP0412(2004)005

Dirac and Majorana Leptonic CP-Violation and Leptogenesis

S. T. Petcov [1]
Scuola Internazionale Superiore di Studi Avanzati and
INFN, I-34014 Trieste, Italy

Abstract. We briefly review the results showing that the CP-violation necessary for the generation of the baryon asymmetry of the Universe in leptogenesis with "hierarchical" heavy Majorana neutrinos can be due exclusively to the Dirac and/or Majorana CP-violating phase(s) in the neutrino mixing matrix U. One can have successful leptogenesis, in particular, even if the only source of CP-violation is the Dirac phase δ in U. For light ν-mass spectrum with normal (inverted) hierarchy, this requires $|\sin\theta_{13}\sin\delta| \geq 0.09$ (0.02), θ_{13} being the CHOOZ angle.

1. Introduction

It was shown recently in [1] that the CP-violation necessary for the generation of the observed baryon asymmetry of the Universe Y_B in the thermal leptogenesis scenario [2] can be due exclusively to the Dirac and/or Majorana CP-violating (CPV) phases in the Pontecorvo-Maki-Nakagawa-Sakata (PMNS) neutrino mixing matrix [3], and thus can be directly related to the low energy CP-violation in the lepton sector (observable, e.g. in neutrino oscillations, etc.). The baryon asymmetry is generated in the regime of large lepton flavour effects in leptogenesis ("flavoured" leptogenesis) [4, 5]. In ref. [6] the baryon asymmetry Y_B produced in "flavoured" leptogenesis, with the requisite CP-violation provided by the PMNS matrix U, was found to depend strongly in certain physically interesting cases on the value of the lightest neutrino mass, $\min(m_j)$, $j = 1, 2, 3$. For specific values of $\min(m_j)$ the asymmetry $|Y_B|$ can be significantly enhanced (by a factor of ~ 100 or more) with respect to that predicted for $\min(m_j) \cong 0$. This enhancement can make the predicted Y_B compatible with the observations even when this is not the case for $\min(m_j) \cong 0$. Here we review briefly some of the results obtained in [1, 6].

Following [1, 6], we consider the the simplest scheme in which leptogenesis can be implemented - the minimal "seesaw"(type I) model of ν-mass generation [7], which includes the Standard Model (SM) plus three heavy right-handed (RH) Majorana neutrinos, N_j. The latter are assumed to have a hierarchical mass spectrum [2], $0 < M_1 \ll M_2 \ll M_3$. We limit our discussion here to the cases of light neutrinos having normal hierarchical (NH), $m_1 \ll m_2 < m_3$, or inverted hierarchical (IH), $m_3 \ll m_1 < m_2$, mass spectrum [3] (see, e.g. [8]). In thermal leptogenesis with "hierarchical" heavy Majorana neutrinos N_j, the CPV lepton asymmetry is produced in out-of-equilibrium lepton number and CP-nonconserving decays of the lightest one, N_1. At later epoch the lepton asymmetry is converted into a baryon asymmetry by $(B - L)$-conserving but $(B + L)$-violating sphaleron interactions [9].

[1] Also at: Institute of Nuclear Research and Nuclear Energy, Bulgarian Academy of Sciences, 1784 Sofia, Bulgaria.
[2] Results for quasi-degenerate in mass heavy Majorana neutrinos were also obtained in [1].
[3] A rather detailed discussion of the case of quasi-degenerate light neutrino mass spectrum is given in [1].

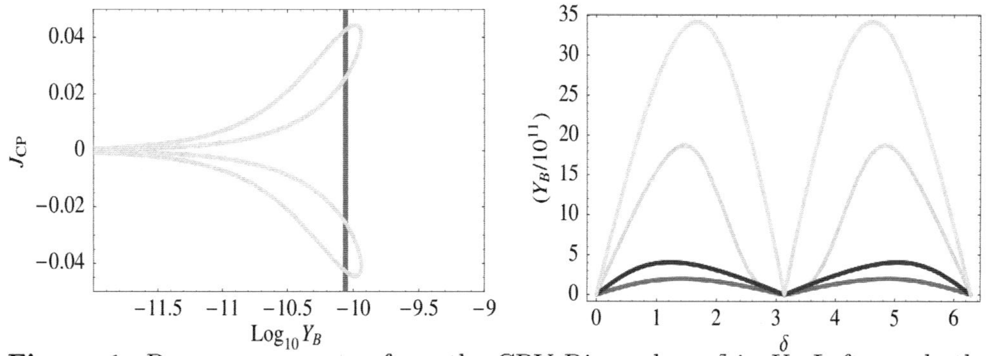

Figure 1. Baryon asymmetry from the CPV Dirac phase δ in U. Left panel: the correlation between $J_{\rm CP}$ and $Y_{\rm B}$ for NH spectrum, s_{13}=0.2, α_{32}=0, R_{12}=0.86, and M_1=5 × 10^{11} GeV. Right panel: $|Y_B|$ as a function of δ for IH spectrum, $R_{11}R_{12} = i\ \kappa\ |R_{11}R_{12}|$ ($|R_{11}|^2 - |R_{12}|^2 = 1$), $\kappa = -1$ (red and dark blue lines), $\kappa = +1$ (light blue and green lines), for $M_1 = 2 \times 10^{11}$ GeV, and $s_{13} = 0.1$ (red and green lines) and $s_{13} = 0.2$ (dark blue and light blue lines). (From [1].)

2. Baryon Asymmetry from CPV Dirac and Majorana Phases in $U_{\rm PMNS}$

The matrix of neutrino Yukawa couplings, λ, together with the Majorana mass matrix of the RH neutrinos, M_R, and the matrix of charged lepton Yukawa couplings, λ^{lep}, plays a crucial role both in the see-saw mechanism and in leptogenesis. In the basis in which M_R and λ^{lep} are diagonal, λ is the only source of CP-violation in the lepton sector. The "orthogonal" parametrisation allows to relate in a rather simple manner λ with the PMNS matrix [10]: $\lambda = v^{-1}\sqrt{M}\, R\, \sqrt{m} U^\dagger$, where R is a complex orthogonal matrix, $R\ R^T = R^T\ R = \mathbf{1}$, $M \equiv {\rm Diag}(M_1, M_2, M_3)$, $m \equiv {\rm Diag}(m_1, m_2, m_3)$, $M_j > 0$, $m_k \geq 0$, and $v = 174$ GeV is the vacuum expectation value of the Higgs doublet field. We are interested in the case when the matrix R conserves the CP-symmetry. Therefore we will assume that R has real and/or purely imaginary elements [1].

Suppose the baryon asymmetry Y_B is produced in the "two-flavour" regime in leptogenesis [4, 5]. This regime is realised at temperatures 10^9 GeV $\lesssim T \sim M_1 \lesssim 10^{12}$ GeV. Under the assumptions made, $|Y_B|$ generated via leptogenesis can be written as [4, 5] $|Y_B| \cong 3 \times 10^{-3}\ |\epsilon_\tau\ \eta|$, where ϵ_τ is the CPV asymmetry in the τ flavour (lepton charge) produced in N_1-decays [4], η is the efficiency factor [5], $|\eta| \cong |\eta(0.71\widetilde{m_2}) - \eta(0.66\widetilde{m_\tau})|$, $\widetilde{m_{2,\tau}}$ being the wash-out mass parameters $\widetilde{m_2} = \widetilde{m_e} + \widetilde{m_\mu}$, $\widetilde{m_l} = |\sum_j m_j\ R_{1j}\ U^*_{lj}|^2$. For real R we have [1]:

$$\epsilon_\tau = -\frac{3M_1}{16\pi v^2}\frac{\sum_k \sum_{j>k}\sqrt{m_k m_j}\,(m_j - m_k)\,\rho_{kj}|R_{1k}R_{1j}|\,{\rm Im}\,(U^*_{\tau k}\,U_{\tau j})}{\sum_i m_i\,|R_{1i}|^2}\,,\ {\rm Im}\,(R_{1k}R_{1j}) = 0\,, \qquad (1)$$

where $\rho_{jk} \equiv {\rm sgn}(R_{1j}R_{1k}) = \pm 1$, $j \neq k$. If $R_{1j}R_{1k} = i\rho_{jk}\ |R_{1j}R_{1k}|$, then ${\rm Im}(R_{1k}R_{1j})$ and $(m_j - m_k)$ in eq. (1) should be replaced [5] by ${\rm Re}(R_{1k}R_{1j})$ and $(m_j + m_k)$ [1].

We shall summarise next briefly some of the results obtained in [1, 6]. Consider the case of NH Spectrum, $m_1 \ll m_2 \ll m_3 \cong (\Delta m^2_{\rm A})^{1/2}$, $\Delta m^2_{\rm A} \cong 2.5 \times 10^{-3}$ eV^2 being the "atmospheric" neutrino mass squared difference. Suppose that $m_1 \cong 0$, $R_{11} \cong 0$ (N_3 decoupling), $R_{12}R_{13}$ is real and the only source of CP-violation is the Dirac phase δ in U. In this case $\epsilon_\tau \propto s_{13}\sin\delta$, $s_{13} \equiv \sin\theta_{13}$. For, e.g. $R_{12}R_{13} > 0$, s_{13}=0.2, δ=$\pi/2$, and $R_{12} \cong 0.86$ (which maximises $|Y_B|$), we have [1]: $|Y_B| \cong 3.6 \times 10^{-13}$ $((\Delta m^2_{\rm A})^{1/2}/0.05$ eV$)(M_1/10^9$ GeV$)$. The observed baryon asymmetry $|Y_B| \cong (8.0 - 9.2) \times 10^{-11}$ can be reproduced for $M_1 \lesssim 5 \times 10^{11}$ GeV if $|\sin\theta_{13}\sin\delta| \gtrsim 0.09$, or $|J_{\rm CP}| \gtrsim 2.0 \times 10^{-2}$, $J_{\rm CP}$ being the rephasing invariant controlling the magnitude of CP-violation in neutrino oscillations [11]. Given the experimental limit

[4] The expression for Y_B we have given is normalised to the entropy density, see, e.g. [1].

[5] Note that real (purely imaginary) $R_{1k}R_{1j}$ and purely imaginary (real) $U^*_{lk}U_{lj}$, $j \neq k$, implies violation of CP-invariance by the matrix R [1]. In order for the CP-symmetry to be broken at low energies, we should have both ${\rm Re}(U^*_{lk}U_{lj}) \neq 0$ and ${\rm Im}(U^*_{lk}U_{lj}) \neq 0$ (see [1] for further details).

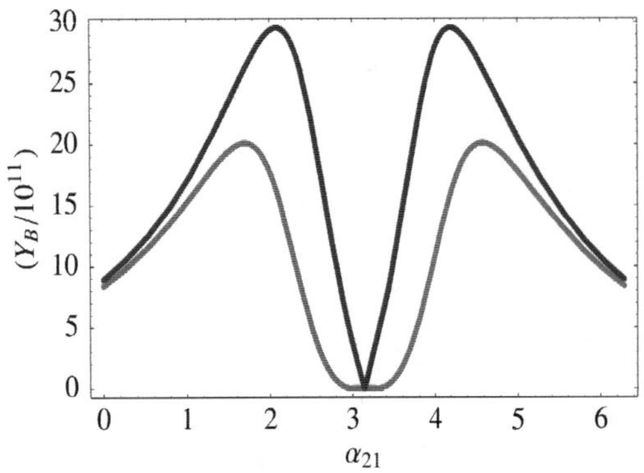

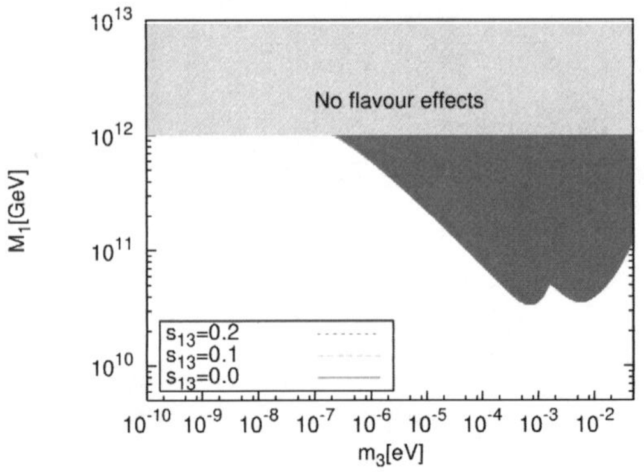

Figure 2. The asymmetry $|Y_B|$ in the case of IH spectrum. Left panel: $|Y_B|$ versus the Majorana phase $\alpha_{21} \equiv \alpha$ of U for purely imaginary $R_{11}R_{12} = -i|R_{11}R_{12}|$, $R_{13} = 0$, $\delta = 0$, $M_1 = 2 \times 10^{11}$ GeV and $s_{13} = 0,\ 0.2$ - black/blue and gray/red lines. Right panel: values of m_3 and M_1 for which the "flavoured" leptogenesis is successful, generating baryon asymmetry $|Y_B| = 8.6 \times 10^{-11}$ (red/dark shaded area) in the case of real R (see [1, 6] for further details).

$\sin\theta_{13} \leq 0.22$, leptogenesis can be successful provided $M_1 \gtrsim 1.9 \times 10^{11}$ GeV. The ranges of values of $\sin\theta_{13}$ and of $|J_{\rm CP}|$ we find are within the sensitivity of the planned θ_{13} reactor neutrino experiments Double CHOOZ and Daya Bay and of the neutrino oscillation experiments on CP- (T-) violation. Since both Y_B and $J_{\rm CP}$ depend on s_{13} and δ, for given values of the other relevant parameters there exists a correlation between the values of $|Y_B|$ and $J_{\rm CP}$ (Fig. 1). In the analogous case of IH spectrum, Dirac CP-violation, $m_3 \cong 0$ and N_3 decoupling ($R_{13} \cong 0$), we find that the observed baryon asymmetry $|Y_B| \cong (8.0 - 9.2) \times 10^{-11}$ can be reproduced if $|\sin\theta_{13}\sin\delta| \gtrsim 0.02$, or $|J_{\rm CP}| \gtrsim 4.6 \times 10^{-3}$ (Fig 1). However, leptogenesis can be successful for $M_1 \lesssim 10^{12}$ GeV only if $R_{11}R_{12}$ is not real [1, 12]. The quoted results correspond to $R_{11}R_{12}{=}i\kappa|R_{11}R_{12}|$, $\kappa{=}\pm 1$. We can still have successful leptogenesis for real R, CP-violation due to the CPV phases in U and $M_1 \lesssim 10^{12}$ GeV, provided m_3 is significantly different from 0 [6]. Typically values of 5×10^{-4} eV $\lesssim m_3 \lesssim 5 \times 10^{-2}$ eV are required. If, however, $R_{1j} \neq 0$, $j = 1, 2, 3$, one can have successful leptogenesis even for $m_3 \sim 10^{-5}$ eV (Fig. 2).

In conclusion, the results presented in the present article underline, in particular, the importance of understanding the status of the CP-symmetry in the lepton sector and, correspondingly, of the experiments aiming to measure the CHOOZ angle θ_{13} and of the experimental searches for Dirac and/or Majorana leptonic CP-violation at low energies.

Acknowledgments

This work was supported by the Italian INFN program on "Fisica Astroparticellare". I would like to thank S. Pascoli and A. Riotto, as well as E. Molinaro, T. Shindou and Y. Takanishi for the fruitful and very enjoyable collaborations on the topics discussed in the present article.

References

[1] S. Pascoli, S.T. Petcov and A. Riotto, Phys. Rev. D **68** (2003) 093007; Nucl. Phys. B **739** (2006) 208.
[2] M. Fukugita and T. Yanagida, Phys. Lett. B **174** (1986) 45.
[3] B. Pontecorvo, Zh. Eksp. Teor. Fiz. **33** (1957) 549, **34** (1958) 247 and **53** (1967) 1717;
Z. Maki, M. Nakagawa and S. Sakata, Prog. Theor. Phys. **28** (1962) 870.
[4] A. Abada *et al.*, JCAP **0604** (2006) 004; E. Nardi *et al.*, JHEP **0601** (2006) 164.
[5] A. Abada *et al.*, JHEP **0609** (2006) 010.
[6] E. Molinaro *et al.*, arXiv:0709.041 (to be published in Nucl. Phys. B).
[7] P. Minkowski, Phys. Lett. B **67** (1977) 421.
[8] S.T. Petcov, Nucl. Phys. B (Proc. Suppl.) **143** (2005) 159 (hep-ph/0412410).
[9] V.A. Kuzmin, V.A. Rubakov and M.E. Shaposhnikov, Phys. Lett. B **155** (1985) 36.
[10] J. A. Casas and A. Ibarra, Nucl. Phys. B **618** (2001) 171.
[11] P.I. Krastev and S.T. Petcov, *Phys. Lett.* B **205** (1988) 84; M. Freund *et al.*, *Nucl. Phys.* B **578** (2000) 27.
[12] S.T. Petcov *et al.*, *Nucl. Phys.* B **739** (2006) 208.

Dirac phase leptogenesis

Steve Blanchet

Max-Planck-Institut für Physik (Werner-Heisenberg-Institut)
Föhringer Ring 6, 80805 München, Germany

E-mail: blanchet@mppmu.mpg.de

Abstract. I present here a concise summary of the preprint arXiv:0707.3024, written in collaboration with A. Anisimov and P. Di Bari. There we discuss leptogenesis when *CP* violation stems exlusively from the Dirac phase in the PMNS mixing matrix. Under this assumption it turns out that the situation is very constrained when a hierarchical heavy right-handed (RH) neutrino spectrum is considered: the allowed regions are small and the final asymmetry depends on the initial conditions. On the other hand, for a quasi-degenerate spectrum of RH neutrinos, the *CP* asymmetry can be enhanced and the situation becomes much more favorable, with no dependence on the initial conditions. Interestingly, in the extreme case of resonant leptogenesis, in order to match the observed baryon asymmetry of the Universe, we obtain a lower bound on $\sin\theta_{13}$ which depends on the lightest active neutrino mass m_1.

1. Introduction

With the understanding of the important role of flavor in leptogenesis [1, 2], it has become progressively clear that the link between low-energy *CP* violation and the size of the baryon asymmetry predicted by thermal leptogenesis [3] is considerably tighter than in the unflavored (or one-flavor) situation. In particular it was shown in [4] for a hierarchical RH neutrino spectrum that a non-zero Majorana phase can be the unique source of *CP* violation contributing to the generation of the baryon asymmetry of the Universe, although the allowed region in the parameter space was small, and mainly in the weak wash-out. The Dirac phase can also do the job, provided the angle θ_{13} is not too small [5].

Within the see-saw model [6], a nice way to see the different contributions to the *CP* violation required for leptogenesis is to choose the following parametrization of the Dirac mass matrix [7]:

$$m_D = U\sqrt{D_m}\,\Omega\sqrt{D_M}, \tag{1}$$

where m_i and M_i ($i = 1,2,3$) denote the light and heavy neutrino masses, respectively, and $D_X \equiv \mathrm{diag}(X_1, X_2, X_3)$. Among the 18 supplementary parameters brought by the see-saw, 6 are *CP* violating phases: 1 Dirac and 2 Majorana phases in the PMNS mixing matrix U, as well as 3 unobservable phases in the Ω matrix. The Ω matrix can be conveniently parametrized as a product of three complex rotations, $\Omega = R_{12}(\omega_{21})R_{13}(\omega_{31})R_{23}(\omega_{32})$.

In [8] we analysed in detail the role of the Dirac phase as the *unique* source of *CP* violation required for leptogenesis. The aim is here to summarize some important results obtained there.

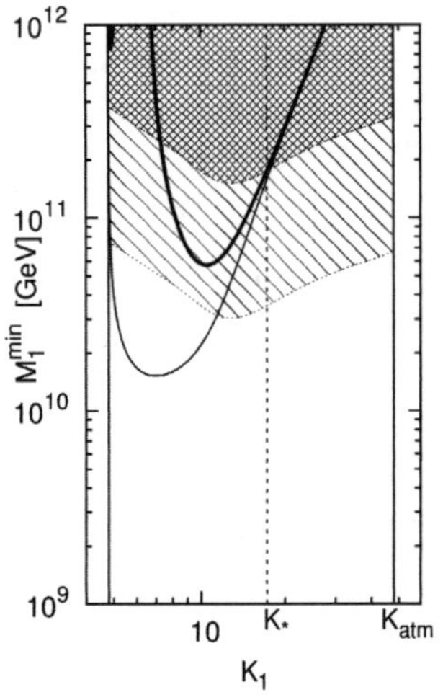

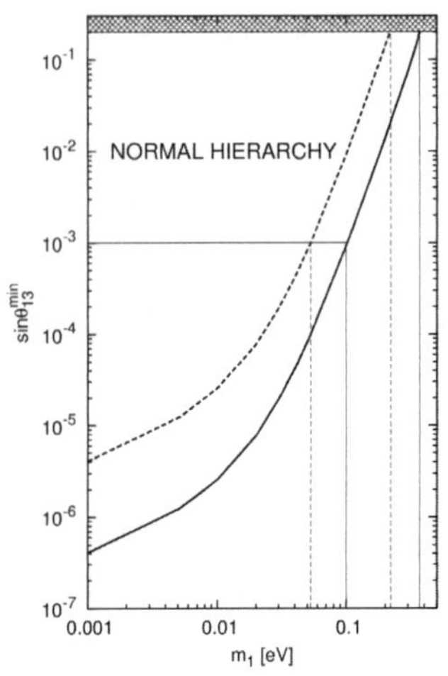

Figure 1. HL, $\Omega = R_{13}$. Lower bound $M_1^{\rm min}$ vs. K_1 for $m_1/m_{\rm atm} = 0.1$, $s_{13} = 0.2$, $\delta = -\pi/2$ and $\omega_{31} < 0$. The thin and thick solid lines denote thermal and vanishing initial abundance of RH neutrinos, respectively.

Figure 2. DL, $\Omega = R_{13}$. Lower bound on $\sin\theta_{13}$ vs. m_1 for $d = 1$ (solid line) and $d = 10$ (short-dashed line). Values $\sin\theta_{13} > 0.20$ are excluded at $3\,\sigma$ by current experimental data.

2. Hierarchical spectrum for the heavy RH neutrinos

In the hierarchical limit (HL), $M_1 \ll M_2 \ll M_3$, one can safely study only the decay of the lightest RH neutrino [1], N_1, and we anticipate that successful leptogenesis will occur for $10^9{\rm GeV} \lesssim M_1 \lesssim 10^{12}{\rm GeV}$, where a two-flavor regime applies, with flavors denoted $\alpha = e+\mu, \tau$.

Since here we consider leptogenesis exclusively from one low-energy phase, namely the Dirac phase, we have that the total *CP* asymmetry $\varepsilon_1 = \varepsilon_{1\tau} + \varepsilon_{1,e+\mu} = 0$, so that the final asymmetry can be written as

$$N_{B-L}^{\rm f} \simeq (\kappa_{1\tau}^{\rm f} - \kappa_{1,e+\mu}^{\rm f})\varepsilon_{1\tau}, \tag{2}$$

where $\kappa_{1\alpha}^{\rm f}$ are the final efficiency factors, which include the effects of the wash-out and the production of the asymmetry.

Since the *CP* asymmetry parameter $\varepsilon_{1\tau} \propto M_1$, if one requires the final asymmetry, including the sphaleron conversion, to match the observed baryon asymmetry of the Universe, $\eta_B^{\rm CMB} = (6.1 \pm 0.2) \times 10^{-10}$ [9], one obtains a lower bound on the lightest RH neutrino mass, M_1.

We show in Fig. 1 an example of lower bound $M_1^{\rm min}$ vs. $K_1 \equiv \sum_k m_k |\Omega_{k1}|^2/m_\star$ ($m_\star \simeq 1.08 \times 10^{-3}$ eV) for the particular choice $\Omega = R_{13}$, with $m_1/m_{\rm atm} = 0.1$, $\sin\theta_{13} = 0.2$ and $\delta = -\pi/2$. When the condition of validity for the fully flavored Boltzmann equations is used (see [10] for a discussion), the region allowed is only outside the squared region. In the dashed region it is not exluded but one expects some modification of the results shown. It can be seen from Fig. 1 that the allowed region is very marginal, and only in the weak wash-out regime, where there is a dependence on the initial conditions. Moreover, in the example shown, we took the angle θ_{13} at its maximal allowed value, so if the upper bound on θ_{13} becomes one order of magnitude more stringent in the future, it will essentially exclude Dirac phase leptogenesis in the hierarchical limit [2].

[1] It was explicitly checked in [8] that, for all cases under study, the contribution from the heavier RH neutrinos was negligible.

[2] This rigorously applies to the example shown in Fig. 1, but this statement turns out to apply for all cases studied in [8].

One way-out of this unfavorable situation is to go to a quasi-degenerate RH neutrino spectrum, where the CP asymmetry can be enhanced and where a strong wash-out is ensured.

3. Quasi-degenerate spectrum for the heavy RH neutrinos

In the degenerate limit (DL), $M_1 \simeq M_2 \simeq M_3$, we anticipate the mass M_1 to be below 10^9 GeV, so that a three-flavor problem arises, with flavors $\alpha = e, \mu, \tau$. Contrary to the HL, the CP asymmetries, as well as the wash-outs, must be now summed over each RH neutrino:

$$N^f_{B-L} \simeq \sum_\alpha (\varepsilon_{1\alpha} + \varepsilon_{2\alpha} + \varepsilon_{3\alpha}) \kappa^{\rm f}_\alpha (K_{1\alpha} + K_{2\alpha} + K_{3\alpha}). \tag{3}$$

In the DL, the CP asymmetries are inversely proportional to the degeneracy parameter $\delta_{ji} \equiv (M_j - M_i)/M_i$. This implies a dependence on δ_{ji} of the lower bound on M_1 for successful leptogenesis. For $\Omega = R_{13}$ we obtained the following 3σ bounds for normal and inverted hierarchy, respectively:

$$M_1 \gtrsim 5.5 \times 10^9 \mathrm{GeV} \frac{\delta_{31}}{|\sin\theta_{13} \sin\delta|} \quad \text{and} \quad M_1 \gtrsim 5 \times 10^{11} \mathrm{GeV} \frac{\delta_{31}}{|\sin\theta_{13} \sin\delta|}. \tag{4}$$

It is interesting to notice how the degeneracy parameter is linked to the size of $\sin\theta_{13}$ for successful leptogenesis at fixed M_1.

The enhancement for close RH neutrino masses is of course limited. Indeed, when $\delta^{\rm res}_{ij} \simeq d M_i m_{\rm atm}/(16\pi v^2)$, one hits a resonance [11, 12], where $d = 1 \div 10$ accounts for the present uncertainty in the location of the resonance. It is interesting to notice that in this extreme situation we obtained a lower bound on θ_{13} which depends on the lightest active neutrino mass, m_1. This is shown in Fig. 2, for $\Omega = R_{13}$ and a normal hierarchy. The corresponding constraint for an inverted hierarchy as well as for other Ω matrices can be found in [8]. Note finally that for a vanishing m_1, successful Dirac phase leptogenesis demands $\sin\theta_{13} \gtrsim 2.3 \times 10^{-7}$ and $\sin\theta_{13} \gtrsim 3 \times 10^{-6}$ for normal and inverted hierarchy, respectively.

4. Conclusion

The possibility that the Dirac phase, which one hopes to measure in the next-generation neutrino experiments, might be the source of CP violation required for leptogenesis is interesting. We obtained that a degenerate RH neutrino spectrum is preferred and that in the extreme case of resonant leptogenesis a nice link between low-energy parameters, such as $\sin\theta_{13}$, m_1 and the mass hierarchy, is present.

Acknowledgments

I would like to thank A. Anisimov and P. Di Bari for collaboration on this project.

References

[1] Nardi E, Nir Y, Roulet E and Racker J 2006 *J. High Energy Phys.* JHEP0601(2006)164

[2] Abada A, Davidson S, Josse-Michaux F-X, Losada M and Riotto A 2006 *J. Cosmol. Astropart. Phys.* JCAP0604(2006)004

[3] Fukugita M and Yanagida T 1986 *Phys. Lett.* B **174** 45

[4] Blanchet S and Di Bari P 2007 *J. Cosmol. Astropart. Phys.* JCAP0703(2007)018

[5] Pascoli S, Petcov S T and Riotto A 2007 *Phys. Rev.* D **75** 083511

[6] Minkowski P 1977 *Phys. Lett.* B **67** 421

[7] Casas J A and Ibarra A 2001 *Nucl. Phys.* B **618** 171

[8] Anisimov A, Blanchet S and Di Bari P 2007 *Preprint* arXiv:0707.3024.

[9] Spergel D N *et al.* [WMAP Collaboration] 2006 *Astrophys. J. Suppl.* **170** 377

[10] Blanchet S, Di Bari P and Raffelt G G 2007 *J. Cosmol. Astropart. Phys.* JCAP0703(2007)012

[11] Pilaftsis A 1999 *Int. J. Mod. Phys.* A **14** 1811

[12] Anisimov A, Broncano A and Plumacher M 2006 *Nucl. Phys.* B **737** 176

Quintessential Kination and Leptogenesis

Eung Jin Chun [1,2] **and Stefano Scopel** [1,*]
[1] Korea Institute for Advanced Study
207-43 Cheongryangri-dong Dongdaemun-gu
Seoul 130-722, Korea
[2] Michigan Center for Theoretical Physics
Department of Physics, University of Michigan
Ann Arbor, MI 48109, USA
[*] Paper presented at the conference by Stefano Scopel

E-mail: `ejchun@kias.re.kr, scopel@kias.re.kr`

Abstract. We show how thermal leptogenesis induced by the CP-violating decay of a right-handed neutrino (RHN) can be compatible to the assumption of quintessential kination, i.e., to a cosmological background model where the energy density of the early Universe is assumed to be dominated by the kinetic term of a quintessence field during some epoch of its evolution. The phenomenology depends on the temperature T_r above which kination dominates over radiation. For instance, when kination stops to dominate when $M/100 \lesssim T_r \lesssim M$ (where M is the RHN mass) the efficiency of the process η, defined as the ratio between the produced lepton asymmetry and the amount of CP violation in the RHN decay, can be even larger than in the standard scenario of radiation domination. On the other hand, a super–weak wash–out regime is obtained ($\eta \ll 1$) for $T_r << M/100$. In this latter situation the lower bound $T_r \gtrsim 5$ MeV $\times(M/\mathrm{TeV})(0.05\,\mathrm{eV}/\tilde{m})$ can be found, where $\tilde{m}$ is the effective neutrino mass scale.

The dark energy component in the present universe can be explained by modifying the standard cosmology with the introduction of a slowly evolving scalar field called quintessence [1]. In this case, an open possibility is the existence of an early era of kination domination, during which the Universe is dominated by the kinetic energy of the quintessence field. During this era, the expansion rate of the Universe is larger compared to the usual radiation domination case, modifying the predictions of the standard cosmological picture. Kination has been discussed in different contexts, such as dark matter and inflation[2].

In this talk we investigate the impact of the quintessential kination scenario on the properties of the thermal leptogenesis induced by the CP-violating decay of a right-handed neutrino (RHN), N [3]. As discussed in detal in [4] and summarized in the following, leptogenesis and kination can indeed be compatible with each other, although the faster expansion rate in the early Universe due to kination dominance can modify the predictions of the standard leptogenesis scenario even by several orders of magnitude.

The kination regime is attained when, in the energy–momentum tensor of the quintessence field ϕ (assumed here as spatially constant): $T_{\mu\nu} = \partial_\mu \phi \frac{\partial \mathcal{L}}{\partial \partial^\nu \phi} - g_{\mu\nu}\mathcal{L}$, the kinetic term $\dot{\phi}^2/2$ dominates over the potential term $V(\phi)$, so that

$$w \equiv \frac{p}{\rho} = \frac{\frac{\dot{\phi}^2}{2} - V(\phi)}{\frac{\dot{\phi}^2}{2} + V(\phi)} \rightarrow 1. \quad (1)$$

In this case the energy density of the Universe, which scales like $\rho \propto a^{-3(1+w)}$, is given by $\rho_{kin} \propto a^{-6}$, instead of $\rho_{rad} \propto a^{-4}$ for the case of radiation domination. In the following, we will assume that, in the epoch after reheating which is relevant to thermal leptogenesis, the energy density of the Universe is dominated by the sum of these two components, $\rho = \rho_{rad} + \rho_{kin}$, with the boundary condition $\rho_{kin}(T_r) = \rho_{rad}(T_r)$. The kination–radiation equality temperature T_r is in principle a free parameter, with the only constraint: $T_r \gtrsim 1$ MeV, in order not to spoil big-bang nucleosynthesis. This implies that:

$$H(z) = \sqrt{\frac{z^2 + z_r^2}{1 + z_r^2}} \frac{H_1}{z^3}, \tag{2}$$

where $z \equiv M/T$ with M the mass of the RHN, $H_1 \equiv H(z = 1)$, $z_r \equiv \sqrt{\frac{g_*}{g_{*r}}} M/T_r$, g_* is the number of relativistic degrees of freedom, and $g_{*r} = g_*(T_r)$.

Let's discuss now thermal leptogenesis in a Minimal Supersymmetric extension of the Standard Model (MSSM) supplemented by right–handed neutrino supermultiplets, i.e. the model described by the following superpotential:

$$\mathcal{W} = \mathcal{W}_{MSSM} + \frac{1}{2} N^c M N^c + y H_2 L N^c. \tag{3}$$

In this case the decay rate of the RHN is given by $\Gamma_d = |y|^2 M/4\pi$ where y is the neutrino Yukawa coupling, with $g_{*r} = 10.75$ and $g_*(T) = 228.75$. This scenario has been extensively studied in the literature [5] in a conventional cosmological setup where the energy density of the early Universe is dominated by radiation.

In our case, introducing as usual the effective neutrino mass scale given by $\tilde{m} \equiv |y|^2 \langle H_2 \rangle^2 / M$, we can get the wash–out parameter, given by the ratio $K \equiv \Gamma_d / H(z = 1)$:

$$K = \frac{63.78}{\sqrt{1 + z_r^2}} \left(\frac{\tilde{m}}{0.05 \text{ ev}} \right). \tag{4}$$

The above equation shows that, depending on z_r, one can have $K >> 1$ (strong wash–out) or $K << 1$ (super–weak wash–out) at fixed $\tilde{m}$. In the general case ($z_r << 1$ corresponds to radiation domination and $z_r >> 1$ to kination domination) the Boltzmann equations that drive leptogenesis are given by:

$$\frac{d\hat{N}}{dz}(z) = -K \sqrt{\frac{1 + z_r^2}{z^2 + z_r^2}} z^2 (\hat{N} - \hat{N}_{eq}) \left[\gamma_d(z) + 2\gamma_s(z) + 4\gamma_t(z) \right] \tag{5}$$

$$\begin{aligned} \frac{d\hat{L}}{dz}(z) &= K \sqrt{\frac{1 + z_r^2}{z^2 + z_r^2}} z^2 \Big[(\gamma_d(z)\epsilon + 2\gamma_s(z)\epsilon_s + 4\gamma_t(z)\epsilon_t)(\hat{N} - \hat{N}_{eq}) \\ &\quad - \frac{\gamma_d(z)\hat{N}_{eq}\hat{L}}{4} - \frac{1}{2}\gamma_s(z)\hat{L}\hat{N} - \gamma_t(z)\hat{L}\hat{N}_{eq} \Big], \end{aligned} \tag{6}$$

where $\hat{N}(z)$ is the combined comoving number density, normalized to its equilibrium value at $z = 0$, of RH neutrinos and sneutrinos, while $\hat{L}(z)$ is the combined asymmetry of leptons and sleptons normalized in the same way. In the above equations we include the dominant $\Delta L = 1$ lepton–number violating scattering amplitudes $\gamma_{s,t}$ proportional to the top Yukawa coupling λ_t and driven by Higgs exchange in the s and t channel [5]. Moreover, the ϵ parameter is the amount of CP–violating in decays, while $\epsilon_{s,t}$ is the CP violation for $\gamma_{s,t}$.

The result of our calculation is shown in Fig. 1, where the efficiency $\eta \equiv \hat{L}(z = \infty)/\epsilon$ is calculated with the boundary condition $\hat{N}(0) = 0$ (vanishing initial RHN density) as a function

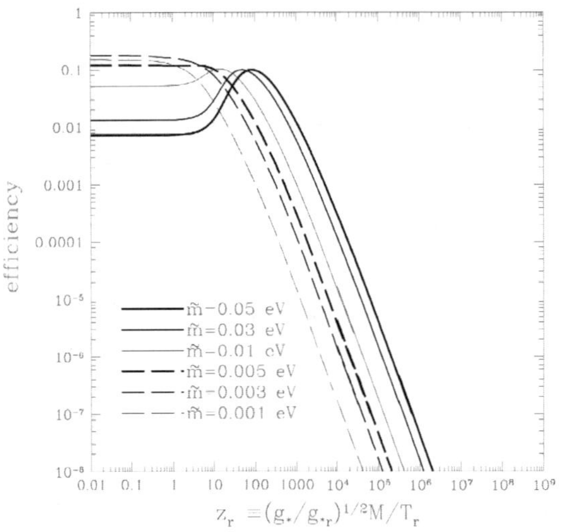

Figure 1. The efficiency as a function of $z_r \equiv \sqrt{g_*/g_{*r}} M/T_r$, for $N(0) = 0$ and for several values of $\tilde{m}$.

of z_r. In this plot a smooth transition between radiation domination (the plateau at $z_r \lesssim 1$) and kination-domination ($z_r \gtrsim 100$) is clearly visible. In particular, one can see than when $z_r >> 1$ and $K << 1$ the efficiency is strongly suppressed, so that a resonant CP violation $\epsilon \simeq 1$ is necessary in order to achieve succesful leptogenesis. For this reason in the calculation we have taken $\epsilon_{s,t} = \epsilon$. In fact, while $\epsilon \neq \epsilon_{s,t}$ is possible[6], in the case of resonant CP violation $\epsilon_{s,t}/\epsilon \rightarrow 1$ is expected to a very high level of accuracy. On the other the values of $\epsilon_{s,t}$ are irrelevant when $K \gtrsim 1$ and $z_r \lesssim 50$ since in this case scattering can be neglected in the first place. Including CP violation in scattering implies that in the super–weak wash–out regime, $K << 1$, one has aproximately $\eta \simeq 10.5K^2$. So the condition for a successful leptogenesis $Y_{\hat{L}} = 4 \times 10^{-3}\epsilon\eta \approx 10^{-10}$, which requires $\eta \gtrsim 5 \times 10^{-8}$ when $\epsilon \simeq 1$, implies the lower bound $T_r \gtrsim 5\ \mathrm{MeV} \left(\frac{M}{\mathrm{TeV}}\right) \left(\frac{0.05\ \mathrm{eV}}{\tilde{m}}\right) \left(\frac{g_*/g_{*r}}{228.75/10.75}\right)^{1/2}$. This means that, depending on the parameters, T_r can be as low as the nucleosynthesis temperature. On the other hand, for $M/100 \lesssim T_r \lesssim M$ the efficiency η can be even larger than in the standard scenario of radiation domination. In this case thermal leptogenesis can proceed for a hierarchical spectrum of heavy neutrino masses, and for this scenario the usual lower bound on the RHN mass M can be relaxed up to a factor of 30 compared to the standard case.

References

[1] Caldwell R R, Dave R and Steinhardt P J, 1998 *Phys. Rev. Lett.* **80** 1582 [arXiv:astro-ph/9708069]

[2] Kamionkowski M and Turner M S, 1990 *Phys. Rev.* D **42** 3310; Salati P, 2003 *Phys. Lett.* B **571** 121 [arXiv:astro-ph/0207396]; Chung D J H, Everett L L, Kong K and Matchev K T *Preprint* 0706.2375 [arXiv:hep-ph]; Chung D J H, Everett L L and Matchev K T *Preprint* 0704.3285 [arXiv:hep-ph].

[3] Fukugita M and Yanagida T, 1986 *Phys. Lett.* B **174** 45; Covi L, Roulet E and Vissani F, 1996 *Phys. Lett.* B **384** 169 [arXiv:hep-ph/9605319].

[4] Chun E J and Scopel S, 2007 *JCAP* **10** 011 [arXiv:0707.1544].

[5] Plümacher M, 1998 *Nucl. Phys.* B **530** 207 [arXiv:hep-ph/9704231]; Giudice G F, Notari A, Raidal M, Riotto A and Strumia A, 2004 *Nucl. Phys.* B **685** 89 [arXiv:hep-ph/0310123]; Buchmuller W, Di Bari P and Plümacher M, 2005 *Annals Phys.* **315** 305 [arXiv:hep-ph/0401240].

[6] Nardi E, Racker J and Roulet E, 2007 *JHEP* **0709** 090 [arXiv:0707.0378]; Pilaftsis A and Underwood T E J, 2004 *Nucl. Phys.* B **692** 303 [arXiv:hep-ph/0309342].

Indirect Searches for Dark Matter with IceCube

Carsten Rott for the IceCube Collaboration[1]
Pennsylvania State University, Department of Physics, 104 Davey Lab, University Park, PA 16802, USA
E-mail: carott@phys.psu.edu

Abstract.
The existence of dark matter can be inferred from a number of observations, among them rotational profiles of galaxies, large scale structures, and WMAP's anisotropy measurement on the cosmic microwave background. Weakly Interacting Massive Particles (WIMPs), 'cold' thermal relics of the Big Bang, are leading dark matter candidates. They are expected to be gravitationally trapped by massive bodies, such as the Sun and the Earth, or objects at the Galactic Center, where they could then annihilate and produce neutrinos, which could be detected by neutrino telescopes. We describe searches performed in AMANDA for these events and the prospects for IceCube.

1. Introduction

In the concordance model of cosmology, only about 4% of the universe's mass-energy content is described by standard model particles. The remaining fraction is predicted to be 23% non-baryonic cold dark matter (DM) and 73% dark energy. Understanding the dark sector of the universe is one of the most challenging problems in modern physics. Neutrino telescopes such as IceCube can search indirectly for DM by detecting neutrinos created in DM annihilation processes. IceCube will instrument a volume of 1 km^3 by 2011 with 80 strings, of which 22 have been deployed, each containing 60 Digital Optical Modules (DOMs). The AMANDA detector, which has been taking data in its final configuration since 2000 has been integrated into IceCube and is combinedly operated [1].

2. WIMP Searches with IceCube and AMANDA

Massive objects could gravitationally capture cold DM, resulting in an enhanced DM density at the center. Neutrinos could then be produced in annihilation processes and would escape from the source, providing a potential signal for kilometer scale neutrino telescopes. Indirect searches for such DM signals are competitive to direct searches for models with significant spin-dependent neutralino-proton couplings [2].

We have searched for such signals (from the Sun and center of the Earth) with AMANDA and upper limits were placed on the neutrino flux produced by neutralino annihilation processes.

In the search for solar WIMPs [3] the background was estimated from off-source data in the same declination band as the Sun. A 90% CL exclusion limit on the neutrino flux was set for a neutralino mass range from 100 GeV to 5 TeV as seen in Fig. 1. Two annihilation channels

[1] IceCube web site at http://www.icecube.wisc.edu

were compared: W^+W^-, which produces a hard neutrino energy spectrum, and $b\bar{b}$ yielding a soft spectrum. The sensitivity for solar WIMPs has also been evaluated using the combined IceCube/AMANDA detector [1] and a considerable improvement compared to AMANDA was achieved. The improvement is due to a reduced trigger threshold, a larger effective volume, and the detector configuration, in which AMANDA forms a more densely instrumented subarray inside the IceCube detector, which is especially important for detection of sub-TeV neutrinos. Peripheral strings can be used as a veto region to more effectively reduce the atmospheric muon background. The detector configuration allows for an analysis of fully and partially contained events yielding additional sensitivity.

We have also searched for vertically up-going muon neutrinos as expected from neutralino annihilations in the Earth's core using the AMANDA dataset and placed a 90% CL upper limit on the flux [3] as shown in Fig. 2 and compared to other experiments.

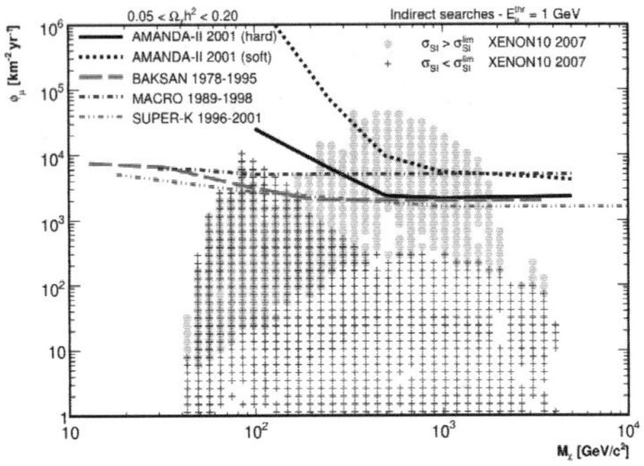

Figure 1. 90% CL upper limit on the muon flux from neutralino annihilations in the Sun compared to other experiments.

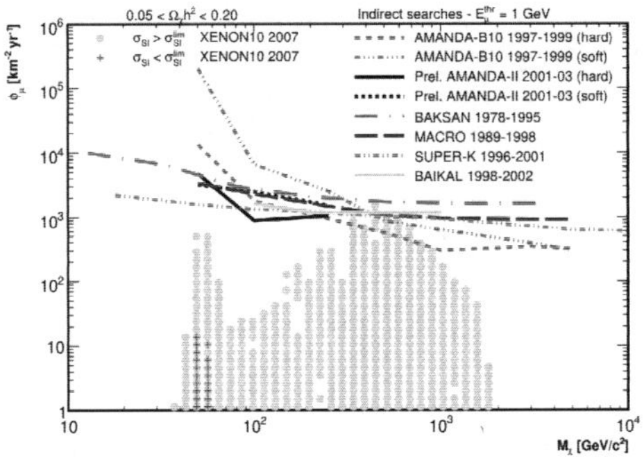

Figure 2. 90% CL upper limit on the muon flux from neutralino annihilations in the Earth compared to other experiments.

The region around the Galactic Center is another good potential candidate for sources of neutrinos from DM annihilations. Located in the Southern sky, it was so far inaccessible to AMANDA due to an irreducible down-going atmospheric muon background. However, the much larger IceCube detector might offer the potential to observe these neutrinos as fully or partially contained events. A dedicated filter for these contained events has been implemented [4].

3. Conclusions and Outlook

The combined IceCube/AMANDA detector provides a significant improvement in sensitivity for searches of DM. Dedicated filters for these events have been implemented and the analysis effort is ongoing. Extensions of IceCube to construct a more densly instrumented detector array near the bottom inside IceCube are also being studied. Such a configuration would further improve the sensitifity to sub-TeV neutrinos, as the deep ice is clearer, the deeper location provides an improved shielding from cosmic ray backgrounds, and surrounding IceCube strings can be used to reject down-going atmospheric muons.

References

[1] A. Achterberg *et al.* [IceCube Collaboration] Contributions to the 30th International Cosmic Ray Conference (ICRC 2007), astro-ph 0711.0353

[2] F. Halzen and D. Hooper, Phys. Rev. D **73**, 123507 (2006)

[3] M. Ackermann *et al.* [AMANDA Collaboration] Astropart. Phys. **24**, 459 (2006); A. Achterberg *et al.* [AMANDA Collaboration], Astropart. Phys. **26**, 129 (2006).

[4] C. Rott, TeV Astroparticle Physics III, Venice 2007

A cosmological bound on radiative neutrino lifetime

A. Mirizzi[1] *, D. Montanino[2] and P. D. Serpico[3]

[1]Max-Planck-Institut für Physik (Werner-Heisenberg-Institut), Föhringer Ring 6, 80805 Munich, Germany
[2]Dipartimento di Fisica and Sezione INFN di Lecce, Via Arnesano, 73100 Lecce, Italy
[3]Center for Particle Astrophysics, Fermi National Accelerator Laboratory, Batavia, IL 60510-0500 USA
* Paper presented at the conference by A. Mirizzi
E-mail: amirizzi@mppmu.mpg.de

Abstract. Neutrino oscillation experiments and direct bounds on absolute masses constrain neutrino mass differences to fall into the microwave energy range, for most of the allowed parameter space. As a consequence of these recent phenomenological advances, older constraints on radiative neutrino decays based on diffuse background radiations and assuming strongly hierarchical masses in the eV range are now outdated. We thus derive new bounds on the radiative neutrino lifetime using the high precision cosmic microwave background spectral data collected by the FIRAS instrument on board of COBE. The lower bound on neutrino lifetime is between a few $\times 10^{19}$ s and $\sim 5 \times 10^{20}$ s, depending on the neutrino mass ordering and on the absolute neutrino mass scale. However, due to phase space limitations, the upper bound on the effective magnetic moment mediating the decay is not better than $\sim 10^{-8}\,\mu_B$. We also comment about possible improvements of these limits, by means of recent diffuse infrared photon background data.

The standard cosmological scenario predicts the existence of a diffuse background of low-energy neutrinos, which has been often investigated in the past to probe non-standard neutrino properties. In this talk we use these cosmological neutrinos to revisit the bounds on neutrino radiative lifetime. The results of our study has been published in [1], to which we address the interested reader for further details.

The radiative decay $\nu_i \to \nu_j + \gamma$ can be thought of as arising from an *effective* interaction Lagrangian of the form

$$\mathcal{L}_{\rm int} = \frac{1}{2}\bar{\nu}^i \sigma_{\alpha\beta}(\mu_{ij} + \epsilon_{ij}\gamma_5)\nu^j F^{\alpha\beta} + \text{h.c.} \tag{1}$$

where ν_i is the neutrino field of mass m_i, and μ_{ij} and ϵ_{ij} are the magnetic and electric transition moments usually expressed in units of the Bohr magneton μ_B. In general, μ_{ij} and ϵ_{ij} are functions of the transferred momentum squared q^2, so that constraints obtained at a different q^2 are independent. In the following, we shall quote the bounds in terms of $\kappa_{ij}^2 = |\mu_{ij}|^2 + |\epsilon_{ij}|^2$.

From simple kinematical considerations it follows that in a decay $\nu_i \to \nu_j + \gamma$ from a state of mass m_i into one of mass $m_j < m_i$, the photon in the rest frame of the decaying neutrinos is thus monochromatic (two-body decay), with an energy $\varepsilon_{ij} = (m_i^2 - m_j^2)/2\,m_i$. At present, the neutrino mass spectrum is constrained by the well-known values of the two squared mass splittings for the atmospheric (Δm_H^2) and the solar (Δm_L^2) neutrino problems. We take their best-fit values and 2σ ranges from [2]. The remaining unknowns in the neutrino spectrum are the absolute mass scale (equivalently, the mass of the lightest eigenstate $m_1 < 2$ eV) and the mass hierarchy. We have discussed the limits on κ_{ij}^2 as a function of m_1 for the two cases of normal and inverted mass hierarchy.

The neutrino mass splittings squared provided by oscillation experiments and present upper bounds on the neutrino mass scale constrain neutrino mass differences to fall in the microwave energy range ($E \sim 10^{-3}$ eV), in most of the allowed parameter space. So, we have derived our cosmological bounds using the high precision cosmic microwave background (CMB) data collected by the FIRAS instrument on board of COBE, which tested the blackbody nature of the spectrum at better than 1 part in 10^4 [3, 4]. In the presence of neutrino decay, the original radiance (energy flux per unit of solid angle) of the "theoretical blackbody" at temperature T would gain an additional term. We have thus performed a chi-squared analysis, comparing the experimental measured CMB spectrum, with the theoretical predictions in presence of a radiative neutrino decay.

We obtain a lower bound on neutrino lifetime between a few $\times 10^{19}$ s and $\sim 5 \times 10^{20}$ s, depending on the neutrino mass ordering and on the absolute neutrino mass scale. Since it is customary to parameterize the neutrino electromagnetic decay via an effective operator of the kind reported in Eq. (1), it makes sense to translate the bounds on the lifetime into bounds on the parameter κ. However, due to phase space limitations, the constraints are not better than $\kappa \sim 10^{-8}\,\mu_B$, which at first sight do not appear competitive with astrophysical limits neither with most of the laboratory bounds. Nevertheless, every experimental measure and every cosmological and astrophysical constraint has its own systematic uncertainties and its own recognized or un-recognized loop-holes. In this sense it is certainly important to use many different approaches to constrain fundamental neutrino properties. In particular, the cosmological bound is based on the appearance of the daughter photons, and thus is very direct. Each decay model must face with our direct bounds on τ which are the strongest attainable with direct livetime measurements. Moreover, it also probes the energy scale $q^2 \sim 10^{-3}\,\mathrm{eV}^2$, unaccessible to both laboratory experiments and stellar arguments.

Possible improvements of our limits could be obtained using Cosmic Infrared Background (CIB) data [5]. Although these bounds appear stronger than those obtained by the FIRAS data in the same range of m_1, we emphasize that at present they should be considered only as indicative. In fact, the CIB flux have still strong uncertainties and the precise spectral shape is essentially unknown. If in the future a better measurement of the CIB flux and a more detailed knowledge of the astrophysical sources contributing to it would be available, stronger and more reliable bounds would be settled.

References

[1] A. Mirizzi, D. Montanino and P. D. Serpico, Phys. Rev. D **76**, 053007 (2007).

[2] G. L. Fogli *et al.*, Phys. Rev. D **75**, 053001 (2007).

[3] D. J. Fixsen, E. S. Cheng, J. M. Gales, J. C. Mather, R. A. Shafer and E. L. Wright, Astrophys. J. **473**, 576 (1996)

[4] J. C. Mather, D. J. Fixsen, R. A. Shafer, C. Mosier and D. T. Wilkinson, Astrophys. J. **512**, 511 (1999)

[5] H. Dole *et al.*, "The Cosmic Infrared Background Resolved by Spitzer. Contributions of Mid-Infrared Galaxies to the Far-Infrared Background," astro-ph/0603208.

Gravitational waves from extreme mass-ratio inspirals

Takahiro Tanaka

Department of Physics, Kyoto University, Kyoto 606-8502, Japan

E-mail: tama@scphys.kyoto-u.ac.jp

Abstract. Extreme mass-ratio inspiral (EMRI) is the main target of Laser Interferometer Space Antenna (LISA). We give a short review on this system as test-bench of general relativity. We briefly mention the recent progress in our study to obtain sufficiently accurate gravitational waveforms from this type of inspiralling binaries.

1. Introduction

There are various types of sources of gravitational waves. Among them, inspiralling binaries are the most promising candidate. This is because reasonably high event rate is expected and because the wave form, which is called a chirp signal, can be presumably theoretically predicted in advance. Besides the inspiralling binaries, there are semi-periodic sources such as binaries with large separation or pulsars, sources possibly correlated with optical counter part such as supernovae or γ-ray bursts, and stochastic background.

We expect that the observation of gravitational waves from inspiralling binaries in general will bring various information; event rate, binary parameters and possible deviation from general relativity. Binaries composed of stellar mass black holes and/or neutron stars are targets of ground based gravitational wave detectors [1, 2, 3, 4]. From these binaries we may be able to study the equation of state of neutron stars, possible correlation with the short γ-ray bursts and the abundance of primordial black hole binaries if they exist. Massive or intermediate mass black hole binaries are the target of the future space missions [5, 6, 7]. From the observation of gravitational waves from these sources, we will obtain information about the formation history of super massive black holes at the galactic centers. The last that we mention is the extreme (or intermediate) mass-ratio inspirals (EMRI), which is the main topic of my talk. This target is expected to be a good probe of the spacetime geometry around the black hole [8, 9].

Binary evolution is divided into three phases. The first is the inspiral phase, where the separation of binary is large compared with the horizon radius. In this phase, one can neglect the effect of internal structure of each object, and hence the evolution of binary can be approximated by that of two point particles [10]. To extract the maximum information from the observation, accurate prediction of the wave form is requested. First of all, detection of weak gravitational wave signal is possible only when we know the possible pattern of wave form in advance. To extract the binary parameters as accurately as possible, the correspondence between the wave form and the binary parameters must be also determined very precisely. When we use the observation of gravitational waves for the test of general relativity [11], we will use the event with the largest signal to noise ratio. In this case more precise theoretical prediction might be required. The inspiral phase is followed by the merging phase, where strongly non-linear dynamics can be understood only by numerical approach. Recently, there has been big progress in the area of numerical relativity, and we are now ready to simulate what happens at this merging phase. After the merging phase, there is the phase of ringing tail corresponding to the quasi-normal oscillations of the black hole formed after the coalescence.

2. Extreme mass ratio inspirals

The main frequency band of LISA (the Laser Interferometer Space Antenna) [5] is between 0.003Hz to 0.03 Hz. These frequencies correspond to the typical time scale of the super-massive black holes with mass between $10^5 M_\odot$ and $5 \times 10^6 M_\odot$. When various kinds of objects like white dwarfs, neutron stars and stellar mass black holes orbit such a super-massive black hole, gravitational waves in LISA band will be generated. Such a binary system with extreme mass-ratio is called EMRI. The basic scenario of the formation of such binaries is as follows. Most of galaxies are thought to have a central black hole, and there is a star cluster around the central black hole. Occasionally a stellar encounter which leads to a large angle scattering occurs in the star cluster. Sometimes the orbit after the scattering comes into an extremely eccentric orbit which passes the very vicinity of the central black hole. If the orbit is close enough to the central black hole, the gravitational radiation reaction is efficient before the star is rescattered by the surrounding star cluster. Then, the orbit is gradually circularized and the star is captured by the central black hole. Although the orbit is very eccentric ($e \sim 1$) at the beginning of the formation of this system, the eccentricity is reduced to a moderate value ($1 - e = O(1)$), say, at the last three years before the coalescence. The event rate is estimated and a few $\times 10^2$ events for 3 year observation by LISA are expected[12, 13], although the estimate of the event rate is still quite uncertain.

Since EMRI has a large mass ratio, the reduced mass μ is much smaller than the total mass M. In this case the small mass star, which we shall call satellite, follows very close to the geodesic in the spacetime governed by the large central black hole. The deviation from the geodesic motion is caused by the radiation reaction due to gravitational wave emission. This effect is suppressed by the mass ratio μ/M and is weak for EMRI. This means that a large number of cycles N are spent before plunge in the strong field region near the black hole horizon. From the comparison with the observational date, roughly speaking, difference in the number of cycles ΔN greater than 1 is detectable. Hence, a large number of cycles are advantageous for precise comparison between the theoretical templates and the observational data. The waveform depends on the metric of the background spacetime governed by the central super-massive black hole. Therefore it is said that gravitational waves from EMRI map the strong field region of the black hole spacetime. It is unlikely that the rotation of the central black hole is slow enough to be able to neglect its effect. Hence, we need to understand the waveform generated by a satellite orbiting a rotating super-massive black hole.

The precise theoretical prediction is possible for a binary system in which the effects other than the gravitational interaction can be neglected. If some other interaction gives a significant effect on the waveform, we cannot use this system as an ideal source for testing general relativity. Here "a significant effect" means that the induced phase shift in the gravitational waveform is greater than $O(1)$ after a large number of cycles over the whole observational period. Hence, even if the effect is a tiny minor correction to the equation of motion of the satellite, it can be harmful.

The super-massive black hole will be associated with the accretion disk. The disk-satellite interaction may cause a significant effect. This was studied by Narayan [14]. Roughly speaking, the interaction time scale of the satellite with the disk surrounding the central black hole is estimated using the usual formula for the dynamical friction time scale:

$$t_{df} = \frac{v_{rel}^3}{4\pi G^2 \mu\rho \log\Lambda}. \tag{1}$$

Here $\log\Lambda$ is the coulomb logarithm of $O(1)$ and G is the Newton's constant. The relative velocity between the satellite and the disk gas is almost at the Keplerian velocity in the case of advection dominated accretion flow (ADAF). The assumption of ADAF will be valid for the most of galactic centers which are not very active like AGN. In the case of ADAF, the accretion flow is almost spherically symmetric. Thus the density of the disk gas ρ will be estimated by using the accretion rate $\dot{M}$ as $\rho = \dot{M}/4\pi r^2 v_r$ where the radial velocity v_r is estimated as $v_r \approx \alpha v_k$ with $\alpha \approx 0.1$ using the phenomenological α-viscosity prescription. Substituting these estimates, we have

$$t_{df} \approx 4.5 \times 10^{12} \frac{M}{10^6 M_\odot} \left(\frac{\mu}{10 M_\odot}\right)^{-1} \left(\frac{\dot{M}}{10^{-2}\dot{M}_{Edd}}\right)^{-1} \text{yr}, \tag{2}$$

where $\dot{M}_{Edd}$ is the Eddington accretion rate.

Using the above formula we can estimate the shift of the orbital frequency Δf due to the dynamical friction as

$$\frac{\Delta f}{f} \approx \frac{T_{obs}}{t_{df}}. \tag{3}$$

T_{obs} is the duration of the observation, which is typically $O(1 \text{ year})$. This shift of the orbital frequency propagates to the change in the number of cycles as

$$\Delta N \approx \Delta f T_{obs} \approx f T_{obs} \frac{T_{obs}}{t_{df}}. \tag{4}$$

The observational frequency band f is $\approx 10^{-3}$Hz, and hence fT_{obs} is at most 10^5 or so. On the other hand, the last factor T_{obs}/t_{df} is $\approx 10^{-12}$ from Eq. (2). Therefore the effect of dynamical friction would be safely neglected in most cases.

3. Templates and test of general relativity

The theoretical prediction of the gravitational wave form from a binary in a quasi-circular orbit takes the form in Fourier space as

$$\tilde{h}(f) \approx A f^{-7/6} e^{i\Psi(f)}, \tag{5}$$

with $A = \mathcal{M}^{5/6}/\sqrt{20\pi^3} D_L$, $\mathcal{M} = \mu^{3/5} M^{2/5}$ and D_L is the luminosity distance. The phase $\Psi(f)$ is obtained in the post-Newtonian expansion as

$$\Psi = 2\pi f t_c - \phi_c + \frac{3}{128}(\pi\mathcal{M}f)^{-5/3}\left[1 + \frac{20}{9}\left(\frac{743}{331} + \frac{11}{4}\frac{\mu}{M}\right)(\pi M f)^{2/3} - (16\pi - \beta)(\pi M f) + \cdots\right], \tag{6}$$

where t_c is the coalescence time and ϕ_c is the phase at $t = t_c$. β is a parameter related to the spins of the component stars. The first term in the square brackets is the Newtonian term. The next terms are post-Newtonian corrections at 1PN order and 1.5PN order, respectively. We know how higher order expansion formally continues although the explicit coefficients in general cases are known only up to 3.5PN order. Only for detection, one may introduce a new search parameter at each order of expansion. In this sense the detailed knowledge about the higher order corrections may not be necessary for detection. However, for the precise extraction of the binary parameters or for the test of general relativity, we need higher order templates which are accurate within the observational error. Since the observational error is inversely proportional to the signal to noise ratio, better templates are requested when we observe by chance a golden event whose signal to noise ratio is extra-ordinary high.

For the test of general relativity. so far two possibilities of modified gravity are most investigated. One is the scalar tensor theory and the other is massive gravity. In these two cases the waveform is modified as

$$\Psi = \cdots + \frac{3}{128}(\pi \mathcal{M} f)^{-5/3}\left[\alpha u^{-2/3} + 1 + \left(\frac{3715}{756} + \frac{55}{9}\frac{\mu}{M} + \frac{128}{3}\frac{\mu}{M}\beta_g\right) u^{2/3} - (16\pi - \beta)\, u + \cdots\right], \tag{7}$$

where we have introduced $u = \pi M f$.

α is the parameter related to the scalar degree of freedom. α is inversely proportional to the so-called Brans-Dicke parameter ω_{BD}. The current bound on the Brans-Dicke parameter coming from the dipole radiation is $\omega_{BD} > 140$ for 4U 1820-30 (Neutron star + white dwarf binary in NGC6624) [15]. Constraint from future observation by LISA for $1.4M_\odot$ neutron star + $400M_\odot$ black hole can be $\omega_{BD} > 2 \times 10^4$ [11]. If we can use higher frequency band (~ 0.1Hz) for observation which is the target of DECIGO project [6], much stronger constraint is obtained like $\omega_{BD} > 5 \times 10^9$ by observing $1.4M_\odot$ neutron star + $10M_\odot$ black hole binary at a cosmological distance. $\beta_g = (\pi^2 M/\lambda_g^2 \int a^3 dt)$ is the parameter related to the graviton mass, where λ_g is the Compton wavelength of the massive graviton, a is the scale factor and η is the conformal time. The expected constraint obtained by future observations by LISA is $\lambda_g > 1$kpc, which is brought by observing the merger of $10^7 M_\odot$ black holes [11].

We said that EMRI is best suited as a probe of the black hole spacetime. The spacetime governed by the central black hole is expected to be given by Kerr geometry. In this case the quadrupole moment Q (and also higher order moment) of the spacetime is related to the spin S like $Q = -S^2/M$. However, in modified gravity theory some deviation from the Kerr geometry may arise. In Ref. [9] it was shown that Q/M^3 can be measured to within $\sim 10^{-2} - 10^{-4}$ based on the last year of data assuming $SNR = 100$.

4. Black hole perturbation

As we mentioned earlier, the template of gravitational wave in the inspiralling phase is obtained by using the standard post-Newtonian expansion. However, the higher order expansion is not easy in this method. In the case of EMRI we may need a very precise template due to the large number of cycles spent near the black hole. To obtain more accurate templates for EMRI, there is a more suitable method, which is the black hole perturbation. When the mass-ratio is large, we can treat the satellite as a small perturber of the central black hole. Then one can solve the Einstein equations

$$G^{\mu\nu}[\mathbf{g}] = 8\pi G T^{\mu\nu},$$

by expanding the metric as

$$g_{\mu\nu} = g^{BH}_{\mu\nu} + h_{\mu\nu}.$$

The method to compute $\mathbf{h}$ at the lowest order of expansion is well-established. Here we do not need to assume the slow motion of the satellite $v/c \ll 1$ as in the case of standard post-Newtonian expansion. At linear order, the equations $G^{\mu\nu}[\mathbf{g}^{(BH)} + \mathbf{h}] = 8\pi G T^{\mu\nu}$ can be reduced to a single master equation $L\psi = 4\pi\sqrt{-g}T$ for a single complex master variable ψ. L is a second order differential operator. For a non-rotating central black hole, the procedure is known as Regge-Wheeler-Zerilli formalism. (There are two real variables in this case.) For a rotating case, one can use the Teukolsky formalism. Recently, a systematic post-Newtonian expansion method was developed for these master equations [16]. Hence, we can systematically derive higher order analytic formulae in the post-Newtonian expansion.

The source term T is a function composed of the energy momentum tensor of the satellite. Schematically, $T^{(1)} = \tau_{\mu\nu}T^{\mu\nu}$ in the case of Teukolsky formalism, and $\tau_{\mu\nu}$ is a second order differential operator. The master variable ψ is a projected component of Weyl curvature. The master equation is usually solved by means of Green function method. First we solve the homogeneous equation $L\Omega = 0$. This equation allows separation of variables by setting $\Omega = \sum R_\Lambda(r)Y_\Lambda(\theta,\varphi)e^{-i\omega t}$, where $\Lambda = \{\ell, m, \omega\}$ represents a set of separation constants. $Y_\Lambda(\theta,\varphi)$ is the angular harmonic function. Substituting this decomposition, we obtain a radial equation $\left[\partial_r^2 + \cdots\right] R(r) = 0$. Using thus obtained solutions of the homogeneous equation, we have $\psi = \sum \Omega_\Lambda^{up} Z_\Lambda$, and Z_Λ is obtained by the convolution of the appropriately normalized mode function and the source term. Since the master variable ψ reduces to the usual two polarizations of gravitational waves as $\psi \sim (\ddot{h}_+ - i\ddot{h}_\times)/2$ in the limit of large radius, the energy and the angular momentum loss rates due to gravitational wave emission are written in term of the amplitude of gravitational waves, Z_Λ. The formulas are

$$\frac{dE}{dt} = -\sum_\Lambda |Z_\Lambda|^2, \qquad \frac{dL_z}{dt} = -\sum_\Lambda \frac{m}{\omega}|Z_\Lambda|^2.$$

5. Adiabatic evolution of orbits and wave form

What we want to obtain is the phase evolution of gravitational waves. As a simple example, let's consider quasi-circular orbits. Basically the evolution of the frequency of gravitational waves is determined by the evolution of the satellite orbital period. Equating the energy lost by the gravitational wave emission with the minus of the change in the binding energy, the change rate of the orbital frequency f is evaluated as

$$\frac{df}{dt} = -\frac{d\mathcal{E}_{GW}}{dt}\left(\frac{d\mathcal{E}_{orbit}}{df}\right)^{-1}. \tag{8}$$

The leading order of $d\mathcal{E}_{GW}/dt$ starts with $O(\mu^2)$. This is because the energy was exactly conserved, if the radiation reaction were turned off. On the other hand, the leading order of the relation between the binding energy and the orbital frequency, $d\mathcal{E}_{orbit}/df$, is determined by the geodesic motion on a given background black hole spacetime, and it is $O(\mu)$. The effect of the self-force, which is the force acting on a particle due to the metric perturbation caused by the particle itself, is higher order correction of $O(\mu^2)$ in $d\mathcal{E}_{orbit}/df$. Roughly speaking, to obtain the leading order correction to the waveform, we therefore have only to know the effect of the self-force on $d\mathcal{E}_{GW}/dt$.

The extension to more general orbits is straight forward in the case of Schwarzschild background. In this case one can assume that the orbit is on the equatorial plane without loss of generality. Then, the geodesics are completely specified by the energy $\mathcal{E}$ and the z-component of the angular momentum $\mathcal{L}$. Both "constants of motion" have the associated timelike and rotational Killing vectors, $\eta^\mu_{(t)} \equiv (\partial_t)^\mu$ and $\eta^\mu_{(\varphi)} \equiv (\partial_\varphi)^\mu$. Therefore we can define conserved current by $\int d\Sigma^\mu \eta^\nu t_{\mu\nu}$ using the effective energy momentum tensor of gravitational waves $t_{\mu\nu}$. Here $d\Sigma^\mu$ is the volume element of a constant time hypersurface.

In the case of Kerr background, we do not have spherical symmetry. Hence, one cannot say that the orbits are on the equatorial plane in general. To specify geodesics off the equatorial plane, we need to consider another "constant of motion". It is well known that the geodesics on the Kerr background possess the third "constant of motion" called Carter constant. However, Carter constant is not associated with any Killing vector field. Instead, it is related to a rank two Killing tensor. Killing tensor satisfies the equation

$$K_{(\mu\nu;\rho)} = 0. \tag{9}$$

Using this Killing tensor, the Carter constant is defined as

$$\mathcal{Q} \equiv \mu K_{\alpha\beta} u^\alpha u^\beta, \tag{10}$$

where $u^\alpha := dz^\alpha/d\tau$ is the four velocity of the satellite.

To evaluate $d\mathcal{Q}/dt$, we need to compute the self-force acting on the particle directly [17]. Though the prescription to calculate the self-force is formally established [18, 19, 20], performing explicit calculation is not so straight forward. However, it was found to be much easier to compute the averaged value of $d\mathcal{Q}/dt$. As an approximation, we may use the averaged values of the change rates for the "constants of motion" instead of those evaluated by using the instantaneous self-force. We call it adiabatic approximation. This approximation will be as good as an estimate using the balance argument mentioned above, and moreover is also applicable to $d\mathcal{Q}/dt$.

Gal'tsov has shown that the radiative field correctly reproduces the results obtained by using the balance argument for $d\mathcal{E}/dt$ and $d\mathcal{L}/dt$ when they are averaged over an infinitely long time interval [21]. Here the radiative field is defined by half retarded field minus half advanced one. Since the radiative field is free from divergence, we do not have to worry about how to regularize the self-force.

However, there had been no justification for applying the same scheme to $d\mathcal{Q}/dt$ until very recently. The breakthrough was brought by Mino, who gave a justification for applying the same scheme to $d\mathcal{Q}/dt$ [22].

Finally a simplified formula for the averaged value of dQ/dt was obtained by [23]. The self-force is explicitly written in term of the metric perturbation as

$$f^\mu[\mathbf{h}] := -\frac{\mu}{2}(g^{\mu\nu} + u^\mu u^\nu)(h_{\nu\rho;\sigma} + h_{\nu\sigma;\rho} - h_{\rho\sigma;\nu})u^\rho u^\sigma.$$

Using this force, the evolution of Carter constant is given by

$$\frac{d\mathcal{Q}}{d\tau} = 2K_\mu^\nu u^\mu f_\nu \tag{11}$$

Some complicated operation is necessary to compute all the metric component from the master variable ψ, although the prescription is given by Chrzanowski [24]. If we naively combine all the formulas, the final expression becomes very messy and hard to handle. But we found that, after several non-trivial manipulations, the final expression is greatly simplified. It turned out to be as easy to evaluate as the averaged values of $d\mathcal{E}/dt$ and $d\mathcal{L}/dt$. Some demonstrative calculations have been performed in [25, 26].

6. Summary

Among the various gravitational sources, extreme mass-ratio binaries are the most suitable ones to prove the geometry of a black hole spacetime. For high-precision test of general relativity, we need both space-borne gravitational antenna and accurate theoretical templates.

As a tool for computing the gravitational waveform, black hole perturbation theory and the radiation reaction due to the emission of gravitational waves have been developed. Here we reported the recent progress in the area of the adiabatic approximation to the radiation reaction. Within this approximation, we are now ready to compute the evolution of the satellite orbit taking into account the gravitational reaction. In this approximation we treat only the averaged rates of change of the "constants of motion". So far, it is not clear if we need to calculate not the averaged effect but the instantaneous self-force for the use of precision test of general relativity in reality. This must be clarified soon. If the computation of instantaneous self-force is really necessary, it is an urgent problem which must be solved before the launch of LISA.

[1] LIGO web page: http://www.ligo.caltech.edu/
[2] TAMA300 web page: http://tamago.mtk.nao.ac.jp/
[3] GEO600 web page: http://www.geo600.uni-hannover.de/
[4] VIRGO web page: http://www.virgo.infn.it/
[5] LISA web page: http://lisa.jpl.nasa.gov/
[6] N. Seto, S. Kawamura and T. Nakamura, Phys. Rev. Lett. **87**, 221103 (2001) [arXiv:astro-ph/0108011].
[7] BBO web page: http://universe.nasa.gov/program/bbo.html
[8] F. D. Ryan, Phys. Rev. D **56**, 1845 (1997).
[9] L. Barack and C. Cutler, Phys. Rev. D **75**, 042003 (2007) [arXiv:gr-qc/0612029].
[10] C. Cutler *et al.*, Phys. Rev. Lett. **70**, 2984 (1993) [arXiv:astro-ph/9208005].
[11] E. Berti, A. Buonanno and C. M. Will, Phys. Rev. D **71**, 084025 (2005) [arXiv:gr-qc/0411129].
[12] J. R. Gair, L. Barack, T. Creighton, C. Cutler, S. L. Larson, E. S. Phinney and M. Vallisneri, Class. Quant. Grav. **21**, S1595 (2004) [arXiv:gr-qc/0405137].
[13] P. Amaro-Seoane, J. R. Gair, M. Freitag, M. Coleman Miller, I. Mandel, C. J. Cutler and S. Babak, Class. Quant. Grav. **24**, R113 (2007) [arXiv:astro-ph/0703495].
[14] R. Narayan, arXiv:astro-ph/9907328.
[15] C. M. Will and H. W. Zaglauer, Astrophys. J. **346**, 366 (1989).
[16] S. Mano, H. Suzuki and E. Takasugi, Prog. Theor. Phys. **95**, 1079 (1996) [arXiv:gr-qc/9603020]; Prog. Theor. Phys. **96**, 549 (1996) [arXiv:gr-qc/9605057];
[17] A. Ori, Phys. Lett. A 202, 347 (1995); Phys. Rev. D 55, 3444 (1997).
[18] B. S. DeWitt and R. W. Brehme, Annals Phys. **9**, 220 (1960).
[19] Y. Mino, M. Sasaki and T. Tanaka, Phys. Rev. D **55**, 3457 (1997) [arXiv:gr-qc/9606018]; Prog. Theor. Phys. Suppl. **128**, 373 (1997) [arXiv:gr-qc/9712056].
[20] T. C. Quinn and R. M. Wald, Phys. Rev. D **56**, 3381 (1997) [arXiv:gr-qc/9610053].
[21] D.V. Gal'tsov, J.Phys. A 15, 3737 (1982).
[22] Y. Mino, Phys. Rev. D **67**, 084027 (2003) [arXiv:gr-qc/0302075].
[23] N. Sago, T. Tanaka, W. Hikida and H. Nakano, Prog. Theor. Phys. **114**, 509 (2005) [arXiv:gr-qc/0506092].
[24] P. L. Chrzanowski, Phys. Rev. D **11**, 2042 (1975).
[25] N. Sago, T. Tanaka, W. Hikida, K. Ganz and H. Nakano, Prog. Theor. Phys. **115**, 873 (2006) [arXiv:gr-qc/0511151].
[26] K. Ganz, W. Hikida, H. Nakano, N. Sago and T. Tanaka, arXiv:gr-qc/0702054.

Status of Gravitational Wave detection

Adalberto Giazotto-INFN Pisa-EGO
adalberto.giazotto@pi.infn.it

1) The discovery

Taylor and Hulse[1] discovery of Gravitational Wave[2] (GW) emission by the compact binary system composed by two Neutron Stars PSR1913+16 has been, for the experimental physicists working in this field, the definitive thrust allowing to reach the extremely sophisticated technology needed for investigating in this field of research.

J. Taylor

R.Hulse

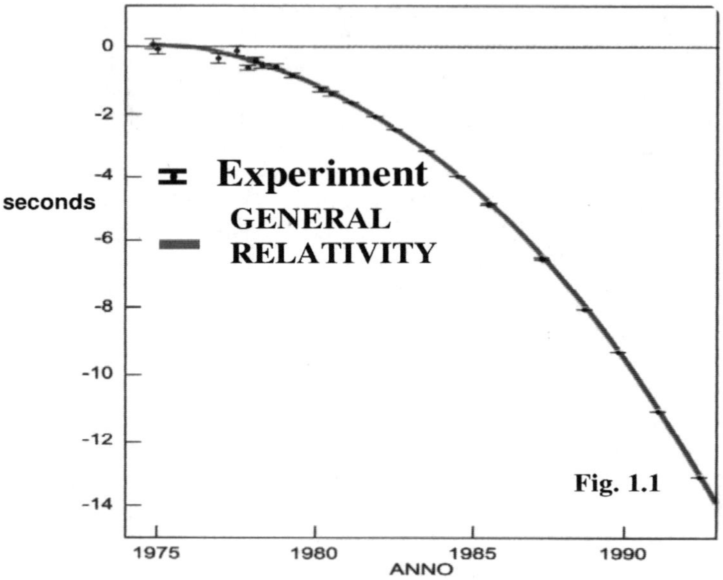

Fig. 1.1

In Fig.1.1 the famous histogram showing the decreasing of periaster passage time produced by the associated emission of GW together with the prediction evaluated through General Relativity (GR), is shown.Today at least other 6 similar systems are investigated and energy loss remarkably agrees with the prediction of GR.

2) The generation and detection of gravitational waves[3]

Gravitational Waves are a consequence of Einstein's GR; they are shown to be ripples in the space -time curvature traveling at light speed. It is easy to show that GW amplitude has two polarizations h^+ and h^x:

$$\begin{cases} h^{TT}_{\mu\nu} = h^{TT}_{11}\begin{pmatrix}0&0&0&0\\0&1&0&0\\0&0&-1&0\\0&0&0&0\end{pmatrix} + h^{TT}_{12}\begin{pmatrix}0&0&0&0\\0&0&1&0\\0&1&0&0\\0&0&0&0\end{pmatrix} = h^+ e^+_{ik} + h^X e^X_{ik} \\ GW\ Along\ X_3 \end{cases} \quad \textbf{Eq. 1}$$

The wave amplitude in the Transverse Traceless (TT) system satisfy the following equation:

$$\ddot{h}^{TT}_{\alpha\beta} = -\frac{2G}{c^4 R_0}\frac{d^2}{dt^2}\int \rho\left(x_\alpha x_\beta\right)^{TT} dV \qquad \textbf{Eq. 2}$$

Where R_0 is the source distance, G is Newton constant, and the integral is extended to source volume. Eq. 2 shows that only asymmetric sources can emit GW; most efficients are coalescing

binaries sources, while pulsar can rely only on ellipticities, usually very small. Supernovae could have relevant asymmetries but this is very unlikely.
In his fundamental papers on Gravitational Waves, A.Einstein shows how a mechanical system is creating GW. For the detection of GW he shows that GW couple to matter and increase the energy of a suited mechanical system but it was F.A.E.Pirani[4] in 1956 who proposed the geodesic deviation equation as a tool for designing a practical GW receiver. The result is that the relative acceleration of two freely falling bodies, due to GW action is: $\frac{d^2\xi_\alpha}{d\tau^2} = \frac{1}{2}\ddot{h}^{TT}_{\alpha\beta}\xi^\beta$

3) Early detectors
It is relevant to start this paragraph by quoting very briefly some highlights of Weber life, the part more connected with GW detection, because I believe that not enough appreciation from the scientific community has been given to this great scientist. As it can be read he was also dealing with interferometric detectors too.

Joseph Weber
-Born on 17 May 1919 in Paterson, New Jersey,
-In the early 1960s, J.Weber[5] turned his attention to testing the general relativistic prediction of gravitational waves from strong gravity collisions.
-Then,in 1959, he studied a detector that might be able to measure displacements smaller than the size of the nucleus.
-He was alone in charting these unknown waters: His first paper on how to build a gravity wave detector was published in 1959.
-Weber developed an experiment using a large suspended bar of aluminum, with a high resonant *Q* at a frequency of about 1 kHz.
-In 1969 he was claiming evidence for observation of gravitational waves (*Physical Review Letters* (1969)) based on coincident signals from two bars separated by 1000 km.Those events, later, have been demonstrated to be consistent with cosmic rays.
-Weber also proposed the idea of doing an experiment to detect gravitational waves using laser interferometric techniques.
-Robert Forward, Weber's student and postdoctoral associate, started in 1970 the first laser interferometric gravitational wave experiments at Hughes Research Labs.
-Joseph Weber, professor emeritus of physics at the University of Maryland, College Park, died on 30 September 2000 in Pittsburgh, Pennsylvania.

JOSEPH WEBER WITH GRAVITATIONAL WAVE DETECTOR

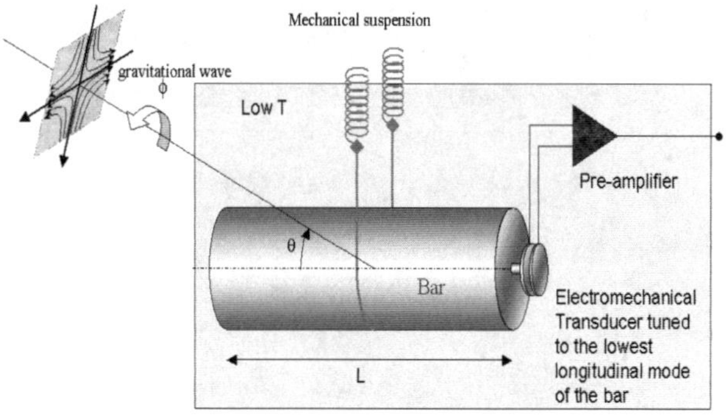

Fig.3.1- A schematic diagram showing a bar detector with the transducer. The GW Riemann forces excite longitudinal modes of the Bar. Bar elongation is detected by the transducer and transformed to an electrical signal. The length L is tipically 2-3 m.
Picture courtesy M. Cerdonio

4) Modern Bar Detectors: Cryogenic Bars[6].

All these modern detectors were originate from early Weber's detectors.

In the world 5 cryogenic bar detectors have been built for working at temperatures <4K: Explorer at CERN (see Fig. 4.1) , Nautilus at Frascati INFN National Laboratory (see Fig. 4.2), Auriga at Legnaro INFN National Laboratory (see Fig. 4.3), Allegro at Luisiana State University (see Fig. 4.4) and Niobe in Perth. Instrumental details can be found in an extensive literature. I report the sensitivity and stability of performances of only four of them since Niobe stopped operation in 1997.

Fig. 4.1

Fig. 4.2

Fig. 4.4

Fig.4.3

IGEC (Resonant Bar Detectors International Gravitational Event Collaboration) was established in 1997 in Perth. IGEC has been the first GW Detector network. During IGEC-1 (1997-2000) the following coincidences where registered:
29 days of four-fold coinc.
178 days of three-fold coinc.
713 days of two-fold coinc.
Sensitivity and bandwidth are shown in Fig. 4.5.

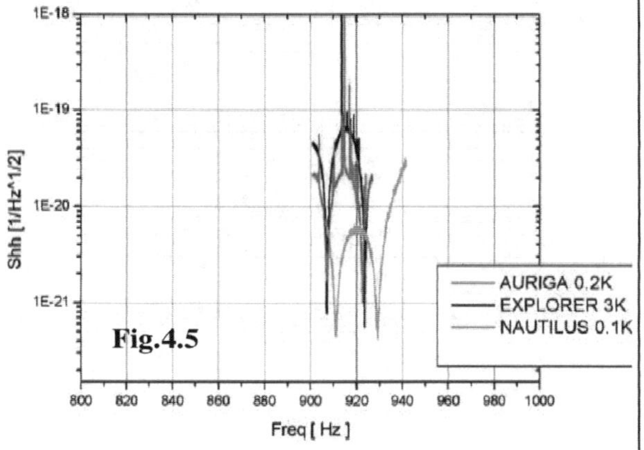

Fig.4.5

Followed by a series of upgrades, resumed operations EXPLORER in 2000, AURIGA and NAUTILUS in 2003, ALLEGRO in 2004 the Collaboration started to take data.
With IGEC-2 (2005--) first data analyzed covered May-November 2005 when no other observatory was operating. Sensitivity and operation stability are shown in Fig. 4.6 and 4.7 respectively Bar Detectors situation at present (oct. 2007) is the following: NIOBE (Perth) stopped operation and did not join IGEC-2, ALLEGRO (LSU) stopped operation in 2007. In 2006 INFN acted as follows:
a)stopped R&D on Spherical Detectors
b)left on the R&D on DUAL.
c)left Auriga, Nautilus and Explorer running on an annual evaluation.
These three detectors are currently running with very good sensitivity and duty cycle. Their role is particularly relevant when Virgo and LIGO will be shut down for upgradings to Virgo Plus and Henanced LIGO; their running is aimed to a coincidence with GEO 600 and Tama 300 interferometers for a kind of observation which has been named "Astro watch".

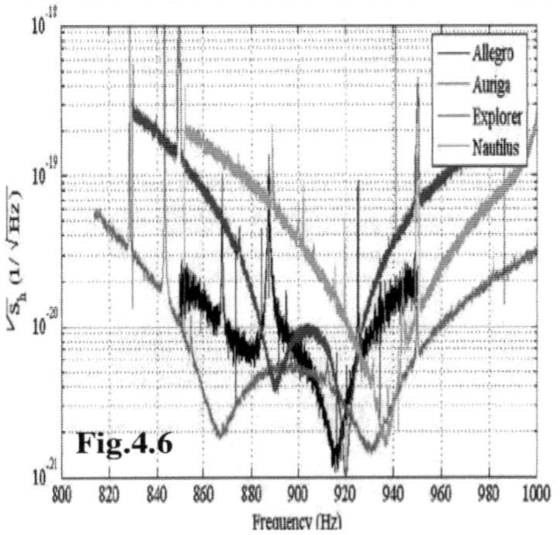

Fig.4.6

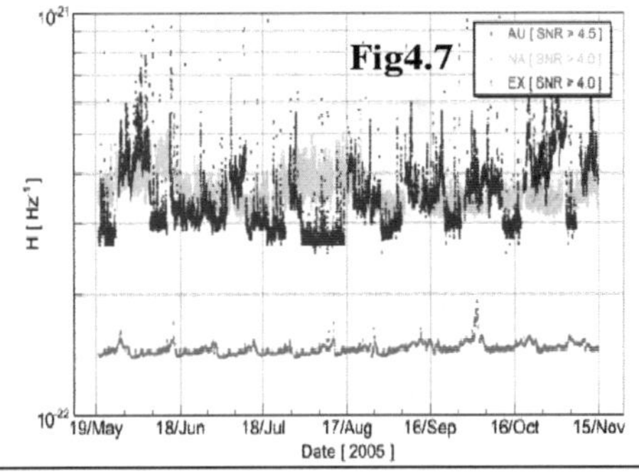

Fig4.7

5) Spherical detectors[7]
A spherical antenna will be able to determine the wave polarization and localize its astrophysical source on the sky. Another very important advantage of this kind of detector comes exactly from the fact that it has many transducers monitoring many quadrupolar modes.
Mario Schenberg: this detector (see Fig.5.1) is built in San Paolo (Brazil) and has a 65cm-diameter CuAl6% sphere weighing about 1.15 ton; sensitivity is shown in Fig. 5.2. In this detector each one of the five quadrupole modes of the sphere has an effective mass for oscillation of 287 kg.

Fig. 5.1

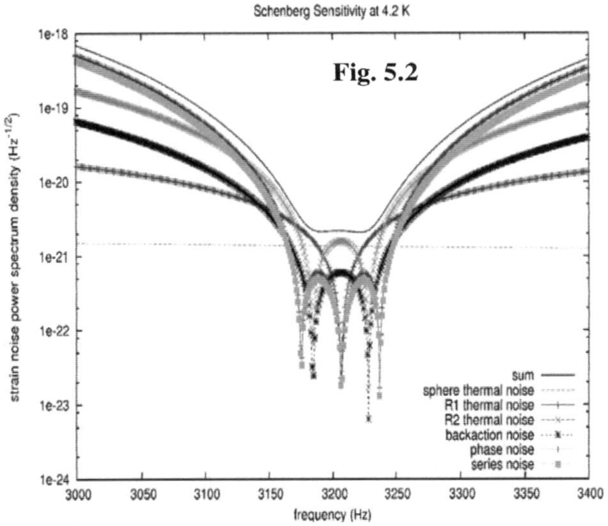

Fig. 5.2

The MiniGRAIL detector is a cryogenic 68 cm diameter spherical gravitational wave antenna (see Fig.5.3) built at Kamerlingh Onnes Laboratory of Leiden University, made of CuAl(6%) alloy with a mass of 1400 Kg, a resonance frequency of 2.9 kHz and a bandwidth around 230 Hz, possibly higher. The quantum-limited strain sensitivity dL/L would be ~$4x10^{-21}$(see Fig. 5.4). The antenna will operate at a temperature of 20 mK.)

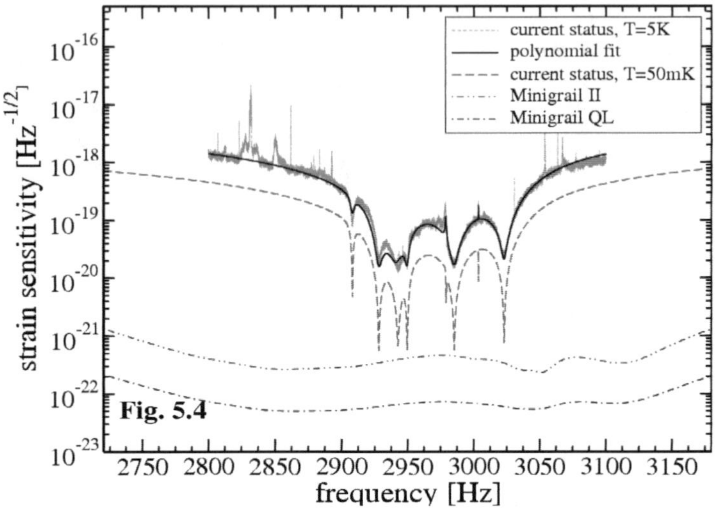

6) Interferometric detectors[8]

In this case free falling masses are interferometer mirrors which can be separated km far apart (3km for Virgo, 4km for LIGO), consequently GW Riemann force can be several order of magnitude larger than in bar detectors. Accompanying large sensitivity is the very large Bandwidth (10-10000 Hz) due to the fact that mirrors are suspended to pendula having resonance in the Hz region; above resonance frequency mirrors are freely falling masses in the horizontal plane.

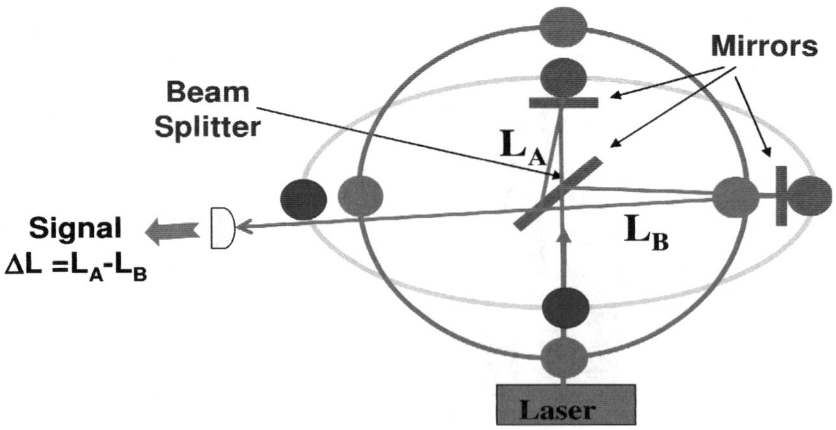

Fig.6.1-GW deform a test masses circular ring (red) in an elliptical one (green) and the red masses displace in the blue position. If two mirrors are attached to the test masses the arm unbalance $\Delta L = L_A - L_B$ can be measured interferometrically. Interferomer's displacement sensitivity can reach ~10^{-19}-10^{-20} m, then, for measuring $h=\Delta L/L \sim 10^{-22}$ L_A and L_B should be km long.

Due to quadrupolar nature of GW a circle of test masses, in red in Fig. 6.1, is deformed in the green ellipse. Always with reference to Fig. 6.1, by connecting mirrors to the freely falling test masses, we can interferometrically measure GW intensity by the relative unbalance $\Delta(L_A-L_B)/L$ of the two ellipse radii; this relative length unbalance is equal to GW amplitude h.
Interferomer's displacement sensitivity can reach $\sim 10^{-19}$-10^{-20} m, then, for measuring $h=\Delta(L_A-L_B)/L \sim 10^{-22}$, L_A and L_B should be km long.
With reference to Fig. 6.3 the two polarizations h_x and h_+ gives contribution to total mirror displacement according to Eq. 3.

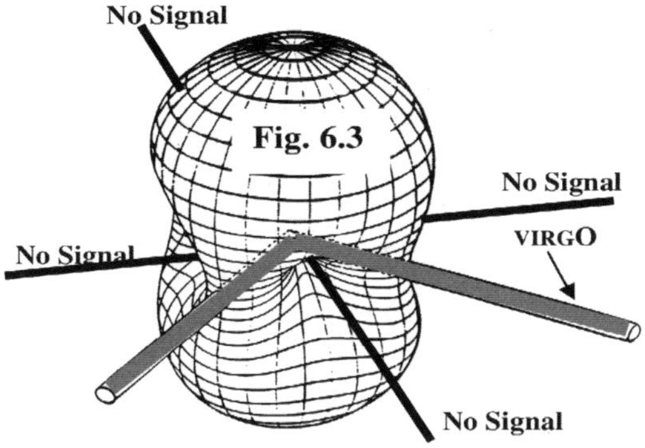

Fig. 6.3

In Fig. 6.2 the Coordinate system having ξ and η axis coinciding with interferometer arms is shown; n is GW versor and θ and φ are his polar and azimuthal angles. From Eq. 3 it is easy to evaluate the sensitivity pattern summed on GW polarizations for an interferometer having arms at 90^0; the antenna pattern is shown in Fig. 6.3. The effect of the two GW polarizations on a circular ring (red) of masses for a wave travelling normally to the circle plane, is shown in Fig. 6.4

Fig. 6.4

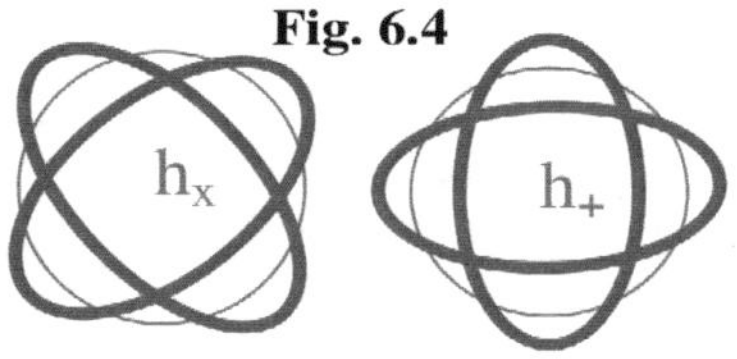

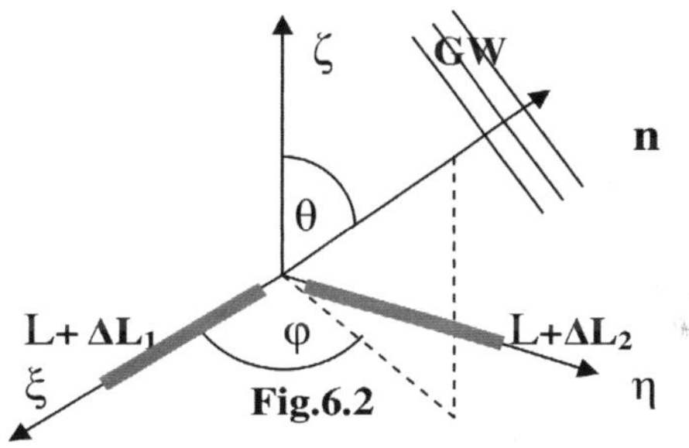

Fig.6.2

$$\frac{\Delta L_1+\Delta L_2}{L}=\frac{1+(n.\zeta)^2}{2}\frac{(n.\xi)^2-(n.\eta)^2}{1-(n.\zeta)^2}h_+ + 2\frac{(n.\xi)(n.\eta)(n.\zeta)}{1-(n.\zeta)^2}h_x \quad \text{Eq. 3}$$

7) Interferometer noises[9]

It is easy to imagine that a plethora of noises is affecting interferometers for GW detection; infact this kind of interferometers has mirrors supended and free to move in the horizontal plane. We should imagine how small is the displacement to be measured by these machines i.e. $<10^{-19}$m/sqrt(Hz) in the frequency band 4-10000 Hz. Just for making a mere list we have (see Fig. 7.1, 7.2):
Optical Noises: Shot Noise, Laser frequency fluctuations, Laser power fluctuations, Diffused and Scattered light hitting vacuum vessels, where collects vibrations, and partially reflected back in the main optical path, Acoustic noise…..
Force Noises : Radiation pressure fluctuations, Thermal noise in the mirrors and in the mirror suspensions, Electromagnetic noises producing forces on the mirrors…..
A pictorial view of noises affecting a suspended mirror interferometer is shown in Fig.7.1. In Fig. 7.2 the accent is given on two kinds of fundamental noises: Thermal noise and Optical Noises i.e. Shot noise, Radiation pressure fluctuation noise and Standard Quantum Limit.Optical noises can be reduced by Quantum Non-Demolition (QND) techniques Thermal noise is likely to be reduced by keeping the Interferometer mirrors close to 4 K.

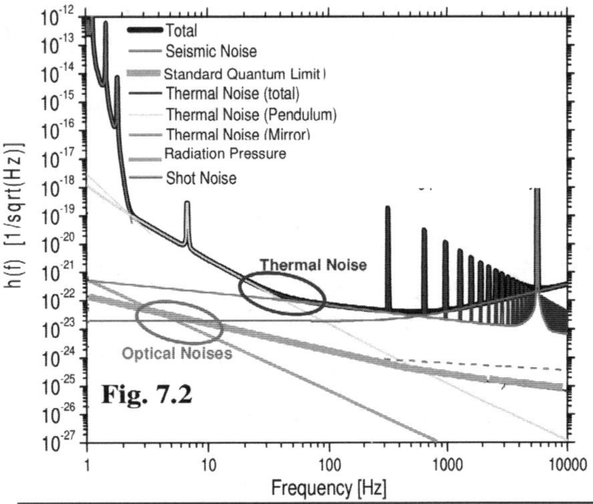

Fig. 7.2

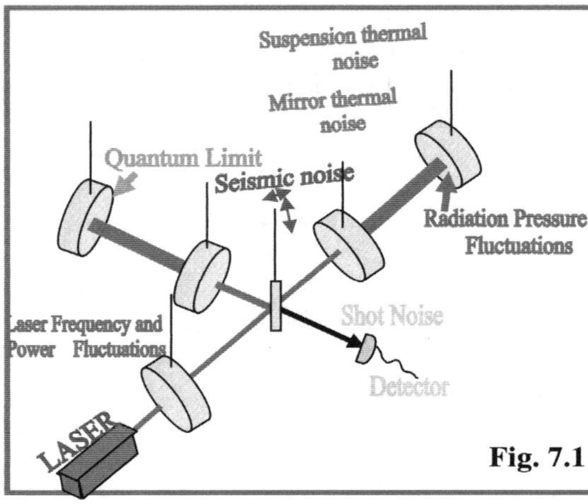

Fig. 7.1

If we define W the laser power, F and L the cavity finesse and length respectively, ν_0 the laser frequency, Ω the observation frequency and M the mirror mass, the optical noises will give the following limit to $\tilde{h}$: In Eq. 4 the first term on the right hand side is due to to radiation pressure fluctuation while the second is due to shot noise. By minimizing $\tilde{h}$ with respect WF^2 we obtain the Standard Quantum Limit (SQL) Eq. 5. Today ways[11]exist, by using squeezed light tecniques, able to cancel Radiation pressure fluctuations in interferometers and then overcoming SQL.

$$\tilde{h}^2 > \left(\frac{2}{LM\Omega^2 c}\right)^2 h\nu_0 WF^2 + \quad \textbf{Radiation press. fluctuations}$$

$$+\left(\frac{\lambda}{4\pi}\right)^2 \frac{h\nu_0}{WF^2}\,\frac{1+\left(\frac{\Omega F L}{c}\right)^2}{L^2} \quad \textbf{Shot noise} \qquad \text{Eq. 4}$$

$$\tilde{\mathbf{h}}_{\mathrm{SQL}} \approx \frac{\mathbf{1}}{\mathbf{L\Omega}}\sqrt{\frac{\hbar}{\mathbf{M}}} \qquad \text{Eq. 5}$$

8)The Network of interferometric GW detectors

Some relevant reasons for starting network coherent analysis are:
-Sensitivity increase.
-Source direction determination from time of flight differences.
-Sources polarizations measurement.
-Test of GW Theory and GW physical properties.

Some relevant astrophysical targets are:
- Far Universe expansion rate measurement.
-GW energy density in the Universe.
-Knowledge of Universe at times close to Planck's time.

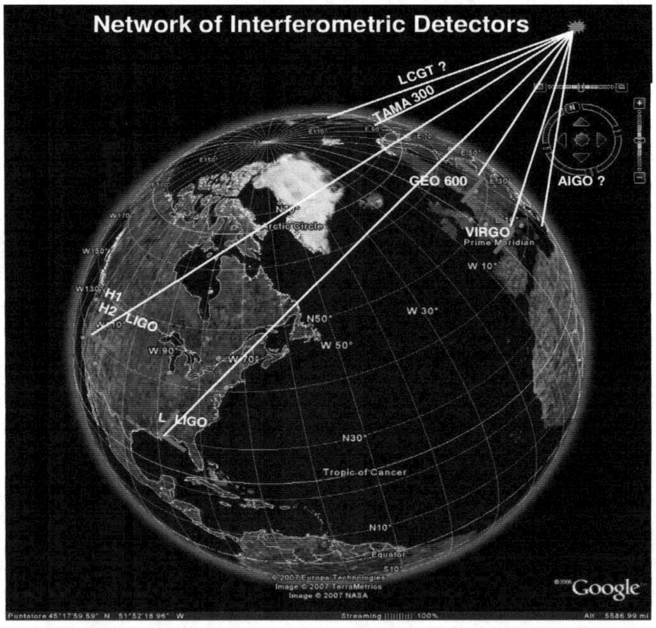

In Fig 8.1 the optical diagram of Virgo, a typical large base GW detector, is shown.
With reference to Fig. 11.1, the laser beam is prestabilized up to 2 Hz, at a frequency stability **$\Delta\upsilon=10^{-4}Hz^{1/2}$** by means of a Ultra Low Expansion cavity and then, by using the 3+3 km cavities operated in common mode, for frequencies in the range 2-10000 Hz, to a frequency stability level of **$\Delta\upsilon=10^{-6}Hz^{1/2}$.** The power recycling mirror PR allows to increase by a factor 20 the power on the beam splitter mirror BS; the two 3 km Fabry-Perot cavities have a Finesse of 30 and, in Virgo, cavities mirrors are aligned by using Anderson's technique.

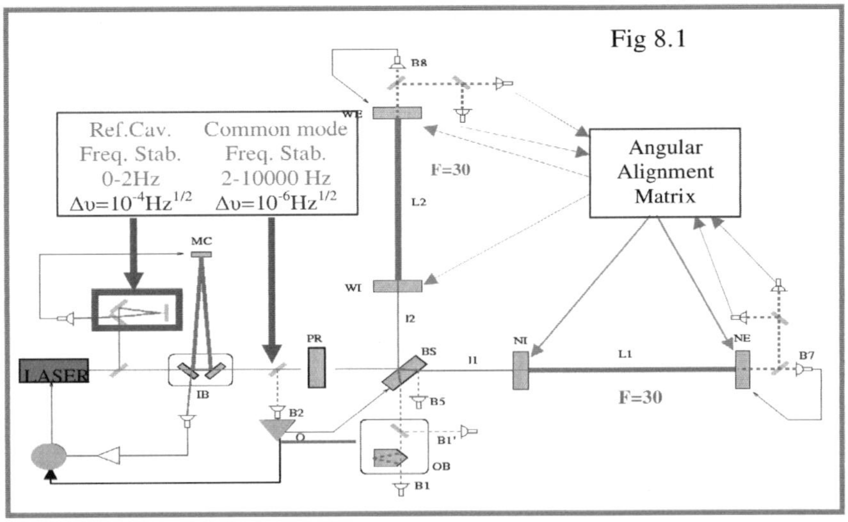

Fig 8.1

In the following a brief description of GW interferometers performances is presented.

LIGO[10]
LIGO project, run by Caltech and MIT, is the largest existing GW detector, composed by 3 interferometers: a 4 km+ 2 km interferometers inside the same vacuum pipe in Hanford (Washington state) (see Fig 8.2) and a 4 km detector in Livingston (Louisiana state) (see Fig. 8.3).
The distance between Hanford and Livingston is about 3000 km.
The sensitivity of Hanford machine, really outstanding, is shown in Fig. 8.4; the graph shows only the sensitivity of Hanford machine, but a graph showing sensitivities of all three LIGO machines will be shown later; the sensitivity progress of Hanford machine is shown in Fig.8.5 while operation stability is shown in Fig.8.6. The science run time progression is shown in Fig.8.7. Fig. 8.6 shows LIGO operation stability; in the ordinate axis of this plot is the distance in Mpc at which a 1.4-1.4 M_0 NS-NS coalescing binary system, averaged on orbital plane directions , is detectable. Sensitivity uniformity and operation stability are remarkable.

Fig 8.2

Fig 8.3

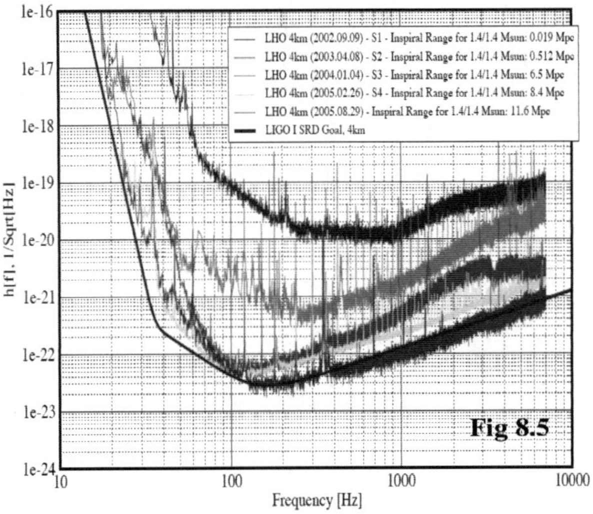

Fig 8.5

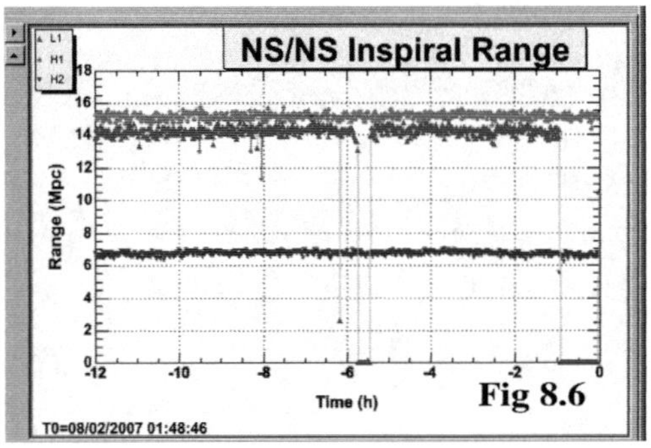

Fig 8.6

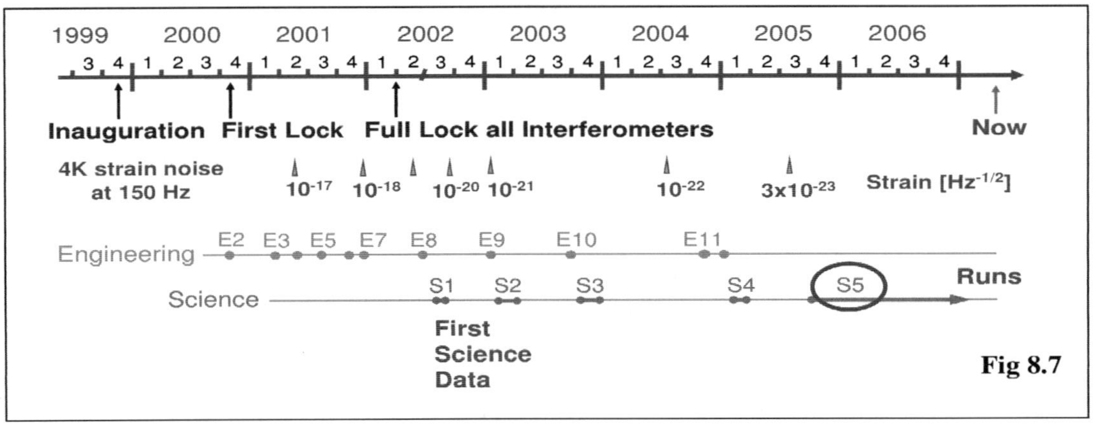

Fig 8.7

Virgo[11]

Virgo is collaboration, initially between INFN (Italy), CNRS (France) at which later added NIKHEF (Holland), has built a 3 km interferometer (see Fig. 11.8) located in Cascina, about 20 km from Pisa.

Fig. 8.8-The Virgo 3 km interferometer.

Virgo has a very sophisticate antiseismic system whose basic elements are called Superattenuators (SA) (see Fig. 8.9).These devices, thoroughly described elsewere[12], are capable to attenuate seismic noise below expected thermal noise down to 4 Hz. In Fig. 8.10 the measured SA attenuation transfer function, measured stage by stage, while the upper limit point at 4 Hz has been otained by shaking SA suspension point.

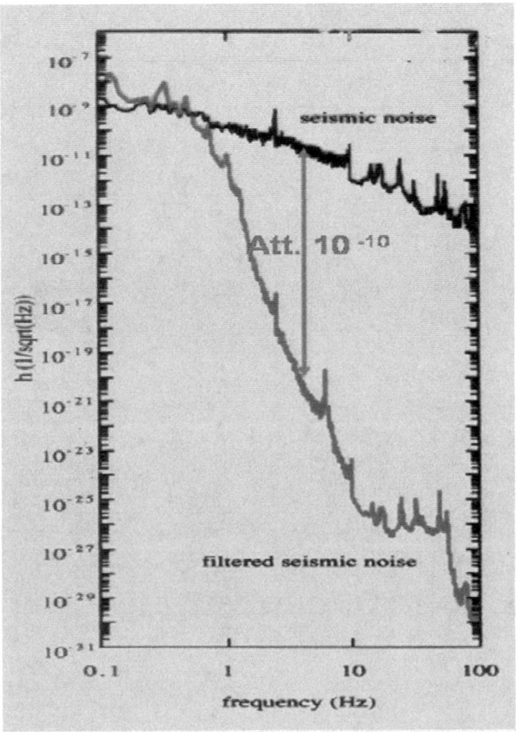

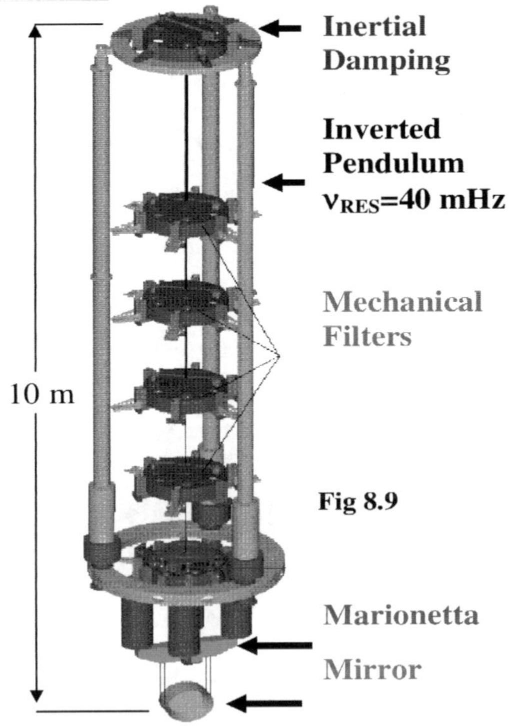

Fig. 8.9-The Virgo Superattenuator.

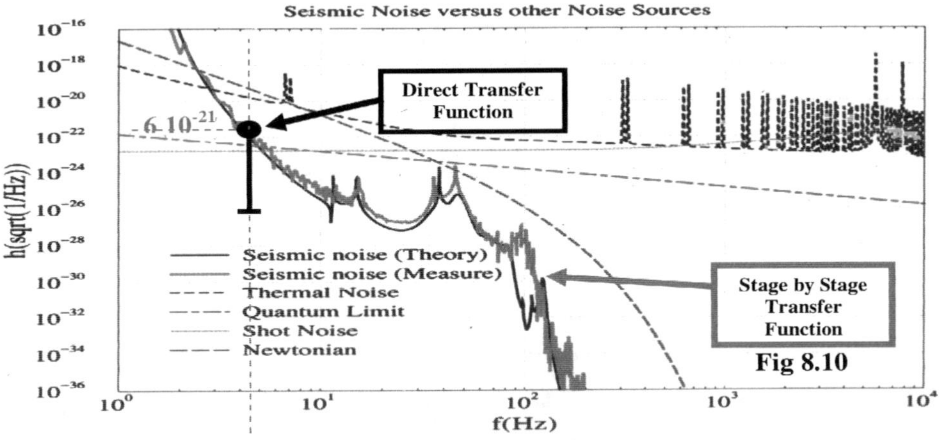

Fig.8.10-The seismic noise remnant amplitude having assumed a seismic spectral displacement $10^{-7}/f^2$ m/sqrt(Hz). The Superattenuator transfer function measured stage by stage (red curve) and directly by shaking SA suspension point (black point).

In Fig. 8.11 Virgo sensitivity progress is shown; it is noticeable the impressive low frequency sensitivity improvements spanning about 8 orders of magnitude. At high frequency design sensitivity has not yet been reached since Virgo is running at low power due to mirror lensing effects. Lensing corrective devices are being tested.

-Virgo operation stability over 8 hours of running is shown in Fig. 8.12; in the ordinate axis of this plot is the distance in Mpc at which a 1.4-1.4 M_0 NS-NS coalescing binary system, averaged on orbital plane directions , is detectable. In Fig. 8.13 Virgo operation duty cycle over 1 month of running is shown. .

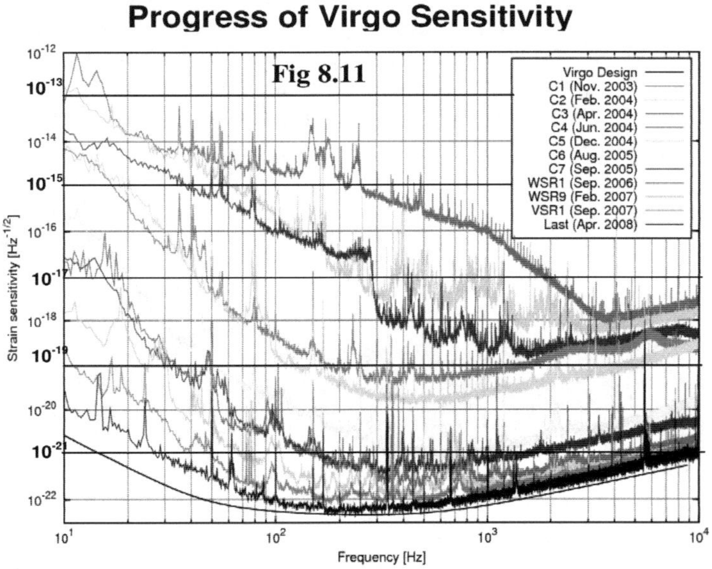

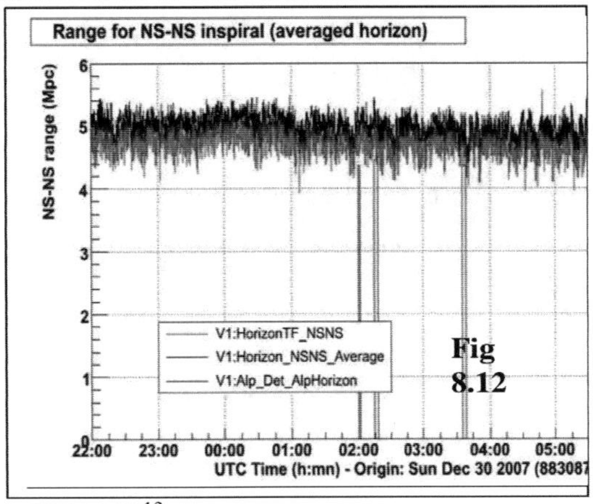

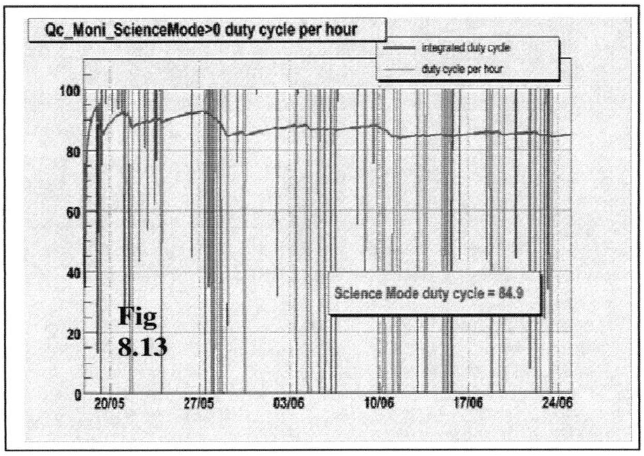

GEO600[13]

GEO600, a collaboration between Great Britain and Germany, is a 600 m long arms signal and power recycled interferometer located near Hannover (see Fig.8.14)

Fig 8.14

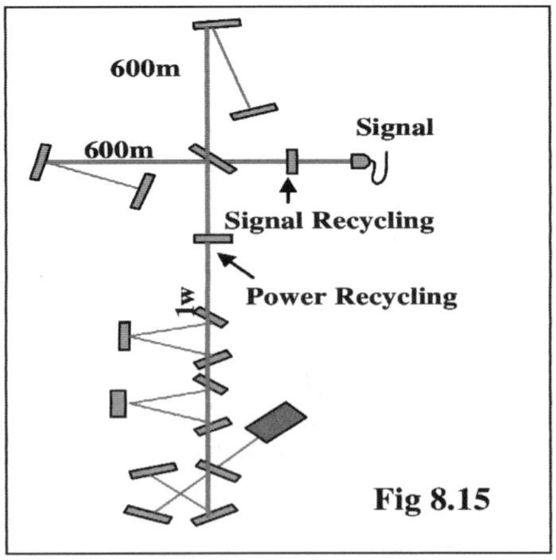

Fig 8.15

The optical diagram is shown in Fig. 8.15; the interferometer has two single pass arms and is equipped with both signal and power recycling mirrors. Mirrors are suspended by means of Silica fibers for reducing thermal noise.

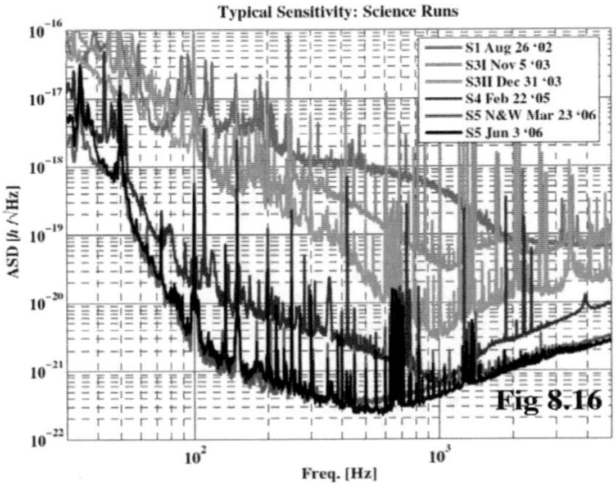

Fig 8.16

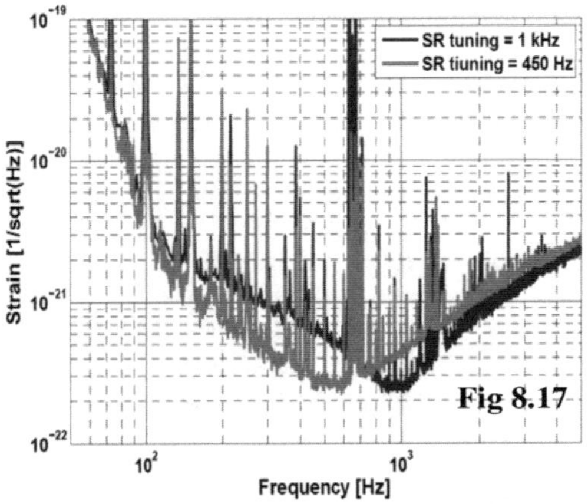

Fig 8.17

Signal recycling is an important addition since may create radiation pressure fluctuation noise reduction in some frequency range. In Fig. 8.16 sensitivity progression is shown, while in Fig.8.17 the effect of quadrature rotation, produced by signal recycling mirror is shown.

TAMA 300[14]

TAMA 300 m interferometer (see Fig. 8.18), located at National Astronomical Observatory of Japan (Tokyo) has been the first interferometric detector reaching sensitivity good enough for starting a realistic gravitational event search. TAMA construction started in 1995, and , even if the project was run by a relatively small number of scientists, it was immediately supplying the world best performances; during the period 2000-2002 attained the world best sensitivity.

Fig 8.18

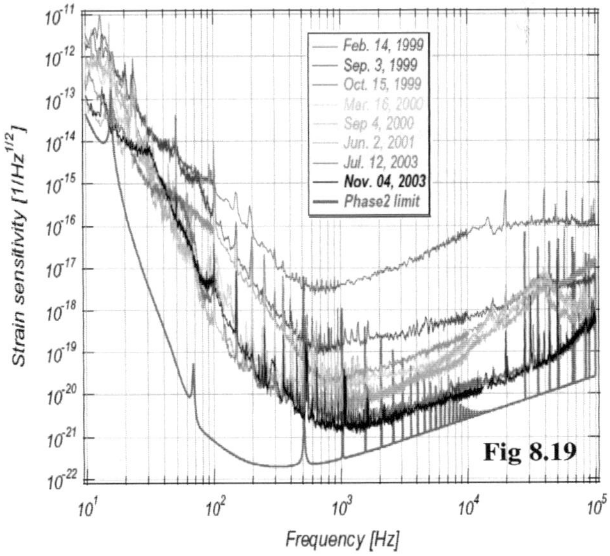

Fig 8.19

In Table 1 TAMA milestone progress is presented:

With DT1 In 1999 TAMA started to make observations, in DT4 attained the world best sensitivity, in DT6 performed 1000 hours of data taking and with DT7 started operating with power recycling. Joint observation with LIGO was performed during DT7 to DT9.
In Fig. 8.19 the sensitivity progress is presented; best strain sensitivity attained was $1.710^{-21}Hz^{-1/2}$ with a recycling factor of 4.5

Table 2. TAMA data-taking runs including long-term observations

Run	Term	Year	Live Time (Hour)
DT1	6-Aug → 7-Aug	1999	7
DT2	17-Sept → 20-Sept	1999	31
DT3	20-Apr → 23-Apr	2000	13
DT4	21-Aug → 4-Sept	2000	161
DT5	2-Mar → 8-Mar	2001	111
DT6	15-Aug → 20-Sept	2001	1038
DT7	31-Aug → 2-Sept	2002	25
DT8	14-Feb → 14-Apr	2003	1158
DT9	28-Nov → 10-Jan	2004	558

Table 1- TAMA data taking runs progress and running times. *Courtesy Kazuaki Kuroda for the TAMA/CLIO/LCGT Collaboration*

In Fig. 8.20 sensitivities of all kinds of operating GW detectors are displayed; original sensitivity graphs for bar detectors can be seen in Fig. 4.6 and for TAMA in fig. 8.19.

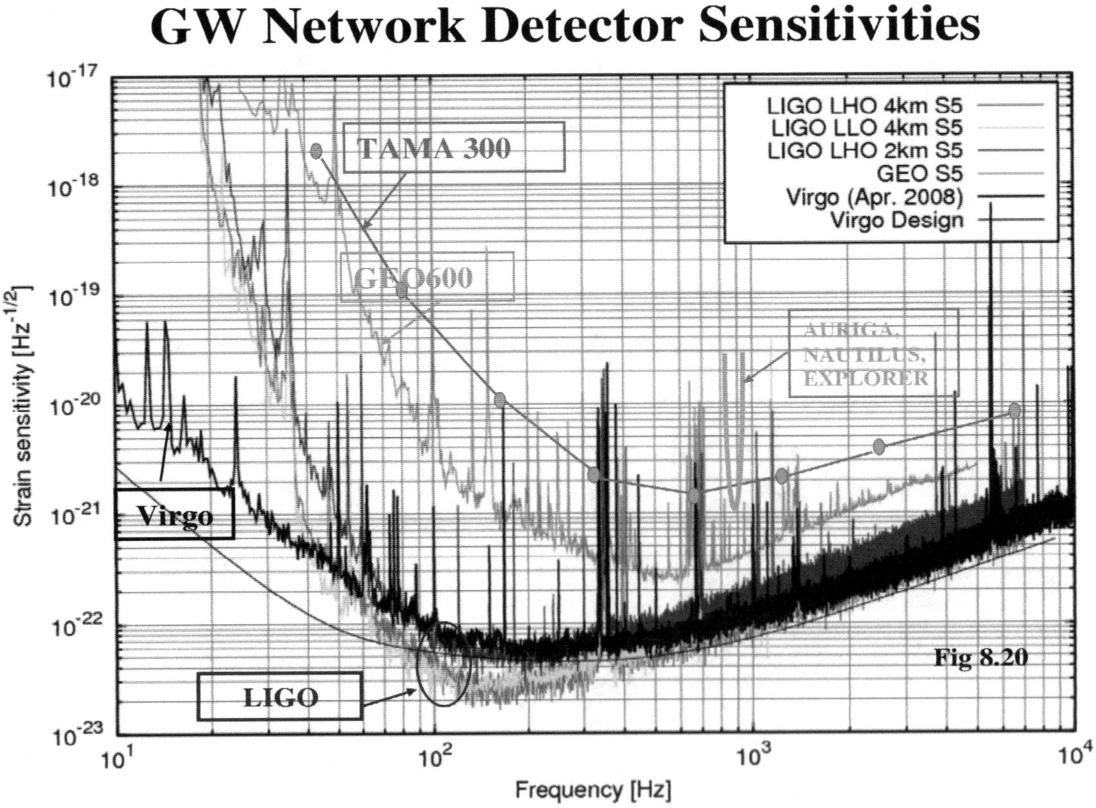

9) Resumè of status of GW detection.

IGEC: Bar Detector Network (Auriga, Explorer, Nautilus, Allegro) started 1997 for detecting impulsive GW[15].

No evidence of any significant signal

Periodic[16]Sources:
LIGO-GEO600: OG da Pulsar (28 known)-
$\varepsilon < 10^{-5} - 10^{-6}$
(no mountains > 10 cm)-
Upper Limits h :
2.10^{-24}@200Hz,
5.10^{-24}@400Hz,
10^{-23}@1KHz. See Fig. 9.1 and 9.2.

Fig. 9.1-Periodic sources upper limits: this is the all sky upper limit for periodic sources strain (95% confidence level), obtained for the Hanford observatory. The plot compares several search method, documented in the S4 paper LIGO-P060010-05-Z.

Fig. 9.2-Periodic sources upper limits: the same of the previous figure, for the Livingston observatory.

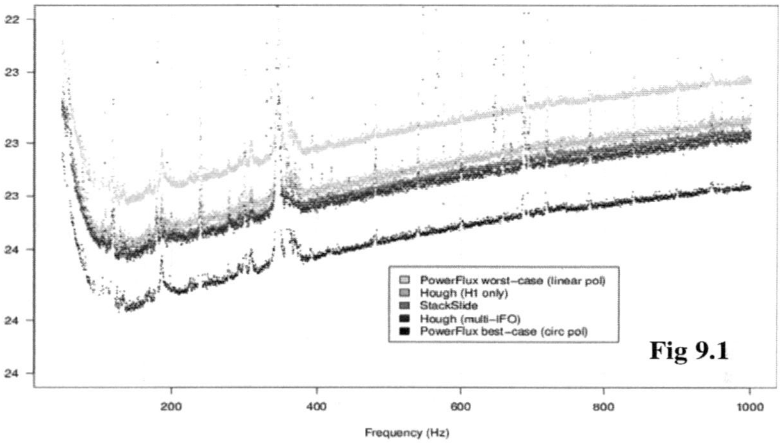

Fig 9.1

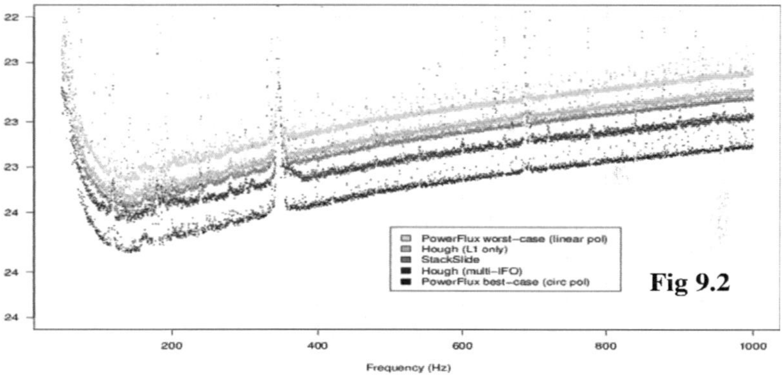

Fig 9.2

Coalescence[17]:
LIGO,GEO600,TAMA (see LIGO paper arXiv:0704.3368) upper limits (see Fig.9.3, 9.4, 9.5): - Fig.9.3-Upper limits on the binary inspiral coalescence rate per year and per L10 as a function of total mass of the binary, for Primordial Black Hole binaries. The darker area shows the excluded region after accounting for marginalization over estimated systematic errors. The lighter area shows the additional excluded region if systematic errors are ignored.

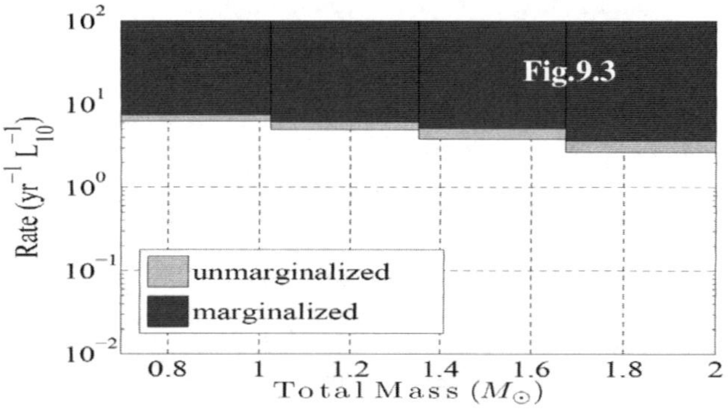

Fig.9.3

Coalescence[17] cont.:

Fig. 9.4-Upper limits inspirals: Same as the previous figure for Binary Black Holes.
Fig. 9.5-Upper limits inspirals: Same as the previous figure for Binary Neutron Stars
Upp. Lim. Coalescing
NS-NS <1 event/(galaxy.year) $2 < M_0 < 6$,
Coalescing BH-BH <1 event/(galaxy.year) $10 < M_0 < 80$

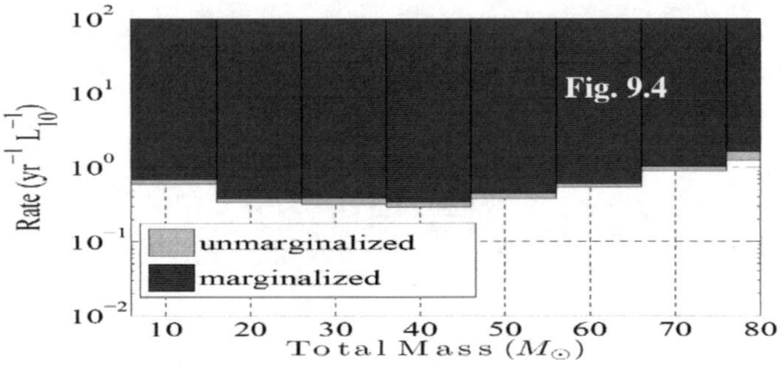

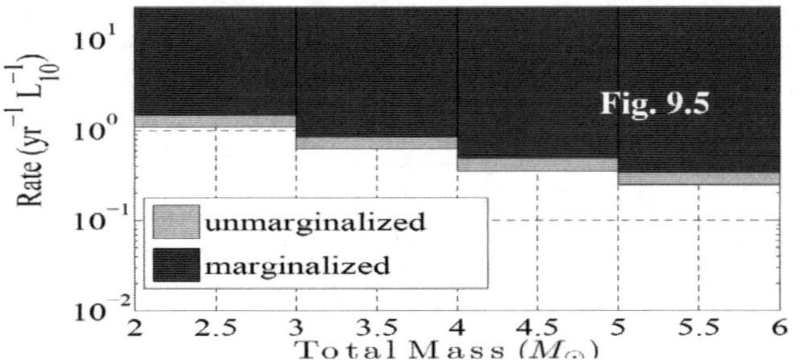

Impulsive[18]:
LIGO and LIGO-Tama upper limits for S1, S2, S4 scientific runs are shown in Fig. 9.6 and 9.7.
Figure 9.6 and 9.7. Exclusion diagrams (rate limit at 90% confidence level, as a function of signal amplitude) for (a) sine-Gaussian and (b) Gaussian simulated waveforms for this S4 analysis compared to the S1 and S2 analyses (the S3 analysis did not state a rate limit). These curves incorporate conservative systematic uncertainties from the fits to the efficiency curves and from the interferometer response calibration. The 849 Hz curve labeled "LIGO-TAMA" is from the joint burst search using LIGO S2 with TAMA DT8 data [18], which included data subsets with different combinations of operating detectors with a total observation time of 19.7 days and thereby achieved a lower rate limit. The hrss sensitivity of the LIGO-TAMA search was nearly constant for sine-Gaussians over the frequency range 700–1600 Hz.

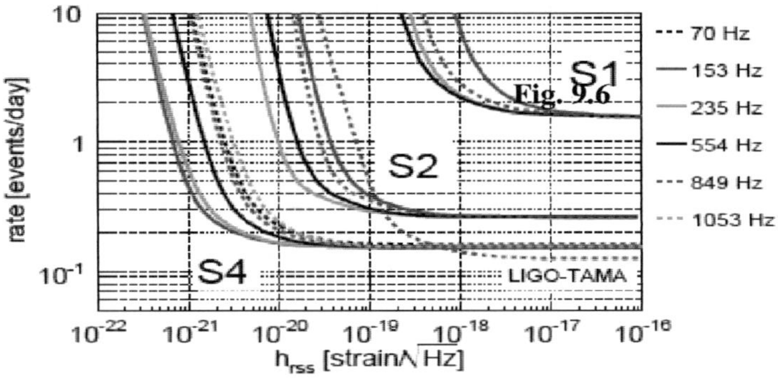

(a) Sine-Gaussians with Q=8.9

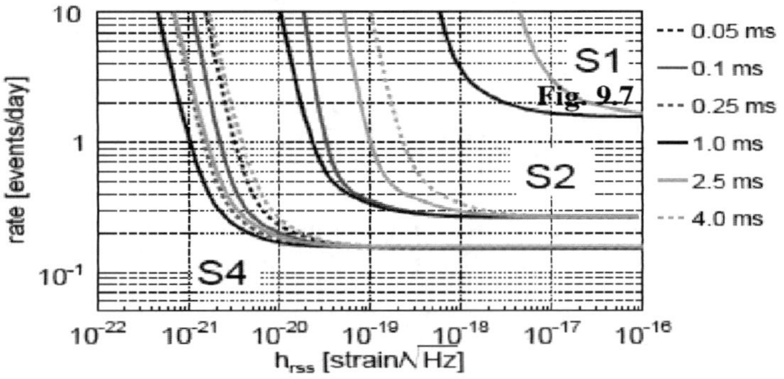

(b) Gaussians

Upper bounds on Stochastic Background[19]: 90% Upper Limit on GW spectrum at 100 Hz (see the model on the right) as a function of α for S3 H1L1 and S4 H1L1+H2L1 combined, and expected final sensitivities of LIGO H1L1 and H1H2 pairs, assuming LIGO design sensitivity and one year of exposure, are shown in Fig. 9.8.

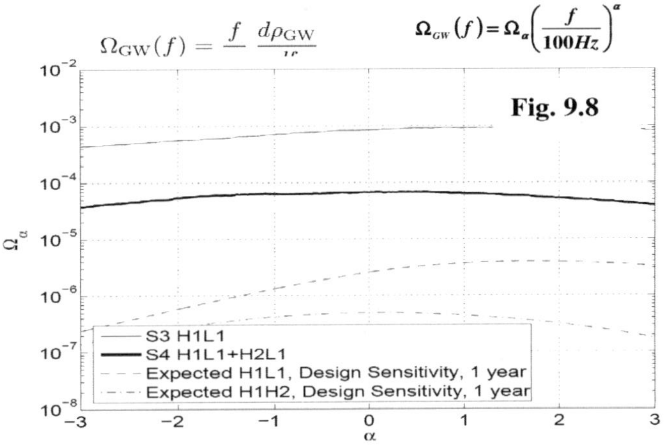

Virgo, LIGO, GEO 600 on May 18, 2007 started common and coherent data taking analysis; main target impulsive events.

10)The Future

In Table 2 the totality of future GW experiments is presented, grouped by status of approval level with a brief explanation of forecast upgradings.

Table 2- **The Future**

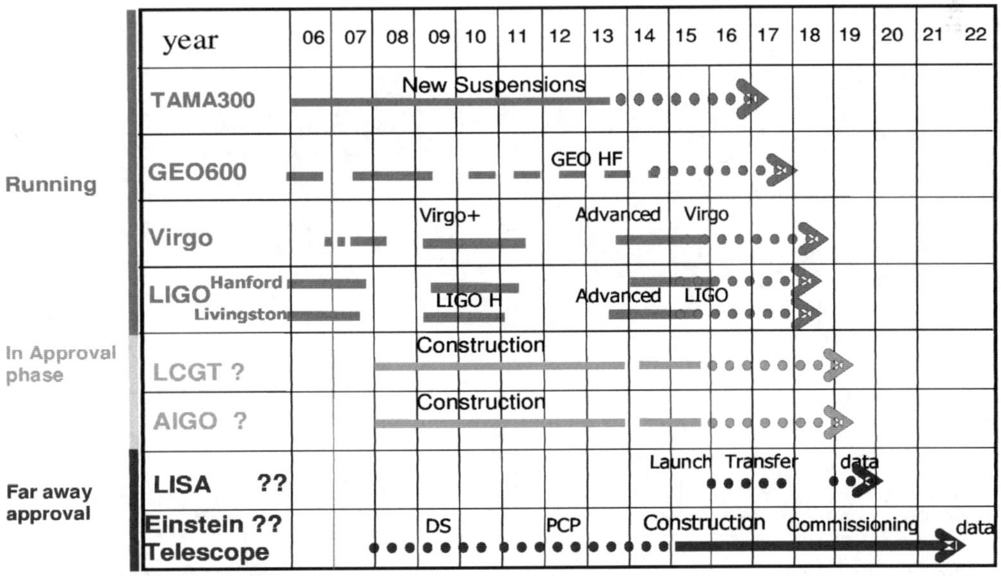

Runnig Interferometers

TAMA intend to operate with new Superattenuators mounted for better control and seismic isolation; timing is outlined in Table 2.

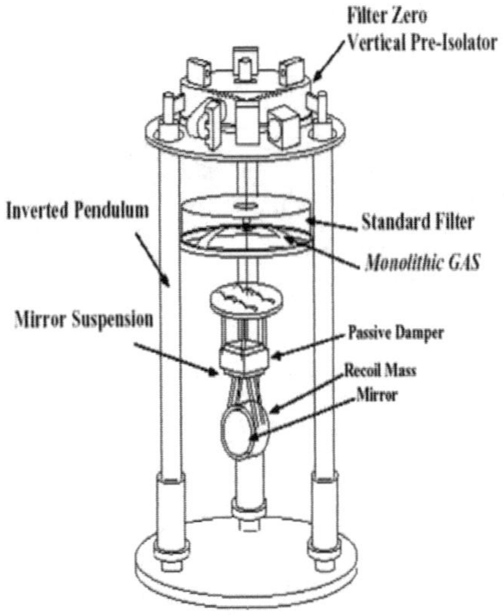

GEO600 intend to push high frequency sensitivity (new name GEO HF) to much higher level as shown in Fig. 10.1. Fig. 13.2-GEO HF high frequency sensitivity is strongly improved by increasing Intra Cavity Power and by optimizing squeezing. Purpose of GEO HF is also the study of advanced techniques to be applied in future large interferometers with particular reference to signal recycling and squeezing. Timing is outlined in Table 2.

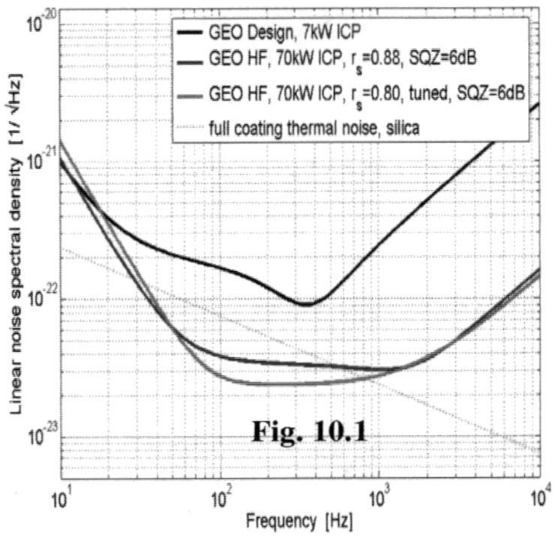

Fig. 10.1

Virgo[20] Plus: The expected sensitivity of the upgrading to Virgo Plus is presented in Fig. 10.2; the main items are the power increase to 50 w, implementation of fused silica suspensions (very much depending on how Virgo removes low frequency noise) and increase of Fabry-Perot cavities finesse to about 150 in such a way to produce firm sensitivity increase in the 100-400 Hz region.
The sensitivity increase in this frequency interval will improve a lot coalescing binary system detection sensitivity range up to 28 Mpc. Expected to start runing in june 2009.

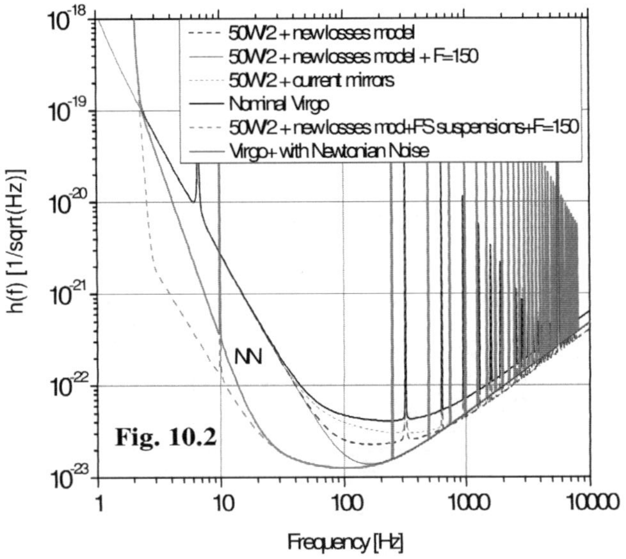

Fig. 10.2

Virgo Advanced:
Optical configuration will be changed in such a way to have the optical waist displaced to 1500m, in this way thermal noise of the cavity entrance mirrors will be reduced. Advanced Virgo expected sensitivity is shown in Fig. 10.3.
A further reduction of thermal noise will, hopefully, come from coatings losses reduction by using other less lossy metals than Titanium in the coating recepies. Laser power will also be increased and optical read out changed from heterodyne to DC.
Virgo Advanced will also be equipped with signal recycling mirror for implementing radiation pressure fluctuations noise reduction.

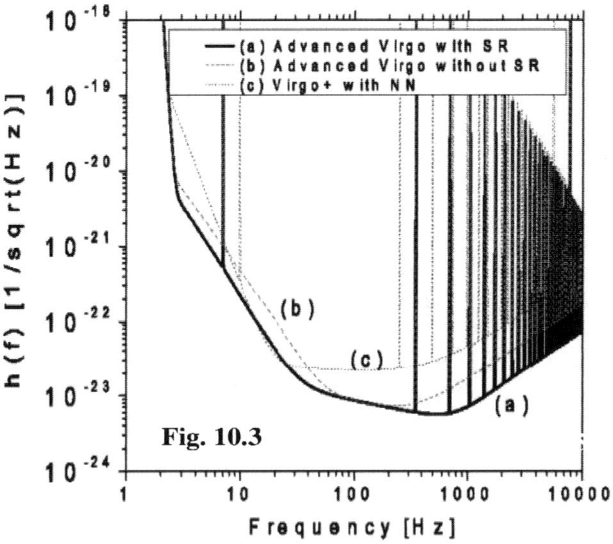

Fig. 10.3

LIGO[21] Henanced.
The first set of improvements will be set in the LIGO Henanced configuration for starting scientific runs together with Virgo Plus in june 2009. They are Higher laser power, DC readout and insertion of an Output modecleaner; LIGO H will also serv as test bench of Advanced LIGO hardware and techniques. Target is a factor of 2 improvement in sensitivity (factor of 8 in event rate).

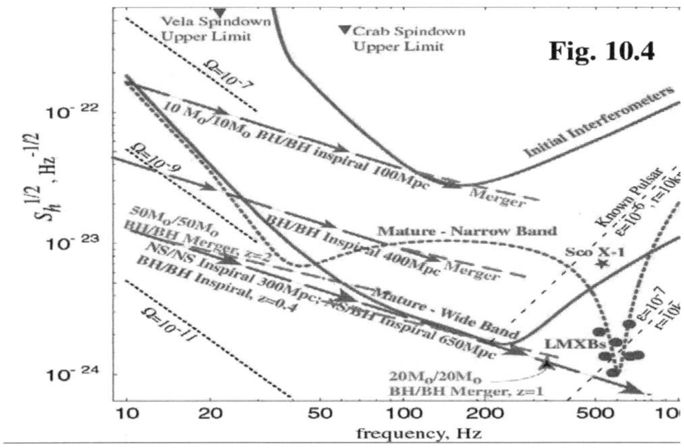

Fig. 10.4

LIGO Advanced:
will implement active anti-seismic system operating to lower frequencies, lower thermal noise suspensions and optics, higher laser power and more sensitive and more flexible optical configuration; expected sensitivity is shown in Fig. 10.4.
In table 3 main technical improvements with respect to initial LIGO, are presented.

Table 3

Parameter	*LIGO*	*Advanced LIGO*
Input Laser Power	10 W	180 W
Mirror Mass	10 kg	40 kg
Interferometer Topology	Power-recycled Fabry-Perot arm cavity Michelson	Dual-recycled Fabry-Perot arm cavity Michelson
GW Readout Method	RF heterodyne	DC homodyne
Optimal Strain Sensitivity	3 x 10^{-23} / rHz	Tunable, better than 5 x 10^{-24} / rHz in broadband
Seismic Isolation Performance	f_{low} ~ 50 Hz	f_{low} ~ 10 Hz
Mirror Suspensions	Single Pendulum	Quadruple pendulum

Interferometers in approval phase

LCGT

Large-scale Cryogenic Gravitational wave Telescope (LCGT)[22] This detector has a very relevant impact in the technology of next generation interferometers for GW detection, infact it is the first one to be designed with Fabry Perot cavity mirrors brought to 20 K. Cold mirror technology seems to be the only one able to win, at lest in part, thermal noise limits.

As it was shown by Numata[22] coating losses seems to be quite insensitive to temperature, infact a loss angle of about $4\ 10^{-4}$ was measured to remain constant at various temperature and down to 4K. This mirror coating anomalous and unfortunate behavior is creating a problem for next generation machines where thermal noise limits will be met. As far as we know, at present, the only known possibilities for reducing mirror thermal noise are coating loss angle reduction by using new coating materials and cooling mirrors. The latter benefits from the fact that thermal stochastic force is decreasing with $T^{1/2}$ (T is the temperature), hence bringing mirrors to 4K there is a factor about 10 expected reduction of mirror thermal noise.

LCGT, expected to work at 20 K, is a 3 km interferometer with cryogenic mirror.

Constructed underground in the depth of the Kamioka mine, has two beams (see Fig. 10.5) in the same pipe and the upper beam is used for a suspension point interferometer; the optical diagram is shown in Fig 10.6 and the expected sensitivity in Fig. 10.7.

Courtesy Kazuaki Kuroda for the TAMA/CLIO/LCGT Collaboration

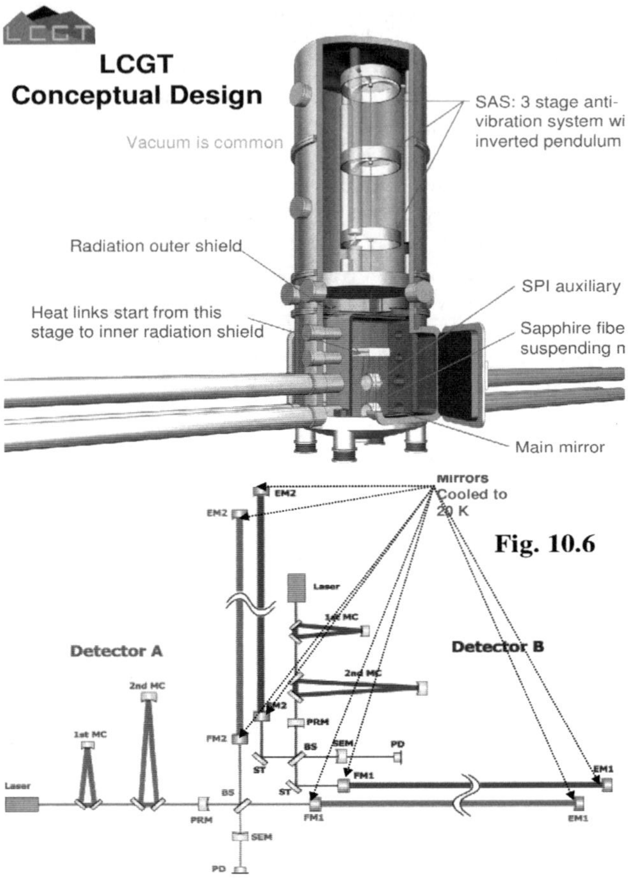

Fig. 10.6

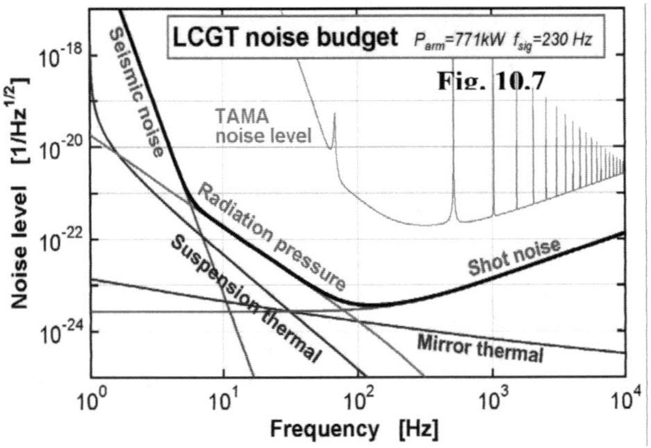

Fig. 10.7

AIGO[24]

AIGO is a project for the construction of a 5+5 km interferometer in Gin Gin, 80 km east of Perth (see Fig 10.8) ; it a very long time since Australian Consortium for Interferometric Gravitational Astronomy (ACIGA) is trying to get approval from Minister for the construction of AIGO. The importance of AIGO, beyond the very relevant fact that a large GW detector is adding to the existing ones, is its geographical position on Earth; infact a very sensitive GW detector in the southern emisphere will allow to measure GW sources angular positions with greater accuracy than the one obtained by Earth northern hemisphere detectors, as shown in Fig.s 10.9. The detector, designed for being sensitive to very low frequency GW sources, is equipped with superattenuators, in some way similar to the Virgo ones. **GW detector community is really looking forward for AIGO approval.**

Plots courtesy David Coward and David Blair

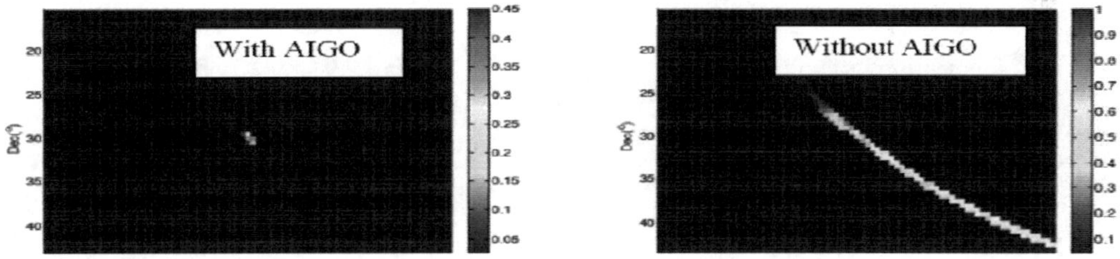

GW detectors whose final approval phase is still far

LISA[25]

LISA consists of three spacecraft mutually separated by 510^6 km. It is aimed to detect GW in the range 10^{-4}-10^{-3} Hz, as shown in Fig. 10.10. One of the of LISA keypoints is the possibility to measure GW emitted by galactic "Standard Candles", such as cataclysmic variables , per ex. AmCvn at 2 mHz and by other binary systems whose GW amplitude is evaluated by the system physical parmeters. If GW will not be detected before LISA being in air, then LISA can give an inconfutable check of General Relativity validity. Let us hope this will not be the case. Lisa description is so complex that we leave it to more specialized literature.

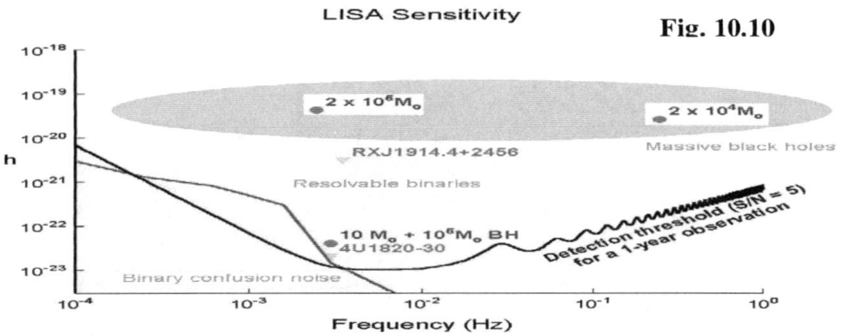

Fig. 10.10

Einstein Telescope[26] (ET)

ET will be the future challenge for GW detection since it should attain a factor 100 more in sensitivity with respect to Virgo and LIGO (see Fig. 13.15). One day of data taking of ET is equivalent to 10^6 days of data taking of Virgo and LIGO machines

It is at least a 10+10 km machine underground mainly for reducing seismic noise and Newtonian noise, but also for avoiding strong surface environmental impact. It is possible that the lower frequency should be less than 10 Hz; this is a real challenge since noises at very low frequency are very rugged and hard to eliminate.

Optical noises will be reduced either by using signal recycling or by injecting squeezed vacuum in beam splitter black port; stiil to understand how to get rid of vacuums intrduced by photodiode inefficiencies. ET will be a cryogenic detector with mirrors kept at 4 K; it is not obvious if it is necessary to cool also power recycling, signal recycling and beam splitter mirrors. ET will be formed by at least 4 detectors separated by some thousand km for solving the inverse problem i.e. determination of source 2 polarizations and 2 angles.

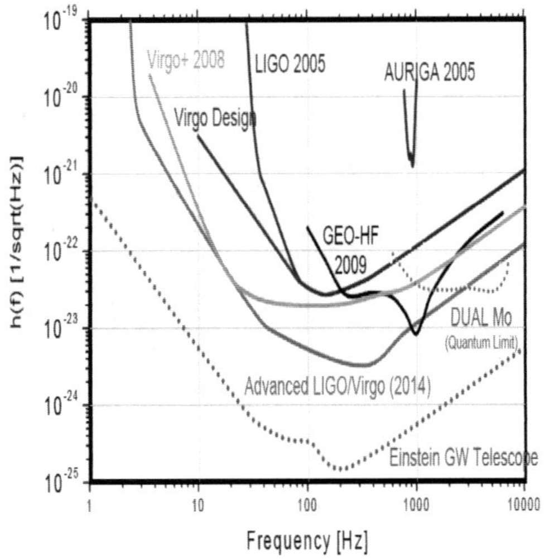

It is likely that the best configuration should contain at least one detector in the southern emisphere. After 2020 ET will be the only operative project; it is likely that Virgo and LIGO will be obsolete. GW detector remnant noises, when sensitivity reaches 10^{-22}-10^{-23}, are almost non-modelizable and in general can not be studied by means of an R&D:

a) Diffused light is one of major sources of noise and can not be easily predicted.

b) Line and power supply e.m. fields are not predictable.

c) Ground Loops are not predictable ....

Since in ET noise should be 100 times smaller than the one in the existing machines, the only way to reach design sensitivity in 2020-2024 is to start as soon as possible the construction of a one-to-one scale prototype machine which will become the first detector of ET array.

Final Considerations

Cryogenic Bar Detectors have reached high sensitivity ($h>10^{-21}$) and a duty cycle so high that the first network of GW detectors has been created i.e.IGEC.

A big step forward in the last 10 years has been in the Interferometer technology. Design sensitivity, above 100 Hz, has been reached ($h>5\ 10^{-23}$) and stability is so good that a very efficient detector network has been set up. The big challenge will be in the operation of Advanced Interferometers for which sensitivity should reach $h>4\ 10^{-24}$

I believe, after what we have learned from Virgo, LIGO, GEO600 and TAMA construction and operation, that class Einstein detectors (third generation) having sensitivity $h>10^{-25}$, seems fesible with very high probability of success. This seems to be the right way for starting GW astrophysics.

Acknowledgments

I am deeply grateful to Stan Withcomb whose presentation at Amaldi 7 has been very much of help for the preparation of this paper; I am also deeply grateful to Giancarlo Cella for carefully reading the manuscript.

References

(1) -R.A.Hulse and J.H. Taylor, **Astrophys. J. Lett**. 195, 151 (1975)-H.V.Shapiro, S.A.Teukolsky, 1984, "*The physics of compact objects*" (J.Wiley and Sons, Eds.), New York.
(2) -A.Einstein, Sitz. Ber. Kon.Preus. **Ak. Wiss**, 688(1916); id. 154(1918)
(3) -C.W.Misner, K.S.Thorne, J.Wheeler, 1973, "*Gravitation*" (W.H.Freeman Eds.), S.Francisco.
-K.S.Thorne, 1987, "Gravitational radiation" in "*300 years of gravitation*" (Edited by S.W.Hawking and W.Israel, Cambridge University Press), New York (9) 330-458.
(4) - Pirani, F.A.E., 1956, Acta Physica Polonica 15, 389..
(5)-J. Weber, **Phys. Rev**.117,306 (1960)
-J. Weber, **Phys. Rev**. Lett. 22,1302(1969)
- H.B.Callen, R.F.Greene, **Phys. Rev**. 86 (1952) 702.
(6) - Eugenio Coccia- STATUS OF THE BAR DETECTORS EXPLORER AND NAUTILUS- Edoardo Amaldi Conference on Gravitational Waves-8 – 14 July 2007, Sydney.
-Luca Taffarello- STATUS REPORT OF AURIGA DETECTOR- Edoardo Amaldi Conference on Gravitational Waves-8 – 14 July 2007, Sydney.
- Status of ALLEGRO- W. Johnson- Edoardo Amaldi Conference on Gravitational Waves- June 20-24, 2005, Okinawa.
-D. Blair et al. "Niobe: improved noise temperature and Background Noise Suppression" in *Gravitational Waves,* third Amaldi Conference, Pasadena Ca. 12-16 july 1999 , AIP Conference Proceedings # 523, Melville, New York).
- J.J. Johnson et al. in *Gravitational Wave Experiments*, Proceedings of the First Edoardo Amaldi Conference, Frascati 1994,(World Scientific, Singapore, 1995)"
-M. Cerdonio *et al*-Wideband dual sphere detector of gravitational waves- Phys. Rev. Lett. 87 (2001) 031101, also in gr-qc/0011002.
7)- Oleksandr Usenko-MINIGRAIL PROGRESS; TEST OF COMPLETE TRANSDUCER CHAINS DOWN TO mK TEMPERATURE- Edoardo Amaldi Conference on Gravitational Waves-8 – 14 July 2007, Sydney. Odylio D. Aguiar-THE SCHENBERG SPHERICAL GRAVITATIONAL WAVEDETECTOR:THE FIRST COMMISSIONING RUNS- Edoardo Amaldi Conference on Gravitational Waves-8 – 14 July 2007, Sydney. -L. Gottardi, A. de Waard, O. Usenko, G. Frossati, M. Podt, J. Flokstra, M. Bassan, V. Fafone, Y. Minenkov, and A. Rocchi- Sensitivity of the spherical gravitational wave detector MiniGRAIL operating at 5K"-Phys. Rev. D 76, 102005 (2007)
(8) - A. Giazotto **-** "Interferometric Detection of of Gravitational Waves", **Physics Rept**, 182, 6, 1989.-
J.Y.Vinet, B.Meers, C.N.Man, A.Brillet**, Phys. Rev**. *D* 38 (1988) 433.Brillet, *Ann. de Phys.* 10 (1985) 219.
(9)-Two books give a lot of important informations in quantum optics research; they are:
Marlon O. Scully and M. Suhail Zubary-Quantum Optics-Cambridge University Press (Sept. 4,1997)
S. M. Barnett and P. M. Radmore-Methods in theoretical quantum optics Oxford University Press (Nov. 1,1997)
The following papers are also instrumental in quantum optics:
-C. M. Caves ,1980,Phys. Rev. Lett.45,75.
-C. M. Caves, B.L. Schumaker:-New formalism for two-photon quantum optics.I. Quadrature phases and squeezed states.- Phys. Rev. A 31,5, may 1985, 3086.
-C. M. Caves, B.L. Schumaker:-New formalism for two-photon quantum optics.II. Mathematical foundations and compact notations- Phys. Rev. A 31,5, may 1985, 3093
-J. M. Courty, A. Heidmann and M. Pinard- Back action cancellation in interferometers by quantum locking.-gr-cq/0301068 v1, jan 2003
-H. J. Kimble et al.-Conversion of conventional gravitational-wave interferometers into QND interferometers by modifying their input and/or output optics.-gr-qc/0008026 v2, 3 sept. 2001

(10)-Daniel Sigg -STATUS OF THE LIGO DETECTORS -Edoardo Amaldi Conference on Gravitational Waves-8 – 14 July 2007, Sydney.
(11)- Frederique Marion - STATUS OF VIRGO - Edoardo Amaldi Conference on Gravitational Waves-8 – 14 July 2007, Sydney.
(12)) - R. Del Fabbro et al., 1988,Phys. Lett. A132,237
-S.Braccini et al., 1997, *"Design of the superattenuators for the VIRGO construction"* VIRGO Internal Report (VIR-TRE-PIS-4600-134).
(13)-Hartmut Grote- THE STATUS OF GEO 600- Edoardo Amaldi Conference on Gravitational Waves-8 – 14 July 2007, Sydney.
(14)- Ryutaro Takahashi -OPERATIONAL STATUS OF TAMA300 WITH THE SEISMIC ATTENUATION SYSTEM (SAS) - Edoardo Amaldi Conference on Gravitational Waves-8 – 14 July 2007, Sydney.
(15)-Francesco Salemi-RESULTS OF THE SEARCH FOR GRAVITATIONAL WAVE BURSTS BY THE NETWORK OF RESONANT BAR DETECTORS. - Edoardo Amaldi Conference on Gravitational Waves-8 – 14 July 2007, Sydney
(16)-LIGO Scientific Collaboration: B. Abbott, et al-All-sky search for periodic gravitational waves in LIGO S4 data-arXiv:0708.3818v1 [gr-qc], submitted 28 Aug 2007
(17)- LIGO Scientific Collaboration: B. Abbott, et al -Search for gravitational waves from binary inspirals in S3 and S4 LIGO data-**s**- arXiv:0704.3368v2 [gr-qc],submitted 25 Apr 2007
(18)- LIGO Scientific Collaboration-Search for gravitational-wave bursts in LIGO data from the fourth science run-Classical and Quantum Gravity 24, 5343-5369 (2007).
-LIGO Scientific Collaboration, TAMA Collaboration-Upper limits from the LIGO and TAMA detectors on the rate of gravitational-wave bursts-arXiv:gr-qc/0507081v3
(19)LIGO Scientific Collaboration-Searching for a Stochastic Background of Gravitational Waves with LIGO-arXiv:astro-ph/0608606v2
(20)- Michele Punturo-VIRGO UPGRADE PLANS- Edoardo Amaldi Conference on Gravitational Waves-8 – 14 July 2007, Sydney.
(21)-Carol Wilkinson- STATUS AND PLANS FOR ADVANCED LIGO INSTRUMENTS- Edoardo Amaldi Conference on Gravitational Waves-8 – 14 July 2007, Sydney.
(22)- Kazauki Kuroda-CURRENT STATUS OF LCGT PROJECT- Edoardo Amaldi Conference on Gravitational Waves-8 – 14 July 2007, Sydney.
(23)-Kazuhiro Yamamoto, Shinji Miyoki, Takashi Uchiyama, Hideki Ishitsuka, Masatake Ohashi, Kazuaki Kuroda, Takayuki Tomaru, Nobuaki Sato, Toshikazu Suzuki, Tomiyoshi Haruyama, Akira Yamamoto, Takakazu Shintomi, Kenji Numata, Koichi Waseda, Kazuhiko Ito, Koji Watanabe- Measurement of the mechanical loss of a cooled reflective coating for gravitational wave detection-gr-qc/0601031 (January 2006)
(24)- David Blair-THE AIGO PROJECT- Edoardo Amaldi Conference on Gravitational Waves-8 – 14 July 2007, Sydney.
(25)- -Scott Hughes-ASTROPHYSICS AND RELATIVITY USING LISA-Edoardo Amaldi Conference on Gravitational Waves-8–14 July 2007, Sydney.
(26)-Harald Lueck PLANS FOR E.T. (Einstein Telescope), A EUROPEAN 3rd GENERATION GRAVITATIONAL WAVE OBSERVATORY-Edoardo Amaldi Conference on Gravitational Waves-8–14 July 2007, Sydney.

Status of LIGO

Keita Kawabe for the LIGO Scientific Collaboration
LIGO Hanford Observatory, PO Box 159, Richland, WA 99352-0159, USA
E-mail: kawabe_k@ligo-wa.caltech.edu

Abstract. LIGO successfully acquired more than one year of three-way coincident observation data using all three detectors during its fifth science run from November 2005 to the end of September 2007. All detectors reached sensitivity better than the design. For the two 4 km detectors, the all-sky averaged detection range exceeded 15 Mpc with a signal-to-noise ratio of 8 for inspiral binary neutron stars of 1.4 solar masses each. The latest sensitivities of the detectors, results from the past science runs, and future prospects of LIGO are presented.

1. Introduction

Laser Interferometer Gravitational-wave Observatory (LIGO) is a project aiming at the detection of gravitational waves (GWs) from astronomical sources using large laser interferometers[1][2]. LIGO has a total of three detectors at two sites. In LIGO Hanford Observatory, an instrument with 4 km arm length (H1) and a 2 km instrument (H2) are operating. LIGO Livingston Observatory is approximately 3000 km away from Hanford Observatory and has a 4 km instrument (L1). The physical separation of two sites allows us to exclude false events caused by local disturbances such as seismic motion, and upon successful detection of GWs would allow us to limit the direction of the source.

All LIGO detectors are Michelson interferometers with some enhancements and suspended optics (Fig. 1). In place of two simple reflectors at the end of orthogonal paths in a Michelson interferometer, each LIGO detector has two "arms" comprising Fabry-Perot resonators of either 4 km or 2 km length. On entering into the resonator, light is in effect bounced back and forth approximately 100 times before coming out of the resonator, enhancing the apparent length of travel the light experiences and thus enhancing the phase change exerted on the light by GWs. The position of the mirrors is controlled in such a way that two beams coming back from orthogonal paths interfere constructively on the beam splitter in the direction going back to the laser. Any differential phase change in the two arms caused by GWs makes a small amount of light leak into the other direction, and this is detected by a photo detector.

LIGO also uses a technique called power recycling[3]. The light going back in the direction of the laser is reflected back into the interferometer by another mirror placed between the beam splitter and the laser, and constructively interferes with the light from the laser that passes through the mirror. This increases the effective input power that impinges on the beam splitter, and thus improves the sensitivity of the interferometer limited by the photon shot noise, the photon counting noise of the light measured.

To date, LIGO has successfully performed 5 science data runs as shown in Tab. 1. The fifth run (S5) was the first long observational run with all instruments typically running at or better than the design sensitivity. S5 started in November 2005 with the mission of acquiring

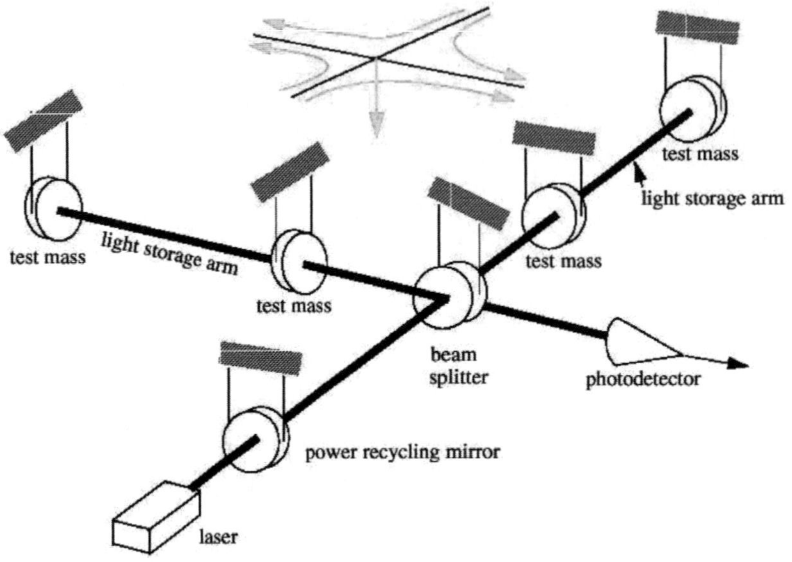

Figure 1. Basic configuration of LIGO instruments.

one year's worth of triple coincident observational data. At the end of September 2007, after completing the mission with more than one year's data, LIGO ended S5. It's worth mentioning that historically LIGO has always collaborated with other projects during science runs. In S5, due to participation of GEO and VIRGO, sometimes in total 5 large interferometers were running at the same time.

Table 1. LIGO runs to date. The "Collaboration" column lists the name of projects that took part in the coincident observation with LIGO.

Run	Started	Period	Collaboration
S1	8/2002	17 days	GEO, TAMA
S2	2/2003	59 days	TAMA
S3	11/2003	70 days	Allegro, GEO, TAMA
S4	2/2005	30 days	Allegro, AURIGA, GEO
S5	11/2005	23 months	GEO, VIRGO

2. Performance of the LIGO Instruments in S5

Figure 2 shows the sensitivity of the LIGO instruments during S5 in terms of linear spectral density of the strain noise. The design strain sensitivity (10^{-21} RMS integrated over a 100 Hz bandwidth centered at the minimum noise region [4], in other words $10^{-22}/\sqrt{\text{Hz}}$ over 100 Hz at around the most sensitive frequency) is represented by a horizontal arrow. Clearly the sensitivity of all of LIGO instruments became better than the design, even though the 2 km instrument is about a factor of two less sensitive than the others due to its smaller length.

The number shown as "Binary Inspiral Range" is the theoretical all-sky averaged detection range for inspiral binary neutron stars, each with 1.4 solar masses, with signal to noise ratio of

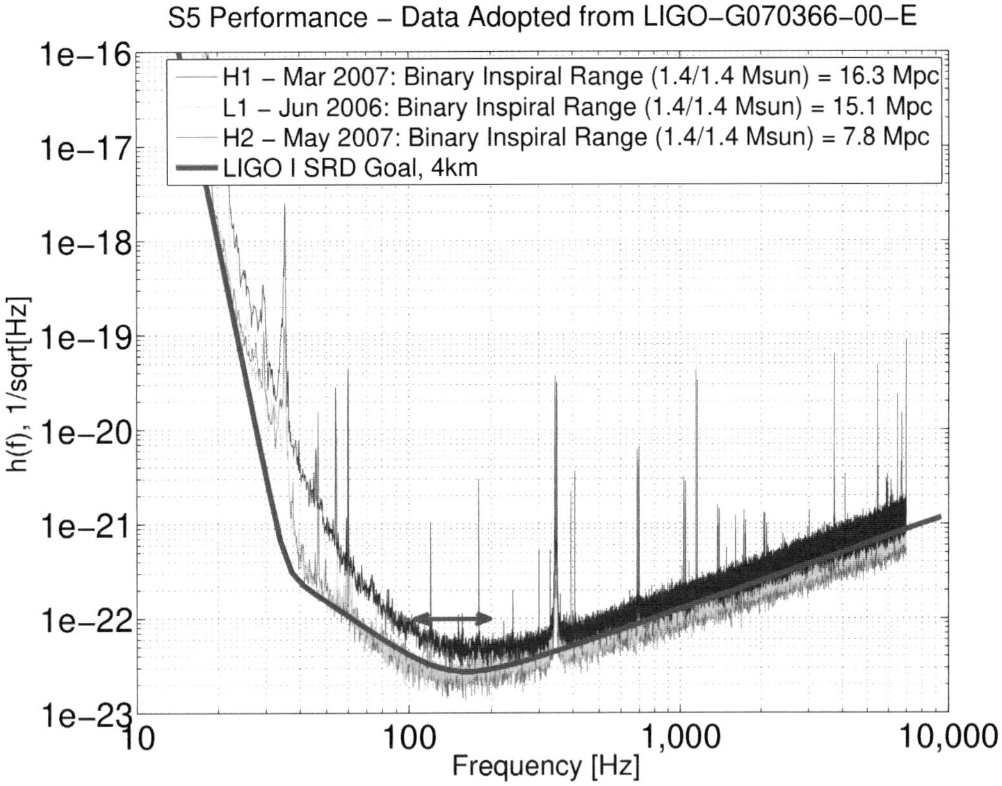

Figure 2. Performance of the Hanford 4 km instrument (red), Livingston 4 km instrument (green) and Hanford 2 km instrument (blue) in S5, plotted as linear spectral density of the strain noise. The arrow represents the design sensitivity, and "LIGO I SRD" is from the initial design parameters for 4 km instruments to realize the design sensitivity.

8 or more. This means that a part of the Virgo cluster is well within our view for sources like the coalescence of compact binary systems such as neutron star and/or black hole binaries.

LIGO performed intensive studies to identify the coupling of various noise contributions into the GW channel of the instruments. Figure 3 is what is called a "noise budget" resulting from such studies. Though this plot is for the H1 detector, the other detectors show quite similar noise budgets. The black trace denoted "DARM" is the measured displacement noise of the interferometer, and the thick dashed line "SRD" is the same as the curve "LIGO I SRD Goal, 4 km" in Fig. 2. Everything else shows the amplitude of corresponding noise sources projected into the displacement noise. It is not within the scope of this paper to discuss these in detail, but the most important part of this plot is that the detector noise is fairly well understood. Above 200 Hz, the noise is dominated by the shot noise (blue broken line, "Shot"). At frequencies lower than 50 Hz, there is no single dominant noise source.

There is a small frequency range between 50 and 100 Hz where the detector noise is not totally accounted for, which shows as a discrepancy between the "DARM" and "total" lines in the figure. There are several hypotheses about the origin of this unknown additional noise which is still under study.

Figure 4 is the histogram of the binary inspiral range of LIGO detectors over the entire run. There were several "commissioning breaks" during the run to improve the sensitivity and

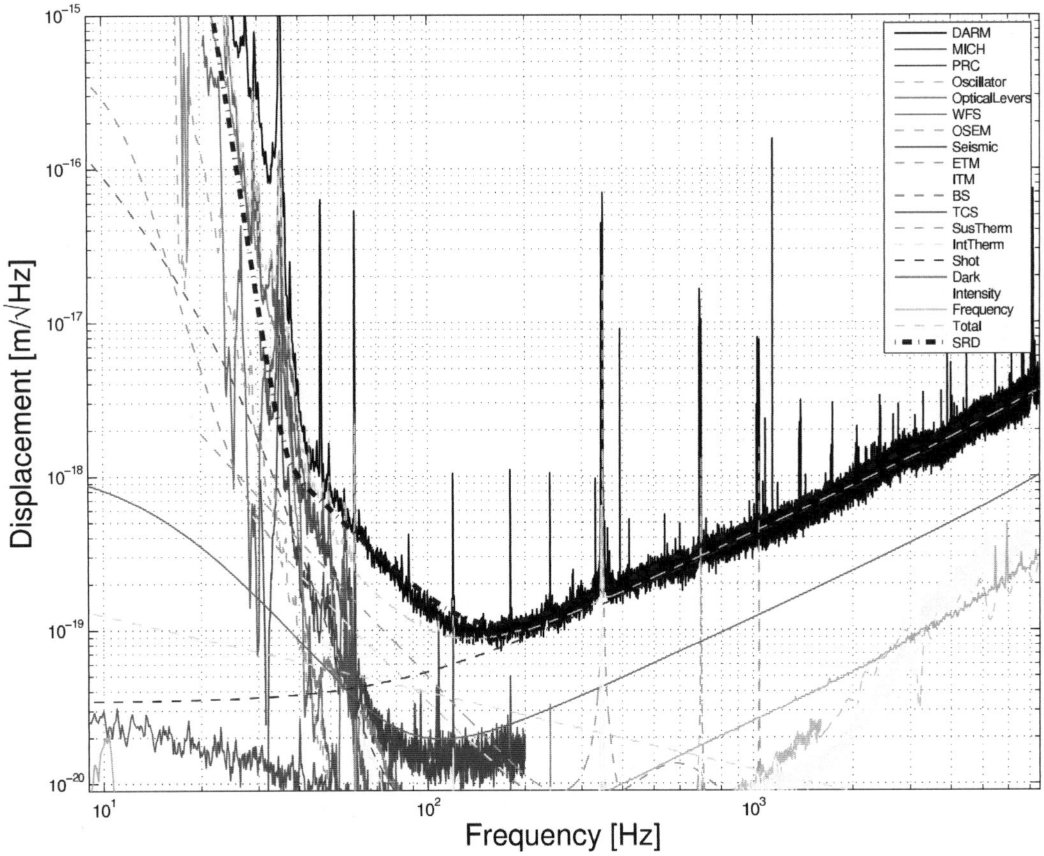

Figure 3. Projection of various noise sources coupling to the displacement sensitivity of the H1 detector. DARM: Displacement noise of the H1 instrument. MICH and PRC: Noise of the auxiliary servo controlling the Michelson part and the power recycling cavity part of the interferometer. Oscillator: Local oscillator noise. OpticalLevers and OSEM: Noise of servos locally damping the angle and position of the mirrors. WFS: Noise of the wave front sensors controlling the angle of the mirrors. Seismic: Noise caused by seismic excitation. ETM, ITM and BS: Actuator noise of End Test Masses, Input Test Masses and Beam Splitter. TCS: Noise coming from the thermal compensation system. SusTherm and IntTherm: Thermally excited motion of the wires suspending the mirrors and the internal resonant modes of the mirrors. Shot: Shot noise. Dark: Noise of the main detection electronics chain. Intensity and Frequency: Intensity and frequency noise of the laser. Total: Root sum square of all of the noise contributions. SRD: See Fig. 2.

reliability of the instruments. After each of these improvements, typical binary inspiral ranges of the instruments shifted to larger values, and this can be seen on the figure as the multiple peaks in each detector's histogram.

Figure 5 shows the weekly duty factor chart for S5. As one can clearly see, LIGO's reliability improved over time during the run. Although there are day-to-day fluctuations due to various reasons, all in all a triple coincidence duty factor (red in the figure) of about 0.6 was maintained except during the early part of the run and during commissioning breaks.

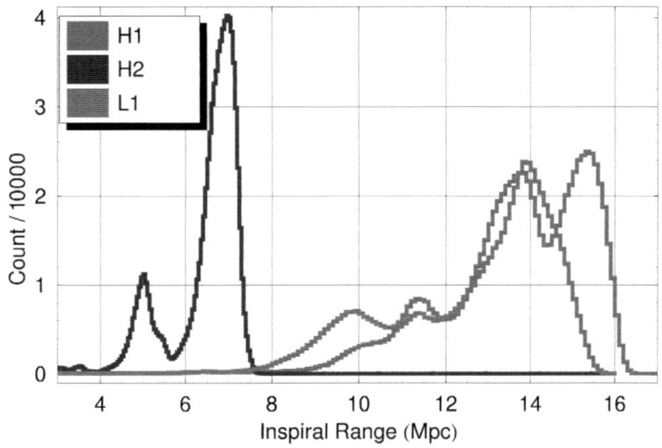

Figure 4. Histogram of binary inspiral range for the entire S5 run. Different peaks represent different periods of the run separated by several "commissioning breaks" to improve the detector performance.

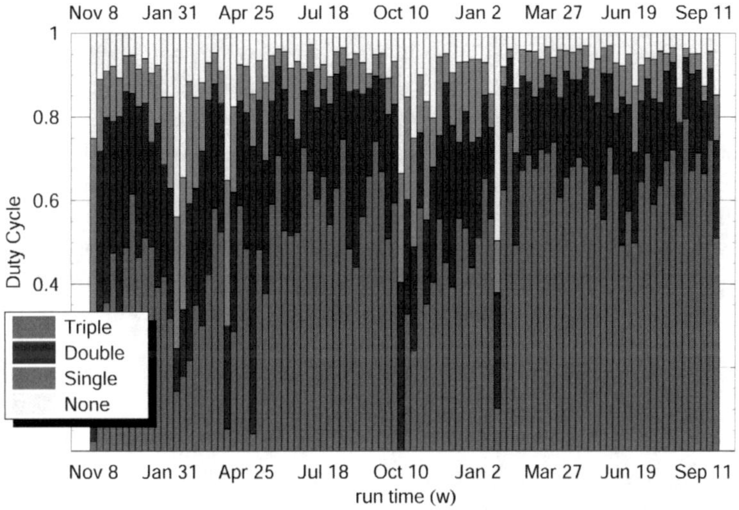

Figure 5. Duty factor chart for S5. Each vertical bin corresponds to a week. Each color shows the time LIGO was operating with three detectors (red), two (blue), only one (green) or none (yellow).

3. Science Results

The data obtained in S5 is still being analyzed by search groups in the LIGO Scientific Collaboration (LSC). Though no GW has been detected yet, upper limits for various sources have been set using the data from S4 and earlier runs.

In the targeted search of continuous GWs from 78 known radio pulsars in S3 and S4, strain upper limits as small as 2.6×10^{-25} and ellipticities as small as 10^{-6} were established with 95 % confidence level[5]. Also, a non-targeted all-sky continuous wave search was conducted for S4 with the best upper limit of 4.23×10^{-24} near 140 Hz with 95 % confidence level[6].

For binary inspiral events, upper limit rates for primordial black hole binaries in the mass range of 0.35 to 1 solar masses, neutron star binaries in the range of 1 to 3 solar masses, and stellar black hole binaries in the range of 3 to 80 solar masses were derived from S4 and S3 data. They were 4.9, 1.2 and $0.5/\text{yr}/L_{10}$ respectively with 90 % confidence level, where L_{10} is 10^{10} times the blue luminosity of the Sun.

For short bursts with unknown waveform, the sensitivity of LIGO in S4 for root-sum-squared amplitude of the gravitational strain was from 10^{-21} to $10^{-20}/\sqrt{\text{Hz}}$ for 50 % detection efficiency[7]. Also a targeted search for burst GWs associated with the SGR 1806-20 hyperflare was conducted[8], resulting in the upper limit $h_{\text{rss}} = 4.5 \times 10^{-22}/\sqrt{\text{Hz}}$ with 90 % confidence.

For a stochastic background, the 90 % upper limit of $\Omega_{\text{GW}} < 6.5 \times 10^{-5}$ was obtained using S4 data in the frequency range of 51–150 Hz assuming a flat spectrum[9], where Ω_{GW} is defined as the energy density spectrum of GW normalized by the critical energy density of the universe as $\Omega_{\text{GW}} \equiv (\text{d}\rho_{\text{GW}}/\text{d}\ln f)\,\rho_{\text{c}}^{-1}$. For the frequency range 850–950 Hz, $\Omega_{\text{GW}} < 1.02$ was obtained from an S4 correlation analysis of L1 and the resonant bar antenna ALLEGRO[10]. Also all-sky upper-limit maps for two different source strain power spectrum models were generated using S4 data[11]. The upper limit for a flat strain power spectrum was between $8.5 \times 10^{-49}\text{Hz}^{-1}$ and $6.1 \times 10^{-48}\text{Hz}^{-1}$, depending on the sky location, with 90 % confidence level.

For a complete list of publications, readers are encouraged to visit the LSC web page[12] and follow the "Observational Results" link.

4. Enhanced LIGO and Advanced LIGO

After S5, the LIGO 4 km detectors are to be upgraded to improve the sensitivity by about a factor of two, which would cover about a factor of 8 larger volume than initial LIGO within the detection range. Some of the key elements in this program called Enhanced LIGO[13] are early adoption of the technologies developed for an even more ambitious program called Advanced LIGO[14, 15]. Such technologies include a high power laser module with output power of more than 30 W, and what is called a "DC" readout scheme (as opposed to the radio frequency modulation-demodulation technique of initial LIGO), combined with an optical resonator called an output mode cleaner. Commissioning of Enhanced LIGO starts in winter 2007. Upon successful commissioning of Enhanced LIGO, another science run (S6) is anticipated in 2009.

Advanced LIGO is a program to achieve roughly a factor of 10 improvement in strain sensitivity or a factor of 10^3 improvement in volume coverage over initial LIGO. We expect to realize this by increasing the laser power further, using larger mirrors with smaller absorption and smaller scattering, better seismic isolation, and new optical schemes, among other things. Advanced LIGO commissioning is planned to start in 2011.

5. Conclusion

During its fifth science run, LIGO acquired more than one year of triple coincidence data. The sensitivity of all instruments reached and exceeded design. Despite the fact that S5 data is still under study and no gravitational wave has been detected yet, LIGO has already set many upper limits for various sources. Enhanced LIGO is already in the commissioning phase. A factor of two improvement in sensitivity is expected, with roughly a factor of eight larger volume in the detection range than initial LIGO, before the Advanced LIGO starts.

Acknowledgements

The authors gratefully acknowledge the support of the United States National Science Foundation for the construction and operation of the LIGO Laboratory and the Science and Technology Facilities Council of the United Kingdom, the Max-Planck-Society, and the State of Niedersachsen/Germany for support of the construction and operation of the GEO600 detector. The authors also gratefully acknowledge the support of the research by these agencies and by the Australian Research Council, the Council of Scientific and Industrial Research of India, the

Istituto Nazionale di Fisica Nucleare of Italy, the Spanish Ministerio de Educacion y Ciencia, the Conselleria d'Economia Hisenda i Innovacio of the Govern de les Illes Balears, the Scottish Funding Council, the Scottish Universities Physics Alliance, The National Aeronautics and Space Administration, the Carnegie Trust, the Leverhulme Trust, the David and Lucile Packard Foundation, the Research Corporation, and the Alfred P. Sloan Foundation. This paper has LIGO Document Number LIGO-P070137-01.

References

[1] Abramovici A *et al.* 1992 *Science* **256** 325–333
[2] Barish B and Weiss R 1999 *Physics Today* **52** 44
[3] Drever R W P *et al.* 1983 *Quantum Optics, Experimental Gravity and Measurement Theory* ed Meystre P and Scully M (New York: Plenum) pp 503–524
[4] Lazzarini A and Weiss R 1996 Tech. Rep. LIGO-E950018-02 LIGO
[5] Abbott B *et al.* 2007 *Phys. Rev. D* **76** 042001
[6] Abbott B *et al. to appear in Phys. Rev. D, arXiv:0708.3818*
[7] Abbott B *et al.* 2007 *Class. Quantum Grav.* **24** 5343–5369
[8] Abbott B *et al.* 2007 *Phys. Rev. D* **76** 062003
[9] Abbott B *et al.* 2007 *ApJ* **659** 918
[10] Abbott B *et al.* 2007 *Phys. Rev. D* **76** 022001
[11] Abbott B *et al.* 2007 *Phys. Rev. D* **76** 082003
[12] URL `http://www.ligo.org/`
[13] Adhikari R, Fritschel P and Waldman S 2006 Tech. Rep. LIGO-T060156-01 LIGO
[14] Gustafson E, Shoemaker D, Strain K and RWeiss 1999 Tech. Rep. LIGO-T990080-00-D LIGO
[15] Barish B and Sanders G 1999 Tech. Rep. LIGO-L990267-00-M LIGO

The Japanese Space Gravitational Wave Antenna; DECIGO

Seiji Kawamura[1], Masaki Ando[2], Takashi Nakamura[3], Kimio Tsubono[2], Takahiro Tanaka[3], Ikkoh Funaki[4], Naoki Seto[1], Kenji Numata[5], Shuichi Sato[1], Kunihito Ioka[6], Nobuyuki Kanda[7], Takeshi Takashima[4], Kazuhiro Agatsuma[2], Tomotada Akutsu[2], Tomomi Akutsu[2], Koh-suke Aoyanagi[8], Koji Arai[1], Yuta Arase[2], Akito Araya[9], Hideki Asada[10], Yoichi Aso[11], Takeshi Chiba[12], Toshikazu Ebisuzaki[13], Motohiro Enoki[14], Yoshiharu Eriguchi[15], Masa-Katsu Fujimoto[1], Ryuichi Fujita[16], Mitsuhiro Fukushima[1], Toshifumi Futamase[17], Katsuhiko Ganzu[3], Tomohiro Harada[18], Tatsuaki Hashimoto[4], Kazuhiro Hayama[19], Wataru Hikida[16], Yoshiaki Himemoto[20], Hisashi Hirabayashi[21], Takashi Hiramatsu[2], Feng-Lei Hong[22], Hideyuki Horisawa[23], Mizuhiko Hosokawa[24], Kiyotomo Ichiki[2], Takeshi Ikegami[22], Kaiki T. Inoue[25], Koji Ishidoshiro[2], Hideki Ishihara[7], Takehiko Ishikawa[26], Hideharu Ishizaki[1], Hiroyuki Ito[24], Yousuke Itoh[27], Shogo Kamagasako[2], Nobuki Kawashima[25], Fumiko Kawazoe[28], Hiroyuki Kirihara[2], Naoko Kishimoto[4], Kenta Kiuchi[8], Shiho Kobayashi[29], Kazunori Kohri[30], Hiroyuki Koizumi[2], Yasufumi Kojima[31], Keiko Kokeyama[28], Wataru Kokuyama[2], Kei Kotake[1], Yoshihide Kozai[32], Hideaki Kudoh[2], Hiroo Kunimori[33], Hitoshi Kuninaka[4], Kazuaki Kuroda[34], Kei-ichi Maeda[8], Hideo Matsuhara[4], Yasushi Mino[35], Osamu Miyakawa[35], Shinji Miyoki[34], Mutsuko Y. Morimoto[4], Tomoko Morioka[2], Toshiyuki Morisawa[3], Shigenori Moriwaki[36], Shinji Mukohyama[2], Mitsuru Musha[37], Shigeo Nagano[24], Isao Naito[38], Noriyasu Nakagawa[2], Kouji Nakamura[1], Hiroyuki Nakano[39], Kenichi Nakao[7], Shinichi Nakasuka[2], Yoshinori Nakayama[40], Erina Nishida[28], Kazutaka Nishiyama[4], Atsushi Nishizawa[41], Yoshito Niwa[41], Masatake Ohashi[34], Naoko Ohishi[1], Masashi Ohkawa[42], Akira Okutomi[2], Kouji Onozato[2], Kenichi Oohara[42], Norichika Sago[43], Motoyuki Saijo[43], Masaaki Sakagami[41], Shin-ichiro Sakai[4], Shihori Sakata[28], Misao Sasaki[44], Takashi Sato[42], Masaru Shibata[15], Hisaaki Shinkai[45], Kentaro Somiya[46], Hajime Sotani[47], Naoshi Sugiyama[48], Yudai Suwa[2], Hideyuki Tagoshi[16], Kakeru Takahashi[2], Keitaro Takahashi[44], Tadayuki Takahashi[4], Hirotaka Takahashi[49], Ryuichi Takahashi[48], Ryutaro Takahashi[1], Akiteru Takamori[9], Tadashi Takano[4], Keisuke Taniguchi[50], Atsushi Taruya[2], Hiroyuki Tashiro[3], Mitsuru Tokuda[7], Masao Tokunari[2], Morio Toyoshima[24], Shinji Tsujikawa[51], Yoshiki Tsunesada[52], Ken-ichi Ueda[37], Masayoshi Utashima[53], Hiroshi Yamakawa[54], Kazuhiro Yamamoto[1], Toshitaka Yamazaki[1], Jun'ichi Yokoyama[2], Chul-Moon Yoo[44], Shijun Yoshida[17], Taizoh Yoshino[55]

[1] National Astronomical Observatory of Japan, Mitaka, Tokyo, 181-8588, Japan
[2] The University of Tokyo, Bunkyo, Tokyo, 113-0033, Japan
[3] Kyoto University, Kyoto, Kyoto, 606-8502, Japan
[4] Institute of Space and Astronautical Science, Japan Aerospace Exploration Agency, Sagamihara, Kanagawa, 229-8510, Japan
[5] NASA Goddard Space Flight Center, Greenbelt, MD 20771, USA
[6] High Energy Accelerator Research Organization, Tsukuba, Ibaraki, 305-0801, Japan

[7] Osaka City University, Osaka, Osaka, 558-8585, Japan
[8] Waseda University, Shinjuku, Tokyo, 169-8555, Japan
[9] Earthquake Research Institute, The University of Tokyo, Bunkyo, Tokyo, 113-0032, Japan
[10] Hirosaki University, Hirosaki , Aomori, 036-8560, Japan
[11] Columbia University, New York, NY 10027, USA
[12] Nihon University, Setagaya, Tokyo, 156-8550, Japan
[13] RIKEN, Wako, Saitama, 351-0198, Japan
[14] Tokyo Keizai University, Kokubunji, Tokyo, 185-8502, Japan
[15] The University of Tokyo, Meguro, Tokyo, 153-8902, Japan
[16] Osaka University, Toyonaka, Osaka, 560-0043, Japan
[17] Tohoku University, Sendai, Miyagi, 980-8578, Japan
[18] Rikkyo University, Toshima, Tokyo, 171-8501, Japan
[19] University of Texas, Brownsville, Texas, 78520, USA
[20] Shibaura Institute of Technology, Saitama, Saitama, 337-8570, Japan
[21] Space Educations Center, Japan Aerospace Exploration Agency, Sagamihara, Kanagawa, 229-8510, Japan
[22] National Institute of Advanced Industrial Science and Technology, Tsukuba, Ibaraki, 305-8563, Japan
[23] Tokai University, Hiratsuka, Kanagawa, 259-1292, Japan
[24] National Institute of Information and Communications Technology, Koganei, Tokyo, 184-8795, Japan
[25] Kinki University, Higashi-Osaka, Osaka, 577-8502, Japan
[26] Institute of Space and Astronautical Science, Japan Aerospace Exploration Agency, Tsukuba, Ibaraki, 305-8505, Japan
[27] University of Wisconsin - Milwaukee, Milwaukee, WI 53201-0413, USA
[28] Ochanomizu University, Bunkyo, Tokyo, 112-8610, Japan
[29] Astrophysics Research Institute, Liverpool John Moores University, Egerton Wharf, Birkenhead L41 1LD, UK
[30] Lancaster University, LA1 4YB, UK
[31] Hiroshima University, Higashi-hiroshima, Hiroshima, 739-8526, Japan
[32] Gunma Astronomical Observatory, Agatsuma, Gunma, 377-0702, Japan
[33] National Institute of Information and Communications Technology, Bunkyo, Tokyo, 113-0001, Japan
[34] Institute for Cosmic Ray Research, The University of Tokyo, Kashiwa, Chiba, 277-8582, Japan
[35] California Institute of Technology, Pasadena, CA 91125, USA
[36] The University of Tokyo, Kashiwa, Chiba, 277-8561, Japan
[37] Institute for Laser Science, The University of Electro-Communications, Chofu, Tokyo, 182-8585, Japan
[38] Numakage, Saitama, Saitama, 336-0027, Japan
[39] Rochester Institute of Technology, Rochester, NY 14623, USA
[40] National Defense Academy, Yokosuka, Kanagawa, 239-8686, Japan
[41] Kyoto University, Kyoto, Kyoto, 606-8501, Japan
[42] Niigata University, Niigata, Niigata, 950-2181, Japan
[43] University of Southampton , Southampton SO17 1BJ, UK
[44] Yukawa Institute for Theoretical Physics, Kyoto University, Kyoto, Kyoto, 606-8502, Japan
[45] Osaka Institute of Technology, Hirakata, Osaka, 573-0196, Japan
[46] Albert Einstein Institute, Max Planck Institute for Gravitational Physics, D-14476 Potsdam, Germany

[47] Aristotle University of Thessaloniki, Thessaloniki, 54124, Greece
[48] Nagoya University, Nagoya, Aichi, 464-8602, Japan
[49] Nagaoka University of Technology, Nagaoka, Niigata, 940-2188, Japan
[50] University of Illinois at Urbana-Champaign, Urbana, IL 61801, USA
[51] Gunma National College of Technology, Maebashi, Gunma, 371-8530, Japan
[52] Tokyo Institute of Technology, Ookayama, Meguro-ku, Tokyo, 152-8550, Japan
[53] Japan Aerospace Exploration Agency, Tsukuba, Ibaraki, 305-8505, Japan
[54] Research Institute for Sustainable Humanosphere, Kyoto University, Uji, Kyoto, 611-0011
[55] Nakamura-minami, Nerima, Tokyo, 176-0025, Japan
[*] Paper presented at the conference by Seiji Kawamura

E-mail address: seiji.kawamura@nao.ac.jp

Abstract. DECi-hertz Interferometer Gravitational wave Observatory (DECIGO) is the future Japanese space gravitational wave antenna. DECIGO is expected to open a new window of observation for gravitational wave astronomy especially between 0.1 Hz and 10 Hz, revealing various mysteries of the universe such as dark energy, formation mechanism of supermassive black holes, and inflation of the universe. The pre-conceptual design of DECIGO consists of three drag-free spacecraft, whose relative displacements are measured by a differential Fabry–Perot Michelson interferometer. We plan to launch two missions, DECIGO pathfinder and pre-DECIGO first and finally DECIGO in 2024.

1. Space gravitational wave antenna DECIGO

DECIGO is the future Japanese space gravitational wave antenna. It stands for DECi-hertz Interferometer Gravitational wave Observatory [1][2]. The objectives of DECIGO are to detect gravitational waves from various kinds of sources mainly between 0.1 Hz and 10 Hz and open a new window of observation for gravitational wave astronomy.

DECIGO will bridge the frequency gap between LISA [3] and terrestrial detectors such as LCGT [4]. It can play a role of follow-up for LISA by observing inspiral sources that have moved above the LISA band, and can also play a role of predictor for terrestrial detectors by observing inspiral sources that have not yet moved into the terrestrial detector band.

DECIGO can reach an extremely good sensitivity. This is because the confusion limiting noise caused by irresolvable gravitational wave signals from many compact binaries is expected to be very low above 0.1 Hz [5].

2. Pre-conceptual design of DECIGO

As shown in Fig. 1, the pre-conceptual design of DECIGO consists of three drag-free spacecraft, whose relative displacements are measured by a differential Fabry–Perot (FP) Michelson interferometer. The arm length was chosen to be 1,000 km in order to realize a finesse of 10 with a 1 m diameter mirror and 0.5 μm laser light. The mass of the mirror is 100 kg and the laser power is 10 W. Three sets of such interferometers sharing the mirrors as arm cavities comprise one cluster of DECIGO. The constellation of DECIGO is composed of four clusters of DECIGO located separately in the heliocentric orbit with two of them nearly at the same position.

The FP configuration requires the distance between two mirrors, thus, the distance between two spacecraft to be constant during continuous operations. This makes DECIGO very different from a possible counterpart with the transponder-type detector (e.g. LISA), where the spacecraft, which are much farther apart, are freely falling according to their local gravitational field. We adopted the FP configuration because it can provide a better shot-noise-limited sensitivity than the transponder configuration due to the enhanced gravitational wave signals.

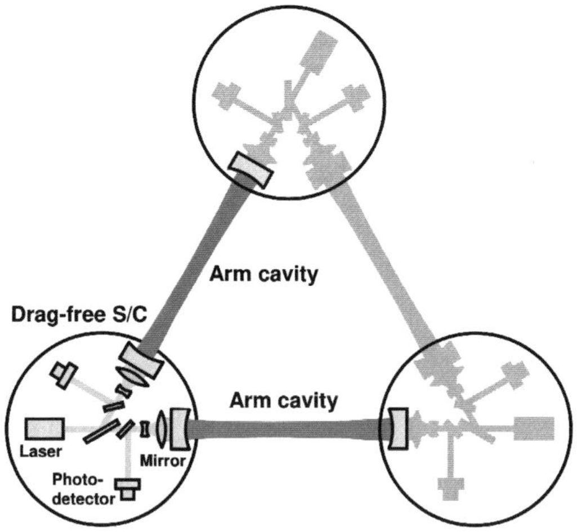

Fig.1. Pre-conceptual design of DECIGO.

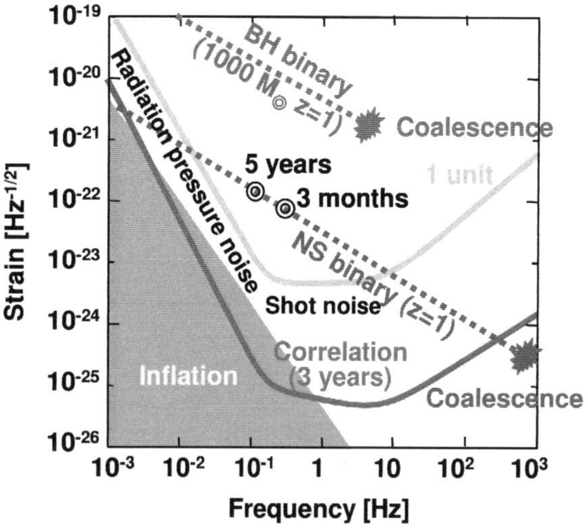

Fig. 2. Sensitivity goal of DECIGO and expected gravitational wave signals.

The FP configuration requires an additional system beyond that of an ordinary drag-free system. In the ordinary drag-free system, the outer spacecraft simply follows the motion of the mirror inside. However, in the FP configuration, the distance between the mirrors should be kept constant, which requires that one of the two mirrors should be controlled with the other mirror as a reference. As a result, the reference mirror dictates the motion of the other mirror by the FP interferometer control system as well as both spacecraft by the drag-free control system. It should be also noted that the FP interferometer control signals, which include gravitational wave signals, are not contaminated by the noisy drag-free control signals.

The lock acquisition of the FP arm cavity is a challenging task. In a ground-based interferometer, the relative motion of the two mirrors of a cavity is small enough to acquire lock of the cavity without much difficulty because the suspension systems of the mirrors are virtually connected to the ground at zero frequency. However, the relative motion of the two spacecraft in space is expected to be much higher. Therefore, we need an additional system that detects the relative motion of the mirrors, and gradually reduces it by actuating the mirrors. Once the relative motion of the mirrors is suppressed well enough, the lock acquisition of the FP cavity will be straightforward.

3. Sensitivity goal of DECIGO and science obtained by DECIGO

The ideal sensitivity of DECIGO is limited only by quantum noise, as shown in Fig. 2. The sensitivity is limited by the radiation pressure noise below 0.15 Hz, and it has an f^{2} frequency dependence. The shot noise limits the sensitivity above 0.15 Hz. It is flat up to 7.5 Hz, and above 7.5 Hz it increases in proportion with frequency because of the signal cancellation in the arm cavities.

In order to realize the sensitivity goal of DECIGO, all the practical noise should be suppressed well below this level. This imposes stringent requirements for the subsystems of DECIGO. We anticipate that extremely rigorous investigations are required to attain the requirements especially in the acceleration noise and frequency noise.

Nevertheless, accomplishing the sensitivity goal of DECIGO will ensure a variety of fruitful sciences to be obtained.

(1) Characterization of dark energy

DECIGO can detect gravitational waves coming from neutron star binaries at z=1 for five years prior to coalescences. It is expected that within this range about 7,000 neutron star binaries will coalesce every year. Therefore, DECIGO will detect gravitational waves coming from a large number of neutron star binaries at the same time. By analyzing the waveforms of these gravitational wave signals precisely, it is possible to determine the acceleration of the expansion of the universe [1]. The acceleration of the expansion of the universe can be also measured by finding host galaxies of each binary and determining their red shifts optically [6]. This will lead to better characterization of dark energy.

(2) Formation mechanism of supermassive black holes in the center of galaxies

DECIGO can detect gravitational waves coming from coalescences of intermediate-mass black hole binaries with an extremely high fidelity. For example the coalescences of black hole binaries of 1,000 solar masses at z=1 give a signal to noise ratio of 6,000. This will make it possible to collect numerous data about the relationship between the mass of the black holes and the frequency of the coalescences, which will reveal the formation mechanism of supermassive black holes in the center of galaxies.

(3) Verification and characterization of inflation

DECIGO can detect stochastic background corresponding to $\Omega_{GW}=2\times10^{-16}$ by correlating the data from the two clusters of DECIGO for three years. According to the standard inflation model, it is expected that we could detect gravitational waves produced at the inflation period of the universe with DECIGO. This is extremely significant because gravitational waves are the only means which make it possible to directly observe the inflation of the universe.

While the inflation background is the primary target for the correlation analysis with the two clusters, it would be important to carefully design the system so that we can disclose various aspects of stochastic gravitational wave backgrounds. One of the interesting measures from fundamental physics is the Stokes V parameter. This parameter characterizes the asymmetry of the amplitudes of the right- and left-handed waves, and it is a powerful measure to probe violation of parity symmetry that interchanges the two circular-polarization modes. By slightly adjusting the relative configuration of the two clusters, we can set sensitivity to the Stokes V parameter [7].

4. Roadmap to DECIGO

DECIGO pathfinder (DPF) and pre-DECIGO will be launched before DECIGO. DPF will test the key technologies with one spacecraft. We expect that it will be launched in 2012. Pre-DECIGO is supposed to detect gravitational waves with minimum specifications. We hope that it will be launched in 2018. Finally it is expected that DECIGO will be launched in 2024 to open a new window of observation for gravitational wave astronomy.

Acknowledgment

This research was partially supported by the Ministry of Education, Science, Sports and Culture, Grant-in-Aid for Scientific Research.

References

[1] Seto N, Kawamura S and Nakamura T 2001 Possibility of direct measurement of the acceleration of the universe using 0.1 Hz band laser interferometer gravitational wave antenna in space *Phys. Rev. Lett.* **87** 221103

[2] Kawamura S *et al* 2006 The Japanese Space Gravitational Wave Antenna - DECIGO *Class. Quantum Grav.* **23** S125

[3] LISA: System and Technology Study Report, ESA document ESA-SCI (2000)

[4] Kuroda K *et al* 2002 Japanese large-scale interferometers *Class. Quantum Grav.* **19** 1237

[5] Farmer A J and Phinney E S 2003 The gravitational wave background from cosmological compact binaries *Mon. Not. R. Astron. Soc.* **346** 1197

[6] Schutz B F 1986 Determining the Hubble constant from gravitational wave observations *Nature* **323** 310

[7] Seto N 2007 Quest for circular polarization of a gravitational wave background and orbits of laser interferometers in space *Phys. Rev. D* **75** 061302

DECIGO Pathfinder

Masaki Ando[1], Seiji Kawamura[2], Takashi Nakamura[3], Kimio Tsubono[1], Takahiro Tanaka[3], Ikkoh Funaki[4], Naoki Seto[2], Kenji Numata[5], Shuichi Sato[2], Kunihito Ioka[6], Nobuyuki Kanda[7], Takeshi Takashima[4], Kazuhiro Agatsuma[1], Tomotada Akutsu[1], Tomomi Akutsu[1], Koh-suke Aoyanagi[8], Koji Arai[2], Yuta Arase[1], Akito Araya[9], Hideki Asada[20], Yoichi Aso[22], Takeshi Chiba[21], Toshikazu Ebisuzaki[23], Motohiro Enoki[24], Yoshiharu Eriguchi[25], Masa-Katsu Fujimoto[2], Ryuichi Fujita[26], Mitsuhiro Fukushima[2], Toshifumi Futamase[27], Katsuhiko Ganzu[3], Tomohiro Harada[28], Tatsuaki Hashimoto[4], Kazuhiro Hayama[29], Wataru Hikida[26], Yoshiaki Himemoto[10], Hisashi Hirabayashi[12], Takashi Hiramatsu[1], Feng-Lei Hong[11], Hideyuki Horisawa[13], Mizuhiko Hosokawa[14], Kiyotomo Ichiki[1], Takeshi Ikegami[11], Kaiki T. Inoue[15], Koji Ishidoshiro[1], Hideki Ishihara[7], Takehiko Ishikawa[16], Hideharu Ishizaki[2], Hiroyuki Ito[14], Yousuke Itoh[17], Shogo Kamagasako[1], Nobuki Kawashima[15], Fumiko Kawazoe[18], Hiroyuki Kirihara[1], Naoko Kishimoto[4], Kenta Kiuchi[8], Shiho Kobayashi[19], Kazunori Kohri[30], Hiroyuki Koizumi[1], Yasufumi Kojima[32], Keiko Kokeyama[18], Wataru Kokuyama[1], Kei Kotake[2], Yoshihide Kozai[31], Hideaki Kudoh[1], Hiroo Kunimori[33], Hitoshi Kuninaka[4], Kazuaki Kuroda[34], Kei-ichi Maeda[8], Hideo Matsuhara[4], Yasushi Mino[35], Osamu Miyakawa[35], Shinji Miyoki[34], Mutsuko Y. Morimoto[4], Tomoko Morioka[1], Toshiyuki Morisawa[3], Shigenori Moriwaki[36], Shinji Mukohyama[1], Mitsuru Musha[37], Shigeo Nagano[14], Isao Naito[38], Noriyasu Nakagawa[1], Kouji Nakamura[2], Hiroyuki Nakano[39], Kenichi Nakao[7], Shinichi Nakasuka[1], Yoshinori Nakayama[40], Erina Nishida[18], Kazutaka Nishiyama[4], Atsushi Nishizawa[42], Yoshito Niwa[42], Masatake Ohashi[34], Naoko Ohishi[2], Masashi Ohkawa[41], Akira Okutomi[1], Kouji Onozato[1], Kenichi Oohara[41], Norichika Sago[43], Motoyuki Saijo[43], Masaaki Sakagami[42], Shin-ichiro Sakai[4], Shihori Sakata[18], Misao Sasaki[44], Takashi Sato[41], Masaru Shibata[25], Hisaaki Shinkai[45], Kentaro Somiya[46], Hajime Sotani[47], Naoshi Sugiyama[48], Yudai Suwa[1], Hideyuki Tagoshi[26], Kakeru Takahashi[1], Keitaro Takahashi[44], Tadayuki Takahashi[4], Hirotaka Takahashi[49], Ryuichi Takahashi[48], Ryutaro Takahashi[2], Akiteru Takamori[9], Tadashi Takano[4], Keisuke Taniguchi[50], Atsushi Taruya[1], Hiroyuki Tashiro[3], Mitsuru Tokuda[7], Masao Tokunari[1], Morio Toyoshima[14], Shinji Tsujikawa[52], Yoshiki Tsunesada[51], Ken-ichi Ueda[37], Masayoshi Utashima[53], Hiroshi Yamakawa[54], Kazuhiro Yamamoto[2], Toshitaka Yamazaki[2], Jun'ichi Yokoyama[1], Chul-Moon Yoo[44], Shijun Yoshida[27], Taizoh Yoshino[55]

[1] The University of Tokyo, Bunkyo, Tokyo, 113-0033, Japan
[2] National Astronomical Observatory of Japan, Mitaka, Tokyo, 181-8588, Japan
[3] Kyoto University, Kyoto, Kyoto, 606-8502, Japan
[4] Institute of Space and Astronautical Science, Japan Aerospace Exploration Agency, Sagamihara, Kanagawa, 229-8510, Japan
[5] NASA Goddard Space Flight Center, Greenbelt, MD 20771, USA
[6] High Energy Accelerator Research Organization, Tsukuba, Ibaraki, 305-0801, Japan

[7] Osaka City University, Osaka, Osaka, 558-8585, Japan
[8] Waseda University, Shinjuku, Tokyo, 169-8555, Japan
[9] Earthquake Research Institute, The University of Tokyo, Bunkyo, Tokyo, 113-0032, Japan
[10] Hirosaki University, Hirosaki , Aomori, 036-8560, Japan
[11] Columbia University, New York, NY 10027, USA
[12] Nihon University, Setagaya, Tokyo, 156-8550, Japan
[13] RIKEN, Wako, Saitama, 351-0198, Japan
[14] Tokyo Keizai University, Kokubunji, Tokyo, 185-8502, Japan
[15] The University of Tokyo, Meguro, Tokyo, 153-8902, Japan
[16] Osaka University, Toyonaka, Osaka, 560-0043, Japan
[17] Tohoku University, Sendai, Miyagi, 980-8578, Japan
[18] Rikkyo University, Toshima, Tokyo, 171-8501, Japan
[19] University of Texas, Brownsville, Texas, 78520, USA
[20] Shibaura Institute of Technology, Saitama, Saitama, 337-8570, Japan
[21] Space Educations Center, Japan Aerospace Exploration Agency, Sagamihara, Kanagawa, 229-8510, Japan
[22] National Institute of Advanced Industrial Science and Technology, Tsukuba, Ibaraki, 305-8563, Japan
[23] Tokai University, Hiratsuka, Kanagawa, 259-1292, Japan
[24] National Institute of Information and Communications Technology, Koganei, Tokyo, 184-8795, Japan
[25] Kinki University, Higashi-Osaka, Osaka, 577-8502, Japan
[26] Institute of Space and Astronautical Science, Japan Aerospace Exploration Agency, Tsukuba, Ibaraki, 305-8505, Japan
[27] University of Wisconsin - Milwaukee, Milwaukee, WI 53201-0413, USA
[28] Ochanomizu University, Bunkyo, Tokyo, 112-8610, Japan
[29] Astrophysics Research Institute, Liverpool John Moores University, Egerton Wharf, Birkenhead L41 1LD, UK
[30] Lancaster University, LA1 4YB, UK
[31] Hiroshima University, Higashi-hiroshima, Hiroshima, 739-8526, Japan
[32] Gunma Astronomical Observatory, Agatsuma, Gunma, 377-0702, Japan
[33] National Institute of Information and Communications Technology, Bunkyo, Tokyo, 113-0001, Japan
[34] Institute for Cosmic Ray Research, The University of Tokyo, Kashiwa, Chiba, 277-8582, Japan
[35] California Institute of Technology, Pasadena, CA 91125, USA
[36] The University of Tokyo, Kashiwa, Chiba, 277-8561, Japan
[37] Institute for Laser Science, The University of Electro-Communications, Chofu, Tokyo, 182-8585, Japan
[38] Numakage, Saitama, Saitama, 336-0027, Japan
[39] Rochester Institute of Technology, Rochester, NY 14623, USA
[40] National Defense Academy, Yokosuka, Kanagawa, 239-8686, Japan
[41] Kyoto University, Kyoto, Kyoto, 606-8501, Japan
[42] Niigata University, Niigata, Niigata, 950-2181, Japan
[43] University of Southampton , Southampton SO17 1BJ, UK
[44] Yukawa Institute for Theoretical Physics, Kyoto University, Kyoto, Kyoto, 606-8502, Japan
[45] Osaka Institute of Technology, Hirakata, Osaka, 573-0196, Japan
[46] Albert Einstein Institute, Max Planck Institute for Gravitational Physics, D-14476 Potsdam, Germany

[47] Aristotle University of Thessaloniki, Thessaloniki, 54124, Greece
[48] Nagoya University, Nagoya, Aichi, 464-8602, Japan
[49] Nagaoka University of Technology, Nagaoka, Niigata, 940-2188, Japan
[50] University of Illinois at Urbana-Champaign, Urbana, IL 61801, USA
[51] Gunma National College of Technology, Maebashi, Gunma, 371-8530, Japan
[52] Tokyo Institute of Technology, Ookayama, Meguro-ku, Tokyo, 152-8550, Japan
[53] Japan Aerospace Exploration Agency, Tsukuba, Ibaraki, 305-8505, Japan
[54] Research Institute for Sustainable Humanosphere, Kyoto University, Uji, Kyoto, 611-0011
[55] Nakamura-minami, Nerima, Tokyo, 176-0025, Japan

Corresponding author e-mail address: ando@granite.phys.s.u-tokyo.ac.jp

Abstract. DECIGO pathfinder (DPF) is a milestone satellite mission for DECIGO (DECi-hertz Interferometer Gravitational wave Observatory) which is a future space gravitational wave antenna. DECIGO is expected to provide us fruitful insights into the universe, in particular about dark energy, a formation mechanism of supermassive black holes, and the inflation of the universe. Since DECIGO will be an extremely large mission which will formed by three drag-free spacecraft with 1000m separation, it is significant to gain the technical feasibility of DECIGO before its planned launch in 2024. Thus, we are planning to launch two milestone missions: DPF and pre-DECIGO. The conceptual design and current status of the first milestone mission, DPF, are reviewed in this article.

1. Introduction

DECIGO, a DECi-hertz Interferometer Gravitational wave Observatory, is a space gravitational wave antenna planned to be launched in 2024 [1][2]. The purpose of DECIGO is to observe gravitational waves at the frequency band mainly between 0.1 Hz and 10 Hz, and thus, to open a new window of gravitational wave astronomy. Since the observation band of DECIGO is between LISA [3] and terrestrial detectors such as LCGT [4], it can be a follow-up of LISA by observing inspiral sources that have moved above the LISA band, and can also be a predictor for terrestrial detectors by observing inspiral sources that have not yet moved into the terrestrial detector band. Moreover, since DECIGO's observation band is free from foreground noises caused by unresolved gravitational-waves from many galactic binaries, it can play an important role in the observation of stochastic background gravitational waves from the early universe.

In the pre-conceptual design, DECIGO will be formed by three drag-free spacecraft which are separated by 1000km to one another. The gravitational-wave signals are detected by measuring the relative displacements with differential Fabry–Perot interferometers. The arm length was chosen so as to realize a finesse of 10 with a 1 m diameter mirror and 0.5 μm laser light. The mass of the mirror is 100 kg and the laser power is 10 W.

Since DECIGO will be an extremely large mission, it is significant to gain the technical feasibility before its planned launch in 2024. Thus, we have a roadmap to launch two milestone missions before DECIGO. DECIGO pathfinder (DPF) is the first milestone mission to test the key technologies with one spacecraft. Pre-DECIGO is supposed to detect gravitational waves with minimum specifications. In this article, we review the conceptual design and current status of DPF.

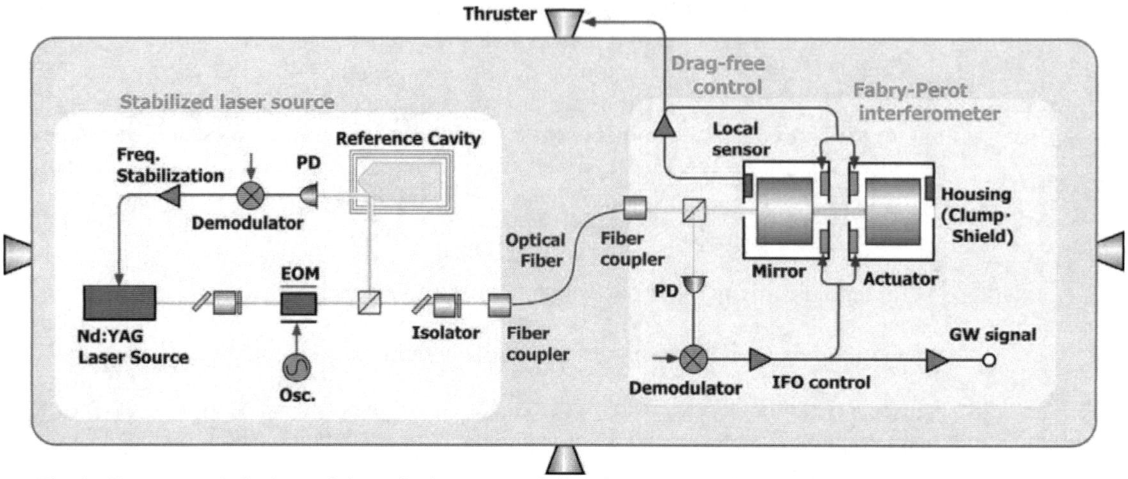

Fig.1. Conceptual design of the mission part of the DECIGO pathfinder (DPF).

2. Purpose and conceptual design of DECIGO pathfinder

The purposes of DPF are to test the key technologies and to make observations at 0.1-1Hz frequency band. The key technologies tested in DPF will be the followings: (1) low-noise operation and observation with a Fabry-Perot interferometer in space (2) operation of a laser source and its stabilization system, (3) demonstration of a drag-free control system (4) demonstration of a launch-lock system for the test-mass mirrors. All of these technologies are critical in the realization of DECIGO. As for the scientific purpose of DPF, DPF targets at 0.1-1Hz gravitational waves from intermediate-mass blackhole inspirals. If merger or ringdown events occur at around the center of our galaxy, DPF will have a sufficient sensitivity to detect them.

DPF will be a small satellite with weight of about 300kg, orbiting the earth along a sun-synchronous orbit with an altitude of 500km. The mission part of DPF is designed to be a prototype of DECIGO, being comprised of a short Fabry-Perot (FP) cavity, a stabilized laser source, and a dreg-free control system (Fig.1). The FP cavity is formed by two mirrors which act as free test masses. Each mirror is placed inside a module called housing. The housing has local sensors and local actuators, which are used to monitor and to control the relative motion between the housing and the mirror. In addition, the housing has a function of launch lock, which clumps the mirror at the launch of the satellite and releases it in the orbit with small initial velocity. The cavity has a baseline length of about 30cm and a finesse of about 100. The length change in the FP cavity, which would be caused by gravitational waves or external disturbances, is measured by means of a stabilized laser beam. In DPF, we use a Nd:YAG laser source in which the frequency is stabilized using an external reference. The requirement for the frequency stabilization is $1\mathrm{Hz/Hz}^{1/2}$. The laser source has an output power of 100mW at a wavelength of 1064nm. The drag-free control of the satellite position works as a shield against external forces such as solar radiations and air drags. The drag-free control is realized by measuring the relative fluctuations between the mirrors and the satellite, and feeding these signals back to the satellite position actuated by low-noise thrusters of the satellite.

With the configuration described above, DPF will have a sensitivity limit of about $h{\sim}10^{-15}$ at around the frequency band of 0.1-1Hz (Fig.2 left). The sensitivity is mainly limited by the laser frequency noise at the high-frequency band and acceleration noises from various kinds of external forces on the test-mass mirrors at the low-frequency band. From this noise curve, we estimated the observable range for blackhole inspirals and ringdowns of a blackhole quasi-normal mode as a function of blackhole mass (Fig.2 right). In this estimation of the observable range of DPF, we used

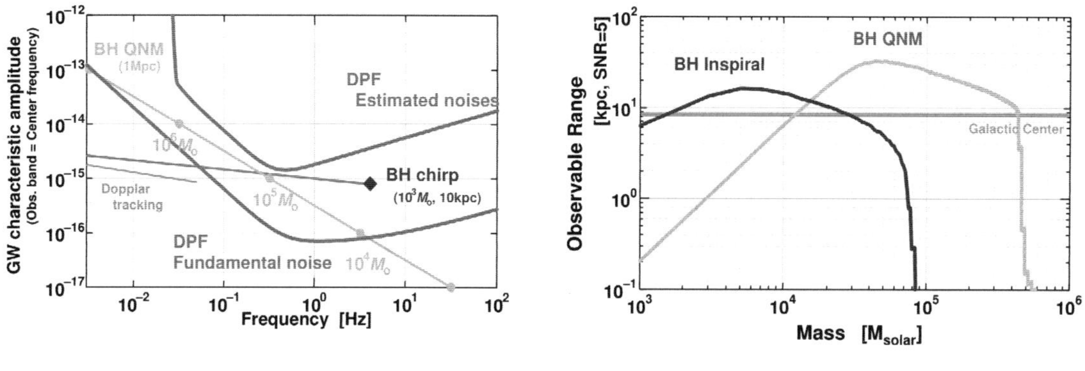

Fig.2. Sensitivity (left) and observable range (right) of DPF. The observable range of DPF for intermediate-mass blackholes will cover the center of our galaxy.

the calculated waveform of gravitational-wave signal by post-Newtonian approximation, and set the detection threshold to be a signal-to-noise ratio of 5. As shown in Fig.2, DPF has a sensitivity to detect gravitational-wave signals, if there is a blackhole inspiral event with 10^3-10^4 solar-masses, or a ringdown event of a quasi-normal mode for 10^4-$4x10^5$ solar-mass blackhole at the center of our galaxy.

3. Current status of DPF

Currently, DPF is selected as one of the candidates of small satellite missions of JAXA (Japan Aerospace Exploration Agency). JAXA have a program to launch at least 3 small satellites in this 5 years, using standard bus systems. The standard bus module has a 200kg weight and a 900mm cubic shape. It will provide a 900W power with 6 solar-cell puddles at maximum, and a 2Mbps downlink telecommunication ability. A Mission module is attached on the upside of the bus module. The bus and mission modules are connected with power lines and communication lines using the SpaceWire standard.

The first mission of the three missions has already decided to be TOPS, which is for the observation of inferior planets. TOPS will be launched in 2011. DPF is now one of the several high-ranked mission candidates for the second or third missions, and will be launched in 2012 in the best and earliest case.

Acknowledgment
This research was partially supported by the Ministry of Education, Science, Sports and Culture, Grant-in-Aid for Scientific Research, and by the Japan Aerospace Exploration Agency.

References

[1] Seto N, Kawamura S and Nakamura T, 2001, Possibility of direct measurement of the acceleration of the universe using 0.1 Hz band laser interferometer gravitational wave antenna in space, *Phys. Rev. Lett.* **87** 221103.
[2] Kawamura S *et al,* 2006, The Japanese Space Gravitational Wave Antenna – DECIGO, *Class. Quantum Grav.* **23** S125. Kawamura S, in this issue.
[3] LISA: System and Technology Study Report, ESA document ESA-SCI (2000).
[4] Kuroda K *et al,* 2002, Japanese large-scale interferometers, *Class. Quantum Grav.* **19** 1237.

Displacement noise free interferometory for gravitational wave detection

S Sato , K Kokeyama , S Kawamura , R L Ward , A Nishizawa , Y Chen[5], A Pai[5] and K Somiya[5]

[1] National Astronomical Observatory of Japan, Tokyo 181-8588, Japan
[2] Ochanomizu University, Tokyo 112-8610, Japan
[3] LIGO Project, California Institute of Technology, Pasadena, California 91125, USA
[4] Kyoto University, Kyoto 606-8501, Japan
[5] Max-Planck-Institut für Gravitationsphysik, 14476 Potsdam, Germany

E-mail: `sato.shuichi@nao.ac.jp`

Abstract. We have demonstrated displacement- and frequency-noise-free laser interferometry (DFI) by partially implementing a recently proposed optical configuration using bidirectional Mach-Zehnder interferometers (MZIs). This partial implementation, the minimum necessary to be called DFI, has confirmed the essential feature of DFI: the combination of two MZI signals can be carried out in a way that cancels displacement noise of the mirrors and beam splitters while maintaining gravitational-wave signals. The attained maximum displacement noise suppression was 45 dB.

1. Introduction

The effect of gravitational wave (GW) and test mass motion are essentially different. The practical application of this fact to the GW antenna was addressed only recently in the references [1, 2]. As was shown theoretically, when each test mass in an N-test-mass array sends and receives light pulses from all other test masses, and if the array contains enough test masses [$N > (d+2)$, where d is the number of spatial dimensions of the array], then there exist combinations of time delays which do not sense test-mass motions or timing noises, but do sense gravitational waves. We will call configurations that cancel both timing and displacement noises displacement-noise-free interferometry (DFI).

Recently, practical optical designs of DFI using laser interferometry in three dimensions have been proposed [3]. In these configurations, the conventional, equal-arm Mach-Zehnder interferometer (MZI) was used as a building block to eliminate laser noise. Four such MZIs were combined, in such a way that they form two pairs of counterpropagating MZIs. Within each pair, the 2 MZIs share the same beam splitters and folding mirrors; subtraction of their outputs balances out displacement noise from motions of the folding mirrors. The two pairs share the same beam splitters, which allows the elimination of beam splitter displacement noise.

In this paper, we study DFI experimentally using a single pair of counterpropagating MZIs and a single pair of independent MZIs, demonstrating the elimination of folding mirror and beam splitter displacement noise.

2. Partial demonstration of 3-D DFI

As described in reference [3], the full configured 3D DFI is composed of two pairs of counterpropagating MZIs, for a total of four MZIs. MZIs of equal-arm length operated on the midfringe are insensitive to laser frequency noise, and by superimposing two counterpropagating MZIs on the same optical path, the displacements of the folding mirrors are sensed redundantly (once by each direction of the MZI) and can thus be unambiguously removed from the signal. An additional pair of counterpropagating MZIs can then be added, which shares beam splitters with the first pair—this second pair allows redundant sensing of beam splitter displacements. As discussed in [3], the signals from the four MZIs can be combined in such a way as to cancel the displacements of all the optics while retaining sensitivity to gravitational radiation.

For the experiment on folding mirror displacement, we have constructed a partial-DFI composed of a single pair of counterpropagating MZIs to demonstrate the cancellation properties of the DFI. The practical experimental setup is shown in Figure 1. The displacement-noise simulator (EOM1) should be located at the exact center of the arm, so the position of EOM1 was carefully tuned. One of the output ports of MZI 1 (output 1a) was monitored with a dc detector (DCPD) and an error signal created by subtracting a static offset; this signal was fed back to a PZT-actuated folding mirror (FM2) after appropriate filtering to give a midfringe locking control. Once the fringe of the MZI 1 is controlled, that of MZI 2 is also automatically controlled because the two MZIs share common optical paths. Other output beams (output 1b and 2a) were received with high-speed photodetectors to monitor differential optical path length variations. The polarization optics (HWP: half wave plate and PBS: polarizing beam splitter) just in front of both high-speed detectors act as an optical attenuator, which serves as a gain compensator for imbalanced outputs of the two detectors. The signals from high-speed photodetectors HPD1 and HPD2 were summed with a power combiner to produce a DFI output signal, which was then monitored with a network analyzer. The DFI features of bi-directional MZIs were demonstrated with a transfer function measurement from noise simulators to DFI output.

The results for displacement-noise suppression are shown in Figure 2. About 30 dB of suppression of displacement noise was attained in a wide frequency region, while the maximum attained suppression was 45 dB at a particular frequency band, achieved when we tuned for maximum suppression in that region. Any imbalance between the two MZI signals determines the suppression ratio; it is believed that the subtly different frequency response of our two photodetectors, which cannot be compensated by simple optical attenuators or path length tuning, was the limiting factor here.

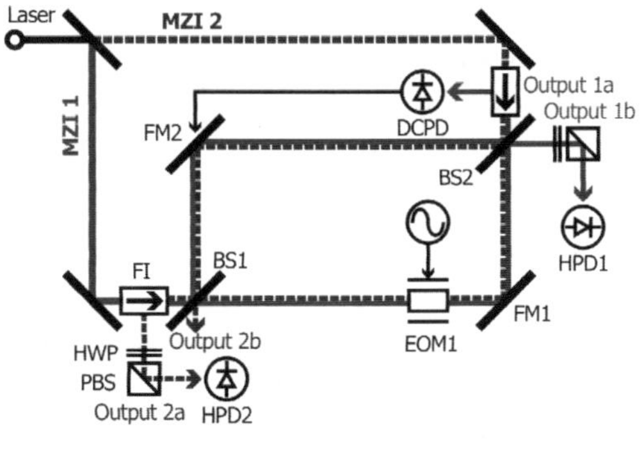

Figure 1. Experimental setup for cancellation of folding mirror displacement.

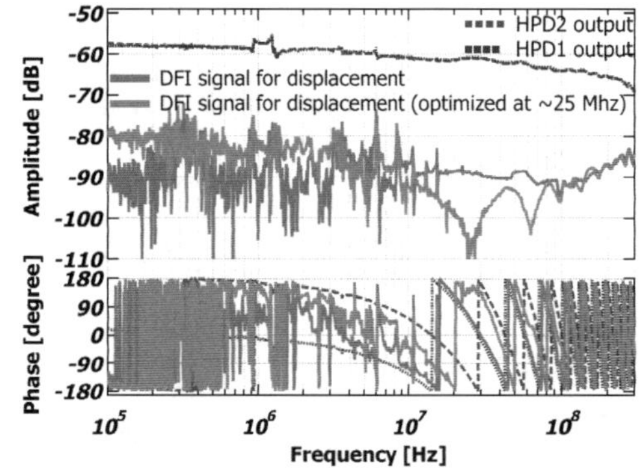

Figure 2. Cancellation of folding mirror displacement.

For the beamsplitter displacement, we have constructed another set of partial-DFI composed of a single pair of MZIs sharing a beamsplitter to demonstrate another cancellation property of the DFI. The practical experimental setup is shown in Figure 3 sharing input beam splitter (BS1). The laser beam is introduced into beamsplitter BS1, then split into two to form independent MZIs. The displacement-noise of the beam splitter was simulated by EOM3, which is located just after BS1. Displacement noise simulation is identical to the previous experiment. One of the output ports of MZI 1 and MZI 2 were monitored with dc detectors (DCPD) and error signals were fed back to a PZT-actuated folding mirrors (FM1 and FM2) independently after appropriate filtering to give midfringe locking control. Other output beams were received with high-speed photodetectors to monitor differential optical path length variations.

The results for displacement-noise suppression are shown in Figure 4. Both amplitude and phase for the two signals were tuned to match each other using optical attenuators and path length adjustments so that the displacement-noise signal disappears in the DFI signal. About 30 dB of suppression of displacement noise was attained in a wide frequency region, while the maximum attained suppression was 45 dB at a particular frequency band, achieved when we tuned for maximum suppression in that region.

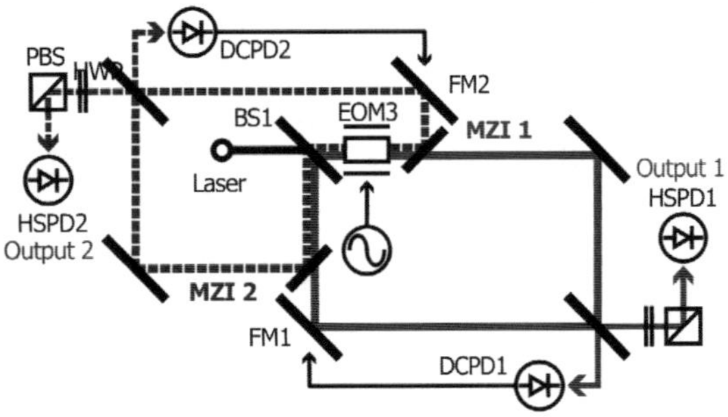

Figure 3. Experimental setup for cancellation of beamsplitter displacement.

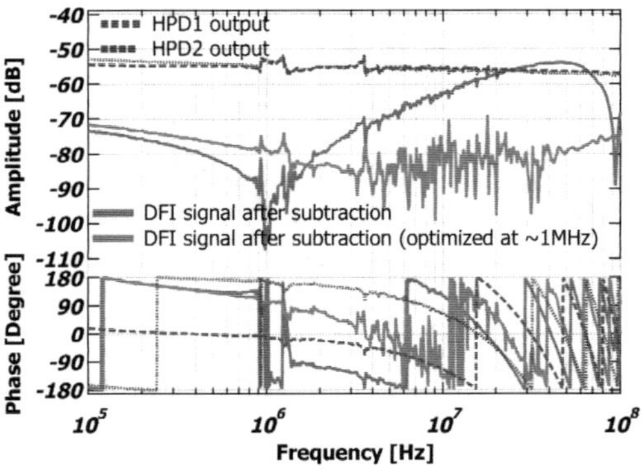

Figure 4. Cancellation of beamsplitter displacement.

3. Summary

We studied DFI experimentally using a single pair of counterpropagating MZIs and a single pair of independent MZIs, demonstrating the elimination of folding mirror and beam splitter displacement noise. The fundamental features of DFI were confirmed with these proof-of-principle experiments. The maximum attained suppression of displacement noise was 45 dB at a particular frequency band; it is believed that the subtly different frequency response of our two photodetectors was the limiting factor.

Acknowledgments

This research was partially supported by the Japanese Ministry of Education, Culture, Sports, Science and Technology under Grant-in-Aid for Scientific Research (B), No. 18340070, 2006. R. L.W. was supported by the U.S. National Science Foundation under Cooperative Agreement No. PHY-0107417. The research of Y. C., A. P., and K. S. is supported by the Alexander von Humboldt Foundation's Sofja Kovalevskaja Programme (funded by the German Ministry of Education and Research).

References

[1] Kawamura S and Chen Y 2004 *Phys. Rev. Lett.* 93, 211103
[2] Chen Y and Kawamura S 2006 *Phys. Rev. Lett.* 96, 231102
[3] Chen Y *et al.* 2006 *Phys. Rev. Lett.* 97, 151103
[4] Sato S *et al.* 2007 *Phys. Rev. Lett.* 98, 141101

VIRGO: a large interferometer for Gravitational Wave detection started its first scientific run

F.Acernese[7,9], M.Alshourbagy[15,16], P.Amico[13,14], F.Antonucci[19], S.Aoudia[10], P.Astone[19], S.Avino[7,8], L.Baggio[1], F.Barone[7,9], L.Barsotti[15,16], M. Barsuglia[11], Th.S.Bauer[21,22], S.Bigotta[15,16], S.Birindelli[15,16], M. Bizouard[11], C.Boccara[12], F.Bondu[10], L.Bosi[13,14], S.Braccini[15], C.Bradaschia[15], A.Brillet[10], V. Brisson[11], D.Buskulic[1], G.Cagnoli[3], E.Calloni[7,8], E.Campagna[3,5], F. Cavalier[11], R.Cavalieri[2], G.Cella[15], E.Cesarini[3,4], E.Chassande-Mottin[10], A.-C.Clapson[11], F.Cleva[10], E.Coccia[23,24], C.Corda[15,16], A.Corsi[19], F.Cottone[13,14], J.-P.Coulon[10], E.Cuoco[2,] S.D'Antonio[23,24], A.Dari[13,14], V.Dattilo[2], M. Davier[11], R.De Rosa[7,8], M.Del Prete[17,16], L.Di Fiore[7], A.Di Lieto[15,16], A.Di Virgilio[15], B.Dujardin[10], M.Evans[2], V.Fafone[23,24], I.Ferrante[15,16], F.Fidecaro[15,16], I.Fiori[2], R.Flaminio[6], J.-D.Fournier[10], S.Frasca[19,20], F.Frasconi[15,*], L.Gammaitoni[13,14], F.Garufi[7,8], E.Genin[2], A.Gennai[15], A.Giazotto[2,15], L.Giordano[7,8], V.Granata[1], C.Greverie[10], D.Grosjean[1], G.Guidi[3,5], S.Hamdani[2], S.Hebri[2], H.Heitmann[10], P. Hello[11], D.Huet[2], S. Kreckelbergh[11], P.La Penna[2], M.Laval[10], N.Leroy[11], N.Letendre[1], B.Lopez[2], M.Lorenzini[3,4], V.Loriette[12], G.Losurdo[3], J.-M.Mackowski[6], E.Majorana[19], M.Mantovani[17,16], F.Marchesoni[13,14], F.Marion[1], J.Marque[2], F.Martelli[3,5], A.Masserot[1], F.Menzinger[2], L.Milano[7,8], Y.Minenkov[23,24], C.Moins[2], J.Moreau[12], N.Morgado[6], S.Mosca[7,8], B.Mours[1], C.N.Man[10], I.Neri[13,14], F.Nocera[2], G.Pagliaroli[23,24], C.Palomba[19], F.Paoletti[2,15], S.Pardi[7,8], A.Pasqualetti[2], R.Passaquieti[15,16], D.Passuello[15], F.Piergiovanni[3,5], L.Pinard[6], R.Poggiani[15,16], M.Punturo[13], P.Puppo[19], P.Rapagnani[19,20], T.Regimbau[12], A.Remillieux[6], F.Ricci[19,20], I.Ricciardi[7,8], A.Rocchi[23,24], L.Rolland[1], R.Romano[7,9], P.Ruggi[2], G.Russo[7,8], S.Solimeno[7,8], A.Spallicci[10], M.Tarallo[15,16], R.Terenzi[23,24], A.Toncelli[15,16], M.Tonelli[15,16], E.Tournefier[1], F.Travasso[13,14], C.Tremola[15,16], G.Vajente[18,16], J.F.J. van der Brand[21,22], S. van der Putten[21,22], D.Verkindt[1], F.Vetrano[3,5], A.Vicerè[3,5], J.-Y.Vinet [10], H.Vocca[13,14], M.Yvert[1]

[1]Laboratoire d'Annecy-le-Vieux de Physique des Particules (LAPP), IN2P3/CNRS, Universitè de Savoie, Annecy-le-Vieux, France
[2]European Gravitational Observatory (EGO), Cascina (Pi), Italia
[3]INFN, Sezione di Firenze, Sesto Fiorentino
[4]Università degli Studi di Firenze, Firenze
[5]Università degli Studi di Urbino "Carlo Bo", Urbino
[6]LMA, Villeurbanne, Lyon, France
[7]INFN, sezione di Napoli
[8]Università di Napoli "Federico II" Complesso Universitario di Monte S.Angelo
[9]Università di Salerno, Fisciano (Sa), Italia
[10]Departement Artemis -- Observatoire de la Cote d'Azur, BP 42209 06304 Nice, Cedex 4, France
[11]LAL, Univ Paris-Sud, IN2P3/CNRS, Orsay, France
[12]ESPCI, Paris, France
[13]INFN, Sezione di Perugia
[14]Università di Perugia, Perugia, Italia
[15]INFN, Sezione di Pisa
[16]Università di Pisa, Pisa, Italia
[17]Università di Siena, Siena, Italia
[18]Scuola Normale Superiore, Pisa, Italia
[19]INFN, Sezione di Roma
[20]Università "La Sapienza", Roma, Italia
[21]NIKHEF, NL-1009 DB Amsterdam, The Netherlands
[22]Vrije Universiteit, NL-1081 HV Amsterdam, The Netherlands
[23]INFN, Sezione di Roma Tor Vergata, Roma, Italia
[24]Università di Roma Tor Vergata, Roma, Italia
[25]Università dell'Aquila, L'Aquila, Italia
* Paper presented at the conference by F. Frasconi
e-mail: franco.frasconi@pi.infn.it

Abstract The VIRGO interferometer is the largest ground based European gravitational wave detector operating at the EGO Laboratory in the Pisa (Italy) countryside. During the last commissioning period relevant progress have been done in approaching its design sensitivity all over the detection bandwidth. Thanks to the effort of the whole Collaboration a long scientific run has been done collecting data for more than 4 months in conjunction with the LIGO detectors. The results obtained from the detector point of view are: a very good stability and a duty-cycle as high as 81% in science mode. In this paper we present the status of the VIRGO interferometer giving an overview of the experimental apparatus together with its most relevant features.

1. Introduction

Nowadays gravitational waves search on Earth is based on broadband detectors able to explore a large fraction of the universe where astrophysical sources are expected to emit signals in the frequency range between a few Hz and a few kHz. The VIRGO interferometer is the largest European gravitational wave detector operating in Italy thanks to the starting effort of the Italian - French Collaboration financed by two founding agencies: the INFN for the Italian side and the CNRS for the French one. Together with similar detectors, like the three LIGO [1] and GEO [2] interferometers based on the same working principle, VIRGO will be a crucial node of a detector network able to perform detailed gravitational field studies. To this purpose the required high sensitivity to differential variation of the two perpendicular arms length of the order of 10^{-18} m, is achievable by using a high power laser source together with the technique to store the light inside the interferometer arms. With the first choice the improvement of the fringe resolution reducing the shot noise is reachable while the light storage into the interferometer arms will increase the phase shift due to a passing gravitational wave. In addition, the detection of gravitational waves on the Earth by using a similar apparatus is related to its capability in isolating the real signals from fake ones. For this reason the optical components of an interferometer should be isolated by a suspension system which is able to reduce noise transmission to the mirror level.

The VIRGO detector consists of a recycled Michelson interferometer with two Fabry-Perot cavities accommodated along the 3 km arms [3]. A laser beam running forward and backward within two long pipes 1.2 m in diameter and the nine towers containing seismic isolation system for the optical components, represent the probe through which a very precise measurement of the optical path length variation along the arms is performed.

2. The VIRGO detector

The VIRGO interferometer is based on a beam generated by a 20 W high-power Nd:YVO4 laser source locked to a 1 W solid state Nd:YAG master laser installed on an optical table (Laser Bench – LB) ground connected and installed within a clean-room laboratory with controlled temperature. As sketched in fig. 1 the beam is sent to a second optical bench (External Injection Bench – EIB) accommodated within the same laser laboratory, to be correctly aligned with respect to the interferometer reference frame before entering into the vacuum system. An optical window separates the laboratory environment from the large vacuum volume of the detector (about 7000 m^3) maintained at a hydrogen partial pressure of 10^{-7} mbar. The first mechanical element under vacuum with an octagonal shape is the Suspended Injection Bench (SIB in fig. 1) hung to a short suspension system conceived to suppress seismic noise transmission to the optical components level [4]. The design of this SIB has been recently revised including, among other sophisticated optical components, a Faraday Isolator system (FI) to reject the back-reflected light. The upper part of this bench is equipped with two flat mirrors optically contacted to a dihedron which form, together with a curved mirror suspended from a short suspension chain about 144 m far away, the Input Mode Cleaner (IMC) triangular cavity. The passage of the beam through this optical cavity is used to spatially filter the beam (selecting the TEM_{00} mode) and for the laser frequency pre-stabilization. In the low frequency region, below 15 Hz, the IMC is stabilized by using a short (30 cm long) rigid reference cavity (RFC) bolted on the bottom part of the SIB.

The beam then enters the main interferometer consisting of a flat Power Recycling (PR) mirror, a Beam Splitter (BS) used to separate the incoming laser beam and two Fabry-Perot cavities each one formed by a flat input mirror (NI and WI) and an end curved one (NE and WE) installed within a vacuum tower 3 km far away. The nominal finesse of these two optical cavities, North and West side respectively oriented, is 50 and they are used to enhance the light path amplifying the beam phase change induced by the optical path length variation.

Since in standard working conditions the interferometer is locked on the dark fringe, all the light (apart some losses and the small amount transmitted through the terminal mirrors) sent by the injection system into the interferometer is back-reflected. For this reason the high reflectivity of the PR mirror (95%) is used to recombine in coherent way this power and creating an additional cavity between the PR mirror and the Michelson interferometer. Thanks to this technique the power impinging onto the BS mirror is amplified by a factor about 35 (recycling factor) reducing the shot noise. All these six mirrors are suspended from a long mechanical isolation system called Superattenuator (SA) and conceived to suppress seismic noise transmission at the mirror level. The SA is the most original element of the entire experimental apparatus. It is a hybrid system combining the passive seismic attenuation performance of a second order mechanical filter chain based on a multi-stage pendulum working principle and an active feed-back control performance applied in the low frequency range (below 4 Hz) where all internal mode of the mechanical structures are confined. This is possible thanks to a three legs soft elastic structure used as a pre-isolation stage designed on the working principle of an Inverted Pendulum (IP). The SA has been built to provide the required passive filtering performance of seismic noise starting from a few Hz in all degrees of freedom. An upper limit of the system has been measured during the commissioning phase obtaining an attenuation factor of 10^{15} at 10 Hz [5].

The top stage of IP is the suspension point of the six filters chain and the inertial platform equipped with sensors and actuators for its active control. Three LVDT (Linear Variable Differential Transformer) positioning sensors together with three horizontal and two vertical accelerometers are able to sense four degree of freedom (3 translations and a rotation around the vertical axis). The combination of these error signals by means of a Digital Signal Processor board (DSP) developed for this purpose, are used to determine the correction signals to be applied at the IP top stage to the coil-magnets actuators damping the normal mode of the chain (Inertial Damping – ID). This is the first active control along the chain able to reduce the residual motion of the suspension point down to a few tenths of micrometer (r.m.s.).

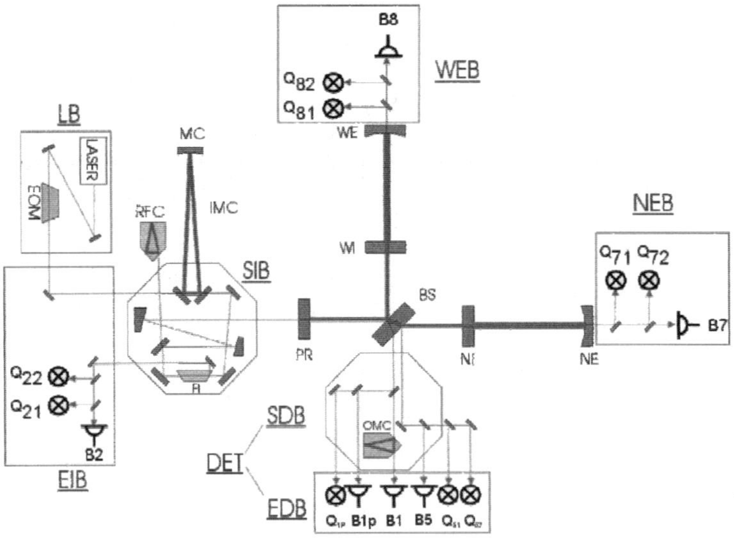

Fig. 1 VIRGO optical layout

The payload suspended from the last filter chain consists of a marionette, a reference mass and a mirror [6]. Four small permanent magnets screwed on the marionette wings together with the same number of coils attached to the end part of the last filter chain legs, represent the second actuation point for the active control of SA. From the marionette four thin wires start. The first pair supports the mirror in a cradle configuration and the second one, with the same technique, supports the reference mass surrounding the mirror itself. On the back side of this optical element four permanent magnets are glued while four coils screwed on the reference mass complete the last actuation point of the chain. This feedback control system (Local Control – LC) is based on optical lever sensors ground connected. Acting on the marionette, it is possible to reduce the mirror angular motion keeping the interferometer aligned within 1 μrad r.m.s. or better. The last coil-magnet pairs (between mirror and reference mass) are used as electronic dampers to recover the position of each mirror after an interferometer un-lock without exciting the mechanical mode of the filter chain.

Finally, the main output signal is reconstructed by a set of high-quantum efficiency InGaAs photodiodes mounted on an optical bench outside of the vacuum vessel (External Detection Bench – EDB). The output beam, before reaching the photodiode B1 (see fig. 1), passes through a monolithic 2.5 cm long cavity (Output Mode Cleaner – OMC) installed on a suspended bench (Suspended Detection Bench – SDB) attached to a SA and maintained in vacuum. The OMC has been included in the VIRGO optical layout to filter the higher order optical modes originating from misalignments and optical defects.

3. Experimental set-up upgrades

After a first commissioning phase, the VIRGO Collaboration decided to perform relevant experimental set-up upgrades devoted to approach the detector design sensitivity. A detector shut-down period started in September 2005 and concluded about four months later has been devoted to fix important problems previously found. Two main changes have been introduced on the interferometer: a new flat monolithic PR mirror replaced the old curved one and a new Suspended Injection Bench (SIB) designed to overcome the annoying problem of the back-reflected light.

3.1. The new Power Recycling mirror

The old suspended PR mirror was designed as a compound optical element consisting of an annular support and a central part with a curved secondary surface (120 mm in diameter) interconnected to the first one by means of a metallic ring. In addition, the PR lens was the final optical component of the input beam mode-matching telescope mounted on the SIB.

The main problem dealing with this composite structure was its inner resonance modes between 100 Hz and 500 Hz disturbing the interferometer controls.

The new flat PR mirror has a monolithic structure and a higher reflectivity (95% instead of 92%) with a diameter of 350 mm. Apart a higher recycling factor due to the increased reflectivity, with the new mirror many improvements have been performed at the longitudinal interferometer control level thanks to the absence of the inner resonance modes.

3.2. The new Suspended Injection Bench

The new SIB has been designed fixing the main problem found during the first commissioning phase and limiting the interferometer performance: the back-reflected light interfering inside IMC. To this purpose the new bench has been equipped with Faraday Isolation (FI) system. The input beam mode- matching telescope has been completely re-designed taking into account the presence of a flat PR mirror. Two large parabolic off-axis mirrors (50 mm and 114 mm in diameter respectively) have been adopted assembling a tunable telescope in two angular degrees of freedom plus the focal length. A rigid metallic support interconnecting the two mirrors has been used to perform a preliminary telescope tuning on bench before installing it on the SIB (see fig. 2). Without changing the suspension geometry but reducing the diameter of the three metallic monolithic wires (1 mm each one instead of 2 mm with an interconnection element at mid length), the SIB has been hung under vacuum to its short SA.

During the commissioning of this new experimental set-up, the residual astigmatism of the parabolic telescope, remotely controllable with pico-metric displacements, has been measured to be of the order of 2-3%. Moreover, even if the two mirrors are mounted on large mechanical supports, no drawback linked to their inner resonance modes has been observed.

4. Preparing the first scientific run

The second phase of the detector commissioning, started in January 2006 after the installation of the up-grades mentioned above, was focused on two main aspects: the continuous improvements of the feedback control loops and the continuous interferometer noise reduction.

To achieve the complete interferometer locking, the VIRGO collaboration has developed the so called "variable finesse" technique [7], locking the interferometer outside the dark fringe working point. In this way a large fraction of the light escapes through the output port and the power buildup in the recycling cavity is slow. As a last step, the interferometer is brought adiabatically on the dark fringe. The technique is called "variable finesse" because the recycling cavity finesse is changed during the lock acquisition path.

Once the interferometer is locked another control loop is necessary to maintain the mirrors perfectly aligned one with respect to the other and with respect to the laser beam. Since the necessary accuracy is not achievable by using the ground based LC, an automatic alignment based on the Anderson technique [8] and using the light transmitted through the end mirrors has been implemented.

At the level of the suspension system, some important control loops have been added to overcome the difficulties in keeping the interferometer locked with bad weather conditions. A particular attention has been devoted to the development of a control strategy to re-allocate the locking force on the marionette and on top of the IP.

This activity has been also driven by the interferometer noise reduction which represented the second important aspect faced during this commissioning phase. Many noise sources limiting the interferometer sensitivity have been identified. In some cases the sensitivity improvement required a fine tuning of the control systems, while a better protection of the interferometer sensors from external disturbances together with environmental noise reduction helped a lot. Acoustic noise source, for instance, have been reduced adding isolating enclosures on in-air optical benches (EIB, EDB, WEB and NEB - see fig. 1).

Fig. 2 The new Suspended Injection Bench (SIB). The parabolic telescope is in foreground while the Faraday Isolator is on right side. On left side the two flat mirrors optically contacted to a dihedron and forming the Input Mode Cleaner cavity are also visible.

All over the second commissioning period a set of Week-end Scientific Run (WSRs) with the interferometer operated at high power in recycled configuration has been organized since September 2006. Those runs have been important for the final tuning of the shifts organization, automatic procedures and for checking detector reliability and data taking process before a long run.

5. The first VIRGO scientific run

The first VIRGO Science Run (VSR1) started on May 18th, 2007 in coincidence with the last period of LIGO S5 run. Even if the detector sensitivity was not the design one all over the bandwidth, in the frequency region 300 Hz-5 kHz the design sensitivity has been reached and it is comparable to the LIGO one (see fig. 3). This result has been considered appropriate to start joint data analysis with the LIGO Scientific Collaboration (LSC). In this context a Memorandum Of Understanding (MOU) has been signed between the two scientific communities that schedules full data exchange, joint data analysis and joint publications.

The VSR1 has been stopped on October 1st, 2007 (in coincidence with the end of S5) after about four months of continuous data taking. The detector duty-cycle in science mode has been as high as 81% thanks to the full locking automation procedure [9] developed during the commissioning phase and the control strategy improvements. The interferometer sensitivity expressed in terms of maximum distance at which a coalescing system formed by two neutron stars with 1.4 solar masses would be detectable, reached a mean value of 4 Mpc.

6. Future perspectives and upgrades

The VIRGO design sensitivity will allow testing some of the present gravitational wave amplitude upper limits. Even if a first detection is possible, the sensitivity of VIRGO and LIGO detectors are not sufficient to open the era of gravitational wave astronomy. Moreover, any potential gravitational wave signal detection should be done by a coincidence of different interferometers having a common sensitivity on a large bandwidth to be considered as a network. For these reasons it is important to prepare an up-grade of the present apparatus improving the design sensitivity of a factor 2 (VIRGO+), so that an increment of a factor 10 or more in the detection rate will be reachable.

The VIRGO collaboration is planning a first set of detector up-grades, VIRGO+ [10], which should be completed (commissioning included) within the first half of 2009. After this phase the detector should be ready for data taking in coincidence with enhanced LIGO (eLIGO).

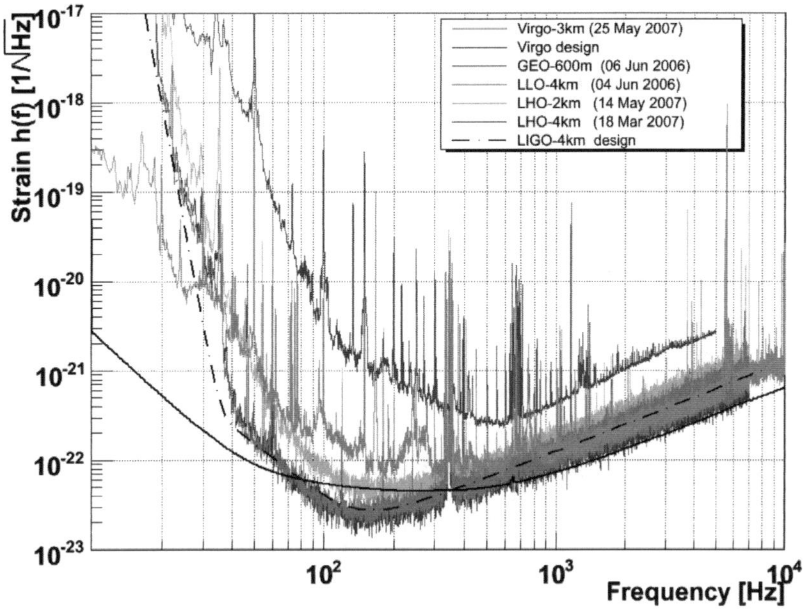

Fig. 3 Sensitivity curves comparison for different detectors (VIRGO, LIGO and GEO-600m)

Since the seismic isolation system (SA) is compliant with the design sensitivity improvement, the scheduled changes will concern:

- the installation of new IMC payload;
- the installation of a system for thermal compensation of the mirrors;
- the improvement of the input laser power (up to 50 watt) by using a new laser amplifier;
- the installation of new control system electronics;
- the installation of new mirrors increasing the Fabry-Perot cavities finesse up to 150.

A more substantial up-grades campaign will take place at the beginning of the next decade, when a second generation detector, called Advanced VIRGO (AdV) [11], will be installed in the present infrastructure. In this case the main goal is to improve the present design sensitivity by one order of magnitude, increasing the event rate by three orders of magnitude. The conceptual design document of AdV is expected to become available by the end of 2007. A dedicated R&D program is in progress with the goal to start the engineering phase in 2010.

7. Conclusion

The first VIRGO Science Run (VSR1) is ended on October 1st, 2007 after more than four months of continuous data taking. The detector duty-cycle in science mode configuration has reached the 81% with about 20 sets of un-interrupted data collection longer than 40 hours very promising for future data analysis. These very good results have been obtained thanks to the locking automation procedure and to the fine tuning of the complex feedback control systems. Even if the design sensitivity is not yet reached all over the bandwidth, the VIRGO interferometer shows the best sensitivity in the frequency range below 35 Hz.

References

[1] http://www.virgo.infn.it
[2] http://www.ligo.caltech.edu
[3] http://www.geo600.uni-hannover.de
[4] G. Ballardin et al., Rev. Sci. Inst., vol. 72, n. 9, (2001) 3643-3652
[5] S. Braccini et al., Astro. Phy. 23 (2005) 557-565
[6] A. Bernardini et al., Rev. Sci. Inst., vol. 70, n. 8, (1999) 3463-3472
[7] F. Acernese et al., Class. Quant. Grav. 23: S85-S89, (2006)
[8] D.Z.Anderson, Appl. Opt., 23, 2944-2949 (1984)
[9] F. Acernese et al., Proceeding of the International Conference on Accelerator and Large Experimental Physics Control System (ICALEPS), Geneva, Switzerland, 10-15 Oct. (2205)
[10] F. Acernese et al., J. Phys. Conf. Ser. 32: 223-229, (2006)
[11] R. Flaminio et. al., 'Advanced Virgo White Paper', VIR-NOT-DIR-1390-304, (2005)

Status of LCGT and CLIO

Masatake Ohashi and the LCGT collaboration
Institute for Cosmic Ray Research, The University of Tokyo, Kashiwa, Chiba, 277-8582, Japan

E-mail: mohashi@icrr.u-tokyo.ac.jp

Abstract. The Japanese gravitational wave proposal, LCGT (Large-scale Cryogenic Gravitational wave Telescope), is to construct a 3km underground interferometer with cryogenic mirrors in the Kamioka Mine. CLIO (Cryogenic Laser Interferometer Observatory) is a 100m-baseline underground cryogenic interferometer built at the same site to serve as a bridge connecting CLIK (7m-baseline prototype cryogenic interferometer) and the planed LCGT. The site of CLIO is near the Super-Kamiokande neutrino detector. The tunnel for CLIO was dug in 2002, and a strain meter for geophysics was installed in the same tunnel in 2003. CLIO was installed in 2005 and the operation started in 2006.

1. Introduction

In 1990's, large laser interferometers were built to detect gravitational waves. There are four kilometer-scale interferometers in the world, three LIGO interferometers in U.S. [1] and one Virgo interferometer in Italy [2], and two sub-kilometer-scale interferometers, GEO600 in Germany [3] and TAMA in Japan [4]. LIGO has just finished the long-term observation named S5 since the end of 2005. The sensitivity of LIGO is good enough to detect coalescing binary neutron star events at 16 Mpc away [5]. But we want to expand the range of observation to 180 Mpc during the next decade. The sensitivity of LIGO will be improved by one order when the next major upgrade is completed (Advanced LIGO [6]). The Japanese plan, LCGT [7], is to construct a 3km underground interferometer with cryogenic mirrors in the Kamioka Mine. The merit of constructing underground is that the seismic motion is much quieter. The seismic noise is expected to be smaller by two orders of magnitude than that in the city area. The mirror cooling is adopted to reduce the thermal noise of mirrors, which is the major noise source in the signal band when the seismic and quantum noises are suppressed enough. We expect that LCGT will detect one event per month.

2. Source of gravitational waves

The target source of LCGT is the coalescence of binary neutron stars at 180Mpc. The theory of general relativity predicts that orbital energy and angular momentum are carried away from the binary star system by gravitational waves radiated from the system. The binary star is one of the most common sources of gravitational wave, and is the simplest case understood. In 1974, Joseph Taylor and Russell Hulse found a binary neutron star in our own galaxy. The period of the orbital motion is 7.75 hours, and is decreasing by about 76 millionths of a second per year. The effect was just as Einstein's theory predicted. The point is that the binary neutron star will merge for the last time and make a black hole with strong gravitational wave radiation. Another source is the continuous gravitational wave radiated from nearby pulsars with rapid rotation. LCGT is expected to have a good sensitivity below 100Hz and suitable for detecting continuous waves.

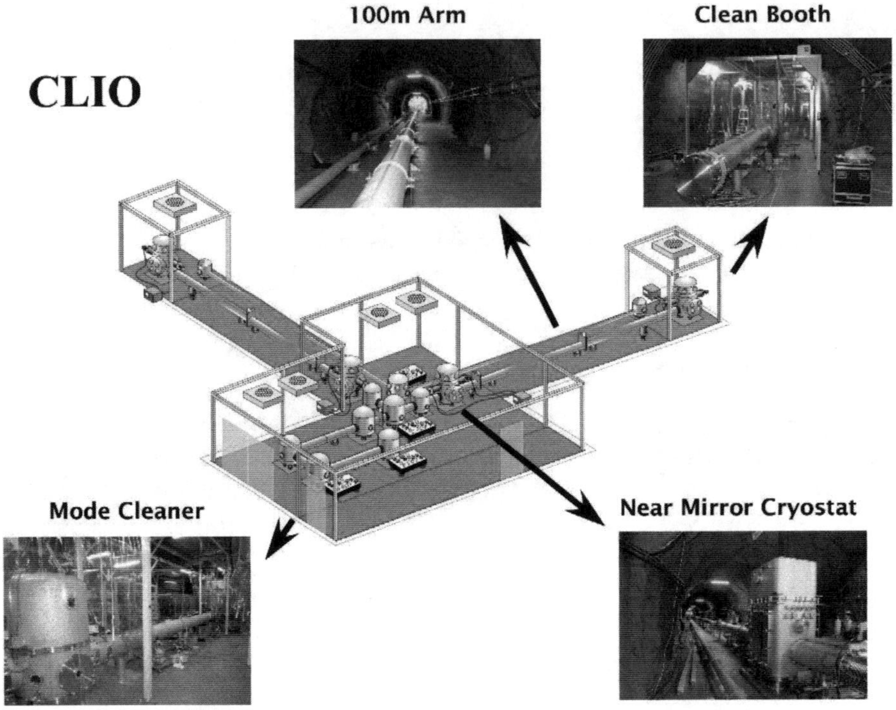

Figure 1. Overview of CLIO with the Mode Cleaner and the Cryostats

3. CLIO

CLIO [8,9] was constructed to demonstrate two unique technologies of LCGT, one to utilize quiet and stable environment in the Kamioka mine and another to adopt sapphire mirrors at cryogenic temperature for thermal noise reduction. CLIO has the locked Fabry-Perot interferometer configuration equipped with a ring mode cleaner for the laser frequency stabilization and the cooling system using refrigerators to cool the sapphire mirrors at 20 K. Whole of the interferometer is housed in tunnels as shown in Fig. 1. The Kamioka mine is located on very hard bedrock and its elastic wave velocity reaches up to 5000 m/sec.

Cryogenic cooling system is shown in Fig.2 [10]. We use pulse-tube-type cryocoolers with low mechanical vibrations. The sapphire mirrors with 10cm in diameter are suspended by pure aluminium wires and cooled down to 20K within a week.

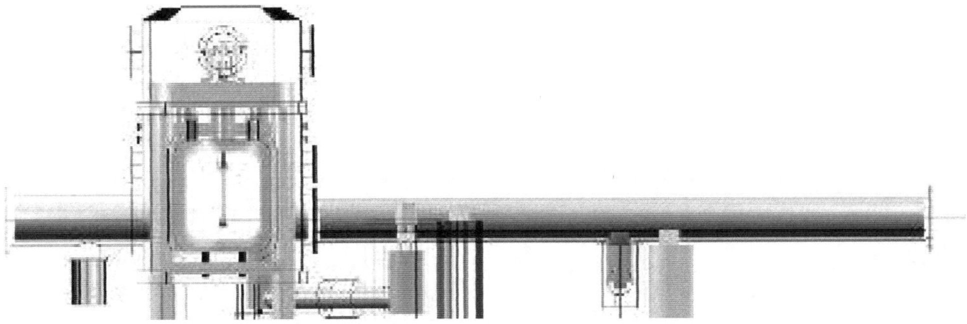

Figure 2. Cryogenic cooling system of CLIO is described. It is connected to several meter vacuum pipes with radiation shield inside.

4. Geophysical strain meter

At the same site, a geophysical strain meter with 100 m baseline [11, 12] was constructed to detect directly the earth core resonance mode in the milli-herz frequency range. It consists of a Michelson interferometer whose mirrors are fixed on the bedrock. The second harmonics wave length (532 nm) of YAG is used for the light source and its frequency is stabilized within an absorption line width of iodine gas [13].

5. Summary

CLIO was constructed in the Kamioka Mine to demonstrate the cryogenic techniques for LCGT. We started the operation of CLIO in 2006. The 57-hours observation data were taken in February 2007 and analyzed for continuous gravitational wave from the pulsar J0835-4510 (Vela pulsar) at the frequency twice of its rotational frequency (~ 22.38 Hz). We use an analysis method based on a matched filtering with a correction of pulsar spin down, a Doppler effect caused by the relative motion between the pulsar and the Earth, and so on [14]. The result of analysis will be submitted soon [15].

Acknowledgment

This research was partially supported by Grant-in-Aid for Scientific Research (A) of Japan Society for the Promotion of Science (18204021).

References

[1] Abramovici A *et al.* 1992 *Science* **256** 325
[2] Bradaschica C *et al.* 1990 *Nucl. Instrum. Methods Phys. Res.* Sect A **289** 518
[3] Willke B *et al.* 2002 *Class. Quantum Grav.* **19** 1377
[4] Ando M *et al.* 2001 *Phys. Rev. Lett.* **86** 3950
[5] Sigg D 2008 *Proc. of 7th Edoardo Amaldi Conference on Gravitational Waves* (Bristol: Institute of Physics Publishing)
[6] Wilkinson C A 2008 *Proc. of 7th Edoardo Amaldi Conference on Gravitational Waves* (Bristol: Institute of Physics Publishing)
[7] Kuroda K 2008 *Proc. of 7th Edoardo Amaldi Conference on Gravitational Waves* (Bristol: Institute of Physics Publishing)
[8] Miyoki S *et al.* 2006 *Class. Quantum Grav.* **23** S231
[9] Yamamoto K 2008 *Proc. of 7th Edoardo Amaldi Conference on Gravitational Waves* (Bristol: Institute of Physics Publishing)
[10] Uchiyama T *et al.* 2006 *J. Phys.: Conf. Ser.* **32** 259
[11] Takemoto S *et al.* 2006 *J. Geodynamics* **41** 23
[12] Araya A *et al.* 2007 *J. Geodetic Soc. of Japan* **53** 81
[13] Araya A *et al.* 2002 *Rev. Sci. Instrum.* **73** 2434
[14] Niebauer T M *et al.* 1993 *Phys. Rev.* D **47** 3106
[15] Akutsu T *et al.* 2008 *Proc. of 12th Gravitational Wave Data Analysis Workshop* in preparation.

Monitoring Sco X-1 for the detection of gravitational waves with networks of gravitational wave detectors

Kazuhiro Hayama , Soumya Mohanty , Malik Rakhmanov , Shantanu Desai , Tiffany Summerscales

[1]Center for Gravitational Wave Astronomy, University of Texas at Brownsville, 80 Fort Brown, Brownsville Texas 78520, US

[2]Southeastern Louisiana University, Hammond, LA 70402, US

[3]Center for Gravitational Wave Physics, Pennsylvania State University, University Park, PA 16802, US

[4]Department of Physics, Andrews University, Berrien Springs, MI 49104, US

E-mail: kazuhiro.hayama@ligo.org

Abstract. Searching for gravitational waves triggered by electromagnetic astronomical observations has several benefits: prior information about the source location enhances detection efficiency, and the use of time coincidence implies analysis of only a small stretch of data which allows sophisticated analysis with high computational costs. During postprocessing, knowledge about the source enables accurate reconstruction of the source parameters, leading to an astrophysical interpretation. In this paper, we present a triggered search pipeline called RIDGE and we demonstrate the search for gravitational waves from Sco X-1, in which X-ray bursts occur frequently, using simulated data. In the analysis, we consider the optimum network combination for Sco X-1.

1. Introduction

Currently, several interferometric gravitational wave (in short, GW) detectors around the world such as LIGO, VIRGO, GEO600, TAMA300 [1] have been taking data. Interferometric gravitational wave (GW) detectors are based on a Michelson interferometer. The quadrupole nature of a GW produces differential changes in the arm lengths of the detectors, being detected as changes in the interference fringe pattern. The data from GW detector preserve both amplitude and phase information. Therefore, it is possible to combine data streams from networks of detectors coherently. The coherent network approach enables us not only to enhance detection efficiency, but also to reconstruct parameters of a GW such as the waveform and the direction to the source [2]. Triggered searches involve looking for gravitational waves from sources such as gamma-ray bursts, magnetars, X-ray bursts, supernovae, blazars etc that also have an observable signature. Having an electromagnetic observation in coincidence with a potential GW signal allows for much higher confidence in the signal and has other important benefits as well: the time coincidence implies that only small data stretches need to be analyzed, which allows us to use algorithms with a high computational cost and to ignore long-term non-stationary behavior of detector noise. Knowing the source location enables accurate recovery of GW signal parameters, which are essential for astrophysical interpretation. Applying the triggered search, we consider the detection of GWs from Sco X-1 which is the low-mass X-ray binary with the

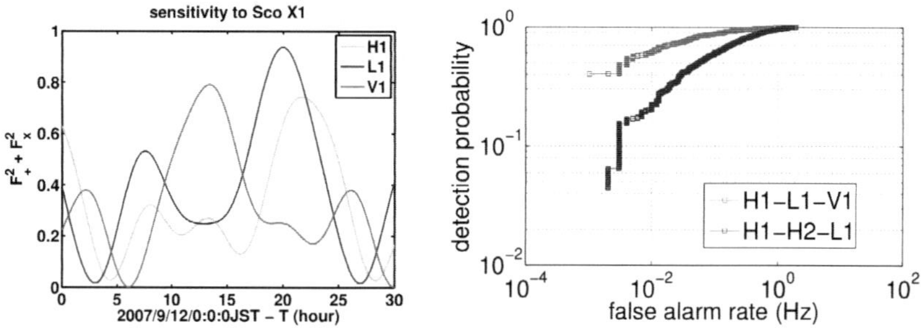

Figure 1. Left:The detector response as a function of time since September 12 (JST), 2007. H1 is the 4 km LIGO Hanford detector, L1 is the LIGO Livingston detector, and V1 is the VIRGO detector. Right: ROC curve at 13 hours. The blue line is for the LIGO only network and the red curve is for the LIGO-VIRGO network.

strongest X-ray emission [3]. In Sco X-1, strong X-ray bursts occur frequently [4], and GWs can be emitted along with the X-ray bursts. Besides searching for GWs associated with the X-ray bursts, we look for bursts even during the quiet phase of Sco X-1 by monitoring Sco X-1. When monitoring Sco X-1, the sensitivity of each GW detector to the direction of Sco X-1 varies in time because of the rotation of the earth. In this paper, we show results of Monte Carlo simulations with some networks of detectors for the detection of GWs from Sco X-1 using the coherent network analysis pipeline "RIDGE", which we have developed, and evaluate the effect of the network quantitatively. We also discuss which networks of detectors are effective for a long-term observation of Sco X-1 so as to provide maximum coverage with enhanced detection efficiency, by minimizing intervals where the network is"blind".

2. Overview of RIDGE pipeline

RIDGE is a fully coherent network analysis pipeline described in Ref. [5]. The pipeline consists of two main components, data conditioning, and generation of detection statistics. The aim of the data conditioning is to whiten the data, so as to remove frequency dependence and any instrumental artifacts in the data. In RIDGE, the data are whitened by estimating the noise floor using a running median [6], and narrow-band noise artifacts arising from sources such as power line interference, and mirror/suspension resonant modes are removed [7]. The resulting conditioned data is passed on to the next step consisting of the generation of a detection statistic. The generation of this statistic is based on maximum likelihood. The GW response of any one detector is a linear combination of the two unknown polarization waveforms $h\ (t)$ and $h_\times(t)$ arriving at the detector from a certain direction on the sky. It was shown that the problem of inverting a set of detector responses to obtain $h\ (t)$ and $h_\times(t)$ is an ill-posed one [8]. In RIDGE, the maximum likelihood is regularized by Tikhonov regularization to solve the problem [8]. The detection statistic is calculated using the regularized maximum likelihood at every sky location.

3. Monitoring Sco X-1

The detector response is a function of time and geographical location of the detector. Because of the rotation of the earth, the detector response is a periodic function with a 24 hour period.

The left plot in figure 1 shows the detector response of LIGO and VIRGO to the direction of Sco X-1 as a function of time since September 12 (JST), 2007. Here, the value of the detector response is defined by summing the squared antenna patterns for the two GW polarizations $h\ $, $h_\times$. From this figure, for the LIGO only network, the value of the squared antenna pattern exceeds 0.5 for about 33% of the time during a day. If VIRGO is added to this network, this network coverage increases to 63% . To study the effect of the addition of VIRGO to the LIGO network quantitatively, we carried out Monte Carlo simulations. For the detector noise PSD, we used the design sensitivity curves for the LIGO and Virgo detectors as given in [1] and kept the locations and orientations the same as the real detectors. Gaussian, stationary noise was generated (2000 sec) at a sampling frequency of 16384 Hz. To simulate instrumental artifacts, we added sinusoids with large amplitudes at (54, 60, 120, 180, 344, 349, 407) Hz. The injected signals corresponded to a single source located at Sco X-1. We assumed that the $h\ (t)$ and $h_{\times}(t)$ waveforms were both sine-Gaussian signals defined by $A\sin(2\pi f_c t)\exp(-t\ /2\sigma\ )$ where the center frequency $f_c = 235$ Hz, $Q := 2\pi f_c \sigma = 9$ and A is a constant value. The $h_{\times}(t)$ waveform is a circularly polarized version of $h\ (t)$ which has been shifted in phase by $\pi/2$. The signal strength is specified in terms of the root-sum-square (hrss) value. We use hrss values of $7 \times 10^{-}$ $\mathrm{Hz}^{-\ /}$. The right plot in the Fig.1 is the Receiver Operating Characteristic (ROC) curve for the simulation at 13 hours after 2007/09/12 00:00 JST. It shows that the detection probability at the false alarm rate of 0.01 Hz is improved from 0.2 to 0.6 by adding VIRGO to the LIGO only network.

4. Discussion

We describe a coherent network analysis based approach to detect GWs associated with electromagnetic observations and study the sensitivity of current detector network of GWs from Sco X-1 with some network combinations. To monitor Sco X-1 with a sensitive detector response for a long period, it is essential to add VIRGO to the LIGO-only network. The LIGO–VIRGO network produces a detector response exceeding 0.5 for about 63% of a day. But even if the LIGO–VIRGO network is used, periods of "blindness" to Sco-X1 still exists. One approach to get rid of this blind periods is to add LCGT [9], which is a proposed cryogenic Japanese detector. We add LCGT-like detector whose sensitivity is 10 times worse than the design sensitivity (so that its sensitivity is similar to LIGO, VIRGO), and carried out Monte Carlo simulations. The simulation shows that the detector response of the LIGO–VIRGO–LCGT-like exceeds 0.5 for about 92% of the time. This shows that the current geographical location and orientation of LCGT is significantly benefitial to this search.

Acknowledgments

K.H. is supported by NASA grant NAG5-13396 to the Center for Gravitational Wave Astronomy at the University of Texas at Brownsville. SDM's work was supported by NSF grant PHY-0555842. The work of S.D. and M.R. was supported by the Center for Gravitational Wave Physics at the Pennsylvania State University and the US National Science Foundation under grants PHY 00-99559, PHY 02-44902, PHY 03-26281, PHY 06-00953 and NSF 428-51 29NV0. The Center for Gravitational Wave Physics is funded by the National Science Foundation under cooperative agreement PHY 01-14375. T. S. is supported by a grant from the Office of Scholarly Research at Andrews University. This paper has been assigned LIGO-P070145-00-Z.

5. References

[1] Abbott, et al. 2004 *Nucl.Instrum.Meth.A* **517** 154, Acernese, et al. 2002 *Class. Quantum Grav.* **19** 1421, Willke, et al. *Class. Quantum Grav.* **21** S417, R. Takahashi, et al. *Class. Quantum Grav.* **S403**

[2] Y. Gürsel and M. Tinto 1989 *Phys. Rev.* **D 40** 3884

[3] Jeremy S. Heyl 2002 *Astrophysical Journal* **574** L74

[4] D. Psaltis et al. 1998 *Astrophysical Journal* **501** L95

[5] K. Hayama, S. Mohanty, M.Rakhmanov, S. Desai 2007 *Class. Quant. Grav.* **24** 681

[6] S. Mukherjee 2003 *Class. Quant. Grav.* **20** 925

[7] S. Mohanty 2002 *Class. Quant. Grav.* **19** 1513

[8] M. Rakhmanov 2006 *Class. Quant. Grav.* **23** S673

[9] K. Kuroda et al. 2006 *Progress of Theoretical Physics Supplement* **163** 54

Recent progress of TAMA300

Koji Arai, on behalf of the TAMA collaboration
National Astronomical Observatory of Japan, Tokyo 181-8588, Japan
E-mail: koji.arai@nao.ac.jp

Abstract. Current effort of the TAMA project is focused on establishment of the interferometer operation with a new vibration isolation system, called "Seismic Attenuation System" (SAS). The SAS employs a multiple stage structure to realize soft spring for all of directions, as well as utilizes active control systems to stabilize mechanical resonances in the low frequency region below 1 Hz. The SASs were installed for four test masses, resulting in improvement of cavity length fluctuation below 150 Hz. We describe the structure of the SAS and its local active control system, and the status of the interferometer.

1. Introduction

TAMA300 [1] is a laser interferometer gravitational wave detector with 300-m cavity arms, being located at the campus of National Astronomical Observatory of Japan, which is about 20 km away from the center of Tokyo. Purposes of TAMA300 are to develop a detector that is able to catch possible gravitational waves from nearby galaxies, and to establish necessary technologies for future km-class interferometers such as LCGT [2]. The construction of TAMA300 started in 1995. The interferometer operated without power recycling from 1999 to 2001, performing six times of observation runs. Power recycling technique was employed in 2001 so as to improve shot noise level by enhancing the internal laser power in the interferometer. From 2003 to 2004, three times of observations were held with improved sensitivity [3]. With those nine times of observation runs, more than 3000 hours data in total was accumulated [4].

Since 2005, we have been working on the new vibration isolation system, called "Seismic Attenuation System" (SAS). The sensitivity of TAMA300 at around 100 Hz was limited by coupling of alignment sensing noise; in order to improve the sensitivity of TAMA300 in the low frequency band, the control bandwidth of the alignment control should be reduced. For this purpose, we replace the previous vibration isolation system [5] [6] to the SAS. The SAS has a multiple stage structure to realize seismic filtration from the low frequency such as 100 mHz. The SAS was installed for all of four test masses and the tuning of the local control systems for the SAS was completed in the summer of 2007.

Our current efforts are focused on interferometer commissioning. We have established operation of the power recycled interferometer with the SAS although tuning of the interferometer subsystems and of the sensitivity is still in progress. In this article, we explain the structure of the SAS and its local active control system. The status of the interferometer is also described.

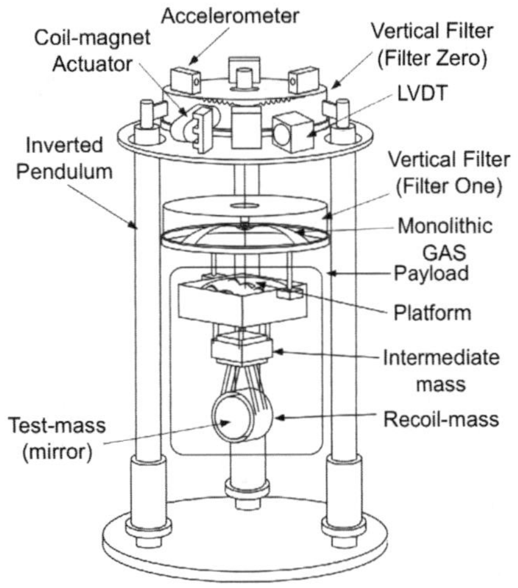

Figure 1. Schematic view of the SAS for TAMA300.

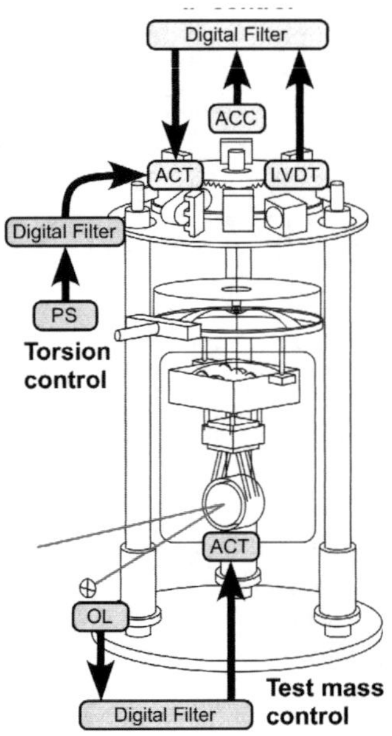

Figure 2. Schematic diagram of the active control systems. "LVDT" and "ACC" are the position sensor and the accelerometer, respectively. "PS" is the photosensor for the torsional motion of Filter-One. "OL" is the optical lever to sense the angular motion of the test mass. "ACT" is the actuator at the IP or the test mass.

2. Seismic Attenuation System

2.1. Mechanical structure

The SAS is a vibration isolation system with a total height of about 2.5 meters, as shown in Figure 1. The SAS was developed by the collaboration of Caltech and TAMA researchers. The design of the structure was based on the prototype experiments [7] [8].

The SAS has a multiple pendulum structure formed by five passive stages: an inverted pendulum (IP) [9], a suspended vertical filter, and a triple pendulum called *payload*. The IP at the first stage provides horizontal vibration isolation. A resonance of the IP at about 30 mHz was achieved by cancellation between elastic restoring force by flex joints and anti-spring force by the gravity. The concept of the IP is based on the superattenuator of VIRGO [10] [11]. The three IP legs are connected to the top part at which a vertical filter, called Filter-Zero, is placed. Another vertical filter, which is the second stage and called Filter-One, is suspended from the top of the IP. Each of those two vertical filters employs a Monolithic Geometrical Anti-Spring Filter (MGASF) [12] for vertical attenuation. The MGASF utilizes compressed blade springs to make the vertical resonant frequency low, such as 500 mHz. The payload [13], which is suspended from Filter-One, is a triple pendulum formed by a platform, an intermediate mass, and the test mass. The intermediate mass is suspended from the platform using miniature MGASFs for additional vertical isolation. The motion of the intermediate mass is damped by eddy current with permanent magnets flexibly supported from the platform [14]. The test mass is actuated by coil-magnet pairs. The coils are attached on a recoil mass that is also suspended from the intermediate mass, so that the actuation can act only on the differential motion between the recoil mass and the test mass and can not excite other undesirable mechanical modes of the multiple pendulum.

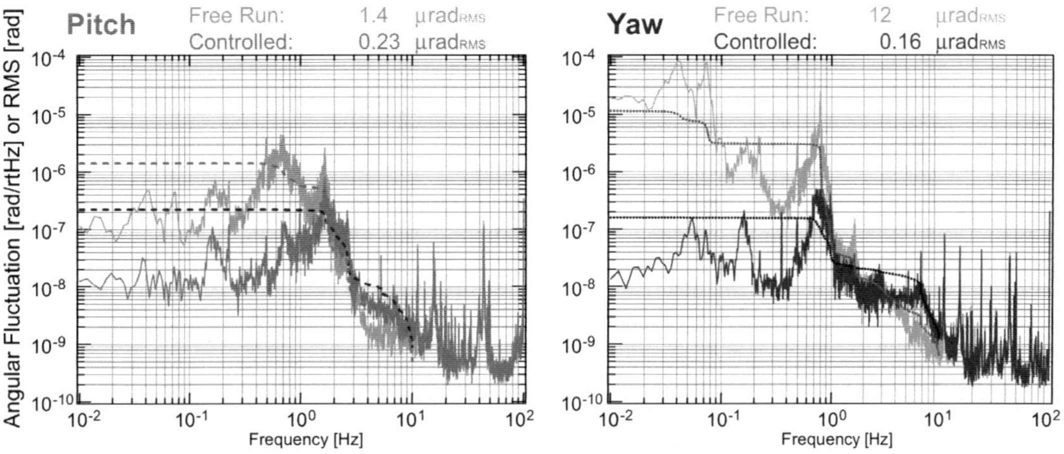

Figure 3. Spectra of the mirror angular fluctuations in the pitch (left) and yaw (right) directions with and without the test mass control. The RMS value of each spectrum integrated from 10 Hz is shown with dashed line. The numbers at the top of each plot show the RMS values of each condition that are derived from the integration from 10 mHz to 10 Hz.

2.2. Local control

In order to ensure the stability of the interferometer, active controls of the SAS is indispensable. The vibration isolation performance of the SAS is essentially provided by low resonant frequencies of the mechanical systems, that are typically below about 1 Hz: The IP and the multiple pendulum have resonant frequency of 0.03–1.6 Hz in the horizontal direction. The MGASFs and the Mini-GASFs has the vertical resonance at 0.5–2 Hz. For the torsional motion, the multiple pendulum has the resonant frequency of 40–500 mHz. It also means that these resonances are easily excited by the seismic motion which has large amplitude in such a low frequency region. In order to suppress the excited motion, three active local controls are applied (Figure 2): an IP control, a torsion control, and a test mass control. All of these control servos are realized by digital control systems based on LabVIEW.

The IP control is the feedback system that senses and suppresses the motion of the IP with the sensors and the actuators placed at the top of the IP. In order to realize the passive isolation performance and the stability at the same time, a two-loop design is used: One loop is an LVDT loop which keeps the position of the IP using position sensors by LVDTs [15] against drift motion. The LVDTs measure the motion of the IP relative to the ground. Therefore, the bandwidth of the LVDT loop is set to be as low as 70 mHz in order to avoid the seismic reinjection from the LVDT loop. The other loop is an ACC loop which damps the inertial motion of the IP using accelerometers [16], thus called *inertial damping*. The control frequency band of the ACC loop is between 70 mHz and 1.9 Hz. Consequently, this loop suppresses microseismic motion and mechanical reaction of the multiple pendulum.

We take linear combinations of the signals from the sensors and to the actuators so that we can treat the system with three horizontal degrees of freedom as three single degree of freedom systems. This technique to simplify the IP control is called *diagonalization* [17]. The sensors and the actuators are tri-symmetrically arranged at the top of the IP. Therefore, the sensing and actuation concern with every mechanical eigenmodes all together, making the mechanical response complicated. On the other hand, each of the eigenmodes is purely sensed and actuated by taking proper linear combinations of the sensor signals, as well as those of the actuator signals. As a result, the mechanical responses of the system are remarkably simplified.

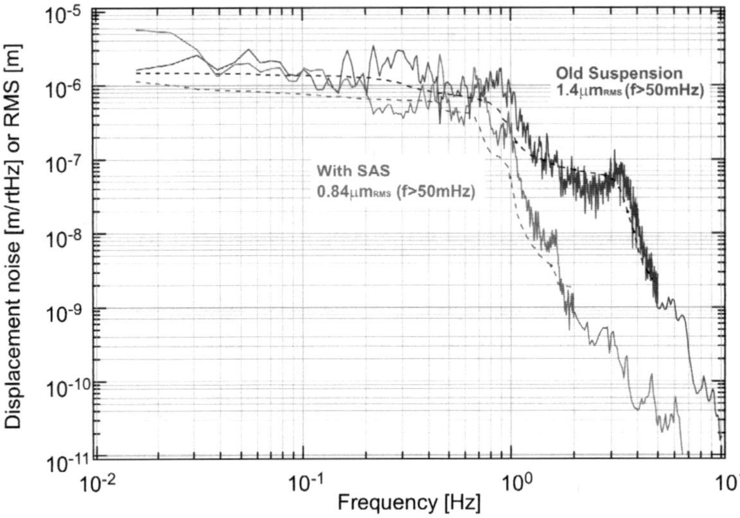

Figure 4. Length fluctuation of the 300-m cavity with the SAS and with the previous suspension system.

The torsion control is to damp the low-frequency torsion resonances of Filter-One and the payload independently from the IP control. It is difficult for the inertial damping to suppress these torsion modes because they do not appear in the top of the IP; the Filter-One is suspended from the IP by a single wire, and therefore the torsional motion of Filter-One is decoupled from that of the IP. The resonant frequencies of the torsion modes are as low as 40 mHz and 80 mHz. Moreover they have high quality factors such as 100. Consequently, these resonances have long decay times of the modes. Once they are excited, the angular motion of the test mass grows up and disturbs interferometer operation about half an hour. The torsion control system senses the angular motion of Filter-One by reflective photosensors, and actuates the rotation of the IP by the IP coil-magnets. The servo filter is designed so that the feedback can be effective only at the mode frequencies so as not to inject unnecessary sensing noise to the system. As a result, the effective quality factors of the modes less than 10 were achieved, which corresponds to the decay time of several minutes.

The test mass control is to stabilize the angular motion of the mirror. The angular sensor is an optical lever that is fixed on optical windows of the vacuum chamber. A visible laser light illuminates the mirror. The position of the reflected beam is detected by a position sensitive detector. The actuator is the coil-magnet pairs of the test mass. Figure 3 shows the angular spectra of the test mass in the pitch and yaw directions. The free-running angular motions in RMS for pitch and yaw are 1.4 $\mu\mathrm{rad}_{\mathrm{RMS}}$ and 12 $\mu\mathrm{rad}_{\mathrm{RMS}}$, respectively. Although these values are too large for the interferometer operation, they are able to be suppressed by control loops with the bandwidth of several Hz, as the fluctuation power is concentrated below 1 Hz. The resultant angular fluctuation with the active control is 0.23 $\mu\mathrm{rad}_{\mathrm{RMS}}$ and 0.16 $\mu\mathrm{rad}_{\mathrm{RMS}}$ for pitch and yaw, respectively, as shown in the same figure. Since these values are about 5 times smaller than that with the previous vibration isolation system, they are considered enough for establishing the interferometer operation. In addition, this measurement let us expect to reduce the alignment noise contribution to the gravitational wave channel. For the optical lever control, the transfer function of the digital servo is configured by precise placement of poles and zeros so that large peak structures in the spectra can be suppressed.

Once these local control systems are configured to be functioning, the interferometer operation

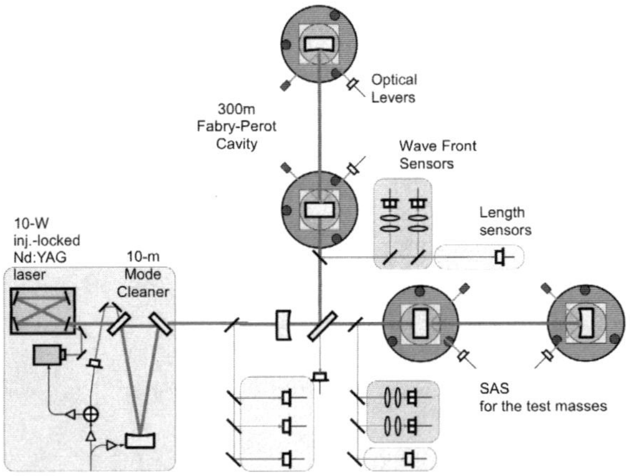

Figure 5. Schematic diagram of the current interferometer configuration of TAMA300

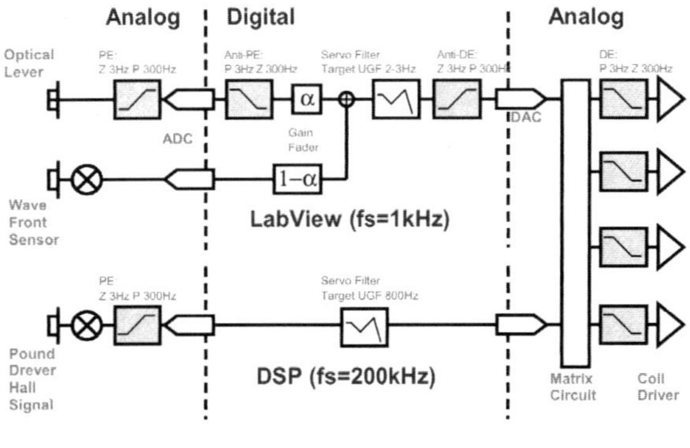

Figure 6. Schematic view of the pre-emphasis/de-emphasis configureation for the alignment and length control loops by digital filters.

come to be available. The longitudinal fluctuation of the 300-m Fabry–Perot cavity was measured by a length correction signal. The displacement spectrum below 10 Hz is shown in Figure 4. By comparison with the spectrum with the previous suspension system, it was confirmed that the length fluctuation with the SAS is improved above 0.1 Hz. Not only the power spectrum, but also the RMS motion of the cavity length is improved from 1.4 μm_{RMS} to 0.84 μm_{RMS}.

3. Interferometer status

After the shaking down of the SAS and its local control systems, the power recycled operation of the 300-m Fabry–Perot Michelson interferometer has become available. Figure 5 shows the current configuration of the TAMA interferometer. The light source is an injection locked Nd:YAG laser with output power of 10 W [18]. The laser beam is injected into a 10-m mode cleaner cavity for spatial filtering and frequency stabilization [19] [20]. The main interferometer is a Fabry–Perot Michelson interferometer with power recycling [21]. For the length sensing, the Pound-Drever-Hall technique [22], Schnupp modulation [23] [24], and third harmonic demodulation [25] are used. For the global alignment sensing, wave front sensors (WFS) are used [26] [27].

The lock acquisition procedure is as follows. First, all of the local controls are turned on and the operating point of the optical lever is adjusted so as to realize a good alignment of the arm cavities. Next, the recycled Michelson interferometer is locked with the third harmonic demodulation signals by actuating the beamsplitter and the recycling mirror. One of the arms is, then, locked with the Pound-Drever-Hall signal by actuating the laser frequency. Finally, the other arm cavity is locked by actuating the corresponding input test mass of the cavity. Once the lock is acquired, the operating points of the optical levers are slowly controlled to fulfill the operating point of the WFSs. In order to obtain gravitational wave signals as low noise as possible, the cavity control signals are switched to the bright and dark port signals. The beamsplitter control signal is also switched to the bright port signal with fundamental frequency demodulation.

For the lock acquisition of the secondary arm, a digital control system is used. The digital filter is a TMS320C6713 DSP based system with a sampling rate of 200 kHz. The achieved control bandwidth was 800 Hz, which was comparable with that of the previously-used analog servo. This system uses three techniques that assist the lock acquisition: Error signal normalization, error signal triggering, and adaptive servo filtering. The error signal of the cavity

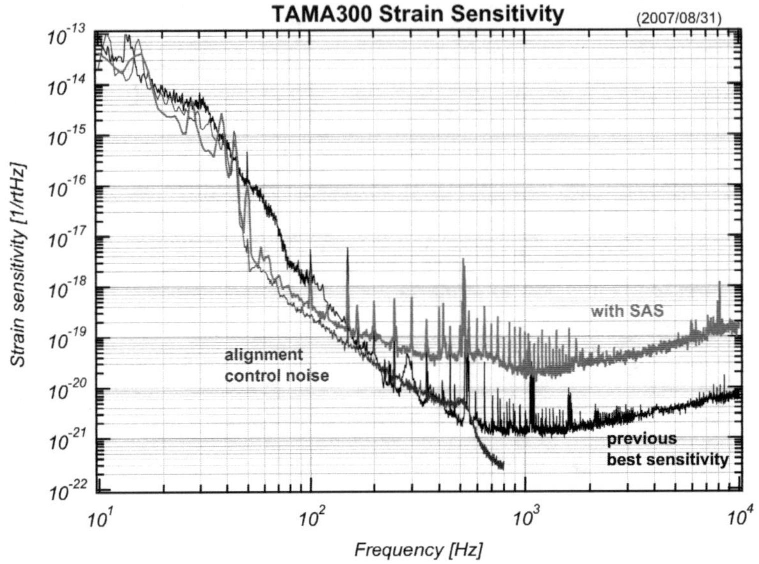

Figure 7. Current strain sensitivity of TAMA300 (red) and the alignment noise level (blue), which is currently limited by DAC noise at around 100Hz. The previous best sensitivity (black) is also shown.

lock is normalized by the transmitted power to extend the linear range of the error signal. This helps the servo system to increase the interaction time by a factor of about three. The error signal triggering turns on and off the servo loops by looking at the light level of the resonance. This is useful to eliminate unwanted feedback by the modulation sidebands and the higher order spatial modes, resulting in reduction of the kick by those spurious signals and allowing us to increase the feedback gain. The adaptive servo filtering is to increase the low frequency control gain when the lock acquisition is detected. We use an adaptive change of the digital filter coefficients in order to realize smooth change of the filter response for glitch-less gain boosting.

Increasing use of digital filters makes the interferometer operation flexible and easier. Digital filters, however, have 100 to 1000 times worse noise levels at the input and the output compared to well-designed analog circuits. In order to avoid coupling of these noises to the interferometer sensitivity, pre-emphasis/de-emphasis (PE/DE) technique is employed (Figure 6). A pair of filters with complementary shapes is inserted into the feedback loop. This enhances the noise immunity, while the total servo shape is preserved. For the input side of the digital system, the PE filter with the gain of $G_{\rm PE}$ is used, while inserting the anti-PE filter into the digital filter with the gain of $G_{\rm antiPE} = 1/G_{\rm PE}$. This reduces the effective noise level of the ADC by a factor of $G_{\rm PE}$. Similarly, for the output, the anti-DE filter which has gain of $G_{\rm antiDE}$ is used, while inserting the DE analog filter with the gain of $G_{\rm DE} = 1/G_{\rm antiDE}$. This reduces the effective noise level of the DAC by a factor of $G_{\rm antiDE}$. Since these filters reduce the sensing and actuation range of the control system, those PE/DE filters are switched off during the lock acquisition, while they are turned on after the lock is acquired.

The current sensitivity of TAMA300 is shown in Figure 7. Sensitivity improvement up to 150 Hz was confirmed although further tuning of the interferometer system is needed for the higher frequency band; not the full of the light is received at the dark port to avoid saturation. We analyzed the level of the alignment control noise, and found the DAC noise of the alignment control loop is very close to the noise level at 100 Hz. Further improvement of the alignment noise is expected by employing additional DA noise reduction and activation of fast WFS servos in stead of the optical lever.

4. Summary

We have been working on the improvement of the vibration isolation system by SAS, Seismic Attenuation System, since 2005. The SASs were installed for the four test masses in 2007. They are now functioning after the shaking down of the local control systems. We have confirmed that the angular fluctuation of the mirror is concentrated below 1 Hz. This leads us to the reduction of the alignment control bandwidth and eventually to the improvement of the sensitivity at about 100 Hz. The improvement of the length fluctuation above 0.1 Hz was confirmed by an arm cavity measurement. The decrease of the RMS mirror motion has also been observed.

With sufficiently stable performance of the SAS, the power recycled Fabry–Perot Michelson interferometer is able to operate. The lock of the arm cavity was established with help of digital filters. The sensitivity of the interferometer was measured, and then improvement up to 150 Hz was confirme. Improvement in the higher frequency band will be achieved by further tuning of the interferometer system such as injection of the full light to the dark port photodetector and refinement of the angular control servos.

Acknowledgments

This work was supported by the Grant-in-Aid for Scientific Research of the Ministry of Education, Culture, Sports, Science and Technology (09NP0801 in Creative Scientific Research and 415 in Priority Areas). Developments of the TAMA SAS were supported by the Advanced Technology Center of National Astronomical Observatory of Japan and the U.S. National Science Foundation under Cooperative Agreement No. PHY-0107417.

References

[1] Ando M and the TAMA Collaboration 2005 *Class. Quantum Grav.* **22** S881-9
[2] Kuroda K and the LCGT Collaboration 2006 *Class. Quantum Grav.* **23** S215-21
[3] Takahashi R and the TAMA Collaboration 2004 *Class. Quantum Grav.* **21** S403-8
[4] Akutsu T et al 2006 *Phys. Rev. D* 122002
[5] Takahashi R, Arai K, and the TAMA collaboration 2002 *Class. Quantum Grav.* **19** 1599-604
[6] Takahashi R et al 2002 *Rev. Sci. Instrum.* **73** 2428-33
[7] Takamori A 2002 Ph.D Thesis (Univ. of Tokyo) http://t-munu.phys.s.u-tokyo.ac.jp/theses/takamori_d.pdf
[8] Márka S et al 2002 *Class. Quantum Grav.* **19** 1605-14
[9] Takamori A et al 2007 *Nucl. Instr. and Meth.* A **582** 683-92
[10] Losurdo G et al 1999 *Rev. Sci. Instrum.* **70** 2507-2515
[11] Ballardin G et al 2001 *Rev. Sci. Instrum.* **72** 3643-52
[12] Cella G et al 2005 *Nucl. Instr. and Meth. A* **540** 502-19
[13] Takamori A et al 2002 *Class. Quantum Grav.* **19** 1615-21
[14] Tsubono K, Araya A, Kawabe K, Moriwaki S, and Mio N 1993 *Rev. Sci. Instrum.* **64** 2237-40
[15] Tariq H et al 2002 *Nucl. Instr. and Meth. A* **489** 570-6
[16] Bertolini A et al 2006 *Nucl. Instr. and Meth. A* **556** 616-23
[17] Losurdo G et al 2001 *Rev. Sci. Instrum.* **72** 3653-61
[18] Yang S, Imai Y, Oka M, Eguchi N, and Kubota S 1996 *Opt. Lett.* **21** 1676-9
[19] Nagano S et al 2002 *Rev. Sci. Instrum.* **73** 2136-42
[20] Nagano S et al 2003 *Rev. Sci. Instrum.* **74** 4176-83
[21] Arai K and the TAMA Collaboration 2002 *Class. Quantum Grav.* **19** 1843-8
[22] Drever R W P, Hall J L, Kowalski F V, Hough J, Ford G M, Munley A J, and Ward H 1983 *Appl. Phys. B* **31** 97-105
[23] Regehr M W, Raab F J, and Whitcomb S E 1995 *Opt. Lett.* **20** 1507-9
[24] Flaminio R and Heitmann H 1996 *Phys. Lett. A* **214** 112-22
[25] Arai K, Ando M, Moriwaki S, Kawabe K, and Tsubono K 2000 *Phys. Lett. A* **273** 15-24
[26] Morrison E, Meers B J, Robertson D I, and Ward H 1994 *Appl. Opt.* 5037-40
[27] Morrison E, Meers B J, Robertson D I, and Ward H 1994 *Appl. Opt.* 5041-9

TAMA300 interferometer development

Daisuke Tatsumi and the TAMA collaboration
2-21-1 Osawa Mitaka, 181-8588 Tokyo, Japan
E-mail: daisuke.tatsumi@nao.ac.jp

Abstract. In recent years, a gravitational wave detector TAMA300 has been upgraded for low frequency noise reduction. Its main subject is to install new seismic attenuation system TAMA SAS for all of four test mass mirrors. To benefit from the excellent performance of TAMA SAS, digital control systems are necessary to damp low frequency resonances. However, the digital control system has a demerit of large ADC and DAC noises. To reduce such digital control noises, pre-emphasis and de-emphasis filters are introduced. By applying these systems, the detector noise reduction below 100 Hz was realized. In this paper, I would like to present the current status of the digital control system and of the new analog filters to reduce digital control noises.

1. Introduction

In recent years, a gravitational wave detector TAMA300 [1] has been upgraded for low frequency noise reduction. Its main subject is to install new seismic attenuation system TAMA SAS [2, 3] for all of four test mass mirrors. To obtain excellent attenuation at 100 Hz region, the system has many low frequency resonances. Actually, horizontal filter of an inverted pendulum has a resonance of 30 mHz and vertical filters of monolithic geometrical anti-spring (MGAS) have resonances of 0.5 Hz. Such a system has large natural motion at low frequencies without any servo systems. Therefore, digital control systems are necessary for stable and excellent performance of TAMA SASs.

In this paper, I would like to present the current status of the digital control system and of new analog filters to reduce digital control noises.

2. Digital control systems

We adopt LabVIEW based digital systems for the inverted-pendulum and test mass alignment controls. About 10 filter banks of pole and zero pair can be applied for each test mass control servo. It is useful for parameter tuning of the servo system. By using such complicating servos, mirror angular motions can be suppressed to a level of sub-micron meter in root-mean-square. It is sufficient for stable interferometer operation.

The digital control system has some demerits. First, delay time of digital processing limits controlled bandwidth. In the case of 1 kHz sampling, the phase delay is 10 degrees at 5 Hz and increases in proportion to the frequency. Therefore, a unity-gain frequency is limited to 5–10 Hz practically. For length control of arm cavities, a unity-gain frequency of 800Hz is needed. A digital signal processor (DSP) of 200 kHz sampling is used for the servo and utilizes for widen linear range of the error signal by a factor of 3. With the help of this digital servo system, the arm cavity is locked stably.

Second, both of ADC and DAC have large electric noises. Our LabVIEW based digital systems have noise levels of about 10^{-5} V/$\sqrt{\text{Hz}}$. In comparison with analog filters, these are 2 or 3 orders of magnitude larger. To benefit from the excellent performance of TAMA SAS, the DAC noises should be reduced by four orders of magnitude at 100Hz.

3. Pairs of pre-emphasis and de-emphasis filters

To reduce ADC and DAC noises, pre-emphasis (PE) and de-emphasis (DE) filters are introduced to the servo loops as shown in Fig. 1. The pre-emphasis filter amplifies sensor output so as to exceed the ADC noise level. Therefore, the noise contribution of ADC could be negligible. The pre-emphasis filter for each sensor should be optimized, since the sensitivities are different. Because an anti pre-emphasis filter compensates for the pre-emphasis one, the servo loop is not influenced in totality. On the other hand, the de-emphasis filter has a role of DAC noise reduction. Similar to the pre-emphasis filters, the anti de-emphasis filter compensates for the de-emphasis one.

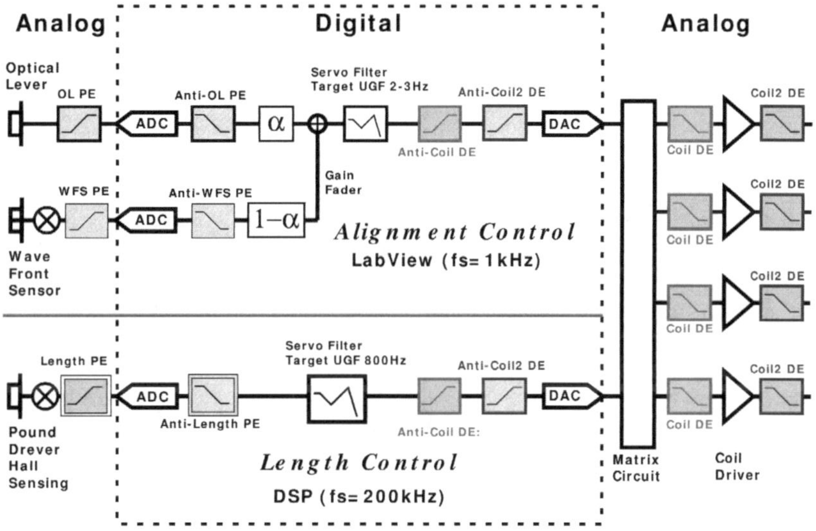

Figure 1. A schematic view of a mirror control servo system including pre-emphasis and de-emphasis filters.

For lock acquisition, high dynamic ranges are necessary for most of electric circuits. Especially for alignment control, optical lever sensors are helpful because of its high dynamic range and robustness for interferometer conditions. Furthermore, to avoid saturation of the signals, some of the filters should be switched to through filter (no-gain buffer amplifier) by external digital control. After the lock acquisition, a simple gain-up filter may cause saturation of the analog signal. Therefore, the filters should have an appropriate frequency dependence so as not to do. To satisfy all of the above requirements, a lot of pre-emphasis and de-emphasis filters with the digital control switch are implemented to the mirror control servo system as shown in Fig. 1.

4. Results

By installation of TAMA SAS and implementation of pre-emphasis and de-emphasis filters, the detector noise reduction below 100 Hz was realized as shown in Fig. 2. The detector noise on 30 August, 2007 is limited not by ADC or DAC noises, but by the sensor noise of optical lever below 50 Hz. The shot noise level of a displacement sensor is worse than that of the previous best sensitivity on 4 November, 2003, since incident laser power is not full.

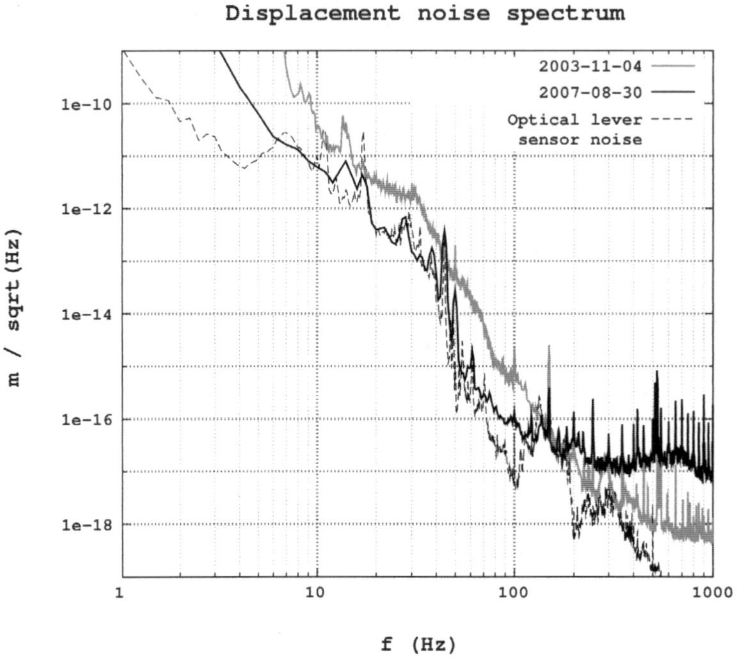

Figure 2. Displacement noise spectrum of TAMA300 detector

To archive the target sensitivity, we need to increase the incident laser power. Laser power is necessary not only for the displacement sensor but also alignment sensors. When laser power becomes to be maximum, wavefront sensors (WFSs) are expected to have about 100 times better noise level than that at the current. By switching the alignment sensor from optical lever to wavefront sensor, further noise reduction will be realized.

5. Summary

To reduce low frequency noises, new seismic attenuation system TAMA SAS was installed. To benefit from the excellent performance of TAMA SAS, digital control system is necessary. Pairs of pre-emphasis and de-emphasis filters are also introduced to reduce the ADC and DAC noises. By applying these systems, the detector noise reduction below 100 Hz is realized. For further noise reduction, detector is under tuning.

Acknowledgments

This research is supported in part by a Grant-in-Aid for Scientific Research on Priority Areas (415) of the Ministry of Education, Culture,Sports, Science and Technology.

References

[1] Tatsumi D *et al* 2007 *Class. Quantum Grav.* **24** S399-403
[2] Márka S *et al* 2002 *Class. Quantum Grav.* **19** 1605-14
[3] Takamori A *et al* 2002 *Class. Quantum Grav.* **19** 1615-21

Summary of the search for inspiraling compact star binaries from TAMA300's observation in 2000-2004

Hideyuki Tagoshi for the TAMA Collaboration

Department of Earth and Space Science, Graduate School of Science, Toyonaka, Osaka 560-0043, Japan

E-mail: tagoshi@vega.ess.sci.osaka-u.ac.jp

Abstract. We summarize the results of the search for inspiraling compact star binaries from TAMA300's observation. In this analysis, 2705 hours of data, taken during the years 2000-2004, are used for the event search. We obtain an upper limit on the rate of the coalescence of compact binaries in our Galaxy of 20 per year at a 90% confidence level.

1. Introduction

In this paper, we present the summary of the data analysis of the TAMA300 detector to search for gravitational waves produced by inspiraling compact star binaries, comprised of non-spinning neutron stars and/or black holes. Inspiraling compact binaries are considered to be one of the most promising sources for ground based laser interferometers.

TAMA300 [1] is a Fabry-Perot-Michelson interferometer with baseline length of 300m located in Mitaka, Tokyo (35°40′N, 139°32′E). TAMA300 has been performed nine observation runs of the detector since 1999. Among them, we use DT4, DT5, DT6, DT8, and DT9 data of TAMA300. The total length of data available for the data analysis, after removing bad quality parts of data, is 2705 hours. A part of DT6 data which was coincident with LISM was already analyzed in [2]. The initial results of the analysis of DT8 data were reported in Ref. [3]. A part of DT8 data which was coincident with LIGO S2 was already analyzed in [4]. In this paper, we analyze these data again together with the other data in a unified way. Until DT6 observation, TAMA300 was the only large scale laser interferometer which was operated. Thus, it is important to analyze such data to search for possible gravitational wave signals. Although all data mentioned above are analyzed, we do not use DT4 and DT5 data to set the upper limit for the event rate, because the length of data from these runs are much shorter than DT6-8-9, and because the quality of data of these runs are much worse than those of DT6-8-9. The total amount of data used for setting the upper limit is 2462.8 hours. We combine the results from DT6, DT8 and DT9 data and obtain a single upper limit on the rate of the coalescence of compact binaries in our Galaxy.

2. Search methods

To analyze the data, we use the matched filtering method, in which we search for the best matched parameters of the theoretical wave form by cross-correlating the data with the theoretical wave form. As the theoretical wave form, we use the non-spinning, restricted post-Newtonian (PN) wave form in which the phase is given to high post-Newtonian order, but only the leading quadrupole term is contained in the amplitude.

The parameters which describe the wave form are the time of coalescence, t_c, the phase of wave at the coalescence, ϕ_c, the total mass $M \equiv m \ \ + m$ and the non-dimensional reduce mass $\eta \equiv m \ m \ /M$ of the binary. We search for the parameters which give the maximum of the output of the matched filter, ρ. The maximization over the phase, ϕ_c, can be taken analytically. The value of parameters, t_c, M and η, which maximize ρ are searched numerically.

The data are divided into subsets of data with length 52.4 seconds. Each subset of data has overlapping data with adjacent data for 4.0 seconds. Each subset of data is Fourier transformed, and the components more than 5kHz are removed. The data are converted to the strain equivalent data by means of the transfer function. The power spectrum density (PSD) of noise is evaluated at neighbor of each subset. Details of the method to evaluate PSD was described in §III.B. of Ref. [2]. With the subset of data, we compute ρ. For each small time interval with length $\Delta t_c = 25.6$msec, we search for t_c, M and η which give the maximum of ρ. The value of ρ at all of t_c can be computed automatically from the inverse FFT with respect to t_c. The search for the best matched M and η is done by introducing the grid points in the two dimensional mass parameter space. The range of masses of each member star of binaries is set to 1 to 3 $M_\odot$. We define by the local maximum of ρ in each small time interval with length, $\Delta t_c = 25.6$msec, and in the whole mass parameter region, together with the parameters, t_c, M and η which realize the local maximum.

The data of TAMA300 contain non-stationary, non-Gaussian noise. Such noise cause many triggers with rate much larger than that expected in the stationary Gaussian noise. In order to distinguish such spurious triggers from triggers caused by real gravitational wave signals, we compute χ value for each trigger with $\rho \geq 7$. We then define $\zeta = \rho/\sqrt{\chi}$ as a new statistic [2]. The statistic, ζ, was used in our previous analysis [2], and was found to be useful to distinguish the spurious triggers from triggers caused by real gravitational wave signals.

3. Results and summary

By using the method described in the previous section, we obtain the trigger lists from each observation run. We find that there are no triggers which deviate from the tail of the distribution of triggers significantly. This fact suggests that there is no candidate trigger which can be interpreted as real gravitational signal.

Next, we evaluate the upper limit to the event rate from each observation run, R_i (i = DT6, DT8, DT9), derived from $R_i = N_i/(T_i\epsilon_i)$ where T_i is the length of data, ϵ_i is the detection probability, and N_i is the upper limit to the number of event which is derived from the distribution of triggers. We evaluate the detection probability of Galactic signals by adding the signals to the real data, and by analyzing the data by the same analysis pipeline used in the real analysis.

We also evaluate the various error sources which affect the detection probability. They include the error of the threshold for the triggers, the error of the Monte Carlo injectioin test, the error of the calibration of the detector, the error of the theoretical wave form, and the error of the binary distribution model.

We obtain the upper limit to the Galactic event rate from each observation; 130_{-}^{5} yr^{-} for DT6, 30_{-} yr^{-} for DT8, and 60_{-} yr^{-} for DT9, at a 90% confidence level. We combine these results and obtain a single upper limit, $17_{-.5}^{5}$yr^{-} at a 90% confidence level. By taking larger value as a conservative upper limit, we obtain 20yr^{-}. This value is much larger than an astrophysically expected value, 8.3×10^{-5}yr^{-} [5] for the coalescence of neutron star binaries. However, this rate is smaller than that obtained by LIGO S2 search [6], 47 yr^{-} MWEG^{-}, or by LIGO-TAMA joint analysis, 49 yr^{-} MWEG^{-} [4]. Here MWEG stand for the Milky Way

	DT6	DT8	DT9
Observation time [hours]	876.6	1100	486.1
Threshold ζ^*	21.8	13.7	17.7
$N_{bg}^{(i)}$	0.1000	0.1255	0.0555
Detection probability	0.18	0.60	0.69
$(\delta R_i)_{\text{fluct}}$ [yr^{-1}]	+20.6	+2.52	+4.04
	−24.0	−2.82	−3.77
$(\delta R_i)_{\text{model}}$ [yr^{-1}]	+55.4	+4.18	+6.84
	−16.6	−1.53	−2.60
R_i [yr^{-1}]	130_{-}^{5}	30_{-} :	60_{-} :
Combined upper limit [yr^{-1}]		$17_{-\,.5}^{\;5}$	

Table 1. Summary of the upper limit to the Galactic event rate. The errors for the upper limit are evaluated in section VI in detail.

Equivalent Galaxy. Main reason for this is that the length of data used in our analysis is much longer than these analyses.

Recently, LIGO reported the results of the analysis from LIGO S3 and S4 data [7]. Their upper limit to the $1.4M_{\odot}$ neutron star binaries can be interpreted to 2.0 yr^{-} MWEG^{-} . LIGO was conducting the 5th science run from November 2005 to September 2007, with its design sensitivity. It can detect the inspiraling binaries up to ~10Mpc distance. They are expected to be able to set a much more stringent upper limit.

Despite of the improvement and long term observation of current detectors, the chance to detect gravitational waves by these first generation detectors will not be very large. We need more sensitive detectors, such like advanced LIGO [8] and LCGT [9].

More details of this work can be found in Ref:[10]

Acknowledgments
This work was supported in part by a Grant-in-Aid for Scientific Research on Priority Area (415) of the Ministry of Education, Culture, Sports, Science and Technology of Japan. This work was also supported in part by JSPS, KAKENHI, No.16540251.

References

[1] M. Ando et al., Phys. Rev. Lett., **86**, 3950 (2001); R. Takahashi et al., Class. Quantum Grav., **21**, S403 (2004).
[2] H. Takahashi et al., Phys. Rev. D **70**, 042003 (2004).
[3] H. Takahashi et al., Class. Quantum Grav. **21**, S697 (2004).
[4] S. Fairhurst and H. Takahashi, Class. Quantum Grav. **22**, S1109 (2005); B. Abbott et al., Phys. Rev. D **73**, 102002 (2006).
[5] V. Kalogera et al., Astrophys. J. **601**, L179 (2004) [Erratum ibid. **614**, L137 (2004).]
[6] B. Abbott et al., Phys. Rev. D **72**, 082001 (2005).
[7] B. Abbott et al., arXive:0704:3368v2.
[8] P. Fritschel, in *Astrophysical Sources for Ground Based Gravitational Wave Detectors*, edited by J.M. Centrella, AIP Conference Proceedings 575, American Institute of Physics, Melville, New York, 2001, pp. 15-23.
[9] K. Kuroda, Class. Quantum Grav., **23**, S215 (2006).
[10] T. Akutsu et al.,Phys. Rev. D **74**, 122002 (2006).

Search for gravitational waves on short duration in TAMA300 data : stellar core collapse and black hole

Ryota Honda , Nobuyuki Kanda , Tomomi Akutsu , Masaki Ando , Yoshiki Tsunesada , and the TAMA collaboration

[1] Graduate School of Science, Osaka City University, Sumiyoshi-ku, Osaka 558-8585, Japan
[2] Institute for Cosmic Ray Research, The University of Tokyo, Kashiwa, Chiba 277-8582, Japan
[3] Department of Physics, The University of Tokyo, Bunkyo-ku, Tokyo 113-0033, Japan
[4] National Astronomical Observatory of Japan, Tokyo 181-8588, Japan[1]
∗ Paper presented at the conference by N. Kanda

E-mail: kanda@sci.osaka-cu.ac.jp

Abstract. We present in the results of TAMA300 data analysis for short duration gravitational waves. The excess power filter, alternative linear filter (ALF) and TF(time-frequency) clustering methods have been employed for burst gravitational waves from stellar-core collapse, and matched filtering method used for the ringdown gravitational waves from black hole quasi-normal oscillations. The observational range of TAMA for the burst gravitational waves is roughly ∼ 1 kpc, and the range for black hole ringdown covers most of our galaxy. We have been developed new method 'time-frequency (TF) clustering' to find the burst waves. This is a TF clustering method on spectrogram (sonogram). Using this method, we can efficiently identify some predicted gravitational wave forms[2] and can exclude typical unstable spike like noises.

1. Introduction

TAMA300 is a 300 m baseline laser interferometric gravitational wave detector in Mitaka, Tokyo, Japan[1]. The interferometer is of Fabry-Perot-Michelson type with a power recycling technique. The sensitivity of the detector is $3\times10^{-}$ $[/\sqrt{\text{Hz}}]$ around 1kHz, and the integral observation time is more than 3000 hours. These observation data of TAMA300 detector is served to search for gravitational waves which has a short time duration. Time duration as Possible sources of such gravitational waves are stellar-core collapse of supernovae, and quasi-normal mode oscillation of black holes, since the typical time scale 10∼100 msec which corresponds to the good sensitive observation band 100Hz∼1kHz of ground-based gravitational wave detectors as TAMA300.

2. Searches for short gravitational waves in TAMA300

In TAMA data analysis, the excess power filter, alternative linear filter (ALF) and TF(time-frequency) clustering method are employed for burst gravitational waves from stellar-core collapses, and matched filtering method is employed for the ringdown gravitational waves from back hole quasi-normal mode oscillations. TAMA's search threshold amplitude for the burst gravitational waves is $h_{rss} \sim 10^{-\ \sim -}$, which is of the order of magnitude in the case of a

[1] present address: Department of Physics, Tokyo Institute of Technology

stellar core collapse at a distance of $\sim$ 1 kpc. We obtain the upper limit the galactic event rate as 5×10 events/day using the excess power filter[3], and 0.55 events/day for $h_{rss} > 10^{-}$ using the ALF trigger[4]. We also have also carried out data analysis to search for ringdown gravitational waves using efficient matched filter templates[5]. The trigger rate of black hole ringdown waves obtained $< 3.4 \times 10^{-}$ events/hour for SNR > 20 [6] .

3. A new approach : Time-frequency clustering method

We have developed a new method by examining 'clustering' of signal powers in time-frequency domain. Comparing with past TF-cluter studies[8] to find efficient power filter window, our study is in order to extract characteristics of gravitational wave signals with a multi-parameter representation[7]. We used predicted burst gravitational waves by Dimmelmeier et al.[2] (DFM waveforms). TAMA real data and DFM waveform software injected on TAMA data are used for following studies.

:

The sampling interval of TAMA's is 50μsec(=20kHz sampling). We split time series data to 3.2 msec chunks. Time shift of next chunk is 0.8 msec, thus the neighbor chunks overlap 2.4 msec. For each of the chunks, signal powers are calculated by using FFT (short-time Fourier transform, STFT) with a Hanning window. The frequency resolution in the STFTs is 312.5Hz. From the signal powers of the chunks at given times, we can construct a 'spectrogram' in time-frequency domain. The figure 1 and 2 are examples of spectrogram of TAMA data around spike like noise, and typical DFM waveform embedded in the TAMA300 data. Each 'pixel' represents the signal power at a given time and a frequency. SFTP spectrum at each frequency band is normalized using average noise.

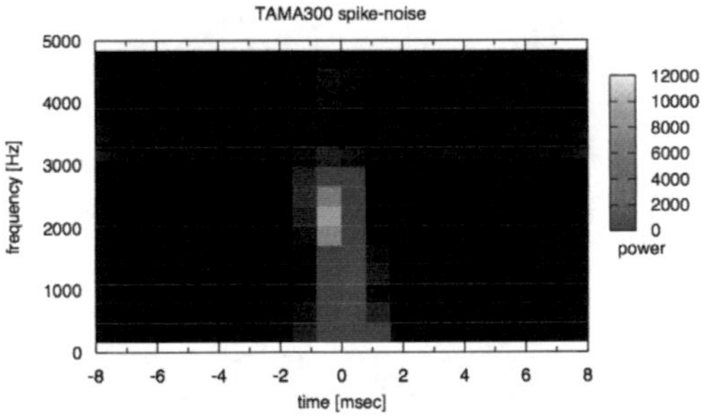

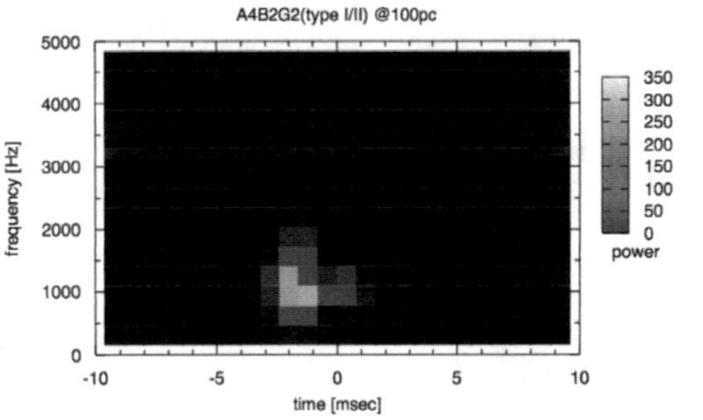

Figure 1. TAMA spike like noise **Figure 2.** burst GW injected TAMA noise
Spectrogram example: The color illuminated pixels connect as 'cluster' of signal.

To identify and define a cluster are as follows: 1) We find all pixel which magnitude exceed signal-to-noise ratio (SNR) = 3. 2) We assume that a local maximum pixel is a 'peak' of the cluster. The peak specified with its time t and frequency f . 3) If one of a 8 pixels around the peak of the cluster exceed the cluster threshold, this pixel is 'connected' pixel. The cluster threshold is settled as square root of SNR of the peak pixel. 4) We check the neighbor of connected pixels, again. 5) If no more connected pixel cannot be found, we define the peak and these connected pixels as one 'TF-cluster' 6) We exclude pixels which belongs to previous cluster, and iterate 2)-5) for other local peak.

The identified TF-cluster can be characterized with these parameters; area S : the number of pixels in the cluster, volume V : the sum of a magnitude of pixels in the cluster, tNs : a

N-th moment of distribution the numbers of pixels along the time axes, tNv : a N-th moment of magnitude distribution along the time axes, fNs : same as a tNs along the frequency axes, fNv : same as a tNv along the frequency axes.

, ffi

Some combinations of the parameters makes possible to separate GW signal from many noises. The figure 3 and 4 show the example of two combination which efficiently separate the gaussian noise from the DFM wave forms. The figure 3 and 4 show the separation using two combinations, $-2.0 \leq f1s \leq 2.0$ and $f2s \leq 5.0$ and $\sqrt{t1s \; + f1s \;}/S \leq 0.15$. One can see that the DFM waveforms can be efficiently separated from spike noises in the figure 3, and from gaussian noises in the figure 4.

The parameters selection for clusters which peak SNR>3 achieved 50% average efficiency for injected DFM waveforms at 350 pc ($h_{rss} \sim 2 \times 10^{-}$) for type I burst[2], and at 170 pc for average. The reduction for noise event clusters by spike signals are more than one order improvement beyond the SNR>100, and factor several for SNR < 80. It means that TF-cluster method is efficient for larger signal case.

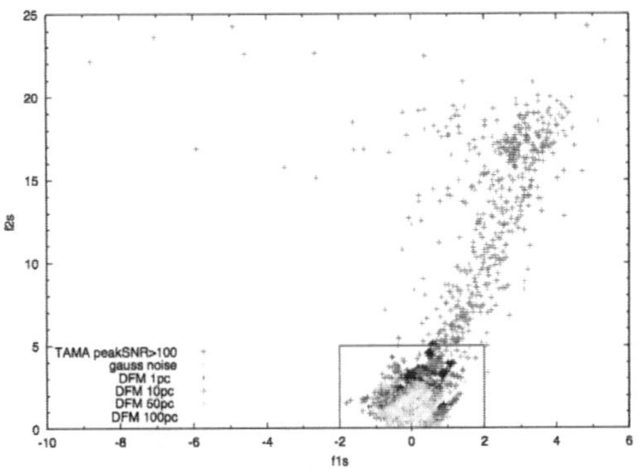

Figure 3. $f1s$ - $f2s$

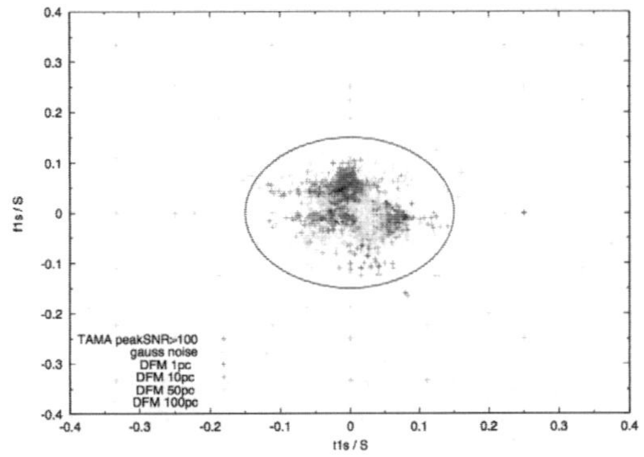

Figure 4. $t1s/S$ - $f1s/S$

4. Summary

TAMA have been developed searched for short duration gravitational waves, for physics target of stelar-core collapse and black hole oscillation. Also we recently practice the newer search algorithm as TF-cluster using many cluster shape parameters. The method is efficient to select gravitational event and reject noise fake as larger SNR case.

5. Acknowledgments

This research is supported in part by a Grant-in-Aid for Scientific Research on Priority Areas (415) of the Ministry of Education, Culture,Sports, Science and Technology.

References

[1] M.Ando, et al.,Physical Review Letters, Volume 86, 3950-3954, (2001)
[2] H.Dimmelmeier, J.A.Font and E.Muller, Astron. Astrophys., 388, 917-935 (2002)
[3] M.Ando, et al., Physical Review D 71, 082002 (2005)
[4] Tomomi Akutsu , et al., Class. Quantum Grav. 23 (2006) S715
[5] Tsunesada, et al. Phys.Rev.D 71, 103005 (2005)
[6] Y Tsunesada,et al., Class. Quantum Grav. 22 No.18 S1129-S1138
[7] R.Honda, Master thesis, 2007
[8] Julien Sylvestre, Phys. Rev. D 66 (2002) 102004

Direct searches for Dark Matter particles (above LN_2 temperature)

P. Belli

INFN, Sezione di Roma Tor Vergata, I-00133 Rome, Italy

E-mail: `pierluigi.belli@roma2.infn.it`

Abstract. Experimental observations and theoretical considerations have shown that a large part of the Universe is made of Dark Matter particles; this has currently motivated a large experimental effort to investigate them by direct detection approaches. In particular, several techniques are developed. A short review of those that use target at temperature above 77 K is summarized, with particular care to the approaches able to offer a suitable model-independent signature. Some perspectives will also be mentioned.

1. Introduction

The problem of the existence of Dark Matter in our Universe dates back to the astrophysical observations at the beginning of the past century. Many observations from the galactic to the cosmological scale support that most of the Universe is dark and large space for Dark Matter (DM) particles in the Universe exists. Most of the DM particles, relics from the Big Bang, were either produced at rest or non relativistic at decoupling time. The DM candidates have to be neutral, stable or quasi-stable (e.g. with a decay time of order of or larger than the age of the Universe) and have to be elusive in the interaction with ordinary matter. These features are respected, for example, by the axions, by axion-like candidates with mass in the keV range, by possible sterile neutrino, by a class of candidates named WIMPs (Weakly Interacting Massive Particles) and by some other candidates which cannot properly be classified as WIMP, but which offer similar phenomenology.

The study of DM candidates implies that a window beyond the Standard Model is investigated. At present, a widely considered candidate for DM is the lightest supersymmetric particle (LSP) that is stable in supersymmetric theories where the R-parity is conserved. The neutralino can be a suitable LSP candidate and is widely considered as DM particle [1]. However, other DM candidates can also be considered, such as e.g. a heavy neutrino of the 4th family or a sneutrino (the spin-0 supersymmetric partner of the neutrino) in a multi-component Dark Matter scenario. Moreover, a dominant contribution due to a sneutrino candidate remains still possible in supersymmetric models with violation of lepton number, where two mass states and a small energy splitting is present [2]. Other possible candidates are the particles from multi-dimensional Kaluza-Klein-like theories, the mirror Dark Matter particles [3], the self-interacting dark matter particles [4], etc. Recently, also a DM candidate interacting only with electrons has been proposed [5]. The case of a light ($\simeq$ keV mass) DM candidate, a bosonic particle either with pseudoscalar or with scalar coupling, has also been considered [6]. The sterile neutrino can also be a good DM candidate.

Several observations have pointed out that our Galaxy should also be embedded in a dark halo with mass at least 10 times larger than that of the luminous matter. The presence of DM particles in this dark halo can be investigated by direct approaches in underground laboratory. In case of WIMP or WIMP-like candidates, their interactions induce recoiling nuclei and/or, in case of inelastic scattering, also the successive de-excitation gamma's. Moreover, it has been pointed out that a recoiling nucleus can also induce ionization and excitation of bound atomic electrons; as a consequence, a certain quantity of electromagnetic radiation can contribute to the signal [7]. Recently, also a DM candidate interacting only with electrons has been proposed [5]. In case the light ($\simeq$ keV mass) bosonic particles are considered [6], the direct detection process is based on the total conversion in the detector of their mass into electromagnetic radiation. In these last processes the target nuclei recoil is negligible and is not involved in the detection process; therefore, because of their electromagnetic nature, these signals are lost in all those experiments using approaches based on discrimination procedures of the electromagnetic component of the measured counting rate.

In the world there are several experiments aiming to perform a direct DM investigation, using different kind of detectors and exploiting different approaches. Some of those that use target at temperature above 77 K will be summarized here, with particular care to the different experimental approaches.

2. Experiments using liquid noble gases

In this section we briefly summarize the experiments using scintillating liquid noble gases as target materials. The DAMA/LXe, started in 1990 [8], was the first experiment that developed liquid Xenon as target material for DM searches. Afterwards, exploiting the pulse shape discrimination techniques, results with a single phase detector have been realized by DAMA/LXe itself [9] and by ZEPLIN-I [10].

In the last years dual phase detectors have been proposed. These experiments exploit some discrimination between nuclear recoils candidates and the electromagnetic component of the measured counting rate through the ratio of the prompt scintillation signal ($S1$) and the delayed signal ($S2$) due to drifted electrons in the gaseous phase.

Recently, some results from prototypes have been published by: i) XENON10 [11], using a fiducial volume of 5.4 kg of liquid natural Xenon and an exposure of 136 kg $\times$ days; ii) WARP [12], using a fiducial volume of 2.3 liters of liquid natural Argon and an exposure of 96.5 kg $\times$ days; iii) ZEPLIN-II [13], using a fiducial volume of 7.2 kg of liquid natural Xenon and an exposure of 225 kg $\times$ days.

These experiments start from sizable counting rate and, in order to try to lower it, need to apply many data selections and handling. Each selection step can introduce systematic errors which can also be variable along the data taking period. After these selections procedures, an analysis based on some discrimination between the electromagnetic radiation and recoiling candidates is applied. In Fig. 1 examples of the published plots – obtained after applying the above mentioned cuts – have been reported for the three experiments, respectively.

The selected events are shown in scatter plots obtained by considering some discrimination variables. The acceptance windows (50% efficiency for XENON10 and ZEPLIN-II) for recoils candidates are also shown there; as it can be seen, several recoils and/or recoil-like candidates survive all the rejection procedures. Generally all of them are interpreted as residual background of various (even "unknown") nature. This appears at the moment a great limitation of this technique.

It is worth to point out some warnings about the present stage of the dual phase detectors, such as: i) the physical energy threshold is not properly evaluated by source calibrations in the energy interval of interest; ii) the detectors suffer from disuniformity (this seems to be an intrinsic limitation of the technique; it needs corrections to be applied and systematics to be accounted for, see in ref. [11, 14, 10, 13]); iii) the used gas is natural Xenon or Argon, with an unavoidable content of Krypton and ^{39}Ar, respectively; iv) the small and not linear light

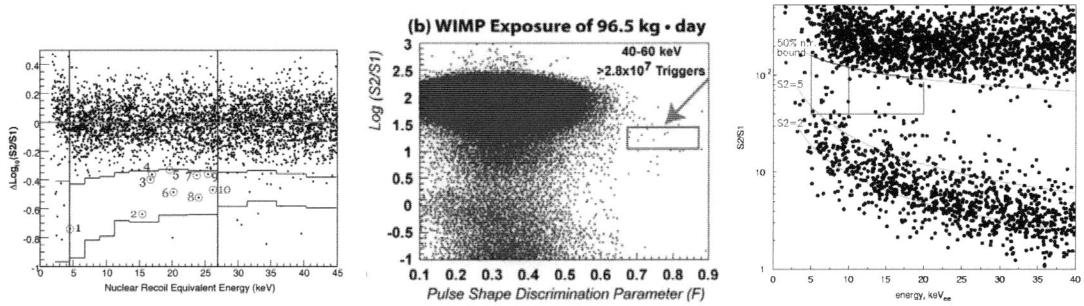

Figure 1. Examples of the results obtained by: XENON10 [11] (*left*), WARP [12] (*center*), ZEPLIN-II [13] (*right*). The events selected by many cuts are reported in these scatter plots of the discrimination variables. The acceptance windows (50% efficiency for XENON10 and ZEPLIN-II) for recoils events are also shown.

responses (e.g. $\simeq 2.2$ and roughly $0.5 \div 1$ photoelectrons/keVee claimed for XENON10 and for WARP, respectively); v) the poor energy resolutions (e.g. $\sigma/E \simeq 13\%$ and 16% at 122 keV γ rays for WARP and ZEPLIN-II, respectively); vi) despite of the small light response, an energy threshold of 2 keVee is claimed by XENON10; vii) the large number of materials involved in the core of the experiments (for example, in XENON10 there are 89 PMTs, all the materials for the electric field, the stainless steel containers, ...) prevents low level counting rate; etc.

To be short, all similar items can affect the robustness of whatever presented conclusion.

Furthermore, it is my opinion that electromagnetic rejection techniques, e.g. double phase read-out in liquid noble gases experiments, heat/ionizing and heat/light signals in bolometers, would require a much deeper and quantitative investigation than those made available up to now. It is worth to note that a "discrimination on an event-by-event base" might be possible only in the ideal case of zero systematics. Moreover, even in an ideal case, the result will not be the identification of the presence of WIMP elastic scatterings as DM signal, because of the well known existing recoil-like indistinguishable background. Finally, the electromagnetic component of the counting rate can contain itself the signal or part of it, as mentioned above, and will be lost by these approaches.

3. Model independent signatures

To obtain a reliable signature for the presence of DM particles in the galactic halo, it is necessary to follow a suitable model independent approach. In principle, two main possibilities exist; they are based on the correlation between the distribution of the events detected in a suitable underground set-up with the galactic motion of the Earth.

The first one – applying only for the WIMPs class of DM candidates – correlates the recoil direction with that of the Earth velocity, but it is practically disfavoured mainly by the technical difficulties in detecting the short recoil track. Few R&D attempts have been carried out so far, while suggestions, based on the use of anisotropic scintillators, were proposed in [15]. Recently, the project DRIFT (see Fig. 2) has collected an exposure of 10.2 kg $\times$ days with a 1 m^3 active volume filled by 40 Torr of CS_2 gas (that is 167 g of target material). A population of nuclear recoil candidates (interpreted as due to the decay of unexpected ^{222}Rn daughter nuclei, present in the chamber) has been observed [16]. Efforts are planned towards the future.

The second possibility, feasible and able to test a large range of cross sections and of DM particle halo densities, is the so-called annual modulation signature [17]. This is the main signature already exploited by the DAMA/NaI set-up ($\simeq$ 100 kg highly radiopure NaI(Tl) detectors) [18, 19]. The annual modulation of the signal rate originates from the Earth revolution around the Sun and offers many peculiarities since a seasonal effect induced by DM particles must simultaneously satisfy all the following requirements: (i) the rate must contain a component

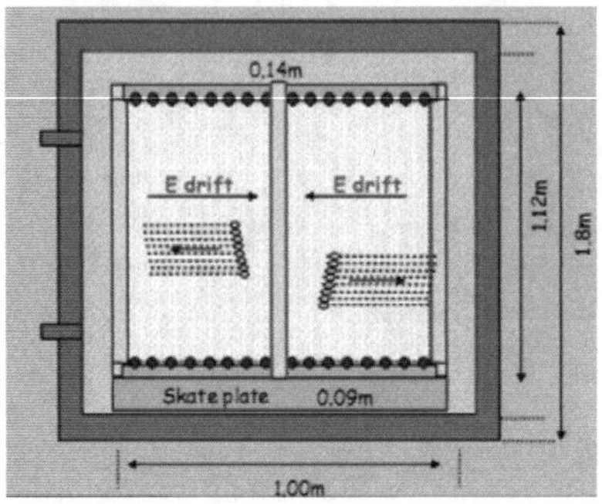

Figure 2. The conceptual schema of the DRIFT project.

modulated according to a cosine function; (ii) with one year period; (iii) with a phase roughly around 2nd June; (iv) the modulation must only be found in a well-defined low energy range, where DM particles can induce signals; (v) it must apply just to those events in which only one detector, in a multi-detectors set-up, actually "fires" (*single-hit* events), since the probability that DM particles would have multiple interactions is negligible; (vi) the modulation amplitude in the region of maximal sensitivity has to be $\leq 7\%$ for usually adopted halo distributions, but it can be significantly larger in case of some possible scenarios such as e.g. those of refs. [2, 6]. To mimic such a signature either spurious effects or side reactions should be able not only to account for the observed modulation amplitude but also to contemporaneously satisfy all the requirements.

The signature approach is a unique tool for a model-independent investigation of DM particles in the galactic halo. In particular, the annual modulation has been exploited – using highly radiopure NaI(Tl) as target material – by the DAMA/NaI experiment, whose model-independent result will be briefly recalled in the following; the corollary model-dependent quests for the DM candidates will be summarized too, since they also allow to offer – at some extent – a view on the complexity of the astrophysical, nuclear and particle physics aspects related to the field.

4. The DAMA/NaI experiment

The low background DAMA/NaI experiment [18, 19, 20] was located in the underground laboratory of Gran Sasso; it has been part of the DAMA project, which is also composed by several other low background set-ups (see Cerulli's paper in these Proceedings, ref. [21] and the references therein).

4.1. The model-independent result

The DAMA/NaI experiment took data over seven annual cycles collecting an exposure of $\simeq 1.1 \times 10^5$ kg $\times$ day. A clear annual modulation has been observed in the measured rate of the *single-hit* events in the lowest energy region, satisfying the many peculiarities of a DM particle induced effect [18, 19]. In Fig. 3 the time behaviour of the residual rate of the *single-hit* events in the (2-6) keV energy interval is reported. Fitting the data with a cosine-like function the presence of annual modulation is favoured at 6.3 σ C.L. with an amplitude equal to (0.0200 ± 0.0032) cpd/kg/keV, a phase $t_0 = (140 \pm 22)$ days and a period $T = (1.00 \pm 0.01)$ year. The period and phase agree with those expected in the case of an effect induced by DM particles in the galactic halo ($T = 1$ year and t_0 roughly at $\simeq 152.5^{th}$ day of the year). For details see refs. [18, 19] and references therein. Note that, while an evidence for a modulated part of the signal with proper features is present in the *single-hit* residuals (events class to which the DM particle-induced signals belong), it is absent in the *multiple-hits* residual rate (event class to which only background events belong) [19]. Since the same identical hardware and the same identical software procedures have been used to analyse the two classes of events, the obtained result offers an additional strong support for the presence of DM particles in the galactic halo, further excluding any side effect either from hardware or from software procedures

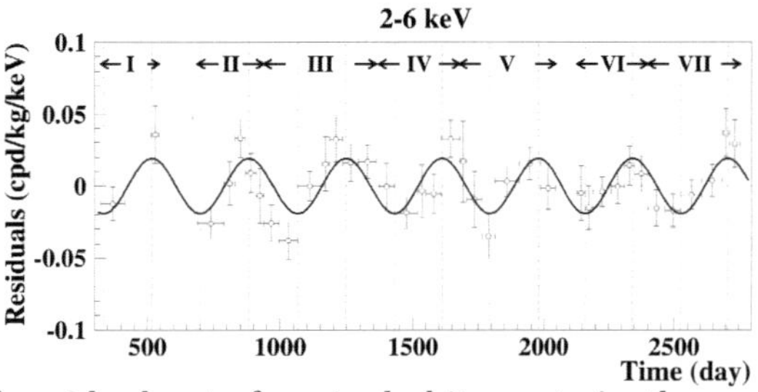

Figure 3. Experimental residual rate for *single-hit* events in the cumulative (2–6) keV energy interval as a function of the time over 7 annual cycles (total exposure 107731 kg × day); end of data taking July 2002. The superimposed curve represents the cosinusoidal function behaviour expected for a DM particle signal with a period equal to 1 year and phase exactly at 2nd June. See ref. [18, 19].

or from background [19].

A careful quantitative investigation of all the known possible sources of systematic and side reactions has been carried out; neither systematic effect nor side reaction able to account for the observed modulation amplitude and to satisfy all the requirements of the signature have been found. For quantitative discussion see ref. [18, 19].

In conclusion, the presence of DM particles in the galactic halo is supported in a model independent way by DAMA/NaI at 6.3 σ C.L.. Except the presently running DAMA/LIBRA, no other experiment whose result can be directly compared with this one is available so far in the field of Dark Matter investigation.

4.2. The corollary model-dependent quests

On the basis of the obtained 6.3 σ C.L. model-independent result, corollary investigations can also be pursued on the nature of the DM particle candidate. This latter investigation is model-dependent and – considering the large uncertainties on the astrophysical, nuclear and particle physics assumptions and on the parameters needed in the calculations – has no general meaning (as it is also the case of exclusion plots and of the DM particle parameters evaluated in indirect detection experiments). Thus, it should be handled in the most general way [18, 19, 5, 6, 7, 22, 23].

For simplicity, the results of the corollary analyses are presented in terms of allowed volumes/regions obtained as superposition of the configurations at given C.L. for the considered model frameworks. The inclusion of other existing uncertainties would further extend the allowed volumes/regions and increase the sets of obtained best fit values. In addition we remind that the results briefly mentioned here and the several other ones available in literature are not exhaustive of the many scenarios possible at present level of knowledge.

For the case of WIMP class candidates, low and high mass (from GeV to many hundreds of GeV) candidates have been considered, interacting with ordinary matter via: i) mixed SI&SD coupling; ii) dominant SI coupling; iii) dominant SD coupling; iv) preferred SI inelastic scattering, in various scenarios [18, 19]. This analysis has also been extended considering a non thermalized DM particle component in the dark halo [22]; other non-thermalized substructures or halo models with caustics can be addressed.

It is worth to note that the model-independent DAMA/NaI evidence is well compatible with theoretical expectations for neutralino in MSSM [1].

It has also been investigated the role of the electromagnetic radiation produced in the interaction of the WIMP with target nuclei [7] (hereafter Migdal effect). In fact, a certain quantity of e.m. radiation (made of escaping electron and of X-rays and/or Auger electrons) arises from the rearrangement of the atomic shells induced by the presence of a recoiling atomic nucleus. This radiation is fully contained in a detector of suitable size and because of its e.m.

nature, this part of the signal is lost in all those approaches based on discrimination procedures of the e.m. component of the measured counting rate. The inclusion of this effect has a sizeable impact in the DM direct searches when interpreted in terms of low mass WIMP candidates. For a discussion see ref. [7].

Recently, the implication of the "channeling" effect in NaI(Tl) crystals has been investigated [23]. Some experimental results, in fact, have shown that ions and recoiling nuclei move in a crystal in a different way than in amorphous materials. In particular, in the case of motion along crystallographic axes and planes, ions manifest an anomalous deep penetration into the lattice of the crystal and their range become much larger than the maximum they would have in case of motion in other directions or in amorphous materials. This is due to the fact that a low-energy ion, entering in the lattice into a channel, transfers its energy mainly to electrons rather than to the nuclei and, thus, its quenching factor approaches the unity. Moreover, it has been demonstrated that such an effect cannot be singled out in the already-collected neutron calibration data [23]. The inclusion of this existing effect gives an appreciable impact in corollary analyses in terms of WIMP (or WIMP-like) candidates [23]. In particular, this implies that lower cross sections are obtained for the allowed volumes/regions in given model frameworks by the DAMA/NaI data. For a detailed discussion see ref. [23]. Similar situation holds for purely ionization detectors, as Ge (HD-Moscow - like experiments), while a loss of sensitivity is expected when pulse shape discrimination is used in crystal scintillators (such e.g. in KIMS) since the channeled events – having $q \simeq 1$ – are probably lost. Moreover, no enhancement can be present either in liquid noble gas experiments (DAMA/LXe, WARP, XENON, ...) or in bolometer experiments; on the contrary some loss of sensitivity can be expected when applying discrimination procedures, based on $q_{ion} << 1$, since some events (those with $q_{ion} \simeq 1$) are lost.

The results further show the role of the existing uncertainties and of the correct description and inclusion of all the involved processes.

The DAMA/NaI model-independent evidence has also been investigated by considering as a DM candidate a light ($\simeq$ keV mass) bosonic particle, either with pseudoscalar or with scalar coupling [6]. For these candidates, the direct detection process is based on the total conversion in the detector's material of the mass of the absorbed bosonic particle into electromagnetic radiation. Thus, in these processes the target nuclei recoil is negligible and is not involved in the detection process; therefore, also signals from these light bosonic DM candidates are lost in experiments applying rejection procedures of the electromagnetic contribution to the counting rate. Large part of the bosonic candidate parameter space allowed by DAMA/NaI are of cosmological interest [6].

Various possibilities for some of the many possible astrophysical, nuclear and particle Physics scenarios can be considered and are also available in literature, such as e.g. refs. [1, 2, 3, 4, 24, 25, 26]. Recently, the DAMA Collaboration have also investigated a DM candidate interacting only with electrons [5]. An investigation on the sterile neutrino as DM candidate is also under consideration.

5. Some perspectives

At present the annual modulation is exploited by the second generation DAMA/LIBRA experiment (a $\sim$ 250 kg highly radiopure NaI(Tl) set-up), in operation [27] at Gran Sasso (see Fig. 4). The first release of results will occur not later than end of 2008.

Two other experiments plan to exploit such a signature: ANAIS [28] at Canfranc tunnel, which is trying to improve the features of old NaI(Tl) detectors by home-made efforts, and KIMS [29] at Yangyang, using CsI(Tl) detectors. The latter experiment has recently presented a preliminary study by means of pulse shape discrimination [29]; considerations similar than those already discussed before about discrimination procedures hold.

Finally, note that some alternative techniques are under investigation, such as those using the

Figure 4. Installating the highly radiopure NaI(Tl) crystals inside the low background shield of the DAMA/LIBRA set-up in HP Nitrogen atmosphere [27].

superheated droplet detectors or bubble chambers (PICASSO [30], SIMPLE [31] and COUPP [32]) and the micro-TPC (e.g. MIMAC [33] and DMTPC [34]).

Moreover, efforts and developments are in progress towards larger mass set-ups by all the experimental activities.

6. Conclusions

Large experimental efforts have been done in the investigation of the DM particles in the galactic halo. In particular, several techniques have been used. Some of them have been summarized here with the particular care to the techniques able to offer model-independent signatures.

References

[1] A. Bottino et al., arXiv:0710.0553; Phys. Rev. D 67 (2003) 063519; Phys. Rev. D 68 (2003) 043506; Phys. Rev. D 69 (2004) 037302.
[2] D. Smith and N. Weiner, Phys. Rev. D 64 (2001) 043502; Phys. Rev. D 72 (2005) 063509.
[3] R. Foot, hep-ph/0308254.
[4] S. Mitra, Phys. Rev. D 71 (2005) 121302.
[5] R. Bernabei et al., in preparation and references therein.
[6] R. Bernabei et al., Int. J. Mod. Phys. A 21 (2006) 1445 and references therein.
[7] R. Bernabei et al., Int. J. Mod. Phys. A 22 (2007) 3155 and references therein.
[8] P. Belli et al., N. Cim. A 103 (1990) 767.
[9] R. Bernabei et al., Phys. Lett. B 436 (1998) 379.
[10] G.J. Alner et al., Astrop. Phys. 23 (2005) 444.
[11] J. Angle et al., arXiv:0706.0039.
[12] P. Benetti et al., astro-ph/0701286.
[13] G.J. Alner et al., astro-ph/0701858.
[14] P. Benetti et al., astro-ph/0603131
[15] P. Belli et al., Il Nuovo Cim. C15 (1992) 475; N.J.C. Spooner at al., IDM Workshop (1997) 481; Y. Shimizu et al., Nucl. Instrum. & Meth. A 496 (2003) 347; R. Bernabei et al., Eur. Phys. J. C 28 (2003) 203.
[16] S. Burgos et al., arXiv:0707.1758; arXiv:0707.1488.
[17] K.A. Drukier et al., Phys. Rev. D 33 (1986) 3495; K. Freese et al., Phys. Rev. D 37 (1988) 3388.
[18] R. Bernabei el al., La Rivista del Nuovo Cimento 26 n.1 (2003) 1-73
[19] R. Bernabei et al., Int. J. Mod. Phys. D 13 (2004) 2127.
[20] R. Bernabei et al., Il Nuovo Cim. A 112 (1999) 545.
[21] DAMA website: http://people.roma2.infn.it/dama
[22] R. Bernabei et al., Eur. Phys. J. C 47 (2006) 263.
[23] R. Bernabei et al., arXiv:0710.0288, to appear on Eur. Phys. J. C
[24] K. Belotsky, D. Fargion, M. Khlopov and R.V. Konoplich, hep-ph/0411093.
[25] E.M. Drobyshevski et al., arXiv:0704.0982; arXiv:0706.3095.
[26] C. Arina and N. Fornengo, arXiv:0709.4477.
[27] R. Bernabei et al., in the volume "Frontier Objects in Astrophysics and Particle Physics", 90 (2004) 581.
[28] J. Amar et al., in the Volume "The Identification of Dark Matter", World Sci. (2007) 237.
[29] H.S. Lee et al., arXiv:0704.0423.
[30] Barnab-Heider et al., Nucl. Instrum. Meth. A555 (2007) 184.
[31] T.A. Girard et al., Phys. Lett B 621 (2005) 233.
[32] W.J. Bolte et al., Journ. of Phys.: Conf. Ser. 39 (2006) 126
[33] D. Santos et al., Journ. of Phys.: Conf. Ser. 65 (2007) 012012
[34] D. Dujmic et al., arXiv:0708.2370.

Dark Matter Detection with Cryogenic Detectors

Dr. Daniel A. Bauer

Fermilab, MS209, PO Box 500, Batavia, IL 60510

bauer@fnal.gov

Abstract. Direct detection of dark matter in the form of Weakly-Interacting Massive Particles (WIMPS) is an active field of research. Cryogenic detectors have been in the forefront of the field, due to their exquisite ability to reject backgrounds from interactions of normal matter. In this paper, I describe the current status and future prospects for such experiments.

1. The landscape of dark matter direct detection

Observations of galaxies, superclusters, distant supernovae and the cosmic microwave background radiation, tell us that ~85% of the matter in the universe is not made of particles that emit or absorb electromagnetic radiation. A leading hypothesis is that it is comprised of Weakly Interacting Massive Particles [1], or WIMPs, that were produced moments after the Big Bang. Particle physics theories provide possible WIMP candidates with masses ~10-1000 GeV/c^2. For example, supersymmetry contains a lightest supersymmetric partner (LSP) that is stable and interacts at roughly the weak-interaction rate, allowing it to decouple from ordinary matter in the early universe with a relic density comparable to the dark matter density [2]. Similarly, many models involving extra dimensions predict the lightest Kaluza Klein excitation is stable, with weak-scale mass and interaction cross sections [3].

If WIMPs are indeed the dark matter, their density in the galactic halo may allow them to be detected via elastic scattering from atomic nuclei in a suitable terrestrial target. The energy depositions and interaction rates are low, requiring that this type of experiment be located deep underground for protection from cosmic rays and requiring the use of radio-pure materials to shield against radioactivity in the environment. The current generation of direct detection experiments is now reaching the level of sensitivity needed to probe theoretical predictions in a way that is quite complementary to accelerator searches. The combination of LHC and WIMP-nucleus elastic-scattering experiments would check the consistency of the models and provide powerful constraints on the parameters. More generally, the low mass of the Higgs inferred from electroweak measurements point to a WIMP-nucleon cross section in the range10^{-45}->10^{-44} cm^2 [4].

The WIMP detection field has grown over the last several years, with the introduction of new technologies. There are now approximately fourteen operating experiments (DAMA/LIBRA, KIMS, CaF2-Kamioka, WARP, XENON-10, ZEPLIN-II, DEAP-I, XMASS, CDMS, CRESST, EDELWEISS, COUPP, PICASSO, DRIFT) and at least another three active R&D programs (ArDM, LUX and DEAP/CLEAN). In this paper, I will focus on the status and prospects for cryogenic direct-detection experiments. A separate contribution to these proceedings will survey non-cryogenic techniques [5].

1.1. WIMP detection

When considering the scattering of WIMPs on nuclei, two types of coupling between the WIMP and a nucleon must be considered: spin-dependent and spin-independent [6]. The balance between the two types of coupling in supersymmetry depends on the flavor composition of the lightest particle and can strongly favor one or the other coupling. However, in the case of spin-dependent couplings, there is a cancellation between opposite-aligned spins in the target nucleus, while in the spin-independent case, all nucleons add coherently. This amplifies the sensitivity of experiments using large-A nuclei to search for spin-independent scattering by a factor A^2. For this reason, the experimental community has focused most attention and resources on attempts to discover WIMPs via spin-independent interactions. Note that this coherence is lost if the WIMP mass becomes much heavier than the mass of the target nucleus. Use of several different target materials can help to distinguish whether any nuclear recoil events seen are due to WIMP interactions, or backgrounds. If no events are seen, upper limits on the cross section are presented as WIMP-nucleon cross sections versus WIMP mass. Even a signal of a few events would constrain the cross section and WIMP mass to within about an order of magnitude.

1.2. Backgrounds

Direct detection dark matter experiments have currently reached a level of sensitivity corresponding to a few events per kilogram of target mass per year. This has already required a large effort to shield the detectors against gammas, electrons and neutrons from radioactivity. In addition, the experiments must be located deep underground to avoid neutrons produced in cosmic ray interactions. Finally, the experiments must have either significant event-by-event discrimination between electron recoils (from electromagnetic background sources) and nuclear recoils (from either WIMP or neutron interactions), or exploit some other expected characteristic of WIMPS such as annual modulation or directionality.

It is crucial that direct detection experiments strive for zero background. This clearly maximizes discovery potential, and allows WIMP sensitivity to improve linearly with increase in target mass and running time. If a background arises, the sensitivity improvement will initially degrade as the square root of exposure (the product of target mass and running time) and then plateau at an irreducible level until the background can be removed or rejected.

1.2.1. Sources of backgrounds

The primary sources of radioactivity in the natural environment are long-lived isotopes of Uranium and Thorium, which yield alpha, beta and gamma particles from their decay chains. Also troublesome are the long-lived isotopes ^{40}K and ^{210}Pb, the latter resulting from decays of ubiquitous Radon gas. It is possible to screen materials for radioactive contamination by detecting either the gamma or alpha particles emitted from surfaces. However, the sensitivity required for direct detection experiments is beginning to challenge the best screening technologies.

A limiting background for all direct detection experiments is the presence of neutrons, since they produce the same nuclear recoil signature as WIMPS. Mounting the experiments far enough underground is sufficient to reduce neutrons from cosmic ray showers. However, neutrons also come from (alpha,n) reactions and fission decays, both resulting from the presence of small residues of radioactivity surrounding the detectors. This background will likely begin to dominate in the next generation of experiments without extreme efforts to further reduce contamination. Use of multiple target materials with different atomic weights can exploit the likelihood that WIMP scattering scales as A^2, whereas neutron cross sections are relatively insensitive to A, to allow statistical rejection of a neutron background. Similarly, the ability to recognize multiple scattering of neutrons, extremely unlikely for WIMPS, can also allow an experiment to recognize a neutron background. However, the presence of a neutron background would still limit discovery potential, since it would require a much larger sample of events to measure and subtract a neutron background, so as to extract a WIMP signal.

1.2.2. Background reduction
Copper, germanium, xenon, neon are among the cleanest materials available to direct detection experiments, with no naturally occurring long-lived isotopes. Argon suffers from the presence of ^{39}Ar, produced by cosmic rays in the atmosphere and having a half-life of 239 years, although it is possible that underground sources of Argon may have much less ^{39}Ar. Lead is one of the best shielding materials against gamma radiation, especially ancient lead which tends to be free of ^{210}Pb ($T_{1/2} = 22$ years). High-density polyethylene and purified water make excellent neutron moderators with which to surround dark matter detectors. It is also important to reduce the presence of Radon gas in the vicinity of detectors, since it tends to plate out on surfaces and leave radioactive decay daughters. There is a substantial literature on these topics [7].

1.2.3. Active background rejection
The most sensitive direct detection experiments have now achieved extremely good active (event by event) discrimination against electromagnetic backgrounds, in some cases exceeding one part in a million. This is most often accomplished by measuring at least two types of signals (e.g. ionization and phonons, ionization and light, or even light and phonons) that differ for nuclear recoil events as compared with electron recoil events. Space does not allow a complete survey of these techniques here, but they have been summarized in recent review articles [8].

2. Cryogenic dark matter detectors
For the purposes of this article, a cryogenic direct detection experiment is defined as one in which the active detector material is maintained at temperatures <77K. In practice, all such experiments use dilution refrigerators and maintain their solid-state detectors at temperatures < 100 mK. This allows the detection of phonons from the interaction of a single particle in a crystal lattice. Comparison to either the collected ionization or scintillation light from the interaction allows excellent discrimination between nuclear recoils and electron recoils. The drawback of the technique is primarily the need for a cryogenic infrastructure, but also the relatively small target mass achieved by placing phonon sensors on solid-state crystals. Nevertheless, cryogenic detectors have been in the forefront of WIMP direct detection experiments for the last decade.

3. Status and prospects for cryogenic experiments
There are three cryogenic direct detection experiments in various stages of operation at present. CDMS II has been operating in the Soudan Underground Laboratory since 2003. EDELWEISS II is commissioning their new apparatus at the Modane Underground Laboratory and CRESST II is at a similar phase in the Gran Sasso Underground Laboratory. There are also a few smaller R&D efforts, as well as plans for future, large cryogenic experiments. I summarize briefly the status and prospects for these experiments in this section.

3.1. CDMS
The Cryogenic Dark Matter Search (CDMS) Collaboration has pioneered the use of low temperature phonon-mediated detectors to search for the rare scattering of WIMPs on nuclei and distinguish them from backgrounds. The CDMS II experiment was constructed over several years (1998-2002) in the Soudan Underground Laboratory (2000 mwe) and has operated successfully since 2003.

3.1.1. Current and pending results
The first year of CDMS II running at Soudan, with a target mass of about 1(0.5) kg Ge(Si) and a raw Ge exposure of 75 kg-days, produced limits on WIMP-nucleus cross sections that were substantially better than achieved by any other experiment at the time. Figure 1 shows the published [9] spin-

independent limits, with a minimum cross section of 1.6 x 10^{-43} cm^2 at 60 GeV/c^2. Spin-dependent limits were also derived for both Ge and Si [10]. Recently, the Xenon 10 experiment [5] has shown improved limits (also shown in Fig 1) based on a raw exposure of 316 kg-days, albeit with a significant background of 10 events in the signal region that limits their discovery potential. Meanwhile, CDMS II has continued to run, now with increased target mass of 4 kg Ge and 1.5 kg Si, and has accumulated a total exposure of ~1200 kg-days as of the end of 2007. The expected sensitivity corresponding to that exposure, assuming no background, is also shown in Fig 1, indicating that CDMS II limits will again be the world's best, reaching cross sections ~2 x 10^{-44} cm^2. Blind analysis of the first 400 kg-days of data is underway and results are expected in early 2008. Analysis of the full data set should follow by summer 2008.

The most difficult background source for CDMS II, as for most solid-state detectors, is due to residual surface contamination, particularly from beta-emitting radioactive isotopes. There is a very thin 'dead layer' on the surface of the crystals, where ionization is not collected efficiently, leading to a potential event signature that mimics that of WIMPs. Fortunately, the phonon signal carries timing information that can distinguish surface events from bulk events, resulting in a 99.9% rejection of surface events in CDMS II. This discrimination can be improved further, with only a gradual loss of nuclear recoil acceptance, allowing the experiment to remain 'background free'.

CDMS II will continue to operate at Soudan at least through 2008, with another data set of ~1200 kg-days expected, and a final cross section goal of ~10^{-44} cm^2 at 60 GeV/c^2. It is important to emphasize that CDMS II is the only direct detection experiment currently running with less than one background event in the signal region. This is crucial for spotting the first few events of a WIMP signal. Indeed, CDMS II may have the reach to discover WIMPs in the coming year or two, before the Tevatron or the LHC see evidence for supersymmetry. Conversely, should supersymmetry be discovered at the Tevatron or LHC, CDMS II may clarify whether the lightest supersymmetric particle constitutes the dark matter.

3.1.2. SuperCDMS 25 kg

The CDMS collaboration has proposed a new SuperCDMS experiment, with 25 kg of Ge target mass. Combined with improved background rejection and location at the very deep (6000 mwe) site at SNOLAB, this experiment would represent another order of magnitude improvement in dark matter sensitivity. Preparation for the SNOLAB phase of the experiment will overlap with continued running of CDMS II detectors at Soudan in 2009, and the full detector payload would operate at SNOLAB in 2011-13.

The SuperCDMS detectors are each 2.5 times more massive than the CDMS II detectors and have both improved rejection capability and decreased surface backgrounds. The projected exposure of 18,000 kg-days should reach a WIMP-nucleon cross-section of ~10^{-45} cm^2 for a WIMP mass around 60 GeV/c^2 as shown in Fig 1.

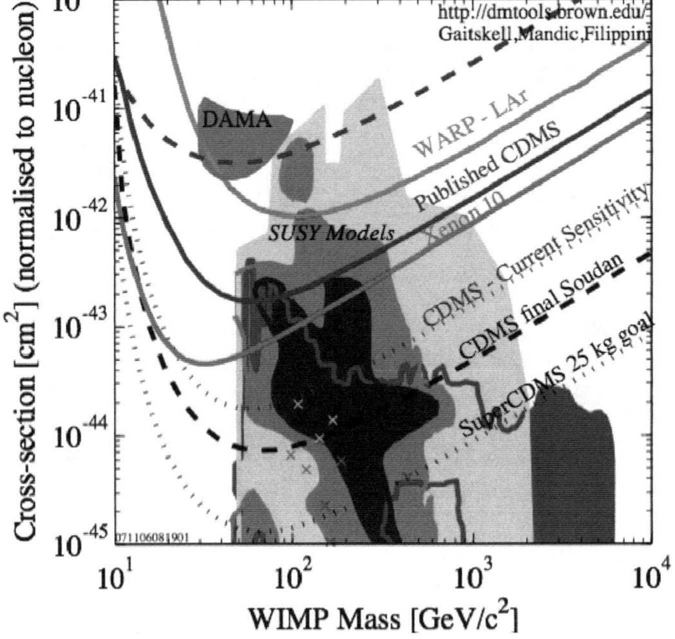

Figure 1. WIMP-nucleon cross section versus WIMP mass for several recent direct detection experiments. In solid blue are the published results of CDMS II, while the recently reported Xenon 10 results are in solid red. The blue dotted curve shows the current CDMS II sensitivity, whereas the black dashed curve shows the expected final sensitivity of CDMS II assuming no background. The black dotted curve is the sensitivity goal for SuperCDMS 25 kg.

3.2. Edelweiss

Edelweiss I operated three, 320 g Ge detectors at the Modane Underground Laboratory (4800 mwe) until 2002, and achieved significant limits on WIMP interactions based on a raw exposure of 62 kg-days [11]. The experiment became limited by backgrounds due to Radon contamination on copper covers surrounding the detectors at the level of ~5 events/kg-day. The Edelweiss I detectors used thermal readout of the phonon signal, which cannot exploit phonon timing to reject surface events. The collaboration is now pursuing Edelweiss II, with a rebuilt apparatus and improved detectors.

3.2.1. Preparations and prospects for Edelweiss II

Edelweiss II has significant improvements compared with the earlier experiment. Much more attention has been devoted to materials screening and the experiment is enclosed in a clean room supplied with air that has reduced Radon concentration. Both gamma (Pb) and neutron (polyethylene) shielding have been made thicker, with more complete coverage, and a muon veto surrounds the experiment. The new cryostat can accommodate up to 40 kg of detectors. Currently available are twenty-one, 320 kg Ge detectors with thermal readout and seven, 400 kg Ge detectors with NbSi thin-film phonon sensors that may ultimately provide surface event rejection similar to that achieved with CDMS II.

The new experiment is currently being commissioned. An engineering run resulted in ~19 kg-days of data with eight detectors, a recoil energy threshold ~30 keV and no events in the nuclear recoil region. Judging from the alpha counting rate, there is still a background from ^{210}Pb, but at a level 2-3 times lower than in the previous experiment. Prototypes of the new NbSi sensor readout indicate 90(99)% surface event rejection, with signal acceptance of 70(50)%. It is expected that physics running will start soon, with a sensitivity goal of 1 x 10^{-43} cm^2 by spring 2008. Thereafter, detectors will be added every few months until the full payload is achieved. Sensitivity of $\sim 10^{-44}$ cm^2 should be possible by 2009-2010 [12].

3.3. CRESST

The CRESST experiment operates $CaWO_4$ detectors with sensors that give simultaneous measurement of phonons and scintillation light for discrimination of nuclear recoils from electromagnetic backgrounds. Oxygen recoils will come mainly from neutrons while W recoils will be preferentially from WIMP interactions. The experiment is located in the Gran Sasso Underground Laboratory (3500 mwe). Results from the earlier CRESST 1 experiment came from a short exposure in 2004 [13], in which it was realized that the shielding surrounding the experiment was not adequate to reject neutron backgrounds.

3.3.1. Preparations and prospects for CRESST II

CRESST II has a substantially improved shield and SQUID-based readout of the phonon signal, as well as a new cryostat and data acquisition system. There is space in the new cryostat for thirty-three, 300 g detectors. A first physics run with three detectors accumulated ~60 kg-days and reached a sensitivity of about 7×10^{-43} cm^2. Backgrounds seem to be understood, and the experiment currently has 3 kg of detector mass deployed and is working on electronic noise and calibration issues [14]. It is expected that a physics run will commence in early 2008. The ultimate goal of CRESST II is to reach a sensitivity $\sim 10^{-44}$ cm^2.

3.4. Eureca

The groups pursuing EDELWEISS and CRESST have tentatively agreed to join forces after their current experiments are complete and pursue a new European facility called EURECA. The goal is to host up to 1000 kg of various cryogenics detector targets, in order to reach spin-independent WIMP-nucleon cross section sensitivity $\sim 10^{-46}$ cm^2. The likely location for the experiment would be Modane and the time scale is probably 2013 or later.

3.5. 3He experiments (Mimac, Ultima)

There has been some R&D work on developing bolometric detection of nuclear recoils from WIMPS in 3He. The technique would employ mechanical wire resonators installed in a 100 μK superfluid 3He cell and driven at resonance via the Lorentz force in a 100 mT uniform B field. Energy depositions would produce quasiparticles that damp the resonator motion and thus shift the resonant frequency. One would also have to sense either ionization or scintillation to provide discrimination against electron recoils.

4. Comparison with collider searches for WIMPs

Collider searches and direct detection are reaching similar WIMP sensitivities at about the same time. The two approaches are complementary. In most models of supersymmetry, if the LHC detects a neutralino, it will also have a sufficiently large elastic scattering cross section with nuclei to be visible in direct detection. Collider experiments are needed to establish WIMP properties, although direct detection can help determine the WIMP mass accurately and indeed have a larger mass reach than at the LHC! Only direct detection can establish convincingly that a WIMP signal detected in collider experiments is a significant part of the dark matter halo of our galaxy.

5. Summary

Cryogenic detectors will continue to be at the forefront of dark matter direct detection over the next few years, with new results expected from CDMS, EDELWEISS and CRESST. Figure 2 shows some sample projections for the next decade, although these should always be treated with caution since history shows they are usually optimistic. The field will benefit greatly from the multiplicity of techniques used if a WIMP signal is detected by one of the experiments. The key to increased sensitivity will be increasing target mass while continuing to reduce and reject backgrounds. Only a background-free experiment has real discovery potential for WIMPs!

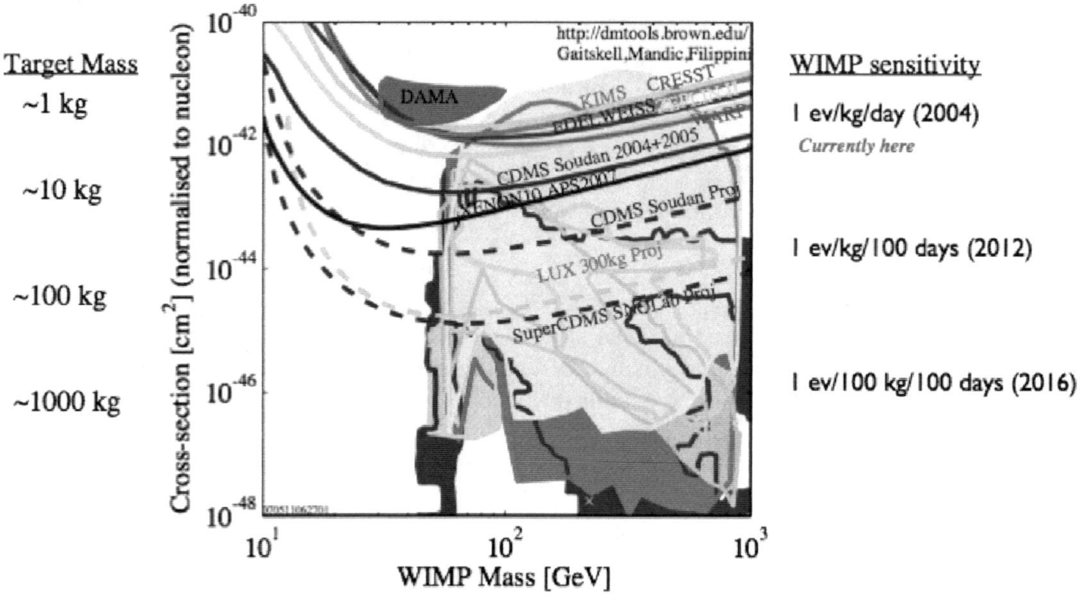

Figure 2. WIMP-nucleon cross section versus WIMP mass projections for upcoming experiments, compared with current limits. Note that as the target mass increases, the background levels must decrease to achieve the sensitivity goals.

References

[1] B.W. Lee and S. Weinberg, *Phys. Rev. Lett.* **39**, 165 (1977).
G. Jungman, M. Kamionkowski, and K. Griest, *Phys. Rep.* **267**, 195 (1996).

[2] D. Hooper, A. Taylor, arXiv:hep-ph/0607086.

[3] K. Agashe and G.~Servant, *Phys. Rev. Lett.* **93**, 231805 (2004); hep-ph/0403143.
G.~Servant and T.M.P.~Tait, *Nucl. Phys.* **B650**, 391 (2003).
H.C.~Cheng, J.L.~Feng and K.T.~Matchev, *Phys. Rev. Lett.* **89**, 211301 (2002).

[4] M. Carena, D. Hooper, A. Vallinotto *Phys.Rev.* **D75**, 055010 (2007).

[5] P. Belli, these proceedings.

[6] J.D.~Lewin and P.F.~Smith, *Astropart. Phys.* **6,** 87 (1996).

[7] See, for example, proceedings of the 2006 Workshop on Low Radioactivity Techniques (LRT2006), Aussois, France, October 2006; AIP Conference Proceedings **897**.

[8] See, for example, R.J. Gaitskell, *Annu. Rev. Nucl. Part. Sci.* **54**, 315 (2004).

[9] D.S. Akerib et al., (CDMS Collaboration) *Phys. Rev. Lett.* **93**, 211301 (2004); *Phys. Rev.* **D72**, 052009 (2005); *Phys. Rev.* **D72**, 052009 (2005).

[10] D.S. Akerib et al., (CDMS Collaboration) *Phys. Rev.* **D73**, 011102 (2006).

[11] S. Fiorucci et al., *Astroparticle Physics* **28**, 143 (2007); A. Benoit et al., *Phys.Lett.* **B545**, 43 (2002).

[12] G. Gerbier, these proceedings.

[13] G. Angloher et al., *Astroparticle Physics* **23** (2005) 325-339.

[14] W. Seidel, these proceedings.

Particle Dark Matter Candidates

Stefano Scopel
Korea Institute for Advanced Study
207-43 Cheongryangri-dong Dongdaemun-gu
Seoul 130-722, Korea

E-mail: `scopel@kias.re.kr`

Abstract. I give a short overview on some of the favorite particle Cold Dark Matter candidates today, focusing on those having detectable interactions: the axion, the KK–photon in Universal Extra Dimensions, the heavy photon in Little Higgs and the neutralino in Supersymmetry. The neutralino is still the most popular, and today is available in different flavours: SUGRA, nuSUGRA, sub–GUT, Mirage mediation, NMSSM, effective MSSM, scenarios with CP violation. Some of these scenarios are already at the level of present sensitivities for direct DM searches.

1. Introduction

About 20% of the energy density of the present Universe appears to be in the form of Cold Dark Matter (CMD) [1], i.e. of particles non–relativistic at the time of their decoupling from the plasma in the Early Universe whose contribution is considered as an indispensable catalyst to the formation of galaxies. CDM clusters at the galactic level, so that it should pervade our solar system and be detectable by direct or indirect techniques. In my talk I will give a short overview on some of the most popular particle CDM candidates today, focusing on those having detectable interactions. So, while Super–Weak DM candidates such as sterile neutrinos and gravitinos represent a viable possibility to explain DM [2], I will not cover them here since they cannot be measured in DM searches. Moreover, I will only discuss direct DM detection, since indirect searches have been discussed elsewhere in this Conference [3].

The properties of a good DM candidate are well known: it should be stable (its decay being forbidden by the conservation of some quantum number), neutrally charged and colourless (in order to be really "dark"), possibly motivated by theory (although "ad hoc" candidates may have the advantage of minimality and simplicity [4]). Moreover, its calculated relic abundance should be compatible to observation, although candidates providing a subdominant contribution to CDM represent a quite reasonable option[1].

In the following $\Omega \equiv \rho/\rho_c$ will indicate a candidate cosmological mass density ρ normalized to the critical one, ρ_c =1.8791 $h^2 \times 10^{-29}$ g cm^{-3}, and $h \equiv H_0/100$ km sec^{-1} Mpc^{-1} will indicate the normalized Hubble constant at present times.

2. The neutrino

The first place to look for a DM candidate is the Standard Model (SM) of interactions. In this case the only viable DM candidate is the neutrino, whose relic abundance as a function

[1] Variety is common in Nature, and a multicomponent CDM might even have larger detection cross sections, and so be easier to detect.

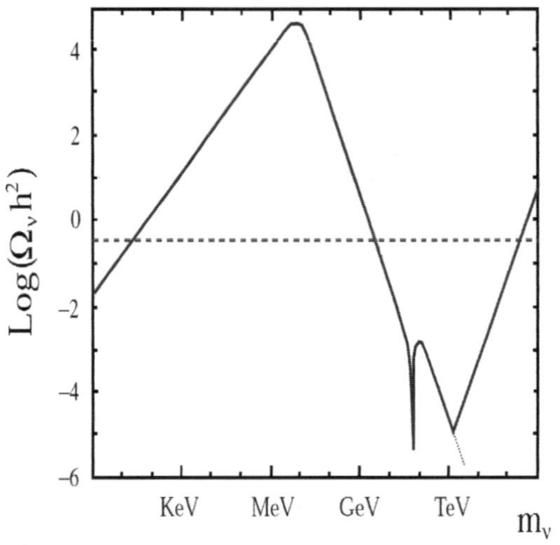

Figure 1. Neutrino relic abundance as a function of the neutrino mass (from[5])

of the mass is shown in Fig. 1, taken from [5]. Since LEP excludes additional neutrinos (with standard weak interactions) with masses $m_\nu \lesssim 45$ GeV one can see from this figure that the mass density of ordinary heavy neutrinos is bound to be very small, $\Omega_\nu h^2 \lesssim 0.001$ for masses up to $m_\nu \sim \mathcal{O}(100)$ TeV. At lower masses the neutrino relic abundance is given by $\Omega_\nu h^2 = \sum m_\nu/91.5$ eV, that, when compared to observation, implies the constraint $\sum m_\nu \lesssim 10$ eV. However a more stringent limit $\sum m_\nu \lesssim 0.66$ eV can be derived combining CMB and Large Scale Structure (LSS) data [1], implying $\Omega_\nu h^2 \lesssim 0.007$: in fact, neutrinos lighter than this bound, which are relativistic (*hot*) at decoupling, erase primordial perturbations due to their free streaming and prevent galaxy formation. So also in this case the neutrino abundance turns out to be small. The bottom line is that, unless neutrinos are mixed with some sterile component [2] they don't work as DM candidates.

The need of a DM candidate outside the SM, when combined with the fact that the SM itself is believed to be an incomplete theory with a cut–off of $\mathcal{O} \sim 1$ TeV, is no doubt intriguing. Moreover, theory is in no shortage of viable DM candidates that have been proposed in the context of Particle Physics in order to solve/alleviate/explain problems that have nothing to do with Cosmology. In the following I will concentrate on some of the most popular.

3. The axion

The axion is the pseudo Goldstone boson of the Peccei–Quinn (PQ) symmetry, introduced in order to explain CP conservation in QCD [6]. Its mass can be considered as a free parameter, spanning several orders of magnitude, and is related to the (unknown) scale of the PQ symmetry breaking f_a. The main production mechanism of relic axions is through misalignment, in which at $T > \Lambda_{QCD}$ the axion field acquires a random initial phase θ_i in its flat potential, while later on, when $T < \Lambda_{QCD}$ and the potential gets "tilted", it starts to oscillate coherently around the minimum, behaving like pressureless non–relativistic DM with density $\Omega_a h^2 \simeq k_a (m_a/10^{-5}\text{eV})^{-1.175}\theta_i^2$ with $0.3 < k_a$ a few. If one assumes no inflation after the PQ phase transition, θ_i is randomly distributed in the Universe, so that making the average, $<\theta_i^2>=\pi^2/3$, the relic abundance falls in the observed range for $m_a \gtrsim 10^{-5}$ eV. Axions can be observed through their conversion to photons when crossing a magnetic field. The present experimental situation is shown in Fig.2, where axion models span the yellow (oblique) band.

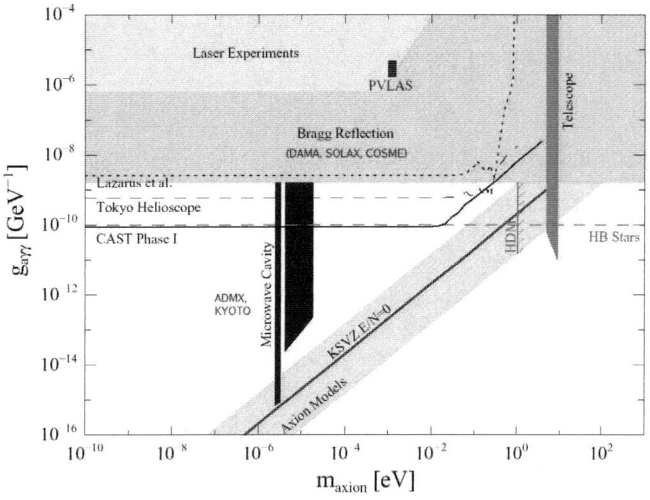

Figure 2. Constraints on the axion–photon coupling as a function of the axion mass.

Note however that the axion–photon coupling is affected by uncertainties on the light quark masses so that it might be suppressed compared to what usually expected[9]. Axions are also produced today in the interior of the Sun and may be measured by making use of resonant Bragg reflection in the same crystal detectors used for direct DM searches[8]. However their sensitivity scales as the exposition to the power 1/8, so that present limits have already saturated their maximal reach and cannot be significantly improved.

4. Weakly Interacting Massive Particles (WIMPs)

The appeal of WIMPs as DM candidates lies in the fact that their thermal cosmological density, given by: $\Omega_{WIMP} \simeq 10^{-38}\text{cm}^3\text{sec}^{-1}/ < \sigma_{ann}v >_{int}$, falls naturally in the correct range if $< \sigma v >_{int}$ is of the order of weak–type interactions. In the last expression $< \sigma v >_{int}$ is the integral over T from the WIMP decoupling temperature T_f to the present one of the thermal average of the WIMP annihilation cross section σ_{ann} to standard particles. This simple picture is modified when the WIMP is close in mass to other particles that may either co–annihilate faster, depleting its density, or slower, enhancing the density by acting as a WIMP reservoir at their decay. Moreover, the cross section may be strongly enhanced (and the relic density suppressed) for particular values of the WIMP mass for which resonant annihilation takes place. Examples of WIMPs are the Heavy Photon in Little Higgs theories, the KK–photon in Universal Extra Dimensions and the neutralino in Supersymmetry.

In the following $\sigma_{\text{scalar}}^{(\text{nucleon})}$ will indicate the coherent WIMP–nucleon cross section.

4.1. Little Higgs (LH)

In LH models [11] the Higgs particle is naturally light because it is the pseudo Goldstone boson from the (collective) symmetry breaking of an effective non–linear σ model. New heavy particles (gauge bosons, singlet and triplet scalars, neutrinos, quarks) are present in the model, and T–parity conservation (where SM particles are even and new ones mostly odd) is introduced in order to forbid dangerous tree level couplings that would put severe constraints from precision Electro–Weak (EW) tests[2]. The Lighest T–odd Particle (LTP) is typically a Heavy Photon B_H,

[2] Note however that it has been pointed out that T–parity is generally broken by anomalies[12]

which is stable and can be a dark matter candidate. Having weak–type interactions the relic abundance of B_H falls naturally in the correct range. On the other hand, its direct detection cross section is expected to be smaller than the present experimental sensitivities (barring some resonant effect), $\sigma^{(\text{nucleon})}_{\text{scalar}} \lesssim 10^{-9}$ pbarn.

4.2. Universal Extra Dimensions (UED)

In the simplest UED model[13] all SM fields propagate in an additional 5th dimension which is compactified on an orbifold S_1/Z_2. The dispersion relation in the 5th dimension, $E^2 = \vec{p}^2 + (p_5^2 + M^2)$ implies for each SM particle an infinite tower of massive states (KK tower) in the effective 4–dimensional theory, with $p_5 = n/R$ $(n = 1, 2, 3, ...)$ while, from electroweak tests constraints, $R^{-1} \gtrsim 300$ GeV. Orbifold compactification, required in order to get rid of unwanted degrees of freedom in the ground (n=0) states, breaks translational invariance in the 5th dimension, but leaves unbroken the residual invariance under discrete πR translations (KK–parity$\equiv (-1)^n$). Typically, the Lighest KK particle is the KK photon $B^{(1)}$ which is stable and a DM candidate. Because of the peculiar evenly–spaced spectrum of KK particles, many modes in UED have similar masses (implying co–annihilations) or integer mass ratios (implying resonant annihilations). In particular, depending on whether the co–annihilating next-to-lighest KK particles are strongly interacting (KK quarks and/or gluons) or not (KK leptons) the relic abundance may be either suppressed or enhanced, so that the range of R^{-1} which corresponds to a correct DM abundance is rather large: 500 GeV $\lesssim R^{-1} \lesssim$ 2 TeV. Direct detection turns out to be below present sensitivities, $\sigma^{(\text{nucleon})}_{\text{scalar}} \lesssim 10^{-9}$ pbarn, and depends on the degree of degeneracy between $B^{(1)}$ and the KK quarks in the propagator.

4.3. Supersymmetry (susy)

Susy is widely considered to be the most natural extension of the SM. In susy every SM particle belongs to a supermultiplet containing partners with opposite statistics (the susy partners of gauge and Higgs bosons being fermionic gauginos and Higgsinos, while those of fermions being scalar fermions or sfermions) in such a way that the total number of fermionic and bosonic degrees of freedom is the same. In general the theory has Yukawa couplings involving squarks and sleptons that violate the baryon and lepton numbers at the tree level and that imply fast proton decay. R–parity is the symmetry that the theory acquires when all these couplings are removed from the superpotential: all SM particles are parity–even, while susy partners are parity–odd, so if R–parity is conserved the Lighest Supersymmetric Particle (LSP) is stable and can be a DM candidate. Moreover, since we know from experiment that particles within supermultiplets are not degenerate in mass, susy must be broken. Unfortunately the mechanism of susy-breaking is model–dependent, so the theory has many free parameters: soft and gaugino masses, the Higgs–mixing parameter μ and the ratio of the two vacuum expectation values $\tan\beta$ (in susy at least two Higgs doublets are required, and the particle content, consisting in 5 states, contains 2 neutral scalars h and H, one neutral pseudoscalar m_A and 2 charged scalars $H^{\pm}$) along with a huge number of flavour–mixing parameters and phases that are usually neglected. In a large region of the susy parameter space the LSP is the neutralino, defined as the superposition of neutral gauginos and Higgsinos: $\chi \equiv a_1\tilde{B} + a_2\tilde{W}_3 + a_3\tilde{H}_1 + a_4\tilde{H}_2$, which clearly has all the properties of an ideal DM candidate. In the following I will discuss the neutralino DM phenomenology in three possible scenarios: SuperGRAvity inspired (SUGRA), Next-to-Minimal SuSy Model (NMSSM) and effective MSSM. Non–Universal SUGRA (nuSUGRA)[14], sub-GUT [15], Mirage mediation [16] and CP violating models [17] are other popular scenarios that can be found in the literature and that I will not cover here.

The SUGRA scenario[18] has just 5 free parameters, all defined at the Grand Unification (GUT) scale: one common soft mass m_0 for scalars, one common soft mass $m_{1/2}$ for gauginos,

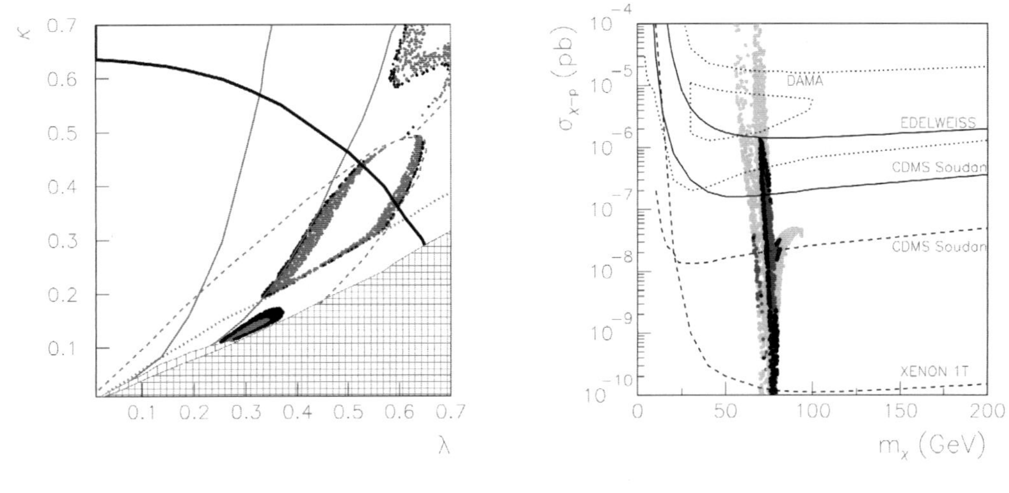

Figure 3. Neutralino phenomenology in the NMSSM scenario. **Left:** Contour plot of the neutralino relic abundance as a function of the superpotential parameters λ and k. **Right:** Neutralino–nucleon cross section as a function of the neutralino mass. Both figures are taken from [20].

$\tan\beta$, the trilinear coupling A_0 and the sign of μ. They are evolved down to the EW scale by making use of renormalization group equations. In this scenario EW symmetry breaking is achieved radiatively, since the large top Yukawa coupling drives one Higgs mass parameter negative in the running from the GUT to the EW scale. This typically implies a large μ parameter with $\chi \simeq \tilde{B}$ and a large value for m_A unless $\tan\beta \simeq 50$. When present experimental limits are taken into account only particular corners of the SUGRA parameter space turn out to be compatible to observation, because the neutralino relic abundance is typically too high. The cosmologically allowed regions correspond to the so–called stau co-annihilation, Higgs funnel and focus point scenarios [18]. The corresponding direct detection cross section is marginally at the level of present sensitivities, $\sigma_{\rm scalar}^{\rm (nucleon)} \lesssim 10^{-7}$ pbarn [19].

The superpotential of the NMSSM [20] contains one additional Higgs singlet S compared to the SUGRA or the MSSM scenarios: $W = W_{MSSM} - \epsilon_{ij}\lambda S H_1^i H_2^j + 1/3 k S^3$. Its vev generates dynamically the μ parameter at the EW scale, explaining why the natural scale of μ, which is the only dimensional parameter in the superpotential, is expected on general grounds to be at the GUT or Planck scale, while it is required to be at the EW scale by phenomenology (the so–called "μ problem" of the MSSM). The main two features of the NSSMS are that one Higgs scalar can be very light because it evades experimental limits by mixing with its singlet component, and that the neutralino has an additional singlino component, $\chi \equiv a_1\tilde{B} + a_2\tilde{W}_3 + a_3\tilde{H}_1 + a_4\tilde{H}_2 + a_5\tilde{S}$. An example of relic neutralino phenomenology in the NMSSM is shown in Fig.3, where on the left-hand side the cosmologically allowed regions are shown in gray in the λ and k plane. In this scenario the presence of additional light Higgs bosons implies more decay channels and resonant decays for the neutralino, which tends to be relatively light and mostly singlino. Thanks to the light Higgs boson exchanged in the propagator the direct detection cross section can be strongly enhanced, as can be seen from the right–hand panel in Fig.3.

In the effective MSSM all soft parameters are defined at the EW scale. Usually in this scenario gaugino mass parameters are assumed to be linked by the relation $M_1 \simeq M_2/2$, which originates from embedding the model in a GUT theory and by assuming that M_1 and M_2 unify at the

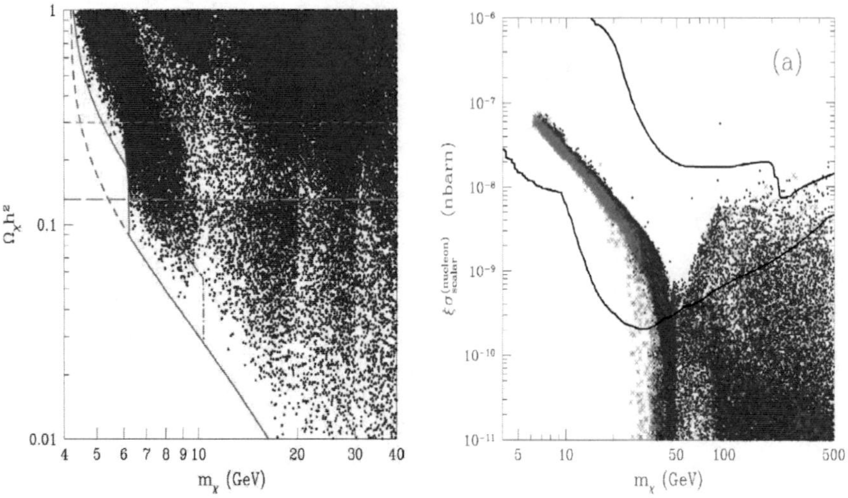

Figure 4. Neutralino phenomenology in an effective MSSM scenario with $M_1 << M_2$. **Left:** Neutralino relic abundance as a function of the neutralino mass[21]. **Right:** Neutralino–nucleon cross section as a function of the neutralino mass [22].

GUT scale. This relation implies the bound $m_\chi \gtrsim m_{\chi^\pm}/2$, where the chargino mass $m_{\chi^\pm}$ has been constrained by accelerator searches to be heavier than about 100 GeV, so that $m_\chi \gtrsim 50$ GeV. The latter limit, which is usually quoted for the neutralino mass, can however be evaded if $M_1 << M_2$. In fact LSP searches at accelerators through the production and decay of heavier neutralinos or the measurement of the invisible width of the Z boson are not sensitive enough to put a direct lower bound on m_χ. In this case, when standard assumptions are made about the history of the Early Universe, a lower bound $m_\chi \gtrsim$ 6 GeV can be derived by making use of the relic abundance [21], as shown in the left panel of Fig.4. Moreover, since Ω_χ and $\sigma^{(\mathrm{nucleon})}_{\mathrm{scalar}}$ are strongly anticorrelated, an upper bound on the former implies a lower bound on the latter. This explains why the cross section $\sigma^{(\mathrm{nucleon})}_{\mathrm{scalar}}$ shown in the right–hand panel of Fig. 4 presents a typical low–mass funnel [22]. The theoretical points overlap nicely with the region enclosed by the solid black line, which is compatible to the annual modulation effect observed by the DAMA experiment[23]. This region is appropriately enlarged in order to take into account the astrophysical uncertainties on the WIMP velocity distribution function as well as the WIMP local density that can strongly affect rate predictions [24], especially at low WIMP masses. This is due to the fact that in direct searches lighter WIMPs need to kick a nucleus above the experimental recoil energy threshold with a higher incoming velocity, so that very light WIMPs probe the high–velocity tail of the distribution, which is more sensitive to the details of the halo modeling. Note, however, that exclusion plots calculated by experimental collaborations are usually obtained by making use of a simplified isothermal sphere model for the velocity distribution, so that some care is required in order to assess in a consistent way how they affect low–mass WIMPs [25]. Moreover, recently the modulation effect was re–analyzed by the DAMA Collaboration in light of a possible effect of channeling in the NaI crystals [26]. This effect may lead to a significant enhancement of the detector response at low recoil energies, displacing the DAMA region to lower WIMP masses, and with direct consequences for the phenomenology of the susy configurations that can explain the annual modulation [27].

5. Conclusions

WIMPs at the TeV scale can be realized in different well–motivated scenarios, like the KK–photon in UED, the heavy photon in Little Higgs and the neutralino in Supersymmetry. All these scenarios can provide CDM with the correct abundance. Neutralino is still the most popular, and today is available in different flavours: SUGRA, nuSUGRA, sub–GUT, Mirage mediation, NMSSM, effective MSSM, scenarios with CP violation. Interestingly enough, some of these scenarios are already at the level of present sensitivities for direct DM searches.

References

[1] Spergel D N *et al.* [WMAP Collaboration], 2007 *Astrophys. J. Suppl.* **170** 377 [arXiv:astro-ph/0603449].

[2] Asaka T, *these Proceedings*; Yamanaka M, *ibidem*;

[3] Bergstrom L, *these Proceedings*; Picozza P, *ibidem*

[4] Tytgat M, *these Proceedings*; Cirelli M, Fornengo N and Strumia A, 2006 *Nucl. Phys.* B **753** 178 [arXiv:hep-ph/0512090].

[5] Kainulainen K and Olive K A, 2002 *Neutrino Mass* (Springer Tracts in Modern Physics), ed G Altarelli and K Winter, *Preprint* arXiv:hep-ph/0206163.

[6] For a Review, see: Kim J E, 1987 *Phys. Rept.* **150** 1.

[7] For a recent review see: Battesti R *et al.*, *Preprint* arXiv:0705.0615 [hep-ex].

[8] Bernabei R *et al.* 2001 *Phys. Lett.* B **515** 6; Cebrian s *et al.*, 1999 *Astropart. Phys.* **10** 397 [arXiv:astro-ph/9811359].

[9] Buckley M R and Murayama H *Preprint* 0705.0542 [arXiv:hep-ph].

[10] Belli P, Cerulli R, Fornengo N and Scopel S, 2002 *Phys. Rev. D* **66** 043503 [arXiv:hep-ph/0203242].

[11] Birkedal A, Noble A, Perelstein M and Spray A, 2006 *Phys. Rev.* D **74** 035002 [arXiv:hep-ph/0603077]; Hubisz J and Meade P, 2005 *Phys. Rev.* D **71** 035016 [arXiv:hep-ph/0411264].

[12] Hill C T and Hill R J, *Preprint* arXiv:hep-ph/0705.0697.

[13] Appelquist T, Cheng H C and Dobrescu B A, 2001 *Phys. Rev.* D **64** 035002 [arXiv:hep-ph/0012100]; Servant G and Tait T M P, 2002 *New J. Phys.* **4**, 99 [arXiv:hep-ph/0209262]; 2003 *Nucl. Phys.* B **650**, 391 [arXiv:hep-ph/0206071]; Kakizaki M, Matsumoto S, Sato Y and Senami M, 2005 *Phys. Rev.* D **71** 123522 [arXiv:hep-ph/0502059]; Kong K and Matchev K T, 2006 *JHEP* **0601** 038 [arXiv:hep-ph/0509119].

[14] Berezinsky V, Bottino A, Ellis J R, Fornengo N , Mignola G and Scopel S, 1996 *Astropart. Phys.* **5** 1 [arXiv:hep-ph/9508249]; Ellis J R, Olive K A, Santoso Y and Spanos V C, 2006 *JHEP* **0605** 063 [arXiv:hep-ph/0603136].

[15] Ellis J R, Olive K A and Sandick P, 2007 *JHEP* **0706** 079 [arXiv:hep-ph/0704.3446].

[16] Choi K, Lee K Y, Shimizu Y, Kim Y G and Okumura K I, 2006 *JCAP* **0612** 017 [arXiv:hep-ph/0609132].

[17] Accomando E*et al.*, *Preprint* arXiv:hep-ph/0608079; Lee J S and Scopel S, 2007 *Phys. Rev.* D **75** 075001 [arXiv:hep-ph/0701221].

[18] Ellis J R, Olive K A, Santoso Y and Spanos V C, 2003 *Phys. Lett.* B **565** 176 [arXiv:hep-ph/0303043]; Battaglia M, De Roeck A, Ellis J R, Gianotti F, Olive K A and Pape L, 2004 *Eur. Phys. J.* C **33** 273 [arXiv:hep-ph/0306219]; Arnowitt R, Dutta B and Hu B, *Preprint* arXiv:hep-ph/0310103; Feng, J L, Matchev K T and Moroi T, 2000 *Phys. Rev.* D **61**, 075005 [arXiv:hep-ph/9909334].

[19] Ellis J R, Olive K A, Santoso Y and Spanos V C, 2005 *Phys. Rev.* D **71** 095007 [arXiv:hep-ph/0502001].

[20] Cerdeno D G, Gabrielli E, Lopez-Fogliani D E, Munoz C and Teixeira A M, 2007 *JCAP* **0706**, 008 [arXiv:hep-ph/0701271]; Cerdeno D G, Hugonie C, Lopez-Fogliani D E, Munoz C and Teixeira A M, 2004 *JHEP* **0412** 048 [arXiv:hep-ph/0408102].

[21] Bottino A, Fornengo N and Scopel S, 2003 *Phys. Rev.* D **67** 063519 [arXiv:hep-ph/0212379]; Bottino A, Donato F, Fornengo N and Scopel S, 2003 *Phys. Rev.* D **68** 043506 [arXiv:hep-ph/0304080].

[22] Bottino A, Donato F, Fornengo N and Scopel S, 2004 *Phys. Rev.* D **69** 037302 [arXiv:hep-ph/0307303]; 2005 *Phys. Rev.* D **72** 083521 [arXiv:hep-ph/0508270].

[23] Bernabei R *et al.* (DAMA Collaboration), 2003 *Riv. Nuovo Cim.* **26N1** 1 [arXiv:astro-ph/0307403]; 2004 *Int. J. Mod. Phys.* D **13** 2127 [arXiv:astro-ph/0501412]; 2006 *Int. J. Mod. Phys.* A **21** 1445 [arXiv:astro-ph/0511262]; 2006 *Eur. Phys. J.* C **47** 263 [arXiv:astro-ph/0604303]; 2007 *Int. J. Mod. Phys. A* **22** 3155 [arXiv:astro-ph/0706.1421].

[24] Belli P, Cerulli R, Fornengo N and Scopel S, 2002 *Phys. Rev.* D **66** 043503 [arXiv:hep-ph/0203242].

[25] Bottino A, Donato F, Fornengo N and Scopel S, 2005 *Phys. Rev.* D **72** 083521 [arXiv:hep-ph/0508270].

[26] Bernabei R *et al.* (DAMA Collaboration), 2007 *Preprint* arXiv:astro-ph/0710.0288.

[27] Bottino A, Donato F, Fornengo N and Scopel S, 2007 *Preprint* arXiv:hep-ph/0710.0553.

Observation of Antimatter in Our Galaxy

Piergiorgio Picozza and Aldo Morselli

INFN and University of Roma Tor Vergata, Rome, Italy

E-mail: `piergiorgio.picozza@roma2.infn.it, aldo.morselli@roma2.infn.it`

Abstract.

Two of the most compelling issues facing astrophysics and cosmology today are to understand the nature of the dark matter that pervades the universe and to understand the apparent absence of cosmological antimatter. For both issues, sensitive measurements of cosmic-ray antiprotons and positrons, in a wide energy range, are crucial.

Many different mechanisms can contribute to antiprotons and positrons production, ranging from conventional reactions like $p + p \rightarrow p + \overline{p}$+*anything* up to exotic processes like neutralino annihilation. The open problems are so fundamental that experiments in this field are of the greatest interest.

Here we will summarize the present situation of the space experiments already flying or ready to fly or in preparation.

1. Introduction

The idea of exploiting cosmic antiprotons positrons and gammas measurements to probe unconventional particle physics and astrophysics scenarios has a long history ([1] and moved the cosmologists for several decades. After the discovery of the CP violation in the weak interactions, Sakharov in '67 formulated the three well known hypotheses that were assumed to be a reasonable starting point to explain the apparent contradiction between the fundamental laws of the nature and the observations.

The search for evidence of antimatter domains in the Universe can be undertaken with two main methods, an indirect one, looking for the spectrum of the Cosmic Diffuse Gamma, and a direct one by searching for heavy Antinuclei and by measuring $\overline{p}$ and e^+ energy spectra.

Concerning the gamma observation in the 100 MeV range, many physical processes in cosmic space can produce γ rays : in particular from interactions of cosmic protons and nuclei with the interstellar medium via π^0 decay. If we assume gamma produced also in proton-antiproton annihilation via π^0 decay, we can obtain an upper limit on the possible amount of annihilating antimatter and on the minimum separation among matter and antimatter domains. From the experimental data on the cosmic diffuse gamma spectrum [2] we obtain that the antimatter/matter fraction is $\leq 10^{-15}$ in the Galactic molecular clouds, $\leq 10^{-10}$ in the Galactic Halo, and $\leq 10^{-5}$ at the level of a cluster; so that we can conclude that if antimatter exists, it is separated from matter at least at the level of 50 Megaparsec.

A wide program of direct antimatter research (see Table 1) has been triggered by the results obtained in the 70's by the teams of R. Golden in USA [3] and of E. Bogomolov in Russia [4] (see figure 1) that identified the first antiprotons in cosmic rays. In fact the antiproton spectrum largely exceeded the expected antiprotons flux produced in the interactions of CR's with the interstellar matter [5]. Many exotic models ranging from antiprotons coming from antigalaxies [6] (blu line) to annihilation of supersymmetric dark matter (red curve) [7], was then promptly formulated to account for the excess.

Even the positron ratio measurements performed in the same years gave a too high flux of positron at energies greater than 10 GeV, explained by some exotic production, like again the annihilation of WIMPs.

1979	First Observation (Golden et al.)
1979	Russian PM (Bogomolov et al.)
1981	Excess reported (Buffington et al)
1985	ASTROMAG Study Started
1987	LEAP, PBAR (upper limits)
1991	MASS
1992	IMAX
1993	TS93, BESS
1994	CAPRICE94, HEAT93
1997	BESS
1998	CAPRICE98, AMS-01
1999	BESS
2000	HEAT-pbar, BESS
2004	BESS Polar I

Table 1. Antimatter in Cosmic Rays

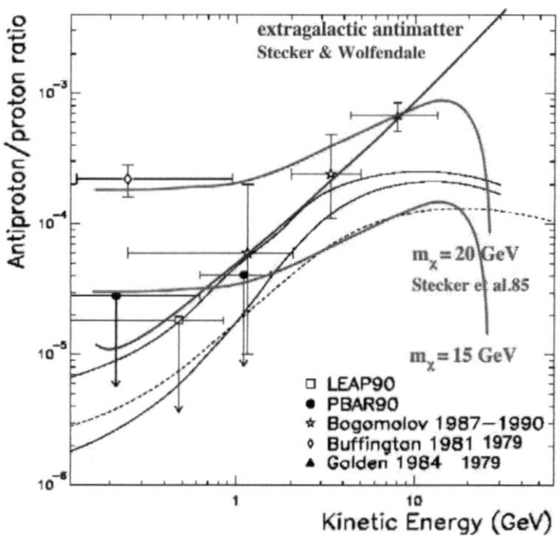

Figure 1. Antiproton/proton ratio: experimental situation before 1990.

In the last 20 years several balloon borne experiments have been performed using novel techniques developed for accelerator physics, mainly by WiZard (MASS89-91, TS93, CAPRICE94-98), HEAT and BESS Collaborations (see Table 1). In 1998 AMS01 flew onboard the Space Shuttle. The core part of all these instruments was a magnetic spectrometer associated with some detectors for hadronic and electromagnetic separation (for a review see [8]) .

In figure 2 there are the present experimental limits for the antihelium/helium ratio with a best value of the order of $3 \cdot 10^{-7}$ obtained combining all the BESS flights data.

In figure 3 are shown the experimental available results for the antiproton flux with different theoretical calculations which account for a pure secondary component, [9, 10, 11] and for a possible contribution from neutralino annihilation (dotted line, from [11]).

In figure 4 the experimental data for the $e^+/(e^+ + e^-)$ fraction are reported, with calculations of a pure secondary component (black solid line), a possible contribution from neutralino annihilation (dotted line, from[12]) and the total flux (grey solid line).

To disentangle these possible contributions of exotic components from the standard production, several space missions has been conceived for measurements at higher energies, with high statistics and during different solar modulation phases. These missions are shown in figure 5.

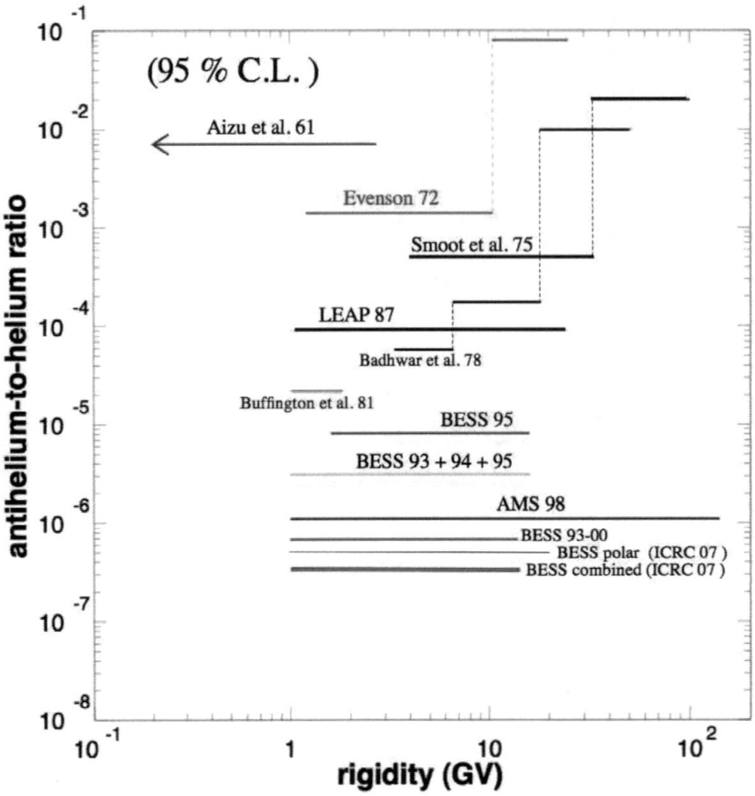

Figure 2. Present experimental limits for the antihelium/helium ratio

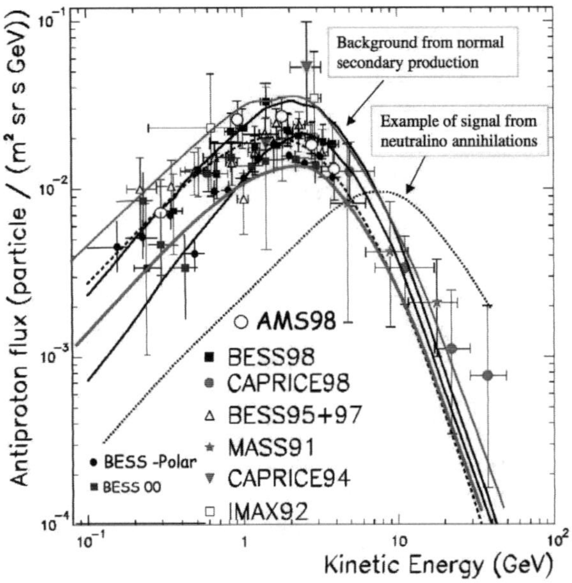

Figure 3. Antiproton flux: experimental situation and theoretical predictions

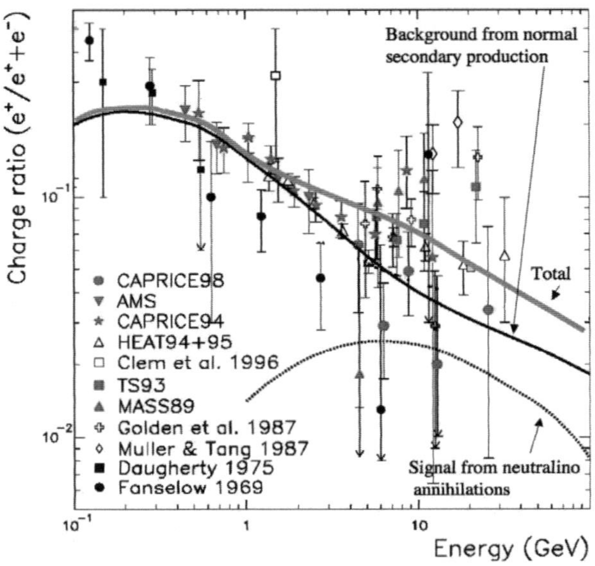

Figure 4. Positron fraction experimental situation and theoretical predictions

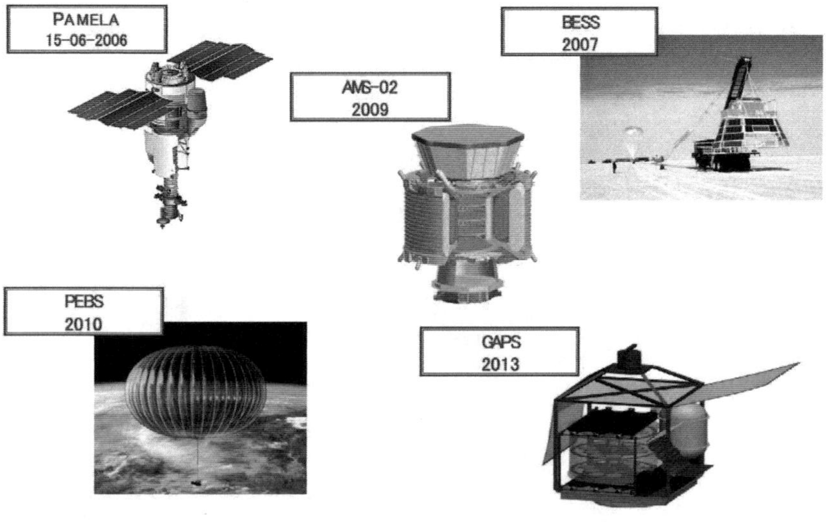

Figure 5. Antimatter and Dark Matter Space Missions

2. The PAMELA and AMS-02 Space Missions

The PAMELA and AMS-02 scientific primary goal is the search for heavy anti-nuclei and non baryonic particles outside the Standard Model, for the understanding of the formation and evolution of our Galaxy and the Universe and for the exploring of the cycles of matter and energy in the Universe. Additional objectives are the studies of galactic cosmic rays in the heliosphere, Solar flares, solar modulation, stationary and disturbed fluxes of high energy particles in the Earth's magnetosphere, anomalous components of cosmic rays. PAMELA (A Payload for Antimatter Matter Exploration and Light-nuclei Astrophysics) [13] was launched into space on board of the Russian satellite DK1 by a Soyuz-U rocket from the Baikonur cosmodrome in Kazakhstan on June 15th 2006. The satellite orbit is elliptical and semi-polar, with an altitude varying between 350 km and 610 km, at an inclination of 70°.

The PAMELA collaboration is composed by six Italian Universities and INFN structures, three Russian research institutions and German and Swedish Universities. The instrument is constituted by a permanent magnet with a microstrip silicon tracker, a plastic scintillators time-of-flight system, a silicon-tungsten electromagnetic imaging calorimeter, an anticoincidence system, a shower tail catcher scintillator and a neutron detector. The combination of these devices allows antiparticles to be reliably identified from a large background of other charged particles. The weight is 470 Kg, the height 1.3 m and both the lateral dimensions 0.7 m. A scheme of the apparatus with flight events displayed is shown in figure 6.

PAMELA is in a continuous data-taking mode since July 2006. All systems are working as expected and scientific data analysis is now on-going.

More precisely, PAMELA is measuring with high statistics:

- Positron flux from 50 MeV to 270 GeV (present limits 0.7 - 30 GeV)
- Antiproton flux from 80 MeV to 190 GeV (present limits 0.4 - 50 GeV)
- Limit on antinuclei $\sim 10^{-8}(\overline{He}/He)$ (present limit about 10^{-6})
- Electron flux from 50 MeV to 2TeV
- Proton flux from 80 MeV to 700 GeV
- Light nuclei flux (up to oxygen) from 100 MeV/n to 200 GeV/n

In addition, it is assuring a continuous monitoring of the cosmic rays solar modulation.

A solar impulsive event proton spectrum is shown in figure 7 compared with the proton spectrum in a solar quite period.

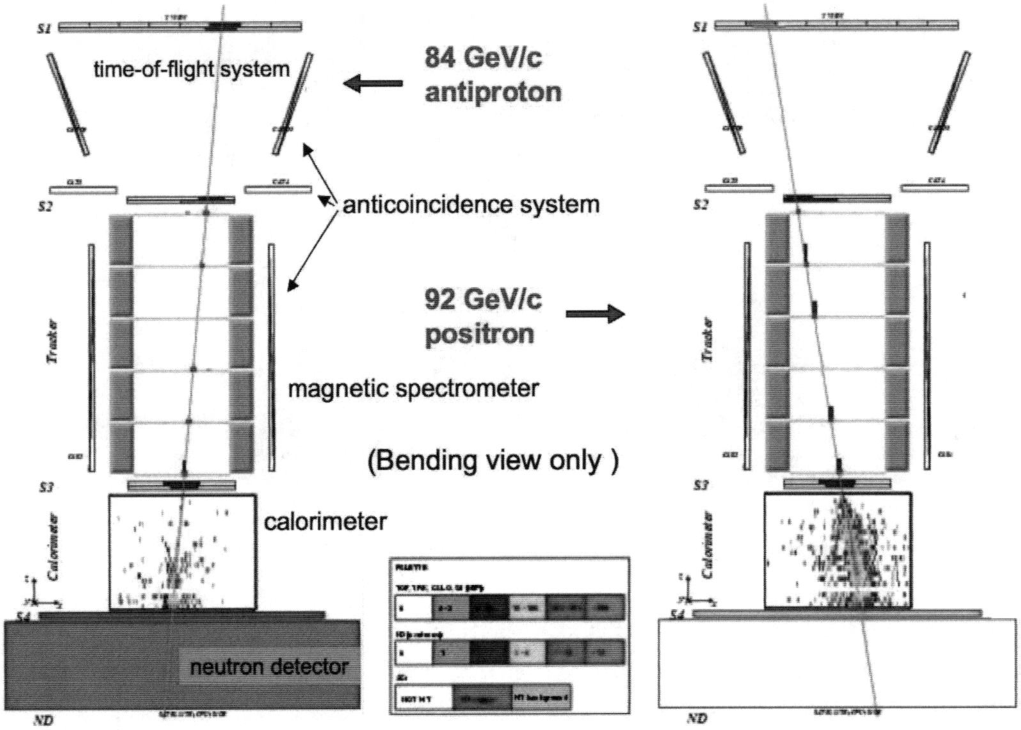

Figure 6. PAMELA Flight events. Bending views only are shown.

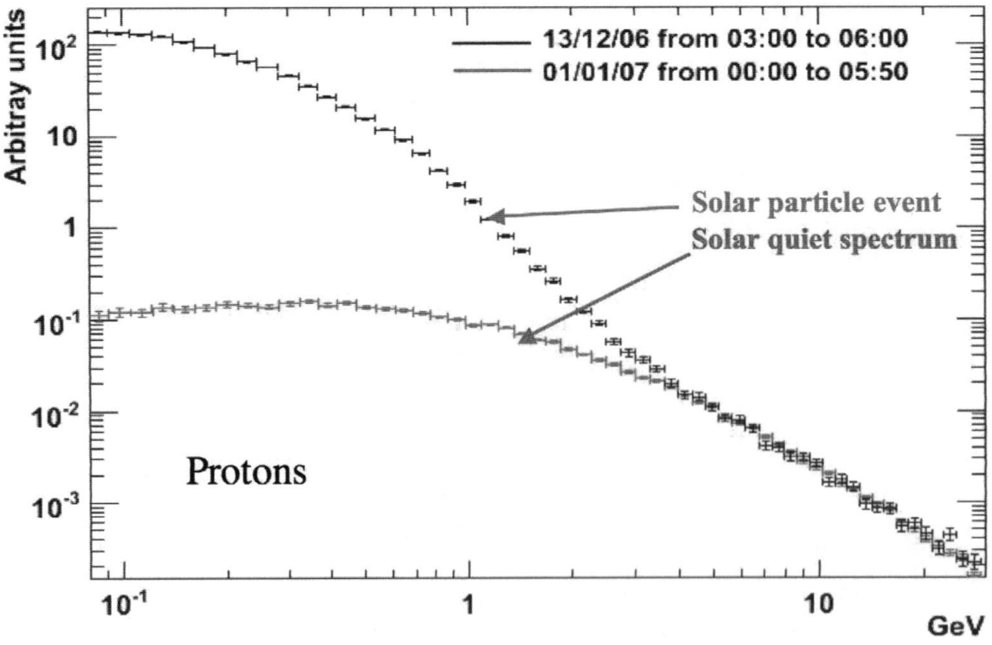

Figure 7. 13 December 2006 - Solar impulsive event.

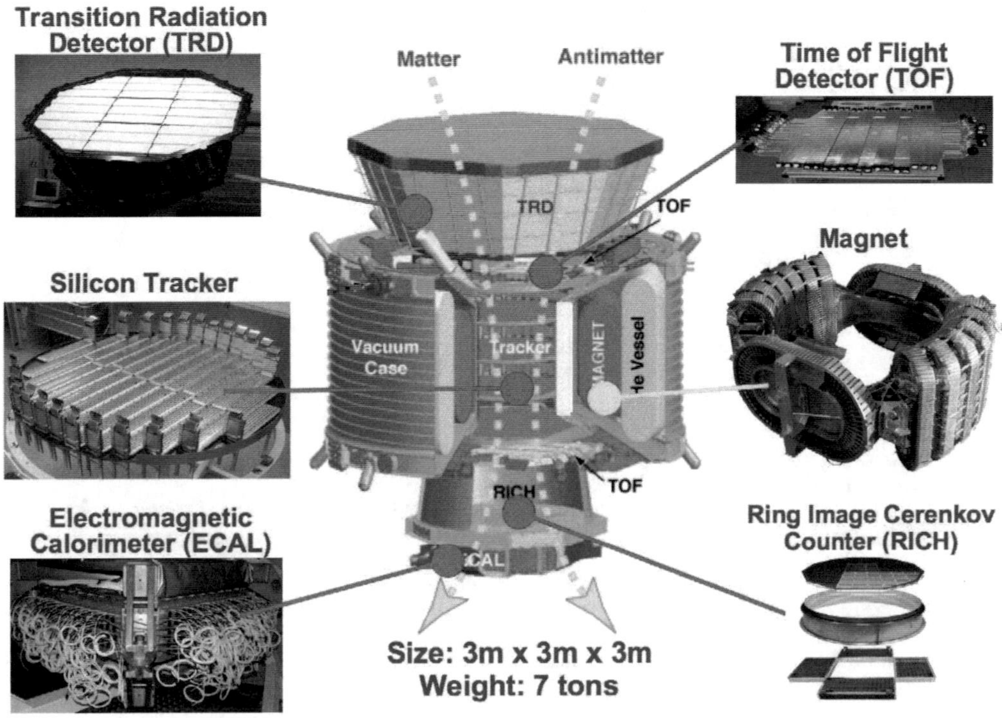

Figure 8. AMS-02 Detector

AMS-02 is a world wide collaboration. The detector in its assembling phase at CERN is shown in figure 8.

It is composed by a superconducting magnet with a microstrip silicon tracker, a strawtube TDR, a RICH, a scintillating fibers electromagnetic imaging calorimeter and a plastic scintillators time-of-flight system. The size is 3 m^3 and the weight is about 7 tons. It is foreseen to be launched in 2009. The large geometric factor, the high magnetic field and the number of different detectors will allow to enlarge the energy range explored by PAMELA, to improve the statistics and to do measurements of antideuterons and gammas.

3. Polar Balloon flights

New opportunities are given to the cosmic ray research at lower cost than space missions by long duration polar flights.

Most of the Antartic flights have flown around South Pole from $\sim$ 8 days up to 42 days.

In December 2007 BESS, an American-Japanese collaboration, will perform the second polar flight with a foreseen duration of $\sim$ 20 days. The Polar BESS long duration experiment features a instrument on the top of a payload with big omni directional solar panels in order to have enough power during the scheduled flight. It is composed by a superconducting solenoid with inside drift chambers as a tracker system. The time-of-flight is composed by plastic scintillators placed outside the magnet. An aereogel counter decreases the contamination of the electron particles.

It will have a good sensitivity to the low energy part of the CR spectrum and will perform a precise measurement of low energy antiprotons, antideuterons and antihelium.

An interesting possibility will be the contemporaneous measurements of BESS and PAMELA

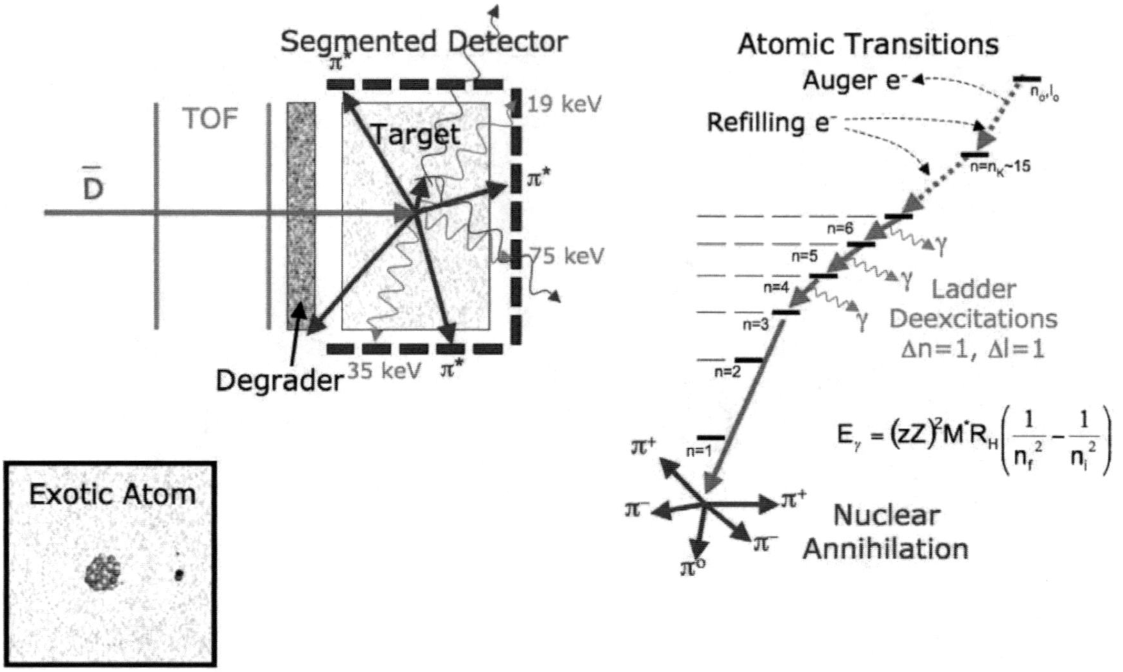

Figure 9. General Antiparticle Spectrometer (GAPS) general scheme.

instruments in the South Pole region.

The best measurements of antideuterons will be performed by the GAPS experiment [14] that will flight as a prototype in 2009 and in its final configuration in 2013. The method is shown in figure 9. The velocity of an antideuteron is measured by a TOF system and a degrader reduces the velocity by energy loss. In the active target the antideuteriun nearly at rest forms with a nucleus an exotic atom in an excited state. A precise measurement of the atomic transitions and of the nuclear annihilation products (mainly pions) allows for achieving comparable sensitivity to third-generation underground dark matter experiments.

References

[1] G.Basini, A.Morselli, M.Ricci, La Rivista del Nuovo Cimento, vol.12, n.4, (1989)

[2] P. Sreekumar, et al., *Astrophys. Journal* 494 (1998) 523

[3] R. Golden et al., *Phys. Rev. Lett.* **43** (1979) 1196

[4] E. Bogomolov et al., *XX Int. Cosmic Ray Conf.*, 2, 72 (1987), Moscow
— *XXI Int. Cosmic Ray Conf.*, 3, 288 (1990), Adelaide

[5] T.K. Gaisser and B.G. Mauger, *Phys.Lett.* **30** (1973) 1264
T.K. Gaisser and R.K. Schaefer, *Astrophys. Journal* **394** (1992) 174
M. Simon and U. Heinbach, *Astrophys. Journal* **456** (1996) 519
T. Mitsui et al., *Phys. Lett.* **B389** (1996) 169

[6] F.W. Stecker and A.W. Wolfendale, *XIX Int. Cosmic Ray Conf.* OG 6.1-8, 354 (1985) La Jolla

[7] F.W. Stecker, S. Rudaz and T.F. Walsh, *Phys.Rev.Lett.* **35** (1985) 1

[8] P.Picozza and A.Morselli, Journal of Physics G: Nucl.Part.Phys.29 (2003) 903

[9] M. Simon, A. Molnar and S. Roesler, *Astrophys. Journal* **499** (1998) 250

[10] L. Bergström, J. Edsjö and P. Ullio, *Astrophys. Journal* **526** (1999) 215

[11] A.Lionetto, A.Morselli, V.Zdravkovic , J. Cosmol. Astropart. Phys. JCAP09(2005)010 [astro-ph/0502406]

[12] E.Baltz and J. Edsjö, Phys. Rev. **D 59** (1999) 023511.

[13] P. Picozza et al., Astroparticle Physics 27 (2007) 296

[14] K.Mori et al., *Astrophys. Journal* 566 (202) 604

Antimatter and Gamma-rays from Dark Matter Annihilation

Lars Bergström

Department of Physics, Stockholm University
AlbaNova University Centre, SE-10691 Stockholm, Sweden

E-mail: `lbe@physto.se`

Abstract. A brief review of the indirect detection signatures of dark matter is given. In particular, detection methods of dark matter particle annihilation to antimatter and γ-rays are reviewed. With the GLAST satellite soon to be launched, a crucial window in the energy range of a few GeV up to 300 GeV will open. The good angular and energy resolution of the instrument means that structures predicted by cold dark matter models can be searched for. Large, currently planned ground-based imaging Cherenkov telescope arrays, may further improve the limits, or discover a signal, if the current understanding of halo dark matter structure is correct.

1. Introduction

The dark matter problem is one of the most obvious unsolved problems of present-day cosmology. A wealth of data support the existence of non-baryonic dark matter, contributing on the order of a quarter of the energy density of the Universe [1]. A particularly interesting class of dark matter models, with basis in particle physics are the so-called WIMPs - weakly interacting massive particles. The idea that something new related to electroweak symmetry breaking could play a role for dark matter is a particularly interesting one, since it gives a relic density of the required order of magnitude, and it may with its electroweak interactions be detectable.

2. Methods of detection

Essentially, there are three methods of detecting WIMPs. First, indications of their existence may be obtained at accelerators, in particular at CERN's upcoming LHC, where in favourable cases also the mass and some couplings to other particles could be measured. However, in most cases it will be difficult to verify that new particles discovered at LHC are indeed the dark matter. Thus, detection by some other methods may be crucial for identification. This problem has recently been studied in some detail for supersymmetric models [2]. Although supersymmetric models are by no means the only contenders for dark matter they are so well motivated by other reasons (such as, e.g., solving the hierarchy problem of particle physics, i.e., to explain why the mass hierarchy $m_W << m_{Pl}$ is stable against radiative corrections), that the lightest supersymmetric particle, most probably the lightest neutralino, has become something of a template for dark matter. One of the prime objectives of LHC is to search for supersymmetry and if found to determine as much as possible about its parameters.

The second way to search for dark matter employs sensitive detectors, typically consisting of sizable amounts of Na I or Ge crystal [3, 4] or liquid noble gas [5]. After many years of steady but slow progress, the direct detection technique now delivers quite impressive data, with experiments now putting limits deep into the supersymmetric parameter plane (for a since several years claimed, but unconfirmed, detection, see [3]). The present generation of detectors, of $50 - 100$ kg of detector material, is presently being enlarged towards the 1000 kg mass scale, which will test WIMPs models, including supersymmetric ones, quite strongly.

Both accelerator searches and direct detection experiments may give indications of the properties of the particle making up the dark matter (given that it is a WIMP, which is the scenario we treat here). However, the third method, indirect detection of the final states produced in dark matter particle annihilation, would in some sense be the most convincing way to determine what makes up dark matter halos, like the one where the Milky Way is embedded. The types of final state particles that may be detected are antimatter (positrons and antiprotons, which are usually a very small part of the ordinary cosmic rays), neutrinos and γ-rays. For supersymmetric dark matter, one of the theoretically favoured templates of a WIMP, there exist extensive computer packages like DarkSUSY [6] or Micromegas [7] which allows to compute expected detection rates for most of the processes of direct and indirect detection discussed here.

3. Neutrinos

Fluxes of neutrinos that are in principle measurable can be produced by annihilations of WIMPs trapped gravitationally at the center of the Earth or the Sun [8]. The mechanism of trapping, however, involves scattering on nuclei which means that the capture rate and therefore in most cases the annihilation rate from the Earth is limited by direct detection bounds. These rates have to be equal in steady state; an exception may occur if the scattering rate is so small that the distribution has not yet equilibrated. Usually one then has unobservable neutrino rates, so generally it is the elastic scattering cross section, not the annihilation cross section, that determines the annihilation rate [9]. This means that the neutrino rate from the Earth is most likely too small to be detected in presently built neutrino telescopes (IceCUBE and Antares), given the present rapidly improving bounds on the spin-independent scattering cross sections [4, 5], which are dominant given the chemical composition of the Earth.

The Sun, however, contains mostly protons, so there the spin-dependent part of scattering is more important. The spin-dependent scattering cross section is notoriously difficult to bound experimentally, so therefore a possibility for a signal from the interior of the Sun remains. In fact, even if the bounds on spin-dependent scattering would improve significantly, the spin-dependent scattering rate is so poorly correlated with annihilation rate that the possibility will still exist [10]. The rate predicted for minimal supersymmetric models are reasonably high. For Kaluza-Klein dark matter particles in universal extra dimensions the predicited rates are, however, probably too small to be measurable at IceCUBE [11].

4. Antimatter

4.1. Positrons

Positrons usually do not give a very competitive method of discovery, as very large factors, "boost factors" are needed [12] to contribute considerably to the measured positron spectrum [13]. This boost factor is an arbitrary factor used to enhanced the rate of positrons over the flux one would get using standard halo models and propagation properties. It could, for instance, be caused by small-scale clumping of the halo, although most estimates would tend to maximize this to $f_B \sim 2 - 8$ [14], perhaps being possible to push by another order of magnitude at most [15]. The other argument against using positrons as a tools for dark matter detection is that they are easily produced in large quantities by all acceleration processes, such as in supernova winds or black hole accretion. Any spectral feature is likewise mostly erased due to the large and rapid energy losses of positrons propagating through the Galaxy. An exception may be for very massive particles annihilating directly into an electron-positron pair, such as happens in Kaluza-Klein models with universal extra dimensions [16]. Also there large boost factors are needed, however.

In priciple, large boost factors may be obtained if there happens to be a clump of dark matter near the solar system [17]. The probability for this to happen is, however, extremely small for realistic models of the distribution of substructure in the galaxy [18].

4.2. Antiprotons

Antiprotons would give a more robust detection rate, although a rather featureless modification of the spectrum from what is expected by cosmic-ray induced antiprotons [19].

The propagation of antiprotons is similar to that of positrons, a main difference being that the synchrotron and inverse Compton energy losses are not important. Instead losses caused by elastic and inelastic scattering on gas and dust in the galaxy become more important. This can be relatively well modeled as the amount of material encountered during an energy loss time can be estimated, and the propagation parameters can be tuned to give a ratio of secondary to primary nuclei (like B/C) that agrees with the observed value.

Early on it was thought that a unique feature of antiprotons produced in dark matter annihilations would be that they populate all of the available phase space down to very low energies as seemed to be indicated by experimental data, whereas the background produced by antiproton production by ordinary cosmic ray protons hitting hydrogen or helium would for kinematical reasons populate higher energies only. However, it was later shown that several effects, in particular energy loss by elastic scattering of antiprotons on nucleons, tend to populate the low-energy part of the spectrum [20], giving a spectrum in excellent agreement with the measured one.

Antiprotons probe a larger part of the galaxy surrounding the observation point than does positrons (for instance, roughly half of the antiprotons produced near the galactic center make it out to the solar radius). Their diffusion properties are also well known, which means that the agreement between the observed flux and the cosmic-ray induced flux can be used to limit "exotic" contributions from dark matter annihilation in models with large boost factors. One example is the model of de Boer [21] which claims to fit well the γ-ray flux measured by the EGRET experiment, by introducing a substantial contribution from dark matter annihilation. This is done by having a rather unusual distribution of dark matter (disk-concentrated with rings) in the galaxy, and in addition working with large boost factors. It seems that this model taken at face value would produce an antiproton flux which is an order of magnitude too large to agree with data [22].

It is possible, though, that high-mass models of dark matter could give a measurable disturbance on the antiproton distribution at higher energies, something which will be interesting to see PAMELA [23] investigate.

4.3. Antinuclei

It is very unlikely that other antinuclei other than antiprotons are produced by cosmic ray interactions. Therefore, even a handful of antideuterons would point to some exotic source. It was suggested [24] that neutralino annihilations could be such a source of antideuterons, produced by the accidental overlap of the wave functions of an antiproton and an antineutron produced at low relative velocity in the quark fragmentation process. This idea is very interesting, but the uncertainties in the theoretical estimates of the formation rate are hard to estimate (for an attempt, see [25]).

5. Gamma-rays

We finally come to one of the most actively pursued indirect detection methods, that of γ-rays created in WIMP annihilations [26]. It is especially important for the new large imaging air Cherenkov telescopes currently operating (e.g., VERITAS, HESS or CANGAROO [27]) or being planned [28], and for the GLAST (γ-ray large area space telescope) satellite planned to be launched in 2008 [29]. There are several recent reviews in this field ([30, 31]) - here we only point out some of the recent highlights.

The annihilation rate towards a direction making the angle Ψ with respect to the galactic centre is conveniently given by the factorized expression [30]

$$\begin{aligned}\Phi_\gamma(\psi) &\simeq 0.94\cdot 10^{-13}\left(\frac{N_\gamma\, v\sigma}{10^{-29}\ \mathrm{cm^3s^{-1}}}\right)\left(\frac{100\,\mathrm{GeV}}{M_\chi}\right)^2\cdot \\ & \quad J\,(\psi)\ \mathrm{cm}^{-2}\ \mathrm{s}^{-1}\ \mathrm{sr}^{-1} \qquad (1)\end{aligned}$$

where we have defined the dimensionless function

$$J\,(\psi) = \frac{1}{8.5\,\mathrm{kpc}}\cdot\left(\frac{1}{0.3\,\mathrm{GeV/cm^3}}\right)^2\int_{line\ of\ sight}\rho^2(l)\,d\,l(\psi)\,, \qquad (2)$$

with $\rho(l)$ being the dark matter density along the line of sight $l(\Psi)$. (Note the numerical factor in Eq. (1) differs by a factor 1/2 from that given in [30]; this takes into account the fact that the annihilating particles are identical, as is the case for supersymmetric neutralinos. See the footnote in connection to Eq. (21) of the publication [32] for a detailed explanation.)

The particle physics factor $N_\gamma v\sigma$, which is the annihilation rate times the number of photons created per annihilation, can usually be rather accurately computed for a given dark matter candidate. DarkSUSY uses PYTHIA [33] to estimate the number of photons with continuum energy distribution created by quark fragmentation for given standard model final states. Special calculations are needed for higher order QED corrections, which may be very important as we will see.

There have been many improvements of the treatment that was first performed in [30], and many interesting new promising features keep appearing, of which can be mentioned:

- The existence of subhalos within dark matter halos [14]. As N-body simulations have become more refined (see, e.g., [34] for a recent example) it seems clear that the hierarchical formation of smaller structures that merge into larger ones will leave intact a population of subhalos which are not destroyed as far as they can be followed by the simulations. As the smaller subhalos formed at an earlier epoch, when the average density of the universe was larger, this may boost annihilation rates by a rather significant factor [34].
- The possible existence of intermediate mass black holes. The black hole at the galactic centre may cause an enhancement of the dark matter density around it (a "spike") [35], although this is not mandatory, and may depend sensitively on the merger history of the black hole [36]. If there would exist a population of intermediate mass black holes, the boosting of density in their vicinity may be operative however, and may give observational γ-ray annihilation rates to be searched for by GLAST [37]. These objects would appear as point sources without optical counterpart, and may be studied further by air Cherenkov telescopes once their location has been established by GLAST. In fact, a few of the very recently discovered very faint dwarf galaxies may give detectable signals if their dominant mass is indeed composed of WIMPs [38].
- The presence of dark matter streams [39, 40, 41]. The microscopic structure of the dark matter halo may also contain dark matter streams, created by late infall of dark matter

particles on an existing halo (for an early, simplified model, see [39]), or by tidal stripping of dark matter particles from dwarf galaxies [40]. N-body simulations are presently not accurate enough to follow these details, that generally appear below the resolution limit. It can be argued, however, that especially for direct dark matter detection, and indirect detection through neutrinos and γ-rays, which all depend sensitively on the dark matter density at specific positions in the halo, these effects can be very important. Recently, a new interesting method to compute these and similar effects has been proposed [41].

- The possibility of an extragalactic signal [42]. The fact that the combined structure formation in the universe would cause an increase of the extragalactic rate (i.e., integrating along the line of sight back to arbitrary large redshift, and combining with the estimated absorption of γ-rays an optical and IR photons), by a factor of order 10^5 to 10^6, was only recently realized [42, 43, 32, 44]. This is an area where observations by GLAST will be crucial.
- "Explosive" annihilation. This phenomenon, discovered in [45], and verified in [46] gives the possibility of very strong γ-ray signals for particular masses (usually in the TeV region). Of course, TeV particles may have difficulty to give the required relic density, unless one tolerates some fine-tuning, as is explicitly done in "split SUSY" models [47].
- The discovery of models where the 2γ and $Z\gamma$ lines would actually be the dominant features of the γ-ray spectrum [48]. These are "inert" Higgs models with an extra Higgs doublet protected by a Z_2 symmetry. Besides other intersting phenomelogy, these models may also break electroweak symmetry radiatively [49].
- Large contributions at high energy from internal QED bremsstrahlung. This interesting phenomenon, originally proposed in [50], which one has to go beyond leading order to see, has recently been proposed as a method to detect an otherwise undetectable dark matter candidate [51]. It has also been applied to MSSM and mSUGRA models, and found to be very important [52], causing sometimes large boosts to the highest energy end of the γ-ray spectrum. The idea, for fermion final states, is that a Majorana fermion (as many dark matter candidates are) suffers a helicity suppression for S-wave annihilation [53]. However, by emitting a photon from an internal charged leg, which only costs a factor of α_{em}/π, the helicity suppression may be avoided. The effect will be that these radiative corrections, instead of as usual being a percent of the lowest order process, may instead give enhancement *factors* of several thousands to millions times the suppressed lowest order, low-velocity, rate [50]. The resulting spectra will have a characteristic very sharp drop at the endpoint $E_\gamma = m_\chi$ of the γ-ray spectrum, see Fig. 1.

6. Connection between antimatter and γ-ray signal

Recently, W. de Boer and colleagues have argued that the spectral distortion, the so-called "GeV anomaly" of EGRET's measurements of galactic diffuse γ-rays can be explained by supersymmetric WIMP annihilation [21]. As explained earlier, this is done by making a very particular model of the distribution of dark matter, which is largely concentrated to the disk, and in addition has two very massive rings of dark matter in the plane of the disk. It is claimed that by using this dark matter distribution, and adjusting the annihilation rate by a rather large "boost factor", the measured γ-ray angular and energy distributions can be reproduced with a canonical supersymmetric WIMP, of mass below 100 GeV.

It is actually still not known whether the GeV excess of EGRET is real, or just an (unknown) experimental artifact as has been recently claimed by some members of the collaboration [54]. In any case, we mentioned that the proposal of de Boer et al. has the problem of a severe overproduction of antiprotons [55]. Given the already somewhat contrived nature of the dark matter distribution, it is difficult to consider very seriously attempts to change the diffusion

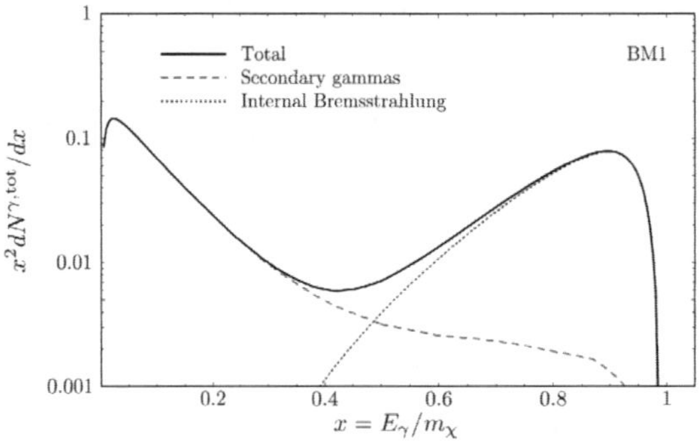

Figure 1. The γ-ray spectrum expected for one of the benchmark models (BM1) defined in [52]. Here $x = E_\gamma/m_\chi$, and the distribution without the bremsstrahlung is shown by the dashed line. There should also be a 2γ and a $Z\gamma$ contribution, not shown here.

properties of antiprotons, as maybe could be a way out [56]. We rather note that the rates for continuum gamma rays and antiprotons are strongly correlated for dark matter candidates that annihilate into standard model particles as they are both given by the fragmentation of quark jets, and are reasonably well understood after a decade of LEP observations. On the other hand, this also means that if one restricts oneself to standard propagation models, and dark matter models which fulfill the constraints from present antiproton measurements (we of course soon anticipate new interesting data from the PAMELA satellite [23] which has been in orbit for well over a year), indications are that the detection of a γ-ray signal has to await GLAST, and possibly the next generation of imaging air Cherenkov telescope arrays.

Acknowledgement
This work was supported by the Swedish Research Council (VR).

References

[1] S. Scopel, these Proceedings.
[2] E. A. Baltz, M. Battaglia, M. E. Peskin and T. Wizansky, Phys. Rev. D **74**, 103521 (2006) [arXiv:hep-ph/0602187].
[3] P. Belli, these Proceedings.
[4] D. S. Akerib *et al.* [CDMS Collaboration], Phys. Rev. Lett. **96**, 011302 (2006) [arXiv:astro-ph/0509259].
[5] J. Angle *et al.* [XENON Collaboration], arXiv:0706.0039 [astro-ph].
[6] P. Gondolo, J. Edsjö, P. Ullio, L. Bergström, M. Schelke and E. A. Baltz, JCAP **0407**, 008 (2004) [arXiv:astro-ph/0406204].
[7] G. Belanger, F. Boudjema, A. Pukhov and A. Semenov, Comput. Phys. Commun. **176**, 367 (2007) [arXiv:hep-ph/0607059].
[8] T. K. Gaisser, G. Steigman and S. Tilav, Phys. Rev. D **34**, 2206 (1986); A. Gould, Astrophys. J. **321**, 571 (1987); L. M. Krauss, M. Srednicki and F. Wilczek, Phys. Rev. D **33**, 2079 (1986).
[9] G. Jungman, M. Kamionkowski and K. Griest, Phys. Rept. **267**, 195 (1996) [arXiv:hep-ph/9506380];
[10] L. Bergström, J. Edsjö and P. Gondolo, arXiv:astro-ph/9906033.
[11] D. Hooper and G. D. Kribs, Phys. Rev. D **67**, 055003 (2003) [arXiv:hep-ph/0208261].
[12] E. A. Baltz, J. Edsjö, K. Freese and P. Gondolo, Phys. Rev. D **65**, 063511 (2002) [arXiv:astro-ph/0109318]; G. L. Kane, L. T. Wang and T. T. Wang, Phys. Lett. B **536**, 263 (2002) [arXiv:hep-ph/0202156].
[13] S. W. Barwick *et al.* [HEAT Collaboration], Astrophys. J. **482**, L191 (1997) [arXiv:astro-ph/9703192].
[14] S. Hofmann, D. J. Schwarz and H. Stoecker, Phys. Rev. D **64**, 083507 (2001) [arXiv:astro-ph/0104173]; V. Berezinsky, V. Dokuchaev and Y. Eroshenko, Phys. Rev. D **73**, 063504 (2006) [arXiv:astro-ph/0511494].
[15] J. Diemand, these Proceedings.
[16] D. Hooper and J. Silk, Phys. Rev. D **71**, 083503 (2005) [arXiv:hep-ph/0409104].

[17] D. T. Cumberbatch and J. Silk, Mon. Not. Roy. Astron. Soc. **374**, 455 (2007) [arXiv:astro-ph/0602320].
[18] J. Lavalle, J. Pochon, P. Salati and R. Taillet, arXiv:astro-ph/0603796.
[19] J. Silk and M. Srednicki, Phys. Rev. Lett. **53**, 624 (1984); F. W. Stecker, S. Rudaz and T. F. Walsh, Phys. Rev. Lett. **55**, 2622 (1985). A. Bottino, F. Donato, N. Fornengo and P. Salati, Phys. Rev. D **58**, 123503 (1998) [arXiv:astro-ph/9804137].
[20] L. Bergström, J. Edsjö and P. Ullio, Astrophys. J. **526**, 215 (1999) [arXiv:astro-ph/9902012]; F. Donato, D. Maurin, P. Salati, A. Barrau, G. Boudoul and R. Taillet, Astrophys. J. **536**, 172 (2001) [arXiv:astro-ph/0103150]; J. W. Bieber, R. A. Burger, R. Engel, T. K. Gaisser, S. Roesler and T. Stanev, Phys. Rev. Lett. **83**, 674 (1999) [arXiv:astro-ph/9903163].
[21] W. de Boer, C. Sander, V. Zhukov, A. V. Gladyshev and D. I. Kazakov, Phys. Rev. Lett. **95**, 209001 (2005) [arXiv:astro-ph/0602325].
[22] L. Bergström, J. Edsjö, M. Gustafsson and P. Salati, JCAP **0605**, 006 (2006) [arXiv:astro-ph/0602632].
[23] M. Casolino *et al.*, arXiv:0708.1808 [astro-ph].
[24] F. Donato, N. Fornengo and P. Salati, Phys. Rev. D **62**, 043003 (2000) [arXiv:hep-ph/9904481].
[25] H. Baer and S. Profumo, JCAP **0512**, 008 (2005) [arXiv:astro-ph/0510722].
[26] D. W. Sciama, Phys. Lett. B **137** (1984) 169; J. Silk and M. Srednicki, Phys. Rev. Lett. **53**, 624 (1984); L. Bergström and H. Snellman, Phys. Rev. D **37**, 3737 (1988); F. W. Stecker and A. J. Tylka, Astrophys. J. **343**, 169 (1989); H. U. Bengtsson, P. Salati and J. Silk, Nucl. Phys. B **346**, 129 (1990).
[27] W. Hofmann, these Proceedings.
[28] V. V. Bugaev, J. H. Buckley and H. Krawczynski, arXiv:0710.0100 [astro-ph].
[29] C. Cecci, these Proceedings.
[30] L. Bergström, P. Ullio and J. H. Buckley, Astropart. Phys. **9**, 137 (1998) [arXiv:astro-ph/9712318].
[31] L. Bergström, Rept. Prog. Phys. **63**, 793 (2000) [arXiv:hep-ph/0002126]; G. Bertone, D. Hooper and J. Silk, Phys. Rept. **405**, 279 (2005) [arXiv:hep-ph/0404175].
[32] P. Ullio, L. Bergström, J. Edsjö and C. G. Lacey, Phys. Rev. D **66**, 123502 (2002) [arXiv:astro-ph/0207125].
[33] T. Sjöstrand, S. Mrenna and P. Skands, JHEP **0605**, 026 (2006) [arXiv:hep-ph/0603175].
[34] J. Diemand, M. Kuhlen and P. Madau, Astrophys. J. **657**, 262 (2007) [arXiv:astro-ph/0611370].
[35] P. Gondolo and J. Silk, Phys. Rev. Lett. **83**, 1719 (1999) [arXiv:astro-ph/9906391].
[36] P. Ullio, H. Zhao and M. Kamionkowski, Phys. Rev. D **64**, 043504 (2001) [arXiv:astro-ph/0101481].
[37] G. Bertone, A. R. Zentner and J. Silk, Phys. Rev. D **72**, 103517 (2005) [arXiv:astro-ph/0509565].
[38] L. E. Strigari, S. M. Koushiappas, J. S. Bullock, M. Kaplinghat, J. D. Simon, M. Geha and B. Willman, arXiv:0709.1510 [astro-ph].
[39] P. Sikivie, Phys. Rev. D **60**, 063501 (1999) [arXiv:astro-ph/9902210].
[40] K. Freese, P. Gondolo and H. J. Newberg, Phys. Rev. D **71**, 043516 (2005) [arXiv:astro-ph/0309279].
[41] M. Vogelsberger, S. D. M. White, A. Helmi and V. Springel, arXiv:0711.1105 [astro-ph].
[42] L. Bergström, J. Edsjö and P. Ullio, Phys. Rev. Lett. **87**, 251301 (2001) [arXiv:astro-ph/0105048].
[43] J. E. Taylor and J. Silk, Mon. Not. Roy. Astron. Soc. **339**, 505 (2003) [arXiv:astro-ph/0207299].
[44] D. Elsaesser and K. Mannheim, Phys. Rev. Lett. **94**, 171302 (2005) [arXiv:astro-ph/0405235].
[45] J. Hisano, S. Matsumoto and M. M. Nojiri, Phys. Rev. Lett. **92**, 031303 (2004) [arXiv:hep-ph/0307216].
[46] F. Boudjema, A. Semenov and D. Temes, Phys. Rev. D **72**, 055024 (2005) [arXiv:hep-ph/0507127]; M. Cirelli, A. Strumia and M. Tamburini, Nucl. Phys. B **787**, 152 (2007) [arXiv:0706.4071 [hep-ph]].
[47] N. Arkani-Hamed, S. Dimopoulos, G. F. Giudice and A. Romanino, Nucl. Phys. B **709**, 3 (2005) [arXiv:hep-ph/0409232].
[48] R. Barbieri, L. J. Hall and V. S. Rychkov, Phys. Rev. D **74**, 015007 (2006) [arXiv:hep-ph/0603188];
N. G. Deshpande and E. Ma, Phys. Rev. D **18**, 2574 (1978);
L. Lopez Honorez, E. Nezri, J. F. Oliver and M. H. G. Tytgat, JCAP **0702**, 028 (2007) [arXiv:hep-ph/0612275]; M. Gustafsson, E. Lundström, L. Bergström and J. Edsjö, Phys. Rev. Lett. **99**, 041301 (2007) [arXiv:astro-ph/0703512].
[49] T. Hambye and M. H. G. Tytgat, arXiv:0707.0633 [hep-ph].
[50] L. Bergström, Phys. Lett. B **225**, 372 (1989).
[51] E. A. Baltz and L. Bergström, Phys. Rev. D **67**, 043516 (2003) [arXiv:hep-ph/0211325].
[52] T. Bringmann, L. Bergström and J. Edsjö, arXiv:0710.3169 [hep-ph].
[53] H. Goldberg, Phys. Rev. Lett. **50**, 1419 (1983).
[54] F. W. Stecker, S. D. Hunter and D. A. Kniffen, pp arXiv:0705.4311 [astro-ph].
[55] L. Bergström, J. Edsjö, M. Gustafsson and P. Salati, JCAP **0605**, 006 (2006) [arXiv:astro-ph/0602632].
[56] W. de Boer, I. Gebauer, M. Weber, C. Sander, V. Zhukov and D. Kazakov, arXiv:0710.5106 [astro-ph].

Right-handed sneutrino as cold dark matter of the universe

Takehiko Asaka

Department of Physics, Niigata University, Niigata 950-2181, Japan

Abstract. We consider the minimal supersymmetric standard model (MSSM) extended by three right-handed neutrinos. Assuming that neutrino masses are purely Dirac-type, the lightest right-handed sneutrino $\tilde{\nu}_R$ can be the lightest superparticle (LSP). We discuss the possibility that the LSP $\tilde{\nu}_R$ becomes dark matter of the universe, paying attention to the production of $\tilde{\nu}_R$ in the early universe. This work is based on collaboration with K. Ishiwata and T. Moroi [1, 2].

1. Introduction

The origin of dark matter (DM) is one of serious problems in particle physics and cosmology. Recently, the relic density of DM has been precisely determined by WMAP observations [3]

$$\Omega_{\rm DM} h^2 = 0.105^{+0.007}_{-0.013}\,, \tag{1}$$

where h is the present Hubble constant in units of 100km/sec/Mpc. Since there is no candidate of DM in the Standard Model (SM), we need new physics beyond the SM. It has been widely discussed that supersymmetric model can provide a viable candidate of DM. The lightest superparticle (LSP) becomes completely stable with R-parity conservation, and can play a role of DM if it is neutral. The well-known candidate of the LSP DM is the lightest neutralino, and others include gravitino and axino.

In [1] we have proposed a possibility that the lightest right-handed sneutrino $\tilde{\nu}_R$ is the LSP DM. (Such a possibility has also been discussed recently in different context. See [4, 5].) This is motivated by very small but non-vanishing neutrino masses strongly suggested by the experimental evidences of neutrino oscillations. The non-zero neutrino masses require physics beyond the SM as well as the MSSM. The simplest way to generate neutrino masses is probably to introduce right-handed neutrinos (as well as sneutrinos in supersymmetric theories).

It has been widely discussed that right-handed (s)neutrinos are introduced together with their super-heavy Majorana masses which are much larger than the electroweak scale $\sim$ 100 GeV. Then the smallness of neutrino masses is naturally explained by the so-called seesaw mechanism [6]. In this case, the right-handed sneutrino cannot be the LSP. On the other hand, if the neutrino masses are purely Dirac-type, the mass of right-handed sneutrino is $\sim$ 100 GeV and can be the LSP. Moreover, as we will explain below, the relic density of the LSP $\tilde{\nu}_R$ can be consistent with (1), and $\tilde{\nu}_R$ is a viable candidate of DM of the universe.

2. Model

We consider the MSSM with three right-handed neutrinos assuming that neutrino masses are purely Dirac-type. The superpotential is then given by

$$W = W_{\rm MSSM} + y_\nu \, \hat{H}_u \, \hat{L} \, \hat{\nu}_R^c \,, \tag{2}$$

where $W_{\rm MSSM}$ is the MSSM superpotential and y_ν denotes the neutrino Yukawa couplings. Here and hereafter, the generation indices are implicit. Further, we introduce tri-linear couplings of $\tilde{\nu}_R$ which will be important in the discussion below: $\mathcal{L} = a_\nu \, y_\nu \, m_{\tilde{L}} \, H_u \, \tilde{L} \, \tilde{\nu}_R^c + h.c.$, where $m_{\tilde{L}}$ is the soft mass for slepton doublet.

In this model, neutrino masses come only from electroweak symmetry breaking as $m_\nu = y_\nu \langle H_u^0 \rangle = y_\nu v \sin\beta$, where $v \simeq 174$ GeV and $\tan\beta = \langle H_u^0 \rangle / \langle H_d^0 \rangle$. Thus, y_ν is determined once m_ν is fixed:

$$y_\nu \sin\beta \simeq 3.0 \times 10^{-13} \times \left(\frac{m_\nu^2}{2.8 \times 10^{-3} \ {\rm eV}^2} \right)^{1/2} \,. \tag{3}$$

We can see that neutrino mass scales in oscillation experiments leads to very small y_ν. One might think that such small couplings are unnatural. However, it is indeed natural in the 't Hooft sense [7], since chiral symmetry of neutrino is restored in the limit of vanishing Yukawa couplings.

In the considering model, right-handed sneutrinos receive masses dominantly from the effects of supersymmetry breaking. (Here we consider gravity-mediation models of supersymmetry breaking.) Then, one should note that the lightest right-handed sneutrino $\tilde{\nu}_R$ can be the LSP. Since $\tilde{\nu}_R$ is neutral, is stable under the R-parity conservation, and also is very weakly interacting, it can be a good candidate of DM provided that its relic density $\Omega_{\tilde{\nu}_R}$ is the right amount (1).

3. Right-handed sneutrino as cold dark matter

Now we discuss the production of $\tilde{\nu}_R$ in the early universe. We should note that $\tilde{\nu}_R$ is not thermalized due to the smallness of y_ν (3). It has been found that $\tilde{\nu}_R$ is effectively produced by decays of superparticles [1, 2], and that there are two distinct contributions: $\Omega_{\tilde{\nu}_R} = \Omega_{\tilde{\nu}_R}^{\rm CE} + \Omega_{\tilde{\nu}_R}^{\rm FO}$ where $\Omega_{\tilde{\nu}_R}^{\rm CE}$ is due to decays of various superparticles in chemical equilibrium, while $\Omega_{\tilde{\nu}_R}^{\rm FO}$ is due to decay of the NLSP after freeze-out.

We first consider $\Omega_{\tilde{\nu}_R}^{\rm CE}$. As shown in [1], it can be estimated by solving the Boltzmann equation for the number density of $\tilde{\nu}_R$. Since the rates of superparticle decays into $\tilde{\nu}_R$ are proportional to y_ν^2, and hence $\Omega_{\tilde{\nu}_R}^{\rm CE} \propto y_\nu^2$. The dominant production occurs when the temperature is comparable to the superparticle mass. There are various processes which produce $\tilde{\nu}_R$. Among them, decay of Higgsino dominates when the effects of the tri-linear scalar couplings are negligible. In this case, we have found that $\Omega_{\tilde{\nu}_R}^{\rm CE} \sim 10^{-2} \, \Omega_{\rm DM}$ and is too small to explain the DM relic density. However, we have also found that the production of $\tilde{\nu}_R$ is enhanced in some cases.

One possibility occurs when the mass of left-handed sneutrino is close to $m_{\tilde{\nu}_R}$. In this case, the left-right mixing angle of sneutrinos becomes larger, which enhance the decay of wino into $\tilde{\nu}_R$. In Fig. 1 we show $\Omega_{\tilde{\nu}_R}^{\rm CE} h^2$ as a function of $m_{\tilde{\nu}_L}$ when $m_{\tilde{\nu}_R} = 100$ GeV, $m_{\tilde{W}} = 300$ GeV and $\mu_H = 150$ GeV. We can see that $\Omega_{\tilde{\nu}_R}^{\rm CE} \simeq \Omega_{\rm DM}$ is obtained with the mass degeneracy of $\sim 10\%$ for $|a_\nu| \sim 3$. Therefore, if this is the case, $\tilde{\nu}_R$ dark matter suggests the light $\tilde{\nu}_L$, which will be a good target of future collider experiments. Even without mass degeneracy between $\tilde{\nu}_R$ and $\tilde{\nu}_L$, $\Omega_{\tilde{\nu}_R}^{\rm CE} \simeq \Omega_{\rm DM}$ is realized when $|a_\nu| \gg 1$. Although such a large value of $|a_\nu|$ may not be obtained in simple supergravity models, it is phenomenologically viable.

Another possibility is due to degenerate neutrinos which are allowed from the data of oscillation experiments. As mentioned above, the larger neutrino masses enhance $\Omega_{\tilde{\nu}_R}^{\rm CE}$. Indeed,

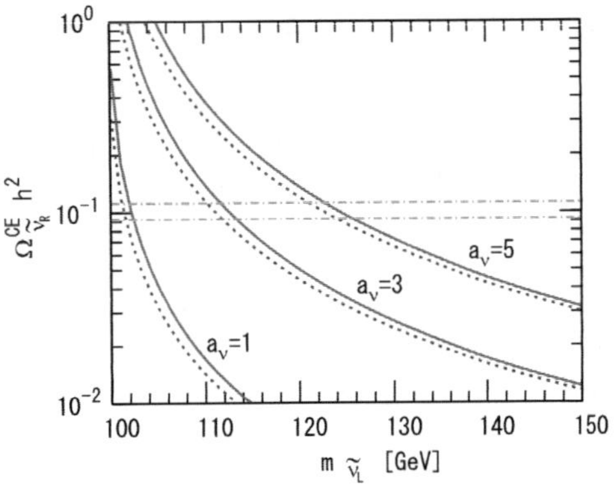

Figure 1. The relic density $\Omega^{CE}_{\tilde{\nu}_R}h^2$ as a function of $m_{\tilde{\nu}_L}$. The solid and dashed lines correspond to the cases with and without the thermal mass of $\tilde{\nu}_L$, respectively. We take $a_\nu = 1, 3, 5$ from left to right. Here $m_{\tilde{\nu}_R} = 100$ GeV, $m_{\tilde{W}} = 300$ GeV and $\mu_H = 150$ GeV. The horizontal dot-dashed lines correspond to the dark-matter density (1).

$\Omega^{CE}_{\tilde{\nu}_R} \simeq \Omega_{DM}$ is obtained when $m_\nu \sim 0.1$ eV. Such a range of m_ν might be verified in future astrophysical observations of cosmic structures.

Next, we turn to consider $\Omega^{FO}_{\tilde{\nu}_R}$. The NLSP, which is assumed to be the LSP in the MSSM sector in this analysis, is very long-lived because of the smallness of y_ν (3). Then, the NLSP decays after its freeze-out time, which gives alternative source of $\tilde{\nu}_R$. The contribution to the relic density via this mechanism is given by $\Omega^{FO}_{\tilde{\nu}_R} = (m_{\tilde{\nu}_R}/m_{\rm NLSP})\Omega_{\rm NLSP}$. Here $\Omega_{\rm NLSP}$ is the "would-be" present density of the NLSP for the case when the NLSP is stable. Since $\Omega_{\rm NLSP}$ is determined by physics around the freeze-out time, $\Omega^{FO}_{\tilde{\nu}_R}$ is also insensitive physics for $T \gg 100$ GeV. Although $\Omega_{\rm NLSP}$ is very sensitive to the MSSM parameters, it can be larger than Ω_{DM} leading to $\Omega^{FO}_{\tilde{\nu}_R} \simeq \Omega_{DM}$. In such a parameter space, the $\tilde{\nu}_R$ DM can be realized. Indeed, we have illustrated this possibility by using the minimal supergravity model in [2]. We should note that the late decay of the NLSP might destroy the success of the BBN. As explained in [2], although the decay of the bino-like NLSP is almost harmless, while that of the stau NLSP is severely restricted. To avoid the BBN constraint the NLSP lifetime must be short enough, which may suggest the degenerate neutrinos and/or the large left-right mixing of sneutrinos.

Finally, we notice that, even if Majorana masses for right-handed neutrinos are present, $\tilde{\nu}_R$ can be the LSP. In this case, however, y_ν becomes larger which results in the universe overclosed by $\tilde{\nu}_R$. To avoid this difficulty, the reheating temperature should be so low that superparticles cannot be thermalized. (See also [4].)

4. Summary

We have discussed the MSSM with three right-handed (s)neutrinos assuming that neutrino masses are purely Dirac-type. We have shown that the lightest right-handed sneutrino $\tilde{\nu}_R$ with mass $m_{\tilde{\nu}_R} \sim 100$ GeV can be the LSP, and that the relic density of $\tilde{\nu}_R$ can be consistent with Ω_{DM}. Thus, the LSP $\tilde{\nu}_R$ is a good candidate for cold dark matter of the universe.

References

[1] Asaka T, Ishiwata K and Moroi T 2006 *Phys. Rev.* D **73** 051301
[2] Asaka T, Ishiwata K and Moroi T 2007 *Phys. Rev.* D **75** 065001
[3] Spergel D N *et al.* [WMAP Collaboration] 2007 *Astrophys. J. Suppl.* **170** 377
[4] Gopalakrishna S, de Gouvea A and Porod W 2006 *JCAP* **0605** 005
[5] McDonald J 2007 *JCAP* **0701** 001
[6] Minkowski P 1977 *Phys. Lett.* B **67** 421; Yanagida T 1979 *Proceedings of the Workshop on Unified Theory and Baryon Number of the Universe* (KEK, Tsukuba) p 95; Gell-Mann M, Ramond P and Slansky R 1979 *Supergravity* (North Holland, Amsterdam); Glashow S L *Proceedings of the Cargése Summer Institute on Quarks and Leptons* (Plenum, new York) p 707
[7] 't Hooft G 1980 *Recent Developments in Gauge Theories* (Plenum) p 135

The neutrino masses and the change of allowed parameter region in universal extra dimension models

Shigeki Matsumoto[a], Joe Sato[b], Masato Senami[c], and Masato Yamanaka[b]

[a] *Institute for International Advanced Interdisciplinary Research, Tohoku University, Sendai, Miyagi 980-8578, Japan*
[b] *Department of Physics, Saitama University, Shimo-okubo, Sakura-ku, Saitama, 338-8570, Japan*
[c] *ICRR, University of Tokyo, Kashiwa, Chiba 277-8582, Japan*
* *Paper presented at the conference by M. Yamanaka*

E-mail: masa@krishna.phy.saitama-u.ac.jp

Abstract. Relic abundance of dark matter is investigated in the framework of universal extra dimension models with right-handed neutrinos. These models are free from the serious Kaluza-Klein (KK) graviton problem that the original universal extra dimension model possesses. The first KK particle of the right-handed neutrino is a candidate for dark matter in this framework. When ordinary neutrino masses are large enough such as the degenerate mass spectrum case, the dark matter relic abundance can change significantly. The scale of the extra dimension consistent with cosmological observations can be 500 GeV in the minimal setup of universal extra dimension models with right-handed neutrinos.

1. Introduction and UED models

As is well known, there is dark matter in this universe, and many models beyond the Standard Model (SM) have been proposed to explain the dark matter. Among those, Universal Extra Dimension (UED) models are one of interesting candidates for new physics. However, because only a short term passes since UED model was proposed, many subject remains which should be investigated. We have solved problems inherent in these models, and calculated the allowed parameter for UED models by estimating the dark matter relic abundance. This proceeding is based on our works [1] and [2].

First, we briefly review UED model. This model is the simplest extension of the SM in the five-dimensional space-time, where the extra dimension is compactified on an S^1/Z_2 orbifold with the radius R. By expanding SM fields on fifth-dimensional space time, we can show that there are infinite tower of excited states. These excited states are called Kaluza-Klein (KK) particles. The SM particles and their KK particles have identical gauge charges. All interactions relevant to KK particles are determined by the SM Lagrangian.

Though the original UED model is phenomenologically good model, the model has two shortcomings. The first one is called the KK graviton problem. In the parameter region where (extra dimension scale $1/R$) $<$ 800 GeV, the KK graviton is the lightest KK particle (LKP), and the next LKP (NLKP) is the KK photon. Hence, the KK photon decays into a KK graviton and a photon at very late time due to gravitational interactions. This fact leads to the serious problem; KK photons produced in the early universe decay into photons in the late universe, and these photons distort the CMB spectrum or the diffuse photon spectrum.

The second problem is the absence of neutrino masses. Since UED model has been constructed as minimal extension of the SM, neutrinos are treated as massless particles. However we know that it is not true. Therefore we must introduce neutrino masses into UED models.

2. Solving the problems

In order to solve these problems, we introduce the right-handed neutrinos into UED models, and assume that they form Dirac mass with ordinary neutrinos. The neutrino masses are expressed as

$$\mathcal{L}_{\nu-\text{mass}} = y_\nu \bar{N} L \Phi + \text{h.c.} \;, \tag{1}$$

where N is the right-handed neutrino, L is the left-handed (doublet) lepton. Thus the second problem, the absence of the neutrino masses, are clearly solved. Once we introduce right-handed neutrinos in the MUED model, their KK particles automatically appear in the spectrum. The mass of the first KK particle of the right-handed neutrino, $N^{(1)}$, is estimated as

$$m_{N^{(1)}} \simeq \frac{1}{R} + O\left(\frac{m_\nu^2}{1/R}\right) \;. \tag{2}$$

By comparing the mass of $N^{(1)}$ with the mass of other KK particles, we can see that $N^{(1)}$ is the NLKP, and the KK photon is the next to next lightest KK particle.

The existence of the $N^{(1)}$ NLKP changes the late time decay of the KK photon. In the models with the KK right-handed neutrino, the KK photon dominantly decays into $N^{(1)}$ and SM left-handed neutrino at tree level. Calculating the decay rate of KK photon decay modes, we can show that the decay of the KK photon is governed by the process, $\gamma^{(1)} \to N^{(1)}\bar{\nu}$. On the other hand, the dominant decay mode associated with a photon, $\gamma^{(1)} \to G^{(1)}\gamma$, comes from the Planck suppressed process, and its branching ratio is

$$\text{Br}_{X\gamma} = \frac{\Gamma(\gamma^{(1)} \to G^{(1)}\gamma)}{\Gamma(\gamma^{(1)} \to N^{(1)}\bar{\nu})} = 5 \times 10^{-7} \left(\frac{1/R}{500\text{GeV}}\right)^3 \left(\frac{0.1\text{eV}}{m_\nu}\right)^2 \left(\frac{\delta m}{1\text{GeV}}\right) \;. \tag{3}$$

As a result, by introducing the right-handed neutrino into UED models, neutrino masses are introduced, and problematic high energy photon emission is highly suppressed. Therefore, two problems in UED models have been solved simultaneously.

3. $N^{(1)}$ dark matter and parameter determination

$N^{(1)}$ cannot decay because it is forbidden by the kinematics. Since $N^{(1)}$ is neutral,massive, and stable, $N^{(1)}$ can be dark matter candidate. Hence, if we introduce the right-handed neutrinos into UED models, dark matter changes from the LKP KK graviton to the NLKP KK right-handed neutrino. The KK graviton dark matter is produced from only decoupled KK photon decay, while $N^{(1)}$ dark matter is produced from decoupled KK photon decay and from the high temperature thermal bath. As a result, in our model, there are additional contribution to the dark matter relic abundance. As total dark matter number density becomes large, the dark matter mass, $\sim 1/R$, becomes small. Therefore, in order to determine the compactification scale $1/R$, we must reevaluate the number density of $N^{(1)}$ dark matter.

Because the number density of the decoupled KK photon has been calculated in previous works [3], we can see the $N^{(1)}$ number density produced from the decoupled KK photon decay. So, we need to calculate the $N^{(1)}$ number density produced from the thermal bath. To do

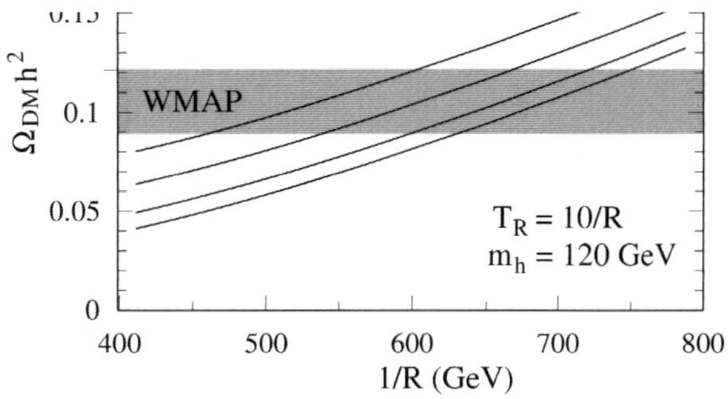

Figure 1. The dependence of the abundance on m_ν with fixed reheating temperature $T_R = 10/R$ and $m_h = 120$ GeV. The solid lines are the relic abundance for $m_\nu = 0, 0.3, 0.5, 0.66$ eV from bottom to top. The gray band represents the allowed region from the WMAP observation at the 2σ level

this, we must take account of the thermal effects. If the thermal bath temperature is high enough, the mass of a particle receives a correction which is proportional to square of thermal bath temperature. For example, the thermal mass of the KK Higgs boson $\Phi^{(n)}$ is given by the following formula,

$$m^2_{\Phi^{(n)}}(T) = m^2_{\Phi^{(n)}}(T=0) + \Big[a(T)\cdot 3\lambda_h + x(T)\cdot 3y_t^2\Big]\frac{T^2}{12}\ . \tag{4}$$

Coefficients $a(T)$ and $x(T)$ are determined by evaluating how many KK modes contribute to the correction at the temperature T. Taking account of such a thermal masses, we calculated the $N^{(1)}$ number density.

Finally, we show the numerical result. In Fig.1, the neutrino mass dependence of the abundance. The horizontal axis is the compactification scale of the extra dimension, and the vertical axis is the relic abundance of the dark matter. The solid lines correspond to the result with $m_\nu = 0, 0.3, 0.5, 0.66$ eV from bottom to top. As shown in this figure, as neutrino mass becomes large, compactification scale can be less than 500 GeV. If compactification scale is less than 500 GeV, in ILC experiment, n=2 KK particles can be produced. Such a particles are very important for discriminating UED from SUSY at collider experiment.

4. Summary

We have solved two problems in UED models (absence of the neutrino mass, energetic photon emission) by introducing the right-handed neutrinos. We have shown that by introducing right-handed neutrino, the dark matter is the KK right-handed neutrino, and we have calculated the relic abundance of the KK right-handed neutrino dark matter. In the UED model with right-handed neutrinos, the compactification scale $1/R$ can be less than 500 GeV. This fact has important consequence on the collider physics, in particular on future linear colliders, because first KK particles can be produced in a pair even if the center of mass energy is around 1 TeV.

References

[1] S. Matsumoto, J. Sato, M. Senami and M. Yamanaka, *Phys. Rev.* D **76**, 043528 (2007)
[2] S. Matsumoto, J. Sato, M. Senami and M. Yamanaka, *Phys. Lett.* B **647**, 466 (2007)
[3] M. Kakizaki, S. Matsumoto and M. Senami, *Phys. Rev.* D **74** (2006) 023504;
S. Matsumoto and M. Senami, *Phys. Lett.* B **633** (2006) 671

Cosmic rays emitted by primordial black holes in a five-dimensional Randall–Sundrum braneworld

Y Sendouda

Yukawa Institute for Theoretical Physics, Kyoto University, Kyoto 606-8502, Japan

E-mail: `sendouda@yukawa.kyoto-u.ac.jp`

Abstract. We probe the Randall–Sundrum braneworld cosmology via the radiative nature of primordial black holes (PBHs). The effects of higher-dimensionality on the cosmic-ray spectra of extragalactic photons and galactic antiprotons from PBHs are investigated. Observational implications for the braneworld geometry as well as the inflationary primordial perturbation as a seed of PBHs are provided.

1. Introduction: Primordial black holes in the Randall–Sundrum cosmology

It is expected that some portion of density contrast in the early universe went into gravitational collapse leading to formation of primordial black holes (PBHs) [1]. PBHs typically have masses much smaller than astrophysical holes thereby acquiring high enough Hawking temperature T_{H} to emit various cosmic rays via the Hawking process [2].

The project here presented is utilisation of PBHs in an inquiry into a five-dimensional braneworld scenario proposed by Randall and Sundrum (RS) [3; 4]. The RS braneworld is modelled out of a four-dimensional (mem)brane, which confines the standard model matter together with positive tension λ, embedded in five dimensional anti-de Sitter space. The model is characterised by a length parameter ℓ below which five-dimensional gravity takes effect, where the experimental upper bound on ℓ is around 0.1 mm [5]. On the brane, cosmological expansion is governed by the modified Friedmann equation

$$H = \frac{8\pi G_{\mathrm{N}}}{3}\rho\left(1+\frac{\rho}{2\lambda}\right). \tag{1}$$

This implies the presence of an exotic early radiation-dominated era in which $\rho\ /\lambda$ term dominates thereby leading to the expansion law $a(t) \propto t^{\ /\ }$. It is in fact what happens when the size of the universe is below ℓ. See e.g. [6] for a review of the model.

Thanks to the following facts, the utility of PBHs to prove such an exotic cosmology is evident. (i) The PBH mass function is sensitive to the expansion history in the early universe since the size of PBHs is closely related to the scale of the Hubble horizon [7; 8]. (ii) Moreover, it has been inferred that in the "ρ " phase, accretion causes power-law growth of the PBH mass as $M_{\mathrm{bh}} \propto t^{\ F/\pi}$, where $0 \le F \lesssim 1$ parameterises the efficiency of accretion [9; 10]. (iii) Finally, Hawking temperature of PBHs smaller than ℓ is lowered. PBHs whose lifetime is the current cosmic age have a temperature as a function of ℓ being

$$T_{\mathrm{H}} \sim 50\left(\frac{\ell}{0.1\ \mathrm{mm}}\right)^{-\ /} \quad \mathrm{keV}. \tag{2}$$

The above characters all bring up quantitative differences from the conventional four-dimensional cosmology. In what follows, we shall see properties of cosmic rays emitted by such PBHs and obtain implications for the RS braneworld.

2. Observational limits on the PBH abundance

First we seek constraints on the abundance of PBH, denoted as α , by calculating two sorts of cosmic-ray spectra emitted by PBHs.

In figure 1, spectra of diffuse photons [11] are shown to be compared with the observations of the extragalactic background [12–15]. The almost black-body spectra directly reflect the Hawking temperature given in equation (2). It will be therefore possible to determine the size of the extra dimension ℓ if the PBH-origin component is identified in the future observations. At this moment, each spectrum determines an upper bound on the PBH abundance for given values of ℓ and F.

The other cosmic ray, primary antiprotons from PBHs in the Milky Way dark halo [16], is shown in figure 2 together with the standard secondary component. Antiprotons from PBHs typically have sub-GeV energies irrespective of ℓ because of the threshold around GeV, the antiproton mass. The PBH-origin antiprotons in the sub-GeV region actually improve the fit to the corresponding observations [17–19], however, here we conservatively find upper limits on the PBH abundance. An investigation reveals that the antiproton flux from PBHs is a decreasing function of the size of the extra dimension ℓ [16].

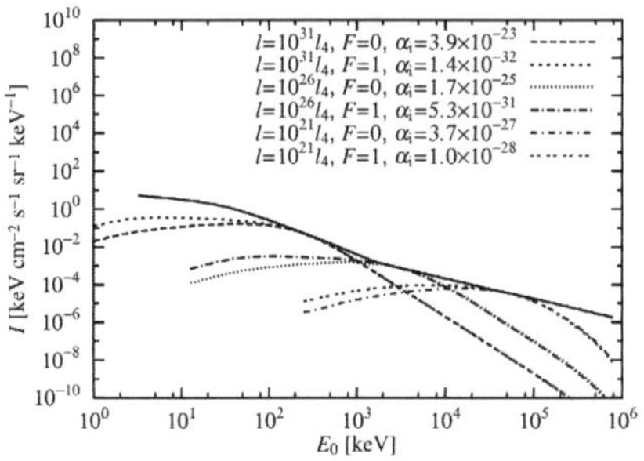

Figure 1. Figure taken from [11]. Each spectrum corresponds to maximal PBH abundance for different values of ℓ and F, where ℓ is the Planck length.

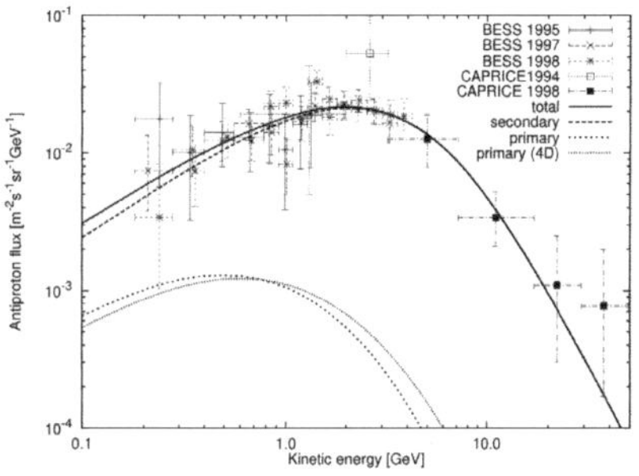

Figure 2. Figure taken from [16]. Sub-GeV antiproton component near its upper limit for parameters $\ell = 10 \quad \ell \quad \sim 0.1$ mm and $F = 1.0$ is presented.

3. Constraints on the RS braneworld

From the above considerations, we find the allowed region in the parameter space $(\ell, F, \alpha\ )$ [11; 16]. Its projection onto the (ℓ, F)-plane are shown in figure 3, where the left-hand side is allowed for the lines each corresponding to a fixed α . We find that when $\alpha \ \sim 10^{-} \ -10^{-}$, constraints from the two cosmic-rays cooperate. In particular, they imply that if the apparent sub-GeV excess of the antiproton observations is really due to PBH evaporation, then an upper bound on the size of the extra dimension is set around 10^{-} m [16].

Finally the upper limits on the PBH abundance α are translated into those on the power spectrum of the inflationary perturbation. The PBH mass function is related to the comoving curvature perturbation spectrum $\mathcal{P}_{\mathcal{R}_c}$ [8]. The upper bounds on the perturbation amplitude are summarised in figure 4 [20]. We can read off from the figure that the comoving length scale of concern approximately ranges between 1–1000 km depending on the size of the extra dimension and the efficiency of accretion.

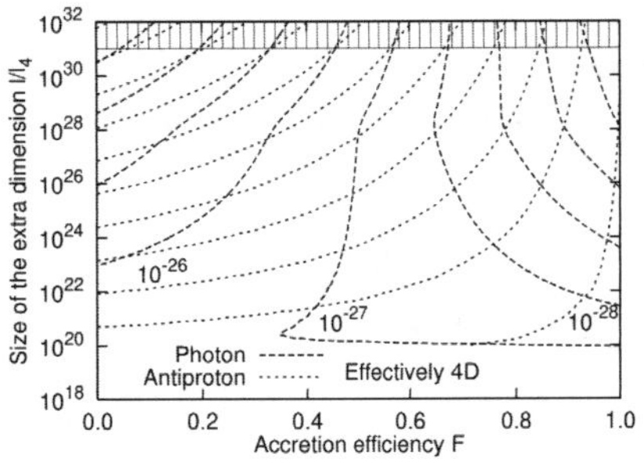

Figure 3. Figure taken from [16]. For each fixed α , only parameters in the left-hand side of corresponding boundary are allowed.

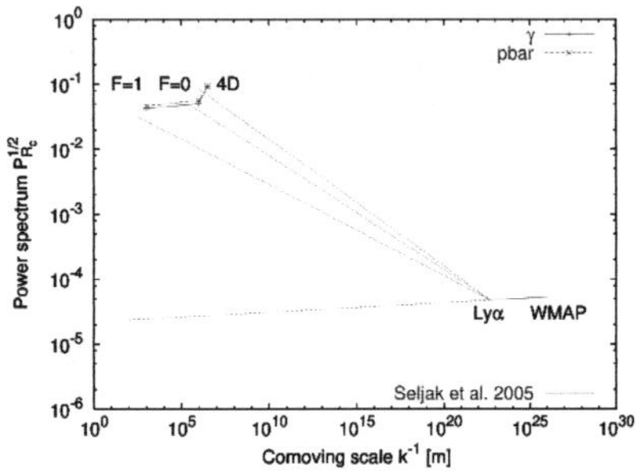

Figure 4. Figure taken from [20]. Upper bounds on the inflationary comoving curvature perturbation are plotted.

Acknowledgments

The author is grateful to Katsuhiko Sato, Shigehiro Nagataki and Kazunori Kohri. The author is in part supported by JSPS through Grant-in-Aid for JSPS Fellows No. 19-7852.

References

[1] Carr B J 1975 *J* **201** 1–19

[2] Hawking S W 1975 **43** 199–220

[3] Randall L and Sundrum R 1999 **83** 3370–3373

[4] Randall L and Sundrum R 1999 **83** 4690–4693

[5] Kapner D J 2007 **98** 021101

[6] Maartens R 2004 **7** 7

[7] Guedens R, Clancy D and Liddle A R 2002 D **66** 043513

[8] Sendouda Y, Nagataki S and Sato K 2006 *J* JCAP06(2006)003

[9] Majumdar A S 2003 **90** 031303

[10] Guedens R, Clancy D and Liddle A R 2002 D **66** 083509

[11] Sendouda Y, Nagataki S and Sato K 2003 D **68** 103510

[12] Gruber D E, Matteson J L, Peterson L E and Jung G V 1999 *J* **520** 124–129

[13] Kinzer R L, Jung G V, Gruber D E, Matteson J L and Peterson L E 1997 *J* **475** 361–372

[14] Sreekumar P (EGRET) 1998 *J* **494** 523–534

[15] Strong A W, Moskalenko I V and Reimer O 2004 *J* **613** 956–961

[16] Sendouda Y, Kohri K, Nagataki S and Sato K 2005 D **71** 063512

[17] Matsunaga H (BESS) 1998 **81** 4052–4055

[18] Orito S (BESS) 2000 **84** 1078–1081

[19] Maeno T (BESS) 2001 **16** 121–128

[20] Sendouda Y 2007 Ph.D. thesis University of Tokyo

Dark matter annihilation in the Milky Way halo

Shunsaku Horiuchi

Department of Physics, School of Science, University of Tokyo, Tokyo 113-0033, Japan

E-mail: horiuchi@utap.phys.s.u-tokyo.ac.jp

Abstract. To constrain the particle properties of dark matter using their annihilation products, robust predictions of the annihilation rates are critical. Since the annihilation rate scales as the dark matter density squared, a precise knowledge of the distribution of dark matter is required. Given the popular galactic center contains large uncertainties in the dark matter density, we focus on the signal arising from the whole Milky Way halo. This is less sensitive to uncertainties in the dark matter distribution, and in terms of flux it is larger than the cosmic signal for halos flatter than NFW. We apply this to a dark matter model in which the principal annihilation products are neutrinos. Being the least detectable Standard Model particle, the neutrino defines a strong and general upper bound on the total annihilation cross section. We show that using the Milky Way halo signal improves previous limits by 1–2 orders of magnitude.

(This article is based on H. Yuksel, S. Horiuchi, J. F. Beacom, S. Ando, "Neutrino Constraints on the Dark Matter Total Annihilation Cross Section," arXiv:0707.0196 [1].)

1. Introduction

High-energy particles from the annihilation of dark matter provide indirect means to probe the particle properties of dark matter. Since the annihilation rate scales as the dark matter density squared, annihilation signatures including gamma rays and neutrinos, which point to the site of production, have been searched for in regions where the dark matter density is expected to be high. Among the many possibilities, the galactic center (GC) remains an attractive site for its closeness and for the expected dark matter cusp. However, the precise annihilation rate strongly depends on the assumed dark matter halo profile; this uncertainty manifests itself as an uncertainty of a few orders of magnitude in size in the enhancement of the annihilation rate (often called the "boost factor"). Given that one places constraints on dark matter properties by assuming a particular dark matter particle make up and a particular dark matter density profile, uncertainties in the boost factor are carried directly to uncertainties in the particle constraints.

Regarding the choice of dark matter particle, there is an entire "zoo" of possible candidates; see a recent review by Bertone, Hooper, and Silk [2]. If the branching ratio to a specific final state were known, a bound on the appearance rate of that state would yield a bound on the total annihilation cross section. However, branching ratios are dependent on the choice of dark matter particle, and for each candidate, there can be different models producing different branching ratios. Any limit on the total annihilation cross section is therefore very weak (specific).

Recently, Beacom, Bell, and Mack have shown that neutrinos can be used as probes to place a strong upper limit on the total dark matter annihilation cross section provided dark matter doesn't annihilate dominantly to new and truly dark particles [3]. We extend on this so-called "neutrino bound", both in terms of robustness and magnitude, by keeping to a conservative

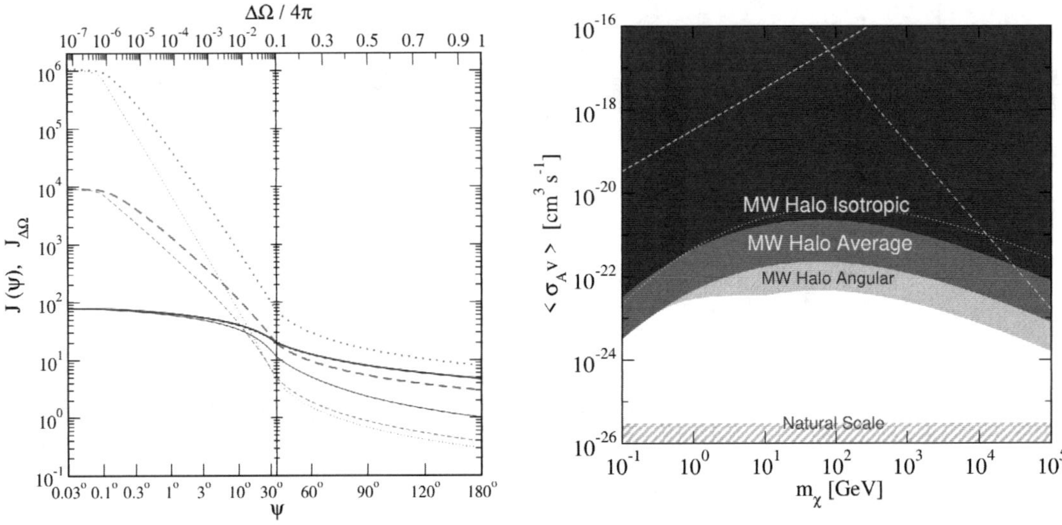

Figure 1. *Left.*—$J(\psi)$ (thin lines) and $J_{\Delta\Omega}$ (thick lines), plotted against ψ and $\Delta\Omega/4\pi = (1-\cos\psi)/2$ respectively, for three halo profiles (Moore, NFW and Kravtsov in order of dotted, dashed and solid lines). Note that the left (right) half is presented in log (linear) scale. *Right.*—Constraints on the dark matter total self-annihilation cross section from components of the Milky Way halo as defined in the text (shaded regions excluded). The constraint from the cosmic signal (dotted line) and the unitarity bound [4] (dot-dashed) are shown.

analysis of the Milky Way halo. We first show that a large field of view centered around the GC dramatically reduces the spread in boost factor between different profiles, and highlight an isotropic component of the galactic halo that is as large as the truly cosmic component for halos flatter than NFW. We then use the halo components for the neutrino bound, showing that previous bounds are improved upon by 1-2 orders of magnitude.

2. Milky Way halo components

We first discuss the annihilation signal flux for a generic dark matter particle. Detailed derivations can be found in [5, 6]; a summary of appropriate material is also presented in [7, 1].

Due to the cusp, the average boost factor within a cone of half-angle ψ centered on the GC is very high for small angles. The left of Fig. 1 shows the boost factor $J(\psi)$ (thin lines) and its average $J_{\Delta\Omega}$ (thick lines) as functions of ψ and $\Delta\Omega/4\pi = (1-\cos\psi)/2$ respectively, for three halo profiles – in order of cuspiness, the Moore [8], NFW [9], and Kravtsov [10] profiles. The density normalization has been chosen consistently with observed data. We define three components, (i) the *Halo Angular* corresponding to a 30° half-cone, (ii) the *Halo Average* corresponding to the entire halo, and (iii) the *Halo Isotropic* corresponding to $J(\psi = \pi)$. The *Halo Isotropic* is the minimum boost factor that arises from our immediate neighbourhood, and yields an isotropic flux. As the figure indicates, large fields of view dramatically decrease the spread in boost factors between different profiles. For example, the *Halo Average* shows a variation of only a factor of a few, $(J_{Ave}^{Moore}, J_{Ave}^{NFW}, J_{Ave}^{Kra}) = (8, 3, 2.6)$, instead of the $\sim 10^3$ spread for 1°.

Interestingly, the boost factor of the *Halo Isotropic* is essentially profile independent - $(J_{Iso}^{Moore}, J_{Iso}^{NFW}, J_{Iso}^{Kra}) = (0.3, 0.4, 0.55)$. In addition, for halo profiles that are at least as flat as the NFW, the predicted flux of annihilation products in this isotropic component is as large as the truly cosmic flux from all dark matter halos, under the assumption of halo universality. For example, the ratio of the *Halo Isotropic* to truly cosmic flux for a Kravtsov profile is ≈ 2.

3. Bounds on the total annihilation cross section from neutrinos

We assume that dark matter annihilation proceeds to Standard Model (SM) particles, and explain the neutrino bound in this section. Of the many possible high energy particles produced in dark matter annihilation, neutrinos, being the hardest to detect of the SM particles, provide the weakest bound on the total annihilation cross section. One may naively expect that they cannot therefore produce any meaningful constraints. However, it has recently been shown that this is not the case, and surprisingly strong flux limits can be derived from recent high-statistics data [3]. Due to the difficultly of neutrino detection, we may interpret the neutrinos as bounding all SM final states, and hence constraining the total annihilation cross section. This is because dark matter with even the slightest branching ratio to particles other than the neutrino will yield a stronger, but particle specific, constraint through that particle (or secondary particle, most often photons). Note that the neutrino bound can be evaded if the final states are dominated by new and truly dark non-SM particles, in which case all dark matter annihilation searches will be more challenging.

On the right of Fig. 1 we show our neutrino bounds, placed considering the Milky Way halo components. We assume a NFW profile, and a signal detection criteria of double the atmospheric neutrino background flux. The null detection over background then yields the bounds shown in the figure. Energy bins of $\Delta \log_{10} E = 0.3$ and $\Delta \log_{10} E = 0.5$ were used for the halo and cosmic neutrino signals, respectively. The already strong previous bound is improved on by 1–2 orders of magnitude, depending on which Milky Way component is utilized.

4. Discussion

Can neutrinos be the only final state? We do not stress this is a realistic assumption, but that it enables us to place robust limits on the total annihilation cross section, which are at the same time strong. The upper limits shown on the right of Fig. 1 have been placed using components of the Milky Way halo, and are 1–2 orders stronger than those placed using the cosmic signal from all extragalactic halos. Advantages of the halo components include: (i) the *Halo Isotropic* is robust to uncertainties in the halo profile, (ii) for profiles cuspier than NFW the *Halo Angular* becomes stronger than in Fig. 1 (iii) for profiles flatter than NFW the *Halo Isotropic* becomes more important relatively, (iv) high energy photons from cosmic dark matter are attenuated, and thus neutrinos may not necessarily be the hardest to detect Standard Model particle (the halo components do not have this concern).

Here we showed the neutrino bound using large energy bins and a conservative detection criteria of 100% as large as the atmospheric neutrino background. The atmospheric neutrino background has been measured with high statistics over six orders of magnitude in energy, and shows zenith dependency. A detailed analysis, including zenith dependency (see also [11]), tighter energy bins and detection criteria, and inclusion of electron neutrinos, should be performed. A complementary study [12] has in fact discussed angular dependence, as well as gamma-rays from Z bremsstrahlung [13]. Charged-lepton pairs accompanying $\nu\bar{\nu}$ pairs may also provide competitive and complementary constraints. A comprehensive study including the points above will yield further stringent limits.

5. References

[1] H. Yuksel, S. Horiuchi, J. F. Beacom and S. Ando, Phys. Rev. D **76**, 123506 (2007) [arXiv:0707.0196].
[2] G. Bertone, D. Hooper and J. Silk, Phys. Rept. **405**, 279 (2005).
[3] J. F. Beacom, N. F. Bell and G. D. Mack, Phys. Rev. Lett. **99**, 231301 (2007) [arXiv:astro-ph/0608090].
[4] K. Griest and M. Kamionkowski, Phys. Rev. Lett. **64**, 615 (1990); L. Hui, Phys. Rev. Lett. **86**, 3467 (2001).
[5] L. Bergstrom, P. Ullio and J. H. Buckley, Astropart. Phys. **9**, 137 (1998).
[6] P. Ullio, L. Bergstrom, J. Edsjo and C. G. Lacey, Phys. Rev. D **66**, 123502 (2002).
[7] S. Ando, Phys. Rev. Lett. **94**, 171303 (2005).
[8] B. Moore, T. Quinn, F. Governato, J. Stadel and G. Lake, Mon. Not. Roy. Astron. Soc. **310**, 1147 (1999);

[9] J. F. Navarro, C. S. Frenk and S. D. M. White, Astrophys. J. **462**, 563 (1996);
[10] A. V. Kravtsov, A. A. Klypin, J. S. Bullock and J. R. Primack, Astrophys. J. **502**, 48 (1998).
[11] D. Hooper and P. D. Serpico, JCAP **0706**, 013 (2007).
[12] M. Kachelriess and P. D. Serpico, Phys. Rev. D **76**, 063516 (2007) [arXiv:0707.0209 [hep-ph]].
[13] V. Berezinsky, M. Kachelriess and S. Ostapchenko, Phys. Rev. Lett. **89**, 171802 (2002).

Rare Components in Cosmic Rays with AMS-02

A. Sabellek, on behalf of the AMS Collaboration

Universität Karlsruhe, IEKP, Postfach 6980, 76128 Karlsruhe, Germany

E-mail: sabellek@iekp.fzk.de

Abstract. The Alpha Magnetic Spectrometer 02 (AMS-02), like the AMS-01 experiment flown in 1998 on board the Space Shuttle, offers a unique opportunity for exploration of cosmic rays. AMS-02 will be installed to the International Space Station (ISS) providing precise data on cosmic radiation spectra in a wide energy range for at least three years. Although the major part of cosmic ray particles is of astrophysical origin, a small component may originate from interaction of dark matter particles. Redundant detection in several sub-detectors with accurate particle identification allows precision measurements and identification of up to now concealed or undetected deviation from Standard Model prediction. After an introduction to detector design, an overview of physics goals of AMS-02 is given.

1. Introduction

The Alpha Magnetic Spectrometer 02 (AMS-02) is a large acceptance spectrometer for high energy charged particles and gamma rays in cosmic radiation. It is designed for maintenance-free operation on board the International Space Station (ISS) for at least three years, comprising the biggest super-conducting magnet in space.

In June 1998 the prototype version AMS-01 has successfully flown for ten days on board the Space Shuttle Discovery. This precursor flight has provided essential information about detector operation and performance under space-conditions and interesting physics data, which results are summarized in [1]. About 1000-times higher statistics of AMS-02 data will widely improve our knowledge about propagation and sources of cosmic rays. AMS-02 also will allow searches for heavy antimatter nuclii, dark matter signatures, which profits from improved proton rejection in the order of 10^6, or even exotic new types of matter [2].

2. AMS-02 Detector Design

Accurate particle identification in cosmic rays requires measurement of energy, rigidity, velocity and electric charge. To gain precision AMS-02 performs a redundant measurement of a particle's properties in different sub-detectors. Core of the experiment is a large super-conducting magnet with a large area silicon strip detector. Figure 1 shows an artist model view of the detector with Transition Radiation Detector (TRD), Time of Flight (ToF), Anti-coincidence Counter (ACC), Ring Image Cherenkov Detector (RICH) and Electromagnetic Calorimeter (ECAL) sub-detectors. The assembled detector has a weight of about 7 tons with an acceptance of 0.5 m^2sr.

Space qualified electronics running with 650 microprocessors control and read out 300,000 physics channels, GPS, Startracker Camera and heat control systems. At an expected trigger rate of up to 2000Hz data is reduced by fast DAQ electronics to a rate of about 2 MBit/s.

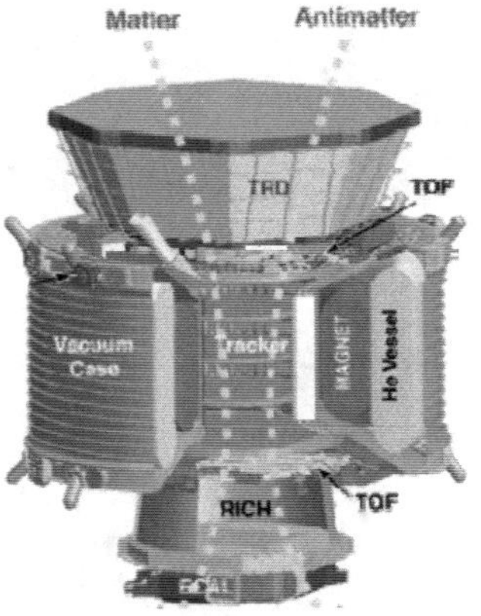

Figure 1

• The super-conducting magnet provides a magnetic field with a bending power of $BL^2 = 0.862\ Tm^2$ for momentum measurement with tracker. It consists of two dipole coils and two set of smaller coils to minimize the stray field outside the magnet. It will be cooled to 1.8 K by 2,500 l of superfluid Helium, which gives a constraint to operation of three to five years.

• The Silicon Tracker inside the magnet has a diameter of 1.1 m and consists of eight planes of 300 μm double-sided microstrip detectors with a resolution of 10μm in bending direction. Rigidity can be measured up to few TeV. The Tracker contributes to particle identification by dE/dx measurement, as well as detection of gammas in conversion-mode.

• The TRD uses production of transition radiation in its fleece radiator layers to suppress the proton signal against positrons with a factor > 100 in non-destructive way. It is built out of 5248 Xe/CO_2 proportional counter tubes in 20 layers and contribute to dE/dx measurement.

• The ToF is designed as a fast primary trigger with a time resolution of 130 ps and velocity measurement of $d\beta/\beta < 3\%$. The 2 x 2 scintillator planes are attached to upper and lower side of the tracker. The 16 cylindrical shell paddles of the ACC surrounding the Tracker operate as veto.

• The RICH detector performs velocity measurement of charged particles up to 20 GeV/n and accurate determination of a particle's charge up to Z=26. It consist of combined NaF and Aerogel radiators and PMT array to detect Cherenkov photons.

• The ECAL is located on bottom of the experiment, where the particle crashes into 18 planes of lead with scintillator fiber inserts. Its thickness corresponds to $16X_0$ ($.5\lambda_h$) and contributes to energy measurement with $dE/E < 5\%$ up to TeV and to proton/positron suppression with a factor of >1000. For gamma ray detection ECAL acts as an independently triggered photon detector with angular resolution of < 1°.

3. Cosmic Ray Astrophysics with AMS-02

Three major physics goals will be presented in the next subsections, keeping in mind that AMS-02 is a multipurpose instrument providing relevant data for various fields of cosmology and particle physics. AMS-02 is a long-duration cosmic radiation monitor.

3.1. Cosmic Ray Physics

A precise measurement of galactic cosmic ray spectra is the basis to investigate origins and to improve parameters of current acceleration and propagation models of cosmic ray particles. AMS-02 is capable to determine chemical composition and abundance of cosmic ray nuclii from Helium to Cobalt with 1% accuracy. Furthermore different Isotopes can be distinguished. Measuring different Helium isotopes already AMS-01 found that spectra are unexpectedly separated in space.

Figure 2

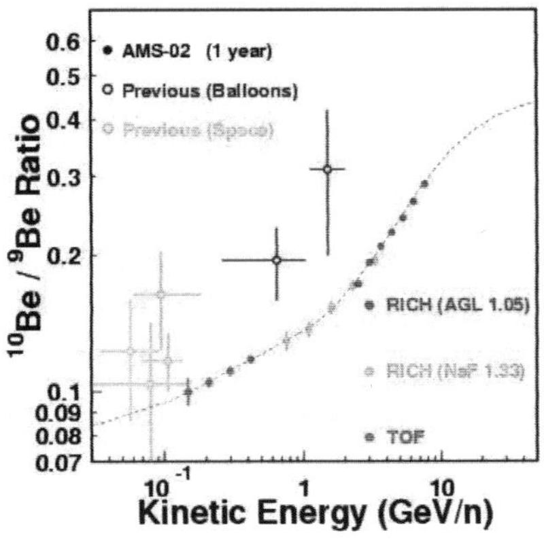

AMS-02 will collect about 10^5 Berillium-10 isotopes in three years and measure isotope ratio shown in figure 2. Since half-life is in the order of confinement time of cosmic ray particles, an exact measurement will improve propagation model parameters like resident time and galactic halo size.

AMS-02 will measure the ratio of Carbon to its spallation secondary Boron up to 1 TeV/n, which gives information about the amount of matter traversed by cosmic ray particles and diffusion processes to understand and fix free parameters of propagation models.

The two complementary modes for gamma detection, either by ECAL or by e^+e^- detection in tracker after conversion, turn AMS-02 also into telescope for high energetic gamma rays from 1 to a few hundreds GeV, complementary to measurement of charged particles with the same instrument. Combined this allows to study cosmic rays interaction with gas as the main source of gamma rays, but also accelerating processes of cosmic rays in 'standard' astrophysics.

3.2. Indirect Search for Dark Matter

The major subject of today's astrophysics certainly is to clarify nature of cold dark matter. Leading idea is some kind of not-yet-seen particle like the Neutralino WIMP candidate as lightest SUSY particle in R-conserving model. Indirect searches are based on quark pair production in annihilation, whose decay then is well known from accelerator experiments. AMS-02 at the same time will measure Antiproton and Positron spectra, which may contain an antimatter excess from annihilating dark matter particles. Furthermore it was shown that one could tune cosmic models from excess in gamma ray flux by fitting a dark matter signal [3].

In standard physics antideuterons in cosmic rays are produced with high energies, whereas antideuterons from Neutralino annihilation can be found blow 1 GeV. Since this excess would be orders of magnitudes, extraction of SUSY signal could be much easier than from antiprotons. A large acceptance spectrometer like AMS-02 is required to measure these still extremely low fluxes.

Low scale quantum gravity predicts bosons with mass, allowing direct annihilation into e^+e^- pairs in a dominant channel. This would result in steep features in the spectrum which is different from Neutralino annihilation signal and could be discover for boson mass of about 300 GeV [4].

3.3. Direct Search for Heavy Antimatter

One of the original goals of AMS-02 is to determine the primordial antimatter content of the Universe, which still is unknown and its existence can not be excluded by standard model on cosmological scale. Up to today no antimatter annihilation signal was found from within our cluster. If there is no heavy antimatter particle found by AMS-02, this would prove there is no antimatter in the Universe, on the other hand a single anticarbon would mean there are antimatter stars.

AMS-02 is designed as a multipurpose instrument and so looking for even new types of matter like stable quark strange matter first proposed in [5]. These Strangelets event signature would simply be an anomalous $Z/A < 0.12$, of which already one candidate was reported by AMS-01.

4. Conclusions

AMS-02 will be a unique and challenging experiment on board the ISS and will be able extend our understanding of the nature of the Universe. AMS-02 perfectly complements current big experiments on ground and in space in exploring new physics.

Individual detector sub-systems are built, successfully tested and passed space qualification procedure. Full detector assembly started in summer 2007 at CERN facilities in cooperation with NASA. Final AMS-02 test for space qualification will be performed in the Large Space Simulator at ESTEC, before shipment to Kennedy Space Center in Florida for launch preparation in 2009.

Acknowledgements

I would like to thank organizers of the TAUP07 Conference in Sendai, Japan.

References

[1] M. Aguilar et al., Physics Reports, vol. 366/6 (Aug.2002), pp.331-404
[2] The AMS collaboration, Design Report, to be published, preprint available at http://ams.cern.ch/AMS/AMS.pdf
[3] W.de Boer, C.Sander, V.Zhukov, A.Gladyshev, D.Kazakov, astro-ph/0508617, 444 (2005) 51
[4] H.C.Cheng, J.L.Feng and K.T.Matchev, Phys.Rev.Lett V89, N21 (2002) 211301-1
[5] E. Witten, Phys. Rev. D,272-285 (1984)

Antideuterons as an Indirect Dark Matter Signature: Design and Preparation for a Balloon-born GAPS Experiment

J. E. Koglin[1], T. Aramaki[1], S. E. Boggs[2], W. W. Craig[3], H. Fuke[4],F. Gahbauer[1,5], C. J. Hailey[1], N. Madden[1], K. Mori[1], R. A. Ong[6], T. Yoshida[4], H. T. Yu[1], and K. P. Ziock[7]

[1]Columbia Astrophysics Laboratory, Columbia University, New York, NY, USA
[2]Space Sciences Laboratory, University of California, Berkeley, CA, USA
[3]Institute of Space and Astronautical Science, Japan Aerospace Exploration Agency (ISAS/JAXA), Sagamihara, Japan
[4]Lawrence Livermore National Laboratory, Livermore, CA, USA
[5]University of Latvia, Riga, Latvia
[6]Department of Physics and Astronomy, University of California, Los Angeles, CA, USA
[7]Oak Ridge National Laboratory, Oak Ridge, TN, USA

E-mail: koglin@astro.columbia.edu

Abstract. The General Antiparticle Spectrometer (GAPS) exploits low energy antideuterons produced in neutralino-neutralino annihilations as an indirect dark matter (DM) signature that is effectively free from background. When an antiparticle is captured by a target material, it forms an exotic atom in an excited state which quickly decays by emitting X-rays of precisely defined energy and a correlated pion signature from nuclear annihilation. We have successfully demonstrated the GAPS method in an accelerator environment and are currently planning a prototype flight from Japan for 2009. This will lead to a long duration balloon (LDB) mission that will complement existing and planned direct DM searches as well as other indirect techniques, probing a different, and often unique, region of parameter space in a variety of proposed DM models. Planes of coarsely pixellated Si(Li) detectors form the heart of the GAPS flight detector, providing both high X-ray energy resolution and good particle tracking. We will describe the proto-flight mission that will verify the performance of our Si(Li) detectors and cooling system in a flight-like configuration. We also will outline the LDB science payload design.

1. Introduction

Many extensions to the standard model predict weakly interacting, massive particles ('WIMPS') that are stable and thus ideal dark matter (DM) candidates. Underground direct detection experiments detect the nuclear recoil of WIMPS that scatter off target nuclei. WIMPS can also annihilate with other WIMPS in the galactic halo producing various debris such as gamma-rays, neutrinos, positrons, antiprotons and antideuterons. The General Antiparticle Spectrometer (GAPS) is specifically designed to uniquely identify low-energy, low-mass antiparticles which allows us to : execute deep searches for the WIMPS predicted by supersymmetric (SUSY) and universal extra dimension (UED) theories using antideuterons; search for evaporating primordial black holes using antideuterons, setting the best limits on primordial black hole density; and perform low energy cosmic-ray antiproton spectroscopy with several orders of magnitude better statistics than current satellite and balloon experiments. The detection rate for antideuterons yields direct information on the particle properties of the DM. When combined with observations from underground experiments, ground-based experiments such as VERITAS, space-based experiments such as GLAST and PAMELA, and accelerator-based experiments, an extremely comprehensive picture of the nature of DM can be obtained.

More than 80 recent papers discuss aspects of antideuteron DM searches. DM searches are highly model dependent, and there are lots of models! To illustrate the key features of a DM search with GAPS, the antideuteron flux expected from three different benchmark models calculated by Baer and Profumo [1] as representative of the most popular candidates for DM are plotted in Figure 1. GAPS is optimized for operation below 0.3 GeV/n, where the DM signal is largest. Also shown in the plot is the anticipated secondary and tertiary background of antideuterons [2]. Finally, Figure 1 shows the sensitivity to antideuterons for long-duration balloon (LDB) campaigns from Antarctica (60 days total over three flights) and the potential sensitivity from an ultra-long duration balloon (ULDB) flight (300 days total) that might be realized by our projected 2013 launch date. For comparison, the upper limit from the BESS experiment is also plotted. Figure 1 illustrates three important points: 1) GAPS is essentially a background free experiment; 2) GAPS has outstanding DM discovery potential for a wide variety of DM models, and 3) GAPS represents a major improvement over the state of the art.

2. GAPS Operating Principle

The goal of GAPS is to uniquely identify antiparticles using a method outlined in a 2002 concept paper [3]. An antiparticle that has been slowed by the atmosphere passes through a TOF system (which measures particle velocity) and is slowed down by dE/dx losses in the target/detector. After stopping in the target, the antiparticle forms an exotic atom in an excited state with

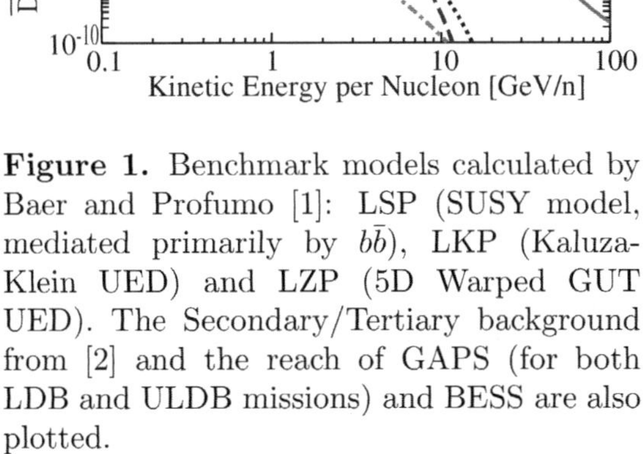

Figure 1. Benchmark models calculated by Baer and Profumo [1]: LSP (SUSY model, mediated primarily by $b\bar{b}$), LKP (Kaluza-Klein UED) and LZP (5D Warped GUT UED). The Secondary/Tertiary background from [2] and the reach of GAPS (for both LDB and ULDB missions) and BESS are also plotted.

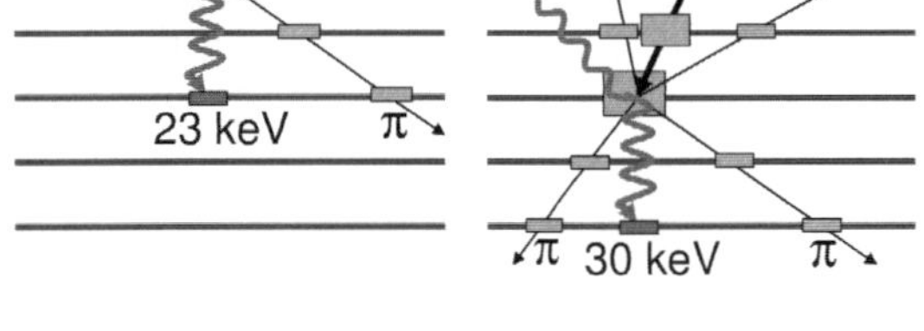

Figure 2. GAPS method of antiparticle identification. For the same measured TOF and angle, an antideuteron (right) will penetrate deeper, typically emit twice as many annihilation pions as well as emit X-rays of different energies than an antiproton (left).

near unity probability. The exotic atom de-excites through both autoionizing and radiative transitions. Through proper target selection, 3-4 X-rays detectable hard X-rays (i.e., in the range ∼10-100 keV) will be produced as the exotic atom de-excites down to the ground state. After X-ray emission, the antiparticle annihilates in the nucleus producing a pion shower (star) (∼5 pions per antiparticle nucleon). The X-ray/pion emission occurs within nanoseconds. In this way, GAPS relies on three techniques to uniquely identify antiparticles as illustrated in the cartoon in Figure 2: 1) time of flight (TOF) and depth sensing to distinguish heavier antideuterons from the lighter antiprotons and protons, 2) simultaneous detection of X-rays emitted as the captured antiparticle makes atomic transitions from an excited state, and 3) multiplicity of pions emitted from nuclear annihilation – on average, roughly proportional to the antiparticle nucleon number.

The X-ray energies, which depend only on mass and charge and are precisely known from quantum theory, uniquely identify the antiparticle. The simultaneous occurrence in a narrow time window of X-rays of the correct energies, along with a pion star, provide an enormously constraining signature with which to suppress background. GAPS is ideally suited to low energy antideuteron searches (<0.3 GeV/n), because it is easy to range out low energy particles. Our designs also allow large grasp compared to superconducting magnets. A GAPS prototype was tested at the KEK accelerator in Japan in 2004 and 2005 [4]. All open physics issues relevant to the design of a flight instrument, and amenable to ground-based testing, were resolved.

3. Balloon Experiment Design

GAPS is amenable to balloon-based experiments since the relevant science can be done with an instrument package of ∼1000 kg. We have designed a complete balloon-based GAPS experiment (bGAPS), as well as a prototype balloon experiment (pGAPS). The X-ray detectors are pixellated, high resolution Si(Li) produced from commercial 10 or 12.5 cm diameter wafers. The Si(Li) in bGAPS will be arrayed in a 13 layer tracking geometry, and each layer covers ∼2 m^2. The Si(Li) will be cooled to ∼-40C to provide low noise, high energy resolution performance. Groups of three Si(Li) will be mounted on the carriers and mechanically fixed to a central Al coupling through which a circulating cooling fluid will carry heat to a radiator panel facing away from the Sun. One-time, deployable shades will reduce the direct solar and albedo solar gain. The coolant temperature at the radiator is predicted by our thermal model to be -60C and -40C at the Si(Li) wafers. The detector planes will be surrounded by highly segmented plastic scintillator to provide the TOF trigger.

Using a smaller, three layer array of Si(Li) detectors, pGAPS will flight test all of the critical features of bGAPS. Through the pGAPS flight we will: confirm proper operation of the Si(Li) detectors at float altitude; measure X-ray and particle backgrounds of relevance to determining the overall instrument sensitivity; confirm the thermal model for predicting the Si(Li) operating temperature and verify the concept for cooling the Si(Li) detectors. The pGAPS experiment is anticipated to take place in late 2009 from Hokkaido, Japan.

4. Acknowledgments

This work was supported in part by NASA APRA grants NAG5-5393 and NNG06WC06G.

References

[1] Baer H and Profumo S 2005 *JCAP* **12** 8–+ (*Preprint* arXiv:astro-ph/0510722)
[2] Duperray R *et al.* 2005 *Phys. Rev.* **D71** 083013 (*Preprint* astro-ph/0503544)
[3] Mori K, Hailey C J, Baltz E A, Craig W W, Kamionkowski M, Serber W T and Ullio P 2002 *ApJ* **566** 604–616 (*Preprint* arXiv:astro-ph/0109463)
[4] Hailey C J, Aramaki T, Craig W W, Fabris L, Gahbauer F, Koglin J E, Madden N, Mori K, Yu H T and Ziock K P 2006 *JCAP* **1** 7–+ (*Preprint* arXiv:astro-ph/0509587)

Search of axions from a nuclear power reactor with a high-purity germanium detector

H M Chang[1,2,*], H T Wong[1,7], M H Chou[1], M Deniz[1,3], H X Huang[1,4], F S Lee[1], H B Li[1], J Li[5,6], H Y Liao[1,2], S T Lin[1,2], V Singh[1], S C Wu[1] and B Xin[4]

[1] Institute of Physics, Academia Sinica, Taipei 11529, Taiwan
[2] Department of Physics, National Taiwan University, Taipei 10617, Taiwan
[3] Institute of High Energy Physics, Chinese Academy of Science, Beijing 100039, China
[4] Department of Engineering Physics, Tsing Hua University, Beijing 100084, China
[5] Department of Nuclear Physics, Institute of Atomic Energy, Beijing 102413, China
[6] Department of Physics, Middle East Technical University, Ankara 06531, Turkey
[*] Paper presented at the conference by H M Chang

E-mail: `htwong@phys.sinica.edu.tw`

Abstract. This article reports the first study of possible emissions of axions from power reactors using Primakoff and Compton conversions as the detection mechanisms. The expected experimental signatures are mono-energetic lines produced by their Primakoff or Compton conversions at a high-purity germanium (HPGe) detector. No evidence of axion emissions were observed and constraints on axion couplings versus axion mass within the framework of invisible axion models were placed. This experimental approach provides a unique probe for axion mass at the keV-MeV range not accessible to the other techniques.

This axion research [1] is a by-product of the research program on neutrino magnetic moment [2] pursued by the TEXONO Collaboration at the Kuo-Sheng (KS) Nuclear Power Station in Taiwan. Axions can be emitted in competition with the photons as a result of neutron captures or nuclear de-excitations at reactor. There are six prominent channels of magnetic gamma-transitions at typical nuclear reactors, as listed in Table 1. Thermal neutron captures on the ^{10}B in the control rods and on proton in the cooling water produce α+^{7}Li* and d+γ/a, respectively. The sources of ^{91}Y*, ^{97}Nb*, ^{135}Xe*, and ^{137}Ba* are all fission daughters. The de-excitations of ^{7}Li* and other fission daughters may emit axions. The axion flux (ϕ_a) at a distance L from a reactor core of fission rate R_f can be described by

$$\phi_a(L) = \frac{R_f \cdot \Phi_\gamma}{4\pi L^2} \cdot \frac{\Gamma_a}{\Gamma_\gamma} \cdot P_{dk} \cdot P_{int} \ , \qquad (1)$$

where (Γ_a/Γ_γ) is the branching ratio of axion emissions in the transitions. P_{dk} and P_{int} are the probabilities of the axions surviving the flight from reactor core to detector without decays or interactions.

The search strategies for reactor axions were inspired by a previous experiment [3]. Two interaction mechanisms of axions with matter were studied: Primakoff and Compton conversions.

[7] Corresponding author

Table 1. Summary of magnetic transitions and their estimated fluxes from reactor. The measured reactor ON-OFF event rates and the 90% CL upper limits (S_u) are also listed.

Channel	E_γ (keV)	Mode	Φ_γ ($10^{10} cm^{-1}s^{-1}$)	ON-OFF ($day^{-1}kg^{-1}$)	S_u ($day^{-1}kg^{-1}$)
p(n,γ/a)d	2230	Isovector M1	22.1	-0.04±0.03	<0.02
^{10}B(n,α)^{7}Li*	478	M1 $(\frac{1}{2}^-) \rightarrow (\frac{3}{2}^-)$	24.7	-0.09±0.36	<0.49
^{91}Y*	555	M4 $(\frac{9}{2}^+) \rightarrow (\frac{1}{2}^-)$	2.10	-0.34±0.15	<0.05
^{97}Nb*	743	M4 $(\frac{1}{2}^-) \rightarrow (\frac{9}{2}^+)$	4.81	0.19±0.31	<0.69
^{135}Xe*	526	M4 $(\frac{11}{2}^-) \rightarrow (\frac{3}{2}^+)$	0.85	0.37±0.15	<0.62
^{137}Ba*	662	M4 $(\frac{11}{2}^-) \rightarrow (\frac{3}{2}^+)$	0.37	-0.19±0.39	<0.46

These processes are respectively and independently sensitive to $g_{a\gamma\gamma}$ and g_{aee}. The total energy of the axions can be fully converted by either of the two processes into measurable ionization energy in the HPGe, such that the experimental signatures are the presence of mono-energetic lines at the known axion energies (E_a's) during the Reactor ON periods. Evidence of reactor axions would manifest as peaks at the known energies of Table 1 in the Reactor ON−OFF residual spectra. As illustrations, the ON/OFF background spectra and the ON−OFF residual spectra for the six candidate lines are depicted in Figures 1 and 2, respectively. The count rates and their errors of the various transitions shown in Table 1 were derived by best-fits of the residual spectra to Gaussian lines at fixed E_a's and resolutions. No excess were observed in all channels and upper limits of the signal rates (S_u) at 90% confidence level (CL) were derived and listed in Table 1.

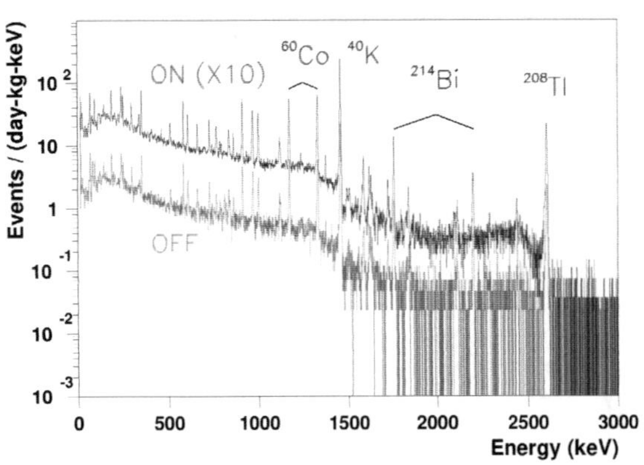

Figure 1. The ON and OFF spectra.

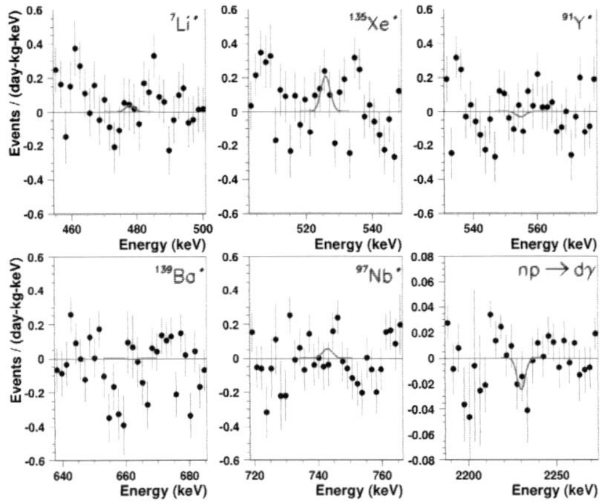

Figure 2. The residual spectra for the six specific channels, with the best-fit Gaussian peaks overlaid.

The experimentally measured upper limits S_u of Table 1 can be translated to bounds among the axion parameters: m_a, $g_{a\gamma\gamma}$, g_{aee}, g^0_{aNN} and g^1_{aNN}. Two approaches were adopted [1]: (a) specific models were used and tested, and (b) model-independent constraints among the parameters were derived.

The model-dependent exclusion plot of $g_{a\gamma\gamma}$ (g_{aee}) versus m_a is depicted in Figure 3 (4). The model-dependence was introduced by fixing $(\Gamma_a/\Gamma_\gamma)_{pn}$ as a function of m_a [1]. The boundaries defined by "Int", "Decay" and "Kine" are constraints due to axions interactions with matter, decays in flights and kinematics, respectively. The bounds from the KS reactor axion searches improve on those of Ref. [3] marked "Zn" in Figures 3 and 4 by two orders of magnitude, owing to enhanced axion flux, lower background and larger data sample. The KS results define the global exclusion boundaries [1] in $g_{a\gamma\gamma}$ for 10^3 eV $\lesssim m_a \lesssim 10^6$ eV and in g_{aee} for 10^2 eV $\lesssim m_a \lesssim 10^6$ eV.

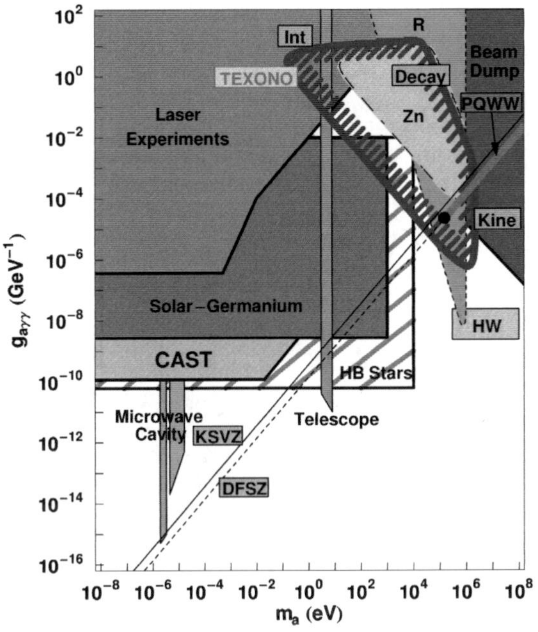

Figure 3. Exclusion plots of $g_{a\gamma\gamma}$ versus m_a for g_{aee}=0. Our model-dependent limits are denoted by "TEXONO".

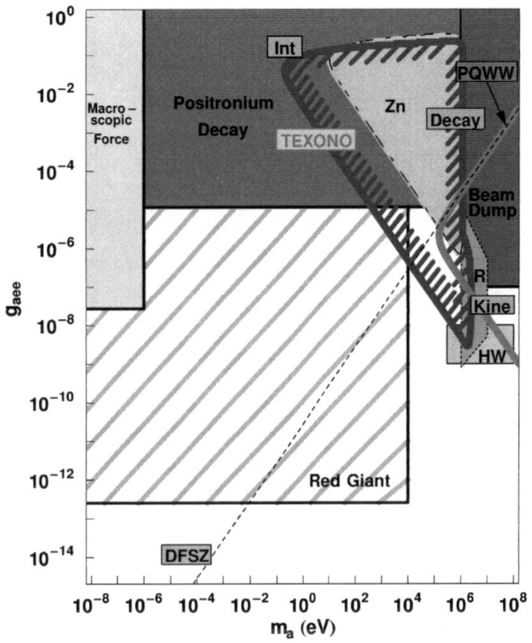

Figure 4. Exclusion plots of g_{aee} versus m_a for $g_{a\gamma\gamma}$=0. Similar conventions as Figure 3 are adopted.

Limit on the Primakoff conversion rate gives rise to the model-independent constraint

$$g_{a\gamma\gamma}^2 \cdot (\frac{\Gamma_a}{\Gamma_\gamma})_{pn} < 5.9 \times 10^{-17}\ \mathrm{GeV}^{-2} \Rightarrow g_{a\gamma\gamma} \cdot g_{aNN}^1 < 7.7 \times 10^{-9}\ \mathrm{GeV}^{-1} \tag{2}$$

which are applicable within the ranges of $m_a \lesssim 10^5$ eV and $g_{a\gamma\gamma} \lesssim 20$ GeV^{-1}, bounded by kinematics and axion interactions in flight, respectively. Similarly, the limit on the Compton conversion rate leads to

$$g_{aee}^2 \cdot (\frac{\Gamma_a}{\Gamma_\gamma})_{pn} < 1.7 \times 10^{-20} \Rightarrow g_{aee} \cdot g_{aNN}^1 < 1.3 \times 10^{-10} \tag{3}$$

for $m_a \lesssim 10^6$ eV and $g_{aee} \lesssim 0.2$.

References

[1] Chang H M et al 2007 *Phys. Rev.* D **75** 052004, and references therein

[2] Li H B et al 2003 *Phys. Rev. Lett.* **90** 131802 Wong H T et al 2007 *Phys. Rev.* D **75** 012001

[3] Avignone III F T et al 1988 *Phys. Rev.* D **37** 618

Prospects of cold dark matter searches with an ultra-low-energy germanium detector

H T Wong[1], M. Deniz[1,2], H B Li[1], S K Lin[1], S T Lin[1], J Li[3], X Li[3], X C Ruan[4], V Singh[5], Q Yue[3], Z Y Zhou[4]

[1] Institute of Physics, Academia Sinica, Taipei 11529, Taiwan.
[2] Department of Physics, Middle East Technical University, Ankara 06531, Turkey.
[3] Department of Engineering Physics, Tsing Hua University, Beijing 100084, China.
[4] Department of Nuclear Physics, Institute of Atomic Energy, Beijing 102413, China.
[5] Department of Physics, Banaras Hindu University, Varanasi 221005, India.
Paper presented at the conference by H T Wong

E-mail: htwong@phys.sinica.edu.tw

Abstract.
The report describes the research program on the development of ultra-low-energy germanium detectors, with emphasis on WIMP dark matter searches. A threshold of 100 eV is achieved with a 20 g detector array, providing a unique probe to the low-mass WIMP. Present data at a surface laboratory is expected to give rise to comparable sensitivities with the existing limits at the $5 - 10$ GeV WIMP-mass range. The projected parameter space to be probed with a full-scale, kilogram mass-range experiment is presented. Such a detector would also allow the studies of neutrino-nucleus coherent scattering and neutrino magnetic moments.

Weakly Interacting Massive Particles (WIMP, denoted by χ) are the leading candidates for cold dark matter (CDM) [1], and supersymmetric (SUSY) particles are the favored WIMP candidates. The popular SUSY models prefer WIMP mass (m_χ) of the range of $\sim$100 GeV, though light neutralinos remain a possibility [2]. Simple extensions of the Standard Model with a singlet scalar favors light WIMPs [3]. Most CDM experiments optimize their design in the high-mass region, and have diminishing sensitivities for $m_\chi < 10$ GeV, where there is an allowed region if the DAMA annual modulation data are interpreted as WIMP signatures [4]. To probe this low-mass region, detectors with sub-keV threshold are necessary. These low-threshold detectors will also open the window to look for WIMPs bound in the solar system [5], as well as non-pointlike SUSY candidates like Q-balls [6]. Sensitivity to sub-keV energy presents a formidable challenge to detector technology and to background control. So far, only the CRESST-I experiment have derived exclusion limits [7] with sapphire(Al_2O_3)-based cryogenic detector at a threshold of 600 eV.

A research program in low energy neutrino and astroparticle physics is pursued [8] by the TEXONO Collaboration at the Kuo-Sheng(KS) Reactor Laboratory. A scientific goal is to develop advanced detectors with kg-size target mass, 100 eV-range threshold and low-background specifications for WIMP searches as well as for the studies of neutrino-nucleus coherent scatterings [9] and neutrino magnetic moments (μ_ν) [10]. The KS laboratory is located 28 m from a 2.9 GW reactor core and has an overburden of about 30 meter-water-equivalence (mwe). Its facilities are described in Ref. [11], where the μ_ν-studies with a 1.06 kg germanium detector (HPGe) at a hardware threshold of 5 keV were reported. This HPGe has also been used for the studies of reactor electron-neutrinos [12] as well as searches for reactor axions [13]. The experimental procedures were well-established and the background above 12 keV were measured. In particular, a background level of about ~ 1 event $kg^{-1}keV^{-1}day^{-1}$(cpd) comparable to those of other underground CDM experiments was achieved.

"Ultra-Low-Energy" germanium (ULEGe) detectors, developed originally for soft X-rays detection, are candidate technologies to meet the challenges of probing into the unexplored sub-keV energy domain [9]. These detectors typically have modular mass of 5-10 grams. Detector array of up to N=30 elements have been successfully built, while there are recent advances in developing single-element ULEGe of kg-size mass [14]. Various prototypes based on this detector technology have been constructed. Depicted in Figure 1 is the measured energy spectrum due to external ^{55}Fe calibration sources (5.90 and 6.49 keV) together with X-rays from Ti (4.51 and 4.93 keV), Ca (4.01 keV), S (2.46 keV) and Al (1.55 keV). Random trigger events uncorrelated to the detector provided the zero-energy pedestals. Pulse shape discrimination (PSD) criteria were applied as illustrated in Figure 2 by correlating two output with different electronics amplifications and shaping times. The electronic noise edge was suppressed by PSD and a threshold of 100 eV was achieved. The deviations of the low energy spectra of the selected events from a flat distribution gave the efficiencies of the PSD cuts.

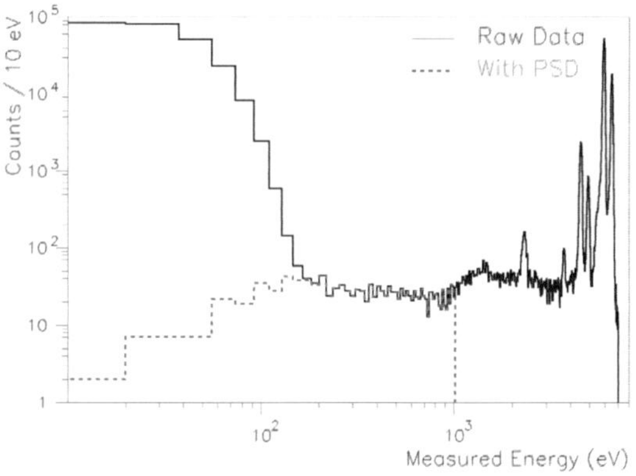

Figure 1. Measured energy spectra with the ULEGe prototype, using ^{55}Fe source producing X-rays from various isotopes. A threshold of 100 eV was achieved, and the electronic noise edge was suppressed by PSD.

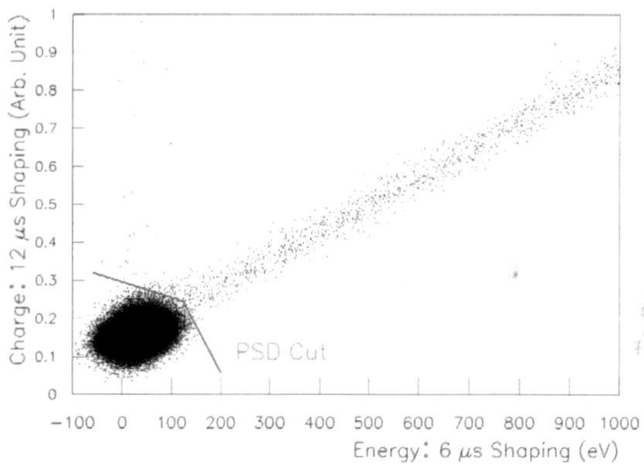

Figure 2. Pulse shape discrimination: correlations of signals with different electronic amplifications and shaping times lead to suppression of the noise edge.

Low-background data were taken at KS under the same shielding configurations as the magnetic moment experiments [11], using a 4-channel ULEGe detector each having an active mass of 5 g. The recorded spectrum with 0.34 kg-day of data after cosmic-rays and anti-Compton vetos and PSD selection is displayed in Figure 3. It can be seen that comparable background level was achieved as the CRESST-I [7] experiment with 1.51 kg-day of data. A summary of the spin-independent exclusion plot is depicted in Figure 4. Comparable limits to CRESST-I can be expected from the current KS results. Intensive efforts on the data analysis are underway. The projected sensitivities with 1 kg-year of data at 100 eV threshold and a 1 cpd background level are shown.

An R&D program towards the realizations of a full-scale experiment is rigorously pursued. Quenching factor measurement for nuclear recoils in Ge with sub-keV ionization energy will be performed at a neutron beam facility. Background studies are conducted at both KS and the

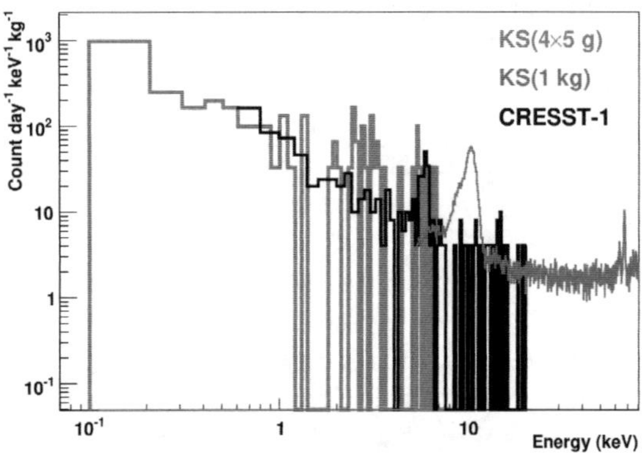

Figure 3. Measured background spectra from 0.34 kg-day of data for the KS 4×5 g ULEGe. Overlaid for comparisons are those of the KS 1 kg HPGe detector [11] and the CRESST-1 experiment [7].

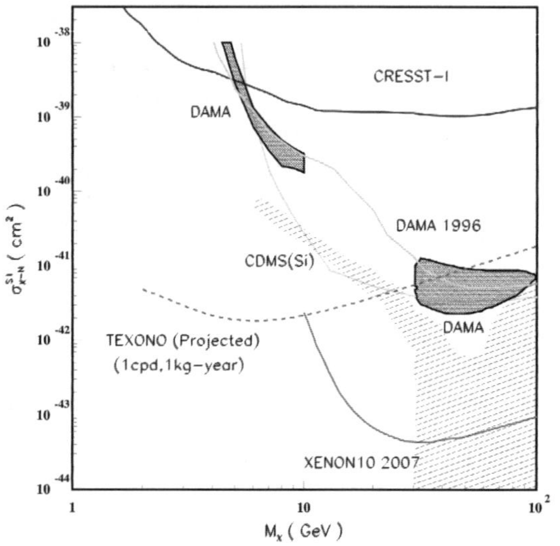

Figure 4. Expected spin-independent sensitivity region for CDM searches using a ULEGe detector with a mass of 1 kg at 1 cpd background level. Also shown are the present exclusion boundaries [1, 15] and the DAMA allowed regions [4]. The striped region is the parameter space favored by the SUSY models [2].

Yang-Yang Underground Laboratory (700 mwe) in South Korea with the various prototypes. Background understanding at the sub-keV range is a challenging and unexplored subject in its own right. External background is expected to be reduced due to self-shielding effects in a kg-mass detector, as well as by additional active veto Ge layers enclosing hermetically the ULEGe inner target [9]. Studies are under way with a 180-g segmented ULEGe prototype equipped with a veto ring and dual-readout channels from both the signal and high-voltage electrodes. A 500-g detector similar to the design demonstrated in Ref. [14] is being constructed.

References

[1] Drees M and Gerbier G 2006 *J. Phys.* **G 33** 233, and references therein.
[2] Bottino A et al. 2005 *Phys. Rev.* **D 72** 083521
[3] He X G et al. 2007 *Mod. Phys. Lett.* **A 22** 2121
[4] Gondolo P and Gelmini G 2005 *Phys. Rev.* **D 71** 123520
[5] Damour T and L.M. Krauss 1998 *Phys. Rev. Lett.* **81** 5726; Collar J I 1999 *Phys. Rev.* **D 59** 063514
[6] Gelmini G, Kusenko A and Nussinov S 2002 *Phys. Rev. Lett.* **89** 101302
[7] Angloher G et al. 2002 *Astropart. Phys.* **18** 43
[8] Wong H T 2004 *Mod. Phys. Lett.* **A 19** 1207
[9] Wong H T et al. 2006 *J. Phys. Conf. Ser.* **39** 266
[10] Wong H T and Li H B 2005 *Mod. Phys. Lett.* **A 20** 1103
[11] Li H B et al. 2003 *Phys. Rev. Lett.* **90** 131802; Wong H T et al. 2007 *Phys. Rev.* **D 75** 012001
[12] Xin B et al. 2005 *Phys. Rev.* **D 72** 012006
[13] Chang H M et al. 2007 *Phys. Rev.* **D 75** 052004
[14] Barbeau P A, Collar J I and Tench O 2007 *JCAP* **09** 009
[15] These Proceedings, for complete references and latest updates.

Status report of the Tokyo axion helioscope experiment

Y Inoue[1], M Minowa[2], Y Akimoto[2], R Ota[2], T Mizumoto[2] and A Yamamoto[3]

[1] International Center for Elementary Particle Physics, University of Tokyo, 7-3-1 Hongo, Bunkyo-ku, Tokyo 113-0033, Japan

[2] Department of Physics, School of Science, University of Tokyo, 7-3-1 Hongo, Bunkyo-ku, Tokyo 113-0033, Japan

[3] High Energy Accelerator Research Organization (KEK), 1-1 Oho, Tsukuba, Ibaraki 305-0801, Japan

E-mail: `berota@icepp.s.u-tokyo.ac.jp`

Abstract. We have searched for solar axions with a detector which consists of a 4 T × 2.3 m superconducting magnet, PIN-photodiode X-ray detectors, and an altazimuth mount to track the sun. The conversion region is filled with cold helium gas which modifies the axion mass at which coherent conversion occurs. In the past measurements, axion mass from 0 to 0.27eV have been scanned. Since no positive evidence was seen, an upper limit to the axion-photon coupling constant was set to be $g_{a\gamma} < 6\text{–}10 \times 10^{-10}$/GeV (95% CL) depending on the axion masses. We are now actively preparing for a new stage of the experiment aiming at one to a few eV solar axions. In this mass region, our detector might be able to check parameter regions which are preferable to the axion models.

1. Introduction

Axions [1] that are thermally produced in the core of the sun through the Primakoff process (Fig. 1) are called the solar axions. Sikivie [2] proposed an ingenious experiment to detect such axions. The detection device called axion helioscope is a system of a strong magnet and an X-ray detector, where the solar axions is converted into X-ray photons through the inverse Primakoff process (Fig. 1) in the magnetic field. Conversion is coherently enhanced even for massive axions by filling the conversion region with light gas. If the axion mass m_a is at around a few eV, detection of the solar axions becomes feasible.

We have constructed an axion helioscope with a dedicated superconducting magnet. Cold helium gas was used as the conversion medium, and the highest axion mass was targeted at 2.6 eV. We report the current status of this experiment.

2. The Sumico V detector

The schematic figure of the axion helioscope is illustrated in Fig. 2. It consists of a superconducting magnet, X-ray detectors, a gas container, and an altazimuth mounting.

The magnet [3] consists of two 2.3-m long race-track shaped superconducting coils running parallel with a 20-mm wide gap between them. The transverse magnetic field in the gap is 4 T. The magnetic field can be maintained without an external power supply with a help

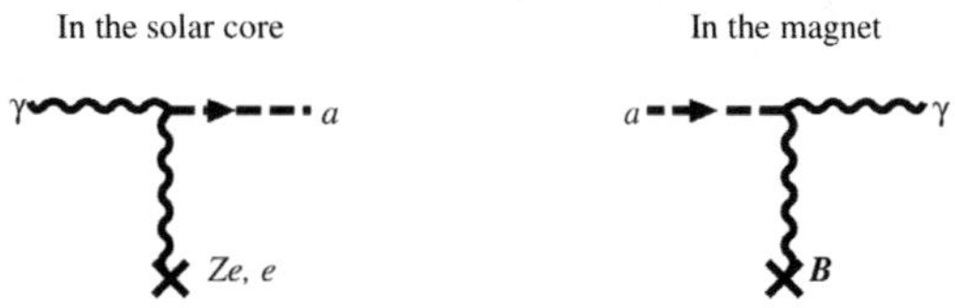

Figure 1. The solar axions produced via the Primakoff process in the solar core are, then, converted into X-rays via the reverse process in the magnet.

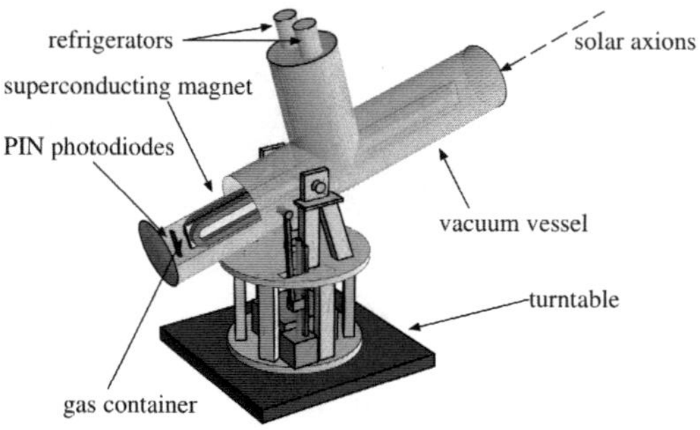

Figure 2. The schematic view of the axion helioscope.

of a persistent current switch. The magnet is kept lower than 6 K by two Gifford-McMahon refrigerators. The container to hold dispersion-matching gas is inserted in the aperture of the magnet. It is made of four 2.3-m long stainless-steel square pipes and 5N high purity aluminium sheets wrapping around them to achieve high uniformity of temperature. The measured thermal conductance between the both ends was 1×10^{-2}W/K at 6 K under 4 T. The "solar" end of the gas container is suspended by three Kevlar cords. The "detector" end at the opposite side is flanged to be fixed to the magnet. This end is terminated with an X-ray window which is transparent above 2 keV and can hold gas up to 0.3 MPa. The gas introducing pipelines are also at this side. The converted X-ray is viewed by sixteen PIN photodiodes. Details on the X-ray detector are given in Ref. [4]. They are constructed in a vacuum vessel which is mounted on an altazimuth mount to track the sun. It can track the sun about a half of a day. During the other half of a day, background spectrum is measured.

3. Past and future measurements

3.1. Phase I — the first measurement

Phase I of the solar observation was performed from 26th till 31st December 1997 without the gas container [5]. Since the conversion region was vacuum, the sensitivity was limited below $m_a < 0.03$ eV. From the absence of an axion signal, an upper limit on the axion-photon coupling is given to be $g_{a\gamma} < 6.0 \times 10^{-10}\text{GeV}^{-1}$ (95% CL).

3.2. Phase II — with low density helium gas

Phase II was performed from July to September 2000, where low density helium was introduced [6]. The sensitive mass region was brought up by an order of magnitude. Again, no signature of axions was seen, restricting the axion-photon coupling to be $g_{a\gamma} < (6.2\text{–}10.4) \times 10^{-10}\text{GeV}^{-1}$ for $0.05 < m_a < 0.27$ eV. The exclusion limit at 95% confidence level obtained from Phase I and II are shown together in Fig. 3.

3.3. Phase III — the next stage

At present, we are preparing for the next stage. In Phase III, high density helium will be introduced aiming at more massive axions up to $m_a \lesssim 2.2\text{–}2.6$ eV. To achieve this goal, several developments have been made. When the superconducting magnet quenches, the high-density helium should be relieved safely before the gas container would burst. The gas introducing pipeline was completely reworked based on the measured pressure change after a forced quench with a safely low pressure helium. In addition, a cryogenic rupture disk was installed as a last resort. Another challenge is that the higher the axion mass becomes, the narrower the resonance

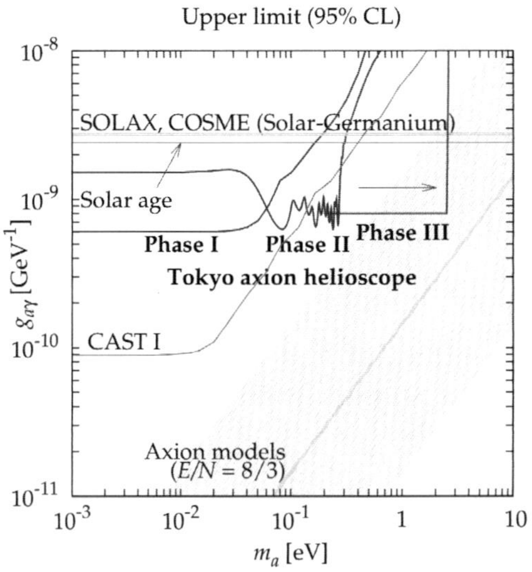

Figure 3. Exclusion limit on $g_{a\gamma}$ versus m_a at 95% confidence level. Phase I and II denote the past results from this experiment, and Phase III shows the area where is aimed at the next stage. The limits inferred from the solar age consideration, the limits from other experiments: SOLAX [7], COSME [8] and CAST [9] are show together. The shaded area shows the preferred axion models.

width of the conversion becomes. The gas density should be controlled with an extreme accuracy typically of the order of 0.1% at $m_a = 2.6\,\mathrm{eV}$, and many data points should be measured to scan a range. For this purpose, an automated gas controlling system was developed. It has successfully demonstrated to stabilize the temperature and to control the pressure of helium gas with a good accuracy.

4. Conclusions

We have searched for solar axions with an axion helioscope, and axion mass from 0 to 0.27eV have been scanned. No evidence of axion signal has been seen so far, implying an exclusion limit to the axion-photon coupling as shown in Fig. 3. In the lower mass region, the above limit has been renewed by CAST. However, we are preparing for the next stage aiming at a higher mass region where our detector might be able to check the parameter regions preferred by theory. In December 2007, a preliminary measurement started at $m_a \simeq 1\,\mathrm{eV}$. Remaing problems are being resolved one by one. The ultimate density will be reached in 2008, although a full range scan will take years.

Acknowledgments

The authors thank the ex-director general of KEK, Professor H. Sugawara, for his support in the beginning of the experiment. The authors are also thankful to the support from the Grant-in-Aid for COE research by the Japanese Ministry of Education, Science, Sports and Culture, and the support from the Matsuo Foundation.

References

[1] Peccei R D and Quinn H R 1977 *Phys. Rev. Lett.* **38** 1440
[2] Sikivie P 1983 *Phys. Rev. Lett.* **51** 1415
[3] Sato Y *et al.* 1998 *Proc. of the 15th International Conference on Magnet Technology (MT-15)* ed Liangzhen L, Guoliao S and Luguang Y (Beijing: Science Press) pp 262–5
[4] Namba T *et al.* 2002 *Nucl. Instr. and Meth.* A **489** 224
[5] Moriyama S *et al.* 1998 *Phys. Lett.* B **434** 147
[6] Inoue Y *et al.* 2002 *Phys. Lett.* B **536** 18
[7] Gattone A O *et al.* (The SOLAX collaboration) 1999 *Nucl. Phys.* B (Proc. Suppl.) **70** 59
[8] Morales A *et al.* 2002 *Astropart. Phys.* **16** 325
[9] Irastorza G I *et al.* (CAST collaboration) 2007 *2007 Electroweak interactions and unified theories* p 301 URL `http://indico.in2p3.fr/conferenceDisplay.py?confId=151`

Discovery of underground argon with low level of radioactive ^{39}Ar and possible applications to WIMP dark matter detectors

C Galbiati[*,1], D Acosta-Kane[1], R Acciarri[2], O Amaize[1], M Antonello[2], B Baibussinov[3], M Baldo Ceolin[3], C J Ballentine[4], R Bansal[5], L Basgall[6], A Bazarko[7], P Benetti[8], J Benziger[9], A Burgers[1], F Calaprice[1], E Calligarich[8], M Cambiaghi[8], N Canci[2], F Carbonara[10], M Cassidy[11], F Cavanna[2], S Centro[3], A Chavarria[1], D Cheng[1], A G Cocco[10], P Collon[12], F Dalnoki-Veress[1], E de Haas[1], F Di Pompeo[2], G Fiorillo[10], F Fitch[13], V Gallo[10], M Gaull[1], S Gazzana[14], L Grandi[14], A Goretti[1], R Highfill[5], T Highfill[5], T Hohman[1], Al Ianni[14], An Ianni[1], A LaCava[15], M Laubenstein[14], H Y Lee[16], M Leung[1], B Loer[1], H. H Loosli[17], B Lyons[1], D Marks[1], K McCarty[1], G Meng[3], C Montanari[8], S Mukhopadhyay[18], A Nelson[1], O Palamara[14], L Pandola[14], F Pietropaolo[3], T Pivonka[5], A Pocar[19], R Purtschert[17], A Rappoldi[8], G Raselli[8], F Resnati[20], D Robertson[12], M Roncadelli[8], M Rossella[8], C Rubbia[14], J Ruderman[1], R Saldanha[1], C Schmitt[12], R Scott[16], E Segreto[14], A Shirley[21], A M Szelc[22,2], R Tartaglia[14], T Tesileanu[1], S Ventura[3], C Vignoli[8], C Visnjic[1], R Vondrasek[16] and A Yushkov[10]

[*] Contribution Presented by C. Galbiati at TAUP 2007. E-mail: `galbiati@Princeton.EDU`

[1] Department of Physics, Princeton University, Princeton, NJ 08544, USA

[2] INFN and Dipartimento di Fisica, University of L'Aquila, L'Aquila 67100, Italy

[3] INFN and Dipartimento di Fisica, University of Padua, Padua 35131, Italy

[4] School of Earth, Atmospheric, and Environmental Sciences, University of Machester, M13 9PL, United Kingdom

[5] AirSep Corporation, Buffalo, NY 14228, USA

[6] Kansas Refined Helium, Otis, KS 67565, USA

[7] Schlumberger Princeton Technology Center, Princeton, NJ 08550, USA

[8] INFN and Dipartimento di Fisica Nucleare e Teorica, University of Pavia, Pavia 27100, Italy

[9] Department of Engineering, Princeton University, Princeton, NJ 08544, USA

[10] INFN and Dipartimento di Scienze Fisiche, University of Naples "Federico II", Naples 80216, Italy

[11] Department of Geosciences, University of Houston, Houston, TX 77204, USA

[12] Department of Physics, Notre Dame University, Notre Dame, IN 46556, USA

[13] Linde Engineering, Murray Hill, NJ 07974, USA

[14] INFN, Laboratori Nazionali del Gran Sasso, Assergi (AQ) 67100, Italy

[15] Felician College, Lodi, NJ 07644, USA

[16] Argonne National Laboratories, Argonne, IL 60439, USA

[17] Physics Institute, University of Bern, Sidlerstrasse 5, 3012 Bern, Switzerland

[18] Department of Earth and Planetary Sciences, Harvard University, Cambridge, MA 02138, USA

[19] Department of Physics, Stanford University, Stanford, CA 94305, USA

[20] INFN and Dipartimento di Fisica, University of Milano Bicocca, Milano 20126, Italy
[21] Linde Gas, Murray Hill, NJ 07974, USA
[22] Instytut Fizyki Jadrowej PAN, 31-342 Krakow, Poland

Abstract. We report on the first measurement of ^{39}Ar in argon from underground natural gas reservoirs. The gas stored in the US National Helium Reserve was found to contain a low level of ^{39}Ar. The ratio of ^{39}Ar to stable argon was found to be $\leq 4\times10^{-17}$ (84% C.L.), less than 5% the value in atmospheric argon (^{39}Ar/Ar=8×10^{-16}). The total quantity of argon currently stored in the National Helium Reserve is estimated at 1000 tons. ^{39}Ar represents one of the most important backgrounds in argon detectors for WIMP dark matter searches. The findings reported demonstrate the possibility of constructing large multi-ton argon detectors with low radioactivity suitable for WIMP dark matter searches.

Argon is an ideal target for WIMPs searches as it allows detection of rare WIMP-induced nuclear recoils down to a few keV by scintillation and/or ionization. Liquid argon, in particular, is an excellent material for use as a detector of ionizing particles. It produces copious scintillation light, allows the drift of the ionization charge over long distances, and it has been used for large detectors [1]. Moreover, the difference in the stopping power between nuclear recoils and β/γ events leads to significant differences in the ratio of ionization charge to scintillation light detected and in the scintillation light pulse shape, providing powerful tools to separate WIMP-induced events from natural radioactivity [2, 3]. Studies with the WARP 3.2-kg liquid argon detector demonstrated that pulse shape alone permits a separation of 1 recoil in 10^8 β's, and the separation by the ratio of scintillation to ionization is 1 in 10^2 [2].

For a liquid argon WIMP detector, the separation of recoil events from β events is particularly important because of the intrinsic background from β decays of ^{39}Ar, present in atmospheric argon. The specific activity of ^{39}Ar (Q = 565 keV, $t_{1/2}$ = 269 y) is $\sim$1 Bq/kg of atmospheric argon [4, 5]. ^{39}Ar is produced by cosmic ray interactions in the atmosphere, principally via the ^{40}Ar(n,2n)^{39}Ar reaction [6, 7].

The WARP 3.2-kg detector [2] published results from a first search for WIMPs obtaining a sensitivity comparable to the best current limits [8]. The high selectivity for argon recoils should permit a sensitive WIMP search with a 140 kg liquid argon detector employing atmospheric argon, currently under construction at Laboratori Nazionali del Gran Sasso [9]. However, in spite of its favorable β/recoil discriminating power, it is highly desirable to use argon with a much lower ^{39}Ar contamination for future, larger detectors. Based on the proven β/recoil discrimination, a 10-fold or more reduction of ^{39}Ar with respect to the atmospheric level would enhance the prospects of future multi-ton argon WIMP detectors.

Argon from natural gas wells is a promising source of ^{39}Ar-depleted argon because ^{39}Ar production induced by cosmic rays is strongly suppressed underground. In the subsurface, ^{39}Ar can be produced by a number of reactions, mainly neutron reactions on potassium, such as ^{39}K(n,p)^{39}Ar [10]. Argon samples collected in groundwater from U and Th rich crystalline rocks revealed an enhanced ^{39}Ar activity, with ^{39}Ar/Ar ratios exceeding atmospheric values by up to 16 times [11]. The enhanced activity is likely due to an abundant local neutron flux, originating from fission or (α,n) reactions induced by decays in the U and Th chains. The ^{39}Ar/Ar ratio in subsurface gases thus depends on the local U, Th, and K concentration, on the porosity of the surrounding rocks, and may vary significantly among geological formations.

We report on the first measurement of the ^{39}Ar/Ar ratio in natural gas wells [12]. We measured argon samples originating from the US National Helium Reserve, located in the Cliffside Storage Facility outside Amarillo, TX [13]. The facility stores crude helium separated by chromatography and/or cryogenic distillation from the nearby helium-rich natural gas fields. The separation processes employed also transfer a limited fraction of the argon contained in the natural gas along with the crude helium stream. All plants producing crude helium are connected to the US National Helium Reserve by the Helium Conservation Pipeline, run by the Bureau of Land Management.

Crude helium samples from the National Helium Reserve were collected from the Helium Conservation Pipeline at the Kansas Refined Helium plant in Otis, Kansas. The composition of the crude helium, as measured by mass spectrometry, can be found in Ref. [12]. The crude helium was processed with a novel Pressure Swing Adsorption plant developed at Princeton to remove non-argon components [12]. Measurements of ^{39}Ar activity were performed at the Low Level Counting Underground Laboratory at the Physics Institute at the University of Bern with a low background proportional counter [4].

The detector background is comparable to the rate due to ^{39}Ar in atmospheric argon and must be determined accurately. Three measurements are performed to measure the ^{39}Ar content in the sample of interest: one with a reference ^{39}Ar-depleted sample to measure intrinsic background, one with standard atmospheric argon for reference, and one with the underground sample itself. The ^{39}Ar activity of the sample under investigation is evaluated by subtracting the detector background from the total measured activity and is then compared to the ^{39}Ar activity in atmospheric argon (background subtracted). More details can be found in Ref. [12].

The best estimate for the ^{39}Ar activity indicates a ^{39}Ar/Ar ratio $\leq 4\times10^{-17}$, which is less than 5% of the ^{39}Ar/Ar ratio in atmospheric argon. The depletion factor from the atmospheric activity of the sample is therefore $\geq$20 at 84.1% C.L. ($\geq$10 at the 97.7% C.L.).

This result is, to the best of our knowledge, the first measurement of the concentration of ^{39}Ar in subsurface gas. It represents an upper limit which is based on the background of the proportional counting system. It therefore motivates the development of new and potentially more sensitive techniques, such as Accelerator Mass Spectrometry (AMS) [14] and ^{39}Ar-decay counting in liquid argon detectors [5], for a more accurate determination of the ^{39}Ar/Ar ratio.

The discovery of low ^{39}Ar/Ar ratio in the US National Helium Reserve is part of a larger program of exploration of several possible underground argon sources. During this investigation, the authors developed the technology required to separate and collect large quantities of argon from natural gas wells. The quantity of argon processed by the Kansas Refined Helium plant is 25 tons per year. Production of a 10 ton target of liquid argon for a dark matter search experiment would take a 1-yr production campaign after construction and commissioning of dedicated separation plants. These findings lead the way for future multi-ton, low background, argon detectors for WIMP dark matter, able to reach sensitivities for the WIMP-nucleon cross section of 10^{-46} cm^2 or smaller.

[1] S. Amerio *et al* (ICARUS Collaboration), Nucl. Instr. Meth. A, **527**, 329 (2004).

[2] P. Benetti *et al* (WARP Collaboration), Astrop. Phys. **28**, 495 (2008).

[3] A. Hitachi *et al* , Phys. Rev. B **27**, 5279 (1983); M.G. Boulay and A. Hime, Astropart. Phys. **25**, 179 (2006).

[4] H.H. Loosli, Earth Plan. Sci. Lett. **63**, 51 (1983).

[5] P. Benetti *et al* (WARP Collaboration), Nucl. Instr. Meth. A **574**, 83 (2007).

[6] B.E. Lehmann and H.H. Loosli, in *Proceedings of the 6th International Symposium on Water-Rock Interaction, 3-6 August 1989, Malvern, UK*, edited by D.L. Miles (A.A. Balkema, 1989), pp. 429432.

[7] B.E. Lehmann and H.H. Loosli, in *Applied Isotope Hydrogeology: A Case Study in Northern Switzerland*, edited by F.J. Pearson *et al* (Elsevier, 1991), pp. 239 296.

[8] D.S. Akerib *et al* (CDMS Collaboration), Phys. Rev. Lett. **96**, 011302 (2006); J. Angle *et al* (XENON Collaboration), `arXiv:0706.0039`.

[9] R. Brunetti *et al* (WARP Collaboration), *WArP: Wimp Argon Programme*, `warp.lngs.infn.it`.

[10] J.T. Fabryka-Martin, Ph.D. thesis, Univ. of Arizona (1988).

[11] H.H. Loosli, B. E. Lehmann, and W. Balderer, Geochim. Cosmochim. Acta **53**, 1825 (1989).

[12] D. Acosta-Kane *et al* "Discovery of underground argon with low level of radioactive 39Ar and possible applications to WIMP dark matter detector", `arXiv:0712.0381`, in press on Nucl. Instrum. Meth. A.

[13] Amarillo Field Office of the Bureau of Land Management, `www.nm.blm.gov/amfo/amfo home.html`.

[14] P. Collon, W. Kutschera, and Z.T. Lu, Ann. Rev. Nucl. Part. Sci. **54**, 39 (2004).

Limits on WIMP-nucleon cross-sections from the ZEPLIN-II data

Vitaly A. Kudryavtsev for the ZEPLIN-II Collaboration
Department of Physics and Astronomy, University of Sheffield, Sheffield S3 7RH, UK
E-mail: v.kudryavtsev@sheffield.ac.uk

Abstract. Results are presented from ZEPLIN-II, the world-first 31 kg two-phase xenon dark matter detector. Discrimination between nuclear recoils and background electron recoils is achieved by recording both the scintillation and ionisation signals generated within the liquid xenon, with the ratio of these signals being different for the two classes of events. This ratio is calibrated for different incident species using neutron and γ-ray sources. Background populations of gamma-induced events and radon progeny events were observed in the data. An acceptance window, defined by the neutron calibration data, of 50% nuclear recoil acceptance between 5 keV and 20 keV electron equivalent energy, had an observed count of 29 events, with a summed expectation of 28.6$\pm$4.3 γ-ray and radon progeny induced background events. These figures provide a 90% CL upper limit to the number of nuclear recoils of 10.4 events in this acceptance window. This corresponds to a WIMP-nucleon spin-independent cross-section with a minimum of 6.6×10^{-7} pb, and to a WIMP neutron spin-dependent cross-section of 7×10^{-2} pb at a WIMP mass of around 65 GeV.

1. Introduction

Search for rare dark matter WIMP events requires a powerful discrimination between nuclear recoils (expected from WIMP and neutron interactions) and electron recoils (from γ-ray background). Event-by-event discrimination in liquid xenon target is based on the ratio of ionisation to scintillation known to be different for the two classes of events. This is implemented in two-phase liquid noble gas detectors where two signals are produced for each event, the 1st one being from the primary scintillation light (S1) and the 2nd one from the secondary scintillation (or electroluminescence) (S2). The latter is caused by electrons produced in the liquid phase, drifted in the electric field through the liquid, extracted into the gas phase and finally accelerated in the high electric field in the gas.

2. ZEPLIN-II experiment

ZEPLIN-II is a two-phase xenon detector with a target mass of 31 kg of liquid xenon, operated at the Boulby Underground Laboratory in the UK at a depth of 2805 m w.e. underground. The target volume is viewed from above by seven 13 cm diameter photomultiplier tubes (PMTs). To protect the target volume from radioactivity in rock (gammas and neutrons), the ZEPLIN-II detector is surrounded by hydrocarbon material and high-purity lead. Part of the hydrocarbon shielding is a liquid scintillator that also acts as an anticoincidence system (active veto) preventing the signals, detected simultaneously in the target and in the veto, being interpreted as WIMP interactions.

The trigger is provided by five-fold coincidences between different PMTs at a single photoelectron level. Either a primary, S1, or a secondary, S2, signal can trigger the data acquisition. The signals from all seven PMTs are recorded with 2 ns sampling time during ± 100 μs around the trigger pulse, covering all possible arrival times for both primary and secondary signals. In an off-line data analysis three-fold coincidences between different PMTs at a single photoelectron level were used to identify and parameterise the primary signal S1.

A number of parameters have been measured for each waveform (Alner et al 2007a, 2007c), the most important being the total areas of the primary and secondary pulses (proportional to the number of photoelectrons), S1 and S2, respectively, the time delay between them and the width of the pulses (that determines whether the pulse is the primary or secondary signal).

Position sensitivity of the experiment in the vertical direction is achieved by considering the time delay of the secondary pulse which is proportional to the drift time of the electrons and thus determines the point of the interaction along the drift field direction. This allows us to exclude events originating on or close to the grid wires that provide the electric field and are contaminated mainly with radon progeny. In the horizontal plane the event position is reconstructed using the relative pulse areas from secondary signals on different PMTs. This method gives bigger uncertainty compared to the drift time due to the large PMT sizes and small photon statistics at low energies.

A daily energy calibration of the detector using a ^{57}Co gamma-ray source allowed monitoring the stability of detector operation. The WIMP/gamma discrimination performance of ZEPLIN-II has been tested by calibrating the detector using high-energy gamma-ray (^{60}Co) and neutron (AmBe) sources. High-energy gamma-rays produce the main electron recoil background, whereas fast neutrons scatter elastically off nuclei producing nuclear recoils in the same way as expected from WIMP scatters. Using the S2/S1 versus S1 plot from neutron calibration run (Alner et al 2007a) the nuclear recoil acceptance box has been defined as retaining 50% of nuclear recoil events at any given energy chosen for analysis (5-20 keV electron equivalent energy). This acceptance box was expected to have a small number of electron recoils due to the tail of S2/S1 distribution observed in the gamma calibration run.

The detector, data acquisition system, analysis procedure and experimental data are described in detail by Alner et al (2007a, 2007b, 2007c).

3. Results

The first data run of the ZEPLIN-II detector had a live time of 31.2 days after time periods with unstable operation conditions were removed from the analysis. A number of software cuts have been applied to the parameterised pulses allowing the selection of single interactions within the fiducial volume of the detector to be made (Alner et al 2007a). For each cut an energy dependent efficiency has been evaluated either from the data or from a combination of data and simulations.

An important cut that reduces significantly the fiducial volume of xenon, is the radial cut. This rejects the events that are reconstructed as being close to the PTFE walls. In reality, due to imperfect position reconstruction of the interaction points in the horizontal plane, a long tail of events assumed to be originated at the walls is reconstructed towards the centre of the detector. The majority of these events is believed to be caused by the alpha decay of radon progeny accumulated on the charged surfaces. To remove most of the 'wall' events the radial cut has been applied reducing the fiducial mass of xenon down to 7.2 kg (Alner et al 2007a).

The analysis of data has revealed 29 events in the nuclear recoil 'acceptance box' defined using neutron calibration data. The expected background rate was estimated by extrapolating distributions of gamma-induced electron recoils and radon progeny events to the acceptance box and found to be 28.6 ± 4.3 (see Alner et al 2007a for full description of the procedure to evaluate the expected background). Applying the procedure described by Feldman and Cousins

(1998) we found that the WIMP signal in the ZEPLIN-II detector is consistent with zero and, hence, the 90% CL upper limit on the number of nuclear recoils was set as 10.4 using central confidence interval. This limit was then converted into an upper limit on the WIMP-nucleon spin-independent (Alner et al 2007a) and WIMP-neutron/proton spin-dependent (Alner et al 2007b) cross-sections. Limit on WIMP-nucleon spin-independent cross-section from ZEPLIN-II is shown in Figure 1 together with the results from some other experiments.

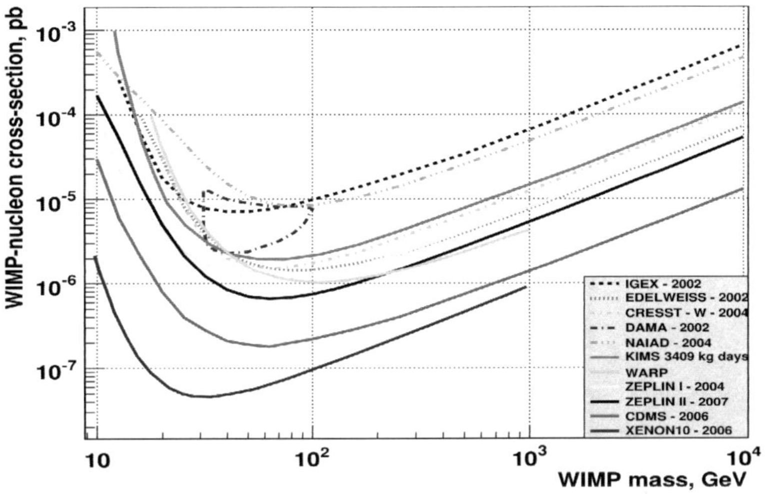

Figure 1. Limits on WIMP-nucleon spin-independent cross-sections from ZEPLIN-II and some other experiments.

4. Summary

The world-first and largest (~ 31 kg) two-phase xenon detector ZEPLIN-II has been operated at the Boulby Underground Laboratory. Data collected during 31.2 days of running did not reveal the presence of WIMP-induced signals. The observed number of events in the nuclear recoil acceptance box, defined using neutron calibrations, was consistent with the expectations of background events based on the extrapolation of the gamma-induced electron recoils and radon progeny events accumulated on the walls, to the acceptance box. The upper limits were set on the WIMP-nucleon spin-independent and spin-dependent cross-sections.

5. Acknowledgments

This work has been funded by the UK Particle Physics And Astronomy Research Council, the US Department of Energy and the US National Science Foundation. We acknowledge support from the Central Laboratories for the Research Councils, the Engineering and Physical Sciences Research Council, the ILIAS integrating activity, Fundação para a Ciência e Tecnologia. We would like to thank Cleveland Potash Ltd., and the underground facility staff for their assistance, and the team from ITEP (Moscow) led by Dr. D. Akimov for their valuable contribution to the ZEPLIN programme.

References

[1] Alner G J et al 2007a *Astroparticle Physics* **28** 287
[2] Alner G J et al 2007b *Physics Letters B* **653** 161
[3] Alner G J et al 2007c Submitted to *Nucl. Instrum. & Meth. in Phys. Res. A*
[4] Feldman G J and Cousins R D 1998 *Phys. Rev. D* **57** 3873.

Status of the EDELWEISS-II experiment

G. Gerbier for the Edelweiss collaboration
CEA Saclay, IRFU, BP 2, 91191 Gif s Yvette, France

E-mail: gilles.gerbier@cea.fr

Abstract. The Edelweiss programme is dedicated to the direct search for Dark Matter as massive weakly interacting particles (WIMPs) with Germanium cryogenic detectors operated in the Laboratoire Souterrain de Modane in the French Alps at a depth of 4800 mwe. After the initial phase Edelweiss I, which involved a total mass of 1 kg, the second step of the programme, Edelweiss II, currently operates 9 kg of detectors and an active shielding of 100 m^2 muon veto detectors and is now in its commissioning phase. The current status and performance of the Edelweiss II set-up in terms of backgrounds will be given, the underground muon flux measured with the muon veto system will be presented.

1. Introduction

Understanding the nature of Dark Matter in the universe is a major challenge for modern cosmology and astrophysics. One of the well-motivated candidates is the generically named Weakly Interacting Massive Particle (WIMP). In the Minimal Supersymetric Standard Model (MSSM) framework, the WIMP could be the Lightest Supersymmetric Particle, which is stable, neutral and massive.

The French-German-Russian EDELWEISS (Expérience pour DEtecter Les Wimps En Site Souterrain) experiment is dedicated to the direct detection of WIMPs trapped in the galactic halo. The detection is performed through the measurement of the recoil energy produced by the elastic scattering of a WIMP off target nuclei.

The main challenges are the extremely low event rate of ~ 1 evt/kg/year and the relatively small deposited energy of < 100 keV.

2. Experimental setup and detectors

2.1. Set up

The EDELWEISS experiment is situated in the Laboratoire Souterrain de Modane (LSM) in the Fréjus Tunnel highway connecting Lyon to Turin behind the French Alps. The 1800 m of rock (4800 mwe) reduces the muon flux down to 4 $m/m^2/day$, a factor 10^6 times less than at the surface. The EDELWEISS-II experiment installation (see figure 1) was completed end of 2005. Specific improvements have been brought to reduce the possible background sources that have limited the sensitivity of EDELWEISS-I [1,2].

The gamma background is screened by a 20 cm thick lead shielding. All the materials used in the vicinity of the detectors have been tested for their radiopurity with a dedicated HPGe detector. A class 10 000 clean room surrounds the whole experiment and a class 100 laminar flow with deradonised air (< 0.1 Bq/m^3) is used when mounting the detectors in the cryostat. In the experimental volume the low energy neutron background in the cavity is attenuated by three orders of magnitude thanks to a 50 cm thick polyethylen shielding. In addition a 100 m^2 plastic scintillator active muon veto surrounding the experiment with a coverage of 99% will tag muons interacting in the lead shielding and producing neutrons. The residual neutron background comes from high energy neutron produced by muon that are not tagged by the veto system and by the neutrons from ^{238}U fission in the lead shielding. Initial Monte Carlo simulation showed that the recoil rate above 10 keV in the detectors is around 0.002 evt/kg/day equivalent to a sensitivity to WIMP-nucleon cross-section of 2.10^{-8} pb for WIMP masses around 100 GeV/c^2. This corresponds to an improvement of a factor 100 compared to EDELWEISS-I sensitivity.

The dilution cryostat used in the experiment is of reversed design, with the experimental chamber on the top of the structure. It is a nitrogen-free system, using three pulse tubes to cool the 100 K and 2 0K screens and a He reliquifier reducing the He consumption. The experimental volume is about 50 liters allowing the installations of 110 detectors of about 330 g each in a compact arrangement for self shielding and multiple interactions identifications.

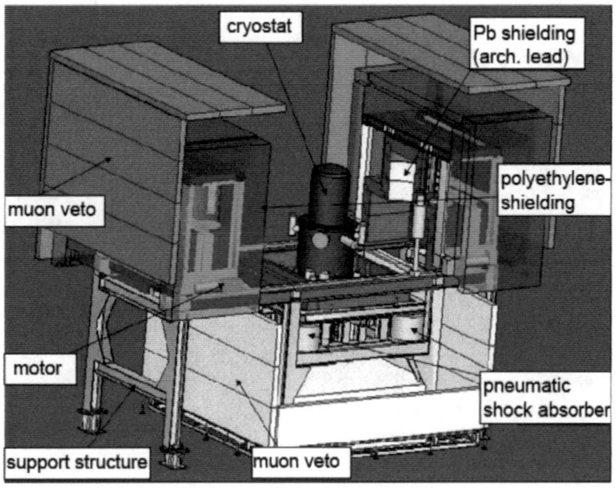

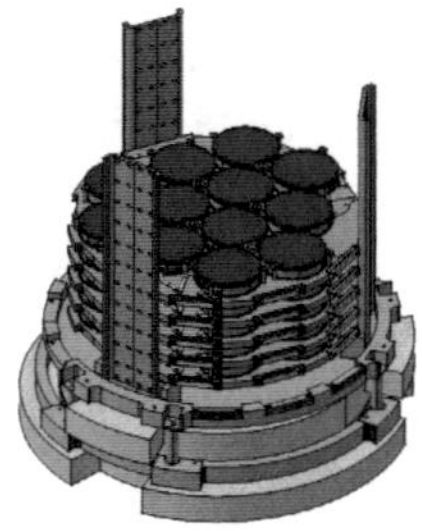

Figure 1. . General scheme of the EDELWEISS-II experiment

2.2. Detectors

EDELWEISS uses high purity Germanium cryogenic detectors with simultaneous measurement of phonon and ionization signals at a temperature of 20-40 mK. A nucleus produces less ionization in a crystal than an electron does, allowing an excellent event-by-event discrimination between nuclear recoils (induced by WIMP or neutron scattering) and electron recoils (induced by α, β and γ radioactivity).

The actual limitation of our present detectors arises from incomplete charge collection for near-surface events. To reach sensitivities to cross-section of interest for SUSY Models it is necessary to improve the rejection capabilities of the detectors in parallel with the active mass.

For this purpose, EDELWEISS-II will run three different types of detectors : (i) the known 320 g Ge/NTD EDELWEISS-I type equipped with new holders, (ii) the 400 g Ge/NbSi [3,4] with two NbSi Anderson insulator thermometric layers for active surface events rejection that have been developed within EDELWEISS collaboration, (iii) a new type of detectors, the 400 g Ge/NTD/INTERDIGIT recently developed [4,5] with the same aim to active surface events rejection thanks to a special interdigitized electrodes sheme. The surface events rejection capabilities of Ge/NbSi and Ge/NTD/INTERDIGIT have been measured in labs and are better than 99%[5,6] for events occuring in the first millimeter under the surfaces.

After the current commissioning phase, data for WIMP search will be acquired with 21 Ge/NTD and 7 Ge/NbSi starting from the fall. Additional prototypes of 400 g Ge/NTD/INTERDIGIT will also be measured. Forty additional detectors (50% Ge/NbSi and 50% Ge/NTD/INTERDIGIT) will be added in the two coming years to enhance progressively the sensitivity.

3. Calibration commissioning runs

In comparison with EDELWEISS-I, EDELWEISS-II is a completely new experiment : new cryostat (up to 110 detectors), new electronics, new acquisition hardware and software and a fully numerical triggering system.

Four cryogenic runs with 8 detectors have been done in 2006 to check and validate the different parts of the setup. This whole year has been dedicated to the tuning of the electronics and improvements on the acquisition and cryogenics, especially for acoustic and mechanical decoupling of the thermal machines. The quasi-total 28 detectors stage (21 Ge/NTD, 4 Ge/NbSi) has been mounted early 2007 for commissioning runs.

During the last run, the mean phonon channel energy resolution (for a charge collection voltage of 5 V) was measured to be 2 keV while best results are at 1.2 keV level for the Ge/NTD detectors. The detectors were operated at 24-26mK with resistances under bias around 1 Mohms. For some detectors, the resolution is still degraded by the noise induced by the pulse tubes and reliquefier and work is in progress to achieve a resolution close to 1 keV for all detectors. Ge/NbSi detectors in operation at this date[4] were showing excess heat capacity of the niobium electrodes that degraded the resolution to around 15 keV for thermal signal and 5 keV for athermal signal. This problem has been solved by using aluminium electrodes and heat resolution of 5 keV and 2 keV for thermal and athermal signal should be obtained for the future detectors. Energy resolutions of the ionization channels were around 1.5-2 keV for both type of detectors.

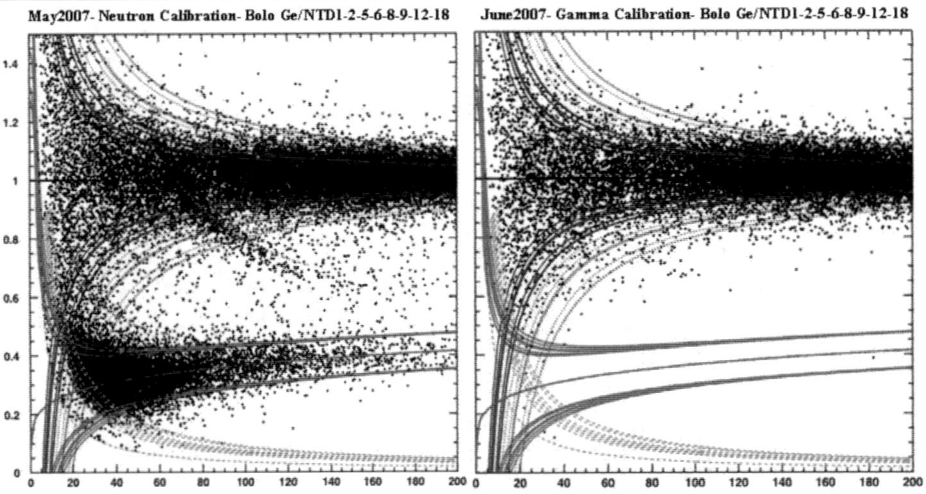

Figure 2. Ionisation/recoil energy ratio versus recoil energy in keV for 8 detectors with recoil thresholds between 20 and 35 keV. On the left, neutron calibration with an Am-Be source, on the right gamma calibration with ^{133}Ba sources. 90 and 99.9% acceptance band are shown.

Neutron and gamma calibration for the 8 Ge/NTD detectors with the best energy resolutions are shown on figure 2. Discrimination capabilities better than 99.9% are obtained, which is similar to EDELWEISS-I performances. The recoil energy threshold varies from 20keV to 35keV.

4. Low background commissioning runs

To quantify the alpha, gamma, and beta backgrounds, low background runs have been performed. The aim of these runs is to measure the rate of miscollected surface events in the detectors in the EDELWEISS-II setup and to extrapolate future sensitivity.

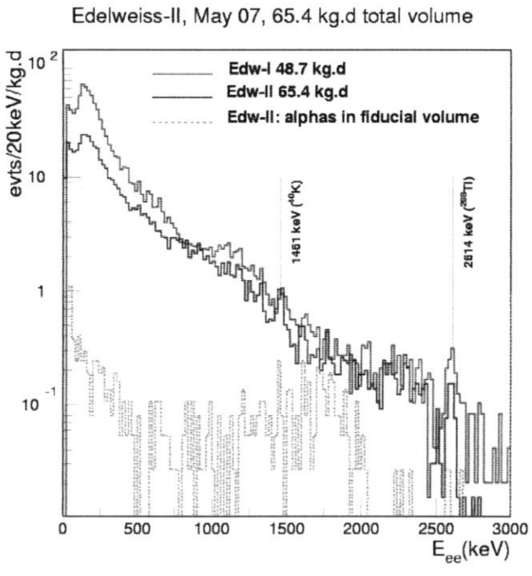

Figure 3. Ionization energy spectrum for low background run for the Ge/NTD detectors (65 kg.d for the total volume). The comparison with EDELWEISS-I gamma background shows an improvement of a factor 2 at low energy. The alpha background are identified by selecting events with a Ionisation/recoil energy ratio less than 0.5.

The electronic recoil background rate is shown on figure 3. For the fiducial volume, this rate is around 0.6 evts/kg/day below 100keV which is a factor 2.5 better than EDELWEISS-I. The alpha background is between 1.6 and 4.4 alpha/kg.day (75-200 alpha/m^2/day) depending of the detector and its near-by environment. The mean alpha rate was 4.2 alpha/kg/day for EDELWEISS-I.

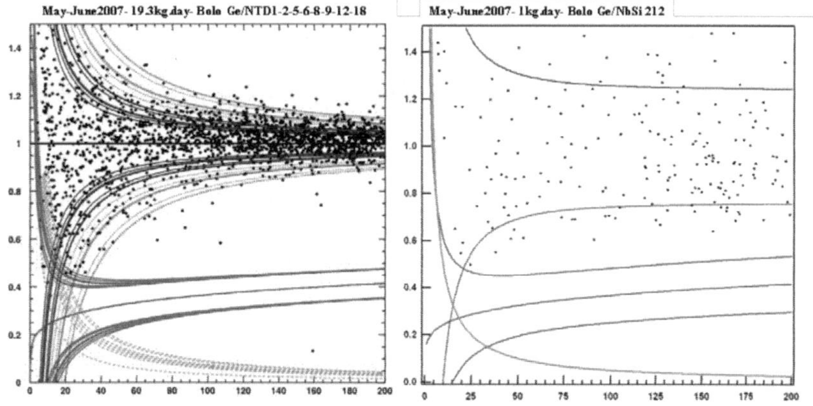

Figure 4. Ionisation/recoil energy ratio vs. recoil energy in keV for background runs. (See text for details).

Figure 4 shows the ionisation/recoil energy ratio as a function of the recoil energy for low energy background runs taken at the end of the commissioning period. Results for an exposure of 19.3 kg.day for the fiducial volume of the 8 Ge/NTD detectors with low thresholds are shown on the left. The rate of bad collected events measured with the method defined in ref[1] is 1.1±0.3 evts/kg/day which is close to EDELWEISS-I results. However, no events are seen in the nuclear recoil band signal, showing a possible improvement relative to EDELWEISS-I, where 3 events were measured with similar exposure. More statistics are obviously needed to draw firm conclusions. Results obtained with a 200g Ge/NbSi detectors is shown on the right for a fiducial exposure of 1 kg.day after the cuts of the surface events. The discrimination parameter is obtained with the athermal phonon signal which causes the rather large dispersion around 1. The Ge/NbSi analysis is explained in details in ref [4].

5. Muon veto measurement and simulations

The veto system consists of 42 plastic scintillator modules of 65 cm width, 5 cm thickness and lengths between 2 m and 4 m. The total surface is about 100 m^2 surrounding almost hermetically the outer polyethylene shielding of the cryostat. Each scintillator module is read out at both ends yielding 84 channels for the muon veto data acquisition system. From a subset of data taken from november 2006 to july 2007, a rate of 11,2 ± 0,3 μ/d has been measured with this set up, while the expected rate taking into account the profile of the mountain [7,8] has been evaluated to be 12,5 ± 0,4 μ/d. Study of coincidences between muon veto and bolometers will be performed in the following months.

In addition, recent and accurate simulations [8] taking into account the deposited energy by muons and associated showers in plastic scintillator detectors show that a significantly larger than anticipated fraction of muon induced showers could be tagged by lowering the threshold in these detectors, thus allowing a potential increased sensitivity of the set up to lower WIMP cross sections. To complete the muon veto installation, a new large (around 1 ton) neutron liquid scintillator detector will be soon operated close to the muon veto and will allow to evaluate the muon induced neutron flux.

6. Tools for background monitoring

We have been developing and are now operating in the environment of Edelweiss set up a ^{3}He neutron monitor sensitive to flux of about 10^{-8} thermal neutron /cm^2/s and a radon monitor with a sensitivity to concentration of radon in air of few mBq/m^3. In addition, a 50 g sapphire scintillation-phonon detector has been successfully operated. The potential for neutron detection with light elements, such as those in sapphire, is being evaluated.

7. Conclusion and prospects

The EDELWEISS-II setup has been validated with calibration and low energy background runs. Energy resolutions and discrimination capabilities close to EDELWEISS-I results have been measured for Ge/NTD detectors. Validation of Ge/NbSi detectors with new aluminium electrodes is in progress and Ge/NTD/INTERDIGIT detectors have shown promising results in surface lab. Low backgrounds physics runs will be taken with the 28 detectors setup with the aim to reach sensitivity to WIMP-nucleon cross-section of ~10^{-7} pb for a WIMP mass of 100 GeV. Additional detectors with active surface rejection capabilities will be installed in the two coming years to enhance progressively the sensitivity to WIMPs. Muon veto and neutron detector will provide a unique tool to understand and limit the ultimate neutron background from muons.

Acknowledgments

This work has been partially supported by the EEC Applied Cryodetector network (Contract HPRN-CT-2002-00322) and the ILIAS integrating activity (Contract RII3-CT-2003-506222)

References

[1] Fiorucci S et al., *Astropart.Phys.* 2007 **28** 143
[2] Sanglard V et al., *Phys. Rev.* 2005 D **71,** 122002
[3] Juillard A et al., *Nucl. Instrum. Methods Phys. Res.* 2006 A **559** 393
[4] Marnieros S et al., *J. Low Temp. Phys.* Proceedings of the 12th international workshop of Low Temperature Detectors, (2007)
[5] Broniatwoski A et al., *J. Low Temp. Phys.* Proceedings of the 12th international workshop of Low Temperature Detectors, (2007)
[6] Defay X et al., *J. Low Temp. Phys.* Proceedings of the 12th international workshop of Low Temperature Detectors, (2007)
[7] Berger Ch at al, *Phys.Rev.* 1989 D **40** 2163--2171
[8] Horn M, PhD Thesis (University of Karlsruhe, to be published)

From DAMA/NaI to DAMA/LIBRA and beyond

R. Bernabei[1], P. Belli[1], F. Montecchia[1], F. Nozzoli[1], F. Cappella[2], A. Incicchitti[2], D. Prosperi[2], R. Cerulli[3], C. J. Dai[4], H. L. He[4], H. H. Kuang[4], J. M. Ma[4], D. Sheng[4] and Z P Ye[4,5]

[1]Dip. di Fisica, Università di Roma "Tor Vergata" and INFN, sez. Roma Tor Vergata, I-00133 Rome, Italy

[2]Dip. di Fisica, Università di Roma "La Sapienza" and INFN, sez. Roma, I-00185 Rome, Italy

[3]INFN, Laboratori Nazionali del Gran Sasso, I-67010 Assergi (AQ), Italy

[4]IHEP, Chinese Academy, P.O. Box 918/3, Beijing 100039, China

E-mail: riccardo.cerulli@lngs.infn.it

Abstract. The low background DAMA/NaI experiment ($\simeq$ 100 kg highly radiopure NaI(Tl)) at the Gran Sasso National Laboratory of the I.N.F.N. had the unique feature to effectively investigate the presence of Dark Matter particle in the galactic halo by exploiting the model independent annual modulation signature. Data collected over seven annual cycles for a total exposure of $\simeq 1.1 \times 10^5$ kg $\times$ day have pointed out at 6.3 σ C.L. a positive result; corollary model dependent quests for the candidate particle have also been carried out in several of the many possible scenarios. At present the second generation DAMA/LIBRA set-up ($\simeq$ 250 kg highly radiopure NaI(Tl)) is in operation. R&D's towards a possible ultimate radiopurity 1 ton NaI(Tl) experiment are also in progress.

DAMA/NaI [1, 2, 3, 4, 5, 6, 7, 8] was proposed in 1990 [9], designed and realized with the main aim to investigate the presence of Dark Matter (DM) particles in the galactic halo by exploiting the model independent annual modulation signature [10]. Other approaches for DM particle investigation have also been exploited and several other rare processes studied [2]. The DAMA project consisting of several low background experimental set-ups, such as: i) DAMA/NaI [1, 2, 3, 4, 5, 6, 7, 8]; ii) DAMA/LXe [11]; iii) DAMA/R&D [12]; iv) DAMA/LIBRA [13]; v) DAMA/Ge for sample measurements, which is located in the low background LNGS Ge facility [14].

DAMA/NaI took data over seven annual cycles collecting an exposure of 107731 kg $\times$ day. A clear annual modulation of the *single-hit* events (i.e. events in which just one detector fires) satisfying the many peculiarities expected for a DM particle induced effect has been observed [1]. The presence of an annual modulation with the proper features is favoured at 6.3 σ C.L. with an amplitude equal to (0.0200 ± 0.0032) cpd/kg/keV, a phase $t_0 = (140 \pm 22)$ days and a period $T = (1.00 \pm 0.01)$ year. Neither systematic effect nor side reaction able to account for the observed modulation amplitude and to satisfy all the requirements of the signature have been found; for details see ref. [1].

A further investigation has been performed on the *multiple-hits* events (i.e. events in which more than one detector fire). The *multiple-hits* events class – on the contrary of the *single-hit* one – does not include events induced by DM particles since the probability that a DM

[5] also: University of Jing Gangshan, Jiangxi, China

particle interacts in more than one detector is negligible. The fitted modulation amplitude is $A = -(3.9 \pm 7.9) \cdot 10^{-4}$ cpd/kg/keV for the *multiple-hits* residual rate [1]. Summarizing, evidence of annual modulation with proper features is present in the *single-hit* residuals (events class to which the DM particle-induced signals belong), while it is absent in the *multiple-hits* residual rate (event class to which only background events belong). Since the same identical hardware and the same identical software procedures have been used to analyse the two classes of events, the obtained result offers an additional strong support for the presence of DM particles in the galactic halo, further excluding any side effect either from hardware or from software procedures or from background. Details can be found in ref. [1].

On the basis of the 6.3 σ C.L. model-independent evidence, several corollary model dependent investigations have also been pursued on the nature of the DM candidate [1, 4, 5, 3, 6, 7, 8].

For the class of the WIMP candidates, low and high mass (from GeV to many hundreds of GeV) candidates have been considered, interacting with ordinary matter via: i) mixed SI&SD coupling; ii) dominant SI coupling; iii) dominant SD coupling; iv) preferred SI inelastic scattering, in various scenarios [1]. This analysis has also been extended considering a non thermalized DM particle component in the dark halo [5]; other non-thermalized substructures or halo models with caustics can be addressed.

It is worth to note that the 6.3 σ C.L. model-independent DAMA/NaI evidence is well compatible with theoretical expectations for neutralino in MSSM [15].

It has also been investigated the role of the electromagnetic (e.m.) contribution produced in the interaction of the WIMP with target nuclei [6], named Migdal effect, not yet considered in the field. In fact, a certain quantity of e.m. radiation (made of escaping electron and of X-rays and/or Auger electrons) arises from the rearrangement of the atomic shells induced by the presence of a recoiling atomic nucleus. This radiation is fully contained in a detector of suitable size and because of its e.m. nature, this part of the signal is lost in all those approaches based on discrimination procedures of the e.m. component of the measured counting rate. The inclusion of this effect has a sizeable impact in the DM direct searches when interpreted in terms of low mass WIMP candidates. For a discussion see ref. [6].

Moreover, recently it has also been investigated the possible implications of the channeling effect in NaI(Tl) crystals [7]. This effect can occur in crystalline materials due to correlated collisions of ions with target atoms, resulting in a range for ions much larger than the maximum range they would have if their motion would be either in other directions or in amorphous materials. When a low-energy ion goes into a channel transfers its energy mainly to electrons rather than to the nuclei in the lattice; thus, its quenching factor (namely the ratio between the detected energy in keV electron equivalent and the kinetic energy of the recoiling nucleus in keV) approaches the unity. The inclusion of this effect gives an appreciable impact in corollary analyses in terms of WIMP (or WIMP-like) candidates [7]. In particular, this implies that lower cross sections are obtained for the allowed volumes/regions in given model frameworks by the DAMA/NaI data. For a detailed discussion see ref. [7]. Also this latter result further shows the role of the existing uncertainties and of the correct description and inclusion of all the involved processes.

As regards other scenarios, the investigation on a DM candidate interacting only with electrons [8] has also been considered, while the sterile neutrino is under analysis.

Finally, the 6.3 σ C.L. model-independent DAMA/NaI evidence has been investigated in terms of light ($\simeq$ keV mass) boson DM candidates [4]. For these candidates, the direct detection process is based on the total conversion in NaI(Tl) crystal of the mass of the absorbed bosonic particle into e.m. radiation, while the target nuclei recoil is negligible and is not involved in the detection process. Also signals from these light bosonic DM candidates are lost in experiments applying rejection procedures of the e.m. contribution to the counting rate. Large part of the bosonic candidate parameter space allowed by DAMA/NaI are of cosmological interest [4].

In conclusion, the DAMA/NaI experiment pointed out at 6.3 σ C.L. the presence of DM particles in the galactic halo by investigating the model independent annual modulation signature over seven annual cycles. Neither systematic effect nor side reaction able to account for the observed effect have been found. In addition, several (of the many possible) corollary quests for the investigation on the nature of the DM candidates have also been pursued. No experiment is available so far – with the exception of DAMA/LIBRA – whose results can be directly compared in a model independent way with that of DAMA/NaI. Thus, claims for contradictions, sometimes reported in the field and in this Conference, have intrinsically no scientific meaning. Some discussion can be found e.g. in ref. [1]. Present positive hints provided by indirect searches are not in conflict with the DAMA/NaI result.

At present the second generation DAMA/LIBRA experiment (a $\sim$ 250 kg highly radiopure NaI(Tl) set-up) is in operation [13]; the first release of results will occur not later than the end of 2008.

A third generation R&D towards a possible ultimate radiopure 1 ton NaI(Tl) detector, proposed by DAMA in 1996, is also in progress.

References

[1] R. Bernabei el al., La Rivista del Nuovo Cimento 26 n.1 (2003) 1-73; R. Bernabei et al., Int. J. Mod. Phys. D 13 (2004) 2127; R. Bernabei et al., Il Nuovo Cim. A 112 (1999) 545.

[2] R. Bernabei et al., Eur. Phys. J. A 24 (2005) 51; Eur. Phys. J. A 23 (2005) 7; Phys. Lett. B 515 (2001) 6; F. Cappella et al., Eur. Phys. J. direct C 14 (2002) 1; R. Bernabei et al., Phys. Rev. Lett. 83 (1999) 4918; P. Belli et al., Phys. Rev. C 60 (1999) 065501; Phys. Lett. B 460 (1999) 236; R. Bernabei et al., Il Nuovo Cim. A 112 (1999) 1541; Phys. Lett. B 408 (1997) 439; Phys. Lett. B 389 (1996) 757.

[3] P. Belli et al., Phys. Rev. D 66 (2002) 043503; R. Bernabei et al., Eur. Phys. J. C 23 (2002) 61; R. Bernabei et al., Phys. Lett. B 509 (2001) 197; P. Belli et al., Phys. Rev. D 61 (2000) 023512; R. Bernabei et al., Phys. Lett. B 480 (2000) 23; R. Bernabei et al., Eur. Phys. J. C 18 (2000) 283. R. Bernabei et al., Phys. Lett. B 450 (1999) 448; R. Bernabei et al., Phys. Lett. B 424 (1998) 195.

[4] R. Bernabei et al., Int. J. Mod. Phys. A 21 (2006) 1445.

[5] R. Bernabei et al., Eur. Phys. J. C 47 (2006) 263.

[6] R. Bernabei et al., Int. J. Mod. Phys. A 22 (2007) 3155.

[7] R. Bernabei et al., arXiv:0710.0288, to appear on Eur. Phys. J. C

[8] Y. Ascasibar et al., Mon. Not. Roy. Astron. Soc. 368 (2006) 1695; C. Jacoby and S. Nussinov, JHEP 05 (2007) 017; D.P. Finkbeiner and N. Weiner, Phys. Rev. D 76 (2007) 083519; M. Pospelov and A. Ritz, Phys. Lett. B 651 (2007) 208, hep-ph/0703128; P. Fayet, Phys. Rev. D 75 (2007) 115017; R. Bernabei et al., in preparation.

[9] P. Belli, R. Bernabei, C. Bacci, A. Incicchitti, R. Marcovaldi, D. Prosperi, DAMA proposal to INFN Scientific Committee II, April 24th 1990.

[10] K.A. Drukier et al., Phys. Rev. D 33 (1986) 3495; K. Freese et al., Phys. Rev. D 37 (1988) 3388.

[11] R. Bernabei et al., Eur. Phys. J. A 27 s01 (2006) 35; Phys. Lett. B 546 (2002) 23; Phys. Lett. B 527 (2002) 182; Nucl. Instrum. & Meth. A 482 (2002) 728; Eur. Phys. J. direct C 11 (2001) 1; Phys. Lett. B 493 (2000) 12; New Journal of Physics 2 (2000) 15.1; Phys. Lett. B 436 (1998) 379; P. Belli et al., Phys. Rev. D 61 (2000) 117301; Phys. Lett. B 465 (1999) 315; Phys. Lett. B 387 (1996) 222; Phys. Lett. B 389 (1996) 783 (err.); Il Nuovo Cim. 19 (1996) 537; Astropart. Phys. 5 (1996) 217.

[12] R. Bernabei et al., ROM2F/2007/17 to appear on Phys. Rev. C; ROM2F/2007/13 to appear on Phys. Lett. B; Nucl. Phys. A 789 (2007) 15; Ukr. J. Phys. 51 (2006) 1037; Nucl. Instrum. & Meth. A 555 (2005) 270; R. Cerulli et al., Nucl. Instrum. & Meth. A 525 (2004) 535; P. Belli et al., Nucl. Instrum. & Meth. A 498 (2003) 352; R. Bernabei et al., Nucl. Phys. A 705 (2002) 29; P. Belli et al., Astropart. Phys. 10 (1999) 115; Nucl. Phys. B 563 (1999) 97; R. Bernabei et al., Astropart. Phys. 7 (1997) 73; Il Nuovo Cim. A 110 (1997) 189.

[13] R. Bernabei et al., in the volume *Frontier Objects in Astrophysics and Particle Physics* Vol. 90 (2004) 581.

[14] P. Belli et al., Nucl. Instrum. & Meth. A 572 (2007) 734; P. Belli et al, in the volume *"Current problems in Nuclear Physics and Atomic energy"*, ed. INR-Kiev, (2006) 479.

[15] A. Bottino et al., arXiv:0710.0553; Phys. Rev. D 67 (2003) 063519; Phys. Rev. D 68 (2003) 043506; Phys. Rev. D 69 (2004) 037302.

Dark matter search with CaF_2 scintillator at Osaka

I Ogawa[1], R Hazama[2], K Mukaida[1], K Kishimoto[1], H Sakai[1], A Katsuki[1], T Itamura[1], S Umehara[1], S Yoshida[3] and T Kishimoto[1]

[1] Graduate School of Science, Osaka University, Toyonaka, Osaka 560-0043, Japan

[2] Graduate School of Engineering, Hiroshima University, Higashi-Hiroshima, Hiroshima 739-8527, Japan

[1] Research Center for Neutrino Science, Tohoku University, Sendai, Miyagi 980-8578, Japan

E-mail: ogawa@phys.sci.osaka-u.ac.jp

Abstract. A CaF_2 scintillator complex system (ELEGANT VI) is developed to search for the axial-vector coupled dark matter (WIMPs) and study the neutrino-less double beta decay of ^{48}Ca. Active light guides of pure CaF_2 crystals which are on both sides of the central CaF_2(Eu) crystal act as a 4π active shield, combined with surrounding CsI(Tl) scintillators. The whole system is at the underground laboratory (Oto Cosmo Observatory) located in Nara. In this article our current status of the investigation are described, especially putting emphasis on the development of flash scaler to do a single photon counting.

1. Introduction

Weakly interacting massive particles (WIMPs) are one of the strongest candidates of the non-baryonic cold dark matter. Search for WIMPs is one of the most challenging issues related to recent particle physics and cosmology. Among them particles that have axial vector coupling with matter (spin coupled dark matter) are unexplored due to small cross sections with matter. The lightest neutral super partners (neutralino) are the most promising candidates since supersymmetry seems to be the theory beyond the standard model.

Our strategy is the direct detection of spin coupled dark matter by elastic nuclear recoils. The dark matter is drifting in a galaxy with a velocity of $\beta \sim 10^{-3}$. The signals are expected to be of low energy and rare, so one has to develop a detector that can sense such weak signals.

2. Detector system

We developed a CaF_2 scintillation detector system to search for WIMPs [1, 2, 3, 4] and study the double beta decay of ^{48}Ca [6, 7, 8]. For the spin coupled dark matter search, ^{19}F nucleus has the best figure of merit, represented by the product of cross section and natural abundance (100 % ^{19}F) [2]. The detector system consists of 23 CaF_2 scintillator complexes surrounded by 38 CsI(Tl) veto scintillators and passive shields. Total weight of the ^{19}F is about 3.2 kg.

The CaF_2 scintillator complex consists of central CaF_2(Eu) cubic crystal ($45 \times 45 \times 45$ mm^3), pure CaF_2 crystals ($45 \times 45 \times 200$ mm^3) on both sides as light guides which are also sensitive to gamma rays and 2 PMTs with quartz window (sensitive to both ultra-violet (UV) and visible light). Scintillation light of the pure CaF_2 is dominantly in the UV region (peak emission at $\sim$ 275 nm), although CaF_2(Eu) emits normal visible light($\sim$ 420 nm). Transmittance of CaF_2(Eu) is limited only for visible light, while pure ones are able to transmit UV light in addition to visible one. The $CaF_2(Eu)$ crystal acts as a filter against the scintillation light from pure CaF_2 crystals. Thus the events occurred at pure crystals, which are mostly generated by the background gamma rays, can be rejected by setting a proper window on the asymmetry parameter (roll-off-ratio) between the signals from 2 PMTs.

WIMPs are thought to give a quite low energy deposit to the detector. Thus it is important to treat properly the signals and to evaluate the detection efficiency in the low energy region. The low energy signal like a few keV from CaF_2 scintillators looks like a bundle of many one photon signals in a long decay time ($\sim 1\ \mu$sec). Further to separate the events occurred at $CaF_2(Eu)$ crystal from CaF_2(pure) ones, we developed flash scalers which measure time distribution of one photon signals and installed them for central 8 CaF_2 scintillator complexes in the ELEGANT VI. In addition to measuring charges from PMTs by ADCs, direct counting of the number of photoelectrons enables us to be free from the fluctuation on the pulse height of one photoelectron. We can define the roll-off-ratio as

$$r = \frac{N_r - N_l}{N_{\text{tot}}}, N_{\text{tot}} = N_r + N_l, \tag{1}$$

where N_l, N_r are number of photoelectrons from left and right PMTs, respectively.

The response of the detector to a nuclear recoil, so-called f-value, was measured by the $^{19}F(n,n')$ reaction for our crystal [5]. The experiment was carried out using a pulsed neutron beam generated from the $d(d,n)^3He$ reaction at the PELLETRON accelerator of Tokyo Institute of Technology. The ~ 3.7 MeV neutron gives similar energy to a ^{19}F nucleus as the scattering by the dark matter. The obtained f-value is a little larger especially at low energy region than that has been measured for other detectors which makes total detection efficiency large.

The schematic drawing of the detector is shown in Figure 1. Segmentation of the CaF_2 detector enables us to reduce the background events which are supposed to be spread in space, while a true event is confined to a small region. Non-hygroscopic features of both CaF_2 and CsI(Tl) crystals enables us to reduce the housing materials near the detectors. This becomes the advantage of low background measurement originate from the housing material and 4π active shield even for X-rays. The materials used in the detector have been carefully selected for low radioactivity. The scintillators are in the airtight box inside of which is purged by pure N_2 gas to remove Rn gas in the air. As conventional passive shields, OFHC copper of 5 cm thick and lead of 10 cm shields are used to reduce the radioactive backgrounds from surroundings. Additional three kinds of shields (LiH-loaded paraffin, Cd sheet and H_3BO_3-loaded water tank moderator) for neutron background are prepared. About factor 3–5 reduction was achieved in the neutron flux with these shields. The whole system is located at the underground laboratory (Oto Cosmo Observatory) located in Nara, Japan which has effectively 1.4 km water equivalent shield to suppress the background originated from cosmic rays.

3. Preliminary Results

Off-line analysis to suppress the background and select the local events occurred at the $CaF_2(Eu)$ crystal are done by setting the conditions; i) no events in CsI(Tl) veto detectors; ii) only one CaF_2 scintillator complex fires; iii) proper selection of central event by roll-off-ratio cut for each N_{tot}. To reject noise events, following cuts were done. For the events which have larger pulse heights than one photoelectron signal, charge information recorded by ADC was used. To reduce the dark current of PMTs, time difference (T_{diff}) between the first one photoelectron signals from two PMTs on both sides were used. The scintillation events are expected to be distribute around $T_{diff} \sim 0$, while the events of dark currents are expected to be constant distribution.

The total analyzed live time is 199.3 days. The energy calibration was done by using ^{241}Am every ~ 3 weeks and in between the gain variation was checked by using LEDs. The preliminary exclusion plot for the WIMP-proton cross section on spin-dependent coupled DM by the present

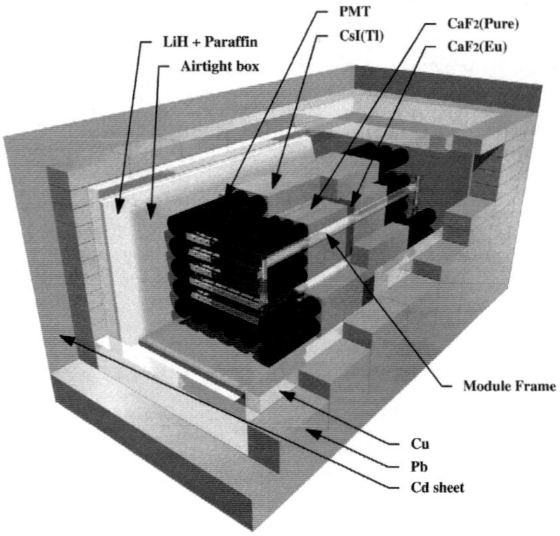

Figure 1. Schematic drawing of detector system.

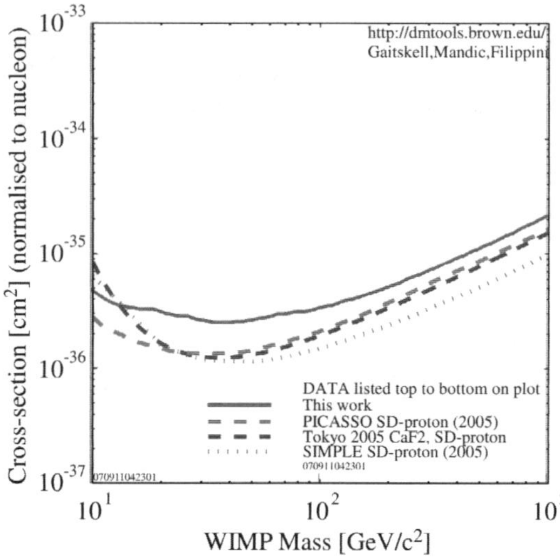

Figure 2. Exclusion plots for the WIMP-proton cross section on spin-dependent coupled DM. (preliminary)

measurement are shown in Figure 2, compared with other experiments which use ^{19}F as a target. Much further reduction is expected by fitting roll-off-ratio (1) with an exact binomial function,

$$F(N_{\rm tot}, N_r) = n_l B^*(N_{\rm tot}, P_l, N_r) + n_{\rm Eu} B^*(N_{\rm tot}, P_{\rm Eu}, N_r) + n_r B^*(N_{\rm tot}, P_r, N_r), \tag{2}$$

using modified binominal distribution function $B^*(N, P, r)$ concerning with trigger efficiency of coincidence,

$$B^*(N, P, r) = \frac{B(N, P, r)}{\sum_{r_i=\frac{N-1}{N}}^{\frac{-N+1}{N}} B(N, P, r_i)}. \tag{3}$$

Thus we can statistically estimate the number of events $(n_{\rm Eu}, n_l, n_r)$ occurred at CaF_2(Eu), left CaF_2(pure) and right CaF_2(pure) crystal separately.

References

[1] Hazama R *et al* 1999 *Proc. 2nd Int. Workshop on the Identification of the Dark Matters (Buxton)* (Singapore: World Scientific) p 329

[2] Hazama R 1998 Doctoral thesis (Osaka University)

[3] Ogawa I *et al* 2000 *Nucl. Phys.* A **663-664** 869c

[4] Kishimoto T *et al* 2001 *Proc. 3rd Int. Workshop on the Identification of the Dark Matters (York)* (Singapore: World Scientific) p 379

[5] Hazama R *et al* 2002 *Nucl. Instr. Meth. in Phys. Res.* A **482** 297

[6] Ogawa I *et al* 2003 *Nucl. Phys.* A **721** 525c

[7] Ogawa I *et al* 2004 *Nucl. Phys.* A **730** 215

[8] Umehara S *et al in this volume*

The CRESST Dark Matter Search

W Seidel[1], G Angloher[1], M Bauer[5], I Bavykina[1], A Bento[1], A Brown[2], C Bucci[3], C Ciemniak[4], C Coppi[4], G Deuter[5], Fv Feilitzsch[4], D Hauff[1], S Henry[2], P Huff[1], J Imber[2], S Ingleby[2], C Isaila[4], J Jochum[5], M Kiefer[1], M Kimmerle[5], H Kraus[2], J-C Lanfranchi[4], R Lang[1], M Malek[2], R McGowan[2], V Mikhailik[2], E Pantic[1], F Petricca[1], S Pfister[4], W Potzel[4], F Pröbst[1], S Roth[4], K Rottler[5], C Sailer[5], K Schäffner[1], J Schmaler[1], S Scholl[5], L Stodolsky[1], B Tolhurst[2], I Usherow-Marshak[5] and W Westphal[4]

[1] Max-Planck-Institut für Physik, Föhringer Ring 6, 80805 München, Germany
[2] University of Oxford, Keble Road, Oxford, OX1 3RH, UK
[3] Laboratori Nazionali del Gran Sasso, Assergi, Italy
[4] Technische Universität München, James-Franck-Straße, 85748 Garching, Germany
[5] Eberhard Karls Universität Tübingen, Auf der Morgenstelle 14, 72076 Tübingen, Germany

e-mail: Seidel@mppmu.mpg.de

Abstract. CRESST is a WIMP dark matter search using scintillating $CaWO_4$ cryogenic detectors with active background suppression. First results obtained in the commissioning run are presented.

1. Introduction

Despite persuasive indirect evidence for the existence of dark matter in the universe and in galaxies, the direct detection of dark matter remains one of the outstanding experimental challenges of present-day physics and cosmology.

A plausible candidate for the dark matter is the Weakly Interacting Massive Particle (WIMP) and it is possible that it can be detected by laboratory experiments, particularly using cryogenic methods, which are well adapted to the small energy deposit anticipated [1]. Supersymmetry provides a well motivated WIMP candidate in the form of the lightest supersymmetric particle. WIMPs [2] are expected to be gravitationally bound in a roughly isothermal halo around the visible part of our galaxy with a density of about 0.3GeV/cm^3 at the position of the Earth [3].

Interaction with ordinary matter is expected via elastic scattering off nuclei. The elastic nuclear scattering can occur via coherent ("spin independent"') or spin-dependent interactions. For the coherent case, a factor A^2 is expected in the cross section, favoring heavy nuclei. We present our WIMP limits in terms of this possibility.

In the CRESST experiment we attempt to detect the WIMP-nucleus scattering using cryogenic methods. For the second phase [4], CRESST-II, we have developed cryogenic detectors based on scintillating $CaWO_4$ crystals [5]. When supplemented with a light detector these provide very efficient discrimination of nuclear recoils from radioactive γ and β backgrounds, down to recoil energies of about 10keV. The heaviest nucleus in our crystal is tungsten, for which the recoil energy due to WIMPsis expected to reach up to about 40keV, with rates below 1 event/(kg day). The mass of each crystal is about 300g.

Passive background suppression is achieved with a low background installation and the deep underground location. The overburden of 3500 meter water equivalent at the LNGS leads to a reduction of the surface muon flux by 10^6 to about $1/(m^2\ h)$, while the detectors themselves are shielded against ambient radioactivity by thicknesses of 14 cm of low background copper and 20 cm of low background lead. A neutron shield consisting of 40cm of Polyethylene and a muon veto, were additionally installed since the last run.

2. Detectors

A single detector module consists of a scintillating $CaWO_4$ "target" crystal, operated as a cryogenic calorimeter (the "phonon channel"), and a nearby but separate cryogenic detector optimized for the detection of scintillation photons (the "light channel"). A schematic representation of a detector module is shown in figure 1.

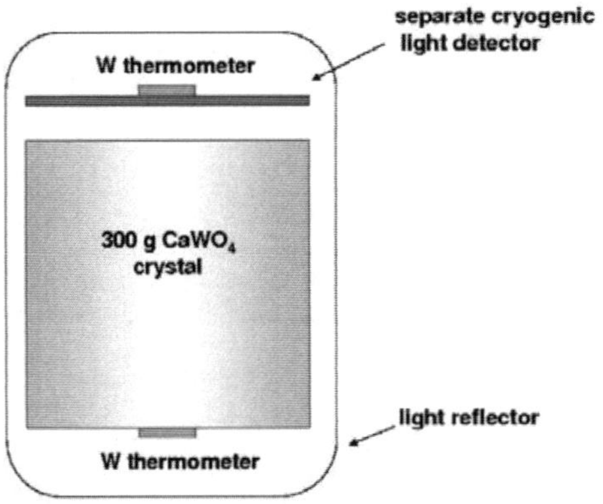

Fig.1 Schematic representation of the detector for coincident phonon light measurements. It consists of two cryogenic detectors enclosed in a highly reflective housing, read out by two tungsten superconducting phase transition thermometers

The phonon channel is designed to measure the energy transfer to a nucleus of the $CaWO_4$ crystal in a WIMP-nucleus elastic scattering. Since a nucleus and an electron or gamma of the same recoil energy differ substantially in the yield of scintillation light, an effective background discrimination against gammas and electrons is obtained by a simultaneous measurement of the phonon and light signals. Among different scintillating crystals, $CaWO_4$ was selected because of its high light yield at low temperatures and the absence of a noticeable degradation of the light yield for events near the crystal surface [6]. Such a degradation, often found in coincident phonon-charge measurements and some scintillators, can cause difficulties as it may lead to a misidentification of electron/photon surface events as nuclear recoils. In addition, the large atomic mass (A=183.86) of tungsten makes $CaWO_4$ a very favorable target for WIMPs with coherent interactions. A more detailed description of the detectors and their operation can be found elsewhere [5].

3. Commissioning run and results

During the last years the CRESST set-up has been upgraded with a neutron shield, a 66 channel SQUID system, new wiring, new bias electronics, a muon veto and a new DAQ. It is now capable to house up to 33 detector modules. After the upgrade a commissioning run was started to optimize the overall performance of the system. This run was also used for a short physics run in which in total 49,8 kg days of data were collected.

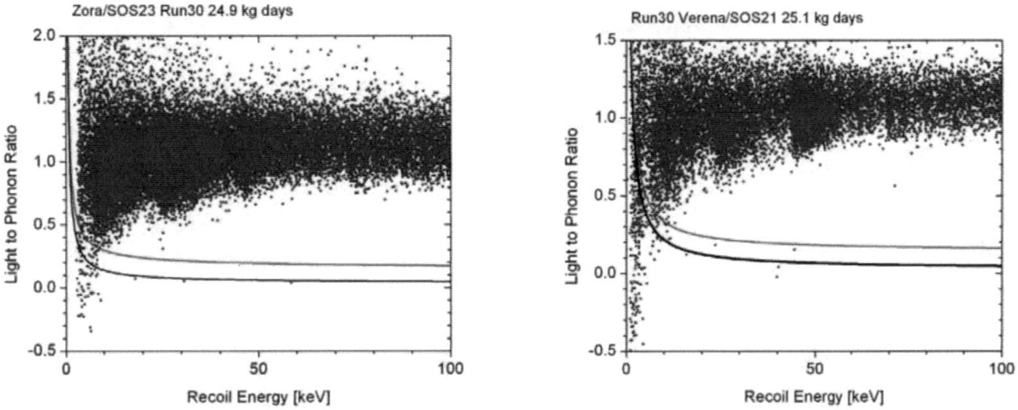

Figure 2.Low energy event distributions for the two modules Zora/SOS23 and Verena/SOS21.The vertical axis represents the light yield expressed as the ratio (energy in the light channel/energy in the phonon channel), and the horizontal axis the energy in the phonon channel. The region below the upper line contains 90% of all recoils while the region below the lower line contains 90% of the tungsten recoils.

Data of two detectors obtained in this run are shown in figure 2. The limits which are obtained from the data in a similar way as described in reference [5] combined with the data of reference [5] are shown in figure 3.

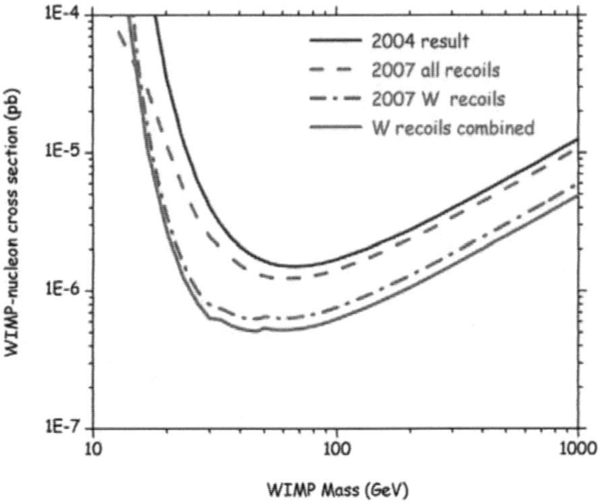

Figure 3. Preliminary limits of the 2007 commissioning run

Since a neutron calibration is still missing these limits should be considered as preliminary. They will be finalized after the neutron calibration in November 2007 and will be published elsewhere.

4. Acknowledgements

This work was partially supported by the DFG SFB 375 “Teilchen-Astrophysik”, Transregio27 “neutrinos and Beyond”, the Munich cluster of Excellence (“Origin and Structure of the Universe”), the EU Network “Cryogenic Detectors” (contract ERBFMRXCT980167), the EU Network HPRN-CT-2002-00322 “Applied Cryodetectors”, The Maier-Leibnitz-Laboratorium (Garching), BMBF, PPARC, two EU Marie Curie Fellowships and the network on direct dark matter detection of the ILIAS integrating activity (R113-CT-2003-506222)

References

[1] M.W.Goodman and E. Witten, Phys. Rev. **D31** (1985) 3059

[2] A general review, including matters related to direct detection, may be found in G.Jungman, M.Kamionkowski, and K.Griest, Phys. Rep. **267** (1996) 195.

[3] J. A. R.Caldwell and J. P.Ostriker, Ap. J. **251** (1981) 61. For a more recent discussion with a wider range of values see E. I.Gates, G.Gyuk, and M. S.Turner,Ap. J. **449** (1995) L123.

[4] Update of the proposal to the LNGS for a second phase of the CRESST dark matter search, MPI-PhE/2001-02.

[5] G. Angloher et al. Astroparticle Physics **23** (2005) 325-339

[6] P.Meunier, M.Bravin, M.Bruckmayer, S.Giordano, M.Loidl, O.Meier, F.Pröbst W.Seidel, M.Sisti, L.Stodolsky, S.Uchaikin and L.Zerle, Appl. Phys. Lett. **75** (1999) 1335.

New Results from the KIMS Experiment

Sun Kee Kim

Department of Physics and Astronomy, Seoul National University, Seoul 151-742

(For the KIMS Collaboration)

skkim@hep1.snu.ac.kr

Abstract. The Korea Invisible Mass Search (KIMS) collaboration has performed experiments on dark matter searches and double beta decay searches at the Yangyang underground laboratory. Recent results of KIMS as well as the status of the R&D on various detectors are presented. For the WIMP search KIMS ruled out the recoil of ^{127}I of DAMA signal region unambiguously and set the most stringent limit on spin dependent WIMP interactions in the case of pure proton coupling. New limits on neutrino-less double beta decays of ^{112}Sn, ^{124}Sn, ^{64}Zn is also reported.

1. The KIMS experiment and Yangynag Undeground Laboratory

Yangyang Undeground Laboratory(Y2L) is located in a tunnel constructed for the Yangyang pumped storage power plant operated by Korea Midland Power Co. Ltd. The vertical rock overburden is 700m. The Korea Invisible Mass Search (KIMS) collaboration has carried out a dark matter search experiment using CsI(Tl) crystal and a development of detectors for the double beta decay search experiment. The temperature and humidity of the experimental hall and the detector chamber have been kept constant over the whole year. The radon level was measured to be 40-80 Bq/m^3. Muon flux is measured using the muon detector, which was used for the neutron shielding as well. The measured muon flux, 2.7×10^{-3} m^{-2}s^{-1}, is consistent with the depth of the laboratory. Neutron flux in the experimental hall is measured with the 1 liter liquid scintillator (BC501A) to be 8×10^{-3} m^{-2}s^{-1} for 1.5 MeV < E_n < 6.0 MeV. The shielding structure for the WIMP search consists of 10 cm thick copper, 5 cm thick polyethylene, 15 cm thick lead and 30 cm thick mineral oil. The mineral oil contains 5% liquid scintillator and it works as a muon detector as well. Inside the innermost copper box, we flow the nitrogen gas continuously to get rid of radon contamination as well as to keep the low level of humidity.

2. WIMP searches using the CsI(Tl) crystal detectors

The KIMS group has developed CsI(Tl) crystals with high radio-purity[1]. The CsI(Tl) crystal is sensitive to both Spin Independent (SI) and Spin Dependent (SD) interaction of WIMP with nucleus. Especially the high sensitivity to the SD interaction in the case of pure proton coupling makes the KIMS experiment complementary to CDMS[2] and XENON[3] experiment who are the most sensitive experiments for the SI interactions and SD interactions in case of pure neutron coupling at the moment.

The source of the major internal background in the crystal, ^{137}Cs has been successfully reduced to below 2mBq/kg by using the highly purified water in the Cs extraction process from the pollucite[1]. Rubidium contamination is reduced to the level of 1 ppb by re-crystallization technique. The contamination of U and Th is estimated to be at the level of 1 ppt. With this level of the impurity level, we expect the counting rate at about few counts/day/keV/kg(cpd) at 10 keV.

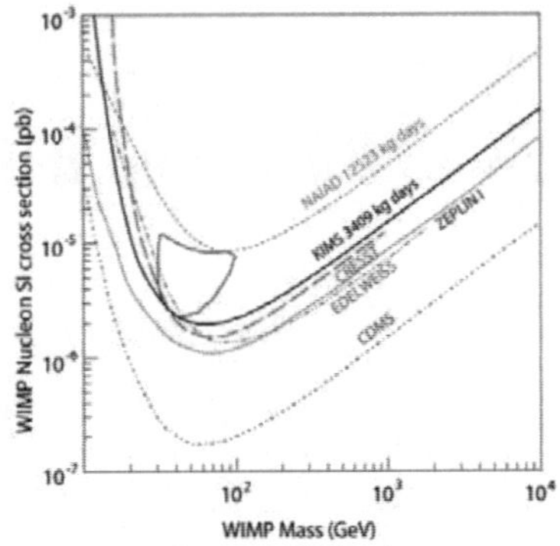

Figure 1 SI limits

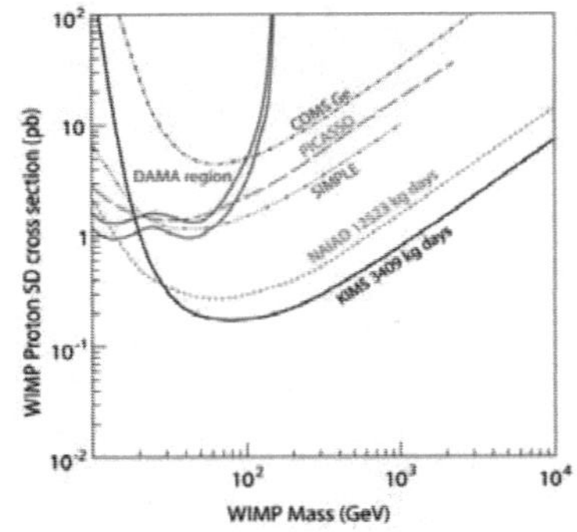

Figure 2 SD limits in case of purely proton coupling

We have carried out a pilot experiment in 2004 with one crystal of 6 kg mass which is smaller than the final crystal. Using this data with amount of 237 kg days, we demonstrated that the CsI(Tl) crystal is highly competitive to other WIMP search experiment[4]. In 2005-2006, we have performed the engineering experiments using four full size CsI(Tl) crystals at the background level of 6 cpd, where we tested various running conditions to optimize the operation for WIMP search. Major part of the data was taken at the temperature T=0°C. The amount of data was 3049 kg days. In order to understand the characteristics of background from photo multiplier tubes (PMT) we have taken data with acrylics boxes in the place of CsI(Tl) crystals mounted with the same PMTs used for WIMP search data. Using this data we developed highly efficient analysis cuts to reduce the PMT background at low energies without introducing bias on WIMP search data. Pulse shape discrimination method is used to extract the nuclear recoil event rates by WIMP interaction from the data where dominant contribution to the event rate came from the gamma ray background. Details of analysis method are described elsewhere[5]. The nuclear recoil event rates at 3 to 11 keV are consistent with null observation of WIMP interaction, we set the upper limit on WIMP-nucleon interaction cross sections as shown in Figures 1 and 2 for the SI interactions and SD interaction in the case of pure proton coupling respectively. DAMA signal region due to the recoil of ^{127}I, the dominant target SI interaction for the DAMA experiment[6], is ruled out. For the SD interactions, KIMS presents the most stringent limit in the case of pure proton coupling.

3. Double beta decay searches

KIMS has developed the experimental techniques for neutrino-less double beta decay search. $CaMoO_4$ crystal can be a good detector candidate for the double beta decay search of ^{100}Mo, once the large sized high efficient scintillation crystals with low background are available. The scintillation properties and background for small crystals and large crystals have been studied in collaboration with Russia and Ukraine groups and the feasibility of the experiment is investigated[7]. One large size crystal was installed in a shield surrounded by CsI(Tl) crystals and lead shielding at Y2L and a long term data taking for the internal background measurement has started.

Metal loaded liquid scintillator is another potential candidate for the double beta decay search. Tin loading to the pseudocumene(PC) based liquid scintillator in terms of the tetramethyltin (TMSN) has been tested. Up to 50% of TMSN (32.6% tin in weight) was successfully loaded into the liquid scintillator. At the 50% TMSN loading, the light yield was about 57% of that of the liquid scintillator without TMSN loading[8].

While developing the new techniques, we have performed an experiment using a setup consisting of a 100% efficiency HPGe detector and a low background CsI(Tl) crystal. We located a sample of one liter tetrabuthyltin (TBSN), which contained 34% by mass of Sn, between the HPGe detector and the CsI(Tl) crystal detectors and measured gamma rays for 140 hours. From the analysis of this data after subtracting background, we did not observe any excess counts over the background. We looked for signals from the double beta decay of ^{124}Sn to the excited states 602.7(2+), 1156.0(0+), and 1325.5(2+) keV and set the 90% CL level lower limits on the half life time of 2.3×10^{18}, 6.7×10^{18}, 7.9×10^{18} years, respectively[9]. Using the same data but by looking for 511 keV gamma ray siganls in HPGe detector and Csi(Tl) crystal detector in coincidence, we searched for the β +/EC of ^{112}Sn. After subtracting background measured with the same amount of de-ionized water sample, we set the 90% CL lower limit of the half life time for β +/EC of ^{112}Sn to be 9.1×10^{17} years[9]. This is five orders of magnitude improvement over the previous limit[10].

A 8.0cm×8cm×1.0cm sample of Zn metal with 99.8 % purity was measured for the period of 332 hours using the same setup used for TBSN measurement. Again applying the same analysis technique that was used for β +/EC of ^{112}Sn search, and subtracting the background measured by the same size oxygen free high conductivity(OFHC) copper sample, we set the 90% CL lower limit of the half life time for β +/EC of ^{64}Zn to be 1.3×10^{20} years[9]. This result is inconsistent with the previous claim of a positive signal with half life of $1.1\pm0.9\times10^{19}$ years[11] and is two order improvement over the previous limit by $ZnWO_4$ crystal[12].

4. Conclusion

The KIMS group reported new limits on WIMP search using the CsI(Tl) crystal detectors. The most stringent limit on SD WIMP interactions for the case of pure proton coupling is obtained. The DAMA signal region is ruled out for both SI and SD WIMP interaction for the WIMP mass greater than 20 GeV/c^2. KIMS has installed 12 crystals (about 100 kg mass) this year and expects to take data more than a year with a stable condition. KIMS is also carrying out detector development for the double beta decay search. A significant progress is made on the improvement of the scintillation properties and of the growth of large size crystals of $CaMoO_4$. Tin loading to the liquid scintillator was successfully done with a reasonable light yield and energy resolution. Measurements of TBSN and Zn samples using an experimnetal setup of HPGe detector and CsI(Tl) crystal, new lower limits on the half life time of neutrinoless double beta decays of ^{124}Sn, ^{112}Sn and ^{64}Zn is established, which are significant improvements over the previous experimental results.

References

[1] H.S. Lee et al., Nucl. Instrum. Meth. A 571, 644 (2007); Y.D. Kim et al., Nucl. Instrum. Meth. A 552, 456 (2005); Y.D. Kim et al., J. Kor. Phys. Soc. 40, 520 (2002)

[2] D.S. Akerib et al., Phys. Rev. Lett. 96, 011302 (2006)

[3] J.Angle et al., arXiv:0706.0039 (2007)

[4] H.S.Lee et al., Phys. Lett. B 633, 201 (2006)

[5] H.S.Lee et al., Phys. Rev. Lett. 99, 091301 (2007)

[6] R.Bernabei et al., Phys. Lett. B 480, 23 (2000); Riv. Nuovo Cim. 25N1, 1 (2003)

[7] H.J. Kim, et al., New View in Particle Physics, Proceeding of VIETNAM2004, 449 (2004); S.Belogurov et al., IEEE Trans. Nucl. Sci., 52, 1131 (2005); F. Danevich et al., Nucl. Instrm. Meth., in print (2007)

[8] M.J.Hwang et al., Nucl. Instrum. Meth. A 570, 454 (2007)

[9] H.J. Kim et al., Nucl. Phys. A, 793, 171 (2007)

[10] J.H.Fremlin, M.C.Walter, Proc. Phys. Soc. A 65, 911 (1952)

[11] I.Bikit, et al., Appl. Radiat. Isot. 46, 455 (1995)

[12] F.A.Danevich et al., Nucl. Instrum. Meth. A 544, 553 (2005)

The XMASS Experiment

Ko Abe[1] (for the XMASS collaboration)

Kamioka observatory, ICRR, University of Tokyo, Higashi-Mozumi, Kamioka, Gifu, 506-1205, Japan

E-mail: abe@suketto.icrr.u-tokyo.ac.jp

Abstract. XMASS is a multi-purpose detector using a large amount of ultra pure liquid xenon. The properties of liquid xenon, such as large scintillation light yield and high atomic number, are well suited for low energy and low background physics targets, including dark matter, solar neutrinos, and double-beta decay. The XMASS collaboration has recently completed a prototype R&D stage. The primary physics goal of the next-stage XMASS 800kg detector will be a dark matter search. Using self shielding an extremely low background rate of 10^{-4}/day/keV/kg and a dark matter sensitivity around 10^{-45} cm^2 in cross section may be achieved. Excavation of the experimental hall in Kamioka started in August, 2007. We plan to finish the construction of the detector and being data taking within the next few years.

1. Introduction

XMASS is a multi-purpose ultra-pure liquid xenon detector. Its physics targets include dark matter searches, low energy solar neutrino measurements and neutrinoless double-beta decay searches [1, 2].

A first-stage 100 kg prototype, to demonstrate the basic detector technology and study the feasibility of a large scale detector has already been completed. We are now constructing an 800 kg detector with the primary physics goal of performing a sensitive dark matter search. Based on the results and experiences obtained with this 800 kg detector, we plan to build a 10-ton class detector with improved dark matter sensitivity, additionally capable of performing low energy solar neutrino measurements.

Liquid xenon has a large scintillation light yield of 42,000 photons/MeV, as good as NaI(Tl). This enables the detection of low-energy signals such as dark matter incident nuclear recoils. Due to its high atomic number (Z=54), the shielding power of liquid xenon for external gamma rays is very large. Moreover, high density liquid xenon (~3g/cm^3) gives a compact detector with relatively small size. Together these properties of liquid xenon realize detector with low backgrounds and a low energy threshold, which are essential for the physics goals of XMASS.

The key background reduction concept for XMASS is "self-shielding", using the active volume of liquid xenon itself as a shield to reduce the gamma-ray background. Absorption in the outer region of liquid xenon results in an extremely low background environment near the detector center. Weakly interacting particles like WIMPs and neutrinos interact throughout the detector, and can be selected by their vertex position to lie within a low-background central fiducial volume.

The vertex position is reconstructed using the pulse-height pattern of hit PMTs. Figure 1 shows the reconstructed vertex position distribution along the direction of an incident gamma ray in the prototype 100 kg detector. The effect of self-shielding can be seen in these figures. 662 keV gamma rays from a ^{137}Cs source are attenuated by a factor 200 over a distance of 24 cm, and there is a factor of 10 attenuation for the 1.17 and 1.33 MeV gamma rays from ^{60}Co. The effect of self-shielding is well reproduced by a Monte-Carlo simulation.

Figure 2 shows the expected sensitivity of the 800 kg detector for spin-independent interactions, assuming 5 years of data with a 100 kg fiducial volume. The detector is sensitive down to ~10^{-45} cm^2, almost two orders of magnitude better than existing experiments.

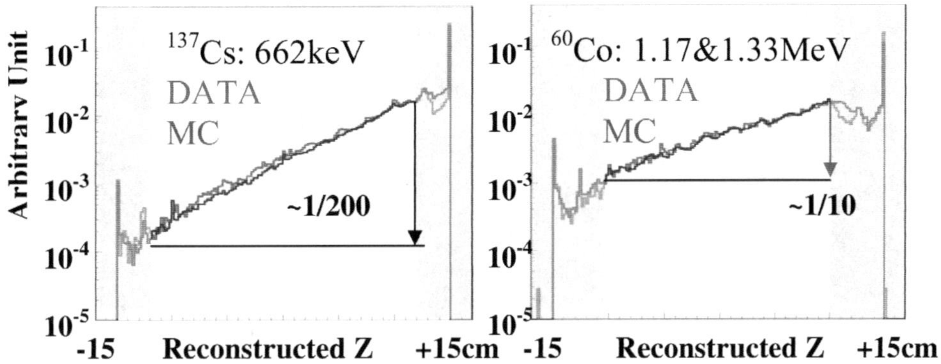

Figure 1. Reconstructed vertex position distributions along the direction of an incident gamma ray in the prototype 100 kg detector. Left: ^{137}Cs source, right: ^{60}Co source. The observed data are drawn in red and are compared to a Monte-Carlo simulation in blue. The shaded regions of the histograms (the right and left ends) are not relevant because the ADCs of the PMTs nearest the vertex position saturate for the events in these regions.

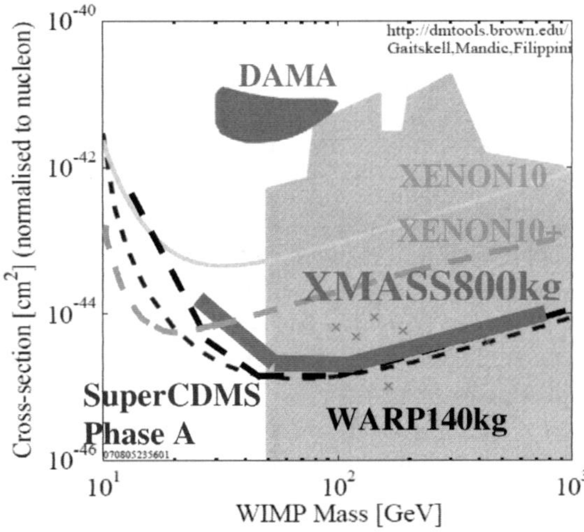

Figure 2. 3-sigma discovery level of the 800 kg detector for a dark matter search assuming spin-independent interactions (thick solid curve). The thin solid curves show current results from other experiments. Dotted and dashed curves show the projected sensitivities of future experiments.

2. Status of the 800kg detector

The 800 kg detector is funded and we plan to construct the detector and commence operations within two or three years. Figure 3 shows a schematic view of the 800 kg detector. There are about 1000 kg of liquid xenon in the volume inside the PMT photocathode plane, which is about 45 cm in radius. The fiducial volume is the region inside a 20 cm radius from the detector center, encompassing 100kg of liquid xenon. A new version of the R8778MOD low background PMT is being developed that has a hexagonal photocathode shape. With 812 hexagonal PMTs, a photocoverage of 67% can be achieved. The scintillation light yield of the detector will be about 4 photoelectrons/keV.

Figure 3. Schematic view of the XMASS 800 kg detector.

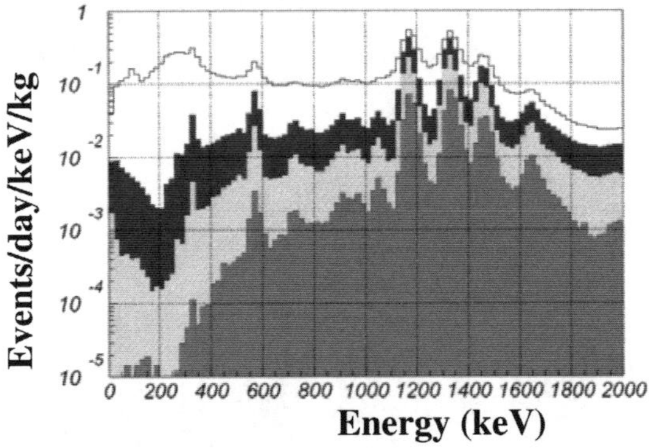

Figure 4. Simulated PMT gamma-ray background. The black line shows a spectrum for observed events in the full detector volume. The blue, green, and red filled histograms show events inside 40, 35, and 25-cm-radius fiducial volumes, respectively.

The target activity level of the new PMTs under development is 1/10 that of the current R8778 PMT. The targets are 1.8×10^{-3}Bq/PMT for the uranium chain, 6.9×10^{-4}Bq/PMT for the thorium chain, 1.4×10^{-2}Bq/PMT for ^{40}K, and 5.5×10^{-3}Bq/PMT for ^{60}Co. Figure 4 shows the PMT gamma-ray background spectrum simulated by Geant4 assuming these activities. The estimated background rate from PMT gamma rays inside a 25-cm-radius fiducial volume is well less than the 10^{-4}/day/keV/kg required to achieve a cross section sensitivity of 10^{-45} cm^2.

The environmental radioactive background, consisting primarily of gamma rays and neutrons, are reduced by a water shield surrounding the detector. Simulations of gamma rays and fast neutrons show that a ~200-cm-thick water layer is enough to shield the environmental background to a level below that of the PMT-induced background.

Requirements on backgrounds originating in the liquid xenon itself are ^{238}U < 1×10^{-14}g/g, ^{232}Th < 2×10^{-14}g/g and ^{85}Kr < 1ppt. These requirements are set by requiring the corresponding contribution to the background to be negligible compared with that from the PMTs. ^{238}U and ^{232}Th daughters were measured for the 100 kg prototype using the Bi-Po coincidence method. Assuming radioactive equilibrium, we measured activity levels of $(9 \pm 6)\times10^{-14}$g/g for ^{238}U and $<23\times10^{-14}$g/g for ^{232}Th. A level of 3.3 ± 1.1ppt was achieved for ^{85}Kr using a distillation system. These radioactive contaminations are within a factor of 10 of the requirements for the 800 kg detector. We believe the requirements can be achieved in the real detector by adding a radon filter and by designing a distillation system with slightly longer tower length.

The full 800 kg detector apparatus will be constructed in the Kamioka mine. Excavation of a new underground area to accommodate the detector started in August, 2007. This area will be ready for detector installation by summer 2008. We expect to finish the construction of the detector and start the dark matter search within the next two to three years.

References

[1] Y. Suzuki et al., "Low energy solar neutrino detection by using liquid xenon", Aug. 2000, hep-ph/0008296.

[2] S. Moriyama, "XMASS EXPERIMENT I", proceedings of the International Workshop on Technique and Application of Xenon Detectors, Kashiwa, Japan, 3-4 Dec. 2001.

The ArDM, a ton-scale liquid argon experiment for direct Dark Matter detection

P. Otyugova, on behalf of ArDM collaboration [1]

Institut für Teilchenphysik, ETHZ, CH-8093 Zürich, Switzerland
Polina.Otiougova@cern.ch

Abstract. The ArDM project aims at developing and operating large noble liquid detectors to search for direct evidence of Weakly Interacting Massive Particles (WIMP) as Dark Matter in the Universe. The initial goal is to design, assemble and operate an approximately 1 ton liquid argon prototype based on the double-phase detection principle to demonstrate the feasibility of a ton-scale experiment with the required performance to efficiently detect and sufficiently discriminate backgrounds for a successful WIMP detection. The detector will independently measure the primary scintillation light and the ionization charge. This paper will mainly describe the concept, R&D results and status of the charge read out system.

1. Introduction

The Argon Dark Matter experiment (ArDM) [1] was initiated in 2004. The goal of the project is to assemble and operate a 1 ton scale double-phase liquid Argon detector with independent readout of the primary scintillation light and the ionization charge. The LAr volume is operated as a TPC, where the charge is readout with fine spatial granularity (few mm resolution) in transverse coordinate and good resolution ($\leq 1\ \mu s$) in longitudinal coordinate, as measured by the time difference between the primary scintillation light and the charge collection time.
A vacuum insulated stainless steel dewar contains a LAr target volume of approximately 850 kg. In order to create a uniform electric field along a vertical drift length of 120 cm, a Greinacher (Cockroft-Walton) chain is used to supply the proper voltages to a cathode (up to $\sim$ 400 kV) placed at the bottom and to a set of field shaping rings positioned along the drift path. Fourteen large area PMTs (8 inches diameter), placed underneath the cathode grid, are used to detect the Ar scintillation light. To maximize the light collection efficiency, the vertical volume delimited by the field shaping rings is enwrapped by light reflector foils. In order to detect VUV Argon scintillation both the surfaces of the PMTs and the reflector foils are covered with a layer of a wavelength shifter Thetra-Phenyl-Butadien(TPB).
In the top part of the dewar, where the Ar vapor is in equilibrium above the liquid surface a charge readout system is positioned. This system will extract the ionization charge from the liquid, multiply it in gas and read it out. Charge multiplication and detection is implemented by a system of Large Electron Multipliers (LEM). Various groups have been developing gas detectors of this type (see for example [2]). This paper will focus on the R&D results on LEM detectors for cryogenic application achieved by our group and their implementation for the ArDM experiment.

[1] http://neutrino.ethz.ch/ArDM/

2. Charge Read out System

LEM is a thick macroscopic hole multiplier manufactured with standard PCB techniques. A 1.5 mm thick, double-sided copper-clad (17.5 mu layer) G-10 plate is drilled with a regular pattern of $500\mu m$ diameter holes, with a distance of $800\mu m$ between centers of neighbouring holes. A $1\mu m$ thick layer of palladium is deposited on copper to avoid oxidization. Charge multiplication is achieved inside the holes by applying a potential difference between the two sides of the plate. The pictures of a single LEM board prototype, a double stage LEM system and simulated electron drift lines are presented on Fig.1. The active area of the prototype is a square of 5.2x5.2 cm^2.

Extensive R&D was performed in order to measure the main characteristics of LEM prototypes.

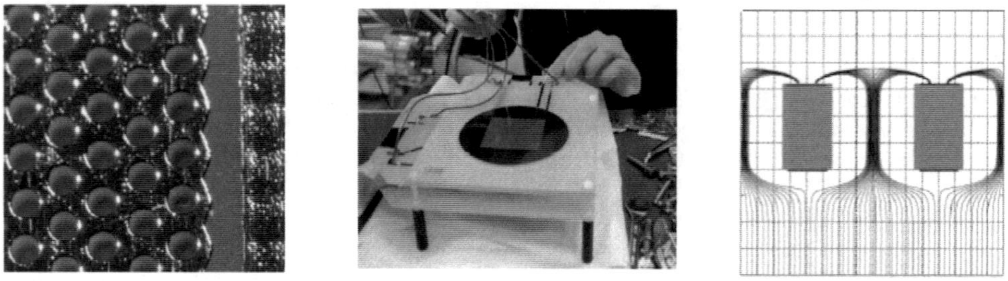

Figure 1. Left: Image of a LEM board prototype. A guard ring is positioned at the side of the sensitive hole region, and it is kept at the same potential. Center: Double-stage LEM assembly. Two LEM plates, spaced by 3 mm, are visible in the upper part of the assembly. The cathode plane is positioned at the bottom. Right: Simulation of electron drift paths in a single stage LEM. Electrons created in the drift gap are focused into LEM holes and collected on a readout electrode.

With a double LEM stage setup we achieved a stable gain of 10^4 in pure Argon low density gas at 87 K and at 0.8 bar. A single stage LEM is not sufficient to obtain this gain. Energy resolution obtained in pure Argon at atmospheric pressure and room temperature with ^{55}Fe r/a source is of the order of 40% FWHM. On Fig.2 signal shapes in pure Ar and in Ar CO_2 (90%/10%) are shown, measured at room temperature and at atmospheric pressure. The signals are recorded at the output of a fast charge preamplifier, connected through a decoupling capacitor directly to the upper electrode of the second LEM stage (Fig.3). There is a well visible difference between the two signal shapes. Signals in pure Ar have an ion-induced component followed by one or more photon feedback components, as expected by secondary avalanches due to photoelectric effect on the cathode or on the electrode of the first LEM stage. The total risetime of the signal is of the order of 20μsec. The slope of the first part of signal is consistent with a signal being generated by ions moving across the holes with a drift velocity of 0.2 mm/μsec. Signals measured in Ar-CO_2 mixture have two visible components. A very fast one O(100)ns, which corresponds to the electron part of the signal and a slow, ion-induced component with a risetime of the order of 5μsec. In Ar-CO_2 mixture we do not expect to observe any photon feedback component since photons are efficiently absorbed by the CO_2 molecules.

The experimental setup for tests in double-phase conditions is shown on Fig.3. The LEM system is positioned inside a stainless steel container cooled down to LAr temperature. The cathode, placed 6 mm below the first LEM stage, provides the drift field to transfer ionization electrons up to the lower electrode of the first LEM. The high voltage is supplied to each electrode of each LEM plate through two independent resistor HV dividers. Pure Argon gas is liquified inside the dewar, covering the cathode, and its level is kept 3 mm below the first LEM stage,

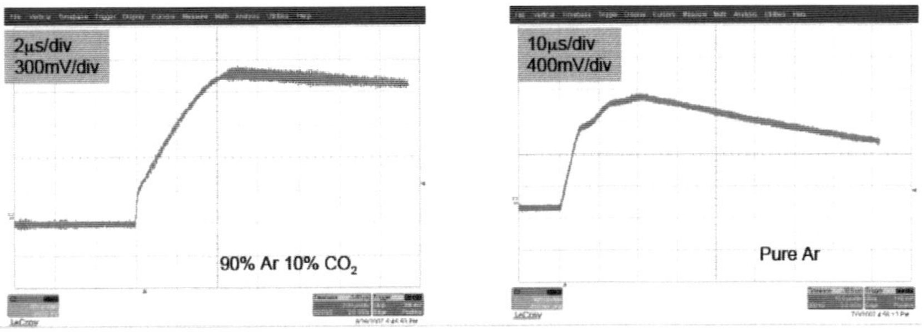

Figure 2. Typical signals measured with ^{55}Fe source. Left: signal in Ar-CO_2 mixture; Right: signal in pure Argon

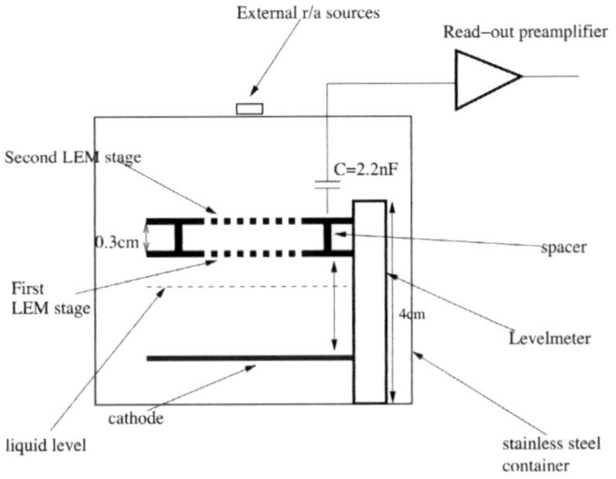

Figure 3. Experimental setup for tests in double phase conditions.

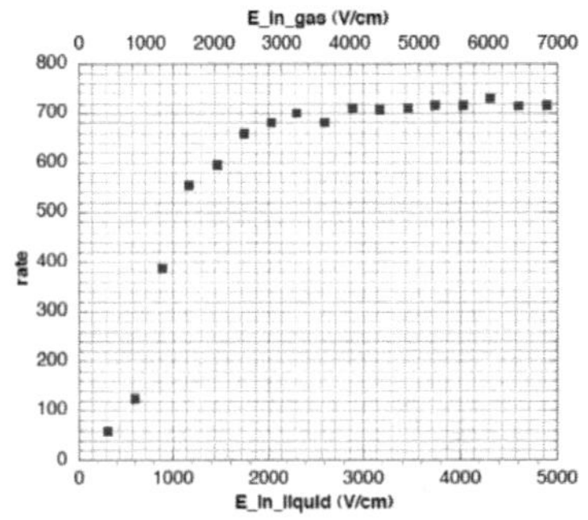

Figure 4. Event rate as a function of electric field in liquid and gas.

as measured by a level-meter with a precision of better than 0.5 mm. The electric field inside the LEM holes is kept at 30 kV/cm, providing a gain of approximately 10^4 in the cold Ar gas in equilibrium with the liquid. We studied the extraction of ionization electrons (generated by an external Co^{60} source) as a function of the potential difference between the cathode and the lower phase of the first LEM stage, as shown on Fig.4. Values of the drift field are given for both liquid and gas Ar phases because of the different electrical permittivity of the two phases. A plateau in the event rate is clearly observed at 3kV/cm drift field in the gas phase.

3. Outlook

In the final setup the charge readout system will cover an area of about 5000 cm^2, subdivided in strips with a granularity down to a few mm. Two planes with orthogonal strips will be readout with electronics of $\sim$ 10 mV/fC sensitivity. This, coupled with a double-stage LEM providing a gain of 10^4, will allow the detection down to a few ionization electrons, as necessary for a Dark Matter experiment. Our plans for the nearest future are first to test a segmented LEM prototype with 16 strips 6 mm wide. In order to get faster rise-time signals, we are also considering the option of adding an anode plane to enhance the fast electron component of the signal.

References

[1] Rubbia A Proc. TAUP 2005, Journal of Phys: conf. Series 36(2006) 133-135.
[2] Chechik R *et al.*, NIM A 2004, vol.535, p.303

PICO-LON Project for WIMPs search

Ken-Ichi Fushimi[1], Kensuke Yasudai[1], Yuuki Kamedai[1], Hiroyasu Ejiri[2,3,4], Ryuta Hazama[5], Kayoko Ichihara[6], Kyoshiro Imagawa[7], Hiroshi Ito[7], Norihiko Koori[1], Hidehito Nakamura[3], Shintaro Nakayama[1], Masaharu Nomachi,[6], Tatsushi Shima[2], Saori Umehara[6], Sei Yoshida[8]

1. Faculty of Integrated Arts and Sciences, The University of Tokushima, Tokushima city Tokushima 770-8502, JAPAN
2. Research Center for Nuclear Physics, Osaka University, Ibaraki city Osaka, 567-0043 JAPAN
3. National Institute of Radiological Sciences, Chiba city Chiba 263-8555, JAPAN
4. Nuclear Science, Czech Technical University, Prague, Czech Republic
5. Faculty of Engineering, Hiroshima University, Higashi Hiroshima city Hiroshima 739-8527, JAPAN
6. Department of Physics, Osaka University, Toyonaka city Osaka, 560-0043 JAPAN
7. Horiba Ltd., Kyoto city Kyoto,601-8510, JAPAN

E-mail: `kfushimi@ias.tokushima-u.ac.jp`
8. Research Center for Neutrino Science, Tohoku University, Sendai city Miyagi 980-8578, JAPAN

Abstract. Highly segmented inorganic crystal has been shown to have good performance for dark matter search. The energy resolution of ultra thin and large area NaI(Tl) scintillator has been developed. The estimated sensitivity for spin-dependent excitation of ^{127}I was discussed. The recent status of low background measurement at Oto Cosmo Observatory is reported

1. Introduction

Dark matter search is one of the most important subjects in nuclear- and astro-particle physics. The particle candidates for cosmic dark matter is a key subject not only for astrophysics but also for particle physics since the particle candidates for dark matter is proposed by various models the beyond standard model.

The components of the universe have been studied by many cosmological observations.. Since the most of the matter in the universe should be the cold dark matter, searching for WIMPs (Weakly Interacting Massive Particles) is quite important. The dark matter in the galaxy has become quite ensuring. One of the most promising candidate for WIMPs is the SUSY neutralino, which interacts with the matter via only weak interaction.

The processes for WIMPs-nucleus interaction are divided into three types; spin-independent elastic scattering (SI), spin-dependent elastic scattering (SD) and spin-dependent inelastic excitation (EX).

In the case of the type-(SI) the scattering amplitude is summed coherently over all the nucleons. Thus the scattering cross section is enhanced by the factor A^2, where A is the mass number of the target nucleus. In the case of the type-(SD), only one nucleon which carries the nuclear spin contributes to the cross section. Thus the cross section is a few orders of magnitude smaller than that of the SI case. The cross section depends on the nuclear spin-matrix element $\lambda^2 J(J+1)$ and has a large ambiguity because this matrix element has a large model dependence for the heavy nuclei [3]. In the case of type-(EX), the target nucleus will be excited to the low lying excited state, which is followed by the gamma ray emission. This process is arisen only if the WIMPs particle has enough kinetic energy. In this case, the matrix element is evaluated from the measured rate of the gamma de-excitation[3, 4]. Thus the model dependence of the cross section is much smaller than the cross section of the SD case.

The segmentation of the detector is shown to be the best way to enhance the position sensitivity[1]. Coincidence measurements of nuclear recoils and γ rays for the inelastic excitation of ^{127}I enhanced the sensitivity using the highly segmented NaI(Tl) detector. Recently, ionized atomic electrons and hard X rays following WIMPs-nuclear interactions have been shown to be useful for the exclusive measurement of nuclear recoils from the elastic scatterings of WIMPs off nuclei [5, 6]. On the other hand, the background events have their own characteristics of timing and spatial profiles. Because the event rate due to the background is reduced by segmentation, a probability of the accidental coincidence of individual background events is vastly reduced.

It has been shown that piling up many thin scintillators enhances the sensitivity for WIMPs search[1]. The highly sensitive detector system PICO-LON (Planar Inorganic Crystals Observatory for LOw-background Neutr(al)ino) has been developed. It consists of many thin NaI(Tl) crystals whose thickness is 0.05cm. Recently, PICO-LON-I which consists of three crystals of thin (0.05cm in thickness) and wide area (6.6cm×6.6cm) NaI(Tl) crystals has been developed. PICO-LON-II which consists of sixteen crystals of thin NaI(Tl) has been also developed. In this report, we describe the excellent performance of a single plate of thin NaI(Tl) which is the foundation of PICO-LON system.

2. Performance test and low background measurement

The energy resolution of the thin NaI(Tl) scintillator was measured for low energy gamma rays. The gamma rays from the sources, ^{241}Am, ^{133}Ba and ^{57}Co were irradiated without any collimator. The typical energy resolution (FWHM) was obtained as small as 18% at 60keV [2].

The energy dependence of the energy resolution was measured. The energy resolution changes inversely proportional to the square root of the energy. The obtained energy dependence agrees with the relationship. From the relationship, the corresponding energy for the single photoelectron was calculated as 0.35keV.

The low background measurement was started in September 2007 at Oto Cosmo Observatory. The observatory is located in the south of Nara prefecture and 150km east from Tokushima. There are three laboratories at the center of the unused railway tunnel whose length is 5km, and the thickness of the overburden rock is 450m. The flux of cosmic rays was reduced by four orders of magnitude compared with the flux at the surface observatory. The fast neutron was reduced by two orders of magnitude.

The concentration of radioactive radon, which is most serious origin of the background against dark matter search, is kept as low as a few Bq/m^3 out of the air-tight container. The PICO-LON system was installed in the air-tight container in which pure nitrogen gas was used. The concentration of radon in the air-tight container is kept as low as 150mBq/m^3. The huge NaI(Tl) scintillators whose dimension of 10.6cm×10.6cm×106cm were used as the active shield. The PICO-LON detector was installed between the huge NaI(Tl) detectors. The air-tight container was covered with 10cm thick pure copper shield and 15cm thick old lead shield to reduce the gamma ray background from the surrounding rock.

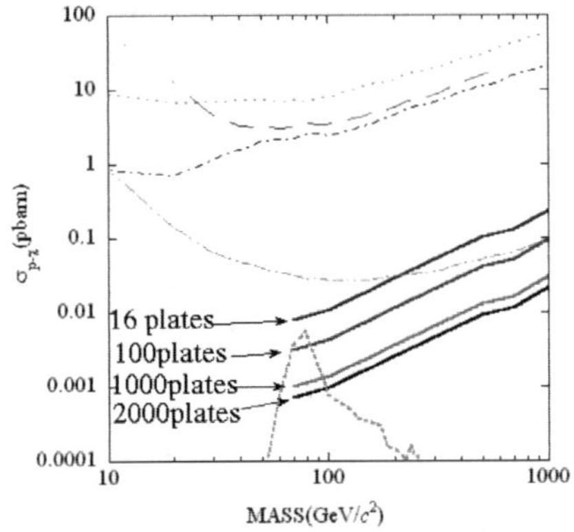

Figure 1. Expected sensitivity for spin-dependent excitation. The cross section was normalized to the proton-WIMPs cross section. The dotted line, dashed line, dash-dotted line and dash-double-dotted lines are the experimental obtained by CRESST, DAMA LXe, ELEGANT-V and NAIAD groups.

3. Prospect

The installation and the tuning of the PICO-LON detector will be done in the winter of 2007. The full measurement will be started in the beginning of 2008. The larger NaI(Tl) plates whose dimension of 15cm×15cm×0.1cm for each plate have been developed and installed in the spring of 2008. The result of high sensitivity measurement will be reported in the summer of 2008.

The PICO-LON system has the ability of extension. The modular plates are piled up and the large volume detector system is constructed. The sensitivity for EX-type of WIMPs interaction depends on the statistics of the energy region of interest (57.6keV). The expected sensitivity by further extension of the PICO-LON system is shown in figure 1. The expected sensitivities by PICO-LON project are calculated by assuming the main background origin is the U-chain contamination which is concentrated in the NaI(Tl) crystal. The concentration of U-chain nuclei are assumed as small as 10μBq/kg for ^{214}Pb and ^{214}Bi, 100μBq/kg for ^{210}Pb [1].

4. Acknowledgment

The authors thank Prof.T.Kishimoto for continuous encouragement of the project. The present work was supported by "Toray Science and Technology Grant" and the University of Tokushima.

References

[1] K.Fushimi et al., Journal of Physical Society of Japan, **74**(2005) 3117.
[2] K.Fushimi et al., Journal of Physical Society of Japan, **75**(2006) 064201.
[3] J.Ellis et al., Physics Letters **B212** (1998) 375.
[4] H.Ejiri, H.Ohsumi and K.Fushimi, Physics Letters **B317** (1993) 14.
[5] Ch.C.Moustakidis, J.D.Vergados and H.Ejiri Nuclear Physics **B727** (2005) 406.
[6] H.Ejiri, Ch.C.Moustakidis and J.D.Vergados, Physics Letters **B639** (2006) 218.

NEWAGE

H.Nishimura[1], K.Hattori[1], C.Ida[1], S.Iwaki[1], S.Kabuki[1], H.Kubo[1], S.Kurosawa[1], K.Miuchi[1], H.Sekiya[2], A.Takada[1], A.Takeda[2], T.Tanimori[1] and K.Ueno[1]

[1] Cosmic-Ray Group, Department of Physics, Graduate School of Science, Kyoto University, Kitashirakawa-oiwakecho, Sakyo-ku, Kyoto 606-8502, Japan
[2] Kamioka Observatory, ICRR, The University of Tokyo Higashi-Mozumi, Kamioka-cho, Hida 506-1205, Japan

E-mail: nisimura@cr.scphys.kyoto-u.ac.jp

Abstract. We developed a three-dimensional gaseous tracking device, and performed a direction-sensitive dark matter search in a surface laboratory. By using 150 Torr carbon-tetrafluoride (CF_4gas), we obtained a sky map drawn with the recoil directions of the carbon and fluorine nuclei, and set the first limit on the spin-dependent WIMP (Weakly Interacting Massive Particles)-proton cross section by a direction-sensitive method.

1. Introduction

Weakly interacting massive particles (WIMPs) are very plausible candidates of dark matter. Most dark matter search experiments are designed to measure only the energy transferred to the nucleus through WIMP-nucleus scatterings [1]-[10]. Because the amplitude of an annual modulation of the energy spectrum, which is distinct signals of WIMPs of these experiment, is only a few % of the rate, positive signals of WIMPs are very difficult to detect. On the other hand, the motion of the solar system with respect to the galactic halo is considered to provide a much larger asymmetry in the direction-distribution of the WIMP velocity observed at Earth. A detection method of a positive signature of WIMPs by measuring the directions of the nuclear recoils have been studied [5]. Gaseous detectors are one of the most appropriate devices to detect this "WIMP-wind" . Since former studies suggest to perform the WIMP-search both via spin-independent (SI) and spin-dependent(SD) interactions [11], we propose a SD-sensitive WIMP-wind search, named NEWAGE (New generation WIMP-search with an Advanced Gaseous tracking device Experiment) [12]. In this paper, we describe the detector performance and report on the direction-sensitive WIMP-search experiment in a surface laboratory.

2. Detector

We developed a small-volumed μ-TPC (NEWAGE 1.0). A μ-TPC is a gaseous time projection chamber with a micro pixel chamber (μ-PIC) readout. A μ-PIC is a gaseous two-dimensional position-sensitive detector manufactured by printed circuit board technology, which is originally developed in our laboratory. We developed a large-sized (31×31cm^2) μ-PICs[13], and subsequently developed a μ-TPC with a detection volume of $23\times28\times30$cm^3[14]. We use the μ-TPC with 150 Torr CF_4 gas (9.0g effective mass) and total gas gain of 2300, which is high enough to maintain the detection efficiency of recoil nucleus. The nuclear detection efficiency was given by $\epsilon = 1.0\times$ erf$((E-45.8)/165.2)$, where erf$(x)=\frac{2}{\sqrt{\pi}}\int_0^x exp(-x^2)dx$ is the error function and E is the energy in keV. The 90% C.L. upper limit of the gamma-ray rejection factor was 1.5×10^{-4}. The position resolution, which is measured using the same method as described in our previous work[14], was 0.8 mm (RMS). The angular resolution of the carbon and fluorine nuclei in the DM energy range was 25° (HWHM). The details of these performance were described in Ref.[15].

3. Surface measurement result

A dark matter search measurement was performed in a surface laboratory at Kyoto University (Lat. 35°2' N, Long. 135°34' E) from November 1st, 2006 to November 27th, 2006. The total live time was 16.71 days and the exposure was 0.151 kg days. The sum of the calibration and maintenance time was 7.51days, and the dead time due to the data acquisition was 1.0days. The obtained spectrum is shown in Fig.1. We initially derived the limits of the WIMP-proton spin-dependent cross section only from the spectrum (conventional method). We next performed a direction-sensitive analysis [15]. We created a sky map of the north sky with the direction of the three dimensional nuclear recoil tracks. Because we did not detect the sense of each track, the map was restricted to half of the sky larger and the southern part of the sky was folded. We plotted every nuclear event in Fig.2. The larger maker, labeled "DM expected", is the direction of the solar motion or Cygnus from which the WIMP-wind is expected. We calculated θ, the angle between the WIMP-wind direction and the recoil direction for each event, and drew the $\cos\theta$ distribution shown in Fig.3. We compared this measured $\cos\theta$ distribution with the expected WIMP-signal. The best fitted function of WIMP-signals of the data was rejected at the 90% confidence level by a χ^2 test. And we set the upper limits shown with thick-collied line (labeled "direction-sensitive") in Fig.4. It was demonstrated that the direction-sensitive method gave a limit slightly better than the conventional spectrum-only method. This is first limit set by a dark matter search experiment with a gaseous tracking device. As for the spin-dependent limit, this is the first result by a direction-sensitive experiment.

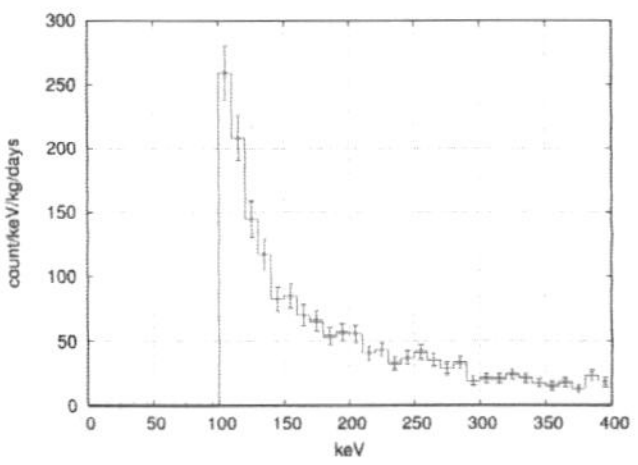

Figure 1. Measured energy spectrum. The detection efficiency is taken into account. [15]

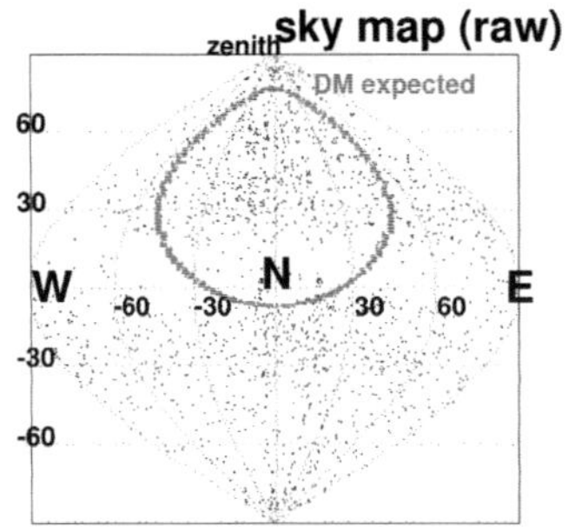

Figure 2. Obtained sky map of north sky. The recoil directions of the fluorine and carbon nuclei are used. The thick line indicates the direction of the solar motion (direction of WIMP-wind). [15]

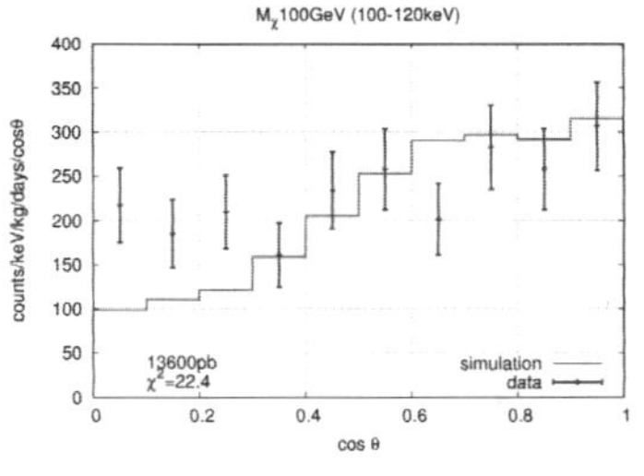

Figure 3. Measured (with error bars) and expected (histogram) distribution of the angle between the recoil direction and the WIMP direction. The expected histogram is that with 100 GeV of WIMPs mass , 100-120keV bin and 1.36×10^4pb. [15]

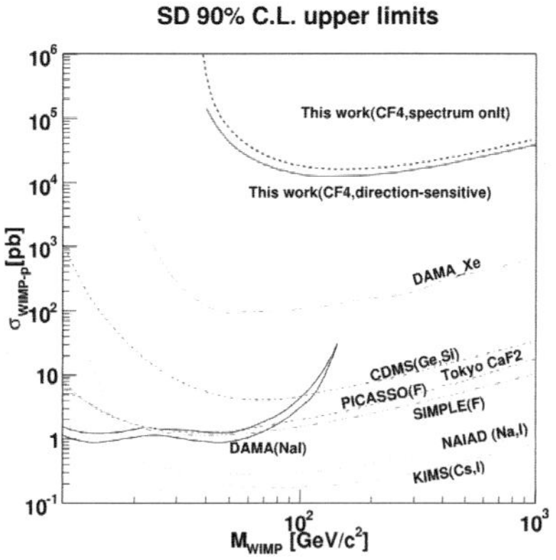

Figure 4. 90% C.L. upper limits on the WIMP-proton spin-dependent cross section The thick solid and dotted lines show the limits obtained with and without the direction information, respectively[15]. Limits from other experiments (DAMA(Xe)[2], DAMA(NaI)[3], NAIAD[4], Tokyo CaF_2[6], SIMPLE[7], PICASSO[8], CDMS[9] and KIMS[10]) are shown for comparison.

4. Future work and Underground experiment

As a next step, we started an underground measurement in Kamioka Observatory located at 2700m.w.e. underground (36°25' N, 137°18' E) from January, 2007 to check the stability of the performance of the detector for a long operation and to measure the background from the detector components. The detector has been operated stably without serious problems from January to August, 2007, while we got the exposure of 0.23 kg days . We estimated internal background caused by the radioactive impurity which must be eliminated. We will improve the purity for low background measurement and make a larger detector for a next generation WIMP search.

Acknowledgments

This work was supported by a Grant-in-Aid in Scientific Research of the Ministry of Education, Culture, Sports, Science, Technology of Japan; research aid program of Toray Science Foundation; and Grant-in-Aid for the 21st Century COE "Center for Diversity and Universality in Physics".

References

[1] P.Beli, R.Bernabei, A.Incicchitti, D.Prosperi, Nuovo Cimento C 15 (1992) 475.
[2] R.Bernabei, et al., Phys. Lett. B 436 (1998) 379.
[3] R.Bernabei, et al., Phys. Lett. B 480 (2000) 23.
[4] G.J.Alner, et al., Phys. Lett. B 616 (2005) 17.
[5] H.Sekiya, et al., in :N.J.C.Spooner, V.Kudryavtsev(Eds.), Proceedings of the Fourth International Workshop on "The Identification of Dark Matte", Edinburgh, UK, 2004, World Scientific, 2005, p. 378.
[6] Y.Shimizu, et al., Phys. Lett. B 633 (2006) 195.
[7] T.A.Girard, et al., Phys. Lett. B 621 (2005) 233.
[8] M.Barnabe-Heider, et al., Phys. Lett. B 624 (2005) 186.
[9] CDMS Collaboration, Phys. Rev. D 73 (2006) 011102.
[10] H.S.Lee, Phys. Rev. Lett. 99 (2007) 091301.
[11] T.A.Girard, F.Giuliani, Phys. Rev. D 75 (2007) 043512.
[12] T.Tanimori, et al., Phys. Lett. B 578 (2004) 241.
[13] A.Takada, et al., Nucl. Instr. Meth. A 573 (2007) 195.
[14] K.Miuchi, et al., Nucl. Instr. Meth A 576 (2007) 43.
[15] K.Miuchi, et al., Phys. Lett. B 654 (2007) 58.

The Inert Doublet Model : a new archetype of WIMP dark matter?

Michel H.G. Tytgat

Service de Physique Théorique, CP225, Université Libre de Bruxelles, Bld du Triomphe 1050 Brussels, Belgium

E-mail: `mtytgat@ulb.ac.be`

Abstract. The Inert Doublet Model (IDM) is a two doublet extension of the Higgs-Brout-Englert sector of the Standard Model with a Z_2 symmetry in order to prevent FCNC. If the Z_2 symmetry is not spontaneously broken, the lightest neutral extra scalar is a dark matter candidate. We briefly review the phenomenology of the model, emphasizing its relevance for the issue of Electroweak Symmetry Breaking (EWSB) and the prospects for detection of dark matter.

1. Introduction

A host of dark matter candidates have been proposed over the years, some more natural than others. While the axion is an old contender, the most acclaimed candidate is the neutralino which, together with the so-called KK-photon and the heavy photon in Little Higgs models, belongs to the category of weakly interacting massive particles (WIMP). (See the review by Scopel at this conference [1]). Other models of dark matter are often looked at with some disdain for, it is argued, they lack motivation. Well, supersymmetry is certainly well motivated, but the fact is that, due to constraints from colliders, considerable fine tuning is required to obtain the right abundance of neutralinos.[1] It is in our opinion fair to consider simpler, albeit phenomenologically interesting, toy models with dark matter. The Inert Doublet Model is such a model [2]. It is a simple extension of the Standard Model, with a Brout-Englert-Higgs sector that could impact electroweak symmetry breaking [3, 4], it is compatible with colliders constraints [5] and has a dark matter candidate that could be observable in the near future [6, 7]. Furthermore, with minor generalization [8], it could establish a link between neutrino physics and dark matter.

2. The IDM and Electroweak Symmetry Breaking

The model we consider is a two Higgs doublet extension of the SM, $H_1 = (h^+ \, (h+iG_0)/\sqrt{2})^T$ and $H_2 = (H^+ \, (H_0 + iA_0)/\sqrt{2})^T$, together with a Z_2 symmetry such that all fields of the Standard Model and H_1 are even under Z_2 while $H_2 \to -H_2$. We assume that Z_2 is not spontaneously broken, *i.e.* that H_2 does not develop a *vev*. As there is no mixing between the doublets, h

[1] By this I mean fine tuning in parameter space on top of the adjustment inherent to the WIMP paradigm. Even the simplest instance of dark matter, a stable heavy neutrino with weak interactions only, has an energy density that spans more than four decades (see figure 1 in [1]) above and below the observed value. Incidentally, why the latter is then of the same order of magnitude as the contribution of baryons is a mystery not addressed by the WIMP paradigm.

plays the role of the usual Higgs particle. Because the extra doublet does not couple to quarks (and leptons in the simplest case) there are no FCNC. The most general renormalisable (CP conserving) potential of the model is

$$V = \mu_1^2|H_1|^2 + \mu_2^2|H_2|^2 + \lambda_1|H_1|^4 + \lambda_2|H_2|^4 \qquad (1)$$

$$+\lambda_3|H_1|^2|H_2|^2 + \lambda_4|H_1^\dagger H_2|^2 + \frac{\lambda_5}{2}\left[(H_1^\dagger H_2)^2 + h.c.\right]$$

with real quartic couplings. The $SU(2) \times U(1)$ symmetry is broken by the vacuum expectation value of H_1, $\langle H_1 \rangle = v/\sqrt{2}$ with $v = -\mu_1^2/\lambda_1 = 246$ GeV while, assuming $\mu_2^2 > 0$, $\langle H_2 \rangle = 0$.

We choose (this is arbitrary) H_0 to be the dark matter candidate. Two H_0 mass ranges are consistent with dark matter observations [3, 6, 9], either $M_{H_0} \lesssim 80$ GeV or $M_{H_0} \gtrsim 500$ GeV. The former is the most interesting from the point of view of phenomenology and observations. The relic abundance of H_0 is shown on Figure 1 below.

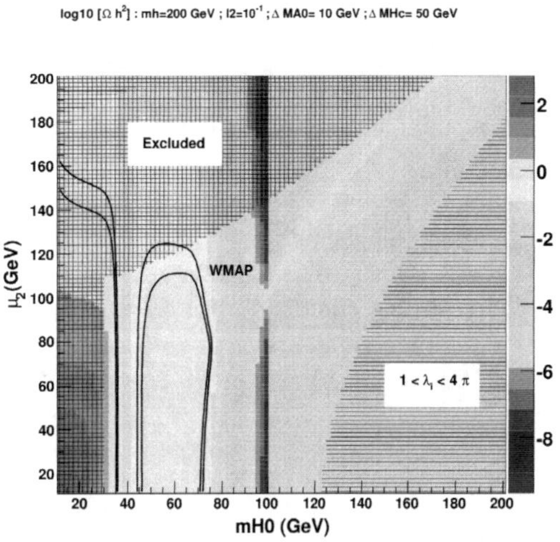

Figure 1. $\log_{10} \Omega h^2$ of H_0 for $M_h = 200$ GeV, computed using micrOMEGAs2.0 [11].

The mass spectrum of the extra scalars depends on the quartic couplings λ_3, λ_4 and λ_5 and on the mass parameter μ_2. Two interesting situations may arise. Firstly, if λ_5 is small, there is an approximate $U(1)_{PQ}$ symmetry and $M_{H_0} \approx M_{A_0}$ while, generically, $M_{H^\pm} \gg M_{H_0}, M_{A_0}$ (large isospin breaking). Less obvious is the custodial symmetry that arises if $M_{H^\pm} \approx M_{H_0}$ (resp. $M_{H^\pm} \approx M_{A_0}$) [10]. Both limits are of phenomenological interest.

First, large isospin breaking leads to large contributions to $\rho = M_W^2/c_\theta^2 M_Z^2$ or, equivalently, to the T parameter [3]

$$\Delta T \approx \frac{1}{24\pi^2 \alpha v^2}(M_{H^\pm} - M_{H_0})(M_{H^\pm} - M_{A_0})$$

This feature has been exploited in [3] to screen the Z and $W^\pm$ gauge boson masses from an equally large (but opposite in sign) contribution from a heavy Higgs (achieving M_h up to $500 GeV$).

Second, we have exploited the custodial symmetry (vanishing contribution to ΔT) in [4] together with the possibility that $M_{H^\pm} \approx M_{H_0} \ll M_{A_0}$ (resp. $M_{H^\pm} \approx M_{A_0} \gg M_{H_0}$) to give large radiative corrections to the effective potential. In particular, the large (negative) radiative effect of the top quark can be compensated by a large contribution from the extra scalars. In the extreme, although very suggestive, limit of vanishing $\mu_{1,2}$ parameters (the Coleman-Weinberg scenario), the electroweak symmetry breaking is literally induced by the extra scalar doublet *i.e.* by WIMP dark matter, a possibility that has been essentially overlooked in the litterature. Some working cases are listed in Table 1.

3. Direct and indirect detection

Being a scalar particle, the cross-section for direct detection is 100 % spin-independent (SI), a specific feature (one could say a limitation) of the model. In Figure 2 we show the prospect for direct detection while Figure 3 displays the integrated flux from the GC into gamma rays

λ_3	λ_4	λ_5	M_h	M_{H_0}	M_{A_0}	$M_{H^\pm}$
5.4	-2.7	-2.7	120	43	395	395
5.4	-2.6	-2.6	120	72	390	390
7.6	-4.1	-4.1	180	12	495	495
7.6	-3.8	-3.8	180	64	470	470
-0.003	4.6	-4.7	120	39	500	55
-0.07	5.5	-5.53	150	54	535	63

Table 1. Instances of mass spectrum from radiative symmetry breaking. The quartic couplings are large, but still in the perturbative regime. All these solutions are compatible with the observed abundance of dark matter.

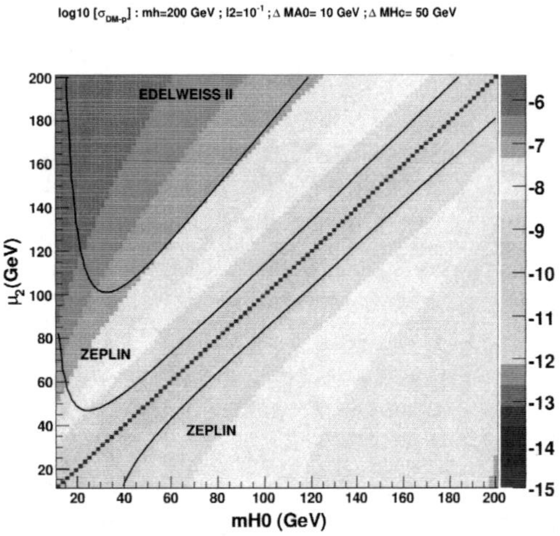

Figure 2. H_0-nuclei cross-section with M_h = 200

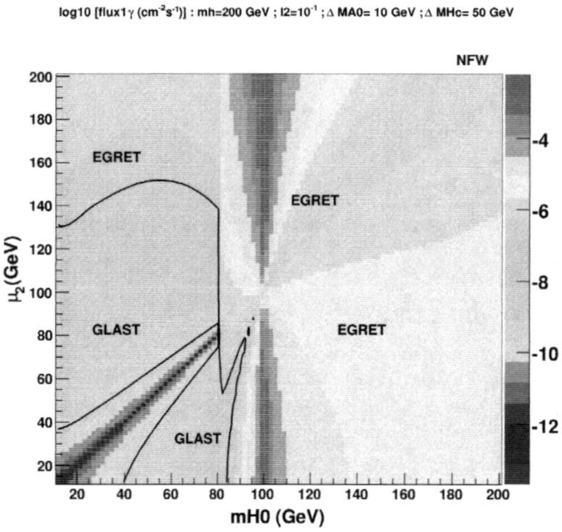

Figure 3. Integrated gamma ray flux from H_0 annihilation with $M_h = 200$ GeV.

[6]. Gustafsson *et al* (see [7] and the talk by Bergstrom at this conference) have shown that annihilation into photon pairs, which occurs at one-loop, could give a significant gamma ray line. For $M_{H_0} \sim 70$ GeV say, this is in the window of opportunity of the forthcoming GLAST detector. As emphasized in [7] this enhancement (*i.e.* compared the neutralino signal) is a distinctive signal, due to the fact that the inert doublet does not couple to quarks. We believe that the lightest scalar of the IDM rightly deserves the title of new dark matter archetype.

Acknowledgements

The author thanks the belgian FNRS and the IAP VI/11 for financial support and Laura Lopez Honorez, Emmanuel Nezri, Josep Oliver and Thomas Hambye for collaboration.

References

[1] S. Scopel, arXiv:0711.2870 [hep-ph].
[2] N. G. Deshpande and E. Ma, Phys. Rev. D **18** (1978) 2574.
[3] R. Barbieri, L. J. Hall and V. S. Rychkov, Phys. Rev. D **74** (2006) 015007 [arXiv:hep-ph/0603188].
[4] T. Hambye and M. H. G. Tytgat, arXiv:0707.0633 [hep-ph].
[5] K. S. Babu and E. Ma, arXiv:0708.3790 [hep-ph].
[6] L. Lopez Honorez, E. Nezri, J. F. Oliver and M. H. G. Tytgat, JCAP **0702** (2007) 028 [arXiv:hep-ph/0612275].
[7] M. Gustafsson, E. Lundstrom, L. Bergstrom and J. Edsjo, Phys. Rev. Lett. **99** (2007) 041301 [arXiv:astro-ph/0703512].
[8] E. Ma, Phys. Rev. D **73** (2006) 077301 [arXiv:hep-ph/0601225].
[9] M. Cirelli, N. Fornengo and A. Strumia, Nucl. Phys. B **753** (2006) 178 [arXiv:hep-ph/0512090].
[10] J. M. Gerard and M. Herquet, Phys. Rev. Lett. **98** (2007) 251802 [arXiv:hep-ph/0703051].
[11] G. Belanger, F. Boudjema, A. Pukhov and A. Semenov, Comput. Phys. Commun. **176** (2007) 367.

Status of the R&D for surface events rejection in the EDELWEISS experiment

X-F.Navick for the EDELWEISS collaboration

CEA DAPNIA/SEDI bat 534 F-91191 Gif Sur Yvette cedex France

navick@cea.fr

Abstract. The first data taking of EDELWEISS-II the Dark Matter search with bolometers, measuring simultaneously ionization and heat, indicate the presence of incomplete charge collection events underneath the electrodes. These events, interpreted as surface interactions limit strongly the sensibility to WIMPs. In this paper, we are describing the status of the different techniques that are under development in our collaboration to identify and reject this type of events.

1. Surface events: the limiting factor

The first data taking of EDELWEISS-II (EDW2) indicates that the WIMPS search with ionization-heat detectors might be affected by the presence of so-called surface events as expected by the last analysis of EDELWEISS-I (EDW1) [1] [2] [3]. These events are due to the energy deposition mainly induced by the interaction of beta particles and X-rays in the first 20 to 100 microns underneath the electrodes for which the collection of free charge carriers is not complete. This is due to efficient trapping and recombination of charges that occurs in the electrodes during the screening of the electric field just after the free charge carrier production. For these events the ionization measurement is lowered in such a way that some electron recoils can mimic the nuclear recoils.

In this paper, we present briefly the status of the investigation of different techniques of reduction, identification and rejection of surface events under development for very low radioactive detectors to improve the physics results of the experiment EDW2.

2. Passive rejection of surface events

2.1. Detectors with amorphous layers

The deposition of a silicon or germanium amorphous layer just underneath the aluminum electrodes strongly reduces the number of surface events [4]. The free carriers can not access to the electrode anymore during the diffusion phase due to the barrier created in the band profile by the amorphous layer. Nevertheless, the surface events are not completely suppressed.

2.2. Ionization-heat detectors with thick electrodes

For electrode thicker than the range of 210Pb betas in aluminum (main source of beta background), no ionization occurs for surface events. Because of the good discriminations between non-ionizing events and low ionization events such as nuclear recoils even at the threshold level (15-20keV), this approach should reduce significantly the number of problematic events. The difficulty to apply this technique is to control the deposition parameters of very thick aluminum electrodes as stability and homogeneity. Moreover, the energetic photons that will interact underneath the electrodes will suffer of incomplete charge collection as the surface events did.

The power of reduction of surface events of both techniques is limited. This is why the more promising techniques based on an active rejection of surface events have been developed.

3. Active rejection of surface events

3.1. Identification of surface events with interleaved-electrode-detectors

The replacement of the standard full electrodes by interleaved electrodes [5] creates two main orientations for the electric fields. The field's lines are perpendicular to the face in the bulk, while they are mainly parallel to the electrodes surface plane near the surfaces. Thus the electrodes involved in the charge carrier collection are different for bulk events and surface events (respectively A and B in Figure 1). This electric field configuration creates some low field areas (event C in the same figure) which might constitute the limiting factor of this technique. A first 200g prototype has been developed in a program called INTERDIGIT.

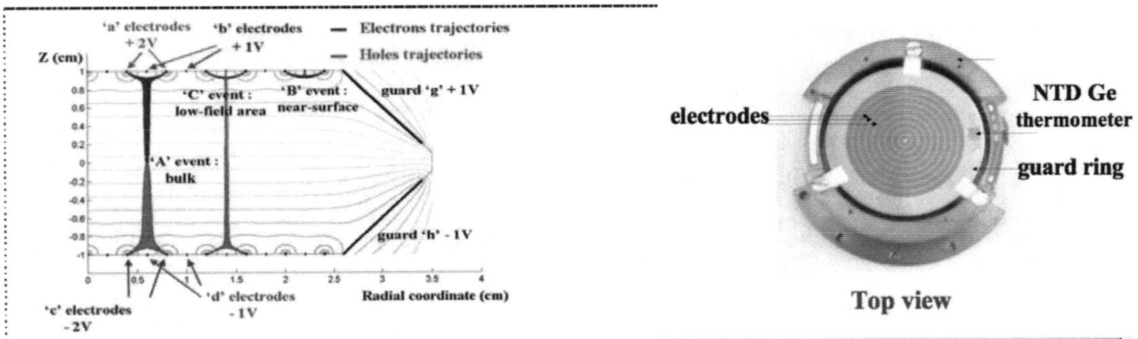

Figure 1. Carrier trajectories and event populations in detectors fitted with interleaved collection electrodes. Picture of the first 200g prototype.

For illustration of present status of test and performance, Figure 2 presents the results of a recent calibration of a 200g prototype with ^{241}Am and ^{252}Cf sources (γ rays and neutrons respectively) before and after the rejection of the surface events. A sizeable fraction of the low penetration 60 keV gamma events at high Q are rejected, most of low Q region ones disappear, while most of the neutron induced interactions in the bulk are kept. Its performances in a low radioactive background will be evaluated before the end of 2007.

3.2. Measurement of the event depth with Pulse shape analysis of the ionization signal

A good resolution in time of the ionization channel (in the 10ns range) allows the extraction of the depth of the energy deposition from the rise of the pulse [6]. A first application of this technique took place in the late run of Edelweiss-1. It showed that the surface event rejection by pulse-shape analysis can only be applied to high energies only (>50keV) due to electronic noise, and that the precision of the event localization is in the order of 1mm. Therefore this technique has a limited interest for low energy surface events rejection which is strongly required for WIMP research. Nevertheless it can be used to characterize the production of a space charge in the volume of the detectors with time. [7][8]

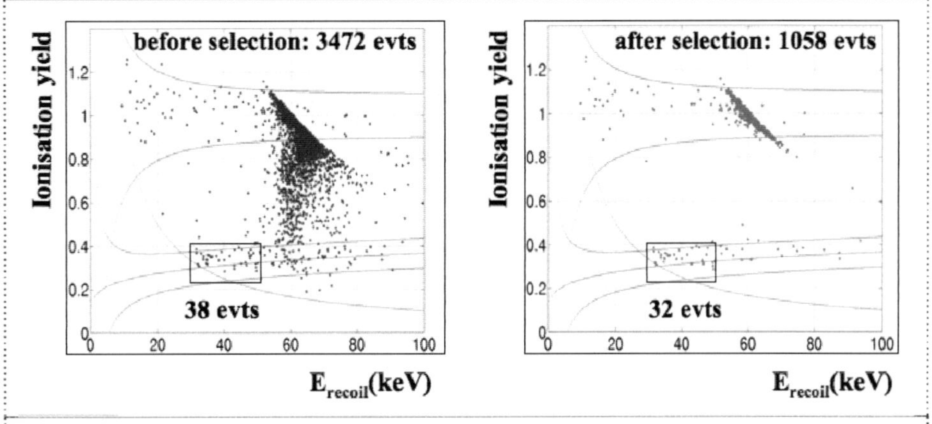

Figure 2. Rejection efficiency of the surface events by the 200g prototype with interdigital electrodes. ^{241}Am and ^{252}Cf radioactive sources

3.3. Athermal phonons measurement with NbSi thin films

This latter technique is based on a transient enhancement of the heat channel signal produced by the direct absorption of athermal phonons in the thermal sensor [9]. When two NbSi films are evaporated on both sides of the detector the enhancement will be larger in the closer film. This allows the determination of the depth of the interaction with a precision better of order of 1mm. Fig 3 shows results of recent beta and neutron calibration performed on a 200g detector. Beta rays from 109Cd source and energy lower than 100 keV are clearly separated from bulk events on the scatter plot of the athermal amplitude asymmetry parameter as a function of ionisation energy (97 % have asymetry higher than 0.3). Right figure shows that neutron recoils, identified as low ionisation yield events as black dots, have indeed low asymmetry, the cut 0.3 keeps most of them. Accurate determination of efficiency and fiducial volumes are under way.

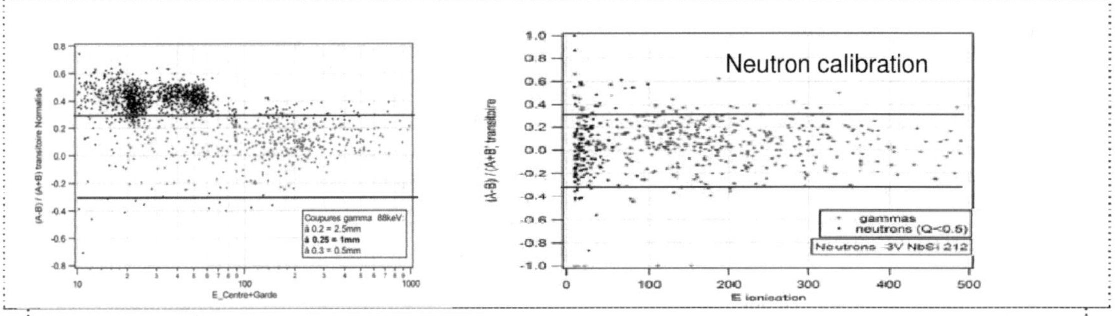

Figure 3. Calibration of the NbSi detectors with a 109Cd electrons source at 18, 62, 84 keV (left), and with a neutron source (right), see text for details.

References

[1] V.Sanglard et al. PhysRevD.71 (2005) 122002

[2] S.Fiorucci et al., *Astropart. Phys.* 28 (2007) 143-153,

[3] T.Shutt et al., NIM A 444 (2000) 340-344

[4] X.-F.Navick et al., NIM A 559 (2006) 483-485

[5] P.L.Brink et al., NIM A 559 (2006) 414-416

[6] A.Broniatowski, et al., NIM A 559 (2006) 378-380

[7] X.-F.Navick et al., NIM A 442 (2000) 267-274

[8] A.Broniatowski, et al., NIM A 559 (2006) 402-404

[9] C.Nones et al. In these proceedings

Neutron background in the Boulby Underground Laboratory

Vitaly A. Kudryavtsev for UKDMC, ZEPLIN-II and ILIAS
Department of Physics and Astronomy, University of Sheffield, Sheffield S3 7RH, UK
E-mail: v.kudryavtsev@sheffield.ac.uk

Abstract. The neutron background in the Boulby Underground Laboratory is reviewed. Measurements of the neutron flux from the rock have been carried out with a small liquid scintillator cell. The neutron flux from rock has been measured as $(1.72 \pm 0.61$ (stat.)± 0.38 (syst.))$\times 10^{-6}$ cm^{-2} s^{-1} above 0.5 MeV. The simulations based on the measurements of the uranium and thorium concentrations in rock predict the flux of 1.21×10^{-6} cm^{-2} s^{-1} above 0.5 MeV, in agreement with the measurements. Muon-induced neutron flux is measured with the large liquid scintillator, also served as an active veto system in the ZEPLIN-II experiment. The rate of gammas depositing more than ≈ 0.7 MeV, from neutron capture was found to be 0.096 ± 0.003 (stat.) events/muon in 20-190 μs time window after the muon signal, about half of the rate predicted by GEANT4 models. This result is preliminary and more detailed simulations are in progress.

1. Introduction

Neutron background can limit the detector sensitivity to rare event searches, such as dark matter WIMPs, double-beta decay, low and intermediate energy neutrinos (solar, reactors, geo-, atmospheric etc.) and proton decay. Neutrons from radioactivity are originated in spontaneous fission (of ^{238}U mainly) and (α,n) reactions on low and intermediate Z isotopes (Z<30). Their energies are limited to about 10 MeV. Neutrons from cosmic-ray muons have spectra extended to GeV energies. Although the flux of muon-induced neutrons deep underground is far below that from radioactivity, they are more penetrating and can be responsible for a significant background component. They can travel far from the muon track or their point of origin, reaching detectors from large distances and reducing the efficiency of an anticoincidence system. Any high-A material, usually used as a shielding against gammas, is also a good target for neutron production by muons or their secondaries.

In this paper we review the measurements and simulations of the neutron flux at the Boulby Underground Laboratory. The Laboratory is located in a salt and potash mine in North Yorkshire (UK) at a depth of 1070 metres. Several dark matter experiments have been and are carried out in the laboratory: NaI, NAIAD, ZEPLIN-I, ZEPLIN-II, ZEPLIN-III, DRIFT-I and DRIFT-II. To investigate the background due to the radioactivity and cosmic rays at Boulby and provide an input for the calculation of detector sensitivity to rare events, several measurements of background have been performed, partly supported by ILIAS (Integrated Large Infrastructures for Astroparticle Science) – an I3 European project, funded within FP6 programme.

One of the early works was the measurement of gamma-ray flux at Boulby using the high purity Ge detector exposed to the gamma radiation from rock (Smith et al., 2005). Measured intensities of various gamma-ray lines were compared to the expectations based on the simulation of the gamma transport and detector efficiencies, and converted into concentrations of uranium, thorium and potassium is rock: U – 67 ± 6 ppb, Th – 127 ± 10 ppb, K – 1130 ± 200 ppm.

2. Neutron background from rock

Neutron background from salt rock in the Boulby was measured using small (6.5 litres) cylindrical liquid scintillator cell (Tziaferi et al, 2007). Liquid scintillator was loaded with Gd to enhance neutron capture. The detector was surrounded by lead and copper which acted as passive shielding against gammas from rock. The volume of the detector was viewed from both ends by the two photomultiplier tubes (PMTs) through quartz windows.

The data acquisition system has been triggered by the delayed coincidences between the two pulses within 195 μs. In case of a neutron event, the first pulse was expected to be from proton recoils whereas the delayed pulses was due to gamma-ray(s) from neutron capture on protons, Gd and other isotopes. The main background in the experiment was caused by random coincidences between two pulses from the gamma-ray background originated in detector components.

Detector was calibrated with a number of gamma-ray sources and ^{252}Cf neutron source. The experiment was running for 4 months collecting about two hundred neutrons with a rate of about 1.5 neutrons per day (for a finite time window of 195 μs between the 1st and the 2nd pulses in an event). Neutrons were discriminated statistically from a much larger background of random coincidences between gamma-ray induced signals. Time delay distribution of two neutron-induced pulses in events (proton recoils and gammas from neutron capture) followed the exponential law whereas that of random coincidences was flat. Measured time delay distribution was a composition of the exponential and a flat background (Tziaferi et al, 2007).

Extensive simulations of neutrons produced in the ^{252}Cf source and in rock, their transport and detection were carried out, allowing the conversion of the measured neutron rate in the detector into the neutron flux in the laboratory: $(1.72 \pm 0.61(\text{stat.}) \pm 0.38(\text{syst.})) \times 10^{-6}$ cm^{-2} s^{-1} above 0.5 MeV (Tziaferi et al, 2007). This agrees well with the predicted neutron flux of 1.21×10^{-6} cm^{-2} s^{-1} above 0.5 MeV based on the U/Th concentrations in rock estimated from the measurements of the gamma-ray intensities with Ge detector (Smith et al 2005).

3. Muon-induced neutrons

The measurements of muon-induced neutrons were carried out at the Boulby Underground Laboratory using the active veto system of the dark matter detector ZEPLIN-II. The veto is filled with about 0.73 tonnes of liquid scintillator and is viewed from above by ten 20 cm ETL hemispherical PMTs. The veto system (together with ZEPLIN-II) is surrounded by a 'castle' made of lead and designed to protect ZEPLIN-II from gamma-rays from rock. The total weight of lead is about 50 metric tonnes making it an excellent target for neutron production. Pure and Gd-impregnated wax and polypropylene were put on top of the veto detector under the castle roof. Part of the vessel of the veto detector was painted with Gd-loaded paint. All materials of and around the veto and their exact locations within the underground laboratory were put into the simulation code based on GEANT4 toolkit (Agostinelli et al, 2003). The principle of neutron detection was based on the delayed coincidences between the first, high-energy, pulse from a muon or muon-induced cascade and the delayed second, low-energy, pulse from neutron capture gamma(s). The electronics and data acquisition system were designed for this purpose.

Energy calibrations with ^{60}Co gamma-ray source were performed at the beginning of the data run and at the end of the experiment. Neutron calibration with an Am-Be source of 0.1 GBq intensity has been carried out before the beginning of the data run. Figure 1 shows measured and simulated time delay distributions between the pulses in the events relative to the trigger pulse in the neutron calibration run. Simulation of this run was carried out using the GEANT4 toolkit taking into account the geometry of the setup, position of the source, neutron interactions and capture. Two free parameters were used to tune the simulated distribution to match the data: (i) the normalisation constant; (ii) the flat component due to random coincidences between either proton recoil pulses from two neutrons or gamma-ray pulses from two neutron captures occurring because of the high event rate. The good agreement between measured and simulated time delay distributions proves the reliability of the GEANT4 toolkit in modeling neutron interactions and capture at MeV and sub-MeV energies. The data on muon-induced neutrons were collected from

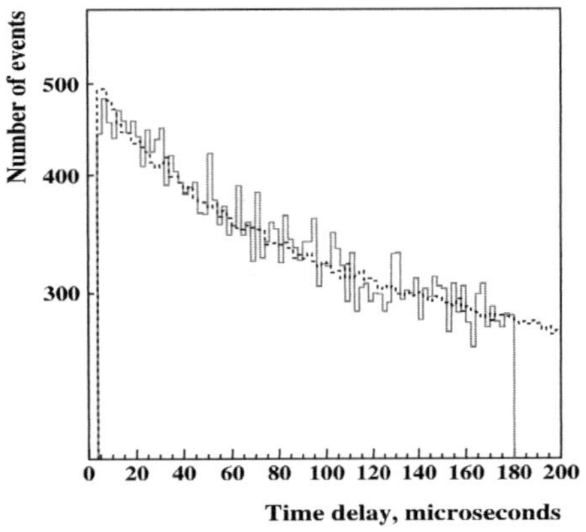

Figure 1. Time delay distribution of pulses in the neutron calibration run.

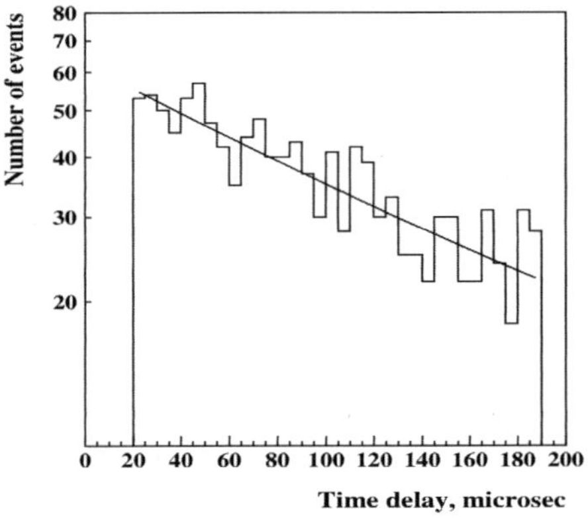

Figure 2. Time delay distribution of pulses in the data run.

August 2006 until April 2007. The live time of the experiment was 204.8 days. 10832 muons depositing more than 25 MeV in the detector were selected for data analysis. The rate of muons was 52.9 ± 0.5 per day in agreement with previous measurements (Robinson et al 2003) giving the value of the total muon flux at Boulby as $(3.92 \pm 0.04$ (stat.)± 0.07 (syst.)$) \times 10^{-8}$ cm^{-2} s^{-1} through a sphere with unit cross-sectional area.

The time delay distribution of pulses in the events relative to trigger (muon) pulses is presented in Figure 2 together with the fit (solid curve) to a combination of an exponential and a flat background. The number of delayed pulses above ≈ 0.7 MeV in 20-190 μs time window after the muon signal was found to be 1239 ± 35 with an expected background, extrapolated from the rate of delayed events after gamma-induced triggers, of 201 ± 10. Background subtraction gives the rate of neutron events as as 0.096 ± 0.003 (stat.) neutrons/muon (for 20-190 μs neutron capture time window after the muon trigger). Preliminary results from GEANT4 simulations (Lindote et al 2007) show very similar time delay distribution of neutron capture events but predict the rate of 0.186 ± 0.003 (stat.)± 0.020 (syst.) neutrons/muon in the time window of 20-190 μs after the muon signal, about twice the measured rate of muon-induced neutrons.

4. Acknowledgments

The author is grateful to all members of the UKDM and ZEPLIN-II Collaborations involved in the reported work. The work was partly supported by the ILIAS integrating activity (Contract R113-CT-2004-506222) and by the Nuffield Foundation (student bursary URB/34132).

References

[1] Agostinelli S et al. (GEANT4 Collaboration) 2003 *Nucl. Instrum. & Meth. in Phys. Res. A* **506** 250
[2] Lindote A et al 2007, in preparation
[3] Robinson M et al 2003 *Nucl. Instrum. & Meth. in Phys. Res. A* **511** 347
[4] Smith P F et al 2005 *Astroparticle Physics* **22** 409
[5] Tziaferi E et al 2007 *Astroparticle Physics* **27** 326

Methods to identify and eliminate surface background events in bolometric detectors for very rare event searches

Claudia Nones[1,2,3]

[1] Dipartimento di Fisica G. Occhialini, P.zza della Scienza 3, 20126 Milano, Italy

[2] Centre de Spectrométrie Nucléaire et de Spectrométrie de Masse, CSNSM-IN2P3/CNRS, Bat.108, 91405 Orsay, France

E-mail: claudia.nones@mib.infn.it

Abstract. The present limitation for experiments looking for rare events, such as neutrinoless double beta decay ($0\nu\beta\beta$) and direct WIMPs detection, is the radioactive background and in particular its discrimination. When the search is based on the bolometric technique, near-surface events constitute a serious problem, due to full volume sensitivity of the detector; in fact for a $0\nu\beta\beta$ experiment they may cause signals close to the Q-value while for a dark matter (DM) experiment, which exploits ionization signal for particle identification, they originates an incomplete charge collection simulating a nuclear recoil. We will described two recently developed techniques for the active suppression of the surface background.

1. Introduction

Background due to near-surface events is the main limitation for very sensitive searches on key topics in astroparticle physics based on the bolometric technique, such as EDELWEISS [1] (WIMP search) and CUORE [2] ($0\nu\beta\beta$ search). In a $0\nu\beta\beta$ experiment, these events cause counts in the energy spectrum close to the Q-value (2.5 MeV in the case of the candidate ^{130}Te discussed here), where the signal is expected, whereas in a Dark Matter experiment, which exploits also the ionization signal for particle identification, they originate an incomplete charge collection simulating nuclear recoils at low energies ($<$ 100 keV) and mimicking therefore the Dark Matter signature. The main source of this surface background consists of radioactive contaminations in materials facing the detectors, such as ^{210}Pb, ^{226}Ra or directly the natural radioactive chain progenitors ^{232}Th and ^{238}U. It is therefore clear that detectors able to identify energy depositions close to the surface are powerful tools to eliminate or at least control this critical background component. These devices can be thought as a safe completion of extremely sophisticated procedures of surface cleaning.

2. How to fight against the surface background problem

There are at least three different approaches to fight against the surface background problem:

- increase the quality of the surface treatment both of crystals and of all the other materials facing them that can be translate into better cleaning procedures;

[3] on behalf of the CUORE Single Module Detector group and on behalf of the EDELWEISS-CSNSM group.

- review the design geometry of the detector mounting structure, in order to minimize the copper surface and to gain efficiency from anticoincidence between closer detectors;
- realize bolometers able to discriminate the origin of the events providing information on the particle impact point exploiting different properties of solid state physics.

In this work we focus on the third approach illustrating two techniques studied and developed for the active suppression of the surface background.

The first approach [3], pursued in the framework of the EDELWEISS collaboration, concerns the identification of surface events in Ge bolometers (for Dark Matter search) or TeO_2 bolometers (for $0\nu\beta\beta$ search of ^{130}Te) equipped with NbSi thin film thermometers acting as out-of-equilibrium phonons sensors. A particle interaction in the absorber produces athermal phonons diffusing away from the impact zone. In the case of a near-surface event, a significant amount of these out-of-equilibrium phonons are absorbed by the NbSi film and induce a signal with an athermal component much larger than for a bulk event of the same energy. Pulse shape analysis enables an effective rejection procedure. Ge bolometers were operated both aboveground and in the Modane underground laboratory, while the first TeO_2 prototype was tested aboveground at CSNSM-Orsay.

The second approach [4], for the moment applied only to TeO_2 bolometers in the framework of the CUORE collaboration, consists in the realization of surface-sensitive composite bolometers (SSB); auxiliary bolometers of different materials (Si, Ge and TeO_2 itself) were successfully coupled to TeO_2 absorbers, first in small scale prototype devices operated aboveground in the Insubria University and then in real CUORE size detectors operated in the Gran Sasso underground laboratory. The origin of the events can be clearly determined by comparing the amplitude of the pulses from the different detector elements. If an α particle comes from outside the bolometer, it interacts with an active layer releasing there all its energy ("surface" event). As a consequence, the temperature will rise and there will be a signal on both the layer thermistor and the TeO_2 one. Because of the small heat capacity of the layer, its thermistor signal will be much higher and faster than that of the TeO_2 thermistor. On the other hand, an event inside the TeO_2 crystal ("bulk" event) will lead to pulses with similar amplitudes and shapes on both thermistors.

3. Experimental results

We report here the main results obtained with two different detectors:

(i) NbSi/TeO_2_1 [5]: the sample is a 12 g TeO_2 crystal (20x20x5 mm^3) characterized by an under layer of amorphous Ge (500 Å) deposited on the two opposite faces of the crystal to improve its roughness. Then two interdigitized combs of electrodes (400 Å of Nb protected by 40 Å of Ir) are deposited, being 1 mm the distance between two adjacent teeth. Finally a square of 14x14 mm^2, 650 Å thick, 8.65 Nb atomic percentage of NbSi is deposited by coevaporation technique. The NbSi films are stabilized by a 60°C annealing and protected with 250 Å SiO coating. This detector has been tested aboveground at the C.S.N.S.M. Laboratory in Orsay, France.

(ii) SSB_1: it is made of a TeO_2 absorber with rectangular shape ($2\times2\times0.5$ cm^3) thermally coupled by means of four epoxy spots to a Ge single crystal with a 1.5×1.5 cm surface and 500 μm thickness. Neutron transmutation doped (NTD) Ge thermistors were glued at the main absorbers and at the auxiliary Ge bolometers with six epoxy spots. This detector has been tested aboveground at the Cryogenic Laboratory of the Insubria University in Como, Italy.

Fig.1 (a) shows the plot of all the events read in coincidence by the two thermometers of detector NbSi/TeO_2_1; in particular we plot the amplitude of athermal component for pulses

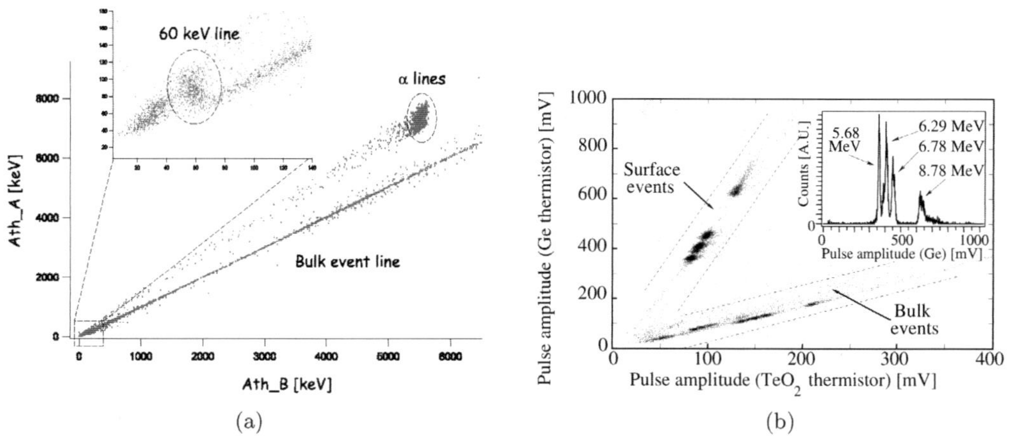

(a) (b)

Figure 1. Main results on surface event discrimination obtained with detectors NbSi/TeO_2_1 and SSB_1. More details in the text.

read by film_A vs. the amplitude of the athermal component for pulses read by film_B. We clearly recognize two different populations of events: a population for which the amplitude of the athermal components of the two thermometers are the same and a population for which the amplitude of the athermal component of film_A is higher than that of film_B. The events "out of the diagonal" are surface events; we clearly identify the α events of the source (used again to test the technique) at high energy but also the 60 keV (zoom), considering that 60 keV γ are absorbed in $\sim$ 300 μm in a TeO_2 crystal and hence are absorbed at the surface of the film_A.

Fig.1 (b) shows the scatter plot of the pulse amplitudes acquired in coincidence from the main absorber (X-axis) and from the active layer (Y-axis) for detector SSB_1. The two classes of events are well separated: the surface events populate a band with a steeper slope with respect to the bulk event band. In particular the four main α lines of the source used to test the technique are clearly appreciable, as also pointed out in the inset where the energy spectrum obtained by selecting surface events is shown.

In both cases, a high rejection efficiency (close to 100% for α particles) has been achieved and demonstrated. It is possible that both techniques can be shared and effectively used as rejection or at least diagnostic methods by the EDELWEISS and CUORE collaborations.

4. Acknowledgments

We would like to thank both the CUORE and EDELWEISS collaborations to have given us the opportunity to develop these innovative detectors. We are pleased to acknowledge that this work has been partially supported by the ILIAS integrating activity (Contract No.RII3-CT-2004-506222) as a part of the EU FP6 programme in Astroparticle Physics.

References

[1] V. Sanglard et al., *Phys. Rev. D* **71**, 122002, (2005).
[2] R. Ardito et al., *hep/ex 0501010.*
[3] A. Juillard et al., *Nucl. Inst. Meth. A* **559**, 393, (2006).
[4] A. Giuliani, L. Foggetta, C. Nones, M. Pedretti and S. Sangiorgio *Nucl. Inst. Meth. A* **559**, 393, (2006).
[5] C. Nones et al., to be published in *Proc. 12th International Workshop on Low Temperature Detectors (Paris).*

Improved measurement of the "head-tail" effect in nuclear recoils

D. Dujmic[3], S. Ahlen[1], P. Fisher[3], S. Henderson[3], A. Kaboth[3], G. Kohse[3], R. Lanza[3], M. Lewandowska[1], J. Monroe[3], A. Roccaro[1], G. Sciolla[3], N. Skvorodnev[2], H. Tomita[1], R. Vanderspek[3], H. Wellenstein[2], R. Yamamoto[3]

[1]Boston University, Boston, MA 02215
[2]Brandeis University, Waltham, MA 02454
[3]Massachusetts Institute of Technology, Cambridge, MA 02139
E-mail: ddujmic@mit.edu

Abstract. We present new results with a prototype detector that is being developed by the DMTPC collaboration for the measurement of the direction tag ("head-tail") of dark matter wind. We use neutrons from a ^{252}Cf source to create low-momentum nuclear recoils in elastic scattering with the residual gas nuclei. The recoil track is imaged in low-pressure time-projection chamber with optical readout. We measure the ionization rate along the recoil trajectory, which allows us to determine the direction tag of the incoming neutrons.

1. Introduction

The non-baryonic dark matter in the form of weakly interacting massive particles (WIMPs) still eludes detection despite recent achievements in the detection technology [1]. Aside from scaling up the size of existing detectors, the improvement in the detection sensitivity is possible by detecting the direction of the incoming dark matter particles. As the Earth moves in the galactic halo, the dark matter particles appear to come from Cygnus constellation. The direction tag of the incoming particle, often referred to as the "head-tail" effect, increases the sensitivity of a directional detector by one order of magnitude [2].

In this paper we present improved results for tagging the direction of low-energy nuclear recoils created by neutrons from a ^{252}Cf source by using a time-projection chamber with optical readout. The neutrons are used in lieu of the dark matter particles because they create similar distributions of recoil energies and angles. The measurement of directionality tag relies on the fact that the ionization rate of recoiling nuclei depends on their residual energy, and therefore the direction of the recoil can be tagged from the light distribution along the track.

2. Experiment and Results

The detector is in more details described in [3]. The chamber utilizes 10×10 cm^2 wire frames. The drift region between the cathode mesh and the ground wire plane is 2.6 cm with average electric field of 580 V/cm, while the amplification region between the ground and the anode wire plane (+2.2 kV) is about 3 mm. The pitch of the wires for the ground (anode) plane is 2 mm (5 mm) and the wire diameter is 15 μm (50 μm). The chamber is filled with CF_4 at 200 Torr.

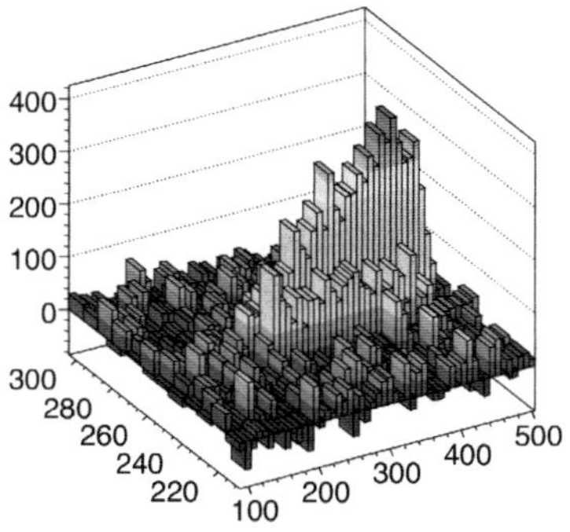

Figure 1. An image of recoil track with neutrons coming from the right. The noticeable asymmetry of the light yield along the wire indicates of the "head-tail" effect.

The scintillation light is recorded with a cooled CCD camera equipped with a photographic lens that images approximately 2 cm^2 of the anode plane. The spread of pixel yields due to ADC noise and dark current is 25 counts. Images are corrected for ADC bias and hot channels are identified and excluded from analysis.

Neutrons are created in the fission of the ^{252}Cf nucleus, which occurs in approximately 3% of all decays and produces 3.8 neutrons per fission [5]. The radioactivity of our ^{252}Cf source is 3.4 mCi and we estimate the total flux of 10^4 neutrons per second into the solid angle (10^{-3} sr) of the detector. The wires of the tracking chamber are aligned with the direction of the neutron beam. The recoil length projected to the wire axis is longer in case of WIMP scattering, therefore, of "head-tail" effect in neutron scattering is expected to be harder. We take sequential 1-second exposures with the CCD camera. We reject images that have segments shorter than 0.7 mm, and recoil tracks that fall close to the boundary of the CCD field of view. The energy of the recoil segment is determined from the projection of the light intensity to the axis perpendicular to the wire. The relation between the light intensity and the energy is determined using alpha particles that travel perpendicular to the wire and deposit a known amount of energy. The range of the recoil segment is calibrated using the known pitch of anode wires and the observed distance between wires in the CCD images.

An image of a nuclear recoil in Figure 1 shows noticeable asymmetry of the light yield along the wire. In order to quantify this effect, we define the skewness γ as the dimensionless ratio between the third and second moments of the light yield along the wire coordinate (x): $\gamma = \frac{\langle (x-\langle x \rangle)^3 \rangle}{\langle (x-\langle x \rangle)^2 \rangle^{3/2}}$. The sign indicates the slope of the light intensity along the track: recoils that travel in the direction of the incoming neutrons have a negative skewness.

A plot of the measured skewness as a function of the segment length is shown in the top plot of Figure 2. The data in this plot corresponds to 3.6 h of live time using 5 mg of CF_4 gas. The head-tail asymmetry is easier to observe for longer tracks that are better aligned with the anode wires and create more scintillation light. The bottom plot in Figure 2 shows the fraction of events with negative skewness as a function of the track length.

3. Discussion and Conclusion

Since the measured light yield is proportional to the energy of the recoil segment and the length is proportional to the track range projected to the wire, these two quantities should be correlated. Figure 3 shows clear correlation between the light yield versus length of the recoil segments.

We collect 1 day of live-time of data without sources and find two events that pass our standard selection cuts. We verify good rejection of gammas by collecting 1/3 day of live-time of data with ^{137}Cs source (8 μCi) placed near the sensitive area of our detector and find zero

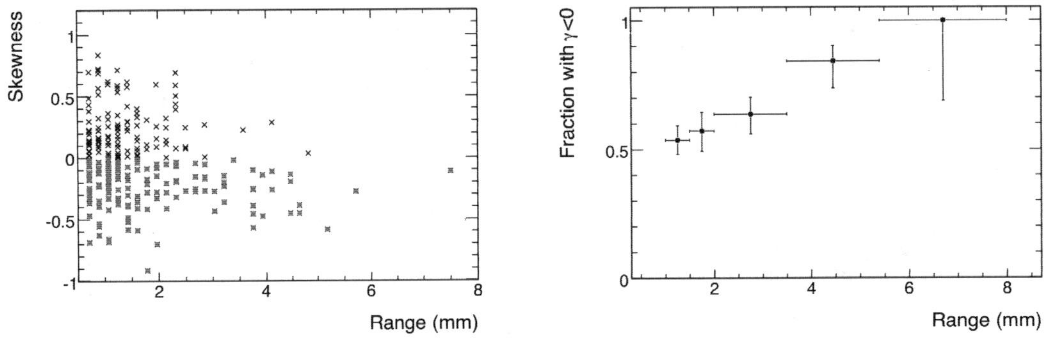

Figure 2. Left: skewness as a function of the track length of the recoil segments. Right: fraction of events with negative skewness as a function of the track length.

events passing the cuts.

We assign a conservative error of 10% to the density of the CF_4 gas. The statistical uncertainty on the energy measurements is about 10%. The systematic error on the energy comes from non-uniformity in wire gain, stability of the gain over time, the pressure measurement and the calibration method that assumes the energy-independent proportionality of the stopping power with the ionization rate. The error on the recoil range comes from the analysis technique that overestimates the range for low-energy recoils with the range close to the diffusion width.

We have presented improved results for tagging the direction of low-momentum nuclear recoils generated in the elastic scattering of low-energy neutrons with CF_4 gas. We have shown that in our current experimental setup the tag of incoming particle can be determined for recoil energies above 200 keV. This threshold can be further reduced with expected improvements in the detector preformance. This study has profound implications for the development of dark matter detectors, as the directionalty will be essential to produce convincing evidence for dark matter particles in the presence of backgrounds.

4. References

[1] R. J. Gaitskell, Ann. Rev. Nucl. Part. Sci. **54**, 315 (2004).
[2] A. M. Green and B. Morgan, Astropart. Phys. **27**, 142 (2007).
[3] D. Dujmic *et al.*, Nucl. Instrum. Meth. A **584**, 327 (2008).
[4] J. F. Ziegler, J. P. Biersack, U. Littmark, The Stopping and Range of Ions in Solids, Pergamon Press, New York, 1985. The code is available online at www.SRIM.org.
[5] M.B. Chadwick et at., Nuclear Data Sheets 107 (2006), 2931-3060

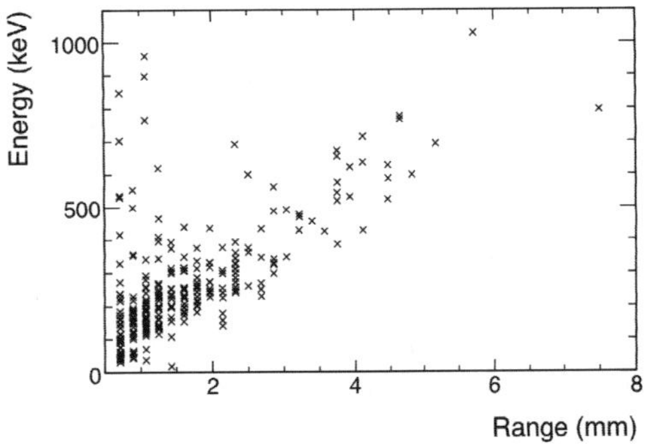

Figure 3. Correlation between the energy and the projected range of the recoil segments in data.

Solar Neutrino Experiments

Mark C Chen

Department of Physics, Queen's University, Kingston, ON, K7L 3N6, Canada

E-mail: mchen@queensu.ca

Abstract. Three solar neutrino experiments are presently collecting data: SAGE, Super-Kamiokande III and Borexino. There is also the KamLAND experiment that is operating and in the process of purifying their liquid scintillator in order to detect ^{7}Be solar neutrinos. The status of those experiments is described in this report. The Sudbury Neutrino Observatory recently completed taking data and the heavy water that was in the SNO detector has been returned. The transition from SNO to SNO+ has begun. SNO+, which is the SNO detector filled with liquid scintillator, aims to measure the flux of *pep* and *CNO* solar neutrinos, amongst other physics goals. The SNO+ experiment is presented here in the context of a synthesis of the results from all solar neutrino experiments. The prospects for other future solar neutrino experiments are also outlined here.

1. Solar neutrino experiments taking data

The field of solar neutrino experiments is now a mature one but one that continues to yield important results. The Chlorine experiment of Ray Davis [1] was the pioneering experiment. It was followed by Kamiokande [2], SAGE [3], Gallex [4], Super-Kamiokande [5], GNO [6] and the Sudbury Neutrino Observatory (SNO) [7]. All of these experiments measured deficits in the flux of electron neutrinos coming from the Sun compared to the predicted fluxes from the Standard Solar Model (SSM) [8]. By detecting neutral-current neutrino-deuteron interactions SNO was able to measure the ^{8}B solar neutrino flux in all active neutrino flavors and in doing so SNO confirmed that the total flux was significantly larger than just the electron neutrino flux (and also consistent with solar model predicted values) [9-10]. This provided conclusive evidence that electron neutrinos produced in the Sun transform to other active flavors. Results from all of these experiments are now well described by including matter-enhanced neutrino oscillations in calculating the survival probability for electron neutrinos propagating from their place of production in the Sun to respective experiments on Earth.

Of the experiments described above only SAGE and the re-built Super-Kamiokande III are still taking data. SAGE is the Russian-American Gallium Experiment. It is a radiochemical experiment that uses 50 tons of metallic gallium and that has been running (performing extractions) since 1990. By counting activity produced by neutrino capture on gallium, experiments like SAGE are sensitivity to the primary reaction *pp* solar neutrinos. At the TAUP2007 conference (for which this report is a proceedings contribution) the SAGE collaboration gave an updated result. After analyzing the results from 157 runs from their starting date of 1990 through 2006, the SAGE collaboration reported a measurement of the solar neutrino rate of $66.2^{+3.3}_{-3.2}\,^{+3.5}_{-3.2}$ SNU (solar neutrino units) [11]. Their measured *pp* solar neutrino signal is in good agreement with SSM calculations which include neutrino oscillations (with the best-fit parameters) – the oscillated, calculated signal is $67.3^{+3.9}_{-3.5}$ SNU [11].

The Super-Kamiokande detector has been reconstructed; this work was completed in April 2006. The full photocathode coverage fraction of 40% has been restored. The detector tank has been filled with water since July 2006 and the current detector is referred to as SK-III. As a ^{8}B solar neutrino detector SK-III has a goal of measuring the upturn in the recoil electron energy spectrum at low energies. This upturn is a prediction of the best-fit oscillation survival probability calculation. In order to see this effect SK-III will need to accomplish three improvements over SK-I. These are illustrated in Figure 1.

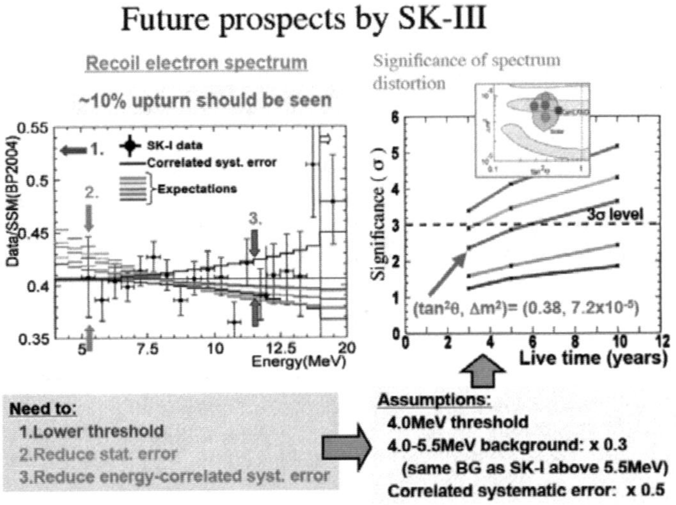

Figure 1. The SK-III goal to demonstrate the upturn in the recoil electron spectrum is shown above [12].

SK-III plans to lower the analysis threshold to 4.0 MeV from the 5.0 MeV used previously. This may require backgrounds in the detector to be reduced. The energy-correlated systematic uncertainties will need to be reduced. Finally statistical errors will be reduced – the plan is for a 10-year run (or longer). By accomplishing these improvements a detection of the spectrum distortion at the 3σ level or greater could be achieved.

SK-III has analyzed 97 live-days of data collected in 2007. The solar neutrino data in SK-III are consistent with previous SK-I results and were reported at the TAUP2007 conference. Noteworthy is the fact that in the central-top region of the detector backgrounds are lower than in SK-I (within the 5.0-5.5 MeV bin); this result is reproduced in Figure 2. By continuing to purify the water in the detector, the SK-III collaboration anticipates that this low background region will be enlarged.

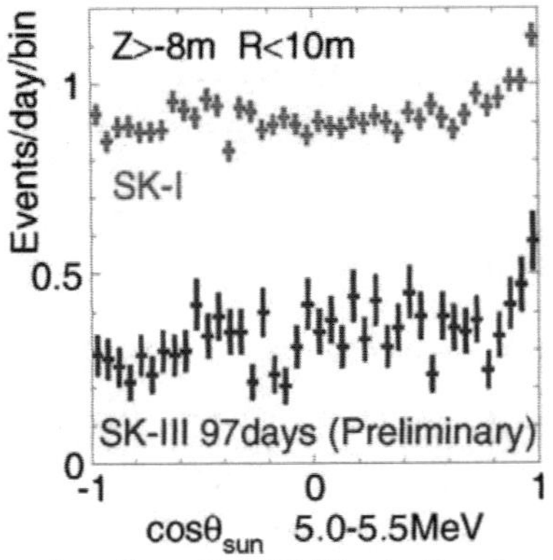

Figure 2. Preliminary data from SK-III illustrating lower backgrounds [12].

In addition to SAGE and Super-Kamiokande III a new solar neutrino experiment is currently taking data. The Borexino experiment was constructed in the late 90's-early 2000's and began taking data in 2007. The Borexino detector contains 300 tons of pseudocumene-based liquid scintillator and detects the scintillation light produced by neutrino-electron scattering. Since scintillator is used as the detection medium (as opposed to water Čerenkov) the high light yield enables detection of the ^{7}Be solar neutrinos with energy 0.86 MeV. In order to detect events at these low energies considerable care is needed, and was taken, in the construction of the Borexino experiment and in the purification of the liquid scintillator in order to eliminate backgrounds from natural radioactivity.

The first results from Borexino were presented in August 2007. The experiment successfully detected ^{7}Be solar neutrinos and determined their rate: 47 ± 7 ± 12 counts/day/100 tons [13]. This rate is consistent with the theoretically predicted event rate including oscillations of 49 ± 4 counts/day, and less than the SSM unoscillated rate which would be 75 ± 4 counts/day [13]. The Borexino results are consistent with other solar neutrino experimental results and the model for oscillations that those data imply. Figure 3 shows the energy spectrum of events in Borexino after a pulse-shape discrimination cut has removed alpha backgrounds from ^{210}Po (other cuts also were applied). The recoil edge spectral feature expected from ^{7}Be solar neutrinos is clearly visible. Backgrounds in this energy region are low compared to the signal.

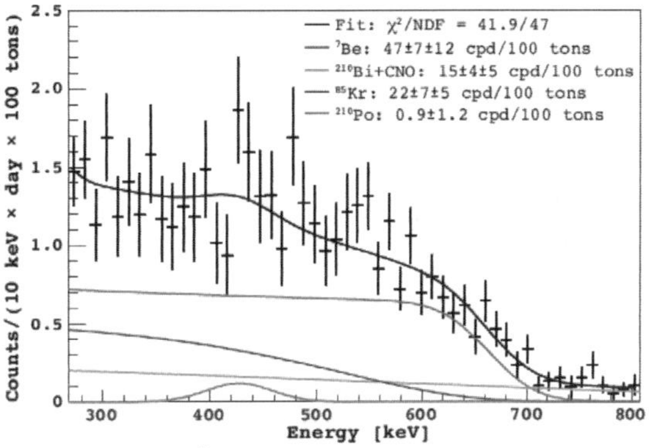

Figure 3. Detection of ^{7}Be solar neutrinos by observing the recoil electron energy spectrum in Borexino [13].

2. Solar neutrino experiments in transition

The KamLAND experiment has been measuring neutrino oscillation parameters by detecting reactor antineutrinos. KamLAND is a large, low background liquid scintillator detector and it would also be capable of detecting ^{7}Be solar neutrinos like Borexino; however singles backgrounds in KamLAND are observed to be too high by a factor of $\sim 10^{5}$. These backgrounds are due to ^{85}Kr and ^{210}Pb contamination in the scintillator and they prevent the observation of ^{7}Be solar neutrinos.

In order to enable the experiment to detect solar neutrinos the KamLAND collaboration built a scintillator purification system. The system uses the techniques of distillation and nitrogen stripping to remove backgrounds from Pb and Kr respectively. The first purification campaign was completed in July 2007. The total quantity of liquid scintillator that was processed was 1699 m^3 (KamLAND contains about 1000 tons of scintillator). Figure 4 shows some data taken during this purification campaign and it can be seen that the purified scintillator re-introduced in the detector (on top) has a lower background. After completing the first purification campaign, the result was that backgrounds had been reduced but not yet by the desired amount. This might have been due to mixing of purified scintillator inside the KamLAND detector after it was re-introduced.

Another experiment that is in transition is the Sudbury Neutrino Observatory. The SNO experiment had 1000 tons of heavy water and detected both charged-current (CC) and neutral-current (NC) neutrino interactions with deuterons in the ultra-pure D_2O. SNO ran in three phases; each phase detected the neutrons produced by neutral-current events differently. The third and final phase of SNO

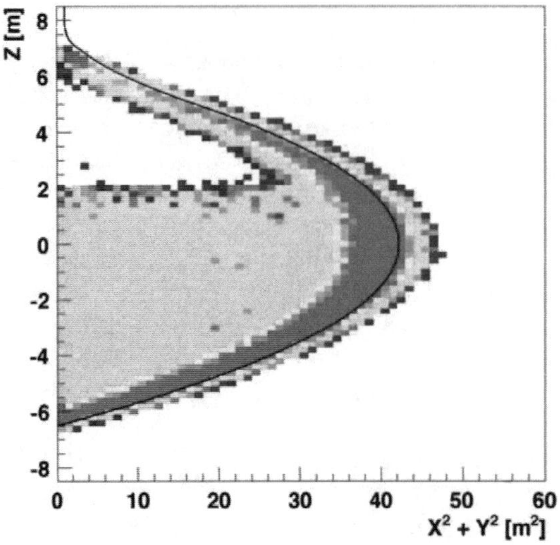

Figure 4. Snapshot of events in the KamLAND detector during the first liquid scintillator purification campaign [14]. The number of events in the top of the detector is much lower showing that the purified scintillator re-introduced in the detector (on top) had much lower backgrounds.

included the deployment of ^{3}He proportional counters distributed in an array inside the heavy water volume. This phase took data from November 2004 until November 2006. The dedicated neutron counters detected NC events in a separate detector subsystem whereas in the previous two phases of SNO neutrons produced capture gamma rays that interacted to produce Čerenkov light detected by photomultiplier tubes. These signals had then to be separated from the Čerenkov light from CC events. The third phase of SNO thus had the potential to break correlations between CC and NC rate measurements, thereby reducing systematic uncertainties.

One important aspect of this event-by-event separation for neutrons (NC signal) is how it might also reduce the uncertainties that affect the measurement of the CC spectral shape. Figure 5 shows the SNO extracted energy spectrum for charged-current neutrino-deuteron events. The large systematic uncertainties at low energy are dominated by the uncertainties related to the separation of CC and NC events in the previous two phases of SNO; but with the event-by-event separation of the third phase the CC-NC correlations reduce to ~0. The will aid in the observation of the predicted upturn in the spectrum at low energies in the data from SNO's third phase.

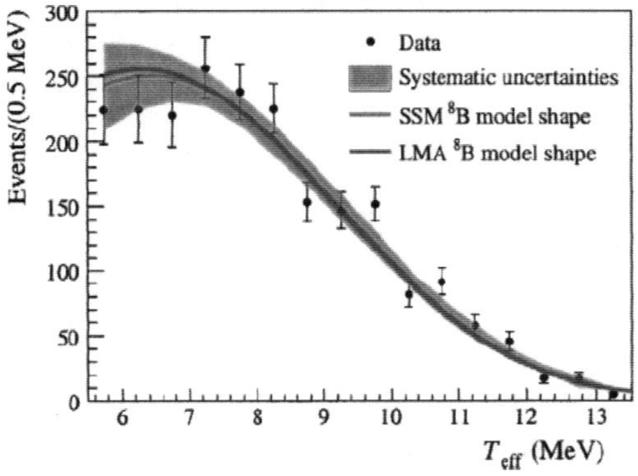

Figure 5. SNO CC energy spectrum extracted from the data from the salt phase [10].

Analysis of data from SNO's NCD (neutral current detectors) phase is taking place. There are also ongoing analyses of past SNO data and there will ultimately be a combined, three-phase analysis. For the near term there are two main analysis efforts. The first blind analysis of the data from SNO's neutron counters is being completed. With the aim of reducing the total uncertainty in SNO's NC measurement down to the ~5% level, a careful understand of the backgrounds in the neutron counters and their uncertainties is required. The second major effort taking place is the lowering of the analysis threshold. Improvements in energy estimators have been developed and they lead to better energy resolution in SNO (and even better understanding of the response function). The result is that tails from low energy backgrounds are smaller. Combining these improvements with improved characterization of these backgrounds allows the analysis threshold to be decreased from 5.5 MeV (electron kinetic energy) down to 4.0 MeV or lower. Figure 6 shows simulated data with a lowered analysis threshold. Backgrounds from Bi (uranium chain) and Tl (thorium chain) will be fitted in this analysis along with neutrino signals.

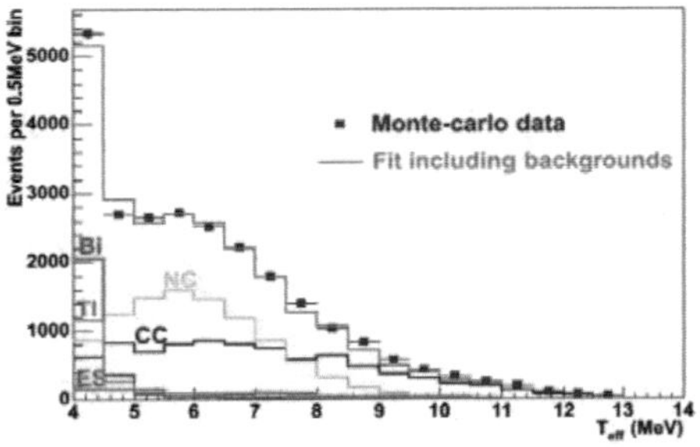

Figure 6. Simulation of a lower energy threshold analysis of SNO data. CC, NC and ES are signals from charged-current, neutral-current and elastic scattering while Bi and Tl are backgrounds.

The SNO experiment stopped taking data at the end of November 2006. The ^{3}He neutron counters were removed by the middle of January 2007 and heavy water removal began in late January 2007. The heavy water that was used by the SNO experiment was borrowed from Atomic Energy of Canada, Limited. On May 28, 2007 the last drops of heavy water were taken out of the SNO acrylic vessel and that event marked the end of the SNO experiment.

There are plans to fill the SNO detector with a liquid scintillator now that the heavy water has been removed. This project is called SNO+ and involves a subset of the SNO collaboration and new collaborators. A liquid scintillator that uses linear alkylbenzene (LAB) as the solvent has been developed by SNO+. It is chemically mild and is compatible with acrylic undiluted, thus has high light yield. LAB is safer and cheaper than pseudocumene. With an LAB-based scintillator in the SNO+ detector, the Monte Carlo simulated light yield in SNO+ is 900-1200 photoelectrons/MeV, depending on the concentration of fluor (e.g. PPO) used.

The density of LAB is similar to that of mineral oil and is 0.86 g/cm^3. With water surrounding the scintillator-filled acrylic vessel, the mechanics in SNO+ need to be reversed from SNO. Whereas the heavy water vessel was held up in SNO, the buoyant scintillator vessel needs to be held down in SNO+. A preliminary engineering evaluation has found feasible solutions for this hold down, likely to be based upon a rope net deployed over the upper hemisphere of the SNO+ acrylic vessel, with hold-down ropes penetrating the photomultiplier support structure and anchored to the cavity floor.

SNO+ aims to detect *pep* and *CNO* solar neutrinos. SNO+ will also detect geo and reactor antineutrinos, serve as a monitor for supernova neutrinos and, with the addition of Nd, plans to search for neutrinoless double beta decay in a separate running phase. This discussion will only focus on the solar neutrino capabilities. The flux of *pep* solar neutrinos is fundamental and related to the flux of *pp* solar neutrinos; it is calculated to ±1.5% uncertainty in the Standard Solar Model [8]. The cross section for neutrino-electron scattering has negligible uncertainty. A measurement of the rate of *pep* solar neutrino interactions can thus be used to determine the survival probability (for $E_\nu = 1.44$ MeV) precisely. It is also interesting to detect ^{7}Be solar neutrinos; but with a solar model uncertainty in the ^{7}Be flux of ±10%, the determination of the survival probability necessarily includes this uncertainty.

Detecting the *pep* solar neutrinos and measuring their survival probability, with precision, is interesting from the point of view of new neutrino physics. The presence of phenomena such as non-standard interactions [15] or mass-varying neutrinos [16] might manifest itself as a different survival probability at the transition between vacuum dominance and matter dominance, in the oscillation of solar neutrinos. The transition region is at neutrino energies between 1-2 MeV. Measuring the survival probability of *pep* solar neutrinos has good sensitivity to new physics.

Detecting the *CNO* neutrinos would be interesting as it could reveal information relevant to solar elemental abundances. A recent problem in solar models is the fact that helioseismology seems incompatible with low metallicity solar models. Solar model sound speeds [8] no longer agree with helioseismology when elemental abundances from [17] are used. The *CNO* neutrino fluxes depend on these abundances; differences of 50-60% in the flux result from using the older metallicity values and the newer ones. A measurement of the *CNO* fluxes would suggest which is compatible.

A background to the *pep* and *CNO* solar neutrinos is the cosmogenic production of ^{11}C. The half-life of ^{11}C is 20 minutes and that makes it difficult to veto or tag these events if the muon rate through a detector is very large. The ^{11}C background scales essentially as the muon rate. The deep location of SNO+ of 6000 mwe depth gives a muon rate that is low enough that this background is not a problem.

3. Synthesis of all solar neutrino data

Across the spectrum of solar neutrinos there have been direct and indirect observations. It is possible to synthesize all of the experimental results to examine what the present state of knowledge is of the survival probability, P_{ee}, of solar neutrinos as a function of neutrino energy.

The CC/NC ratio in SNO is a direct measurement of the average survival probability of the ^8B solar neutrinos that were detected. The mapping of detected recoil electron energy (SNO CC reaction) onto neutrino energy is direct enough (for the purpose of this study) since the response function is peaked and requires just the shift due to the reaction Q-value of 1.44 MeV.

The Chlorine experiment has contributions from ^7Be, *pep*, *CNO* and ^8B solar neutrinos; the latter is known from SNO data, and can be subtracted. The subtraction of a large component leaves the remainder with a large uncertainty; nevertheless, this gives information about the P_{ee} at intermediate energies ~1 MeV. The Borexino experiment has directly detected ^7Be solar neutrinos. The detection in Borexino is with the neutrino-electron scattering reaction; there is a neutral-current component in this signal for neutrinos that have oscillated to mu and tau flavor. Correcting for this gives the electron neutrino survival probability at 0.86 MeV, also at intermediate energies.

The radiochemical gallium experiments have contributions from *pp* solar neutrinos and higher energy neutrinos. One can subtract the ^8B and ^7Be contributions using SNO and Borexino data, respectively. The result gives a glimpse of the survival probability at lower energies.

This synthesis of all data is displayed in Figure 7; the calculated survival probability curve using the best-fit oscillation parameters is also shown in this figure. Also shown in this plot is a hypothetical measurement by SNO+ of the survival probability at 1.44 MeV (from *pep* neutrinos). The SNO+ error bars include estimated statistical, systematic and solar model uncertainties.

4. Future solar neutrino experiments

Several future solar neutrino experiments are being developed (in addition to SNO+ that has already been discussed). There are two noble liquid scintillation detectors under R&D. CLEAN (Cryogenic Low Energy Astrophysics with Neon) would contain 100 tons of liquid neon which is expected to have negligibly low intrinsic radioactive backgrounds. XMASS (Xenon MASSive detector) would contain 10 tons of liquid xenon which benefits from having very effective self-shielding. Both experiments would look for neutrino-electron scattering. CLEAN and XMASS plan to detect *pp* solar neutrinos; rates would be ~1 *pp* event/day per ton, for a 50 keV recoil electron threshold.

Two other experiments are under R&D. They are experiments that plan to use charged-current reactions to detect *pp* solar neutrinos. LENS (Low Energy Neutrino Spectroscopy) would look for neutrino capture on ^{115}In. This reaction can be tagged by subsequent gamma decays and would allow

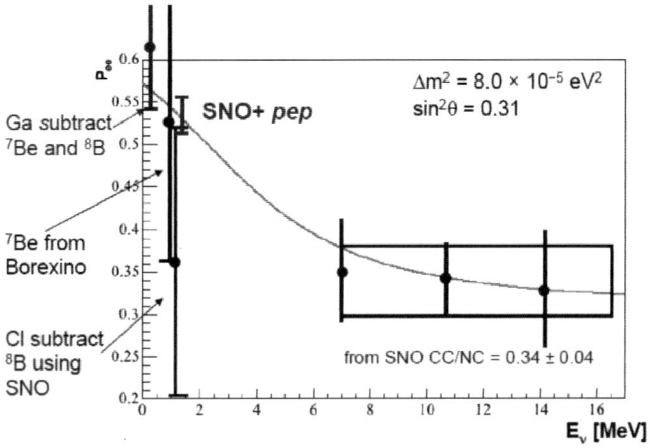

Figure 7. The survival probability for solar neutrinos as a function of energy. The data points are a synthesis of all experimental data interpreted as survival probability. The curve is the calculation using best-fit LMA oscillation parameters.

for 40 *pp* events/year per ton of indium (a rate that includes event tag efficiency). LENS would use an indium-loaded liquid scintillator deployed in a novel scintillation lattice chamber which produces event patterns that allow for discrimination from the large intrinsic ^{115}In beta decay background. MOON (Molybdenum Observatory Of Neutrinos) would look for neutrino capture on ^{100}Mo. This reaction can be tagged by a subsequent beta decay with a rate of 120 *pp* events/year per ton of ^{100}Mo isotope (9.6% natural abundance).

In summary, solar neutrinos will continue to be studied in the future. There will be further analysis of ^{8}B (and *hep*) solar neutrinos by SNO and further measurements by SK-III. The *pep* and *CNO* neutrinos will be detected by the planned SNO+ experiment. Borexino is measuring the flux of ^{7}Be solar neutrinos and KamLAND is lowering backgrounds via scintillator purification in order to detect ^{7}Be neutrinos also. There is R&D underway for four future *pp* solar neutrino experiments including two that would use charged-current reactions.

References

[1] Cleveland B T *et al.* 1998 *Astrophys. J.* **496** 505
[2] Hirata K S *et al.* 1990 *Phys. Rev. Lett.* **65** 1297
[3] Abdurashitov J N *et al.* 1999 *Phys. Rev.* C **60** 0558
[4] Hampel W *et al.* 1999 *Phys. Lett.* B **447** 127
[5] Fukuda Y *et al.* 1996 *Phys. Rev. Lett.* **77** 1683
[6] Altmann M *et al.* 2000 *Phys. Lett.* B **490** 16
[7] Boger J *et al.* 2000 *Nucl. Instrum. Meth. Phys. Res.* A **449** 172
[8] Bahcall J N, Serenelli AM and Basu S 2005 *Astrophys. J.* **621** L85
[9] SNO Collaboration 2002 *Phys. Rev. Lett.* **89** 011301; SNO Collaboration 2002 *Phys. Rev. Lett.* **89** 011302; SNO Collaboration 2004 *Phys. Rev. Lett.* **92** 121301
[10] SNO Collaboration 2005 *Phys. Rev.* C **72** 055502
[11] Gavrin V 2007, *Proc. of TAUP2007*, in this same publication
[12] Takeuchi Y 2007, *Proc. of TAUP2007*, in this same publication
[13] Arpesella C *et al.* astro-ph/0708.2251
[14] Kishimoto Y 2007, *Proc. of TAUP2007*, in this same publication
[15] Friedland A, Lunardini C and Peña-Garay C 2004 *Phys. Lett.* B **594** 347
[16] Barger V, Huber P and Marfatia D 2005 *Phys. Rev. Lett.* **95** 211802
[17] Asplund M, Grevesse N and Sauval J astro-ph/0410214 v2

Atmospheric and Long-Baseline Neutrinos

Kenzo Nakamura

High Energy Accelerator Research Organization (KEK)
Oho, Tsukuba, Ibaraki, 305-0801 Japan

E-mail: kenzo.nakamura@kek.jp

Abstract. Recent results and near future prospects of atmospheric neutrino observations and accelerator-based long baseline neutrino oscillation experiments are presented.

1. Introduction

Since the discovery of atmospheric neutrino oscillations in 1998, remarkable progress has been made in neutrino physics. In particular, accelerator-based long baseline neutrino oscillation experiments play important roles not only for the confirmation of atmospheric neutrino oscillations and determination of their oscillation parameters, but also for the future measurements of θ_{13} and CP violating phase δ in the MNS matrix and neutrino mass hierarchy. This talk mainly addresses the recent results from atmospheric neutrino observations including non-oscillation physics and accelerator-based long baseline neutrino oscillation experiments. Near future prospects of atmospheric and accelerator long baseline experiments are also briefly mentioned.

2. Non-oscillation physics from atmospheric neutrino measurements

The topics for very high-energy neutrinos are covered by F. Halzen [1]. Here we review new results and future prospects for non-oscillation physics from atmospheric neutrino measurements below O(10-100) TeV.

2.1. Muon charge ratio measured by the MINOS far detector

The only new result to mention here is the muon charge ratio measured with the 5.4 kt MINOS far detector (for more details, see Section 3.2) which is the first large magnetic detector located deep underground at a depth of 2070 m.w.e. At this depth, the average charge ratio is measured to be $N_{\mu^+}/N_{\mu^-} = 1.374 \pm 0.004(\text{stat})^{+0.012}_{-0.010}(\text{sys})$. Figure 1 shows N_{μ^+}/N_{μ^-} as a function of $E_{\mu,0}$, where $E_{\mu,0}$ is the muon energy at the surface projected back from the muon momentum as measured underground. The MINOS measurement [2] covers $E_{\mu,0} = 1$ - 7 TeV and offers the first high-statistics data in the TeV energy range. At lower energies of 0.3 - 1 TeV, L3+C data [3] are also plotted. These data show a rise of the muon charge ratio with energy in the TeV range. In Figure 2, other muon charge ratio data are also shown together with the MINOS and L3+C results. The curve shown in Figures 1 and 2 is calculated with a simple "πK" model which assumes Gaisser and Stanev's expression [4] for the muon intensity at the surface, as a function of muon energy and zenith angle, to be applicable indepenently for μ^- and μ^+. According to this qualitative model, the rise of charge ratio as a function of $E_{\mu,0}$ is consistent with an increasing contribution of kaon decays [2].

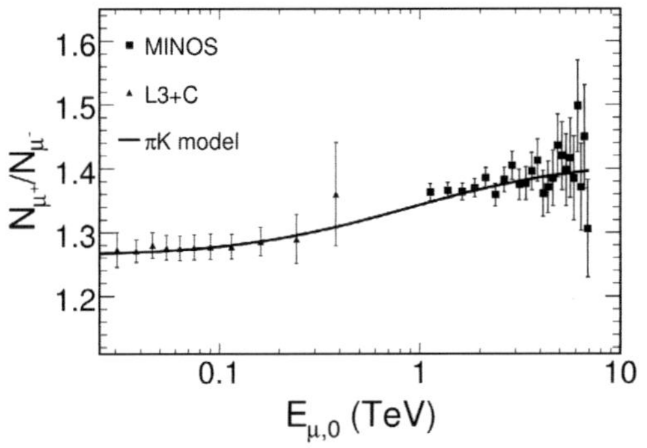

Figure 1. Plotted are the MINOS and L3+C muon charge ratio data. The curve shows the prediction of the πK model.

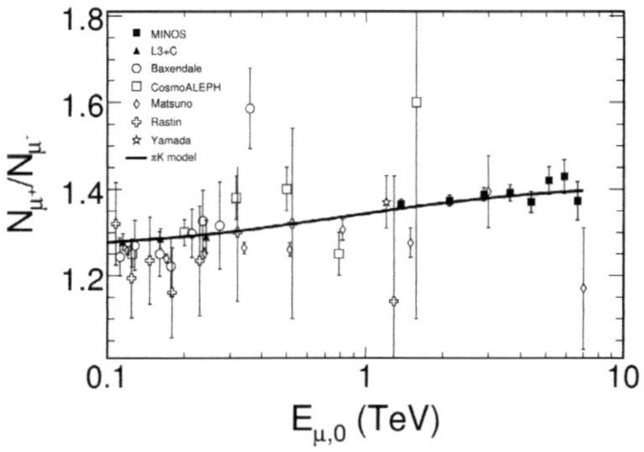

Figure 2. Muon charge ratio data from 0.1 to 7 TeV. The curve shows the prediction of the πK model.

2.2. INO

INO (India-based Neutrino Observatory) is a proposed experiment [5] with a 50 - 100 kt magnetized ICAL (Iron CALorimeter) detector. The active elements of ICAL will be resistive Plate Chambers (RPCs). The INO site recommended by the site selection committee is the one in Nilagiri mountains near Pushep at Tamilnadu, in the Southern part of India. The depth of the underground laboratory will be comparable to that of Gran Sasso (3800 m.w.e.). INO is planning a two-phase approach to scientific programs. In the first phase, INO will be a high-precision atmospheric neutrino experiment with muon charge identification capability. Naturally, it will extend the MINOS measurement of muon charge ratio at the deep underground site. However, the pair meter technique (method based on the energy dependence of the cross section of electron pair production by muons [6]) will allow the measurements of 10 - 300 TeV muons. Of course, INO will have capabilities of measuring atmospheric neutrino oscillations. Measurements of signatures of matter effects and mass hierarchy are among the physics goals aimed at in the first phase. In the second phase, it is hoped that INO would function as a detector for a neutrino factory to be constructed somewhere in the world.

3. Neutrino oscillation results and prospects

The first evidence for the neutrino oscillation was presented by the Super-Kamiokande Collaboration in 1998 [7]. The zenith-angle distributions of the μ-like events (mostly muon-neutrino initiated charged-current interactions) observed in the 50,000 ton water Cherenkov detector showed clear deficit compared to the no-oscillation expectation. This striking feature was explained to be due to almost pure two-neutrino oscillation in vacuum, $\nu_\mu \leftrightarrow \nu_\tau$, with $\Delta m^2 \sim 2-3\times10^{-3}$ eV^2, and a nearly maximal mixing angle, $\sin^2\theta \sim 1$.

The $\Delta m^2 \geq 2\times10^{-3}$ eV^2 region can be explored by accelerator-based long baseline experiments with typically $E \sim 1$ GeV and $L \sim$ several hundred km. With a fixed baseline distance and narrower neutrino energy spectrum, the value of Δm^2, and also with higher statistics, the mixing angle, are potentially better constrained in accelerator experiments than from atmospheric neutrino observations. The first accelerator-based long baseline neutrino oscillation experiment K2K and the subsequent MINOS experiment confirmed the atmospheric neutrino oscillation.

3.1. ν_μ disappearance: atmospheric neutrino results

The results from Super-Kamiokande-I (SK-I for short) measurement (April 1996 - November 2001, total exposure of 92 kt·yr) were reported in [8]. After the unfortunate accident in 2001, the Super-Kamiokande detector was rebuilt within a year with about half of the original number of photomultiplier tubes. The experiment with the rebuilt detector is called SK-II. The Super-Kamiokande collaboration has released the SK-I (92 kt·yr) and SK-II (49 kt·yr) combined preliminary results last year (SK-I+II 2006). The qualitative features of the zenith-angle distributions of various categories of events (fully-contained sub-GeV and multi-GeV e-like/μ-like, partially-contained, and upward-going muon events) look quite consistent with the SK-I results [8]. A two-neutrino oscillation analysis with the hypothesis of $\nu_\mu \leftrightarrow \nu_\tau$ using the SK-I+II 2006 zenith-angle distributions gave the best fit to the data with $\sin^2 2\theta = 1.00$ and $\Delta m^2 = 2.5 \times 10^{-3}$ eV^2. Figure 3 shows the 68%, 90%, and 99% CL allowed regions in the $(\sin^2 2\theta, \Delta m^2)$ plane. At 90% CL, the allowed parameter ranges are 1.9×10^{-3} eV$^2 < \Delta m^2 < 3.1 \times 10^{-3}$ eV^2 and $\sin^2 2\theta > 0.93$.

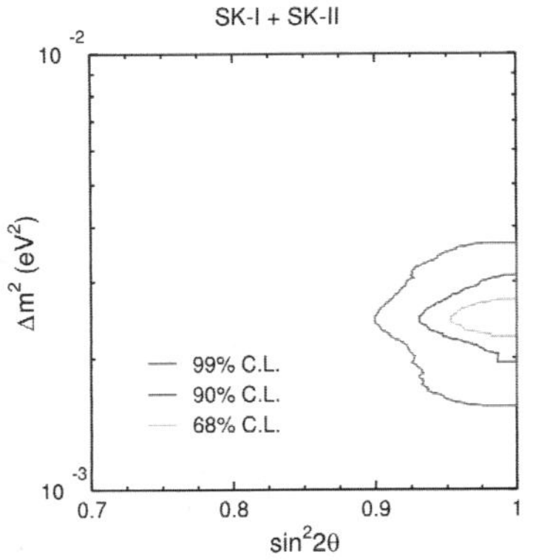

Figure 3. Allowed region for the $\nu_\mu \leftrightarrow \nu_\tau$ oscillation parameters obtained from the zenith-angle distributions of the SK-I+II 2006 atmospheric neutrino data.

Although the Super-Kamiokande atmospheric neutrino observations gave compelling evidence for neutrino flavor conversion $\nu_\mu \leftrightarrow \nu_\tau$ which is consistent with vacuum neutrino oscillations, other exotic explanations such as neutrino decay [9, 10] and quantum decoherence [11] may not be completely ruled out from the zenith-angle distributions alone. The firm evidence for neutrino oscillation is to confirm characteristic sinusoidal behavior of the conversion probability as a function of L/E where L (km) is the distance travelled and E (GeV) is the energy of the neutrino. It should be noted that L and E are reconstructed from the observed quantities. By selecting events with high L/E resolution from the SK-I+II 2006 data, evidence for the dip in the L/E distribution was observed at the right place expected from the interpretation of the data in terms of $\nu_\mu \leftrightarrow \nu_\tau$ oscillations (see Fig. 4). This dip cannot be explained by alternative hypotheses of neutrino decay and neutrino decoherence, and they are excluded at about 5σ in comparison with the neutrino oscillation interpretation. The SK-I+II 2006 results are consistent with, and update, the SK-I L/E results [12]. Figure 5 shows 68%, 90%, and 99% CL allowed regions in the $(\sin^2 2\theta, \Delta m^2)$ plane. The best fit was obtained at $\sin^2 2\theta = 1.00$ and $\Delta m^2 = 2.3 \times 10^{-3}$ eV^2. At 90% CL, the constraints obtained from the L/E analysis are $2.1 \times 10^{-3} < \Delta m^2 < 2.7 \times 10^{-3}$ eV^2 and $\sin^2 2\theta > 0.96$. These results are consistent with the SK-I+II 2006 zenith-angle analysis of fully-contained, partially-contained, and upward-going muon events (see Fig. 3).

3.2. ν_μ disappearance: accelerator-based long baseline experiments

The K2K experiment with the baseline distance of 250 km started data-taking in June 1999 and data were intermittently taken until November 2004. The final ν_μ disappearance results have been reported in [13]. A two-flavor neutrino oscillation analysis using the observed number of

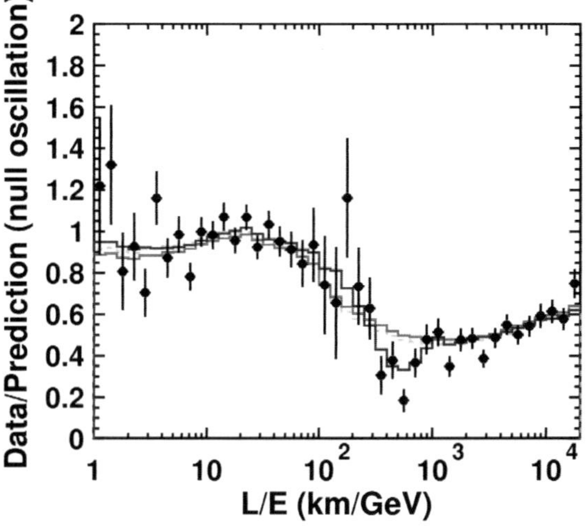

Figure 4. L/E analysis using the SK-I+II 2006 data. The points are the ratio of the data to the Monte Cralo expectation without neutrino oscillation as a function of the reconstructed L/E. Lines are the best-fit expectations for neutrino oscillation (solid line having a dip at L/E $\sim 5 \times 10^2$), neutrino decay (dotted line) and neutrino decoherence (solid line with no dip).

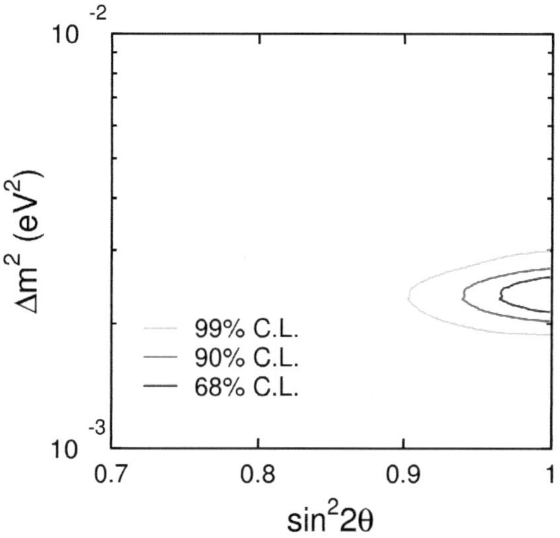

Figure 5. Allowed oscillation parameter regions for 2-flavor $\nu_\mu \leftrightarrow \nu_\tau$ oscillation, obtained from the SK-I+II 2006 L/E analysis.

fully contained events and the shape of the reconstructed E_ν spectrum gave the best fit point in the unphysical region, $(\sin^2 2\theta, \Delta m^2) = (1.2, 2.6 \times 10^{-3}\ \text{eV}^2)$. The allowed Δm^2 region at $\sin^2 2\theta = 1.0$ is between 1.9 and 3.5×10^{-3} eV^2 at the 90% CL with the best-fit value of 2.8×10^{-3} eV^2. The probability that the observations are due to a statistical fluctuation instead of neutrino oscillation is 0.0015% or 4.3 σ.

MINOS is a long baseline neutrino oscillation experiment with near and far detectors. Neutrinos are produced by using 120 GeV protons from the Fermilab Main Injector. The far detector is a 5.4 kt (total mass) iron-scintillator tracking calorimeter with toroidal magnetic field, located underground in the Soudan mine. The baseline distance is 735 km. The near detector is also an iron-scintillator tracking calorimeter with toroidal magnetic field, with a total mass of 0.98 kt. The neutrino beam is a horn-focused wide-band beam. Its energy spectrum can be varied by moving the target position relative to the first horn and changing the horn current.

MINOS started the neutrino-beam run in 2005. Initial results were published [14] using the data taken between May 2005 and February 2006 (Run-I). MINOS also took data from June 2006 through July 2007 (Run-II). Recently, preliminary results on the analysis of ν_μ disappearance using the Run-I and part of the Run-II data, corresponding to a combined exposure of 2.50 $\times 10^{20}$ POT (protons on target) are reported [15]. The data used in this analysis were taken with a "low-energy" option for the spectrum of the neutrino beam (the flux was maximized in the 1-3 GeV energy range). The observed number of candidate events for charged-current ν_μ interaction is 563. This number is compared to 738 ± 30 events expected for no neutrino oscillation. Figure 6 compares the reconstructed ν_μ energy spectrum at the far detector (Run-I and part of Run-II data combined) with the predicted spectrum with and without oscillations. Figure 7 shows the ratio of the NC (neutral-current) background-subtracted spectrum to the predicted

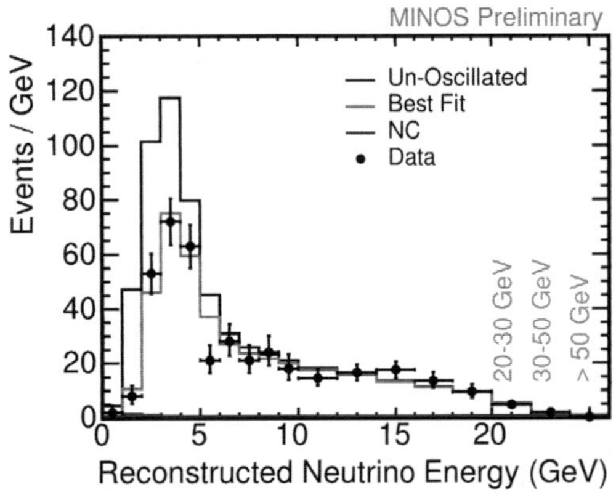

Figure 6. Reconstructed charged-current ν_μ energy spectrum (Run-I and part of Run-II of the MINOS data combined). The spectrum calculated with the best-fit parameters and that for no oscillation are also shown. The amount of the expected neutral-current (NC) background shown by the histogram labelled NC is small.

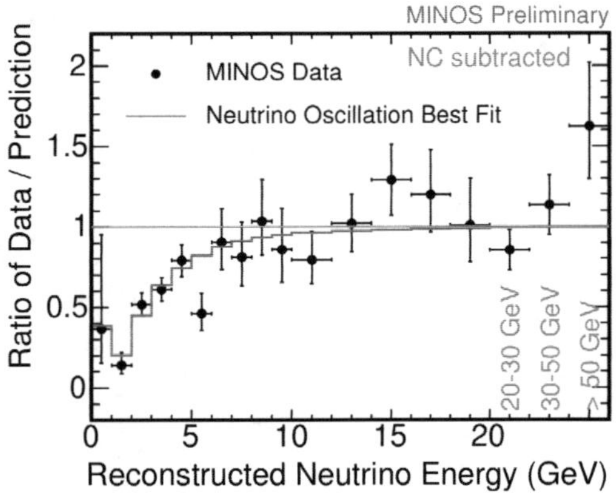

Figure 7. The ratio of the NC-background subtracted spectrum to the predicted spectrum for no oscillation. The histogram shows the expectation with the best-fit parameters.

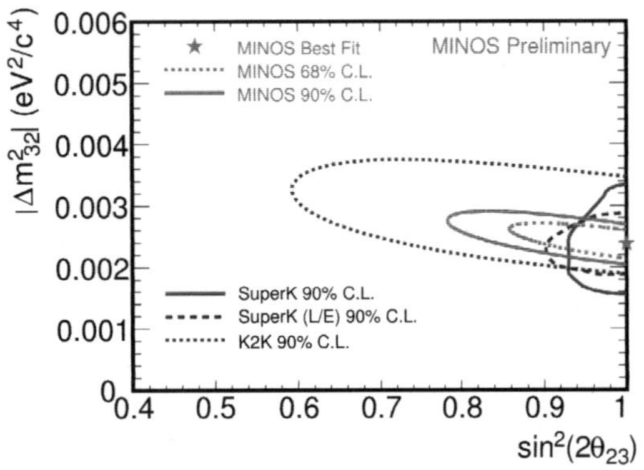

Figure 8. Preliminary MINOS best-fit point (star) and the 68% and 90% CL allowed regions. Also shown are the 90% CL allowed regions from the SK-I zenith [8] and L/E [12] analyses and that from K2K [13].

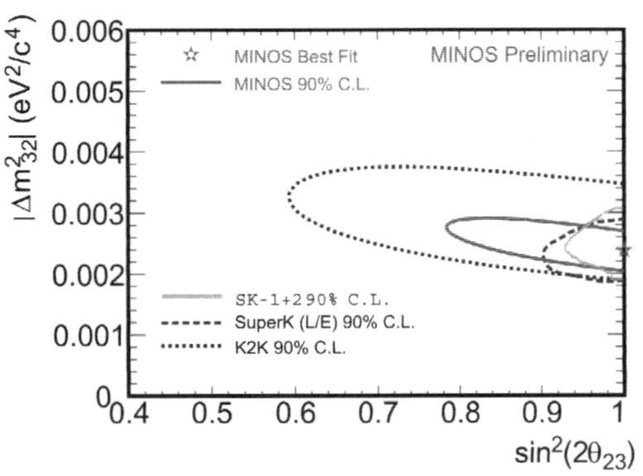

Figure 9. MINOS's preliminary best-fit pioint (star) and 90% CL allowed regions from MINOS [15], SK-I+II 2006 zenith analysis (see Figure 3), and SK-I L/E analysis [12] are shown.

spectrum for no oscillation. A simultaneous fit to the reconstructed ν_μ energy spectra assuming $\nu_\mu \leftrightarrow \nu_\tau$ two-neutrino oscillation gives the best-fit oscillation parameters of $\sin^2 2\theta = 1.00_{-0.08}$ and $\Delta m^2 = (2.3^{+0.20}_{-0.16}) \times 10^{-3}$ eV^2. At 90% CL, $\sin^2 2\theta > 0.84$. Figure 8 shows the 68% and 90% CL allowed regions, compared with the 90% CL allowed regions obtained from the SK-I zenith-angle dependence [8], the SK-I L/E analysis [12], and the K2K [13] results. The preliminary MINOS best fit point is shown by the star. Figure 9 compares the 90% CL allowed regions from MINOS [15], SK-I+II 2006 zenith analysis (see Figure 3), and SK-I L/E analysis [12].

3.3. ν_μ disappearance: future prospects

At Fermilab, significant proton accelerator improvements are expected in 2008, and MINOS envisages an exposure of 6 $\times 10^{20}$ POT by the end of 2008. The uncertainty in the value of Δm^2 would be reduced to about 2/3 of the preliminary MINOS result with 2.5 $\times 10^{20}$ POT. T2K experiment Phase-I now under construction at J-PARC (Japan Proton Accerelator Research Complex) will aim at 5 $\times 10^{21}$ POT in 5 years [16]. T2K will use Super-Kamiokande as a far detector, with a baseline distance of 295 km. The expected accuracies of the two-flavor $\nu_\mu \leftrightarrow \nu_\tau$ oscillation parameters are $\delta(\sin^2 2\theta) \sim 0.01$ and $\delta(\Delta m^2) < 1 \times 10^{-4}$ eV^2.

3.4. ν_τ appearance

The Super-Kamiokande group searched for the appearance of ν_τ from $\nu_\mu \to \nu_\tau$ oscillations in the SK-I atmospheric neutrino data [17]. A characteristic topology of the hadronic decay of τ (more spherical than that of backgrounds) was taken as the signature to search. After a series of cuts to reduce the backgrounds, τ signals were selected by applying the likelihood method and, independently, neural network method. Figure 10 shows the zenith-angle distribution of the selected events. The shaded area shows an excess of τ-like events. Note that these events are seen in the upward-going direction. For the likelihood analysis, the hypothesis of no ν_τ appearance is disfavored by 2.4σ. Similar result is obtained for the neural network analysis.

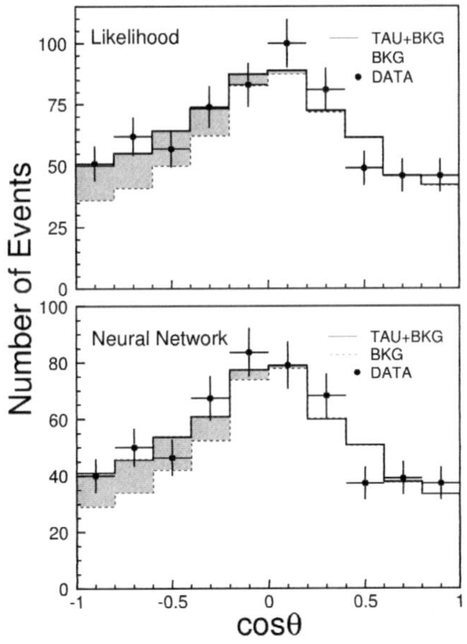

Figure 10. SK-I search for τ appearance. Zenith-angle distributions of the events selected by the likelihood (top) and neural network (bottom) methods are shown. The solid histogram indicates the best-fit including ν_τ and the dashed histogram indicates the background. The shaded area shows a fitted excess of τ-like events.

A promising method to confirm the appearance of ν_τ from $\nu_\mu \to \nu_\tau$ oscillations is an accelerator long baseline experiment using emulsion technique to identify short-lived τ leptons event-by-event. The only experiment of this kind is OPERA [18] with a neutrino source at CERN and a detector at Gran Sasso with the baseline distance of 732 km. The detector is a combination of the "Emulsion Cloud Chamber" and magnetized spectrometer. The CNGS (CERN Neutrinos to Gran Sasso) neutrino beam with $<E_\nu>=$ 17 GeV is produced by high-energy protons from the CERN SPS. With so-called shared SPS operation, 4.5×10^{19} POT/yr is expected. With this beam and 1.35 kt target mass (25% reduction with respect to the proposal), a ν_τ appearance signal of about 10 events is expected in 5 years run. In 2007, a 3 weeks of CNGS commissioning run and another 3 weeks of physics run is scheduled. The installation of emulsion target will be finally completed in 2008.

3.5. ν_e appearance

Measurement of the small mixing angle θ_{13} and the CP violating phase δ in the MNS matrix is among the important goals of future neutrino experiments. In the accelerator neutrino oscillation experiments with conventional neutrino beams, θ_{13} is measured using $\nu_\mu \rightarrow \nu_e$ appearance. Its probability is given by $P(\nu_\mu \rightarrow \nu_e) = \sin^2 2\theta_{13} \cdot \sin^2 \theta_{23} \cdot \sin^2(1.27\Delta m^2_{32} L/E)$, where $|\Delta m^2_{32}| \sim \Delta m^2$ and $\theta_{23} \sim \theta$ with Δm^2 and θ being the atmospheric neutrino oscillation parameters. By examining the exact expression for the oscillation probability, however, it is understood that some of the neglected terms have rather large effects and the unknown CP-violating phase δ causes uncertainties in determining the value of θ_{13}. Actually, from the measurement of $\nu_\mu \rightarrow \nu_e$ appearance, θ_{13} is given as a function of δ for a given sign of Δm^2_{32} (neutrino mass hierarchy). Therefore, a single experiment with fixed L and E (narrow-band ν beam energy) cannot determine the value of θ_{13} though it is possible to establish non-zero θ_{13}.

If θ_{13} is not much less than the CHOOZ limit [20], $\sin^2 2\theta_{13} < 0.19$, the ongoing experiments MINOS and OPERA will have a chance to establish non-zero θ_{13}. T2K Phase-I [16] will have a better sensitivity, $\sin^2 2\theta_{13} < 0.008$ (for the case of $\delta = 0$ or π). An experiment proposed in US, NOνA [19], will have a similar sensitivity.

In order to determine the value of θ_{13} in the accelerator neutrino oscillation experiments, simultaneous measurement of the CP-violating phase δ is necessary. This will be achieved by measuring the appearance probabilities of $\nu_\mu \rightarrow \nu_e$ and $\overline{\nu}_\mu \rightarrow \overline{\nu}_e$. A very high intensity neutrino beam will be required to measure CP violation in this way with a conventional neutrino beam line. T2K Phase-II [16] (with beam power upgrade and a Megaton water Cherenkov detector) and NOνA [19] with upgraded beam power are known as proposals for "superbeam" experiments.

4. Conclusions

Starting with 1998, neutrino oscillation has been established. We now know $|\Delta m^2_{32}|$ and θ_{23} from atmospheric and accelerator long baseline experiments, and we want to measure θ_{13} next. Whether we can measure δ and neutrino mass hierarchy in the future experiments is crucially dependent on the unknown size of θ_{13}.

References

[1] Halzen F 2008 *in these Proceedings*
[2] Adamson P *et al* (MINOS Collab.) 2007 *Phys. Rev.* D **76** 052003
[3] Achard P *et al* (L3+C Collab.) 2004 *Phys. Lett.* B **598** 15
[4] Gaisser T K and Stanev T 2006 *J. Phys.* G **33** 245
[5] Sajjad Athar M *et al* (INO Collab.) 2006 India-based neutrino observatory *Report* INO/2006/01
[6] Alekseev I S and Zatsepin G T 1960 *Proc. Int. Conf. on Cosmic Rays (Moscow)* vol 1 p 324
[7] Fukuda Y *et al* (Super-Kamiokande Collab.) 1998 *Phys. Rev. Lett.* **81** 1562
[8] Ashie Y *et al* (Super-Kamiokande Collab.) 2005 *Phys. Rev.* D **71** 112005
[9] Barger V *et al* 1999 *Phys. Rev. Lett.* **82** 2640
[10] Barger V *et al* 1999 *Phys. Lett.* B **462** 109
[11] Lisi E *et al* 2000 *Phys. Rev. Lett.* **85** 1166
[12] Ashie Y *et al* (Super-Kamiokande Collab.) 2004 *Phys. Rev. Lett.* **93** 101801
[13] Ahn M H *et al* (K2K Collab.) 2006 *Phys. Rev.* D **74** 072003
[14] Michael D G *et al* (MINOS Collab.) 2006 *Phys. Rev. Lett.* **97** 191801
[15] The MINOS Collaboration 2007 Preliminary Results from MINOS on an Exposure of 2.5 $\times 10^{20}$ 120 GeV Protons on the NuMI Target *Preprint* Fermilab-pub-07-413; arXiv:0708.1495 [hep-ex]
[16] Itow Y *et al* 2001 hep-ex/0106019
[17] Abe K *et al* 2006 *Phys. Rev. Lett.* **97** 171801
[18] Guler M *et al* (OPERA Collab.) 2000 OPERA: An appearance experiment to search for $\nu_\mu \leftrightarrow \nu_\tau$ oscillations in the CNGS beam *Preprint* CERN/SPSC 2008-028
[19] Ayres D S *et al* (NOνA Collab.) 2005 NOνA: Proposal to build a 30 kiloton off-axis detector to study $\nu_\mu \rightarrow \nu_e$ oscillations in the NuMI beamline *Preprint* FERMILAB-PROPOSAL-0929; hep-ex/0503053
[20] Apollonio M *et al* (CHOOZ Collab.) 2003 *Eur. Phys. J.* C **27** 331

MiniBooNE Oscillation Results and Implications

Michael H. Shaevitz
for the MiniBooNE Collaboration

Abstract. The MiniBooNE Collaboration has reported first results of a search for ν_e appearance in a ν_μ beam. With two largely independent analyses, no significant excess was observed of events above background for reconstructed neutrino energies above 475 MeV and the data are consistent with no oscillations within a two neutrino appearance-only oscillation model. An excess of events ($186 \pm 27 \pm 33$ events) is observed below the 475 MeV oscillation search cut. This low-energy excess cannot be explained by a two-neutrino oscillation model. This report presents an overview of the MiniBooNE first result and decribes some initial cross checks and investigations associated with the low energy excess.

Columbia University, Department of Physics, New York, NY 10027

E-mail: shaevitz@nevis.columbia.edu

1. Introduction

MiniBooNE was motivated by the result from the Liquid Scintillator Neutrino Detector (LSND) experiment [1], which has presented evidence for $\bar{\nu}_\mu \to \bar{\nu}_e$ oscillations at the $\Delta m^2 \sim 1$ eV^2 scale. LSND observed an approximate 3.8 σ excess of electron antineutrinos which would correspond to a $\bar{\nu}_\mu \to \bar{\nu}_e$ oscillation probability of 0.26 ± 0.08 %. When combined with the positive solar and atmospheric oscillation observations, LSND would require for explanation the addition of one or more sterile neutrinos[2] or further extensions of the Standard Model (*e.g.*, Lorentz Violations[3]). The MiniBooNE experiment is designed to have the same "L/E" as LSND but at higher neutrino energy in order to probe neutrino oscillations with similar oscillation parameter sensitivity but with different systematic uncertainties. The initial MiniBooNE oscillation analysis is performed within a two neutrino appearance-only $\nu_\mu \to \nu_e$ oscillation model where the ν_μ events are used to constrain the predicted ν_e rate. Other than oscillations between these two species, no effects beyond the Standard Model are assumed.

2. The MiniBooNE Experiment

The MiniBooNE experiment was proposed in the summer of 1997 and has been operating since 2002. The experiment uses the Fermilab Booster neutrino beam where 8 GeV protons (typically, 4×10^{12} protons within a ~ 1.6 μs beam spill at a rate of 4 Hz) are incident on a 71 cm long beryllium target which is inside a toroidal magnetic focusing horn. Positively charged pion and kaons are focused into a 50 m long decay region which is 91 cm in radius (See Fig. 1). The data used in the first oscillation result corresponds to 5.579×10^{20} protons on target. The main flux of neutrinos is from pion and kaon decay to muon neutrinos but there is also an intrinsic component of electron neutrino flux from kaon and muon decay with a flux ratio of $\nu_e/\nu_\mu \approx 0.5\%$. Neutrino interactions in the data sample are mostly charged-current quasi-elastic

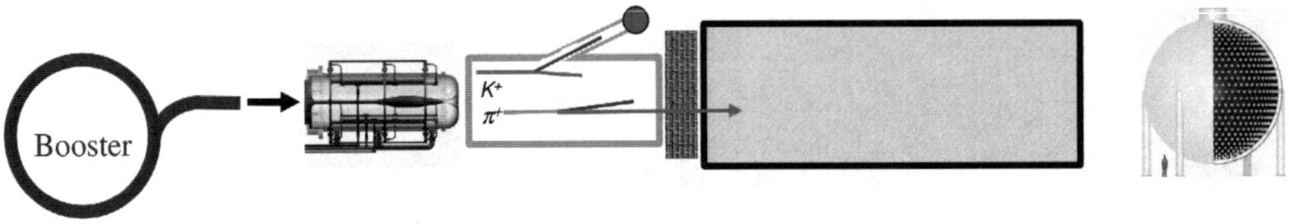

Figure 1. Schematic layout (not to scale) of the MiniBooNE beamline and detector. Protons from the 8 GeV Fermilab booster are directed onto a Be target inside a magnetic focusing horn. Pion and kaons are focused into a 50 m decay region that ends in a steel beam dump. The MiniBooNE neutrino detector is then located after $\sim$ 500 m of dirt shielding.

(CCQE) scattering (39%), neutral-current (NC) elastic scattering (16%), charged-current (CC) single pion production (29%), and NC single pion production (12%).

The MiniBooNE neutrino detector is located 541 m downstream of the beryllium target and 1.9 m above the center of the beam line. The detector is a spherical tank of inner radius 610 cm and filled with 800 tons of pure mineral oil (CH_2) with a density of 0.86 g/cm^3 and an index of refraction of 1.47. Charged particles produce both prompt directional Cherenkov light and longer time constant isotropic scintillation light in a ratio of about 3 to 1 for $\beta \approx 1$ particles. The detector consists of an inner spherical target region of radius 575 cm with 1280 equally-spaced inward-facing 8-inch phototubes (PMT) providing a 10% photocathode coverage; there is also an optically isolated outer veto shield region 35 cm thick with 240 8-inch phototubes. The detector has been designed to detect and measure neutrino events in the energy range from 100 MeV to a few GeV. The system is used to determine the event vertex, outgoing particle angles, and incident neutrino energy as well as separate electron neutrino from muon neutrino induced events.

3. Data Analysis

In the analysis of the data, PMT charge and time information in the 19.2 μs beam window is used to reconstruct neutrino interactions and identify the product particles. The reconstruction uses a detailed model of extended-track light production and propagation in the tank to predict the charge and time of hits in each PMT. Event parameters are varied to maximize the likelihood of the observed hits, yielding the vertex position and time of the event and the direction, energy, and, for photons, the conversion distance of the Cherenkov ring(s). For ν_e events, the event vertex, direction, and energy are reconstructed on average with resolutions of 22 cm, 2.8°, and 11%, respectively, while NC π^0 events are reconstructed with a π^0 mass resolution of 20 MeV/c^2.

The energy calibration for observed muon and electron tracks is accomplished over an energy range from 50 to 1000 MeV using a combination of techniques including muon-decay electrons, gammas from π^0 decay, and stopping and through-going cosmic ray muons. The E_ν^{QE} resolution is about 10% at 800 MeV for both muon and electron neutrino events.

Within the beam window, the data is divided into "subevents", collections of PMT hits clustered in time within $\sim$ 100 ns. Electron neutrino candidates are selected by requiring one subevent (as expected for ν_e CCQE events) with fewer than 6 hits in the veto and more than 200 hits in the main tank (above the muon-decay electron endpoint). This very effectively removes entering cosmic-ray muons and their associated decay electrons

The types of outgoing particles can be identified by their time structure and hit patterns. Muons have a sharp outer Cherenkov ring that is filled in by the muon travel distance, NC π^0 events have two Cherenkov rings, and signal electrons have one ring that is not sharp due to multiple scattering and the electromagnetic shower process. Thus, the PMT hit patterns can be used to identify candidate electron neutrino events.

The final analysis cuts are designed to isolate a sample of ν_e-induced events that are primarily CCQE. Two particle identification algorithms were used for the analysis; one, a likelihood based analysis, is the baseline for the first result and the other a "boosted decision tree" (BDT)[4][5] method is used as a secondary complementary analysis.

For the likelihood based particle identification analysis, the PMT hit patterns in the events are reconstructed under four hypotheses: a single electron-like Cherenkov ring, a single muon-like ring, two photon-like rings with unconstrained kinematics, and two photon-like rings with $M_{\gamma\gamma} = m_{\pi^0}$. To identify ν_e events and reject muon and π^0 events, E_{vis}-dependent cuts are applied on $\log(L_e/L_\mu)$, $\log(L_e/L_{\pi^0})$, and $M_{\gamma\gamma}$, where L_e, L_μ, and L_{π^0} are the likelihoods for each event maximized under the muon 1-ring, electron 1-ring, and fixed-mass 2-ring fits, and $M_{\gamma\gamma}$ is from the unconstrained two-ring fit.

The BDT reconstruction uses a simpler model of light emission and propagation. The particle identification uses 172 quantities such as charge and time likelihoods in angular bins, $M_{\gamma\gamma}$, and likelihood ratios (electron/ pion and electron/muon) as inputs to boosted decision tree algorithms [4] that are trained on sets of simulated signal events and background events with a cascade-training technique [5].

4. Oscillation Analysis

An oscillation signal in MiniBooNE would correspond to an excess of candidate electron neutrino events over expectation. Understanding the expected events is therefore the key to the oscillation search and the uncertainties in this expectation is a dominant factor in determining the sensitivity of the experiment. The primary uncertainties are associated with the ν fluxes, the ν cross sections, the modeling of the detection and identification efficiencies, and the rates of misidentifications. One needs to know the neutrino fluxes including the intrinsic ν_e flux from μ, K^+, and K^0 decay as well as the ν_μ flux which sets the scale of the signal and various misidentification backgrounds. These fluxes are used in conjunction with the neutrino cross sections and the detector simulation to predict the various types of event samples in the data. For MiniBooNE, the primary backgrounds to an oscillation ν_e signal are the intrinsic ν_e events along with misidentified neutral current ν_μ produced π^0, radiative $\Delta \rightarrow N\gamma$, and externally produced γ events. All of the important backgrounds can be directly constrained in both normalization and spectrum from observed non-background events in MiniBooNE; this procedure significantly reduces most of the systematic uncertainties.

The estimated number of background events with reconstructed neutrino energy, E_ν^{QE}, between 475 MeV and 1250 MeV after the complete event selection cuts is 358 ± 35. (The reconstructed energy, E_ν^{QE}, is determined from the reconstructed lepton energy and angle with respect to the known neutrino direction.) For comparison, the estimated number of ν_e CCQE signal events is 163 ± 21 for the LSND central expectation of 0.26% $\nu_\mu \rightarrow \nu_e$ transmutation. Studies of random triggers have established that no significant backgrounds survive the analysis cuts other than those due to beam related neutrinos (both ν and $\bar{\nu}$), which can be divided into either ν_μ-induced or ν_e-induced backgrounds. The small fraction of ν_e from μ, K, and π decay in the beamline gives a background that is indistinguishable from oscillations except for the energy spectrum. CC ν_μ events are distinguished from ν_e events by the distinct patterns of Cherenkov and scintillation light for muons and electrons, as well as by the observation of a delayed electron from the muon decay. NC π^0 events with only a single electromagnetic shower reconstructed are the main ν_μ-induced background, followed by radiative Δ decays giving a single photon, and then neutrino interactions in the dirt surrounding the detector, which can mimic a signal if a single photon penetrates the veto and converts in the fiducial volume.

The largest misidentification background is from NC π^0 production ($\nu_\mu + N \rightarrow \nu_\mu + N + \pi^0$) where one of the decay photons is undetected. For most of these events, both gammas from the π^0 decay are detected. One can use these detected π^0 events to measure the production rate in bins of π^0 momentum. The Monte Carlo is used to correct the production rate for inefficiency, backgrounds, and resolution. The measured rate is then used to reweight the Monte Carlo and, thus, constrain the misidentified NC π^0 background. In a similar way, the background from gammas entering from outside the detector, "dirt" background, can be constrained from the sample of inward-pointing events at high radius that have a large component of this background.

The MiniBooNE oscillation analysis uses the combined information from both candidate ν_e and ν_μ events. Systematic and statistical uncertainties are included using a fully correlated covariance matrix in E_ν^{QE} bins. The systematic uncertainties come from the analyses of both MiniBooNE and external data. The neutrino flux systematic errors are determined from the uncertainties of particle production measurements, the detector model systematic errors are mostly determined from fits to MiniBooNE data, and the neutrino cross section systematic errors are determined from MiniBooNE data as well as from external sources, both experimental and theoretical. The uncertainties are propagated using the detector simulation Monte Carlo into a covariance matrix in terms of E_ν^{QE} for each group of systematic errors. These matrices are then summed including the statistical errors to form the final uncertainty covariance matrix. This matrix is used for perform a χ^2 fit of the observed data to a Monte Carlo prediction including backgrounds and possible oscillation signals in order to set limits and confidence intervals.

5. First MiniBooNE Oscillation Result

MiniBooNE presented the first oscillation result in April, 2007.[6] The MiniBooNE analysis used a signal-blind analysis technique where candidate ν_e events ($\sim$ 5000) with conservative requirements were sequestered until the reconstruction software and analysis cuts were finalized; all other events (at the few 100,000 level) were available for examination and testing. After the analysis cuts were set and signal-blind tests of the data to MC agreement were performed, an oscillation analysis was performed in the $475 < E_\nu^{QE} < 1250$ MeV energy region. For the likelihood based analysis, $380 \pm 19(stat)$ data events were observed as compared to the no oscillation prediction of $358 \pm 35(syst)$ events; this agreement implies that there is no indication of an oscillation signal in the MiniBooNE data. The best fit oscillation parameters are $(\Delta m^2, \sin^2 2\theta) = (4.0\ \text{eV}^2, 0.001)$ with a 99% probability as compared to a null oscillation hypothesis probability of 93%.

Fig. 2 shows candidate ν_e events as a function of E_ν^{QE}. The vertical dashed line indicates the minimum E_ν^{QE} used in the two-neutrino oscillation analysis. There is no significant excess of events ($22 \pm 19(stat) \pm 35(syst)$ events) for $475 < E_\nu^{QE} < 1250$ MeV; however, an excess of events ($186 \pm 27(stat) \pm 33(syst)$ events is observed below 475 MeV. This low-energy excess cannot be explained by a two-neutrino oscillation model; its source is under investigation and will be discussed more below .

A single-sided raster scan to a two neutrino appearance-only oscillation model is used in the energy range $475 < E_\nu^{QE} < 3000$ MeV to find the 90% CL limit corresponding to $\Delta\chi^2 = \chi^2_{limit} - \chi^2_{bestfit} = 1.64$. As shown in Fig. 3, the LSND 90% CL allowed region is excluded at the 90% CL. The plot also shows the results from the BDT analysis which has similar sensitivity. The two analyses are very complementary, with the BDT analysis having a better signal-to-background ratio, and the likelihood analysis having less sensitivity to systematic errors from detector properties.

6. Going Beyond the First Result

Investigations of the low energy excess are focusing on a broad range of possibilities from detector reconstruction issues to incorrect or new sources of background. Several new physics possibilities have also been proposed. Any explanation that involves extra backgrounds or signal sources could have important consequences for future oscillation experiments such at T2K or Nova. All

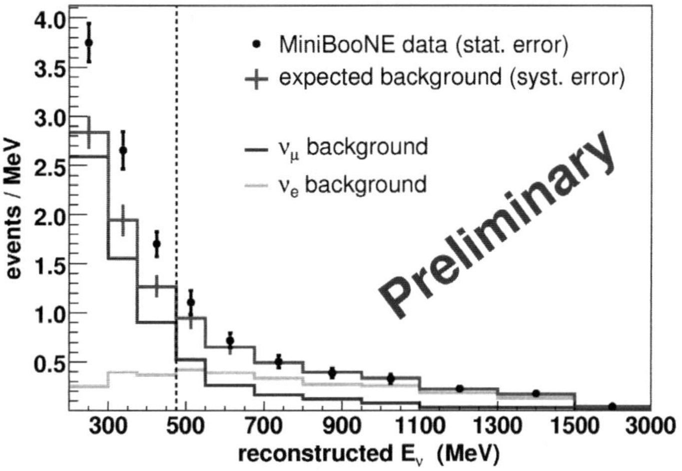

Figure 2. The number of candidate ν_e events as a function of E_ν^{QE}. The points represent the data with statistical error, while the histogram is the expected background with systematic errors from all sources. The vertical dashed line indicates the threshold used in the two-neutrino oscillation analysis. Also shown are the background contributions from ν_μ and ν_e induced events.

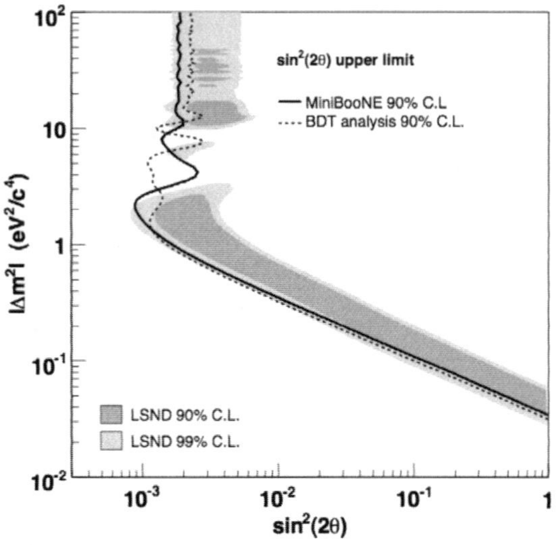

Figure 3. The MiniBooNE 90% CL limit (thick solid curve) from the TBT analysis for events with $475 < E_\nu^{QE} < 3000$ MeV within a two neutrino oscillation model. Also shown is the limit from the boosted decision tree analysis (dashed curve) for events with $300 < E_\nu^{QE} < 3000$ MeV. The shaded areas show the 90% and 99% CL allowed regions from the LSND experiment.

$\mathbf{E}_{\nu}^{QE}$ (MeV)	200 − 300	300 − 475	475 − 1250
Total Background	284 ± 25	274 ± 21	358 ± 35
ν_e intrinsic	26	67	229
ν_μ induced	258	207	129
NC π^0	115	76	62
NC $\Delta \to N\gamma$	20	51	20
Dirt	99	50	17
other	24	30	30
Data	375 ± 19	369 ± 19	380 ± 19
Data - Background	91 ± 31	95 ± 28	22 ± 40

Table 1. Preliminary results for the predicted background and observed data event numbers in three E_{ν}^{QE} bins. The total background is broken down into the intinsic ν_e and ν_μ induced components and the ν_μ induced background is further broken down into its separate components. The indicated uncertainties are systematic for the total background, statistical for the data, and a quadrature combination for the data to Monte Carlo difference.

of the low energy candidate ν_e events have be visually scanned using computer event displays and have been found to be consistent with single-ring, electromagnetic-like events. MiniBooNE cannot distinguish outgoing electrons or photons so the low energy events could be of either type.

A significant excess is seen in all three bins below 475 MeV as shown in Fig. 2. Table 1 lists the event numbers in three E_{ν}^{QE} bins with the background separated into its various components. For the bin corresponding to the oscillation analysis, the main background is intrinsic ν_e events from muon and kaon decay. In contrast, the ν_e background becomes less important in the lower energy bins where ν_μ induced backgrounds from NC π^0 and Δ decay along with "Dirt" become more important. As described above, MiniBooNE constrains these low energy backgrounds using observed events so enhancements of these backgrounds to explain the excess would contradict these observations. On the other hand, these constraints do have systematic and statistical uncertainties which are quantified in the errors presented in Table 1.

Investigation of possible processes that might be improperly modeled or missing in the simulations is another possible change to the predicted background. Since MiniBooNE cannot tell the difference between a single outgoing electron or single outgoing gamma, the background or observed excess could be either. MiniBooNE has been investigating photonuclear processes that are not included in the current simulation. This process could absorb one of the gammas in a NC π^0 event giving a single gamma background. Initial estimates are at the 10-20% level in the two lowest energy bins and work is progressing on more refined calculations. Recently, another standard model process, anomaly-mediated,single-photon production by neutrinos, has been proposed by Harvey, Hill, and Hill[7] as a possible source of low energy events. This process has never been observed and the MiniBooNE excess could possibly be the first observation if the rates and kinematic distributions are shown to be consistent.

Several new physics interpretations of the excess have also been put forward in recent papers. Models with oscillations to sterile neutrinos have been considered (3+1 or 3+2)[8] but these models have trouble describing both the appearance and disappearance oscillation data. Prior to MiniBooNE's first result, it was proposed[9] that sterile neutrinos could take shortcuts in extra dimensions leading to $\nu_\mu \to \nu_e$ oscillations in MiniBooNE mainly at low energy. This model is currently being investigated with respect to the low energy excess. As indicated previously, Lorentz invariance violation[3] might lead to new phenomenology where the oscillation depends on $E \times L$ or possibly on the sidereal time.

Experimentally, new information related to the MiniBooNE excess should become available in the near future. MiniBooNE will run with antineutrinos during the next several years which will give new measurements of possible oscillations and checks on the systematic uncertainties. The SciBooNE detector is now taking data in the MiniBooNE beamline and will provide a cross check of electron neutrinos from kaon decay and be a near detector for a two-detector disappearance measurement. MiniBooNE also sees offaxis neutrinos from the NuMI/MINOS beam. These events will provide an important cross check for the low energy excess since the energy and distance is similar to the Fermilab booster neutrino beam. In addition, the NuMI events are very different in background composition being dominated by ν_e backgrounds in the low energy region. Recently, a new experiment, MicroBooNE[10], has been proposed to put a 70 ton Liquid Argon detector near MiniBooNE to search for the low energy events with a fine-grained detection medium. This detector will have high ν_e efficiency, will be able to separate electron from gamma events, and be nearly free of backgrounds from misidentified particles. If the MiniBooNE low-energy excess is due to a new source of ν_e events, MicroBooNE would see a corresponding excess of 53 ± 9 events and establish a signal at the 9 σ level.

In summary, MiniBooNE has ruled out the LSND result interpreted as two neutrino $\nu_\mu \to \nu_e$ oscillations described by the standard L/E dependence. At low energies outside of the oscillation search region, MiniBooNE observes an excess of ν_e events; studies are currently underway to determine if these events are from unexpected backgrounds or possibly an indication of a new physics source.

Acknowledgements
We acknowledge and thank Fermilab, the Department of Energy, the National Science Foundation and Los Alamos National Laboratory for their support during this project.

[1] C. Athanassopoulos *et al.*, Phys. Rev. Lett. 75, 2650 (1995); C. Athanassopoulos *et al.*, Phys. Rev. Lett. 77, 3082 (1996); C. Athanassopoulos *et al.*, Phys. Rev. Lett. 81, 1774 (1998); A. Aguilar *et al.*, Phys. Rev. D 64, 112007 (2001).
[2] M. Sorel, J. M. Conrad, and M. H. Shaevitz, Phys. Rev. D 70, 073004 (2004).
[3] T. Katori, A. Kostelecky and R. Tayloe, Phys. Rev. D 74, 105009 (2006).
[4] B. P. Roe *et al.*, Nucl. Instrum. Meth. A543, 577 (2005); H. J. Yang, B. P. Roe, and J. Zhu, Nucl. Instrum. Meth. A555, 370 (2005); H. J. Yang, B. P. Roe, and J. Zhu, Nucl. Instrum. Meth. A574, 342 (2007).
[5] Y. Liu and I. Stancu, Nucl.Instrum.Meth.A578:315-321,2007, arXiv:Physics/0611267.
[6] A. Aguilar *et al.*, Phys.Rev.Lett.98:231801,2007, arXiv:0704.1500 [hep-ex]
[7] J. Harvey, C. Hill, and R. Hill, "Anomaly Mediated Neutrino-Photon Interactions at Finite Baryon Density", hep-ph0708.1281.
[8] M. Maltoni and T. Schwetz, "Sterile neutrino oscillaitons after the first MiniBooNE results", hep-ph0705.0107.; G. Karagiorgi, NuFACT07 conference proceedings.
[9] H. Pas, S. Pakvasa, and T. Weiler, Phys. Rev. D 72, 095017 (2005).
[10] http://www-microboone.fnal.gov/

Double beta decay review

Fabrice Piquemal

CENBG Université Bordeaux I and CNRS/IN2P3 BP120 Le Haut Vigneau 33175 Gradignan Cedex

E-mail: piquemal@cenbg.in2p3.fr

Abstract. The neutrinoless double beta decay is the most sensitive process to search the leptonic number violation in the neutrino sector. Its discovery will prove the Majorana nature of neutrino (neutrino and anti-neutrino are the same state). This paper will present the status of the running experiments (10 kg of $\beta\beta$) emitter sources) as well as the projects of detectors for 100 kg of isotopes.

1. Introduction

The recent discovery of the neutrino oscillations coming from various neutrino sources (atmosphere, reactors, sun) have shown that the neutrino is a massive particle. It is the first manifestation of physics beyond the Standard Model of the electroweak interaction. Nevertheless, various properties of the neutrino are still unknown: absolute mass, hierarchy scheme mass, nature (Dirac or Majorana), CP violation, magnetic moment,..... Establishing whether neutrinos are Dirac fermions (different from their antiparticle) or Majorana fermions (spin 1/2 particles identical to their antiparticles) is of fundamental importance for understanding the underlying symmetries of particle interactions and the origin of ν masses. The observation of neutrinoless double beta decay $\beta\beta(0\nu)$ could answer to some of these questions. Independantly of the process leading to the $\beta\beta(0\nu)$ decay, its discovery would imply the non-concervation of the leptonic number and will show that the neutrinos are Majorana fermions.

There is an intensive experimental search for neutrinoless double beta decay with various techniques. Today, the detectors are able to accommodate 10 kg of enriched isotopes. The next step is to built detectors with at least 100 kg of $\beta\beta$ emitters . This article will present the experimental status as well as the main experiments and projects. More complete reviews can be find in [1],[2], [3],

2. Double beta decay

The neutrinoless double beta decay consists to the decay of two neutrons in two protons with emission of two electrons:

$$(A,Z) \rightarrow (A,Z+2) + e^- + e^-$$

This process is forbidden in the Standard Model of the electroweak interaction because of the non-conservation of the leptonic number. The electron sum energy of the $\beta\beta(0\nu)$ decay would be simply a peak at the value of the transition energy. Due to the energy resolution the ultimate background is the allowed double beta decay ($\beta\beta(2\nu)$):

$$(A,Z) \rightarrow (A,Z+2) + e_1^- + e_2^- + \bar{\nu}_1 + \bar{\nu}_2$$

There are several possible processes leading to the $\beta\beta(0\nu)$ process: exchange of light neutrino, exchange of Majoron, right-handed interaction, exchange of supersymetric particles, ... In the case of light neutrino exchange , the expression of the half-live depends on the effective neutrino mass:

$$[T_{\frac{1}{2}}^{0\nu}(0^+ \rightarrow 0^+)]^{-1} = G^{0\nu}.|M^{0\nu}|^2.\langle m_\nu \rangle^2$$

Where $G^{0\nu}$ is the calculable phase space factor, $M^{0\nu}$ is the nuclear matrix element (NME) of the process. The calculation of the NME is a difficult subject, several models (QRPA methods, Shell Model) are used. There are recent progress on these calculations and the discrepencies between models tend to be reduced.

$\langle m_\nu \rangle = ||U_{e1}|^2 m_1 + |U_{e2}|^2 m_2 \exp i\phi_1 ||U_{e3}|^2 m_3 expi\phi_2$ is the effective Majorana neutrino mass m_1, m_2, m_3 are the neutrino mass eigenstates, U_{e1}, U_{e2}, U_{e3} are the coeficients of the neutrino mixing matrix and ϕ_1, ϕ_2 are the Majorana CP pahses (± 1 if CP is conserved).

In this paper, the range for the limit on the effective neutrino mass is calculated taking into account NME calculations from [5], [6], [7]

The effective neutrino mass can be rewritten as a function of the oscillation parameters and the CP phase violation. Three regions corresponding to the three hierarchy mass schemes are observed:

- The quasi-degenerate (QD) scheme where $m_1 \simeq m_2 \simeq m_3$ in this case $m_{\nu_{min}} > \sqrt{|\delta m_{32}^2|}$. In this case the predicted range for the effective neutrino mass is $\langle m_\nu \rangle > 0.05$ eV
- The inverted hierarchy (IH) where $m_3 \ll m_1 < m_2$ ($m_3 = m_{\nu_{min}}$) corresponding to $0.01 < \langle m_{nu} \rangle < 0.05$ eV
- the normal hierachy (NH) where $m_1 < m_2 \ll m_3$ ($m_1 = m_{\nu_{min}}$) would imply $\langle m_\nu \rangle < 0.005$ eV

Detailed discussion can be found in [4].

Today, the experiments using $\tilde{1}0$ kg of enriched isotopes have a sensitivity to explore the QD region. The next generation of double beta decay experiments will have $\tilde{1}00$ kg of enriched isotopes and will start to explore the IH region. To fully explore the IH scheme, it will be necessary to be able to built detector with a mass of 1 ton of isotope.

3. experimental techniques

In case of background the experimental half-life can be written has:

$$T_{1/2}^{0\nu}(y) > \frac{\log 2.N}{k_{C.L.}} \cdot \frac{\epsilon}{A} \cdot \sqrt{\frac{M.t}{N_{bckg}.\Delta E}}$$

Where: M: mass (g), ϵ: efficiency, $k_{C.L.}$: Confidence Level, N: Avogadro number, t: time (y), N_{bckg}: background events ($keV^{-1}.g^{-1}.y^{-1}$), ΔE: energy resolution (keV)

Today, there is no experimental technique able to optimize all the parameters. One can distinguish four main techniques for the $\beta\beta(0\nu)$ decay search:

- Pure calorimeter like semi-conductor detectors or bolometers where the efficiency, mass and good energy resolution are optimized
- Tracko-calo detectors to suppress the background by identification of the electrons and to allow choice of the isotope.
- TPC (Xe) for optimization of the efficiency and mass and also background if the daughter

nucleus can be identified.
- Inorganic scintillators or loaded organic scintillator to optimize efficiency and mass

The main backgrounds come from the natural radioactivity (in particular ^{208}Tl and ^{214}Bi), the neutrons (mainly through (n,γ) reactions), the radon and the cosmics rays (through cosmogenic production and neutron spallation production)

3.1. The present experiments
The best results for the limit on the effective neutrino has been obtained with Ge diodes enriched at 86 % in ^{76}Ge by the Heidelberg-Moscow (11 kg) and IGEX experiments (8.4 kg). There is a very contreversial 4 σ claim about observation of signal from the Heidelberg-Moscow experiment.

Today, two experiments are running Cuoricino (bolometric experiment) and NEMO3 (tracko-calo experiment).

3.2. Cuoricino experiment
Cuoricino is based on the technique of cryogenic detectors. Such detectors are operating at very temperature ($\tilde{8}$ mK). They have a very low heat capacity allowing to measure the small energy released by a radioactive decay event in the crystal. The Cuoricino detector (Figure 1 is tower made of modules of crystals of TeO_2 corresponding to a total mass of 40 kg. There are two sizes of crystals: 5x5x5 cm^3 (44 detectors) and 3x3x6 cm^3 (18 detetors). The energy resolution at 2530 keV (energy of the $\beta\beta(0\nu)$ decay for ^{130}Te) is between 7 and 9 keV (FWHM) depending of the crystal size. The detector is shielded by 20 cm of lead (10 cm internal layer of roman archelogical lead) and 10 cm polyethylen for neutrons. It is running since 2003 in the Laboratori Nazionali del Gran Sasso (LGNS, Italy). Due to the high natural enrichment of Te in ^{130}Te (34%), there is no need to have isotopic enrichment. Nevertheless 2 crystals (3x3x6 cm^3) are enriched in ^{130}Te (75%)and two in ^{128}Te (82.3 %) for the $\beta\beta(2\nu)$ search. The Figure 2 shows the results for 11.83 kg.y of ^{130}Te. A limit for the $\beta\beta(0\nu)$ decay is set to $T_{1/2} > 3.010^{24}y$ (90% C.L.) corresponding to a range for the effective neutrino mass limit : $\langle m_\nu \rangle < 0.19$ - 0.68 eV depending of the nuclear matrix element calculations [10].

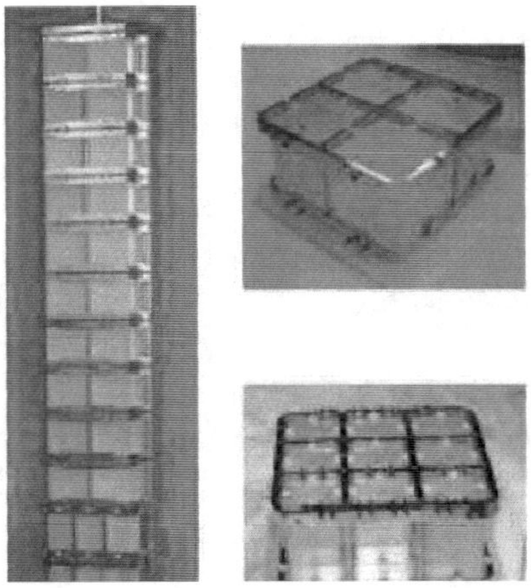

Figure 1. Cuoricino tower and individual 4 and 9 modules.

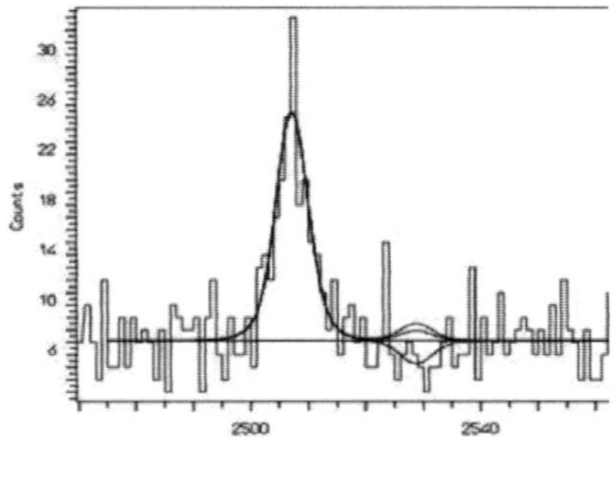

Figure 2. Cuoricino spectrum for 11.83 kg.y.The solid lines are the best fit to the region, and the bounds (68% and 90%) CL on the number of candidate $\beta\beta(0\nu)$-decay events .

3.3. NEMO3 experiment

The NEMO3 detector is looking for $\beta\beta(0\nu)$ decay by the direct detection of the two emitted electrons by the mean of a tracking device associated to a calorimeter. The detection of the electron as well as the measurement of the angular distribution and the individual energy of the electron allow to reduce drastically the background and in case of signal to distinghuish the process leading to the $\beta\beta(0\nu)$ decay. The detector is running since 2003 in the Laboratoire Souterrain de Modane (France). It consists in a thin central source foil (40-60 mg/cm^2) sandwiched by 6180 open drift cells operating in Geiger mode to reconstruct three-dimensionnal tracks and surrounded by a calorimeter made of 1940 plastic scintillator coupled to low radioactive photomultipliers developed by Hamamatsu company. The energy resolution (FWHM) at 1 MeV ranges from (14% to 16%) depending of scintillator size. A solenoid intalled around the detector produce a 25 gauss in order to distinguish electrons and positrons. The detector is shield by 20 cm of iron, water and wood to reduced background from external γ-rays and neutrons. A typical event is shown in Figure 3. The detector contains ^{100}Mo (6914 g) and ^{82}Se (932 g) for $\beta\beta(0\nu)$ searches but also ^{130}Te (454 g),^{116}Cd (405 g), ^{150}Nd (37 g), ^{96}Zr (9.4 g) and ^{48}Ca (6.99 g) for the $\beta\beta(2\nu)$ half-live measurements . The Mo source has been purified to reduce content of ^{214}Bi and ^{208}Tl. The Figure 4 shows the results for ^{100}Mo for 693 days of data taking. The limits obtained for ^{100}Mo are $T_{1/2} > 5.810^{23} y$ (90% C.L.) and $\langle m_\nu \rangle < 0.8$ - 1.3 eV. For ^{8}2Se, the present results are $T_{1/2} > 2.110^{23} y$ (90% C.L.) and $\langle m_\nu \rangle$ ¡ 1;4 - 2.2 eV [11].

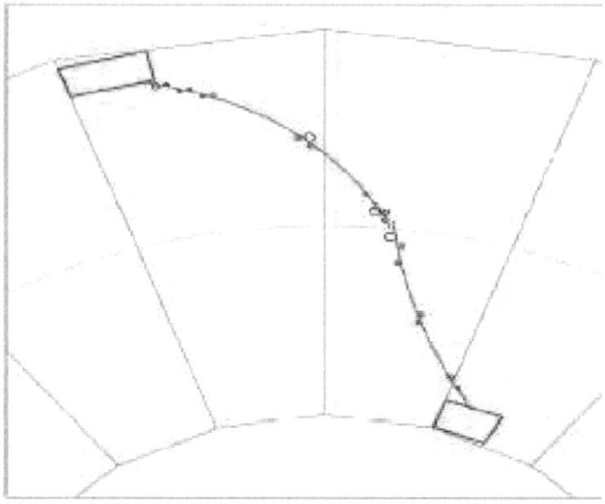

Figure 3. Tranversal view of $\beta\beta$ e selected in NEMO3 data: two tracks (s circles represent the hit drift cells) coming from the same vertex on the source foil and associated each one to a hit scintillator (rectangular boxes) with an internal time of flight..

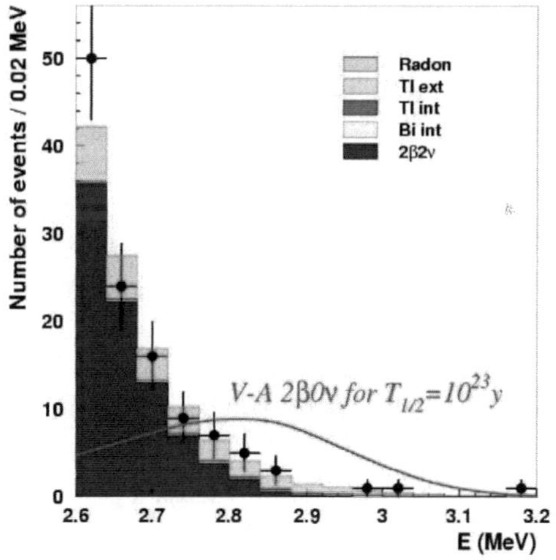

Figure 4. NEMO3 spectrum of the energy of the two electrons above 2.6 MeV of ^{100}Mo. The shape of expected $\beta\beta(0\nu)$ signal corresponds to the pink curve.

3.4. GERDA

The basic concept of GERDA experiment is to operate Ge diodes in liquid nitrogen or argon in order to remove materials to reduce the background. Moreover, an active anticoincidence with the scintillation light from the liquid argon can be used. The GERDA collaboration has scheduled three phases. The first one, starting in 2009 in the Laboratori Nazionali del Gran Sasso, will consist in 17.9 kg of enriched Ge (86% in 76Ge) coming from Heidelberg-Moscow and

IGEX experiments. The aim is decrease the background by a factor 10 (10^{-2} counts/keV/kg/y) compared to HM and IGEX experiment to reach a sensitivity of $T_{1/2} > 3.10^{25} y$ and $\langle m_\nu \rangle$ ¡ 0.25 eV to check the claim of ref. The second phase will correspond to 40 kg of enriched ^{76}Ge diodes with a part of Ge segmented diodes (20 kg). The combination of pulse shape analysis and coincidence method with segmented detectors is a powerwul way to reject the background coming from γ-rays. The expectation for the sensitivity with a background of $10^{-3} counts/keV/kg/y$ and 3 years of data taking is $T_{1/2} > 2.10^{26} y$ $\langle m_\nu \rangle < 0.11$ eV. This phase in planned for 2010. If Phase I and II are successful, the collaboration has a project of 1 ton detector to be sensitive to $T_{1/2} > 5.10^{27} y$ $\langle m_\nu \rangle < 0.02$ eV [8].

3.5. Majorana

The Majorana collaboration proposes to operate 500 kg of 86% enriched Ge detectors divided in modules of 60 kg. The objectives are to use very pure materials, to improve the pulse shape discrimination and to use segmented cristals to reject multi-site events corresponding to background events. The $\beta\beta$ events are single site events. It is plane to use electroformed Copper, free from radioactive contaminants and to form it underground to avoid cosmogenic contaminant production. Presently, segmented Ge crystals are under study to check the impact of the segmentation on signal and background. The R&D phase will start with 30-60 kg of enriched Ge crystals (with some of them segmented). The goal is to reach a background of less than 1 counts/t/y in the region of interest around 2033 keV. The second phase planned for 2011 is to run 30 kg of enriched Ge during three years to reach the following sensitivity sensitivity $T_{1/2} > 1.10^{26} y$ and $\langle m_\nu \rangle < 0.14$ eV to check the claim ref. The ultimate goal is to reach $T_{1/2} > 4.10^{27} y$ with a background less than 1 count/ton/keV/y. To minimize the background from cosmogenic production inside the crystals, the detector will be located in the Sudbury Underground Laboratory (6000 m.w.e.) [9].

3.6. CUORE

CORE is an extrapolation of Cuoricino experiment currently running in the Laboratori Nazionali del Gran Sasso. It consists of an array of 19 towers of 13 modules of 4 TeO_2 crystals each 5x5x5 cm^3. The total mass is 741 kg of TeO_2 and consequently 203 kg of ^{130}Te. The present background of Cuoricino in the energy region of the $\beta\beta(0\nu)$ decay is 0.18 $\pm$ 0.02 count/keV/kg/y. This background is mostly coming from the surface contamination of the crystals and the copper frame. A strategy to reduce these surface events is under study. The expected background with suppression of surface event is 0.01 count/keV/kg/y corresponding to a sensitivity of $T_{1/2} > 2.110^{26} y$ and $\langle m_\nu \rangle < 0.03$ - 0.07 eV for 5 years of data. The data taking is forseen in 2011. With an improvement of background reduction at the level of 0.001 count/keV/kg/y, the sensitivity would be $T_{1/2} > 6.610^{26} y$ and $\langle m_\nu \langle < 0.02$ - 0.04 eV. [10]

3.7. EXO

The EXO experiment is looking for $\beta\beta(0\nu)$ decay of ^{136}Xe to ^{136}Ba with a Time projection chamber (TPC). The main avantages of ^{136} are the ease to enrich it compare to the other $\beta\beta$ candidates and also the long period of the $\beta\beta(2\nu)$ process. The challenge is to attempt to tag the $^{136}Ba^{++}$ ion after its partial neutralization to Ba^+. The identification of the final state of the decay will drastically reduce the background and could be potentially a zero backgorund experiment. The final state ion tagging is possible by the observation of fluorescence light under laser light excitation, a R&D program to extract Ba is in progress. EXO will start by using 200 kg of enriched ^{136}Xe in liquid phase by detecting the scintillation light without Ba ion tagging. The prototype EXO-200 is under construction in the WIPP laboratory. The expected sensititity after 2 years of data taking is $T_{1/2} > 6.410^{25}y$ and $\langle m_\nu \langle < 0.27$ - 0.38 eV. The final goal with Ba tagging is to reach with 1 ton $T_{1/2} > 8.310^{26}y$ and $\langle m_\nu \langle < 0.05$ - 0.14 eV [12]

3.8. SuperNEMO

The SuperNEMO experiment is an extrapolation by a factor ten of the NEMO3 experiment. The detector will be modular and consists in 20 modules with 5 kg each of enriched source based on the NEMO 3 principle: a thin central source foil between two tracking volume surrounded by a calorimeter. The total mass of istope will be 100 kg. The collaboration is in a phase of R&D to improve the energy resolution of the calorimeter. The objective is to reach 4% (FWHM) at 3 MeV instead of 8% for NEMO3. In NEMO 3 the background rate is 0.4 count/kg/y in the region of interest (2.8 -3.2 MeV because of the energy resolution) corresponding to 0.001 count/kg/keV/y for a calorimeter. The origins of the background are to the $\beta\beta(2\nu)$ decay because of the energy resolution and residual pollution of the source foil in ^{214}Bi and ^{208}Tl. A challenge is to reduced by a factor 10 these backgrounds to have activities less than 20 μBq/kg in ^{214}Bi and 2 μBq/kg ^{208}Tl. The gas of the tracking detector must be radon free at the level of 100 μBq/m^3 which impose severe constraint on the activities of the materials use to built the detector. With this technique several isotopes can be studied like ^{82}Se, ^{150}Nd, ^{116}Cd, ^{100}Mo and ^{48}Ca. The baseline is to use 100 kg of ^{82}Se, the expected sensitivity for 5 years of data taking is $T_{1/2} > 1.10^{26}y$ and $\langle m_\nu \langle < 0.07$ - 0.12 eV. The schedule is to start to run the first module in 2011 [11].

3.9. Other projects

There are various other projects for double beta decay searches. A non-exhautive list is:

- CANDLES experiment will use CaF_2 scintillating crystals to look for $\beta\beta(0\nu)$ decay of ^{48}Ca. The very low natural isotopic abundance of ^{48}Ca will be partly compensate by the number of crystals [13].
- SNO+ experiment proposes to dissolve Nd inside the liquid scintillator. Due to the large volume of liquid scintillator, it is possible to dissolve a large mass of natural Nd or enriched ^{150}Nd. Such detecor allow high mass and have high efficiency [14].
- DCBA is a project of gazeous TPC to study $\beta\beta(0\nu)$ decay of ^{150}Nd. The energy will be measure by the curvature of the tracks. This detector could be used with different isotopes [15].
- MOON is a project to study large mass of ^{100}Mo by inserting thin source foil between plates of plastic scintillators [16].
- COBRA would use semi-conductor crystal of ZnCdTe allowing to study several double beta decay istope at the same time (^{150}Cd, ^{130}Te, ^{128}Te,...) with a possible tracking capability [17].

A summary of some projects for double beta decay searches is given in Table 1.

4. Conclusion

The double beta decay search is a very active field. Several experiments are running or will start very soon using different experimental techniques and studying various isotopes. Because of the uncertainties on the nuclear matrix element calculations, it is not possible today to determine the best candidate for $\beta\beta(0\nu)$ search. Since few years, progress have been performed in the NME calculation and the discrepencies between the model are decreasing. In the next years, the effective neutrino mass region corresponding to the claim of $\beta\beta(0\nu)$ signal from HM experiment should be explored. Around 2015, the experiments with isotope mass on the order of 100 kg will provide results starting to test the inverted hierachy for the neutrino mass scheme . This is an essential step before to go to 1 ton detector in order to study and to understand backgrounds.

Table 1. Running experiment and some projects for neutrinoless double beta decay searches.

Experiment	Isotope	Mass (kg)	FWHM at $Q_{\beta\beta}$ (KeV)	Background cts/FWHM.kg.y	$T^{0\nu}_{1/2}(y)$	$\langle m_\nu \rangle$ (eV)	Status
Cuoricino	^{130}Te	10.4	7 - 9	0.18	6.10^{24}	0.1 - 0.7	Running
NEMO3	^{100}Mo	6.9	240	0.1	2.10^{24}	0.3 - 0.7	Running
CUORE	^{130}Te	203	5	0.05	2.110^{26}	0.03 - 0.07	Funded
GERDA Phase I	^{96}Ge	17.9	4	0.01	3.10^{25}	0.2 - 0.5	Funded
GERDA Phase II	^{96}Ge	40	4	0.001	2.10^{26}	0.07 - 0.2	Funded
MAJORANA	^{96}Ge	30 - 60	4	0.003	1.10^{26}	0.1 - 0.3	Funded
EXO-200	^{136}Xe	200	50	0.95	6.410^{25}	0.2 - 0.7	Funded
SuperNEMO	^{82}Se	100	100	0.01	1.10^{26}	0.07 - 0.12	R&D
	^{150}Nd	100	1.10^{26}	0.07	R&D		
CANDLES	^{48}Ca	0.5			1.10^{26}	0.5	Funded
SNO++	^{150}Nd	500	?	?	6.10^{26}	0.03	R&D
DCBA	^{150}Nd	500	100	?	1.10^{26}	0.07	R&D

References

[1] C.R. Physique 6 (2005) 778-788

[2] E. Fiorini, "Double beta decay and Majorana neutrinos," *Prepared for 12th International Workshop on Neutrinos Telescopes: Twenty Years after the Supernova 1987A Neutrino Bursts Discovery, Venice, Italy, 6-9 Mar 2007*

[3] F. T. . Avignone, S. R. Elliott and J. Engel, "Double Beta Decay, Majorana Neutrinos, and Neutrino Mass," arXiv:0708.1033 [nucl-ex].

[4] S. Pascoli and S. T. Petcov, "Majorana Neutrinos, Neutrino Mass Spectrum and the —— 0.001 eV Frontier in Neutrinoless Double Beta Decay," arXiv:0711.4993 [hep-ph]

[5] V. A. Rodin, A. Faessler, F. Simkovic and P. Vogel, "Assessment of uncertainties in QRPA 0nu beta beta-decay nuclear matrix elements," Nucl. Phys. A **766**, 107 (2006) [Erratum-ibid. A **793**, 213 (2007)]

[6] E. Caurier, E. Martinez-Pinedo, F. Nowacki, A. Poves, A. Zuker. Rev. Mod. Phys. **77**, 427 (2005)

[7] M. Koteleinen and J. Suhonen Phys. Rev. C **75** (2007) 051303

[8] S. Schonert *et al.* [GERDA Collaboration], "Status of the Germanium Detector Array (GERDA) in the search of neutrinoless beta beta decays of Ge-76 at LNGS," Phys. Atom. Nucl. **69**, 2101 (2006).

[9] F. T. . Avignone [Majorana Collaboration], "The MAJORANA 76Ge neutrino less double-beta decay project: A brief update," arXiv:0711.4808 [nucl-ex].

[10] C. Arnaboldi *et al.* [CUORICINO Collaboration], "Results from the CUORICINO neutrinoless double beta decay experiment," arXiv:0802.3439 [hep-ex].

[11] S. Soldner-Rembold [NEMO 3 Collaboration], "Search for Neutrinoless Double Beta Decay with NEMO 3 and SuperNEMO," arXiv:0710.4156 [hep-ex].

[12] K. Wamba [EXO Collaboration], "EXO: A Status Report,"

[13] S. Umehara *et al.*, "CANDLES for double beta decay of Ca-48," J. Phys. Conf. Ser. **39**, 356 (2006).

[14] M. C. Chen, "SNO and SNO+," AIP Conf. Proc. **944**, 25 (2007).

[15] N. Ishihara [DCBA Collaboration], "Present status of the DCBA experiment searching for neutrinoless double beta decay," *Prepared for 9th Accelerator and Particle Physics Institute: APPI 2004, Iwate, Japan, 16-20 Feb 2004*

[16] H. Ejiri, "Spectroscopic studies of double beta decays and MOON," AIP Conf. Proc. **942**, 33 (2007).

[17] D. Y. Stewart [COBRA Collaboration], "The COBRA double-beta decay experiment," Nucl. Instrum. Meth. A **580**, 342 (2007).

Reactor neutrino oscillation experiments: recent results and future prospects

Karsten M Heeger
Department of Physics, University of Wisconsin, Madison, WI, 53706, USA
E-mail: heeger@wisc.edu

Abstract.
Reactor neutrino experiments have played a key role in the history of neutrino physics. From the first observation of the free antineutrino by Reines and Cowan to the observation of $\overline{\nu}_e$ disappearance at KamLAND, reactors have been an invaluable tool in the study of neutrinos and helped shape our understanding of neutrinos and neutrino oscillation. Recent data from KamLAND unambiguously reveal the signature of neutrino oscillation in the observed $\overline{\nu}_e$ spectrum and determine the $\overline{\nu}_e$ oscillation parameter Δm^2_{12} to 2.8%. A suite of next-generation reactor experiments are now under construction to determine the yet unknown neutrino mixing angle θ_{13} with a sensitivity of $\sin^2 2\theta_{13} < 0.01$. Stringent control of detector systematics and minimizing backgrounds will be key to the continued success of reactor neutrino oscillation experiments.

1. Neutrino Physics at Reactors

For more than five decades experiments have studied the flux and properties of antineutrinos emitted from reactors. From the first direct observation of the antineutrino by Reines and Cowan in 1956 [1] to the searches for neutrino oscillation at the ILL, Goesgen, Bugey, Krasnoyarsk, Palo Verde, and Chooz reactors [2], experiments at reactors have searched for neutrino oscillation, placed the best laboratory limit on the neutrino magnetic moment [3] and provide the best limit on the yet unknown neutrino mixing angle θ_{13} [4]. By measuring the time-dependent variation of the $\overline{\nu}_e$ flux and spectrum reactor antineutrinos have also been used recently to remotely monitor the fuel composition of nuclear reactors for non-profileration purposes [5].

Nuclear power plants emit copious amounts of $\overline{\nu}_e$. They are produced in the decay of the fission products inside the reactor [6]. Fission products are generally neutron rich nuclei that can undergo successive β-decays each yielding one $\overline{\nu}_e$. Each fission produces on average 6 $\overline{\nu}_e$'s with an accompanying energy release of ~200 MeV. From the energy release and the nuclear fuel composition of a reactor the total number of electron antineutrinos emitted from the nuclear reactor can be predicted. A typical nuclear power plant operating at 3 GW thermal output produces on average 7×10^{20} $\overline{\nu}_e$'s per second. The pure electron flavor content of the antineutrinos produced in the β-decay products of the fission isotopes make nuclear reactors a desirable source for neutrino studies. The Sun and nuclear reactors are two of the very few sources that provide pure flavor beams of neutrinos free of other contamination.

In the last 50 years a series of reactor experiments with increasing baseline have searched for electron antineutrino disappearance as a signature of neutrino oscillation. As early as 1969 Bruno Pontecorvo suggested that lepton flavor may not be conserved and neutrinos might change flavor [7]. Detectors filled with liquid scintillator were placed near nuclear plants at distances from tens of meters up to about one kilometer to measure the flux of $\overline{\nu}_e$ emitted from the nuclear reactor. However, for all experiments up to a distance of 1 km from the reactor the observed reactor $\overline{\nu}_e$ flux was found to be consistent with the expectation from no oscillation.

2. Precision Measurement of Antineutrino Oscillation with KamLAND

Following the observation of atmospheric neutrino oscillation in Super-Kamiokande and the discovery of solar neutrino flavor transformation at SNO the Kamioka Liquid Scintillator Antineutrino Detector (KamLAND) experiment has measured the reactor $\overline{\nu}_e$ flux at a distance of $\mathcal{O}(100$ km), two orders of magnitudes further away from the reactor source than any other experiment. KamLAND is a high-resolution liquid scintillator detector with 1 kton of ultrapure liquid scintillator [8, 9]. Antineutrinos interact with the protons in the scintillator through the inverse beta reaction $\overline{\nu}_e + p \rightarrow e^+ + n$ with a 1.8 MeV energy threshold. With its long baseline KamLAND is uniquely positioned to confirm with terrestrial neutrinos the observed solar neutrino oscillation and provide the best possible measurement of Δm^2_{12}. The principal contribution to the reactor $\overline{\nu}_e$ flux measured by KamLAND comes from 53 power reactors in Japan. Most reactors are located within a distance range of $L = 175 \pm 35$ km from KamLAND. To a good approximation the survival probability for reactor electron antineutrinos is given by the following equation:

$$P(\overline{\nu}_e \rightarrow \overline{\nu}_e) \simeq cos^4\theta_{13}\left(1 - sin^2 2\theta_{12} sin^2\frac{\Delta m^2_{12} L}{4E_\nu}\right) \tag{1}$$

where L is the distance of travel of the neutrinos and E_ν is the antineutrino energy. Reactor neutrinos typically have a mean energy of about ~4 MeV. Matter effects that arise in the interaction of neutrinos with electrons inside matter [10] are usually negligible for reactor neutrinos.

The expected $\overline{\nu}_e$ flux from the Japanese reactors and the resulting interaction rate at KamLAND are calculated from the reactor operation data including thermal power generation, fuel burn-up, exchange and enrichment records. The expected flux is calculated using the fission rates and $\overline{\nu}_e$ spectra [6]. The $\overline{\nu}_e$ contribution from Japanese research reactors and all reactors outside Japan is 4.5%. We assume that these reactors have the same average fuel composition as the Japanese power reactors.

In 2003 KamLAND reported the first observation of reactor $\overline{\nu}_e$ disappearance [8]. With an exposure of 162 ton-yr corresponding to 145.1 days of data KamLAND found the ratio of the number of observed $\overline{\nu}_e$ candidate events minus backgrounds to the expected number of $\overline{\nu}_e$ events to be $N_{observed} - N_{bkgd}/N_{expected} = 0.611 \pm 0.085(stat.) \pm 0.041(syst.)$. This result confirmed with terrestrial neutrinos the neutrino oscillation solution favored by a global analysis of all solar neutrino data [8]. With higher statistics KamLAND reported in 2005 evidence for distortion in the measured energy spectrum, a unique signature of the energy-dependent neutrino oscillation effect [9].

At this conference KamLAND presents a precision measurement of the $\overline{\nu}_e$ oscillation parameters based on an analysis of data collected between March 2002 and May 2007. New features of this analysis include increased livetime, a lower analysis threshold, reduced uncertainty in the $^{13}C(\alpha,n)^{16}O$ backgrounds, reduced systematic in the fiducial volume, and a modified likelihood analysis procedure. To reduce the fiducial volume uncertainty, one of the dominant systematics in the previous analysis, the collaboration deployed a 4π calibration system that allows the calibration of the detector response with passive radioactive sources throughout the entire detector volume. The 4π system consists of a segmented calibration pole with multiple radioactive calibration sources that are used for the calibration of the energy response and vertex reconstruction throughout the detector. This provides a direct test of the vertex and energy reconstruction at radii out to R=5.5 m near the fiducial volume boundary. Figure 1 shows the conceptual design of the 4π calibration system in the KamLAND detector and the vertex distribution of ^{60}Co/^{68}Ge composite source data collected in 4π calibration runs. A suite of radioactive sources including ^{60}Co, ^{68}Ge, and AmBe were used with the 4π system to map out the detector response in θ, ϕ, and r. Using the 4π system the fiducial volume is determined to 1.6% within 5.5 m from the center of the detector. Together with the ^{12}B/^{12}N event ratio which is used to establish the uncertainty between 5.5 m and 6 m a combined 6-m-radius fiducial volume uncertainty of 1.8% is obtained. See Table 2.

With 1491 days of livetime and a fiducial volume of (R_{prompt},$R_{delayed}$) $<$ 6.0 m which corresponds to an exposure of 2.44×10^{32} proton-years KamLAND observed 1609 events over a background of 276 ± 23.5 events. The number of expected events for this time period is 2179 ± 89 events. Figure 2 shows the prompt energy spectrum, the no-oscillation expectation, and the best-fit oscillation scenario. The significance of $\overline{\nu}_e$ disappearance above a 2.6 MeV analysis threshold is 8.5σ and the significance of spectral distortion is $> 5\sigma$. An analysis of the measured prompt spectrum taking into account the $\overline{\nu}_e$ rate, the shape of the spectrum, and the time variation of the reactor flux yields the best-fit oscillation parameters $\tan^2\theta_{12} = 0.56^{+0.14}_{-0.09}$ and $\Delta m^2 = 7.58^{+0.22}_{-0.21} \times 10^{-5}$ eV^2 (Figure 3). The recent KamLAND data nicely reveal the oscillatory signature in the L/E dependence of the neutrino survival probability. See Equation 1 and Figure 3. Given its effective, flux-weighted average baseline of $L_0 = 180$ km to the surrounding reactors KamLAND has little sensitivity to θ_{13} (Figure 3).

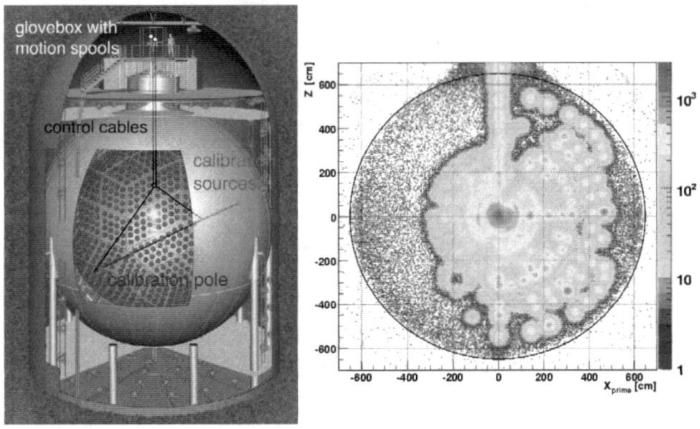

Figure 1. Illustration of the KamLAND detector with 4π calibration system (left) and vertex distribution of ^{60}Co/^{68}Ge composite source in 4π calibration runs (right).

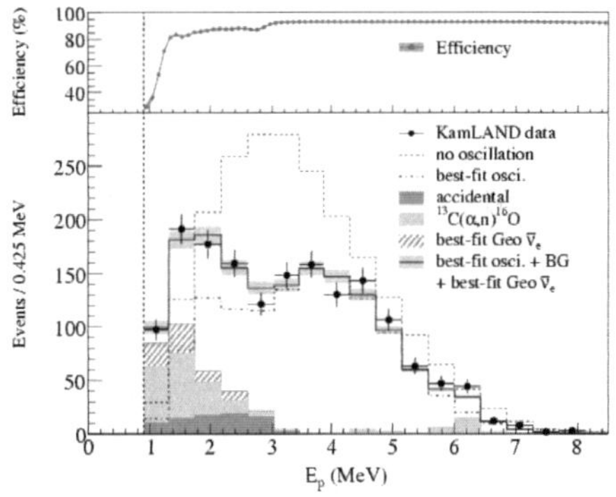

Figure 2. Prompt energy spectrum of reactor $\overline{\nu}_e$ measured in KamLAND.

3. A Precision Measurement of θ_{13} with Reactor Antineutrinos

Recent results from Super-Kamiokande, SNO, and KamLAND, have unambiguously demonstrated that neutrinos have mass and change flavor. Two of the three mixing angles in the U_{MNSP} neutrino mixing matrix have been measured by the solar and atmospheric experiments and were found to be large and maximal respectively. The third neutrino mixing angle, θ_{13}, has not been measured yet but is known to be small. The Chooz experiment has provided the best limit of $\sin^2 2\theta_{13} < 0.15$ [4]. The pattern of the neutrino mixing parameters revealed so far is strikingly different from that of quark mixing and places significant constraints and guidance on the construction of models involving new physics. The measurement of θ_{13} has sparked worldwide interest as it raises important questions about the phenomenology and theory of neutrino mixing

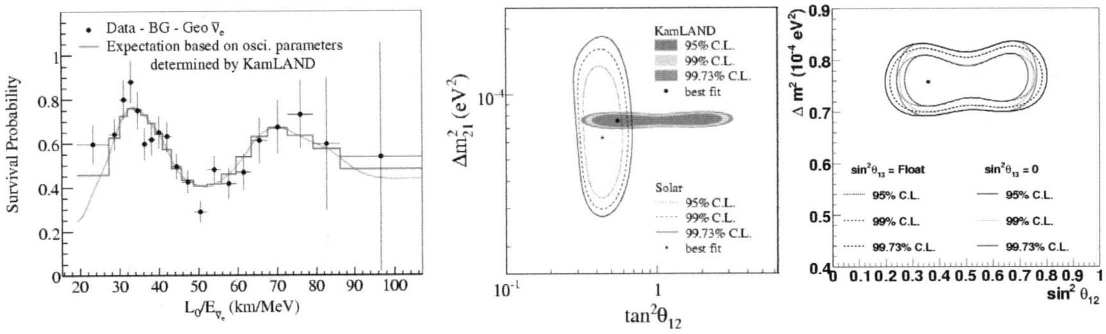

Figure 3. Survival probability as a function of L/E (left) and oscillation parameter space allowed by KamLAND (center and right) for the two-neutrino and three-neutrino oscillation case.

and is key to the future of neutrino oscillation physics. The size of θ_{13} determines the electron component to the third neutrino mass eigenstate. If it turns out to be zero or very small, it would point to some μ-τ symmetry in neutrino mixing. A non-zero θ_{13} is prerequisite to observable CP violation in the lepton sector. CP violation is a key ingredient to leptogenesis which can explain the observed baryon asymmetry in the Universe through out-of-equilibrium L-violating decays of heavy Majorana neutrinos. θ_{13} is also important for determining the nucleosynthesis yields in supernovae and the early Universe where neutrino mixing and neutrino interactions play a key role [12]

There are two approaches to measuring the neutrino mixing angle θ_{13}. One involves the appearance measurement $\nu_\mu \to \nu_e$ with high-energy neutrinos at accelerators over long baselines. In this case the oscillation probability $P(\nu_\mu \to \nu_e)$ is dependent on the mixing angles as well the mass hierarchy and the mass splittings. P is sensitive to matter effects which allow the determination of CP-violating phases. Such a measurement is rich in physics but the parameter degeneracy makes it challenging. On the other hand, the precise measurement of $\overline{\nu}_e$ disappearance with reactor antineutrinos is a clean measurement of the antineutrino oscillation in vacuum in the absence of matter effects. Since $\Delta m^2_{12} \ll \Delta m^2_{23}$, $\Delta m^2_{13} \approx \Delta m^2_{23}$ and we can predict the mass splitting and the wavelength of the unobserved θ_{13} oscillation. The relatively low energy of reactor $\overline{\nu}_e$ allows the observation of the oscillation effect over short distances of typically $\mathcal{O}$(1-2 km). Next-generation reactor experiments maximize their sensitivity to θ_{13} by optimizing detector baselines and using multiple detectors to cancel uncertainties in the reactor $\overline{\nu}_e$ flux prediction (see Table 1).

A relative measurement between two detectors placed near and far from the reactors largely eliminates reactor flux uncertainties and most of the detector-related systematics that have limited absolute $\overline{\nu}_e$ flux measurements in KamLAND and other past experiments. A change in the observed $\overline{\nu}_e$ interaction rate between these two detectors due to the decrease in the $\overline{\nu}_e$ survival probability as a function of distance would be an indication of sub-dominant $\overline{\nu}_e \to \overline{\nu}_{\mu,\tau}$ oscillation and allows the determination of $\sin^2 2\theta_{13}$. Uncertainties related to the absolute target mass and the absolute detection efficiency cancel in this relative measurement. Key to this concept is the use of "identical" detectors near and far to control relative systematic uncertainties. For the expected improvement in the uncertainties of the relative measurement over an absolute $\overline{\nu}_e$ flux measurement see Tables 1, 2, 3, and 4. The goal of next-generation neutrino oscillation experiments is to measure $\mathrm{in}^2 2\theta_{13}$ to < 0.01 at 90% CL. This sets the limit for realistic studies of leptonic CP-violation in the near future. New technologies are needed if $\sin^2 2\theta_{13} \ll 0.01$.

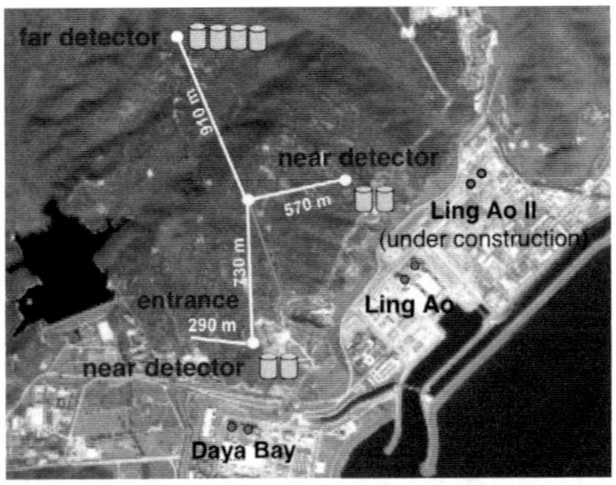

Figure 4. Layout of the Daya Bay reactor θ_{13} experiment. Eight detectors will be located in three experimental halls at different distances from the reactor cores.

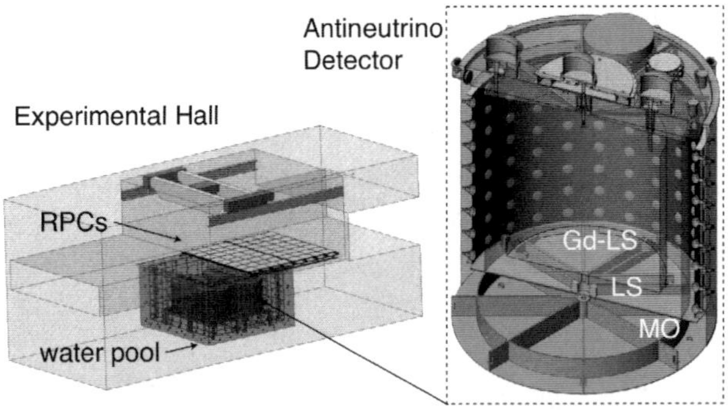

Figure 5. Illustration of the experimental hall with muon water pool in which the antineutrino detectors will be located.

Several collaborations worldwide have undertaken the challenge to measure θ_{13} from the disappearance of reactor $\overline{\nu}_e$. The Double Chooz, Reno, and Daya Bay experiments are currently in various phases of construction and will start taking data in 2009-2010. With a total of eight detectors of 20-ton target mass each installed in three underground experimental halls Daya Bay [13] is the largest experiment under construction and will reach the highest sensitivity of $\sin^2 2\theta_{13}$ to < 0.01 at 90% CL. The Daya Bay nuclear power plant presently operates 4 reactor cores with a total thermal power output of 11.6 GW_{th}. Two more cores are under construction and will increase the total power output to 17.4 GW_{th} by 2011. The layout of the Daya Bay experiment is shown in Figure 4.

Common to all future reactor θ_{13} experiments is the design of the antineutrino detectors. Gd-liquid scintillator is used as the target for the $\overline{\nu}_e$ interactions. With high proton density, good light yield, and a large neutron capture cross-section it provides an easily identifiable $\overline{\nu}_e$ signal and good background discrimination. The inner target volume is surrounded by a liquid scintillator gamma catcher to eliminate the need for position reconstruction and fiducial volume cuts like in KamLAND. A mineral oil buffer region surrounds the target and provides shielding against radioactive backgrounds from the PMTs, the detector tank, and other detector elements. At Daya Bay all antineutrino detectors are placed in water pools instrumented with PMTs and RPCs. The pools act a muon veto and provide shielding against rock and environmental gamma background. All detectors will be located in underground halls to provide shielding against muons and cosmogenic backgrounds. See Figure 5.

Control of the relative detector systematics and the determination of backgrounds will be essential for the success of the upcoming reactor θ_{13} experiments. In the case of Daya Bay the $\overline{\nu}_e$ flux is measured in pairs of "identical" detectors deployed at the different sites. Tests have shown that the relative target mass in these detectors can be measured to $< 0.1\%$. We expect to calibrate and understand the relative antineutrino detection efficiency to $< 0.25\%$. See Table 4. With 4 (or 6) reactor cores Daya Bay will have an additional uncertainty from the multiple baselines between the detectors and reactor cores. This effect, however, is small in the relative measurement between detectors. See Table 3. The main challenge will be to understand cosmogenic backgrounds and in particular beta-delayed neutron emitters such as $^8He/^9Li$ to better than 0.3%. This is likely to be the limiting uncertainty in the experiment. With these projected uncertainties a sensitivity of $\sin^2 2\theta_{13} < 0.01$ at 90% CL is achievable in three years of

Table 1. KamLAND: Reactor-related systematic uncertainties in a flux measurement with a single detector [11].

Reactor related	%
$\overline{\nu}_e$-spectra	2.4
reactor power	2.1
fuel composition	<1.0
long-lived nuclei	0.3
time lag	0.01
total	3.4

Table 2. KamLAND: Detector-related uncertainties in a single detector [11].

Detector related	%
fiducial volume	1.8
energy scale	1.5
efficiency	0.6
OD veto	0.2
cross-section	0.2
livetime	0.03
total	2.4%

Table 3. Daya Bay: Net correlated errors from multiple reactors in a relative flux measurement between multiple detectors [13].

Number of cores	σ(power)	σ(location)	σ(total)
4	0.035%	0.08%	0.087%
6	0.097%	0.08%	0.126%

Table 4. Daya Bay: Detector-related uncertainties in a relative measurement between two detectors [13].

source	baseline (%)	goal (%)
no. of protons	0.3	0.1
energy cuts	0.2	0.1
position cuts	0.0	0.0
time cuts	0.1	0.03
H/Gd ratio	0.1	0.1
n multiplicity	0.05	0.05
trigger	0.01	0.01
live time	<0.01	<0.01
total	0.38%	0.18%

Figure 6. Signal and background event rates in the Daya Bay detectors in the three experimental halls Daya Bay (DYB), Ling Ao (LA), and the far site (Far). All numbers are rates in events/day/module unless stated otherwise [13].

Event Rate	DYB	LA	Far
antineutrino	930	760	90
natural radiation	<50	<50	<50
single neutron	18	12	1.5
β-emission isotopes	210	141	14.6
accidental/signal	<0.2%	< 0.2%	<0.2%
fast neutron/signal	0.1%	0.1%	0.1%
^{8}He-^{9}Li/signal	0.3%	0.2%	0.2%

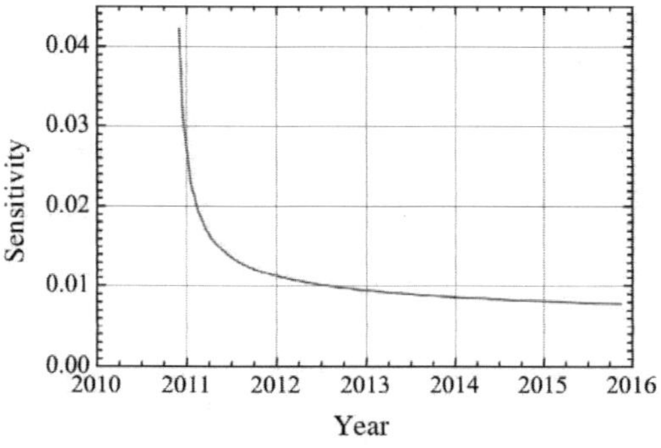

Figure 7. Sensitivity of the Daya Bay experiment to $\sin^2 2\theta_{13}$ [13].

running. Tables 6 and Figure 7 show the expected signal event and background rates and the sensitivity of the Daya Bay experiment to $\sin^2 2\theta_{13}$ as a function of running time.

4. Summary and Conclusions

KamLAND has played a key role in the history of reactor neutrino physics. Designed to observe antineutrinos from nuclear power plants in Japan, KamLAND made the first observation of reactor $\overline{\nu}_e$ disappearance at a flux-weighted average distance of about 180 km, found direct evidence for neutrino oscillation in spectral distortions, and has made a precision measurement of the reactor $\overline{\nu}_e$ oscillation parameters. With the help of the 4π calibration system for the determination of the fiducial volume, reduced uncertainties in the$^{13}C(\alpha,n)^{16}O$ background, and improved analysis techniques a precision of 2.8% on Δm^2_{12} was obtained. Together, solar neutrino experiments and KamLAND provide unambiguous evidence that neutrinos have mass and change flavor, and they provide the most precise determination of the neutrino oscillation parameters Δm^2_{12} and θ_{12}. Recent data from KamLAND illustrate clearly the oscillatory signature of neutrino mixing.

A suite of next-generation reactor antineutrino experiments including Daya Bay, Double Chooz, and Reno are now under construction to make a precision measurement of the reactor $\overline{\nu}_e$ flux over short distances of 0.2-2 km in order to measure the yet unknown mixing angle θ_{13}. With the control of relative detector systematic uncertainties to $< 0.38\%$ Daya Bay will be able to reach a sensitivity of of $\sin^2 2\theta_{13} < 0.01\%$ at 90% CL in three years of data taking. Fifty years after the discovery of the antineutrino, nuclear reactors still play an important role in understanding the mysteries of neutrinos, and the next reactor experiments may hold the clue to opening the path towards the study of leptonic CP violation with neutrinos.

Acknowledgments

KamLAND is an international collaboration from Japan, the United States, and France led by the Research Center for Neutrino Science, Tohoku University, Sendai, Japan [14]. The Daya Bay experiment is an international collaboration of scientists from China, the United States, and Europe [15]. The author is grateful to both collaborations for their support and discussions. This work was supported in part by the University of Wisconsin.

References

[1] Cowan C L, Reines F, Harrison F B, Kruse H W, and McGuire A D 1956 *Science* **124**, 103-104; Reines F, Cowan C L, 1956 *Nature* **178**:446-449; Reines F, Cowan C L, Harrison F B, McGuire A D, and Kruse H W, 1960, Phys. Rev.**117**, 159-173.

[2] Kwon H *et al.*, 1981, Phys.Rev.D**24**, 1097. Zacek G *et al*, 1986, Phys.Rev.D**34**, 2621. Cavaignac J F, *et al.* 1984, Phys.Lett.B**148**:387-394; Declais Y *et al.*, 1994, Phys.Lett.B**338**:383-389; Achkar B *et al.*, 1995, Nucl.Phys.B**434**:503-534; Boehm F *et al.*, (Palo Verde Collaboration), 2001, Phys.Rev.D**64**:112001; Apollonio M *et al.*, (CHOOZ Collaboration), 2003, Eur.Phys.J.C**27**:331-374.

[3] Daraktchieva Z *et al.* (MUNU Collaboration), 2005, Phys.Lett.B**615**: 153-159.

[4] Apollonio M *et al.*, (CHOOZ Collaboration), 1999, Phys. Lett. B **466**, 415; 2003, Eur.Phys.J.C**27**:331-374.

[5] Bernstein A, Wang Y, Gratta G, and West T, 2002, J.Appl.Phys.**91**,4672-4676.

[6] Vogel P and Engel J, 1989, Phys.Rev.D**39**, 3378; Nakajima K *et al.*, 2006, Nucl.Instrum.Meth.**A569**:837-844.

[7] Pontecorvo B, 1958, Sov.Phys.JETP**7**:172-173; 1957, Zh.Eksp.Teor.Fiz.**34**:247; Pontecorvo B, 1968, Sov.Phys.JETP**26**:984-988; 1967, Zh.Eksp.Teor.Fiz.**53**:1717-1725; Gribov V N, Pontecorvo B, 1969, Phys.Lett.B28:493.

[8] Eguchi K *et al.* (KamLAND Collaboration), 2003, Phys.Rev.Lett.**90**:021802.

[9] Araki T *et al.* (KamLAND Collaboration), 2005, Phys.Rev.Lett.**94**:08180.

[10] Wolfenstein L, 1978, Phys.Rev.D**17**: 2369; Mikheyev S P, Smirnov A Yu, 1989, Prog.Part.Nucl.Phys.**23**:41-136.

[11] Abe S *et al.* (KamLAND Collaboration), 2008, submitted to Phys.Rev.Lett., arXiv:0801.4589.

[12] Balantekin B and Yuksel H, 2005, New J. Phys. 7 51; Abazajian K, Beacom J, and Bell N, 2002, Phys.Rev.**D66** 013008.

[13] Guo X *et al* (Daya Bay Collaboration), 2007, Daya Bay Project - Technical Design Report

[14] KamLAND project web sites: http://www.awa.tohoku.ac.jp/KamLAND/ and http://kamland.lbl.gov/.

[15] Daya Bay project web site: http://dayawane.ihep.ac.cn/

First results on ^{7}Be solar neutrinos from the Borexino real time detector

Presented by Gianpaolo Bellini on behalf of the Borexino Collaboration:

G Bellini[2], J Benziger[3], S Bonetti[2], B Caccianiga[2], F Calaprice[4], F Dalnoki-Veress[4], D D'Angelo[2], H de Kerret[6], A Derbin[7], A Etenko[5], K Fomenko[9], R Ford[4], D Franco[2], C Galbiati[4], S Gazzana[1], M Giammarchi[2], M Goeger-Neff[10], A Goretti[4], C Grieb[11], S Hardy[11], G Heusser[15], Aldo Ianni[1], Andrea Ianni[4], M Joyce[11], G Korga[1], D Kryn[6], M Laubenstein[1], M Leung[4], E Litvinovich[5], P Lombardi[2], L Ludhova[2], I Machulin[5], G Manuzio[12], F Masetti[13], K McCarty[4], E Meroni[2], L Miramonti[2], M Misiaszek[14], D Montanari[1], M E Monzani[1], V Muratova[7], L Niedermeier[10], L Oberauer[10], M Obolensky[6], F Ortica[13], M Pallavicini[12], L Papp[1], L Perasso[2], A Pocar[4], R S Raghavan[11], G Ranucci[2], A Razeto[1], A Sabelnikov[5], C Salvo[1], S Schönert[15], H Simgen[15], O Smirnov[9], M Skorokhvatov[5], A Sonnenschein[4], A Sotnikov[9], S Sukhotin[5], Y Suvorov[2,5], V Tarasenkov[5], R Tartaglia[1], G Testera[12], D Vignaud[6], R B Vogelaar[11], F von Feilitzsch[10], M Wojcik[14], O Zaimidoroga[9], S Zavatarelli[12], G Zuzel[15]

[1] I.N.F.N. Laboratori Nazionali del Gran Sasso – Assergi – Italy
[2] Dipartimento di Fisica dell'Universita' degli Studi e INFN – Milano – Italy
[3] Princeton University, Chemical Engineering Department – Princeton, NJ – USA [4]Princeton University, Physics Department – Princeton, NJ – USA
[5] RRC Kurchatov Institute – Moscow – Russia
[6] Laboratoire AstroParticule et Cosmologie APC – Paris – France
[7] St. Petersburg Nuclear Physics Institute – Gatchina – Russia
[8] Massachusetts Institute of Technology, Department of Physics – Cambridge, MA – USA
[9] Joint Institute for Nuclear Research – Dubna – Russia
[10] Technische Universitaet Muenchen – Garching – Germany
[11] Virginia Tech, Physics Department – Blacksburg, VA – USA
[12] Dipartimento di Fisica dell'Universita' e INFN – Genova – Italy
[13] Dipartimento di Chimica dell'Universita' e INFN – Perugia – Italy
[14] Marian Smoluchhowski Institute of Physics, Jagiellonian University – Krakow - Poland
[15] Max-Plank-Institut fuer Kernphysik – Heidelberg – Germany

This paper is dedicated to: C Arpesella, M Deutsch, B Freudiger, A Martemianov, S Vitale, members of the Collaboration, now deceased.

Abstract. The Borexino experiment started the data taking may 15th 2007. It is until now the only experiment able to detect in real time neutrino interactions below 2 MeV. This is due to the very high radio-purity reached by the detector, more than one order of magnitude better than the goal of the design. The detector allows the single fluxes of low energy solar neutrinos to be determined from all sources not previously studied: ^{7}Be, pep, CNO and hopefully pp. Here the first measurement of the ^{7}Be, obtained from the first 47.6 live days of data taking, is presented.

1. A touch of history

In 1990 a group of physicists conceived a project with the goal of studying in real time the solar neutrinos at energies below 2 MeV. For this study, the technology of liquid scintillation has been preferred over water Cherenkov detection, mainly for reasons of higher light yield and greater energy resolution.

The detection in real time of neutrinos below 2 MeV in an underground experiment becomes possible only if the background due to the natural radioactivity is strongly reduced. This is the main reason why the water detectors cannot lower the energy threshold below 5 MeV. A feasibility study revealed that the radioactivity levels of the scintillator should not exceed: 10^{-16} g/g for Th and U equivalent, 10^{-14} g/g for the natural K and ≅ 10^{-18} for the ratio $^{14}C/^{12}C$; in addition all the construction materials were subjected to a severe selection for radiopurity.

Two years of R&D on the purification showed that the measurements of such low background levels were not possible with the existing detectors. For instance, one of the more sensitive tool available, the plasma source mass spectrometer, was only able to reach 10^{-13}-10^{-14} g/g. Consequently the Borexino Collaboration constructed and installed a detector at the Gran Sasso underground Laboratory with 4 tons of scintillator (the Counting Test Facility-CTF), which was able to measure radioactivity levels down to $5 \cdot 10^{-16} g/g$.

In 1995 the CTF measurements showed that, in principle, the Borexino detector was feasible because the design radio-purity goals were reachable [1].

In 1996 the Borexino project has been approved by the Italian INFN and by the German Agencies, and in 1997 by the US NSF. The construction and installation of the detector and the auxiliary plants progressed until August 2002, when the activities have been stopped due to problems between the Gran Sasso Laboratory and the local authorities. Unfortunately, it was only in 2005 that the Collaboration was allowed to resume the operations. A re-commissioning of the set-up was performed and full operations were possible in late spring 2006. The detector was filled with highly purified water by the end of the year.

At the beginning of 2007 the detector has been filled with the scintillator mixture (Pseudocumene plus 1.5 g/l of a fluorescent dye) and with the so-called Buffer liquid (Pseudocumene plus 5.0 g/l of quencher), in addition to the shielding water; all these liquids have been purified before they were inserted into the detector. These operations have been completed on May 15th 2007 and the data taking with the full detector began.

2. The detector

In figure 1 a sketch of the Borexino detector is shown.

Borexino in located in the Hall C of the Gran Sasso underground Laboratory, featuring an overburden of ≅ 4000 m water equivalent. The rate of cosmic muons at the detector location is ≅ $1/m^2 \cdot h$.

The core of the detector consists of 278 tons of liquid scintillator with two components (Pseudocumene (PC) plus PPO, 2,5-diphenyloxazole) contained in a nylon Inner Vessel (IV; 125 μm thick) with a 4.25 m radius. The IV is surrounded by 890 tons of Pseudocumene loaded to a light quencher (DMP, dimethylphtalate). A stainless steel sphere (SSS, having 6.75 m radius) contains both the liquid scintillator and the quencher Pseudocumene. The aim of the buffer liquid is to shield the IV from the radioactive emissions of the phototubes (PMTs) mounted on the sphere, from the stainless steel of the sphere itself, in addition to the residual external radiations (rock and water). The choice of Pseudocumene as buffer liquid aims to reduce the buoyancy on the IV to a negligible value.

The PMTs are 2214, 1843 of which coupled to an optical concentrator designed to detect only the light emitted by the scintillator; the remaining 371 without optical concentrators capture also the photons produced in the buffer region. Between the IV and the SSS an external nylon vessel functions to stop the radon emitted by the PMTs and by the stainless steel.

The SSS is shielded in all directions by, at least, two meters of ultra-pure water (2100 tons in total) contained in a cylindrical dome. Finally 200 PMTs are mounted on the external surface of the SSS

(Outer Detector); they point outwards to detect the Cherenkov light produced in the shielding water by cosmic muons crossing the detector.

In Borexino the ν–e scattering is studied.

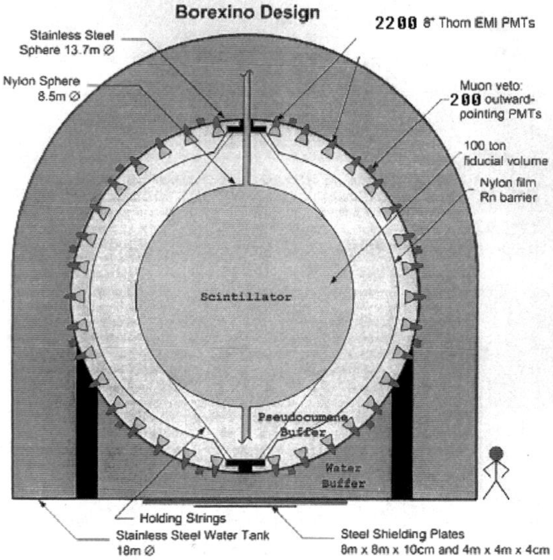

Figure 1. Schematic drawing of the Borexino detector.

3. Tools for a success

The set-up includes a relevant number of subsystems to store the PC, purify the PC, a concentrated solution of PC plus PPO (master solution), and the water, to fill the detector with the scintillator and with buffer liquid, and to perform precise cleaning of all the plants. In addition several safety installations have been integrated into the system.

But even more important are the methods, the procedures, the precautions adopted to construct and install the detector and the auxiliary systems [2]. Here the most important of them are listed:

i) The scintillator was cleaned before the filling of the detector: the PC and separately the master solution were purified via water extraction, distillation (at 80 mbar of pressure and 90-95 ^{0}C of temperature), nitrogen stripping, ultra-fine filtration. The master solution has been added to the pure PC during the filling. The Nitrogen used for the stripping has been purified and selected to obtain the following upper limits: Rn<8 μBq/m^3, Ar<0.012 ppm, Kr<0.02 ppt.

ii) Any water used in the installation has been purified to reduce the contaminants down to: U/Th equivalent<10^{-14} g/g, Rn<1 mBq/m^3, ^{226}Ra<0.8 mBq/m^3.

iii) All construction components: concrete, stainless-steel, gaskets, PMT glass and ceramics, sealing materials, etc. have been subjected to a rigorous solution process to keep the radioactive contents at minimum levels.

iv) All the critical stainless-steel subsystems were electro-polished. Their tightness was tested to be better than 10^{-8} $bar \cdot cm^3 \cdot s$. This is particularly important because the Rn content in the underground air ranges between 40 and 120 Bq/m^3.

v) All operations concerning the detector have been carried out in cleans rooms at classes: 10, 100, 1000, depending on the operation conditions. During the installation the detector itself was a clean room of class 10000. All operations on the auxiliary plants have been made in N_2 atmosphere, whenever possible.

vi) A precision cleaning was applied to the detector, storage vessels, piping and plant components. The cleaning involved acid solutions, detergent and ultra-pure water.

vii) Special care was taken during the PC procurement: a dedicated loading station was constructed at the producer plant (Polimeri Europa in Sardinia), which allowed the loading of clean isotanks in a N_2 atmosphere. The isotanks were then shipped to Gran Sasso and unloaded using a dedicated pumping station.

viii) Extreme precaution was taken in the fabrication and assembly of the nylon vessels. The raw material was selected and the film was extruded in controlled areas. Finally the nylon vessels were constructed in a clean room equipped with a dedicated Rn control system.

4. Data acquisition

The firsts tests of the electronics and DAQ have been carried out with the detector filled with N_2 and water. The data taking with the detector fully filled with scintillator, buffer liquid and water, started may 15th, 2007. An initial period of time was devoted to the tuning of the electronics.
Here the results of the first 47.4 live days of data acquisition are reported.

The main trigger fires when ≥ 30 PMTs detect one photo-electron, at least, within 60 ns. The energy threshold is about 60 keV. Time and charge of each PMT, detected in 7.2 μs, are recorded by means a TDC (≅ 0.5 ns of resolution) and by an 8 bits ADC, respectively. The typical triggering rate is ≅ 11 cps (dominated by the ^{14}C decays).

The Outer Detector gives a muon signal when ≥ 6 PMTs fire (99.8% of probability of muon rejection). In addition, all events within 2 ms after a μ crossing the PC are rejected to discard muon induced activity. The muon rate on the IV is 0.0055±0.0002 s^{-1}.
In the Borexino detector the time and the total charge of each event are measured. The event location is reconstructed from the PMT timing. The absolute time is also provided via a GPS link.

The number of hits and the total collected charge are more or less equivalent because the detector is designed in such a way that, at energies lower than few MeV, in most of the cases the charge of the PMTs corresponds to one photoelectron.

Two fully independent analyses with two independent codes have been carried out. The reconstruction programs as used in the analysis, we are referring to here, have not yet been calibrated with external radioactive sources. The calibrations at present are based upon internal signals.

5. Detector performances

A fiducial volume (FV) is defined within the IV, in order to have a further shielding of 1.25m against the radiations emitted by the vessel nylon and the residual background from the PMTs.

The nominal IV radius is 4.25 m, corresponding to a content of 278 tons of scintillator. The effective IV radius has been reconstructed by means of various methods: the spatial distribution of the ^{14}C decays, the decay of the Thoron (τ=80 s) emitted by the IV nylon and particulate present on the IV surface, the external background γs and the laser light diffused from teflon bowls, located on the IV external surface. All these methods, with the exception of the external background γs, reconstruct the IV radius at about 4.60 m. The external γs give a slightly larger number for the radius. Because a calibration with external sources has not been done yet, the Borexino Collaboration has decided to assume an IV with 4.60 m, with a total very conservative error of 25%.

The FV is defined by considering a volume containing the 35% of the total events, corresponding to the ratio between the nominal FV and IV. In addition a cut on the vertical axis, less than 1.8 m, has been applied in this analysis of the first days, due to a temporary presence in the north hemisphere of Rn, consequence of refilling operations. Then, according with these criteria, the total fiducial volume for the analysis, we are referring here, is 87 tons.

The light yield was evaluated first fitting only the ^{14}C spectra shape (featuring an electron with 156 keV end point [3]) and also by assuming it as free parameter in a global fit on the total energy spectrum and in particular by fitting the ^{14}C distribution, the ^{210}Po width and the ^{7}Be neutrino Compton edge.

The light yield corresponds to about 500 photoelectrons per MeV. This value already takes into account the quenching factor. The high light yield favors a good spatial resolution (16 cm at 500 keV, with a $N_{p.e.}^{-1/2}$ scaling) and a good energy resolution (10% at 200 keV,8% at 400 keV,6% at 1 MeV.

The background levels in the detector are in general better than what expected:

A) The ^{14}C presence fits the design goal. The measured ratio ^{14}C/^{12}C is 2.7±0.6 10^{-18}.

B) The background introduced by the Th family is measured via the sequence of the ^{212}Bi-^{212}Po (τ=236 μs), assuming a secular equilibrium. This method, already checked in CTF, has 93% efficiency. The measured contamination is <6.6 10^{-18} g/g, definitively better that our goal.

C) A very low background level was measured also for the ^{238}U family contamination. In this case the ^{214}Bi-^{214}Po coincidence (τ=236 μs) has been used and a value of ^{238}U equivalent <2 10^{-17}g/g was found. A check of our ability to identify the sequence ^{214}Bi-^{214}Po has been made studying the same correlated decays when, during the filling, a very well detectable Rn and Rn daughters were present in the detector. The decay time of the ^{214}Po has been measured to be 236±4 μs, in perfect agreement with the nominal value.

D) An important presence of ^{210}Po is observed (≅ 90 cpd/1 ton) without evidence of ^{210}Bi. The measurement of its decay time shows a value in agreement with its nominal τ (200 d) and this supports the hypothesis that the contamination of ^{210}Pb should be small. When the ^{210}Po will be reduced, we will able to understand better how negligible is the presence of ^{210}Pb and consequently of its daughters (as ^{210}Bi), For the moment the possible background due to ^{210}Bi has been taken as free parameter in the fits.

E) Another important contaminant can be the ^{85}Kr. Its presence in the scintillator can be studied directly via the well tagged decay ^{85}Kr-^{85m}Rb-^{85}Rb, with the sequence β (173 keV) and γ (514 keV), 1.46 μs apart [4]. The B.R. of this decay is very small (0.43%) and consequently a longer time of data taking is needed. Nevertheless we searched for this decay, finding two events which give an upper limit of 35 cpd/ton (90% c.l.). Until the statistics will be not improved the ^{85}Kr is taken as free parameter in the total fit.

6. Data analysis and results

In figure 2 the photoelectron distribution of the raw data is shown. The lower plot is obtained after the subtraction of the muons and the μ correlated activities, the ^{222}Rn daughters, the α particles (via the α/β discrimination), In addition the FV cut is applied. The shape of the distribution shows a remarkable agreement with what we can expect from the most important signal and unavoidable background: ^{7}Be, ^{14}C,^{11}C (figure 3).

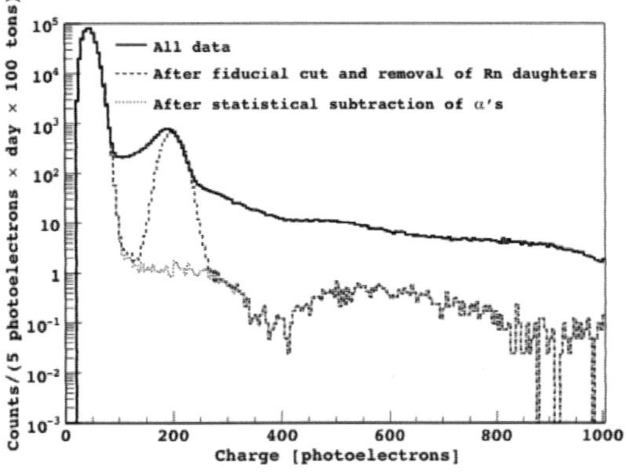

Figure 2. Energy spectrum of the raw data (see text).

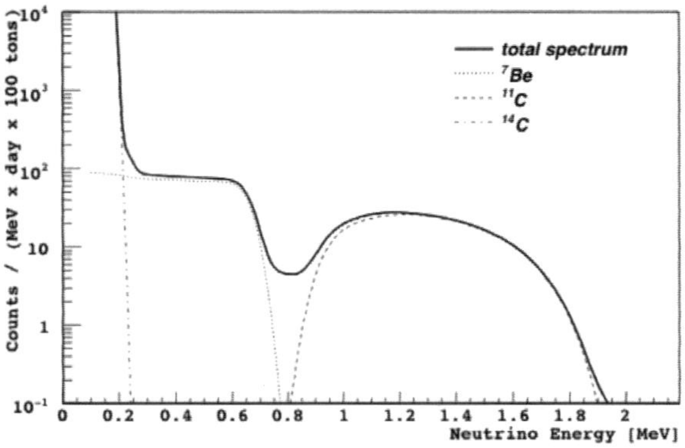

Figure 3. Spectrum of the more important signal and irreducible background (see text).

In these first weeks of data taking and analysis, we have focused our efforts on the ^{7}Be region. The 250-800 keV energy region is being fitted, once subtracted from the spectrum the identified ^{214}Bi, ^{214}Po and a number of ^{214}Pb decays equivalent to the ^{214}Bi-^{214}Po coincidences. We proceeded with two different fits: one without α subtraction and a second one with the α statistical subtraction, via the α/β discrimination, based upon the Gatti Parameter [5].

The statistical errors are determined by the χ^2 profile. The systematic errors are dominated by the uncertainty in the FV definition.

Because in the energy range of the fit the CNO and ^{210}Bi spectra are overlapped, their contributions are fitted together.

In figures 4 and 5 the two fits are presented, without and with α subtraction, respectively.

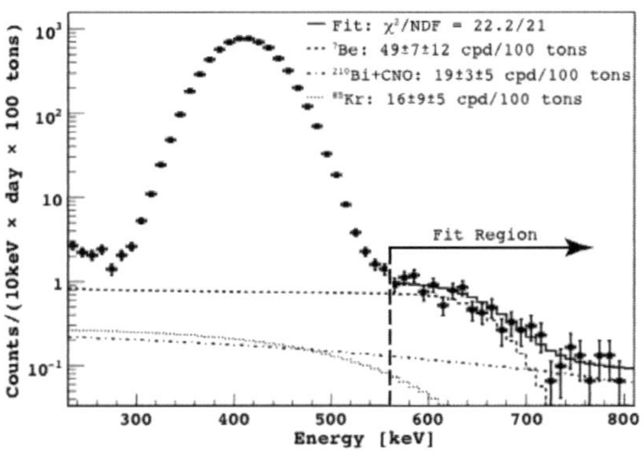

Figure 4. Energy spectrum without alpha subtraction

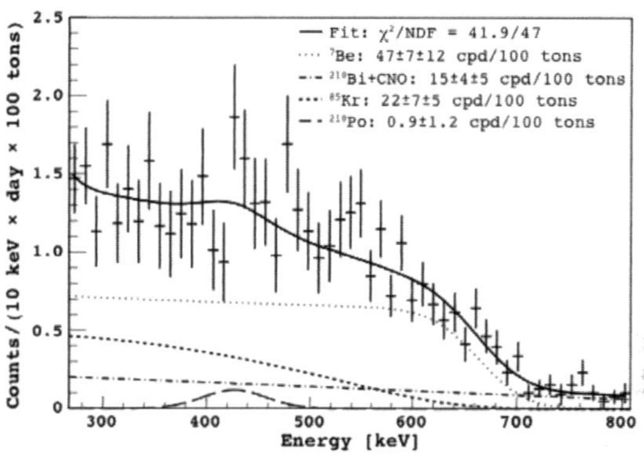

Figure 5. Energy spectrum once subtracted alphas via α/β discrimination.

The first fit has been carried out in two steps. In a first step the energy region 250-800 keV is fitted leaving the following parameters free: Light Yield, ^{210}Po, σ_{210Po}, ^{85}Kr, CNO+^{210}Bi, ^{7}Be. In a second one the fit has been restricted to the range 560-800 keV, with only ^{7}Be and CNO+^{210}Bi as free parameters, while the other ones are kept fixed at the values obtained from the 1st fit. The results are summarized in table 1, in cpd/100 tons, together with the statistical errors.

Table 1. (cpd/100 tons)

^{7}Be	^{85}Kr	CNO+^{210}Bi
49±7	16±9	19±3

The statistical error quoted for ^{7}Be includes, in addition to the contribution obtained from the χ^2 profile, also the change in the ^{7}Be rate best fit if the ^{85}Kr contribution is fixed at its maximum value (rate + 3σ).

The second fit is carried out on the energy spectrum where the alphas have been subtracted using the Gatti parameter [5], as shown in figure 5. This method, already tested in CTF [6] is based on the different time response of the scintillator for β and α particle. The reference curves have been obtained from the study of the ^{214}Bi-^{214}Po decays, present in the scintillator during the filling period, as the ^{222}Rn daughters (the number of photoelectrons for individual shapes within 2 ns is taken into account; the times of the PMT hits are compensated for the travel distances). Then the Gatti parameter distributions for each 20 keV bins in the full range 200-800 keV have been calculated and the alphas have been subtracted from them (CTF tests demonstrated that the pulse shapes do not depend on energy). The efficiency of the method is 98.5%.

The fit on this spectrum has been carried out in the 240-800 keV range, assuming as free parameters: ^{7}Be,CNO+^{210}Bi, ^{85}Kr and the residual ^{219}Po. The results of this fit are summarized in Table 2 .

Table 2. (cpd/100 tons)

^{7}Be	^{85}Kr	CNO+^{210}Bi	^{210}Po (resid.)
47±7	22±7	15±4	0.9±1.2

In both cases the fits show a good degree of confidence; in the first fit we obtain χ^2/NDF=22.2/21; in the second χ^2/NDF=41.9/47.

The results of the fits on the two spectra, with and without alpha subtraction, agree very well each to other. The quoted errors are statistical only; a total systematic error of 25% has to be added on both the results.

In figure 6 the survival probability for ν_e from the Borexino (^{7}Be) and SNO (^{8}B) are compared with the curve expected from the best fit LMA oscillation values ($\Delta m^2_{1-2} = 7.92 \cdot 10^{-5} eV^2, \sin^2\theta_{1-2} = 0.314$) [7] and from the Solar Standard Model values quoted in BPS07 [8] (based upon the metallicity as in GS98). In the same plot the determination of the pp flux before and after Borexino data are also shown.

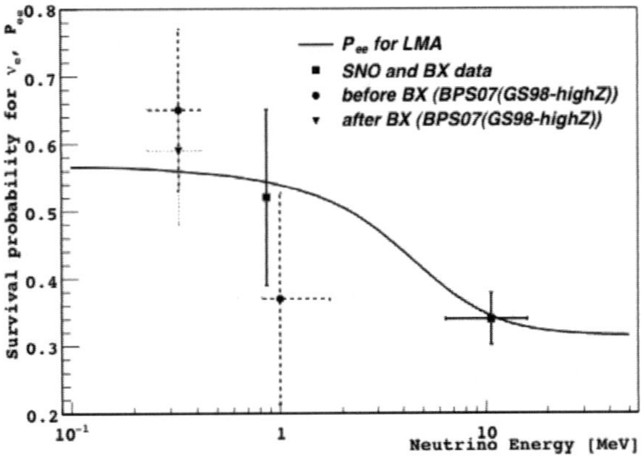

Figure 7. Survival probability for ν_e compared with the model curve (see text).

7. What next

The very good radio-purity achieved in Borexino, definitively higher than expected, opens possibilities beyond the experiment goals. Then in addition to the ^{7}Be flux, we are confident to be able to measure also the pep and CNO fluxes: the main problem for the study of these two fluxes is the production, μ induced, of ^{11}C nuclides within the scintillator. Its production triggers a sequence of events, which provides a very good tagging: $\mu + ^{12}C \rightarrow ^{11}C + n + \mu$, $^{11}C \rightarrow ^{11}Be + e^{+} + \nu_e$, neutron capture after ~300 μm with the emission of a 2.2 γ. With this threefold coincidence we can disentangle the ^{11}C, while missing only 14% of ν events; this method has been already checked in CTF [9].

In addition, due to the very good resolution, the pp flux could be studied because the ^{14}C spectrum ends well below 200 keV.

A direct check of the ν origin can be obtained by studying the seasonal variations of the flux, due to the Earth orbit eccentricity. Taking into account the radio-purity achieved in Borexino, we can reproduce this effect with a C.L. of 2.5 σ in one year and of 3.6 σ in two years.

Another subject for our analysis program concerns the search for $\bar{\nu}$ from the Sun, Earth (geo-neutrinos) and nuclear reactors. We expect 7-17 events per year for geo-neutrinos, depending on the Earth models, to be selected using the Reines-Cowan detection reaction [10]. The background due to the reactor $\bar{\nu}$ is particularly favourable at Gran Sasso, due to the absence of reactors in Italy; a S/N $\cong$1 is expected in the energy range of the geo-neutrinos. The total flux of $\bar{\nu}$ from reactors is $\cong$ 20 events/years with a baseline of 1000 km, on the average. In both cases the whole detector of 300 tons is considered, due to the very good tagging of the $\bar{\nu}$ interactions: prompt positron followed by a n capture after ~300 μs.

Borexino is also a good observatory for a Supernova explosion in our Galaxy. In Table 3 the channels with higher rate are quoted for a Supernova Standard @ 10kpc together with their total fluxes.

Table 3. The higher rates channels in Borexino for a Supenova explosion @ 10 kpc.

Inverse beta decay→79 (E_ν>1.8 MeV)	ν-p elastic scattering→55 (E_ν>0.25 MeV)	$^{12}C(\nu,\nu)^{12}C^*$→17 (E_γ=15.1 MeV)

For the first and the third channels other experiments can collect more statistics. For the second channel Borexino is almost unique because, due to the quenching, the p is pushed toward the low energies and, to detect it, a very low threshold is needed. In addition only the higher energy ν from Supernova are effective, and then this channel is triggered only by ν_μ and ν_τ.

Finally we plane to push the limit of the ν magnetic moment to ~ 5 10^{-11} Bohr Magneton, by means a 2.5 MCi ^{51}Cr source, to be installed in a access pipe designed on purpose below the Borexino at ~ 8.25 m from the detector's centre.

References

[1] Borexino Collaboration 1998 *Phys. Lett.* B **422** 349; 2002 *Astropart. Phys.* **18** 1

[2] Borexino Collaboration 2002 *Astropart. Phys.* **16** 205

[3] Borexino Collaboration 2000 *Nucl. Instr. and Methods* A **440** 360

[4] Borexino Collaboration 1998 *Astropart. Phys.* **8** 141

[5] Gatti F et al. 1996 *Nucl. Instr. and Methods* A **370** 609

[6] Borexino Collaboration in publication on *Nucl. Instr. and Methods* A

[7] Fogli G 2006 *Progr. Part. Nucl. Phys.* **57** 742

[8] Bahcall J et al.; reported by Peña-Garay at Neutrino Telescopes 2007

[9] Borexino Collaboration 2006 *Phys. Rev.* C **74** 045805

[10] Borexino Collaboration 2006 *Eur. Phys. Journal* C **47** 21

Nuclear physics inputs needed for geo-neutrino studies

G Bellini[1,2], G Fiorentini[3,4], A Ianni[5], M Lissia[6], F Mantovani[4,7,9] and O Smirnov[8]

[1] Dipartimento di Fisica, Università degli Studi di Milano, I-20100 Milano, Italy

[2] Istituto Nazionale di Fisica Nucleare, Sezione di Milano, I-20100 Milano, Italy

[3] Dipartimento di Fisica, Università degli Studi di Ferrara, I-44100 Ferrara, Italy

[4] Istituto Nazionale di Fisica Nucleare, Sezione di Ferrara, I-44100 Ferrara, Italy

[5] INFN, Laboratori Nazionali del Gran Sasso, I-67010 Assergi (AQ), Italy

[6] Istituto Nazionale di Fisica Nucleare, Sezione di Cagliari, I-09042 Monserrato (CA), Italy

[7] Centro di GeoTecnologie CGT, I-52027 San Giovanni Valdarno (AR), Italy

[8] Joint Institute for Nuclear Research, 141980, Dubna, Moscow region, Russia

Abstract. Geo-neutrino studies are based on theoretical estimates of geo-neutrino spectra. We propose a method for a direct measurement of the energy distribution of antineutrinos from decays of long-lived radioactive isotopes.

1. Introduction

Geo-neutrinos, the antineutrinos from the progenies of U, Th and ^{40}K decays in the Earth, bring to the surface information from the whole planet, concerning its content of radioactive elements. Their detection can shed light on the sources of the terrestrial heat flow, on the present composition, and on the origins of the Earth.

Although geo-neutrinos were conceived very long ago, only recently they have been considered seriously as a new probe of our planet, as a consequence of two fundamental advances that occurred in the last few years: the development of extremely low background neutrino detectors and the progress on understanding neutrino propagation. From the theoretical point of view, the links between the geo-neutrino signal and the total amount of natural radioactivity in the Earth have been analyzed by several groups; various "reference models" for geo-neutrino production have been presented in the literature; these have been refined with geological and geochemical studies of the region surrounding the detector. KamLAND and Borexino are collecting geo-neutrino data, while several planned

[9] Corresponding author: fabio.mantovani@unisi.it

experiments (e.g. SNO+, LENA, HANOHANO, EARTH...) have geo-neutrino measurements among their primary goals. A recent review is presented in [1].

It seems to us that this most intense and most interesting activity has to be complemented with some deepening of the nuclear physics which is at the basis of geo-neutrino production and which is crucial for interpreting future geo-neutrino data.

All experiments, either running or in preparation, use the well tagged inverse beta decay on free protons

$$\bar{\nu}_e + p \rightarrow e^+ + n - 1.806 \text{ MeV} \quad (1.1)$$

as the reaction for geo-neutrino detection. The signal is estimated from the cross section $\sigma(E_\nu)$ of (1.1) and from the decay spectrum $\mathrm{f}(E_\nu)$ of geo-neutrinos produced in the decay chains of ^{235}U and ^{232}Th, the relevant quantity being the "specific signal" defined as:

$$s = \int_{E_0}^{E_{MAX}} \mathrm{d}E_\nu \sigma(E_\nu) \mathrm{f}(E_\nu) \quad (1.2)$$

where the integral is between the threshold energy ($E_0 = 1.806$ MeV) and the maximal geo-neutrino energy E_{MAX} and the spectrum $\mathrm{f}(E_\nu)$ is normalized – say - to the number N of geo-neutrinos produced in each decay chain, $N = \int_0^{E_{MAX}} \mathrm{d}E_\nu \, \mathrm{f}(E_\nu)$.

It is important to observe that the specific signal is affected by unknown uncertainties. In fact, whereas the cross section $\sigma(E_\nu)$ is known with an accuracy of half of per cent or better, it is difficult to assess the accuracy of $\mathrm{f}(E_\nu)$, which is determined from rather indirect measurements and questionable theoretical assumptions.

The goal of this note is to provide a framework for a direct measurement of $\mathrm{f}(E_\nu)$*, so that the accuracy of the specific signal can be established.*

2. Why should geo-neutrino spectra be measured?

Geo-neutrinos are produced through β and βγ processes:

$$X \rightarrow X' + e + \bar{\nu}_e \quad (1.3)$$

$$X \rightarrow X'^{*} + e + \bar{\nu}_e, \quad X'^{*} \hookrightarrow X' + n\,\gamma \quad (1.4)$$

In order to determine the geo-neutrino decay spectra $\mathrm{f}(E_\nu)$ one has thus to know:

i. the probabilities of populating the different energy states of the final nucleus;
ii. the shape of the neutrino spectrum for each transition.

Probabilities are generally derived by measuring the relative intensities of gamma lines. In particular, the transition to the ground state (pure β) is indirectly determined by a subtraction procedure, which brings with it the propagation of all errors on the measured intensities. We remark that this transition, which is associated with the most energetic geo-neutrinos, gives the largest contribution to the specific signal.

For each transition, the shape of the neutrino spectrum is generally calculated assuming the well known "universal shape". This expression, see [1], corresponds to momentum independent nuclear matrix elements (as for allowed transitions) and includes the effect of the bare Coulomb field of the nucleus through the relativistic Fermi function. Electron screening and nuclear finite size effects are not considered. Furthermore, the same "universal shape" expression is used for forbidden transitions, where momentum dependent nuclear matrix element can appear.

All this suggests to us that some experimental check is needed.

3. Towards a direct measurement of geo-neutrino decay spectra

When the nucleus X decays, whichever is the transition involved, energy conservation provides a connection between the neutrino energy E_{ν}, the kinetic energy of the electron T_{e} and the total energy of the emitted gammas, E_{γ}:

$$Q = E_{\nu} + T_{e} + E_{ã} \qquad (1.5)$$

where $Q = M_{X} - M_{X'} - M_{e}$ is the Q-value for the decay.

In order to measure the geo-neutrino spectrum, one needs a detector which is capable of measuring the "prompt" energy deposited together by the electron and gammas, $E_{prompt} = T_{e} + E_{ã}$. When measured decay events are displayed as a function of E_{prompt}, by a specular reflection one immediately obtains the number of events as a function of neutrino energy, at $E_{\nu} = Q - E_{prompt}$, see Figure 1 for the decay spectrum of ^{214}Bi.

For such a measurement one needs a detector that can collect the energy lost by both electrons and gammas, and which has a similar response to both electrons and gammas. Essentially, this is a calorimetric measurement. In principle, it can be done with bolometers, which have very good energy resolution but long dead times. A sufficiently large liquid scintillator detector is suitable for such measurements. Although energy resolution is limited, nevertheless it can contain both electrons and gammas and a significant statistics can be collected in a short time.

We suggest to exploit the potential of the Counting Test Facility (CTF), which is operational and available in the underground Gran Sasso Laboratory.

Like Borexino, the CTF design [2] is based on the principle of graded shielding, see Figure 2 and 3. The active scintillation liquid in CTF is a 4 ton mass of pseudocumene enclosed in a transparent nylon sphere, the CTF vessel. Outside this vessel there is a volume of ultra-pure water which is enclosed in a second nylon sphere, the CTF shroud. A set of inward-facing PMTs is arrayed outside the shroud. The entire apparatus, surrounded by another volume of water, is contained in a cylindrical stainless tank. The bottom surface of the tank holds 16 upward-facing PMTs used to tag the tracks of muons passing through the detector.

The facility is equipped with a rod system, which can be used to insert a small, cylindrical quartz vial inside the CTF vessel. A suitable source, dissolved in the liquid scintillator, can be placed in the vial. Electrons are stopped inside the vial and the scintillation light is propagated within CTF through the quartz (which is transparent to the near-UV wavelengths of scintillation light and has index of refraction near that of the scintillator) whereas gamma conversion occurs inside the CTF inner vessel. The inward facing PMTs outside the shroud can thus detect light originating from both electrons and gammas.

We think that CTF, in addition to being available and operational, offers a particular advantage. The detection method which we propose for measuring geo-neutrino decay spectra is rather similar to that which will be adopted by all experiments (KamLAND, Borexino, SNO+...) for detection of geo-neutrino interactions. In particular, Borexino has the same liquid scintillator and the same phototubes as CTF. Some systematic errors will cancel when extracting the geo-neutrino interaction signal in Borexino by using the geo-neutrino decay spectrum measured with CTF.

4. What has to be measured?

Geo-neutrinos with energy above the threshold for (1.1) arise only from the chains of ^{238}U and ^{232}Th.

In particular, for ^{238}U only three nuclides (^{234}Pa, ^{214}Bi, ^{210}Tl) contribute to the geo-neutrino signal. The contribution from ^{210}Tl is negligible, due to its small occurrence probability, and the uranium contribution to the geo-neutrino signal comes from five β decays: one from ^{234}Pa and four from ^{214}Bi. Table 1 lists the effective transitions, i.e., those that can produce antineutrinos with energy above the threshold E_0. In fact, 98% of the uranium signal arises from the two transitions to the ground state (in bold in Table 1) and an accuracy better than 1% is achieved by adding the third one.

^{232}Th decays into ^{208}Pb through a chain of six α decays and four β decays. In secular equilibrium the complete network includes five β-decaying nuclei. Only two nuclides (^{228}Ac and ^{212}Bi) yield antineutrinos with energy larger than 1.806 MeV. The thorium contribution to the geo-neutrino signal comes from three β decays: one from ^{212}Bi and two from ^{228}Ac (see Table 2). In fact, 94% of the thorium signal arises from the transition to the ground state of ^{212}Po (in bold in Table 2).

We remind that, assuming the chondritic ratio for the global uranium and thorium mass abundances, $a(Th)/a(U) = 3.9$, one expects that geo-neutrinos from uranium (thorium) contribute about 80% (20%) of the total U+Th geo-neutrino signal.

In summary:

a) *98% of uranium geo-neutrino signal comes from just two transitions, one from ^{214}Bi and the other from ^{234}Pa. They provide 77% of the expected total U+Th signal.*
b) *A single decay of ^{212}Bi accounts for 94% of the thorium signal. It provides 20% of the expected U+Th signal.*

Just three transitions have to be investigated experimentally. In this respect, the following considerations can be useful:

a) ^{222}Rn ($\tau_{1/2}$ = 3.8 days) can be easily dissolved in the scintillator and the decay of ^{214}Bi is uniquely identified by the subsequent decay of ^{214}Po ($\tau_{1/2}$ = 0.16 ms)
b) By dissolving ^{238}U in the scintillator, one can detect the beta decay of ^{234}Pa (superimposed however with that of ^{234}Th). The subsequent decays of the chain are effectively blocked by the long half-live of ^{234}U ($\tau_{1/2}$ = 245.5 Kyr).
c) For the investigation of ^{212}Bi decay one has to start with a ^{224}Ra source ($\tau_{1/2}$ = 3.7 days) or with a ^{232}Th source. The decay of ^{212}Bi can be easily identified by the subsequent α-decay of ^{212}Po ($\tau_{1/2}$ = 0.3 μs).

An experimental program on these lines has already started at Gran Sasso.

Acknowledgments

We are grateful for enlightening discussions and valuable comments to L. Miramonti, B. Ricci and C. Tomei.

This work was partially supported by MIUR (Ministero dell'Istruzione, dell'Università e della Ricerca) under MIUR-PRIN-2006 project "Astroparticle physics".

References

[1] Fiorentini G, Lissia M and Mantovani F 2007. Geo-neutrinos and earth's interior, *Phys. Rep.*, **453/5-6** 117-172.

[2] Borexino Collaboration, 1998. A large-scale low-background liquid scintillation detector: the counting test facility at Gran Sasso, *NIM*, **406** 411.

Table 1. Effective transitions in the ^{238}U chain, from [1]

i	j	$R_{i,j}$	E_{MAX} [keV]	I_k	ΔI_k	Type		S_U [%]	S_{tot} [%]
$^{234}Pa_m$	**^{234}U**	**0.9984**	**2268.92**	**0.9836**	**0.002**	**1st forbidden 0^-**	**0^+**	**39.62**	**31.21**
^{214}Bi	**^{214}Po**	**0.9998**	**3272.00**	**0.182**	**0.006**	**1st forbidden 1^-**	**0^+**	**58.21**	**45.84**
			2662.68	0.017	0.006	1st forbidden 1^-	2^+	1.98	1.55
			1894.32	0.0743	0.0011	1st forbidden 1^-	2^+	0.18	0.14
			1856.51	0.0081	0.0007	1st forbidden 1^-	0^+	0.01	0.01

Table 2. Effective transitions in the ^{232}Th chain, from [1].

i	j	$R_{i,j}$	E_{MAX} [keV]	I_k	ΔI_k	Type		S_{Th} [%]	S_{tot} [%]
^{212}Bi	**^{212}Po**	**0.6406**	**2254**	**0.8658**	**0.0016**	**1st forbidden 1^-**	**0^+**	**94.15**	**20.00**
^{228}Ac	^{228}Th	1.0000	2069.24	0.08	0.06	Allowed 3^+	2^+	5.66	1.21
			1940.18	0.008	0.006	Allowed 3^+	4^+	0.19	0.04

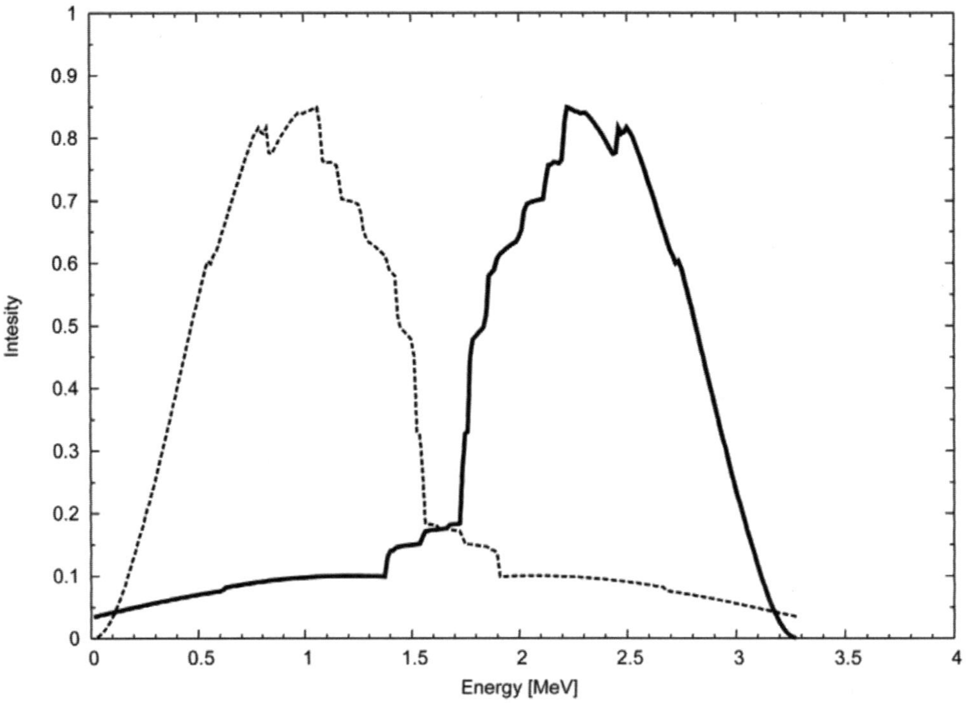

Figure 1. The decay spectrum of ^{214}Bi as a function of the prompt energy E_{prompt} (full line) and of the neutrino energy E_ν (dashed line).

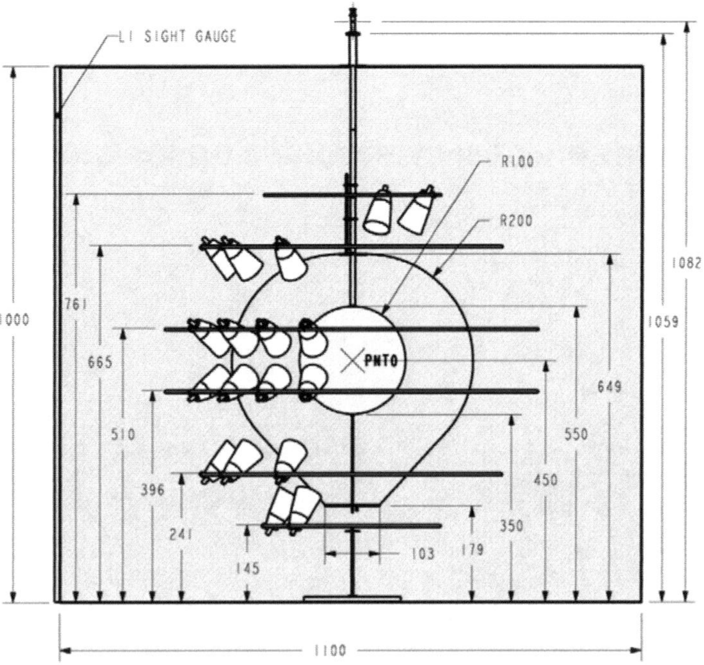

Figure 2. Side view of the design of CTF. The vessel (labelled R100 in this drawing) and shroud (R200) are shown, as well as the six rings of PMTs, the cylindrical tank, and the tubes used for filling and draining the vessel. The point PNT0 is the nominal center of the sphere of PMTs and of the CTF vessel. Dimensions are given in cm. Courtesy of the Borexino collaboration.

Figure 3. A photograph of the CTF viewed from below. Courtesy of the Borexino collaboration..

Low-energy neutrino observation at Super-Kamiokande-III

Yasuo Takeuchi
Kamioka Observatory, ICRR, Univ. of Tokyo,
456 Higashi-Mozumi, Kamioka-cho, Hida-shi, Gifu 506-1205, Japan

E-mail: takeuchi@icrr.u-tokyo.ac.jp

Abstract. Super-Kamiokande-III (SK-III) has been started its observation in July 2006. The main targets of low-energy neutrinos are the solar neutrinos and the diffuse supernova neutrino background. In this paper, the current status of the solar neutrino observation in SK-III is reported.

1. Introduction

Super-Kamiokande-III (SK-III) has been started its observation in July 2006. The photo coverage of the inner detector was recovered to the original design [1], so the energy and the vertex resolutions against the low-energy events came back. In SK-III, we would like to observe the solar neutrinos below 5.0MeV in order to examine the possible distortion of the solar neutrino energy spectrum with high sensitivity. About a 10% increase is expected by the current MSW LMA solution at around 4MeV in SK. Only SK-III could reach to this energy region among existing solar neutrino experiments. The supernova neutrinos are another main subject in low-energy region in SK. We extensively studied the supernova burst neutrinos and the diffuse supernova neutrino background (DSNB) in SK-I and SK-II data. We are also studying on a possible signal tagging for the DSNB events by neutron capture. In this paper, the current status of the low-energy solar neutrino observation in SK-III is reported.

2. The current status of solar neutrino observation in SK-III

In the SK-III detector, water circulation isn't stable yet. So, the data quality isn't enough to measure the solar neutrino flux below 5.0MeV with small enough statistical uncertainties. We have started the data taking with the Super-Low-Energy (SLE) trigger mode [2] since January 24, 2007. The trigger efficiency is almost 100% at 5.0MeV with this SLE mode. In this paper, the first look of the SK-III SLE data is reported. The live time of this SLE dataset is 97 days until May 21, 2007.

As the first step, we have applied preliminary data reduction tools to the SK-III SLE 97days data, then compared the SK-III data with the SK-I final results. In this analysis, the efficiency for the ^{8}B solar neutrino signal of the SK-III final data sample was roughly adjusted to the SK-I analysis. The analysis tools for SK-III are still under improvements. Figure.1 shows the data reduction steps of SK-III and SK-I. In SK-III, the low-energy background from the detector wall, probably due to FRP, is increased. But, they can be easily reduced to the SK-I level by the pre-ambient BG cut, because the quality parameter of the event reconstruction is very low for the increased background events, comparing to the real solar neutrino events. The event rate in the final data sample in the 22.5kton fiducial volume is almost similar between SK-I and SK-III. Figure 2 shows angular distributions of SK-III and SK-I. The back ground event rate in 6.5-8.0MeV in SK-III is higher than SK-I. We think this is due to ^{16}N events which aren't cut in SK-III, while they are reduced in SK-I. The ^{16}N events will be also removed in SK-III. In 22.5 kton fiducial volume, SK-III has a similar data quality as SK-I.

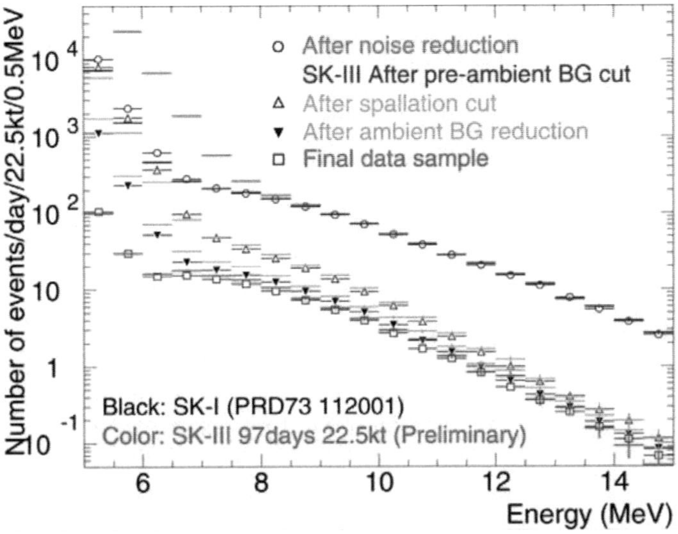

Figure 1. The reduction steps of SK-III (colour) and SK-I (black) solar neutrino data in the 22.5kton fiducial volume. The noise reduction is a simple cut to remove obvious non-physics events. The spallation cut remove the events originated from cosmic-ray muons. The ambient BG reduction removes bad quality events. In the final data sample, external gamma-ray events are removed. After the noise reduction, low-energy event rate of SK-III is higher than SK-I due to the background events from detector wall. These are removed by the pre-ambient BG cut only in SK-III.

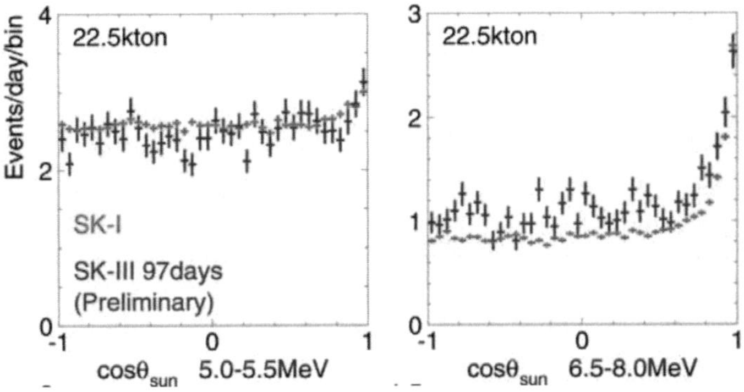

Figure 2. The angular distributions of SK-III (blue) and SK-I (red) in the 22.5kton fiducial volume. The θ_{sun} is defined as the angle between the direction from the Sun and the direction of the event. So, the peak around $\cos\theta_{sun}$~1 is solar neutrinos. The higher event rate in 6.5-8.0MeV in SK-III are probably due to ^{16}N events, which aren't cut yet in SK-III while they are reduced in SK-I

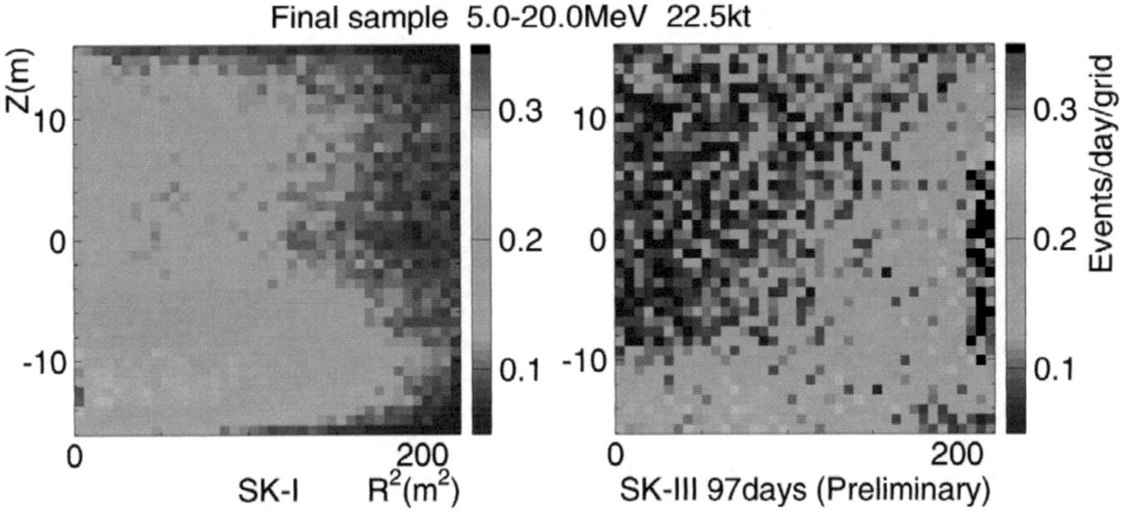

Figure 3. The vertex distribution of the solar neutrino final samples of SK-I and SK-III. See [2] for the definition of the detector $R^2=X^2+Y^2$ (horizontal) and Z (vertical). Both SK-I and SK-III rates in R>~10m are reduced by the same external event cut. The event rate of SK-III in the central-top region is lower than that of SK-I. The high rate around the bottom in SK-I is due to Rn from the water circulation [3].

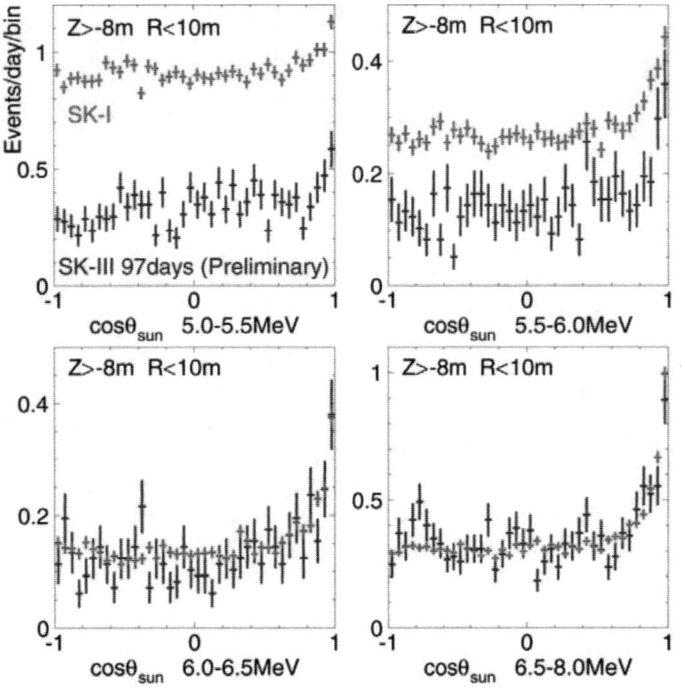

Figure 4. The angular distributions of SK-III (blue) and SK-I (red) in the central-top region. The definition of θ_{sun} is same as Fig. 2. The vertical axis is the observed event rate in the final data sample. In the 5.0-5.5MeV region SK-III background event rate is lower than SK-I, while the signal rate looks similar. We think this is mainly due to improvement of the water quality in SK-III.

Figure 3 shows the vertex distribution of the solar neutrino final data samples. The event rates of SK-III around detector barrel (R^2>~$200m^2$) and bottom (Z<~ -10m) are higher than those of SK-I. We think these are mainly due to the remaining low-energy background events from the detector wall. We are now improving the data reduction tools against these background events in SK-III.

However, the event rate in the central-top region of SK-III is lower than that of SK-I, from Figure. 3. Figure 4 shows angular distributions of this central-top region. From Figure 4, only the background rate of SK-III is lower than SK-I while the solar neutrino signals are similar, especially in 5.0-5.5MeV energy region. This would mean the background of SK-III in low-energy region is reduced. We think this is mainly due to the water quality improvement in SK-III. We have added an additional Reverse Osmosis (RO) device in the main flow of the water purification system during SK-I. In general, RO removes small particles efficiently. So, it also removes Radium in the particles from water. This RO was used for the initial purified water filling of SK-III, but not used for the filling of SK-I. We are now trying to enlarge this low-background region by stabilizing water circulation in the SK detector, applying improved analysis tools, and so on.

3. Conclusion

The Super-Low-Energy data in SK-III has been taken since January 2007, with almost 100% trigger efficiency at 5.0MeV electron total energy threshold. As the first look of the SLE dataset above 5.0MeV in SK-III, the background rate in the lowest energy region in the central-top area is lower than that of SK-I. We are trying to enlarge the low-background region, and then measure the solar neutrino flux below 5.0MeV.

References

[1] Super-Kamiokande collaboration 2003 *Nucl. Instr. and Meth.* **A501**, 418
[2] Super-Kamiokande collaboration 2006 *Phys. Rev.* **D73**, 112001
[3] Super-Kamiokande collaboration 1999 *Phys. Lett.* **B452**, 418

News from the Sudbury Neutrino Observatory

Christine Kraus
Department of Physics, Queen's University, Kingston, ON, K7L3N6, Canada
E-mail: tine@owl.phy.queensu.ca

Abstract. The Sudbury Neutrino Observatory (SNO) [1] is a 1000 tonne heavy water Cerenkov detector, that has finished its data taking period on November 29th 2006. The heavy water was contained in a 12 m diameter acrylic sphere, surrounded by 9500 PMTs. It is located in INCOs Creighton mine in Sudbury, Ontario, Canada - 2039 m (6000 mwe) underground. The so-called third phase of the experiment had in addition to the heavy water an array of 36 ^{3}He and 4 ^{4}He proportional counters installed. In this configuration it can detect neutrons produced by the neutral current interaction of solar neutrinos by neutron capture in these counters (NCDs = neutral current detection array). This is a way to determine the neutral current flux independent of the PMT system with very different systematic uncertainties and is therefore determining the CC and NC reactions in separate streams.

1. Introduction

The SNO experiment detects ^{8}B solar neutrino reactions in three different reactions:

$$\begin{array}{rcll} \nu_e + d & \rightarrow & p + p + e^- & \text{(CC)} \\ \nu_x + d & \rightarrow & p + n + \nu_x & \text{(NC)} \\ \nu_x + e^- & \rightarrow & \nu_x + e^- & \text{(ES)} \end{array} \tag{1}$$

The Charge Current (CC) interactions are only possible for electron type neutrinos, while the Neutral Current (NC) and Elastic Scattering (ES) reactions are sensitive to all flavours. The Elastic Scattering reaction is reduced due to cross section. The three different phases of the SNO experiment use three different ways to determine the NC flux from the ^{8}B neutrinos from the sun or in other words three different ways to detect neutrons. The first phase is just using with pure heavy water, where the neutron captures on deutrium and produces a single gamma, that is detected by the PMTs. The second phase had additional 2000 kg of salt (NaCl) introduced into the detector. The NC reaction therefore took place on ^{35}Cl. This has two advantages, the neutron capture efficiency is significantly larger and multiple gammas are produced at a higher energy. Results from these phases can be found here [2],[3]. For the third and final phase neutrons were detected in proportional counters, filled with ^{3}He as active gas on an event-by-event basis. An array of ^{3}He proportional counters (Neutral Current Detection array, or NCD) consisting of 36 strings was deployed. In additon 4 strings of counters filled with ^{4}He were installed. These strings are insensitive to neutrons and can therefore be used to study backgrounds in form of alpha particles. The detection in the NCDs is based on:

$$^3\text{He} + n \rightarrow p + {}^3\text{H} + 0.76 \text{ MeV} \tag{2}$$

A detailed description of the NCD array can be found here [4]. As in previous phases, the PMT system (9500) is recording signals from CC, NC and ES reactions. Therefore the measurements of the CC signal (PMT) and NC signal (NC) are decoupled. The fraction of NC signal observed by the PMTs is significantly lower than the ones detected with the NCDs.

2. Neutron calibrations

For the calibration of the detector, these 40 discrete counters are a new challenge. The frequency and diversity of neutron calibrations has been increased significantly compared to earlier phases of the SNO experiment. The complexity of neutron calibration analyis for the third phase has also increased and is necessary to ensure, that the neutron detection efficiency and detector response can be determined with great accuracy. Between November 27, 2004 and November 29, 2006 several calibration sessions with various sources were performed. A manipulator system which allows a source to be positioned in two perpendicular planes is used to deploy optical, γ and neutron sources. We used various neutrons sources with different rates and characteristics - ^{252}Cf, 3 different ^{241}Am^{9}Be, ^{232}Th. In addition we introduced distributed soures in form of activated ^{24}Na brine and radon into the detector. Activated ^{24}Na ($t_{1/2}$=14.959 h) mixed into the heavy water produced neutrons via deuterium photodisintegration induced by the 2.754 MeV gammas emitted with the ^{2}4Na beta decay. The following graphs show the distribution of the sodium inside the detector in both systems. These data are used to evaluate the neutron capture efficiency and the related uncertainties. The second path for these studies is closer to previous phases and is based on using point sources. In addition the energy spectrum for neutrons is also taken from the sodium spike data.

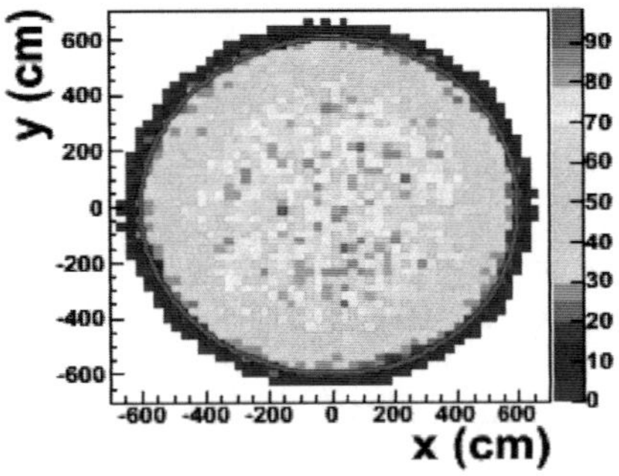

Figure 1. Distribution of ^{24}Na as seen by the PMT system.

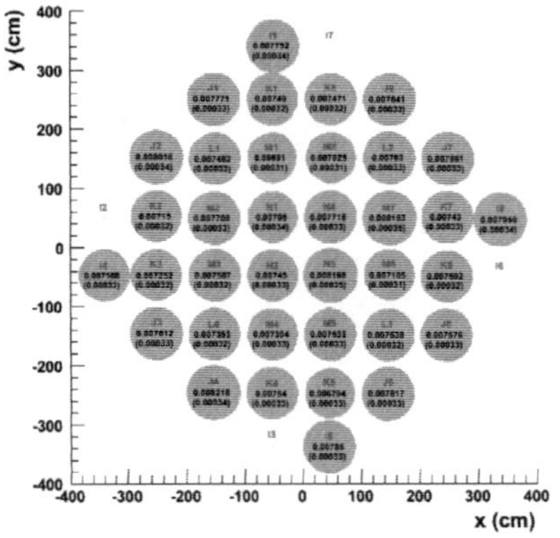

Figure 2. Distribution of ^{24}Na as seen by the NCDs.

3. Analyis method

Signals from the NCDs are digitized and recorded by scopes. The recorded pulse shapes can be used to distinguish between neutrons, alphas and discriminate from instrumental backgrounds, including discharges and oscillatory noise. Several methods have been developed and improve the signal to background ratio significantly. They also alter the remaining alpha energy spectrum especially under the neutron peak, where data are statistically limited by having only 4 ^{4}He strings. The analysis group has investigated several pulse shape discrimination techniques. The analysis is blinded. This means a fraction ($< 30\%$) of the final data set is hidden and in addition an unknown admixture from neutron events following muon interactions is inserted into the data set. The analysis procedures are developped and tested on this blind data set. Using neutron calibration data from the mixed sodium source as the neutron energy spectrum and then performing a fit to this signal plus an underlying background will determine the detected number of neutrons. Detection efficiencies, threshold efficiencies need to be folded in and the known neutron background subtracted to finally derive the number of signal neutrons, that will be used to determine the NC flux. The following figure demonstrates the approach to the analysis.

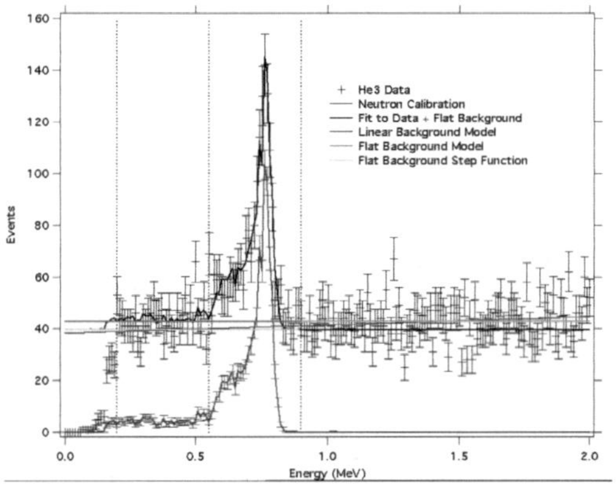

Figure 3. Spectrum for the blind data set. The neutron spectrum from sodium and various approaches for background estimates are shown as well.

4. Outlook

In parallel the analyis group is working on combining the first and second phase with a significantly lower energy threshold - 3.5 MeV instead of 5.5 MeV. This requires a sophisticated understanding of the low energy background events, that had been excluded from the analysis for previous publications. These results will be published soon.

5. References

[1] SNO Collaboration, Nucl. Instr. and Meth.**A449, 172 (2000)**

[2] SNO Collaboration, Phys. Rev. C75, 045502 (2007)

[3] SNO Collaboration, Phys. Rev. C72, 055502 (2005)

[4] J.F. Amsbaugh *et al.* **Nucl. Instr. and Meth. A579, 1054-1080 (2007)**

KamLAND Status for Solar Neutrino Phase

Yasuhiro Kishimoto for KamLAND collaboration

Research Center for Neutrino Science, Tohoku University, Aramaki Aoba-ward Sendai-city, Japan, 980-8578

E-mail: kisimoto@awa.tohoku.ac.jp

Abstract. KamLAND has been trying to reduce backgrounds to observe neutrinos from the Sun. The first activities for the ^{7}Be neutrino observation was performed form the spring to the summer of 2007. We report the status of the purification campaign and also describe studies for CNO neutrino measurement.

1. Introduction

The Sun is a well-known intensive source of neutrinos, which are continually produced in the core of the Sun by nuclear reactions. The studies of the neutrino flux have been one of the most exciting research field in elementary and astroparticle physics and have provided us with evidence for the neutrino flavor oscillation (see, *e.g.* [1], [2] and references therein). And the ^{7}Be neutrino has been observed by Borexino collaboration [3]. So, the solar neutrino has gained more attention. The KamLAND collaboration plans to observe ^{7}Be neutrino with a large liquid scintillator detector, KamLAND, and started to purify the liquid scintillator to reduce backgrounds. R&D for future CNO neutrino measurement in KamLAND are also now going. These status and future prospects are reported in this paper.

2. KamLAND

KamLAND is the large-sized underground detector, containing 1000t of liquid scintillator surrounded with 1879 PMTs[4]. The scintillator is made of pseudo-cumene, dodecane, and PPO. With this detector, solar neutrinos are planed to be measured via electron scattering. But the cross section is so small that lower background levels are required. Main backgrounds against the neutrino measurements are ^{85}Kr, ^{210}Bi(^{210}Pb daughter), and ^{40}K. Other radioactive sources, such as ^{238}U and ^{232}Th, are low enough even before the purification. The background sources and their radioactivities are listed in Table 1.

3. Purification System

To reduce the backgrounds, two methods are adopted: One is distillation for metallic or ionic chemicals, such as ^{40}K. The other is purging for gaseous chemicals, such as ^{85}Kr. These methods ware studied well with small and medium sized glassware. Experimental results showed very good performance for our purpose[5]. With these encouraging results, we have developed purification apparatus in Kamioka mine, which consists of two storage tanks, three sets of distillation columns, and purge columns with an ultra-pure N_2 generator. The three sets of distillation columns are for the three chemical components of KamLAND liquid scintillator.

Table 1. Background activities in KamLAND before and after the purification. We have two layers in KamLAND liquid scintillator after the purification because of our operation as described in the text, so the concentration of the upper ($z > 4$m) and lower part ($z < 2$m) are shown individually. The fiducial volume before the purification is $r < 4$m for ^{210}Bi and ^{85}Kr, and $r < 3.5$m, $\rho > 1.2$m for ^{40}K. That after the purification of is $r < 5$m for ^{210}Bi and ^{85}Kr, and $r < 3.5$m for ^{40}K. No defined value of ^{40}K activity in the upper region is calculated because of small statistics in the small upper volume. ^{222}Rn concentration is measured within $r<$ 6m during the purification. Required reduction factors are also shown.

Isotope	Before [Bq/m^3]	After (Upper) [Bq/m^3]	After (Lower) [Bq/m^3]	Required $[\frac{\text{After}}{\text{Before}}]$
^{210}Bi	$(4.2^{+0.8}_{-0.6}) \times 10^{-2}$	$(2 \pm 1) \times 10^{-4}$	$(1.0 \pm 0.1) \times 10^{-2}$	10^{-6}
^{40}K	$(4.4 \pm 0.4) \times 10^{-5}$	NA	$(1.3 \pm 0.1) \times 10^{-5}$	10^{-6}
^{85}Kr	$(5.1^{+0.2}_{-0.4}) \times 10^{-1}$	$(1.4^{+0.1}_{-0.2}) \times 10^{-2}$	$1.85^{+0.01}_{-0.02} \times 10^{-1}$	$10^{-5\sim-6}$
^{222}Rn	2.8×10^{-8}	1×10^{-4}	1×10^{-4}	$< 10^{-3}$

Each distillation column distills a pure chemical, depending on its boiling point. So the liquid scintillator is distilled separately, then mixed together into liquid scintillator again. The distilled liquid scintillator is purged with ultra pure N_2. The construction of the apparatus was accomplished in 2006, followed by shakedown and parameters' tuning in early 2007.

4. Purification in 2007

In May in 2007, we started the purification campaign. KamLAND liquid scintillator was drawn out from KamLAND bottom plug to be purified with the apparatus. Purified scintillator was then filled into KamLAND from the top. The purification operation was very safe and stable. During the purification, we conducted various quality checks of the liquid scintillator, such as transparency and light yield measurements. Details of these measurements are discussed in elsewhere (see related KamLAND papers in this proceeding). The purification was paused on August 1st because there started blasting in the Kamioka mine to construct new experimental laboratories. The total amount of purified liquid scintillator is 1699m^3, it corresponds to 1.4 times of the whole KamLAND liquid scintillator volume. During the purification, reduction factors of the radioactive concentrations were found to be not so good as expected from the test experimental results. One reason of this was that original and purified scintillator started to be mix together, which was observed by tracing movement of ^{222}Rn daughters through ^{214}Bi-^{214}Po coincidence events. Other possibility was that the reduction factor itself in our apparatus was less than our expectation. To deal with this issue, we changed the density and the temperature of the last 173m^3 of purified liquid scintillator: $\Delta\rho = -0.03\%$ and $\Delta T = +1.0$K, respectively, to have more distinct separation between new and old scintillator with the density and temperature difference. Separation of old and new liquid scintillators became more clear after these changes. The spectrum of the upper and the lower part are presented in Figure 1.

The concentrations are shown in Table 1.

5. Next Purification Campaign

The radioactivities after the first purification is not yet sufficient to observe ^{7}Be solar neutrinos. KamLAND plans to start purification after some upgrade of the apparatus. A new temperature controller will be installed in the pipe line. More stringent temperature control with this device and much more careful adjustment of the liquid scintillator density will keep the new and old liquid scintillator boundary. We also will again search purification parameters to reduce radioactive chemicals much more effectively and use their values in the next purification. We are going to restart the purification activities in the Spring of 2008.

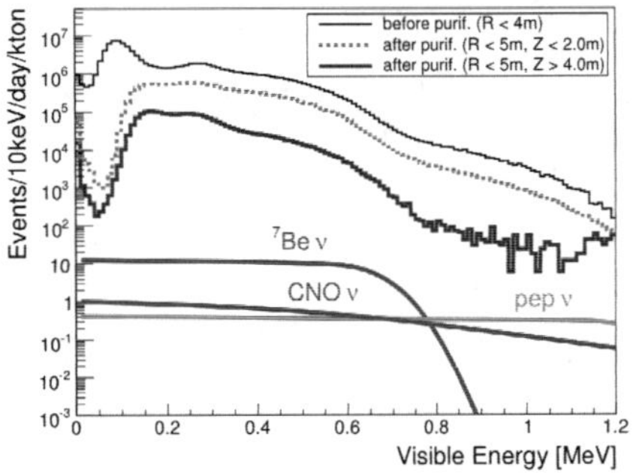

Figure 1. Spectrum in KamLAND before and after the purification. The spectrum before the purification are presented by a thin solid line. The spectrum of the upper (z> 4m) and the lower (z< 2m) part of r < 5m by a solid and broken line, respectively. Expected neutrino signals (^{7}Be, CNO, and *pep*) are also shown.

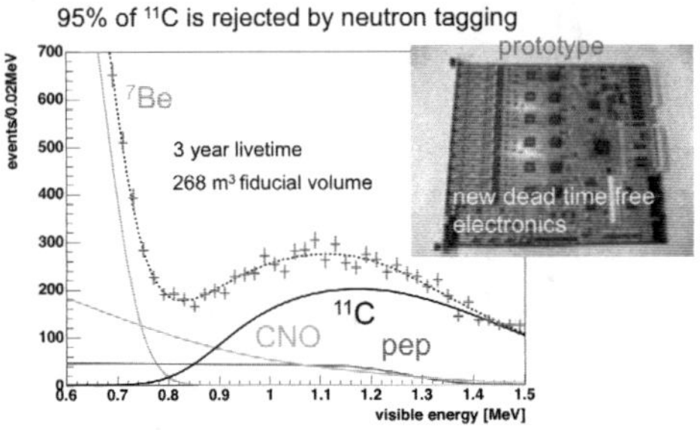

Figure 2. Expected spectrum of CNO neutrino after three years lifetime and 268m^3 target volume. 95% reduction of ^{11}C with the triple coincident are assumed. A snapshot of the prototype of an electronics are also added. The prototype has been developed to record high multiplicity events after μ passing through KamLAND and to utilize the ^{11}C rejection.

6. Toward CNO neutrino measurement

Measurement of CNO neutrinos is one of physics targets of the future KamLAND. Once background level is low enough to observe ^{7}Be neutrinos after the purification, dominant background for CNO neutrinos will be ^{11}C, which is produced by the interaction of a cosmic muon. The interaction of muon in liquid scintillator is well studied[6], and the reaction rate in KamLAND can be estimated. 95% of ^{11}C is produced in the reaction, $^{12}\mathrm{C} + X \rightarrow {}^{11}\mathrm{C} + n + Y$, where $X = \mu$, n, p, π, e, γ, and so on. So we can reject ^{11}C events by triple coincidence of μ, n, and ^{11}C. To utilize this triple coincidence as a rejection method, we have been developing a new front end electronic to measure the high multiplicity reaction with high efficiency. A prototype of this electronics was developed and shows good performance. Expected spectrum is shown in Figure 2. We can expect to measure $\mathrm{CNO}\,\nu + pep\,\nu$ with about 6% statistical error after three years lifetime in 268m^3 fiducial volume, assuming 95% rejection of ^{11}C[7].

7. Summary

The first purification of KamLAND liquid scintillator has been accomplished on August 1st. The reduction in KamLAND radioactivities are determined. We are going to purify KamLAND liquid scintillator in the Spring of 2008 after some upgrade of the apparatus to observe ^{7}Be solar neutrino. The new electronics have also being developed in parallel for CNO neutrino observation.

References

[1] Bahcall J. N. 1986 *Neutrino Astrophysics* (Cambridge U. Press, Cambridge, England)
[2] Fogli G. L. *et al.* 2006 *"Solar neutrinos"*, Proceedings of *NO-VE 2006*, p.69
[3] Arpesella C. et al. 2007 *Phys. Lett. B* PLB-D-07-00772R2
[4] Araki T. *et al.* 2005 *Phys. Rev. Lett.* **94** 081801
[5] Kishimoto Y. 2005 *"Liquid Scintillator Purification"* Proceedings of LRT2004, p.193
[6] Hagner T. *et al.* 2000 *Astroparticle Phys.* **14** 33
[7] Shimizu I. 2006 *"Neutrinos from the Sun: Observation"* Slide in the 4th COE Symposium of Exploring New Science by Bridging Particle-Matter Hierarchy, Tohoku Univ.

The U.S. National Science Foundation's Deep Underground Laboratory at Homestake- DUSEL

Kevin T. Lesko

Institute for Nuclear and Particle Astrophysics, Lawrence Berkeley National Laboratory, 1 Cyclotron Road, MS 50R5008, Berkeley, CA 94720-8158, USA
Email: ktlesko@lbl.gov

The US National Science Foundation has launched a program to establish a world-class, multi-disciplinary deep underground science and engineering laboratory - DUSEL. The NSF has selected as the prime site to be developed for an international world-class research facility. Recently there has been increasing interest in underground physics, astrophysics and other fields of science that require deep underground laboratory and research facilities. I will review the scientific case for DUSEL and the near-term experiments anticipated in Homestake, as well as presenting the efforts to create the Sanford Laboratory at Homestake in advance of DUSEL.

One of the most compelling questions in 21st century science is the mass and make-up of our universe. We now know that its large fraction is made of unknown material. Leading theoretical conjecture points to a weakly interacting massive particle or WIMPs. Recent advances in detector technology highlight the need for appropriately outfitted space in a deep laboratory to pursue the next generation of detectors whose mass will likely surpass 1 ton and which will require exquisite shielding from cosmic rays at depths of over 7000 feet below ground. We need to prepare sites which has to be sheltered by the earth's crust from the background cosmic rays and human activity for exquisitely sensitive particle physics and astrophysics experiments searching for the subtle but unmistakable signatures of a revolutionary new physics of the universe.

The study of neutrinoless double beta decay has made significant advances in technology. Coupling these advances with the increase in knowledge of the neutrino mass states and mixing angles highlights the need for next-generation experiments with detector masses, again, approaching 1 ton. These large-scale experiments will require excellent support and appropriately outfitted laboratory space below 7000 feet below ground.

Both neutrinoless double beta decay and dark matter experiments are being approached in a phased manner, with intermediate mass prototype detectors. Homestake's 4850 Level (feet below ground) and the laboratory at the7400 Level are well suited to this phased approach and in providing the essential experimental requirements for these efforts.

Long baseline neutrinos experiments will require massive detectors, with masses over 100 ktons. These large cavities will present engineering challenges. Producing these at depth enables parallel scientific programs, in the same detector, of nucleon decay searches, supernovae detections, and even atmospheric and solar neutrino detection.

Nuclear astrophysics efforts, centered around a modern high-current accelerator will be a showcase experiment for the nuclear astrophysics community.

Biologists probe the secrets of microbial life at extreme depths, in hot, harsh environments sequestered for millennia from the earth's surface. Geoscientists and engineers research the behavior of subsurface rock, minerals, water and energy sources.

A new Deep Science Initiative, structured around DUSEL would extend the frontiers of particle physics and astrophysics, biology, geoscience and engineering–and foster the synergies among them. This Deep Science Initiative and DUSEL will yield discoveries about the fundamental nature of our own planet, about the life that it harbors and about the universe that is its home. It would contribute strongly to the basic science. It would provide unique opportunities for innovation in underground technology, with immediate and long-term applications. A Deep Science Initiative would address the need to sustain research in fundamental and applied science and to educate, train and inspire the next generation of scientists and engineers. Fig.1 shows conceptual design of DUSEL at Homestake. The need of DUSEL is well documented with many reports and studies. The NSF has adopted a careful thorough review of sites and recently selected Homestake as the site to focus their efforts in creating the facility. The Homestake Collaboration is currently working to developing plans for DUSEL at Homestake in South Dakota.

I shall discuss the multidisciplinary research efforts identified in the nearer term for Homestake Interim Laboratory and the Initial Suite of Experiments being discussed by the NSF and its committees [1].

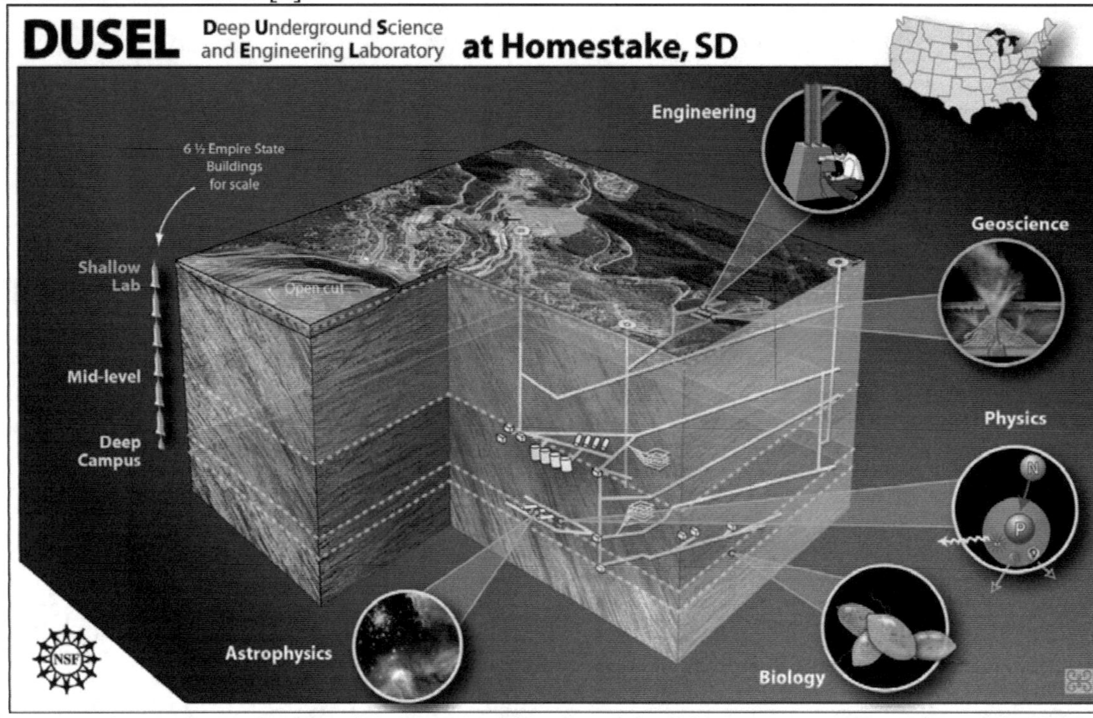

Figure 1. The Deep Underground Science and Engineering Laboratory at Homestake. This figure presents the conceptual design DUSEL in the former Homestake goldmine in South Dakota.

References

[1] http://www.dusel.org/

SAGE: status and future

V N Gavrin
Institute for Nuclear Research RAS,
60th October Anniversary prospect, 7a, Moscow 117312, Russia

E-mail: gavrin@dionis.iasnet.ru

Abstract. Present status, current results, and plans of SAGE for immediate future are discussed.

1. Introduction

The gallium experiments gave a great impact upon a view of neutrino oscillation and have supplied most important motivation for creation of SNO experiment whose result led us to solution of the solar neutrino problem. SAGE is currently the only operating solar neutrino experiment which provides the determination of the fundamental *pp* neutrino flux [1] and a continuous monitoring of the low energy solar neutrino flux with increasing sensitivity over very long time period.

2. Gallium results

SAGE is running with about 50 tons of metal gallium at Baksan Neutrino Observatory INR RAS [2] since January 1990. The global best fit capture rate for all SAGE data from January 1990 through December 2006 (157 runs and 288 separate counting sets) is $66.2^{+4.8}_{-4.5}$ SNU, where the statistical and systematic uncertainties are combined in quadrature.

The final result from 123 runs in the Gallex/GNO experiment is 69.3±5.5 (stat + syst) SNU [3]. The weighted combination of both Ga experiments, SAGE and Gallex/GNO, is thus

67.6±3.7 SNU. Present Ga experiment result.

The good agreement between results of the Ga experiments has led to increased confidence in the measured event rates. It was very good that for many years there were two Ga experiments operating at the same time and it is indeed unfortunate that the GNO experiment was terminated for non-scientific reasons.

3. Source experiments

The experimental procedures of both Ga experiments have been checked by exposing the gallium target to reactor-produced neutrino sources whose activity was close to ~ 1 MCi. Gallex has twice used ^{51}Cr sources [4]; SAGE has irradiated their target with a ^{51}Cr source and a ^{37}Ar source [5, 6]. The published results, expressed as the ratio R of the measured ^{71}Ge production rate to that expected due to the source strength, are shown in Figure 1. The weighted average value of the ratio for the four experiments is R = 0.88±0.05, more than two standard deviations less than unity.

The most likely explanation is that the cross sections for neutrino capture to the lowest two excited states in ^{71}Ge, both of which can be reached using either ^{51}Cr or ^{37}Ar sources, have been overestimated [7]. If the contribution of these two excited states to the predicted rate is set to zero, then R=0.93±0.05, reasonably consistent with unity. Recent reanalyzing of the data of Gallex two source experiments yields for the ratio R = 0.90±0.09 [8]. Since other auxiliary tests, especially the ^{71}As experiment of Gallex, have given great confidence in the knowledge of the various efficiencies in the Ga experiments, it lead us to conclusion that the source experiments on Ga have measured the contribution of the first two excited states to the ^{71}Ga capture cross section and found this contribution to be rather small. A new experiment with considerably higher rate from the neutrino source is discussed now in SAGE to settle this question.

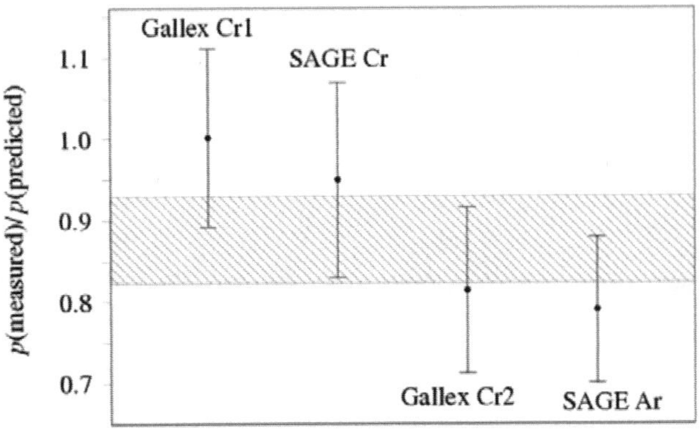

Figure 1. Results of all neutrino source experiments with Ga. The hashed region is the weighted average of the four experiments. See [6] for details.

4. Comparison of Ga result to theory

4.1. The solar neutrino capture rate in ^{71}Ga

For comparison to theory, the fluxes at the Sun from two standard solar models with differing composition [9] have been calculated taking into account neutrino propagation from the Sun to the Earth and the results of the neutrino source experiments with gallium [10]. There is an excellent agreement between the calculated ($67.3^{+3.9}_{-3.5}$ SNU) and observed (67.6±3.7 SNU) capture rate on ^{71}Ga.

4.2. The pp neutrino flux

Using the combined capture rate from the SAGE and Gallex/GNO experiments given above (67.6±3.7 SNU) and the data from Cl, SNO, and SK experiments an electron neutrino *pp* flux of $(3.41^{+0.76}_{-0.77})\times10^{10}/(\text{cm}^2\ \text{s})$ has been calculated. For comparison, the standard solar model calculates the *pp* flux produced in the Sun to be $(5.94\pm0.01)\times10^{10}/(\text{cm}^2\ \text{s})$ [9]. If we multiply this rate by the average survival probability for *pp* neutrinos, which from [10] is 0.555 ($1^{+0.038}_{-0.040}$), we obtain a *pp* flux at the Earth of $(3.30^{+0.13}_{-0.14})\times10^{10}/(\text{cm}^2\ \text{s})$, in excellent agreement with the value determined above from solar neutrino experiments.

5. Is the neutrino capture rate in Ga constant?

The possibility of variability over longer time periods has been considered by several authors [11,12,13,14]. In a plot of the data there appears to be a difference between early and late time periods, which gives a visual hint of a long-term decrease, as illustrated in Figure 2. The Gallex-GNO data are shown on the left and SAGE data on the right of Figure 2. The average rate prior to 1997 is higher in both experiments than in the data after 1997.

If one assumes the rate in Gallex-GNO varies linearly in time then the best fit gives [3] capture rate = 82±10 – (1.7±1.1) × [t(year) – 1990]. These trend lines are plotted for both experiments in Figure 2 and there is reasonably good visual agreement with the measured data.

When examined quantitatively, however, the evidence for long-term variability becomes less convincing. A χ^2 test applied to the Gallex-GNO data with (without) the assumed time variation yields χ^2/dof = 10.8/5 (13.2/6), prob. = 5.6% (4.0%), i.e., the fit to both the time-varying rate and to a constant rate is more or less equally bad. For the SAGE data the fit to a constant rate gives

χ^2/dof = 11.7/16, prob. = 76%, whereas the fit to the central Gallex-GNO trend line yields χ^2/dof = 11.4/17, prob. = 83%, i.e. the fit to both rate hypotheses is quite good.

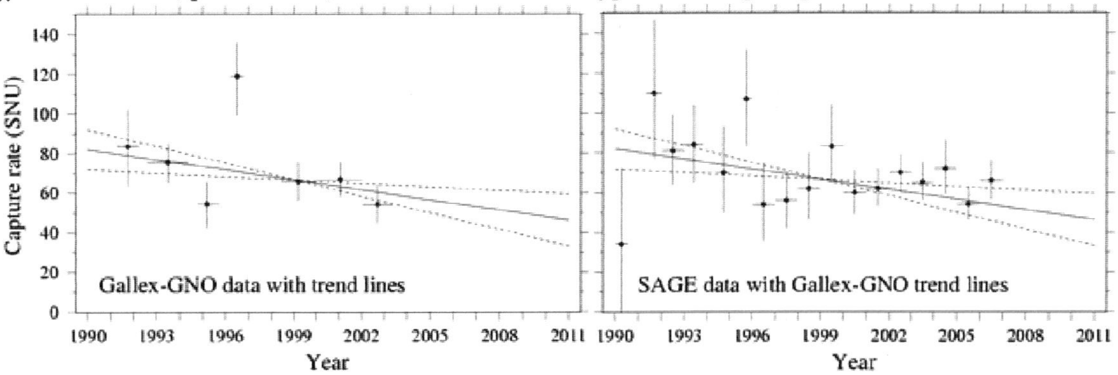

Figure 2. Gallex-GNO results (left panel) and SAGE results (right panel) vs time.

Up to now it is not known if this apparent variability is a statistical fluctuation or an indication of a real effect, such as has been considered by Pulido *et al.* [15]. At the present time we cannot differentiate between these two hypotheses, but it should become possible to do so with considerable additional data.

SAGE plans to continue running for at least nearest three years. We also consider a possibility of creation of a ^{51}Cr neutrino source with activity ~ 2 MCi to measure capture rate of neutrinos from this source on ^{71}Ga with accuracy ~ 4%.

Acknowledgments

It is a pleasure to thank the Organizers of this conference for warm hospitality and support. SAGE Collaboration appreciates the long-term strong support of the Russian Ministry of Science, the Russian Foundation of Basic Research, the U.S. National Science Foundation and the U.S. Department of Energy.

References

[1] SAGE Collaboration, Abdurashitov J N 2002 *J. Exp. Theor.Phys.* **95** 181; *Zh.Eksp.Teor. Fiz.* **122** 211 (*Preprint* astro-ph/0204245).

[2] SAGE Collaboration, Abdurashitov J N 1999 *Phys.Rev.* C **60** 055801 (*Preprint* astro-ph/9907113).

[3] Gallex Collaboration, Altmann M 2005 *Phys.Lett.* B **616** 174 (*Preprint* hep-ex/0504037)

[4] Gallex Collaboration, Hampel W 1998 *Phys.Lett.* B **420** 114

[5] Abdurashitov J N *et al.* 1999 *Phys.Rev.*C **59** 2246

[6] Abdurashitov J N *et al.* 2006 *Phys. Rev.* C **73** 045805 (*Preprint* nucl-ex/0512041)

[7] Haxton W C 1998 *Phys. Lett.* B **431** 110 (*Preprint* nucl-th/9804011)

[8] Hampel W (private communication); Kirsten T (see these Proceedings)

[9] Bahcall J N and Pinsonneault M N 2004 *Phys.Rev.Lett.* **92** 121301 (*Preprint* astro-ph/0402114)

[10] Gavrin V N and Cleveland B T 2006 (*Preprint* nucl-ex/07003012)

[11] Pandola L 2004 *Astropart. Phys.* **22** 219 (*Preprint* hep-ph/0406248)

[12] Sturrock P A, Caldwell D O and Scargle J D 2006 *Astropart.Phys.***26** 174 (*Preprint* hep-ph/0409064)

[13] Strumia A, Cattadori C, Ferrari N and Vissani F 2002 *Phys.Lett.*B **541**, 327 (*Preprint* hep-ph/0205261)

[14] Cattadori C, Ferrari N and Pandola L 2005 *Nucl.Phys.* B **143** 3

[15] Pulido J, Chauhan B and Picariello M 2006 (*Preprint* hep-ph/0611331)

Retrospect of GALLEX/GNO

Till Kirsten

Max-Planck-Institut für Kernphysik, Saupfercheckweg 1, 69117 Heidelberg, Germany

E-mail: till.kirsten@mpi-hd.mpg.de

Abstract. After the completion of the gallium solar neutrino experiments at the Laboratori Nazionali del Gran Sasso (GALLEX, GNO), we shortly summarize the major achievements. Among them are the first observation of solar pp-neutrinos and the recognition of a substantial (40%) deficit of sub-MeV solar neutrinos that called for ν_e transformations enabled by non-vanishing neutrino masses. We also inform about a recent complete re-analysis of the GALLEX data evaluation and reflect on the causes for the termination of GNO. From our gallium data we extract the e-e survival probability P_{ee} for pp-neutrinos after subtraction of the ^{8}B and ^{7}Be contributions based on the experimentally determined ^{8}B- (SNO/SK) and ^{7}Be- (Borexino) neutrino fluxes as P_{ee}(pp only) = 0.52 ± 0.12.

1. Introduction

The Gallium solar neutrino experiments at the Laboratori Nazionali del Gran Sasso have been terminated for external non-scientific reasons. This triggers a short retrospect of the achievements of GALLEX and GNO (Section 2). In Section 3 we report on a recent update that is based on data that were impossible to acquire before completion of the low rate measurement phase (solar runs). After some reflections on the causes for the termination of the gallium experiments at Gran Sasso (Section 4), I give a first quantitative estimate of the separate pp solar neutrino production (Section 5). This ultimate goal of GALLEX can be approached after the first results from BOREXINO allow subtracting the ^{7}Be-neutrino contribution from the gallium results.

2. Retrospect of GALLEX /GNO

GALLEX and GNO have recorded solar neutrinos with energies above 233 keV via the inverse EC reaction $^{71}Ga(\nu_e,e^-)^{71}Ge$ in a 100-ton gallium chloride target (containing 30.3 tons of gallium) between May 1991 and April 2003 [1][2][3]. The key features are summarized in Figure 1. Of utmost importance is the low energy threshold. 98+ % of all solar neutrinos are sub-MeV (pp: ~91%, ^{7}Be: ~7%). The pp-neutrino flux is coupled to the solar luminosity. It is a fundamental parameter that must definitely be measured, as precisely as possible. Stringent limitations (or observation) of departures from the standard solar model can be obtained if the flux of pp-neutrinos is deduced, provided that, as it is now the case, neutrino properties are properly folded in. Below ~1-2 MeV, the vacuum oscillation domain takes over from the matter oscillation domain at > 2 MeV.

The milestones and key references are given in Table 1 (PLB = Physics Letters B). The solar neutrino recordings cover more than a full solar cycle (1991-2003), with a break in 1997. In addition, very important demonstrations of the reliability of the radiochemical method in general and of the GALLEX detector in particular have been provided with the ^{51}Cr source experiments and with the Arsenic 71 tests.

The GALLEX collaboration announced the first observation of solar pp-neutrinos at 'Neutrino 92' in Granada on June 8th, 1992. Subsequently, the GALLEX recordings (till 1997) confirmed both the presence of pp-neutrinos [1] and a significant deficit (≈ 40 %) in the sub-MeV neutrino induced ^{71}Ge production rate [2]. At that time, this was the strongest indication for neutrino transformations on the way between the solar core and the Earth, implying non-zero neutrino mass and non-standard physics [4][5]. The subsequent GNO observations have improved the quality of the data, added important restrictions on the presence of possible time variations, and substantially reduced the total error on the charged current reaction rate for pp-neutrinos as measured by the inverse EC reaction on gallium [3]. Without radiochemical gallium detectors, the majority (92%) of all solar neutrinos would still remain unobserved.

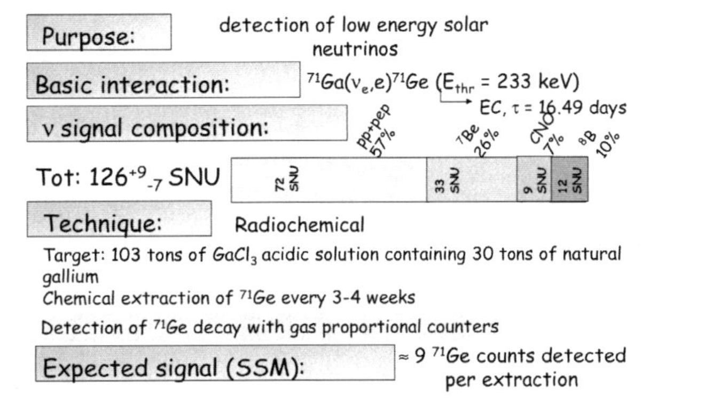

Figure 1. Scheme of the Gallium experiments

1986 - 1990	Construction of the detector	
May 1991 – May 1992	GALLEX I data taking 15 Solar runs, 5 Blanks 83.4 ± 19 SNU	PL B285 (1992) 376 PL B285 (1992) 390
Jun 1994 – Oct 1994	1st ^{51}Cr source experiment	PL B342 (1995) 440
Oct 1995 – Feb 1996	2nd source ^{51}Cr experiment	PL B420 (1998) 114
Feb. 1997	End of Solar Data taking	PL B447 (1999) 127
	GALLEX Final Result 1594 days – 65 runs: 77.5 ± 7.7 SNU	
Feb 1997 – Apr 1997	Test of the detector with ^{71}As	PL B436 (1998) 158
Apr 1998 – Apr 2003	GNO data taking	PL B490 (2000) 16 PL B616(2005) 174

Table 1. Milestones for GALLEX and GNO.

Gallium experiments at LNGS have provided a long time record of low energy solar neutrinos and determined the bulk production rate with an accuracy of ± 5.5 SNU (Table 2). This is based on 123 solar runs, 65 from GALLEX and 58 from GNO (Figure 2). The result and its precision will remain without competition from upcoming low-threshold real-time experiments for many years to come. Recorded is a fundamental astrophysical quantity, the neutrino luminosity of the Sun. In the astrophysical context, the gallium results shed light on the individual contributions of the PPI, PPII and CNO cycles to the solar luminosity and on the agreement of the energy production derived from the photon- and from the neutrino luminosity respectively [3].

Radiochemical experiments are first generation experiments. Opposite to real-time experiments, they provide no spectral or directional information. The statistical errors equal now the intrinsic systematic errors. This success implies their end. However, they have been crucial in paving the way into the excitement of the third millennium!

3. A recent update

The largest components of the systematic errors are the absolute efficiencies of the individual miniaturized low-level gas proportional counters. To maintain the extremely low background rates (~1 count per month) internal calibration that requires high ^{71}Ge activities was prohibited as long as the counters were used in the solar neutrino runs. However, after the completion of the solar runs such calibrations (using ~10^5 Ge-decays per counter) became possible and were done for the old counters

GNO	$62.9 \pm^{6.0}_{5.9}$ SNU
GALLEX	$77.5 \pm^{7.6}_{7.8}$ SNU
GALLEX+GNO	69.3 ± 5.5 SNU
SAGE	$66.9 \pm^{5.3}_{5.0}$ SNU

Range of SSM predicted rates:	
No oscillations:	122 – 131 SNU
With oscillations (global fit)	68 – 72 SNU

Table 2. Final results

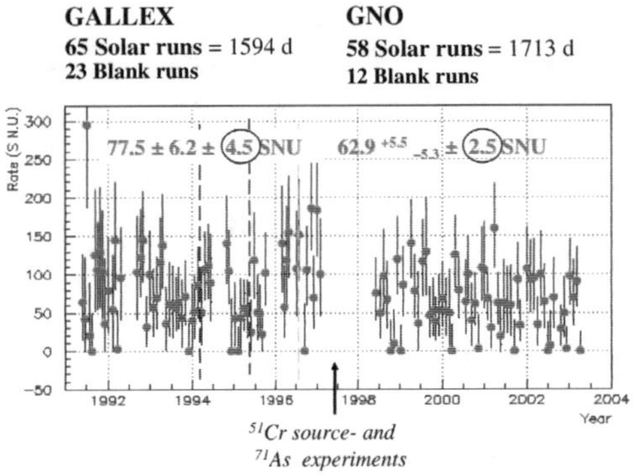

Figure 2. Individual solar runs

used in GALLEX. This triggered a complete re-analysis of the GALLEX data [6]. In addition to the efficiency error reduction, two other effects have been included:

- Improved Rn-cut efficiency after completion of the respective multi-year low-rate measurement.
- Full pulse shape analysis of the counter pulses instead of rise time analysis.

Among the four GALLEX data taking periods, by far the largest change is for GALLEX IV. This is the period that had created some concern because of a +2.5 σ deviation from the mean. Now this has reduced itself to +1σ. The integral updated results are given in Table 3; a full publication is in preparation.

GALLEX	$77.5 \pm^{7.6}_{7.8}$ SNU
re-evaluated	$73.1 \pm^{7.1}_{7.3}$ SNU
GNO (unchanged)	$62.9 \pm^{6.0}_{5.9}$ SNU
GALLEX+GNO	69.3 ± 5.5 SNU
re-evaluated	67.5 ± 5.1 SNU

Table 3. Updated Results

Partitioning of the Gallium Cake

(Experiments, SSM and LMA)

	SNU	P_{ee}	Experim.
measured	68 ± 5		GX/GNO
$^7Be_{exp}$	22 ± 7	0.68 ± 0.20	Borexino
$^8B_{exp}$	4 ± 1	0.33 ± 0.08	SK + SNO
^{13}N, ^{15}O	4.5^{+1}_{-3}	~0.5(assumed)	-
pep	1.5 ± .5	~0.5(assumed)	-
pp only	36 ± 8	0.52 ± 0.12	Gallium

Table 4. Deduction of the pp-flux

Also the Cr-source data have been reanalyzed with new efficiencies and an improved solar subtraction to include also the GNO data (see Table 2). This Cr-source update yields a reduction of the ratio R = $activity_{measured}/activity_{expected}$ from 93 ± 8% [7] to 88 ± 8% (1σ). The expectation value for ground state transitions only is 95 ± 1% [8]. Our ^{71}As experiments exclude any Ge-yield errors above 1%. Consequently, the excited state contribution is probably close to zero, instead of (5±3)% as estimated by Bahcall. This is also supported by the SAGE Cr-source result. If this is true, then the ^{71}Ge production rate prediction on Ga (without neutrino oscillations) must be reduced by ~2 SNU from ^7Be-neutrinos, from 34.8 SNU to 32.7 SNU (for pp-neutrinos, there is no change).

4. The causes for the termination of the Gallium experiments at LNGS

It has never been questioned that GALLEX/GNO has been a great success. This would not automatically justify a continuation after the main goals have been achieved. However, as long as real-time pp-neutrino detection was (and still is) a long way to go, there existed very important reasons to continue GNO:

- Continous pp-neutrino monitoring is an astrophysical necessity.
- pp-neutrino observations simultaneously with Borexino real-time beryllium neutrino observations would help to check on non-standard scenarios (time variations).
- Further neutrino source experiments would improve the knowledge of relevant cross sections.

Still, in 2004 it was decided by INFN that the GNO experiment should be terminated since it constitutes a safety risk which prevents the release of restrictions imposed on the Laboratory Nazionali del Gran Sasso (LNGS) by national safety authorities. This was not true. GALLEX/GNO (in Hall A) never spilled a single drop of liquid. The spill occurred in Hall C (Borexino). Ironically, GALLEX introduced the double spill tray safety conception in 1986, almost 20 years before it now became mandatory for the Laboratory. I quote here 'Physics Today' as an independent observer:

Quote: „The Gallium Neutrino Observatory (GNO) is the price that Italy's underground lab is paying to get back on its feet after a small chemical spill nearly two years ago" [9].

I wish to state with due respect that the GALLEX corpse feeded Borexino for survival.

GNO17, the last regular (semi-annual) GNO meeting was held at LNGS on April 6, 2005. On February 28, 2006 a *Final Celebration Ceremony for GALLEX/GNO* ended a successful fifteen year long period that had started with the *Inauguration Ceremony* on November 30, 1990. In April 2007, the gallium (chloride) was sold to Recapture Metals Inc., Ontario, Canada.

5. Oscillation parameters and the partitioning of the Gallium cake

If the LMA(MSW) solution is the correct explanation of the SNO/SK data, then vacuum oscillations must dominate below 1 MeV and the mixing angle is estimated as $\theta = 32 \pm 1.6$ degrees [10].

The first preliminary results from the BOREXINO experiment on the flux of solar ^{7}Be neutrinos make it now possible to deduce the pp-neutrino flux separately. We extract from our data the e-e survival probability P_{ee} for pp-neutrinos after subtraction of the ^{8}B and ^{7}Be contributions based on the *experimentally* determined ^{8}B-(SNO/Superkamiokande) and ^{7}Be-(BOREXINO) neutrino fluxes as P_{ee}(pp only) = 0.52 ± 0.12 (Table 4).

The results imply the experimental verification of the solar model and of the neutrino oscillation mechanisms at sub-MeV energies that are otherwise inaccessible.

References

[1] GALLEX Collaboration 1992 *Phys.Lett.* B **285** 376

[2] GALLEX Collaboration 1999 *Phys.Lett.* B **447** 127

[3] GNO Collaboration 2005 *Phys.Lett.* B **616** 174

[4] Bahcall J, Krastev P and Smirnov A 1998 *Phys. Rev.* D **58** 096016

[5] Kirsten T 1999 *Rev.Mod.Phys.* **71** 1213

[6] Kaether F 2007 Datenanalyse des Sonnenneutrinoexperiments GALLEX *PhD Thesis* Heidelberg University

[7] GALLEX Collaboration 1998 *Phys.Lett.* B **420** 114

[8] Bahcall J 1997 *Phys.Rev.* C **56** 3391

[9] Feder T 2004 On the mend, Gran Sasso sacrifices low-energy neutrino observatory *Physics Today* **57(#5)** 31

[10] Bahcall J and Pinnsonneault M 2004 *Phys.Rev.Lett.* **92** 121301

[11] BOREXINO Collaboration 2007 First real time detection of ^{7}Be solar neutrinos by Borexino *Preprint arXiv:0708.2251v1 [astro-ph] 16 Aug 2007*

LENS, MINILENS –Status and Outlook

R. S. Raghavan* and the LENS Collaboration
*Virginia Tech, Universities of Indiana, Louisiana State, North Carolina, North Carolina State and Princeton, Brookhaven National Laboratory and INR Moscow
E-mail: raghavan@vt.edu

Abstract

The current status of development of the LENS detector for measuring the luminosity of the sun in neutrinos is outlined. Plans for testing the technology and concepts of background control of LENS via a 250 liter prototype—MiniLENS- are described.

The low-energy solar ν spectrum offers unique access to specific low-energy evidence for the MSW-LMA ν model and new scenarios of non-standard particle physics and astrophysics. A crucial test is the fundamental equality of the L_ν, the solar luminosity inferred from ν fluxes and the *photon* luminosity L. Any inequality $L_\nu/L \neq 1.00$ that survives at *high precision*, implies new discovery in astrophysics or ν physics. A precision measurement of the ν luminosity implies the measurement of the complete set of pp, ^{7}Be, pep, & CNO ν fluxes, which together, make up >99.99% of the ν luminosity. The $L_\nu = L$ test requires the individual fluxes of specific energies so that they can be reverse-engineered to pre-conversion fluxes (needed to derive L_ν,) using a ν model. Energy specific, real-time ν detection is thus essential. The needed technology is available in LENS (Low-Energy Neutrino Spectroscopy), the only charged-current (CC) based real-time detector developed up to now for low energy solar ν's. The neutrino target in LENS is the 96% isotope ^{115}In. The recent success of Borexino in detecting ^{7}Be solar neutrinos (and possibly other low energy neutrinos in the future) provides a major new impetus. Borexino uses ν-electron *scattering* based on both CC and NC interactions. The combination of Borexino and LENS data for the same neutrino source eliminates the need for determining the absorption nuclear matrix element for the In reaction. A separate LENS experiment for this purpose with a MCi radioactive source (planned earlier), may no longer be needed.

Neutrino detection in In-LENS is based on the charged current (CC) *tagged* ν_e- capture in ^{115}In: $\nu_e + {}^{115}In \rightarrow e^- + (\tau = 4.76\ \mu s)\ 2\gamma + {}^{115}Sn$. The detection medium is an organic liquid scintillator containing 8-10wt% of indium (InLS). This reaction is unique because: 1) It provides a prompt e^- signal with the energy $E(e^-) = E(\nu) - Q$ that uniquely specifies the incident ν energy and thus its spectrum. The Q value is only 114 keV, the lowest known for CC solar ν capture. The reaction is thus sensitive to >95.5% of the pp ν continuum (0-420 keV). 2) The prompt electron is followed by a tag of 2 γ's after a mean delay of $\tau = 4.76$ μs. This central feature allows on-line tagging of ν events and powerful discrimination against background.

R&D of the In based LENS has achieved major results in the last few years. The main aspects are, a) Monte Carlo simulations that have revealed new analysis tools to discriminate against the random coincidence background from the decay of the In target. b) Development of In-loaded scintillators with properties suitable for LENS. c) A new type of detector design—the scintillation lattice—that allows digital 3-D location of nuclear events an order of magnitude more precisely than the analog TOF methods of the usual array of longitudinal detector modules.

Background Suppression: The major background arises from the natural beta-activity of ^{115}In. The MINILENS R&D focuses on this problem. Fig. 1 shows the topologies of the In-neutrino tag and the random coincidence background due to the Indium decay. Extensive GEANT4 monte carlo simulations that examine the topology of events. Fig. 1 shows the signal/background ratios and detection efficiencies achievable by the various cuts. The results indicate that the overall In decay background can be suppressed by a factor $6x10^{11}$ at the cost of a signal loss by factor of 1.6, leading to a S/N ratio of ~3, sufficient for achieving the physics objectives of LENS.

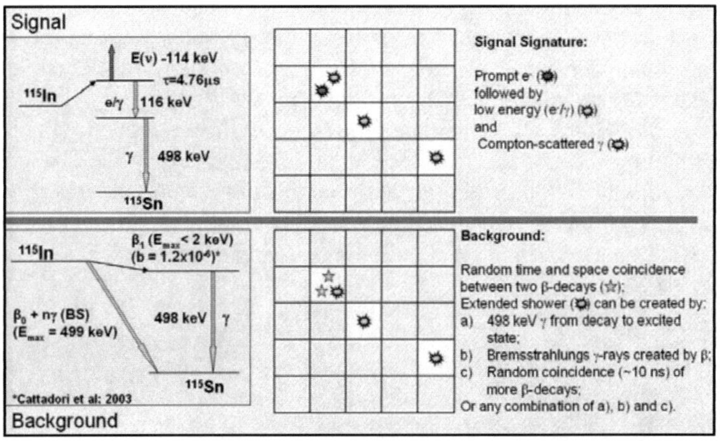

	Signal /y /t In	Bgd tot /y /t In	Bgd A1 /y /t In	Bgd A2 /y /t In	Bgd B /y /t In
RAW	62.5	79×10^{11}			
Valid tag (Energy, Branching, Shower) in Space/Time delayed coinc. with prompt event in vertex	50	"Free" 2.76×10^5	8.3×10^4	2.8×10^3	1.9×10^5
+ ≥3 Hits in tag shower	46	2.96×10^4	2.6×10^4	2.5×10^3	1.4×10^3
+Tag Energy = 613 keV	44	306	0.57	4.5	293
+Tag topology	40	13 ± 0.6	0.57	4.0	8.35

Fig. 1: Signal and background topologies (top) and role of experimental cuts in LENS background analysis by a factor ~1.6.

Indium loaded Liquid Scintillator (InLS): The choice of InLS as the basic technology for LENS was clear early in the historical development of LENS. Thus a viable InLS technology has been the focus if LENS R&D programs for more than a decade. The LENS demands on the InLS are extraordinarily stringent and unprecedented. These have been achieved for the first time in our work in the last 2 years. The following typical figures of merit are observed using the LS solvent pseudocumene (PC).

- Indium loading : ~ 8 wt%
- Scintillation light efficiency relative to pure PC: Y ~8000 hν /MeV
- (1/e) Light attenuation length L_e for 430 nm scintillation light: $L_{1/e}$ ~ 800 cm
- Chemical and Optical Stability (from time history of L_e): > 8 months

Recently we have studied the suitability of a new solvent Linear Alkylbenzene (LAB). The flash point of this solvent is high (~100C vs 45C for PC) thus more compatible with underground operations. It is also considerably less expensive. All the figures of merit above for PC (In loading, scintillation yield, attenuation length, long term stability) have been reproduced with somewhat less favorable figures of merit. The work so far shows that LAB may be suitable as the solvent of choice for LENS and will be field tested in the prototype project MINILENS.

Scintillation Lattice Detector Design: We have developed a new detector technology: the 3-D scintillation lattice chamber (Fig. 2). The basic idea extends the 1-D light piping property of the long module to 3-dimensions using a close packed 3-D assembly of InLS filled cubic cells with air gaps surrounding each cell. In such a device, scintillation light generated in a cell travels only in channels along the 3 coordinate axes intersecting the vertex cell If the cubic array faces are optically coupled with PMT's, the light triggers only the 6 PMT's viewing the channels specific to the event cell. This is seen in a laboratory model (Fig. 2) consisting of a cubic array of acrylic cubes that simulate the optical segmentation. A plastic scintillator somewhere in the array (= a nuclear event) is fluoresced by UV light. The fluoresence light triggers only the PMT's at the end of the channels (the bright squares on the faces) that lead back along the channels to the orignal event. Event location is thus *digital.* Important progress has been made in applying this concept to real detector configurations of optical segmentation using plastic foils in a solvent.

Fig. 2 Concept of a scintillation lattice

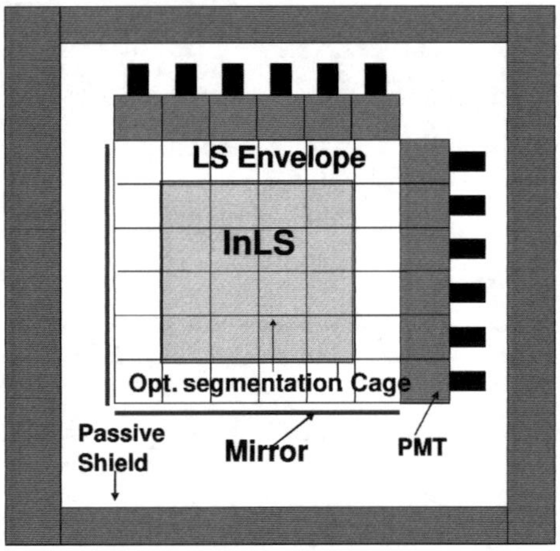

Fig. 3 Schematic Design of MINILENS

MINILENS: The roadmap to the stage of readiness for the full-scale LENS calls for a practical demonstration of the essential elements of LENS. We see two phases in this roadmap. In phase I, the R&D will continue to address the next essential questions. In phase II we plan to globally test all the R&D conclusions, design plans and strategies in a modest, scalable detector –MINILENS that will demonstrate the essential control of background as well as field testing the technology, e.g.,large scale InLS production and handling and construction and operation of the scintillation lattice. We propose to construct MINILENS with In mass of ~20-40 kg in the lattice design as a 6x6x6 array of cubic cells 125 mm on the side (see Fig. 3 for a schematic design. The outer array in Fig. 3 would be filled with non-In LS as an active buffer and the inner volume with ~250 L of InLS containing a minimum of 8% In). The PMT's cover the faces (6 or 3 with mirrors on the other 3 faces), sufficient for full 3D event localization. Passive shielding will be added as needed. MINILENS thus contains all the basic elements of our design for LENS and is scalable to the full experiment. The In singles rate in such a detector will be at least 5 kHz, sufficient to critically test the trigger design, DAQ and the background suppression concepts. MINILENS will be located at the recently commissioned Kimballton Underground Research Facility at a depth of ~1400 mwe, near the VT campus. Thus we can make a direct test of the operability of LENS at these depths.

The LENS work at Virginia Tech is supported in part by the National Science Foundation; Brookhaven National Laboratory is supported by the ONP of in the Department of Energy.

Low-level techniques applied in experiments looking for rare events

G Zuzel
Max-Planck Institutt für Kernphysik, Saupfercheckweg 1, 69117 Heidelberg, Germany

E-mail: Grzegorz.Zuzel@mpi-hd.mpg.de

Abstract. Techniques allowing investigation and reduction of backgrounds are crucial for all experiments looking for rare events. In this paper investigations of ^{222}Rn, germanium spectroscopy and noble gas mass spectrometry will be described. Selected measurements and their applications in the Borexino and GERDA experiments will be discussed.

1. Introduction

Low-level techniques could be defined as techniques allowing investigations of natural and man-made radioisotopes at very low activities (concentrations). One can consider here studies of radioactive noble gases (diffusion, emanation), germanium spectroscopy, mass spectrometry (ICP-MS, noble gas mass spectrometry), analysis of surface contaminations (α- and β-spectroscopy), investigation of purification techniques (gases, liquids), constructing algorithms for background events rejection (time coincidences, pulse shape analysis) and modelling of backgrounds using experimental results and Monte Carlo simulations. In the following chapters as selected topics radon studies, germanium spectroscopy and noble gas mass spectrometry will be discussed. Some measurements performed in the frame of the Borexino [1] and GERDA [2] experiments will be described.

2. Studies of ^{222}Rn

^{222}Rn is one of the most dangerous isotopes for many low-background experiments. It belongs to the ^{238}U chain and as an inert noble gas it has typically high mobility (diffusion coefficient) allowing easy escape from bulk materials and diffusion into active parts of detectors. Together with the short-living daughters ^{222}Rn generates radiations with a wide energy range. Through a broken equilibrium in the chain (^{210}Pb level) surface contamination with long-living heavy metals can also cause problems (e.g. it may appear some years after cleaning as ^{210}Po).

Activity of ^{222}Rn can be measured using for example low-background proportional counters designed originally for the GALLEX/GNO experiment [3]. They are hand-made out of very pure quartz and have an active volume of about 1 cm^3. The background and detection efficiency (for alpha particles) for energies above a 50-keV threshold is ~1 count per day and 50 %, respectively. Counters are filled with radon samples using a special gas filling line.

^{222}Rn emanation can be studied using emanation chambers [4]. Two stainless steel electro-polished and completely metal-sealed vessels of the volume of 20 and 80 l are available. For samples up to 1 l volume small glass vials could be used. The absolute detection limit obtained for emanation tests is at the level of 100 μBq (~50 atoms).

To measure radon concentration in gases a system called MoREx (Mobile Radon Extraction Unit) has been constructed [5]. It takes the advantage of the ^{222}Rn pre-concentration in charcoal traps at low temperatures. In connection with low-background counting it is possible to analyze gases contamination down to ~0.5 μBq/m^3 (STP).
To analyze ^{222}Rn and ^{226}Ra in the water STRAW (System for the ^{222}Rn and ^{226}Ra Assay of Water) can be used [6]. The sensitivities are at the level of 0.1 and 0.8 mBq/m^3 for radon and radium, respectively.
In many cases it is also interesting to know radon diffusion through different materials. This knowledge helps to design proper barriers protecting active parts of the detectors from radon moving from outside. An apparatus for this kind of research was constructed and successfully used in the frame of the Borexino experiment to measure the diffusion and solubility coefficients for radon barrier- and inner vessel foils [7], also as a function of the relative humidity of the foils vicinity [8]. The diffusion measurements could be performed at the 10^{-13} cm^2/s level.

3. Noble gas mass spectrometry

Other noble gases, which can be considered as sources of background e.g. in Borexino are ^{39}Ar or ^{85}Kr present in nitrogen as traces. In principle their concentrations could be measured as that for radon, using low-background proportional counters, however the pre-concentration does not work for them efficiently enough, and achievable detection limits would be here much poorer. Other solution is to look for natural Ar and Kr using a highly sensitive noble gas mass spectrometer. Concentrations of ^{39}Ar and ^{85}Kr can be later reconstructed taking into account their fairly known specific activities. Ar/Kr is detected using a VG3600 mass spectrometer devoted to investigate rare gases at low concentrations in terrestrial and extraterrestrial samples. Besides the mass spectrometer, the system consists of a sample preparation and a sample purification section. In the first one the volume of the investigated sample can be determined (or if necessary divided into smaller fractions), in the second one the sample is purified from reactive gases before it is introduced into the spectrometer. For final purification Al-Ti getter pumps are used. The calibration is performed with air or Ar/Kr standards. Taking into account the intrinsic background of the whole system and measuring 1 cm^3 of nitrogen we could achieve the following detection limits: 1×10^{-9} m^3/m^3 for Ar in N_2 (~1.4 nBq/m^3 for ^{39}Ar in N_2) and 1×10^{-13} m^3/m^3 for Kr in N_2 (~0.1 μBq/m^3 for ^{85}Kr in N_2) [9].

4. Germanium spectroscopy

Germanium spectroscopy is one of the most powerful tools to identify γ emitters (U/Th chain, ^{60}Co, ^{40}K). The strength of this technique comes from the excellent energy resolution of germanium diodes and from their very low intrinsic background. External background can be usually effectively shielded or vetoed. In order to reach high detection sensitivities one should assure also a large sample chamber and proper determination of efficiencies (measurements or simulations).
Max Planck Institute operates several detectors at different locations. The most sensitive world-wide are GeMPI-type spectrometers located at the Gran Sasso laboratory in Italy. Due to the extremely low background und large capacity, for high density materials it is possible to reach the detection limit at the level of 10 μbq/kg (U/Th) [10]. Some detectors are also run in the MPI underground laboratory (15 m w.e.), however because of the shallow depth the lowest measurable concentrations are in the range of 1 mBq/kg [11].

5. Applications in Borexino

5.1. ^{226}Ra in the inner vessel foil

The purity requirements for the construction materials in Borexino, especially for the scintillator and its nylon vessel were very strong. The allowed contamination for the second one with ^{238}U was at the level of 1 ppt, what corresponds to ~12 μBq/kg for ^{226}Ra (in secular equilibrium). This limit could not be reached with any Ge spectrometer (foil is a low density material) and a new technique had to be developed. It is based on the combination of the ^{222}Rn diffusion and emanation measurements and uses the fact that the radon diffusion coefficient in nylon depends very strongly on its water content. Measuring emanation rates for dry and wet material one could determine the surface and bulk ^{226}Ra concentrations at the levels of 0.5 μBq/m^3 and 10 μBq/kg, respectively. In the frame of those investigations a proper material for the vessel was identified [12]. It was also shown that ^{238}U and ^{226}Ra in nylon was in disequilibrium.

5.2. Nitrogen

One of the most important applications of ultra-pure nitrogen in Borexino was the scintillator purification. The purity requirements for N_2 were the following: 7 μBq/m^3 for ^{222}Rn, 0.4 ppm and 0.2 ppt for Ar and Kr, respectively. Such a gas was delivered by the SOL company after thorough purity and transportation tests, where possible (re-)contamination sources were eliminated [13]. Beside mentioned ultra-pure- Borexino uses also ^{222}Rn-free- [5] and boil-off nitrogen (obtained from technical quality LN_2) produced with the rate up to 100 m^3/h.

6. Applications in GERDA

6.1. Germanium spectroscopy of steel samples

As in many other experiments germanium spectroscopy was applied in GERDA to screen construction materials like steel (for the cryostat) or copper (for detector holders and shields). The screening was done in Heidelberg and in Gran Sasso using the GeMPI detectors. One of the most important achievements was finding of steel with surprisingly low U/Th contamination of the order of < 0.1 – 1 mBq/kg [14]. Batches with the lowest concentrations (< 0.1 – < 0.4 mBq/kg) were used to construct the cylindrical part of the cryostat, the others for the top and bottom heads. Low U/Th content in steel allowed thickness reduction of the copper shield installed inside the vessel (keeping the same background index).

6.2. Surface cleaning tests

Cleanliness of surfaces located close to the Ge detectors is one of the key issues in GERDA. This concerns mainly the long-living radon daughters deposited for example on the copper support structure of the diodes. To test different cleaning procedures some Cu disc were loaded with radon daughters and then cleaned applying selected methods. Removal efficiencies for ^{210}Pb, ^{210}Bi and ^{210}Po were investigated using γ-, β- and α-spectroscopy, respectively. The main result from this research is that the electro-polishing usually removes effectively all mentioned isotopes, while etching with acids does not work for ^{210}Po practically at all [15]. Further tests are planned for steel samples.

7. References

[1] Alimonti G et al. 2002, *Astropart. Phys.* **16** 205

[2] Abt I et al. GERDA Proposal 2004 hep-ex/0404039

[3] Wink R et al. 1993 *Nucl. Instr. Meth. A* **329** 541

[4] Rau W and Heusser G 2000 *Appl. Rad. Isot.* **53** 371

[5] Heusser G et al. 2000 *Appl. Rad. Isot.* **52** 691

[6] Simgen H et al. 2003 *Nucl. Instr. Meth. A* **497** 407

[7] Wojcik M et al. 2000 *Nucl. Instr. Meth. A* **449** 158

[8] Wojcik M and Zuzel G 2004 Nucl. Instr. Meth. A **524** 355

[9] Zuzel G et al. 2004 *Appl. Rad. Isot.* **61** 197

[10] Heusser G Laubenstein M and Neder H 2006 *Radioactivity in the Environment* **8** 495

[11] Heusser G 1995 *Ann. Rev. Nucl. Part. Sci.* **45** 543

[12] Zuzel G et al. 2003 *Nucl. Instr. Meth. A* **498** 240

[13] Simgen H and Zuzel G 2007 *AIP Conf. Proc.* **897** 45

[14] Budjas D et al. 2007 in preparation

[15] Wojcik M and Zuzel G 2007 *AIP Conf. Proc.* **897** 53

Low energy neutrino experiments sensitivity to physics beyond the Standard Model

Timur Rashba

Max-Planck-Institut für Physik (Werner-Heisenberg-Institut), Föhringer Ring 6, 80805 München, Germany
and
IZMIRAN, 142190, Troitsk, Moscow region, Russia

E-mail: `timur@mppmu.mpg.de`

Abstract. We discuss the sensitivity of future low energy neutrino experiments to non-standard interactions of neutrinos with quarks and electrons. A new limit on the electron neutrino effective charge radius from a new evaluation of the weak mixing angle by a combined fit of all electron-(anti)neutrino electron elastic scattering measurements is presented.

1. Introduction

Non-standard neutrino interactions (NSI) with quarks and electrons are still allowed to be rather large [1, 2, 3]. It was shown recently that coherent neutrino nuclei scattering is a very sensitive probe to NSI [1, 2]. The sensitivity to some specific NSI models, such as extra heavy neutral gauge bosons, leptoquarks, supersymmetry with broken R-parity, could be better than present constraints and, therefore, could give complementary bounds to future LHC limits [2].

Another topic to be discussed are new limits on the weak mixing angle, $\sin^2\theta_W$, and on the electron neutrino effective charge radius, $\langle r^2_{\nu_e}\rangle$, obtained from the combined analysis of all available low energy (anti)neutrino-electron elastic scattering experiments [4].

2. Non-standard neutrino interactions

It was demonstrated that neutrino coherent scattering off nuclei has a lot of potential to make precision tests of new physics, in particular, to probe NSI [1, 2]. A new experimental proposal is able to test large NSI parameters and therefore may exclude new solutions to the solar neutrino data [5] and constraint NSI parameters that naturally appear in many extensions of the Standard Model. The sample sensitivity is illustrated in Fig. 1. All further details can be found in Refs. [1, 2].

As concrete coherent neutrino-nuclei interaction proposals, we have discussed the TEXONO case, the stopped pion source with a noble gas detector and the beta beams [2]. It was found that a coherent neutrino-nuclei scattering using reactor neutrinos, such as the TEXONO proposal, or a beta-beam neutrino source, could have a high sensitivity to new interactions coming from leptoquarks or R-parity breaking SUSY, and it has been shown that a stopped-pion neutrino source experiment could also improve the current R-parity breaking SUSY constraints. On the other hand, for this kind of experiments an improved constraint to extra heavy neutral gauge bosons seems to be difficult.

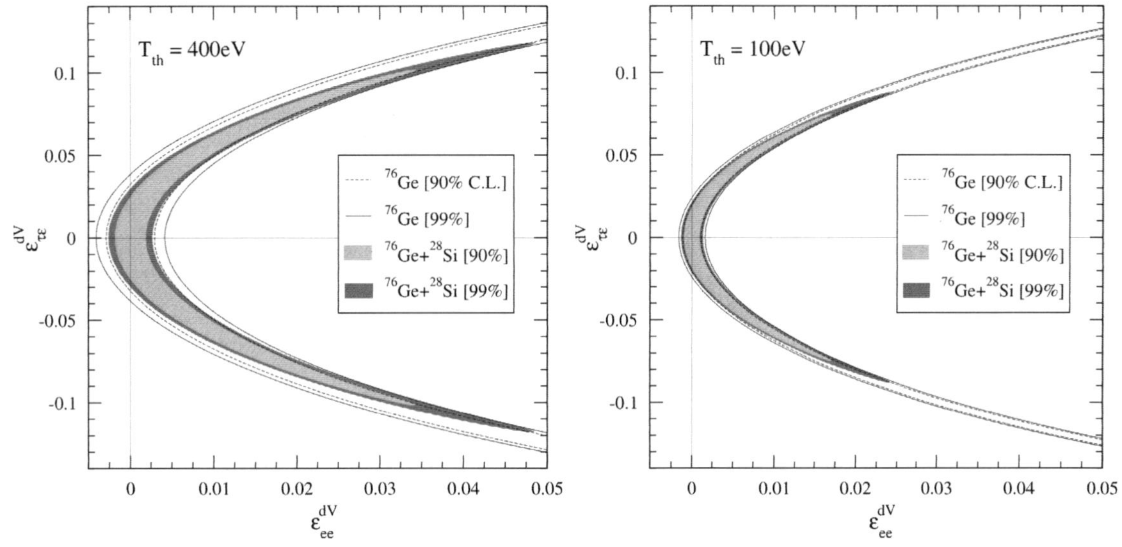

Figure 1. The allowed regions of non-universal NSI parameters ε_{ee}^{uV} and ε_{ee}^{dV} are shown at 90 and 99% C.L. for combined data from two detectors of ^{28}Si and ^{76}Ge (colored regions) and only for ^{76}Ge (solid and dashed lines). It was assumed 1 kg mass and 1 year of data taking for both detectors, only statistical errors are taken into account. The results presented for two values of threshold, $T_{th} = 400$ eV (left) and 100 eV (right panel). Figures are taken from Ref. [1] and all further details can be found there.

For the particular case of leptoquarks, it have been found that all the discussed low energy neutrino experiments have the potential to improve the present bound on leptoquark masses and couplings. In particular, the sensitivity to the vector leptoquark mass is of the order of 800 GeV, assuming an electroweak value of the coupling, $\lambda_q^2/4\pi = 1/137$. For the case of supersymmetry with broken R-parity, the perspectives to improve the constraint on the λ'_{1j1} and the corresponding mass for the $\tilde{d}_L$ are also very promising for all the experimental setups [2].

3. Electron neutrino charge radius

The straight-forward definition of a neutrino charge radius has been proved to be gauge-dependent and, hence, unphysical. There have been recent discussions in the literature on the possibility to define a physically observable neutrino charge radius [6, 7]. All relevant references can be found in Ref. [4]. A more general interpretation of the experimental results is that they are limits on certain nonstandard contributions to neutrino scattering.

All available measurements of the (anti)neutrino-electron scattering from the following reactor and accelerator experiments were analyzed: first measurement of neutrino-electron scattering made by Reines, Gurr and Sobel (Irvine), the Kurchatov institute group at the Krasnoyarsk reactor, the group from Gatchina at the Rovno reactor, MUNU at the Bugey reactor, LAMPF and LSND. We haven't included into our analysis the recent reactor neutrino results from TEXONO and GEMMA experiments, which have put very strong limits on neutrino magnetic moments. The reason is that both experiments are working at very low energy thresholds searching for neutrino magnetic moments and are not yet sensitive to the weak mixing angle and neutrino charge radius. All references to the experimental data used in the analysis are given in Ref. [4].

The data of (anti)neutrino-electron scattering experiments listed above were used to obtain new limit on the weak mixing angle with 10% precision at energies below 100 MeV [4]

$$\sin^2\theta_W = 0.259 \pm 0.025\,. \tag{1}$$

This result is not competitive with the current best measurements at low energies obtained from atomic parity violation [8] and from Moller scattering by SLAC E158 [9], both experiments have a precision better than 1%. However, it is derived from a different channel and it could therefore give new information about other effects, such as the electron neutrino effective charge radius.

This value of the weak mixing angle was used to set a new limit to the electron neutrino effective charge radius squared [4]

$$-0.13\times10^{-32}\ \mathrm{cm}^2 < \langle r^2_{\nu_e}\rangle < 3.32\times10^{-32}\ \mathrm{cm}^2 \quad \text{at } 90\%\ \text{C.L.} \tag{2}$$

This result improves previously published bounds [10].

One can conclude, that from the measurement of the weak mixing angle with 1% resolution it is possible to find a strong evidence for electron neutrino charge radius, which is estimated theoretically to be of the order of 0.4×10^{-32} cm^2 [6].

4. Summary

Neutrino coherent scattering off nuclei has a lot of potential to make precision tests of new physics, not only in the case of a non-zero neutrino magnetic moment as had already been studied in the literature, but also to test non-standard neutrino interactions.

The weak mixing angle with 10% precision at energies below 100 MeV from (anti)neutrino electron scattering off electrons was recently obtained. To get this result the combined set of all available data from accelerator (LSND and LAMPF) and reactor (Irvine, Rovno, Krasnoyarsk and MUNU) experiments was used. This analysis was also applied to set a new limit to the electron neutrino effective charge radius squared which improves previous bounds.

Future reactor experiments with the estimated precision of $\delta(\sin^2\theta_W)\sim 1\%$ will be able to find a strong evidence for theoretically predicted electron neutrino effective charge radius.

Acknowledgments

I would like to thank my collaborators Juan Barranco and Omar Miranda. I am grateful to the organizers of TAUP-2007 for the fruitful atmosphere and very good organization of the conference. The work was supported by the Marie Curie Incoming International Fellowship of the European Community and by the Deutsche Forschungsgemeinschaft under grant No. TR-27.

References

[1] J. Barranco, O. G. Miranda and T. I. Rashba, JHEP **0512** (2005) 021 [arXiv:hep-ph/0508299].

[2] J. Barranco, O. G. Miranda and T. I. Rashba, Phys. Rev. D **76** (2007) 073008 [arXiv:hep-ph/0702175].

[3] J. Barranco, O. G. Miranda, C. A. Moura and J. W. F. Valle, arXiv:0711.0698 [hep-ph].

[4] J. Barranco, O. G. Miranda and T. I. Rashba, arXiv:0707.4319 [hep-ph].

[5] O. G. Miranda, M. A. Tortola and J. W. F. Valle, JHEP **0610** (2006) 008 [arXiv:hep-ph/0406280].

[6] J. Bernabeu, J. Papavassiliou and J. Vidal, Phys. Rev. Lett. **89** (2002) 101802 [Erratum-ibid. **89** (2002) 229902] [arXiv:hep-ph/0206015].

[7] K. Fujikawa and R. Shrock, arXiv:hep-ph/0303188.

[8] S. C. Bennett and C. E. Wieman, Phys. Rev. Lett. **82**, 2484 (1999) [arXiv:hep-ex/9903022].

[9] P. L. Anthony *et al.* [SLAC E158 Collaboration], Phys. Rev. Lett. **95**, 081601 (2005) [arXiv:hep-ex/0504049].

[10] W. M. Yao *et al.* [Particle Data Group], J. Phys. G **33** (2006) 1.

Scintillator purification, detector performance and first results from Borexino

M. Pallavicini

Università di Genova & INFN – via Dodecaneso, 33 I-16146 Genova, Italy

marco.pallavicini@ge.infn.it

On behalf of the Borexino Collaboration

G. Bellini, J. Benziger, S. Bonetti, B. Caccianiga, F. Calaprice,F. Dalnoki-Veress, D. D'Angelo, H. de Kerret, A. Derbin, A. Etenko,K. Fomenko, R. Ford, D. Franco, C. Galbiati, S. Gazzana,M. Giammarchi, M. Goeger-Neff, A. Goretti, C. Grieb, S. Hardy, G. Heusser,Aldo Ianni, Andrea Ianni, M. Joyce, G. Korga, D. Kryn, M. Laubenstein,M. Leung, E. Litvinovich, P. Lombardi, L. Ludhova, I. Machulin, G. Manuzio, F. Masetti, K. McCarty, E. Meroni, L. Miramonti, M. Misiaszek,D. Montanari, M. E. Monzani, V. Muratova, L. Niedermeier, L. Oberauer,M. Obolensky, F. Ortica, M. Pallavicini, L. Papp, L. Perasso, A. Pocar, R. S. Raghavan, G. Ranucci, A. Razeto, A. Sabelnikov, C. Salvo, S. Schönert ,H. Simgen, O. Smirnov, M. Skorokhvatov, A. Sonnenschein, A. Sotnikov, S. Sukhotin, Y. Suvorov, V. Tarasenkov, R. Tartaglia, G. Testera, D. Vignaud, R. B. Vogelaar, F. von Feilitzsch, M.Wojcik, O. Zaimidoroga, S. Zavatarelli, G. Zuzel

Abstract. The Borexino experiment has begun data taking in May 2007 after a long R&D work and preparation. The liquid scintillator purity exceeds even optimistic expectations, and the first detection of ^{7}Be solar neutrinos has been possible after less than two months of data taking. This note shows briefly which are the main issues that were addressed in order to obtain such an extreme radiopurity, the detector performance and a few details concerning this first result.

1. Introduction

The Borexino experiment [1], now operating at the Laboratori Nazionali del Gran Sasso, has been designed and built mainly with the purpose of detecting low energy solar neutrinos, in particular those emitted by the electron capture decay of ^{7}Be in the Sun.

Solar neutrinos are detected by means of their elastic scattering off electrons in a liquid scintillator target made of pseudocumene (PC,1,2,4-trimethylbenzene) and doped with PPO (2,5-diphenyloxazole, a fluorescent dye) at a concentration of 1.5 g/l. The scintillator is contained in a 300 m^3 thin (125 μm) nylon vessel (R=4.25 m). This inner nylon vessel is kept almost buoyant free at the center of a large Stainless Steel Sphere (R=6.85 m) which acts at the same time as a container of the remaining buffer liquid (~1000 m^3 of PC with a small addition DMP to quench the unwanted scintillation light coming from pure PC) and as mechanical support of 2212 EMI PMTs that collect the scintillator light produced in the scintillator. Another outer nylon vessel (R=5.5 m) act as a barrier against Rn emanated from the SSS and the PMTs and against other possible contaminations. Finally, the SSS is installed inside a 3000 m^3 Water Tank which provides the necessary schielding against rock induced external backgrounds and is used as a Cerenkov detector to tag the residual muons that penetrate the Gran Sasso mountain. A schematic view of the Borexino detector is shown in Fig. 1

The detection of solar neutrinos by means of scintillation allows good energy resolution, very low energy threshold (about 50 KeV), good spatial reconstruction, and good discrimination between α and β nuclear decays. However, the direction of the incoming neutrino cannot be measured and each neutrino induced event is intrinsically indistinguishable from any β or even γ decay within the scintillator. This fact makes the experiment feasible only if extreme radiopurity requirements are met for both the scintillator and the materials of which the detector is built.

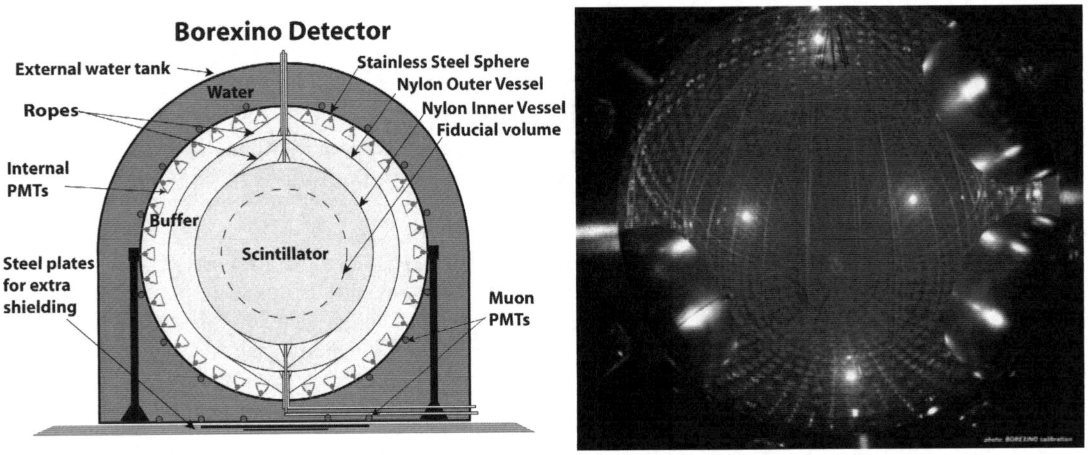

Figure 1 – Left: Schematic view of the Borexino detector; Right: a picture of the detector taken immediately after the end of the filling

2. Scintillator purification and detector filling

In the last 15 years the Borexino collaboration has developed innovative techniques for the measurement of ultra low radioactive backgrounds in materials and for the purification of the scintillator [1]. A set of purification plants have been designed and built, and their effectiveness was tested by means of the Counting Test Facility [2] at Gran Sasso. These plants include, among others, a water plant, an ultrapure nitrogen plant, a PC distillator, and another distillator designed for the purification of PPO dissolved in concentrated solution in PC.

The water plant can produce water at a rate of 1.8 m^3/h with ^{238}U and ^{232}Th content below 10^{-14} g/g, ^{222}Rn content around 1 mBq/m^3 and typical resistivity around 18.2-18.3 MΩ at 20 °C.

Particular attention was devoted to the development of low Argon Krypton nitrogen that was used to purify the scintillator from all air borne contaminations. The content of both ^{39}Ar and ^{85}Kr in this nitrogen is such that the expected background in the 100 t fiducial volume should be below 1 c/d.

The whole amount of PC (~1300 m^3), both in the scintillator and in the buffer, was distilled by means of a 6 stages distillator at 60 mbar pressure and ~90 °C temperature. The PC was then sparged with nitrogen gas to remove air borne impurities (^{222}Rn, ^{39}Ar and ^{85}Kr) and humidified to preserve the mechanical properties of the nylon vessel. Being the PPO solid, the purification of PPO was done preparing a concentrated solution PPO in PC (120 g/l) and then purifying this solution first with several cycles of water extraction and then by means of a single stage distillator.

The filling of the detector was done as following: first, the nylon vessel and the SSS have been filled with ultrapure water. This water filling allowed a final rinsing of the detector and of the nylon vessels. Then, the water was replaced with PC-PPO solution in the inner nylon vessel and with PC-DMP solution in the buffers. At the same time, the water taken out of the SSS was passed through a charcoal filter and used to filled the Water Tank. The filling proceeded smoothly from August 1st, 2006 until May, 2007 when data taking began.

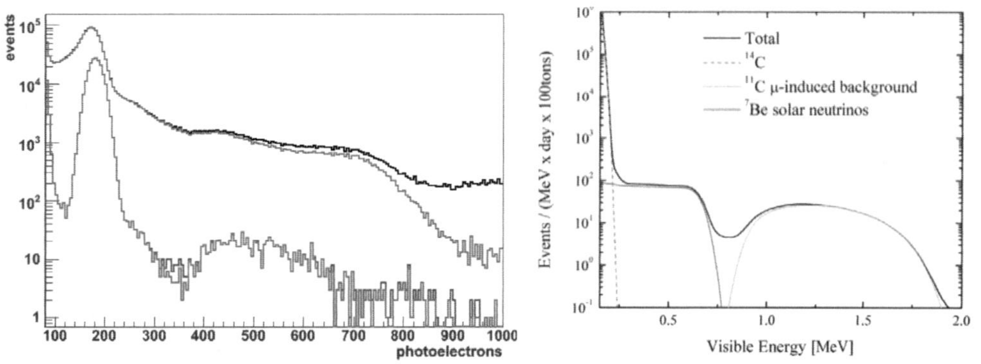

Figure 2 – Left: the effect of different cuts in the Borexino spectrum; the black curve is the raw spectrum; the brown one is after muon cut, the purple one is after fiducial volume cut and the red one is the final after Rn cut. For comparison, on the right the expected rate due to ^{7}Be neutrinos and irreducible background is shown.

3. First detection of ^{7}Be solar neutrinos

The purity of the Borexino scintillator is good enough to allow the clear detection of ^{7}Be solar neutrinos using a couple of months of data. Fig. 2 shows a comparison between the expected signal rate in 100 t of fiducial volume and the corresponding spectrum seen on data. The compton shoulder expected from mono-energetic neutrinos scattering on electrons is evident (red curve).

The analysis does not require many cuts. The red curve in Fig. 2 (left) is obtained just by removing muon induced events by exploiting the capabilities of both the outer Cerenkov detector and the inner detector pulse shape, by removing ^{222}Rn events and its daughers and by removing the external background by means of a fiducial volume cut. The spectrum in Fig. 2 left is not converted into energy. The conversion is approximately linear below 1 MeV with a yield of about 500 pe/MeV.

The very good α/β separation and the very good light yield allow to fit the spectrum both with and without the statistical subtraction of the ^{210}Po peak which is evident in the spectrum. The two approaches yield consistent results. Details of this result are published in [3] and [4].

4. Conclusions

The Borexino experiment has successfully completed the construction of the detector and the data taking has begun. The scintillator is very pure and the first detection of ^{7}Be solar neutrinos has been possible after a few months of data taking. Precision measurement will require better understanding of the detector, more statistics and a calibration campaign. This paper is dedicated to the memory of Cristina Arpesella, Martin Deutsch, Burkhard Freudiger, Andrei Martemianov and Sandro Vitale.

References

[1] Borexino Collaboration – G. Alimonti et.al., Astropart.Phys.16:205-234,2002.
Borexino Collaboration – C. Arpesella et.al, Astropart.Phys.18:1-25,2002.
[2] Borexino Collaboration – G. Alimonti et.al., Astropart. Phys. 8:141, 1998
[3] Borexino Collaboration – C. Arpesella et. al., astro-ph 0708.2251, accepted by PLB
[4] G. Bellini on behalf of Borexino Collaboration – these proceedings

LENA: A multipurpose detector for low energy neutrino astronomy and proton decay

T Marrodán Undagoitia[1,2], F von Feilitzsch[1], M Göger-Neff[1], L Oberauer[1], W Potzel[1], A Ulrich[1], J Winter[1] and M Wurm[1]

[1]Physik-Department, Technische Universität München,
James-Franck-Str., 85748 Garching, Germany

E-mail: tmarroda@ph.tum.de

Abstract.
The status of the feasibility studies for the proposed large liquid-scintillator detector LENA (**L**ow **E**nergy **N**eutrino **A**stronomy) is reported. Recent technical investigations of optical properties of possible scintillator mixtures are described and the physics potential concerning supernova, solar and geophysics as well as proton decay is reviewed.

1. Introduction

Organic liquid-scintillator detectors, as KamLAND [1] or Borexino [2], have already shown the great physics potential of such a technology. The LENA (**L**ow **E**nergy **N**eutrino **A**stronomy) project suggests a 50 kt liquid scintillator detector for further investigations of low-energy neutrinos (from supernovae, the Sun, our Earth or from artificial sources) and for the search for proton decay.

2. Current detector design

The detector design has been described elsewhere [3][4]. It is based on a cylinder with a diameter of 30 m and a length of approximately 100 m. To look at the scintillation light, the inner volume will be instrumented with ~13 000 photomultiplier tubes providing a surface coverage of ~30% which results in a photoelectron (pe) efficiency of at least 10^2 pe/MeV. Thus, sub-MeV neutrino astronomy becomes feasible. The inner volume will be filled with about 50 kt of liquid scintillator, the exact composition being still under discussion. In the outer part, a water Cherenkov detector will tag through-going muons. Most of the research areas mentioned above require a low background event rate. Thus, the detector should be placed underground at a depth of around 4 000 m.w.e.

3. Liquid scintillator feasibility studies

In the last years, intensive R&D on liquid scintillators and photo-sensors has been set up for LENA. Optical properties of various scintillator mixtures have been tested to guarantee good energy resolution, high light collection, fast timing and precise position reconstruction. The

[2] Corresponding author. Tel: +49 89 289 12328

solvents PXE, LAB and Dodecane, and the solutes PPO, pTP, bisMSB and PMP, mixed in different concentrations [5][6] have been under investigation.

Experiments to investigate the light propagation have been performed and it has been shown that absorption and scattering lengths in the range of 20 m can be reached. These values correspond to an one-dimensional attenuation length of ~10 m. Monte Carlo calculations show [9] that for such a scintillator, a light yield of 180 pe/MeV would be obtained.

Furthermore, the light production has been studied in detail. Measurements concerning the fluorescence decay constants show that different scintillator mixtures have a strong influence on the time structure. Using the single-photon counting method, the time evolution of the scintillation-photon emission has been obtained. The spectrum can be fitted by three exponential decay times: for a mixture of PXE with $2\,g/\ell$ PPO, a shortest decay constant of around 3 ns has been found. For LAB with $2\,g/\ell$ PPO, typical values around 6 ns have been observed. Also for a mixture of pure Dodecan with $2\,g/\ell$ PPO a relatively slow time constant of about 5 ns was found. In addition, measurements of the time constants of PXE with different concentrations of PPO have been performed. The shortest time constant has been found for PXE with $6\,g/\ell$ PPO with ~2.3 ns, for smaller fluor concentrations the value increases as the energy transfer between PXE and PPO is less efficient. For larger concentrations of PPO, the time constant increases again as self-absorption processes in PPO start playing a role. Studies of the efficiency of the energy transfer and its dependence on the time constants are on-going. Further details will be given in [7].

To further characterize the mixtures, emission spectra of the scintillator samples have been measured where two excitation methods have been used: excitation by ultraviolet light and by electrons of ~10 keV. For the second method, a device [8] has been used consisting of a small electron accelerator with a thin exit window. This 300 nm silicon-nitride window allows the electrons to escape from the electron gun vacuum with only ~ 10% energy loss. Both excitation methods result in the same scintillation spectra. This leads us to the conclusion that there is no dependence of the spectra on the excitation mechanism.

4. Physics potential

One of the main goals of the project is the search for proton decay. Monte Carlo calculations have shown [9] that a unique sensitivity to the proton decay channel $p \to K^+\overline{\nu}$ is obtained in a liquid-scintillator detector. For this decay mode, which is favored by some Supersymmetry theories, a sensitivity for the proton lifetime $\tau > 4 \cdot 10^{34}$ y at 90% C.L. after 10 y measuring time has been derived. This value gives an order of magnitude improvement with respect to the current best limit which has been achieved by the Superkamiokande detector [10].

In case of a supernova explosion, LENA will provide valuable information on spectra and fluxes of the arriving neutrinos. Especially $\overline{\nu}_e$ will be measured with high accuracy due to the large cross section for the reaction $\overline{\nu}_e p \to n e^+$. For an 8 $M_\odot$ supernova exploding in the center of our galaxy (10 kpc distance) a signal of ~15 000 events is expected. Recently, a study of the neutrino rates for different supernova models and for different oscillation scenarios (including possible values of θ_{13} and ν-mass hierarchy) has been performed [11]. It is shown that separation of different explosion models is possible due to the detection via the neutral current reaction on ^{12}C which is not affected by neutrino oscillations.

Even in the absence of a galactic supernova, core-collapse explosion mechanisms can be investigated by detecting the diffuse background of supernova neutrinos from explosions that happened throughout the history of the Universe. As all flavours of neutrinos and antineutrinos are produced, $\overline{\nu}_e$ can be used to measure this neutrino background by their capture on protons. In our calculations it has been shown [12], that within 10 y of exposure, various supernova models can be discriminated and information on the supernova rate in the near Universe can be gained which is closely linked to the star formation rate.

The LENA detector will be capable to detect solar neutrinos via neutrino electron scattering. Due to the high statistics on ^{7}Be neutrinos (~5400 events/day), time fluctuations, e.g. the day-night counting asymmetry which is caused by Earth matter effects, could be studied. The experience gained with the Borexino experiment [2] will significantly help to recognize and discriminate various types of background events. If the ^{11}C cosmogenic background can be successfully identified, neutrinos from the pep reaction and from the CNO cycle could be detected. Spectroscopy in this 1-2 MeV region will allow to look at the solar matter effect resonance and eventually to search for non-standard interactions [13].

KamLAND has shown [14] that $\overline{\nu}_e$ from beta decay of radioactive elements inside the Earth can be tagged. For LENA at a location in an already existing mine close to Pyhäsalmi (Finland) [16] about 1 000 events per year are expected [15]. Decays from Uranium and Thorium could be separated as the produced neutrinos have different characteristic energy spectra.

In addition, neutrino oscillation parameters could be studied with LENA. For a location close to power reactors, a good sensitivity to θ_{12} has been predicted [17]. Furthermore, the possibility to use LENA for a beta beam experiment is being investigated.

5. Summary and Outlook

A large liquid-scintillation detector of the scale of 50 kt constitutes a unique tool for neutrino astrophysics, geophysics and elementary particle physics. The good energy resolution, low energy threshold and high light yield make it a good choice for a future observatory.

Recently, LENA has joined the European initiative LAGUNA [18] (**L**arge **A**pparatus for **G**rand **U**nification and **N**eutrino **A**strophysics). In this framework, three different detector technologies with common physics interests are being studied: A 500 kt water Cherenkov detector, MEMPHYS (**ME**gaton **M**ass **PHYS**ics) [19], a 100 kt liquid Argon device, GLACIER (**G**iant **L**iquid **A**rgon **C**harge **I**maging **E**xpe**R**iment) [20] and the already described 50 kt liquid scintillator, LENA. Not only the physics potential but also common technical requirements as well as questions concerning possible underground locations are discussed within a European context.

Acknowledgments

This work has been supported by funds of the Maier-Leibnitz-Laboratorium (Garching), the Virtual Institute for Dark Matter and Neutrinos (VIDMAN, HGF), the Deutsche Forschungsgemeinschaft DFG (SFB Transregio TR27) and by the Cluster of Excellence *Origin and structure of the Universe.*

References

[1] Eguchi K et al. (KamLAND Collaboration) 2003 *Phys. Rev. Lett.* **90** 021802
[2] Arpesella C et al. (BOREXINO Collaboration) 2007 Submitted to *Phys. Lett. B (0708.2251[astro-ph])*
[3] Oberauer L, von Feilitzsch F, and Potzel W 2005 *Nucl. Phys.* B *(Proc. Suppl.)* **138** 108
[4] Marrodán Undagoitia T et al. 2006 *Prog. Part. Nucl. Phys.* **57** 283 *(hep-ph/0605229)*
[5] Wurm M Diploma thesis 2005 Technische Universität München, Physik-Department, Garching
[6] Chen M Talk at *"2nd Topical Workshop in Low Radioactivity Techniques"*, October 1-4, 2006 Aussois, France
[7] Marrodán Undagoitia T et al. To be published
[8] Morozov A et al. 2005 *Eur. Phys. J.* **33** 207 and references therein
[9] Marrodán Undagoitia T et al. 2005 *Phys. Rev.* D **72** 075014 *(hep-ph/0511230)*
[10] Kobayashi K et al. (Super-Kamiokande Collaboration) 2005 *Phys. Rev.* D **72** 052007
[11] Winter J Diploma thesis 2007 Technische Universität München, Physik-Department, Garching
[12] Wurm M et al. 2007 *Phys. Rev.* D **75** 023007 *(astro-ph/0701305)*
[13] Miranda O G, Tortola M A and Valle J W F 2006 *JHEP* **0610** 008 *(hep-ph/0406280)*
[14] Araki T et al. (KamLAND Collaboration) 2005 *Nature* **436** 499
[15] Hochmuth K et al. 2007 *Astropart. Phys.* **27** 21 *(hep-ph/0509136)*
[16] Peltoniemi J 2001 *Phys. Scripta* **T93** 102
[17] Petcov S T and Schwetz T 2006 *Phys. Lett* **B642** 487
[18] Autiero D et al. 2007 *(0705.0116 [hep-ph])*
[19] de Bellefon A et al. 2006 *(hep-ex/0607026)*
[20] Eridato A and Rubbia A 2005 *Nucl. Phys. (Proc. Suppl.)* **139** 301

Collective neutrino flavor transitions in Supernovae: Analytical and numerical aspects

G.L. Fogli[1,2], **E. Lisi**[2*], **A. Marrone**[1,2], **A. Mirizzi**[1,2,3]

[1] Dipartimento Interateneo di Fisica "Michelangelo Merlin", Via Orabona 4, 70126 Bari, Italy
[2] Istituto Nazionale di Fisica Nucleare, Sezione di Bari, Via Orabona 4, 70126 Bari, Italy
[3] Max-Planck-Institut für Physik, Föhringer Ring 6, 80805 München, Germany
[*] Paper presented at the Conference by E. Lisi

E-mail: `eligio.lisi@ba.infn.it`

Abstract. Non-linear effects on supernova neutrino oscillations, associated with neutrino-neutrino interactions, are known to induce collective flavor transformations near the supernova core for $\theta_{13} \neq 0$. For typical electron density profiles (as taken from shock-wave simulations at a few seconds after bounce) these transformations precede ordinary matter effects, and become more amenable to both numerical computations and analytical interpretations in inverted hierarchy—while they basically vanish in normal hierarchy. We numerically evolve the neutrino density matrix in the region relevant for self-interaction effects, using thermal spectra and a representative value $\sin^2\theta_{13} = 10^{-4}$. Our results neatly show the collective phenomena of synchronization, bipolar oscillations, and spectral split, with analytically understandable features. They also suggest that averaging over neutrino trajectories plays a minor role in the final outcome. The split/swap of (anti)neutrino spectra emerges as an unmistakable signature of the inverted neutrino hierarchy.

1. Introduction

Neutrinos emitted by core-collapse supernovae (SN) represent an important laboratory for both particle physics and astrophysics. While propagating in the dense SN environment, they can feel not only the presence of background matter (via ordinary Mikheev-Smirnov-Wolfenstein, MSW, effects) but also of the gas of neutrino and antineutrinos (via self-interaction effects). The self-interactions appear to modify the flavor evolution of SN neutrinos in a collective way, completely different from ordinary MSW effects. However, the analysis is made complicated by the non-linear nature of the flavor evolution equations, and by the fact that neutrino-neutrino interactions depend on the intersection angle ϑ_{ij} of their trajectories [1].

We have recently contributed to the analysis of such collective neutrino flavor transitions in SN in the work [2], to which we refer the reader for further details and references. In particular, we have shown that, in the context of typical SN explosion simulations, collective effects (which precede MSW effects) are not strongly affected by a detailed (multi-angle) description of the ϑ_{ij} distribution. Indeed, one of the main observable effects arising for nonzero mixing angle θ_{13} in inverted neutrino mass hierarchy—the so-called spectral swap and split—is already captured by the (single-angle) approximation where the ϑ_{ij}'s are replaced by an average angle. These results will be briefly illustrated in the following. See also the talk [3] at this Conference for a study of single-angle *vs* multi-angle simulations with variable neutrino-antineutrino asymmetry.

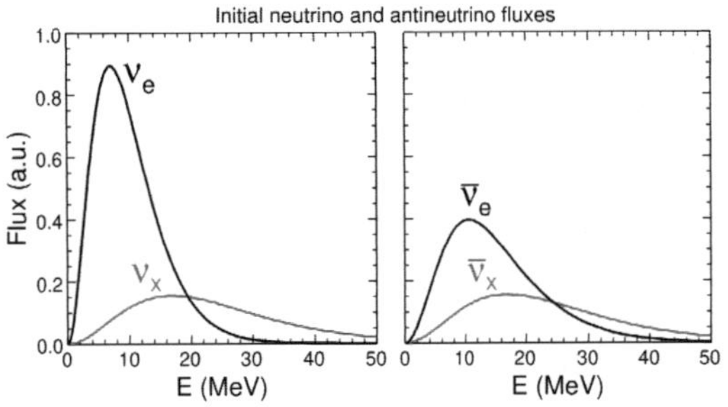

Figure 1. Initial neutrino and antineutrino fluxes at the neutrinosphere ($r \simeq 10$ km) in arbitrary units.

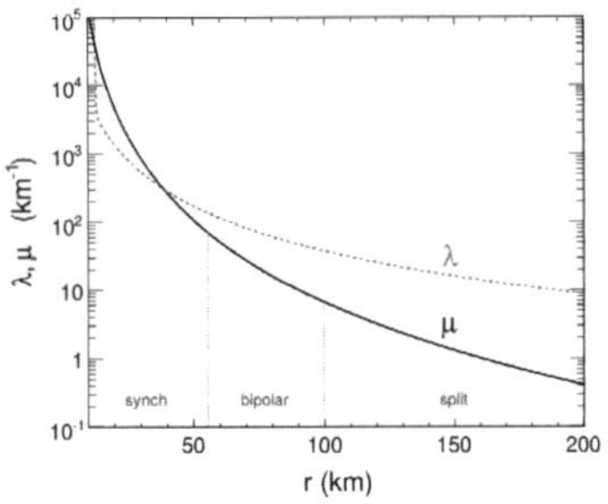

Figure 2. Radial profile of the λ and μ parameters defined in the text.

2. Supernova neutrino input

Three main quantities with dimension $[T^{-1}]$ ("frequencies") characterize the flavor evolution of SN neutrinos along the radial coordinate r, namely, the energy E or the associated vacuum oscillation frequency ω,

$$\omega = \frac{\Delta m^2}{2E} \ , \tag{1}$$

(in terms of the largest squared mass difference $\Delta m^2 = m_3^2 - m_{1,2}^2$), the interaction energy difference λ in matter between ν_e and ν_x (where $x = \mu, \tau$),

$$\lambda(r) = \sqrt{2}G_F\, N_{e^-}(r) \ , \tag{2}$$

$N_{e^-}(r)$ being the net electron number density at the point r, and the self-interaction strength parameter $\mu(r)$,

$$\mu(r) = \sqrt{2}G_F\, [N(r) + \overline{N}(r)] \ , \tag{3}$$

where N and $\overline{N}$ are the effective neutrino and antineutrino densities, respectively.

We fix the above frequency parameters according to SN shock-wave simulations (a few seconds after bounce) as shown in Fig. 1 (energy spectra) and in Fig. 2 (interaction energies). Figure 2 also shows the expected ranges where the so-called synchronized oscillations (joint flavor evolution of neutrinos and antineutrinos at high density), bipolar oscillations (partially disjoint flavor evolution of neutrinos and antineutrinos at intermediate densities) and spectral split effects (with freeze-out of energy spectra at low density) are analytically expected to take place. Ordinary MSW effects (if any) occur only later, beyond the radial range of Fig. 2.

3. Equations of motion (EOM)

In a two-family (ν_e, ν_x) scenario, the 2×2 neutrino and antineutrino density matrices in flavor basis can be decomposed onto Pauli matrices via "polarization" (Bloch) vectors $\mathbf{P}$ and $\overline{\mathbf{P}}$, respectively, at any fixed energy. In single-angle approximation, the individual polarization vectors obey the following coupled equations of motion in flavor space,

$$\dot{\mathbf{P}} = (+\omega\mathbf{B} + \lambda\mathbf{z} + \mu\mathbf{D}) \times \mathbf{P} \ , \tag{4}$$

$$\dot{\overline{\mathbf{P}}} = (-\omega\mathbf{B} + \lambda\mathbf{z} + \mu\mathbf{D}) \times \overline{\mathbf{P}} \ , \tag{5}$$

where $\mathbf{D} = \mathbf{J} - \overline{\mathbf{J}}$, and $\mathbf{J}$, $\overline{\mathbf{J}}$ are the integral (energy-averaged) polarization vectors. The z-component of the polarization vectors is directly linked to the ν_e survival probability (e.g., inversion of z-component corresponds to complete flavor change). In the multi-angle case, a further dependence on the neutrino trajectory crossing angle arises in the above equations.

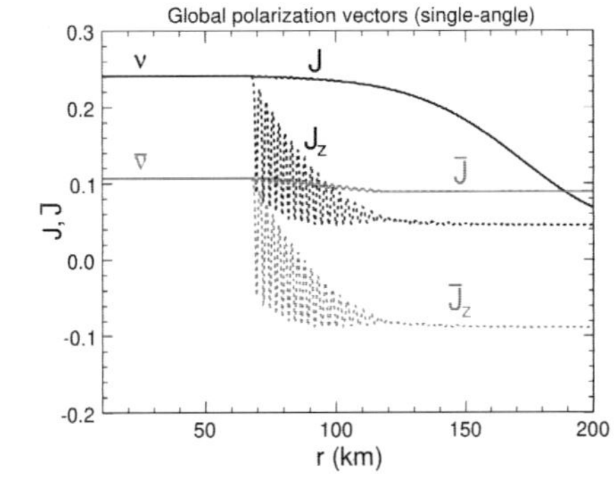

Figure 3. Global polarization vectors in single-angle approximation.

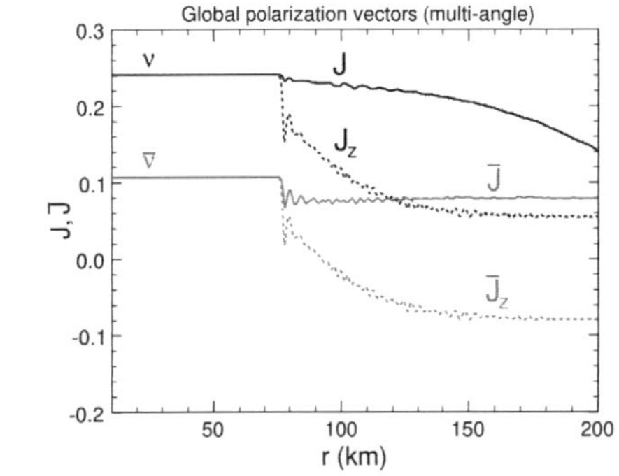

Figure 4. Global polarization vectors in multi-angle calculation.

4. Results and conclusions

We have solved numerically the EOM in the single-angle and multi-angle cases. Many of the observed features can be understood analytically. The main results are the following.

Figures 3 and 4 show the behavior of the norm and of the z-component of the global polarization vectors for neutrinos and antineutrinos, assuming inverted neutrino hierarchy (i.e., $\Delta m^2 < 0$). We have fixed the mixing angle $\theta_{13} = 10^{-2}$, but its exact value has little relevance (as far as it is nonzero). The multi-angle case appears to be a "smeared" version of the single-angle one, but the average behavior and the final states at the end of collective effects ($r \sim 200$ km) are similar. Figures 5 and 6 display the energy spectra at the end of collective effects in single-angle and multi-angle calculations. In both cases, a comparison with Fig. 1 shows that the ν_e and ν_x spectra are split and swapped above a critical neutrino energy, while the $\overline{\nu}_e$ and $\overline{\nu}_x$ spectra are basically interchanged. This peculiar feature (spectral split/swap) appears to be a robust, important signature of self-interaction effects in inverted hierarchy, which cannot be mimicked by ordinary MSW effects. See [2] for further details, explanations and references.

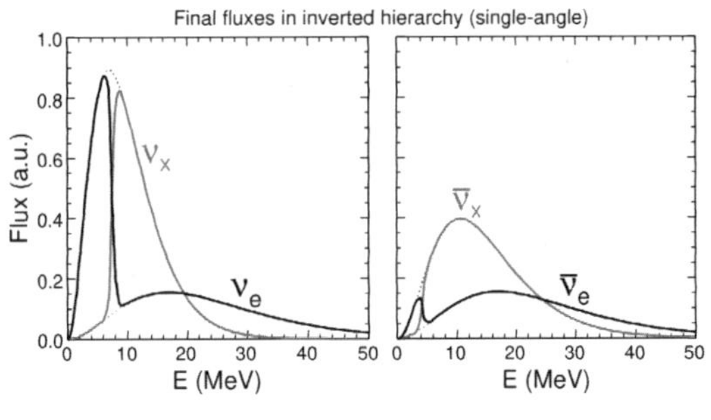

Figure 5. Final fluxes in single-angle approxim.

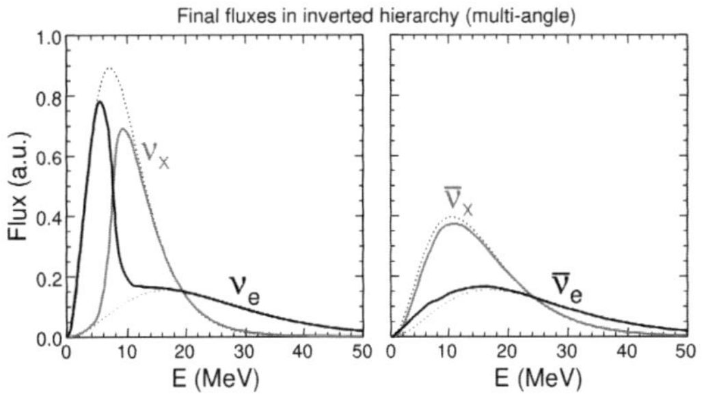

Figure 6. Final fluxes in multi-angle calculation.

Acknowledgments

This work is supported in part by the Italian INFN and MIUR through the "Astroparticle Physics" research project. In Germany, A.M. is also supported by an Alexander von Humboldt fellowship grant.

References

[1] H. Duan, G. M. Fuller, J. Carlson and Y. Z. Qian, Phys. Rev. D **74**, 105014 (2006).
[2] G.L. Fogli, E. Lisi, A. Marrone, and A. Mirizzi, JCAP **12** (2007) 010.
[3] Talk by G.G. Raffelt, these Proceedings.

Constraints on oscillation parameters through the MSW effect of supernova neutrinos

Shiou Kawagoe[1], T Kajino[1], H Suzuki[2], K Sumiyoshi[3] and S Yamada[4]

[1]Department of Astronomical Science, School of Physical Sciences, The Graduate University for Advanced Studies(SOKENDAI), National Astronomical Observatory of Japan, 2-21-1 Osawa, Mitaka-shi, Tokyo 181-8588, Japan

[2]Department of Physics, Faculty of Science and Technology, Tokyo University of Science, 2641 Yamazaki, Noda-shi, Chiba 278-8510, Japan

[3]Numazu College of Technology, 3600 Ooka, Numazu-shi, Shizuoka 410-8501, Japan

[4]Department of Physics, Faculty of Science and Engineering, Waseda University, 3-4-1 Okubo, Shinjuku-ku, Tokyo 169-8555, Japan

E-mail: shiou@th.nao.ac.jp

Abstract. We try to constrain the neutrino oscillation parameters by studying the MSW matter effect on supernova neutrinos. Using an implicit Lagrangian code for general relativistic spherical hydrodynamics (Yamada,1997), we calculate the propagation of shock waves passing through the stellar envelopes over dozens of seconds, which is based on modeling the adiabatic collapse of iron core.

We study how quantitatively the influence of the shock propagation appears in the deformation of neutrino energy spectrum. In addition, we are able to examine the θ_{13} dependence of time evolution of the neutrino detection rates at SK. Using our supernova model, we found that the MSW effect appears, at first, from low-energy region and moves toward high-energy region as the shock wave propagates. We find that the shock propagation has strong influence on the supernova neutrino oscillation through the evolution of density profile induced by the shock propagation.

1. Introduction

There are still a lot of mysteries on the mechanism of the core-collapsed supernovae and the nature of supernova neutrinos. 99 % of the gravitational energy for the core collapse is released as neutrinos. The neutrino oscillation has been studied in various experiments [1] But, we cannot determine the remaining three neutrino oscillation parameters, i.e. the sign of mass difference (Δm_{13}) , the mixing angle (θ_{13}) and the CP violating phase (δ).

It has recently been found that the shock wave changes the density profile around the resonance point in a few seconds after the core bounce. The effects of the shock wave appear as a decrease in average energy of ν_e in the case of normal mass hierarchy (or $\bar{\nu_e}$ in the case of inverted mass hierarchy) at stellar surface [2]. If $\sin^2 2\theta_{13}$ is large, the average neutrino energy as a function of time decreases according to shock propagation [3]. The neutrino event number is found to be different according to the θ_{13} value (large or small)[4]. The time dependence of the event rates was calculated in various cases of θ_{13} [5] and they discussed the influence of the shock wave on the neutrino event rate for a specific energy. They however used simplification of parameterized shock-wave profiles.

In contrast, we followed the shock wave propagation for a long time numerically, which is generated by a model of adiabatic collapse adopted in our hydrodynamics calculation, and obtained more realistic density profile. Then we apply the profile to the neutrino oscillation phenomena in order to examine the θ_{13} dependence of time evolution of the neutrino event rate in all energy ranges from 5MeV to 60MeV.

2. Numerical Method

As the first step, we intend to construct an approximate models of prompt explosion using an implicit Lagrangian code for general relativistic spherical hydrodynamics [6], we perform simplified calculations of core collapse and bounce by following adiabatic collapse with fixed electron fraction. We adopt the presupernova model of $15M_{\odot}$ star provided by Woosley and Weaver [8]. The numerical tables of Shen's relativistic equation of state (EOS) [9] and Timmes's EOS [10] are adopted for the high and low density matters, respectively. We succeeded in calculating propagation of shock waves which are generated by adiabatic collapse of iron cores and pass into the stellar envelopes[14]. In our simulation, the density behind the shock wave hardly decreases because we neglected neutrino cooling of the proto-neutron star. Detailed study of the cooling effect of the proto-neutron star is underway.

In order to calculate neutrino oscillation effect on the spectrum, we have to solve the time evolution of the neutrino wave function along the density profile of our supernova interior [11]. We here put the CP violating phase δ equal to zero in the CKM matrix. The neutrino oscillation parameters are taken from the analys of the various observations [12], except for θ_{13}; $\sin^2 2\theta_{12}$=0.84, $\sin^2 2\theta_{23}$=1.00, Δm^2_{12}=8.1×10^{-5}eV and Δm^2_{23}=2.2×10^{-3}eV. Using four values of $\sin^2 2\theta_{13} = 10^{-2}, 10^{-3}, 10^{-4}$ and 10^{-5}, we calculate the neutrino survival probabilities. The neutrino energy spectra from supernova are obtained by multiplying the survival probability by original neutrino spectra that change in time [13], We assume detection at the Super-Kamiokande (SK) and Sudbury Neutrino Observatory (SNO). SK is filled with 32,000 ton pure water, and SNO is filled with 1,000 ton heavy water. The finite energy resolution of the detector was neglected here. The event number is obtained by integrating over the angular distribution of the events. We assumed a supernova at the center of the Milky Way (10[kpc]) and neglected the Earth effect.

3. Result and Discussion

Figure 1 show the expected event rates of $\bar{\nu_e}$ in the case of inverted mass hierarchy. The upper part shows the expected event rates in the SK as a function of time. This figure shows that the behavior of event rate is roughly divided into two features. When $\sin^2 2\theta_{13}$ is large, large $\bar{\nu_e}$ event rate is expected. When θ_{13} is large, the $\bar{\nu_e}$ neutrinos pass into the H-resonance in adiabatic condition, and $\bar{\nu_e}$ and $\bar{\nu_x}$ completely convert to each other. The average energy of neutrino is different ($\bar{E_{\nu_e}} < \bar{E_{\bar{\nu_e}}} < \bar{E_{\nu_x}}$). Therefore, the average energy of $\bar{\nu_e}$ becomes higher, and the event rate of $\bar{\nu_e}$ of large θ_{13} is larger than the event rate of $\bar{\nu_e}$ of small θ_{13}. The lower part of Figure 1 shows the ratio of the event rates of $\bar{\nu_e}$ with and without shock in the case of the inverted mass hierarchy. The shock front does not reach H-resonance point before 3 seconds. The event rate with shock is smaller than that without shock after 3 seconds. This is because the influence of the shock wave appears mainly for high-energy neutrinos as the shock propagates, and the number of high energy $\bar{\nu_e}$ decreases. On the other hand, in the case of small θ_{13}, the difference is not clearly seen at any time. We hence confirm that the influence of the shock wave appears at late times only if $\sin^2 2\theta_{13}$ is large. In the case of the inverted mass hierarchy, the adiabaticity of $\bar{\nu_e}$ is influenced by the shock wave, because $\bar{\nu_e}$ is related to the H-resonance [15]. On the other hand, the expected event rates of ν_e in the SK hardly change in the case of the inverted mass hierarchy.

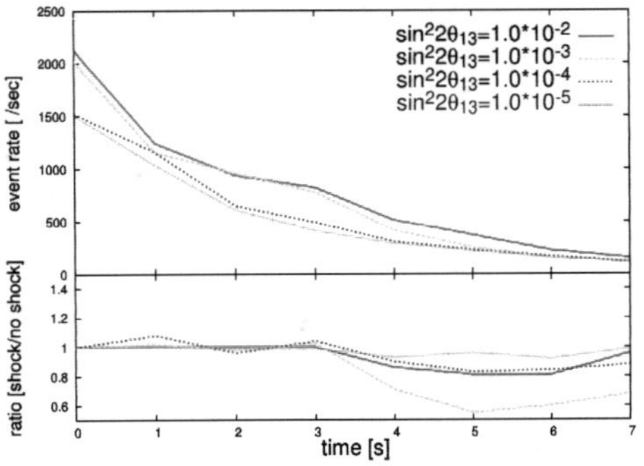

Figure 1. The upper part shows the expected event rates of $\bar{\nu_e}$ in the case of inverted mass hierarchy in the SK as a function of time, and the lower part shows the ratio of event rate with shock and without shock.

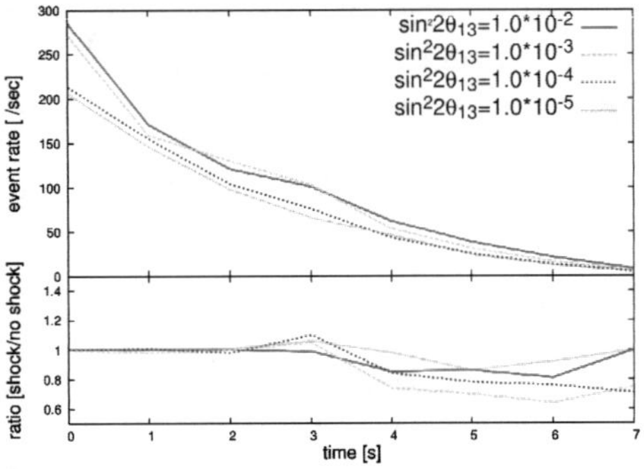

Figure 2. The upper part shows the expected event rates of ν_e in the case of normal mass hierarchy in the SNO as a function of time, and the lower part shows the ratio of event rate with shock and without shock.

Figure 2 is the same as Figure 1, but the expected event rates of ν_e in the case of normal mass hierarchy in the SNO. The upper part shows the expected event rates of ν_e as a function of time and the lower part shows the ratio of event rate with shock and without shock. This figure shows similar behavior to Figure 1. In the case of normal mass hierarchy, the adiabaticity of ν_e is influenced by the shock wave, because ν_e is related to the H-resonance. Therefore, the behavior of Figure 2 is almost the same as the Figure 1.

We point out that the constraint on θ_{13} can be inferred by the integrated event rate over the whole energy range. When $\sin^2 2\theta_{13}$ is large, the influence of the shock wave appears after a few seconds in the observation. It is time that the shock wave reached H-resonance. Therefore, observing the time evolution of the event, we would limit $\sin^2 2\theta_{13}$ and know the propagation of shock wave inside the star for large $\sin^2 2\theta_{13}$. If the detailed information of supernova neutrinos are obtained in future supernova events, it may reveal the appearance of the propagation of the shock wave by comparing the observation with our theoretical predication. Moreover, it is possible to feed back to the construction of the theoretical model of supernova explosion and to tie with clarification of more detailed explosion mechanism.

References

[1] Kajita T and Super-KAMIOKANDE Collaboration 2002 *nmgm.meet.* Part**A** 203
[2] Takahashi K, Sato K, Dalhed H E and Wilson J R 2003 *Astropart. Phys.* **20** 189
[3] Tomàs R, Kachelrieß M, Raffelt G, Dighe A, Janka H T and Scheck L 2004 *J. Cosmol. Astropart. Phys.* JCAP09(2004)015
[4] Takahashi K and Sato K 2003 *Nucl. Phys.* A **718** 455
[5] Fogli G, Lisi E, Mirizzi A and Montanino D 2006 *J. Cosmol. Astropart. Phys.* JCAP06(2006)012
[6] Yamada S 1997 *Astrophys. J.* **475** 720
[7] Sumiyoshi K, Suzuki H, Yamada S and Toki H 2004 *Nucle. Phys.* A **730** 227
[8] Woosley S E and Weaver T 1995 *Asrtophys. J. S.* **101** 181
[9] Shen H, Toki H, Oyamatsu K and Sumiyoshi K 1998 *Nucl. Phys.* A **637** 435
[10] Timmes F X and Arnett D 1999 *Astrophys. J. S.*. **125** 277
[11] Takahashi K and Sato K 2003 *Prog. Theor. Phys.* **109** (2003) 919
[12] Maltoni M, Schwetz T, Tortola M A and Valle J W F 2004 *New. J. Phys.* **06** 122
[13] Dighe A S, Smirnov A Y and Alexei Yu 2000 *Phys. Rev.* D **62** 3007
[14] Kawagoe S, Kajino T, Suzuki H, Sumiyoshi K and Yamada S 2006 *J. Phys.: Conf. Series* **39** 294
[15] Fukugita M and Yanagida T 2003 *Physics of Neutrinos and Applications to Astrophysics* (Berlin: Springer)

Multi-angle effects in collective supernova neutrino oscillations

A Esteban-Pretel[1], S Pastor[1], R Tomàs[1], G G Raffelt[2,*], G Sigl[3]

[1] AHEP Group, Institut de Física Corpuscular, CSIC/Universitat de València, Edifici Instituts d'Investigació, Apt. 22085, 46071 València, Spain

[2] Max-Planck-Institut für Physik (Werner-Heisenberg-Institut), Föhringer Ring 6, 80805 München, Germany

[3] II. Institut für Theoretische Physik, Universität Hamburg, Luruper Chaussee 149, 22761 Hamburg, Germany

[*] Paper presented at the conference by G G Raffelt

E-mail: raffelt@mppmu.mpg.de

Abstract. We study two-flavor collective neutrino oscillations in the dense-neutrino region above the neutrino sphere in a supernova (SN). The angular dependence of the neutrino-neutrino interaction potential causes "multi-angle" effects that can lead either to complete kinematical decoherence in flavor space or only to small differences between different trajectories. This nonlinear system switches abruptly between "self-maintained coherence" and "self-induced decoherence" among the angular modes, depending on the strength of the deleptonization flux. For a realistic SN the quasi single-angle behavior is probably typical, simplifying the numerical treatment and probably allowing for the survival of observational features of flavor oscillations.

The neutrinos streaming off a collapsed supernova (SN) core are so dense near the neutrino sphere that they produce a significant refractive effect for each other, leading to collective oscillation effects. The practical importance of these nonlinear phenomena was only recently recognized and studied in a series of papers [1, 2, 3, 4, 5, 6, 7, 8, 9, 10, 11, 12, 13, 14, 15]. An introduction was given at this conference by Eligio Lisi [16].

The SN neutrino fluxes are thought to obey the hierarchy $F_{\nu_e} > F_{\bar{\nu}_e} > F_{\nu_x}$ where ν_x stands for any of ν_μ, ν_τ, $\bar{\nu}_\mu$, and $\bar{\nu}_\tau$. In other words, there is an excess of $\nu_e\bar{\nu}_e$ pairs relative to $\nu_x\bar{\nu}_x$. In an inverted hierarchy situation, this pair excess converts collectively into $\nu_x\bar{\nu}_x$ pairs, a process that does not violate flavor-lepton number and thus does not require mixing: it could also proceed as an ordinary pair annihilation process. Neutrino refraction causes this pair process to proceed collectively and very fast, almost independently of the mixing angle, i.e., we have to do with a collective "speed-up effect" [1, 2]. The unpaired ν_e excess flux from deleptonization is conserved [5]. In the adiabatic limit, it is the low-energy part of the ν_e spectrum that survives, leading to a step-like feature in the ν_e spectrum (a "spectral split" [9, 10] caused by a "step-wise spectral swapping" [4, 12]). All of these effects happen in the dense-neutrino region that typically extends from the neutrino sphere out to a few hundred kilometers. If the ordinary MSW resonances occur at larger radii, the collective phenomena and subsequent MSW transformations are independent, the former producing the initial condition for the latter. For an inverted hierarchy the only effect of matter in the collective-transformation region is a decrease of the effective mixing angle [5, 8]. Conversely, if the matter density profile is very shallow, the ordinary MSW effect can prepare the initial condition for the collective effects [14].

Mixed neutrinos are described by matrices of density $\rho_{\mathbf{p}}$ and $\bar\rho_{\mathbf{p}}$ for each (anti)neutrino mode. The diagonal entries are the usual occupation numbers whereas the off-diagonal terms encode phase information. The equations of motion are $\mathrm{i}\partial_t \varrho_{\mathbf{p}} = [\mathsf{H}_{\mathbf{p}}, \varrho_{\mathbf{p}}]$, where the Hamiltonian is [17]

$$\mathsf{H}_{\mathbf{p}} = \Omega_{\mathbf{p}} + \mathsf{V} + \sqrt{2}\,G_{\mathrm{F}} \int \frac{\mathrm{d}^3\mathbf{q}}{(2\pi)^3} \left(\varrho_{\mathbf{q}} - \bar\varrho_{\mathbf{q}}\right)\left(1 - \mathbf{v_q}\cdot\mathbf{v_p}\right), \tag{1}$$

$\mathbf{v_p}$ being the velocity. In the mass basis, the matrix of vacuum oscillation frequencies is $\Omega_{\mathbf{p}} = \mathrm{diag}(m_1^2, m_2^2, m_3^2)/2E$. The matter effect is represented, in the weak interaction basis, by $\mathsf{V} = \sqrt{2}\,G_{\mathrm{F}} n_B\,\mathrm{diag}(Y_e, 0, Y_\tau^{\mathrm{eff}})$. For antineutrinos the only difference is $\Omega_{\mathbf{p}} \to -\Omega_{\mathbf{p}}$. The effective tau-lepton density $Y_\tau^{\mathrm{eff}} \approx 10^{-5}$ arises from radiative corrections [18] and can be important in a genuine three-flavor treatment of ordinary [19] or collective [15] SN neutrino oscillations. For the latter, the influence of Y_τ^{eff} can be rather sensitive to deviations from maximal 23-mixing.

The angular factor $(1-\mathbf{v_q}\cdot\mathbf{v_p})$ in Eq. (1) derives from the current-current nature of the weak-interaction Hamiltonian. In an isotropic ensemble it averages to unity, whereas the neutrinos streaming off a SN core are strongly non-isotropic so that different angular modes experience a different strength of the neutrino-neutrino interaction potential. One may expect that this effect leads to kinematical decoherence among angular modes and thus to flavor equilibrium [1, 2]. In a symmetric ensemble with equal densities of neutrinos and antineutrinos, this is indeed the case. Such a system is highly unstable in that an infinitesimal deviation from exact isotropy is enough to trigger an exponential run-away towards flavor equilibrium [7]. On the other hand, a numerical simulation of the flavor evolution of SN neutrinos revealed that multi-angle effects were small: All angular modes evolved nearly collectively, very similar to an isotropic ensemble [4]. Likewise, in a multi-energy system every energy mode feels a different Hamiltonian, yet the evolution is collective, i.e., multi-energy effects do not lead to kinematical decoherence.

We have performed a numerical exploration of multi-angle effects in a spherically symmetric system where the neutrinos are emitted from a "neutrino sphere" [8]. We have identified the asymmetry between the ν_e and $\bar\nu_e$ flux as the crucial parameter. Realistic deleptonization fluxes in SNe seem sufficient to suppress multi-angle decoherence. Therefore, in practice multi-angle effects seem to be a subdominant feature of collective SN transformations.

A better understanding of collective oscillations can be developed in the two-flavor case in terms of the usual flavor polarization vectors $\mathbf{P_p}$ to express the matrices $\rho_{\mathbf{p}}$. The equations of motion are $\partial_t \mathbf{P_p} = \mathbf{H_p} \times \mathbf{P_p}$. Ignoring the ordinary matter effect, the "Hamiltonian" is

$$\mathbf{H_p} = \omega\mathbf{B} + \sqrt{2}\,G_{\mathrm{F}} \int \frac{\mathrm{d}^3\mathbf{q}}{(2\pi)^3} \left(\mathbf{P_q} - \bar{\mathbf{P}}_{\mathbf{q}}\right)\left(1 - \mathbf{v_q}\cdot\mathbf{v_p}\right), \tag{2}$$

where $\mathbf{B}$ is a unit vector in flavor space in the "mass direction" and $\omega = \Delta m^2/2E$ the vacuum oscillation frequency. In an isotropic situation where the velocity terms average to zero, the second term is of the form $\mu\,\mathbf{D}$ where the vector $\mathbf{D}$ is the difference between the total neutrino and antineutrino polarization vectors and $\mu = \sqrt{2}G_{\mathrm{F}}\,n_\nu$ the neutrino-neutrino interaction strength. The individual $\mathbf{H_p} = \omega\mathbf{B} + \mu\mathbf{D}$ all lie in a single plane. In the adiabatic limit the polarization vectors follow the Hamiltonians so that they also lie in this co-rotating plane. This observation explains the collective nature of the multi-energy evolution (all polarization vectors stay in the same plane that rotates around $\mathbf{B}$ with a certain frequency ω_{c}) and explains the spectral splits in that the final $\mathbf{H_p}$ in the co-rotating plane are either aligned or anti-aligned with $\mathbf{B}$ [9, 10]. This picture also illustrates a crucial difference to the symmetric system where initially $\mathbf{D} = 0$ and the overall evolution is purely pendular [5]—there is no co-rotating plane.

In a spherically symmetric situation every mode is characterized by its vacuum oscillation frequency ω and its angle θ relative to the radial direction, providing $v = \cos\theta$ as the radial velocity. Now the individual Hamiltonians are $\mathbf{H_p} = \omega\mathbf{B}+\mu(\mathbf{D}-v_\mathbf{p}\mathbf{F})$ where $\mathbf{F}$ is the flux term of the neutrino ensemble, i.e., the same as $\mathbf{D}$, but every mode weighted with $v_\mathbf{p}$. If all polarization vectors lie in a single plane, then also all $\mathbf{H_p}$ are in that plane so that an evolution in a co-rotating plane is self-consistently possible. Our numerical simulations show that those cases with little kinematical decoherence correspond to the polarization vectors essentially staying in a co-rotating plane with some zenith-angle spread. On the other hand, the vectors $\mathbf{B}$, $\mathbf{D}$ and $\mathbf{F}$ do not have to stay in a single plane and in fact, this appears to be an unstable arrangement. Kinematical multi-angle decoherence corresponds to strong deviations from this coplanar situation. In a spherically symmetric situation, the neutrino-neutrino interaction strength decreases with r^{-4}. In a toy model with an artificially slow decrease of $\mu(r)$ one can make the evolution arbitrarily slow and adiabatic. In this case multi-angle decoherence appears to be unavoidable.

It appears that in a realistic SN the collective evolution is slow enough to be essentially adiabatic with regard to the development of a spectral split, but fast enough that the co-planar arrangement of the polarization vectors survives. This picture may provide the key to an analytic understanding of the conditions for multi-angle kinematical decoherence.

The assumption of spherical symmetry of the overall system severely restricts possible solutions. The evolution is a one-dimensional problem along the radial direction. It remains to be studied if this symmetry has an important impact on kinematical decoherence, i.e., if a system with fewer symmetries would decohere more easily.

Acknowledgments

This work was partly supported by the Deutsche Forschungsgemeinschaft (TR-27 "Neutrinos and Beyond"), by the Cluster of Excellence "Origin and Structure of the Universe" (Garching and Munich), by the European Union (contracts No. RII3-CT-2004-506222 and MRTN-CT-2004-503369), and by the Spanish grants FPA2005-01269 (MEC) and ACOMP07-270 (Generalitat Valenciana). AE was supported by an FPU grant from the Spanish Government. SP and RT were supported by MEC contracts (*Ramón y Cajal* and *Juan de la Cierva*, respectively).

References

[1] R. F. Sawyer, hep-ph/0408265.
[2] R. F. Sawyer, Phys. Rev. D **72**, 045003 (2005) [hep-ph/0503013].
[3] H. Duan, G. M. Fuller and Y. Z. Qian, Phys. Rev. D **74**, 123004 (2006) [astro-ph/0511275].
[4] H. Duan, G. M. Fuller, J. Carlson and Y. Z. Qian, Phys. Rev. D **74**, 105014 (2006) [astro-ph/0606616].
[5] S. Hannestad, G. G. Raffelt, G. Sigl and Y. Y. Y. Wong, Phys. Rev. D **74**, 105010 (2006) [astro-ph/0608695].
[6] H. Duan, G. M. Fuller, J. Carlson and Y. Z. Qian, Phys. Rev. D **75**, 125005 (2007) [astro-ph/0703776].
[7] G. G. Raffelt and G. Sigl, Phys. Rev. D **75**, 083002 (2007) [hep-ph/0701182].
[8] A. Esteban-Pretel, S. Pastor, R. Tomàs, G. G. Raffelt and G. Sigl, Phys. Rev. D, in press (2007) [astro-ph/0706.2498].
[9] G. G. Raffelt and A. Yu. Smirnov, Phys. Rev. D **76**, 081301 (2007) [arXiv:0705.1830].
[10] G. G. Raffelt and A. Yu. Smirnov, Phys. Rev. D **76**, 125008 (2007) [arXiv:0709.4641].
[11] H. Duan, G. M. Fuller and Y. Z. Qian, Phys. Rev. D **76**, 085013 (2007) [arXiv:0706.4293].
[12] H. Duan, G. M. Fuller, J. Carlson and Y. Z. Qian, Phys. Rev. Lett., in press [arXiv:0707.0290].
[13] G. L. Fogli, E. Lisi, A. Marrone and A. Mirizzi, JCAP, in press [hep-ph/0707.1998].
[14] H. Duan, G. M. Fuller, J. Carlson and Y. Z. Qian, arXiv:0710.1271.
[15] A. Esteban-Pretel, S. Pastor, R. Tomàs, G. G. Raffelt and G. Sigl, arXiv:0712.1137.
[16] E. Lisi, "Collective neutrino flavor transitions in supernovae: Analytical and numerical aspects", talk at TAUP 2007, 11–15 Sept. 2007, Sendai, Japan.
[17] G. Sigl and G. Raffelt, Nucl. Phys. B **406**, 423 (1993).
[18] F. J. Botella, C. S. Lim and W. J. Marciano, Phys. Rev. D **35**, 896 (1987).
[19] E. K. Akhmedov, C. Lunardini and A. Yu. Smirnov, Nucl. Phys. B **643**, 339 (2002) [hep-ph/0204091].

KamLAND (Anti-Neutrino Status)

Itaru Shimizu for the KamLAND Collaboration

Research Center for Neutrino Science, Tohoku University,
Aramaki Aoba, Aoba, Sendai, Miyagi 980-8578, Japan

E-mail: `shimizu@awa.tohoku.ac.jp`

Abstract. The KamLAND experiment is a long baseline experiment, designed to explore the neutrino oscillation of $\bar{\nu}_e$'s from reactors, at a flux-weighted average distance of 180 km. The sensitivities to oscillation parameters are much improved with enhanced statistics of 2881 ton-yr data set, added data lowering the energy threshold, and suppressed systematic uncertainties. As a result, we firmly established the spectral distortion of reactor $\bar{\nu}_e$'s at $>5\sigma$, and other hypotheses for neutrino disappearance, neutrino decay and decoherence, are strongly disfavored.

1. Introduction

The KamLAND experiment searched for the neutrino oscillation of $\bar{\nu}_e$'s emitted from distant power reactors. Previously, KamLAND revealed a significant deficit of $\bar{\nu}_e$'s and excluded all solar neutrino solutions but the LMA solution assuming CPT invariance [1], followed by direct evidence of spectral distortion of the reactor $\bar{\nu}_e$'s above the neutrino energy 3.4 MeV [2]. In this report, we have extended the analysis down to the energy threshold of inverse β-decay ($\bar{\nu}_e + p \rightarrow e \;\; + n$), and suppressed the systematic and background uncertainties. The exposure to reactor $\bar{\nu}_e$'s is almost 4 times over previous results owe to the more data collection and the enlarged fiducial volume radius from 5.5 m to 6 m. These improvements contributed the precise measurement of neutrino oscillation parameters.

2. Efficiency and systematic uncertainties

Electron anti-neutrino events are selected by the delayed coincidence. Prompt scintillation light from the e gives information on the incident $\bar{\nu}_e$ energy, $E_{\bar{\nu}_e} \simeq E_{\rm p} + \overline{E}_n + 0.8$ MeV, where $E_{\rm p}$ is the prompt event energy including the positron kinetic energy and the annihilation energy, and $\overline{E}_n$ is the average neutron recoil energy. Neutron capture on hydrogen emits a 2.2 MeV γ-ray with $\sim$ 200 μs delay and is a powerful tool for reducing backgrounds.

The detection efficiency and its uncertainties are evaluated with MC simulations. The anti-neutrino selection is based on the discriminator $L = \frac{f_{\bar{\nu}_e}}{f_{\bar{\nu}_e} \;\; f_{acc}}$, where $f_{\bar{\nu}_e}$ and f_{acc} are probability density function (PDF) for $\bar{\nu}_e$ signals and accidental backgrounds, as a function of prompt and delayed energies, space and time correlation, radial distances from the detector center ($E_{\rm p}$, $E_{\rm d}$, ΔR, ΔT, $R_{\rm p}$, $R_{\rm d}$). For the discrimination of accidental backgrounds, we determined a selection value $L_{cut}(E_{\rm p})$ to get the maximal figure of merit $\frac{S}{\sqrt{S \;\; B_{acc}}}$ for each prompt energy interval of 0.1 MeV. We evaluated the efficiency uncertainties by comparing MC simulations with the Ge and Am Be calibration data. Table 1 summarizes the systematic uncertainties on the expected event rate of reactor $\bar{\nu}_e$'s, and the overall uncertainty is 4.1%. The fiducial volume uncertainty

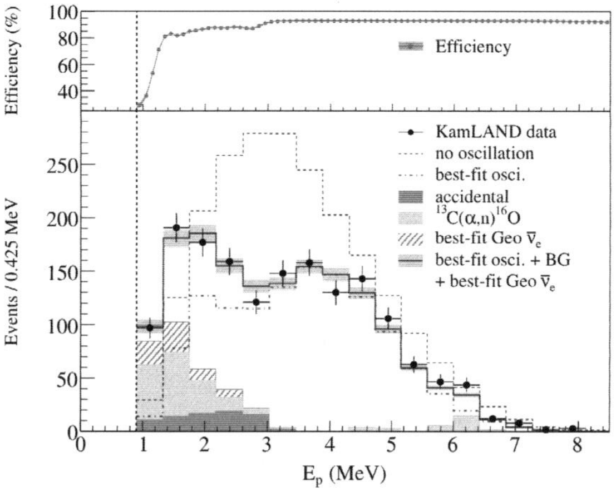

Figure 1. Energy spectrum of the observed prompt events, along with the expected no oscillation spectrum and best-fit including neutrino oscillations. The detection efficiency has an energy dependence as shown in the top panel.

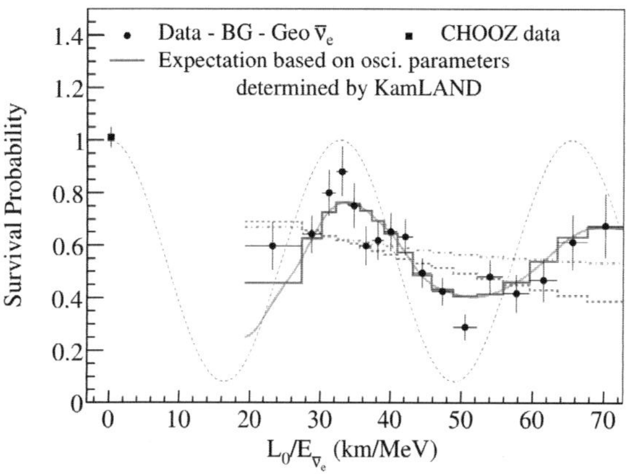

Figure 2. Ratio of the observed $\bar{\nu}_e$ spectrum to the expectation for no oscillation versus L $/E$. Observed energies are converted with a fixed L $=$ 180 km. The curves show the corresponding expectation for the best-fit oscillation, best-fit decay (dashed), and best-fit decoherence (dot-dashed) models.

1.8% was significantly improved over the previous result [2], because we confirmed the vertex reconstruction bias was less than 3 cm established by a new "off-axis" calibration system.

Table 1. Estimated systematic uncertainties on the expected event rate of reactor $\bar{\nu}_e$'s.

Detector-related (%)		Reactor-related (%)	
Fiducial volume	1.8	$\bar{\nu}_e$-spectra	2.4
Energy scale	1.5	Reactor power	2.1
L-selection eff.	0.6	Fuel composition	1.0
Cross section	0.2	Long-lived nuclei	0.3
total	2.4	total	3.3

3. Results

O

The data presented in this report were collected from March 9, 2002 through May 12, 2007, including the reanalysis of the data used in earlier results [1, 2]. In the absence of anti-neutrino disappearance, we expect 2179 ± 89(syst) events from all reactors and 276.1 ± 23.5 events from backgrounds [3], and we observed 1609 events. The oscillatory energy spectrum of the data is inconsistent with the expected no oscillation shape. From the χ test in $0.9 < E_{\rm p} < 8.5$ MeV, the no spectral distortion is excluded at more than 5σ considering the background and systematic uncertainties.

The expected no oscillation spectrum and the best fit of oscillation spectrum are shown in Figure 1. The oscillation evaluation is based on the maximum likelihood method considering background and systematic uncertainties. The geo neutrino contributions from U and Th-decays are treated as free parameters to avoid uncertainties from geology. The best-fit for Δm and tan θ are 7.58_ : $\times 10^{-5}$ eV and 0.56_ : if tan θ < 1. To illustrate oscillatory behavior of the data, the L $/E$ distribution is shown in Figure 2. The two alternate hypotheses for neutrino disappearance, neutrino decay [4] and neutrino decoherence [5], give different L $/E$ dependence. The minimum χ of decay and decoherence are 34.5 and 45.1 larger than neutrino

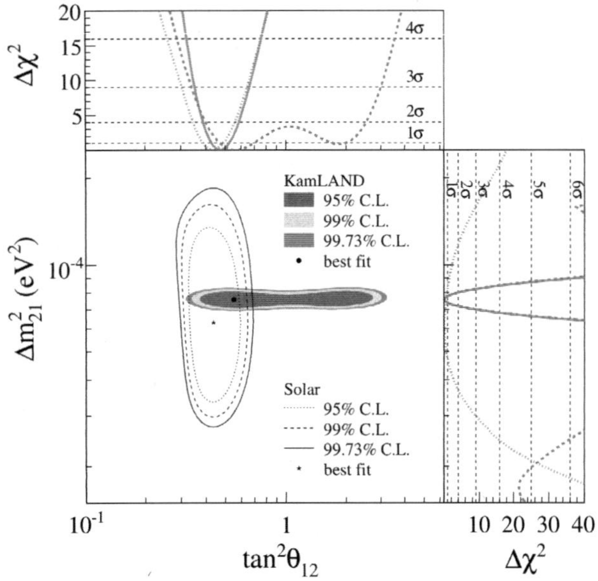

Figure 3. Allowed region for neutrino oscillation parameters from KamLAND and Solar. The side-panels show the marginalized $\Delta\chi$ -projections for KamLAND (dashed), Solar (dotted), and KamLAND + Solar (solid).

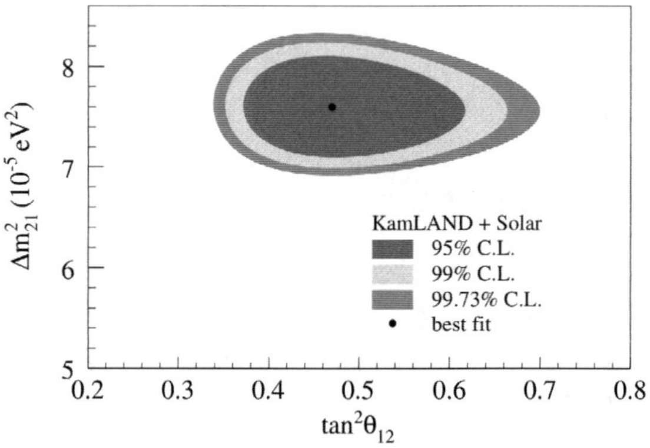

Figure 4. Result of a combined two-neutrino oscillation analysis of KamLAND and Solar under the assumption of CPT invariance. The best-fit gives Δm $= 7.59_{-}$: $\times 10^{-5}$ eV and tan θ $= 0.47_{-}$: $_{5}$ with marginalized errors.

oscillation, strongly indicating the neutrino oscillation is the most favored hypothesis. The data covers 2 cycles of oscillatory shape expected from neutrino oscillation.

The allowed region of neutrino oscillation parameters are shown in Figure 3, including the $\Delta\chi$ profiles projected onto each parameter. From the KamLAND data, the most favored solution is the so-called LMA I (at Δm $\sim 7.6 \times 10^{-5}$ eV), and other solutions at higher or lower Δm are disfavored at more than 4σ. Assuming CPT invariance, the allowed region, especially for tan θ , is more constrained by combining the solar data [6, 7] as shown in Figure 4.

The calculated neutrino fluxes from a reference earth model [8] for U and Th are 29 TNU and 8 TNU, corresponding to 57 events and 13 events after considering the detection efficiencies and neutrino oscillation. From the anti-neutrino analysis including the oscillation constraint on reactor $\bar{\nu}_e$'s by the solar data, the measured geo neutrino flux is 39 ± 14 TNU fixing the Th/U mass ratio to 3.9. The obtained result is consistent with a earth model prediction.

References

[1] K. Eguchi *et al.* (KamLAND Collaboration), Phys. Rev. Lett. **90**, 021802 (2003).
[2] T. Araki *et al.* (KamLAND Collaboration), Phys. Rev. Lett. **94**, 081801 (2005).
[3] K. Ichimura and Y. Minekawa, Background Study for KamLAND Reactor Neutrino experiment, poster session at tenth international conference on Topics in Astroparticle and Underground Physics (TAUP 2007), Sendai, Sep. 14 (2007).
[4] V.D. Barger *et al.*, Phys. Rev. Lett. **82**, 2640 (1999).
[5] E. Lisi *et al.*, Phys. Rev. Lett. **85**, 1166 (2000).
[6] B. Aharmim *et al.* (SNO Collaboration), Phys. Rev. **C72**, 055502 (2005).
[7] J. N. Bahcall, A. M. Serenelli and S. Basu, Astrophys. J. **621**, L85 (2005).
[8] S. Enomoto, E. Ohtani, K. Inoue and A. Suzuki, Earth Planet Sci. Lett. **258**, 147 (2007).

Double Chooz - A reactor θ_{13} experiment

Masakazu Motoki for the Double Chooz Collaboration

Tohoku University, 6-3 Aramaki-aza Aoba, Aoba-ku, Sendai, 980-8578, Japan

E-mail: motoki@hbs.tohoku.ac.jp

Abstract. The Double Chooz experiment is a next generation reactor experiment to measure θ_{13}. It is being installed in France. The Double Chooz R&D's are in final stages, and detector construction just started. The experiment will start data taking in 2009, which will be the quickest to explore θ_{13} below the current CHOOZ limit ($\sin^2(2\theta_{13}) < 0.2$), by reaching a sensitivity down to $\sin^2(2\theta_{13}) \sim 0.03$. We will know or set a strong limit on θ_{13} within a few years.

1. Introduction

The neutrino mass and weak eigenstates are related through the Pontecorvo-Maki-Nakagawa-Sakata (PMNS) neutrino mixing matrix [1, 2, 3]. A synthesis of solar, reactor, atmospheric and accelerator neutrino oscillation data requires the existence of at least three-neutrino mixing. The PMNS mixing matrix can be parameterized by three mixing angles θ_{sol}, θ_{atm}, θ_{13} and one or three CP-violating phases depending on if it is Majorana or not [4]. Concerning the determination of the PMNS mixing parameters, the measurement of the angle θ_{13} is the next experimental step to accomplish.

A reactor antineutrino disappearance experiment with multi-detectors is a cost-effective way to extend our reach in sensitivity for the neutrino mixing angle θ_{13} free from ambiguities from CP violation and matter effects.

Reactor neutrino experiments measure the survival probability $P_{\bar{\nu_e}\to\bar{\nu_e}}$ of the electron antineutrinos emitted from the nuclear reactor. This survival probability does not depend on the δ-CP phase. Furthermore, because of the low energy as well as the short baseline considered, matter effects are negligible [5]. Assuming a normal mass hierarchy scenario, $m_1 < m_2 < m_3$, the $\bar{\nu_e}$ survival probability can be written [6, 7]

$$\begin{aligned} P_{\bar{\nu_e}\to\bar{\nu_e}} &= 1 - 2\sin^2\theta_{13}\cos^2\theta_{13}\sin^2\left(\tfrac{\Delta m^2_{31}L}{4E}\right) \\ &-\tfrac{1}{2}\cos^4\theta_{13}\sin^2(2\theta_{12})\sin^2\left(\tfrac{\Delta m^2_{21}L}{4E}\right) \\ &+2\sin^2\theta_{13}\cos^2\theta_{13}\sin^2\theta_{12}\left(\cos\left(\tfrac{\Delta m^2_{31}L}{2E} - \tfrac{\Delta m^2_{21}L}{2E}\right) - \cos\left(\tfrac{\Delta m^2_{31}L}{2E}\right)\right) \end{aligned}$$

CHOOZ provides the upper bound $\sin^2(2\theta_{13}) < 0.20$ (90 % C.L.), assuming $\Delta m^2_{atm} = 2.0 \times 10^{-3} eV^2$ (this upper limit is strongly correlated with the assumed value of Δm^2_{atm}) [8, 9].

2. The Double Chooz experiment

2.1. Overview

High precision neutrino oscillation measurements can be achieved by a multi-detector experiment with detectors located at different distances. The sensitivity of the experiment is then nearly unaffected by the dominant uncertainties related to neutrino production and interaction.

The antineutrinos used in the experiment are those produced by the pair of reactors located at the Chooz-B nuclear power station, located in the Ardennes region, in the northeast of France. The maximum operating thermal power of each core amounts to 4.27 GW.

The Double Chooz experiment [10, 11, 12, 13] will employ two almost identical detectors of medium size, each containing 10.3 m^3 of liquid scintillator target doped with 0.1% of gadolinium. The far detector will look for $\bar{\nu_e}$ disappearance at a distance of 1.05 km from the reactor cores; it will be located in the existing site of the previous CHOOZ experiment, screened from cosmic ray muons by a natural rock overburden of 300 m.w.e. on average. A near detector will be installed at ~400 m from the cores, to measure the $\bar{\nu_e}$ flux and spectrum from the reactors before the oscillation. The main uncertainties at CHOOZ came from the uncertainty in the knowledge of the antineutrino flux coming from the reactor. This systematic error vanishes by normalizing to the near detector data.

Reactor antineutrinos are detected through the inverse beta decay (threshold of 1.806 MeV)

$$\bar{\nu_e} + p \rightarrow e^+ + n.$$

Experimentally one takes advantage of the coincidence signal of the prompt positron followed in space and time by the delayed neutron capture. This very clear signature allows us to reject most of the accidental backgrounds.

2.2. Detector

The fiducial neutrino target consists of 10.3 m^3 of Gd-loaded scintillator (20% PXE + 80% dodecane), contained in an acrylic vessel, a cylinder of 246 cm height, 230 cm diameter, and 8 mm thickness. The acrylic vessel provides a hardware definition of the fiducial volume.

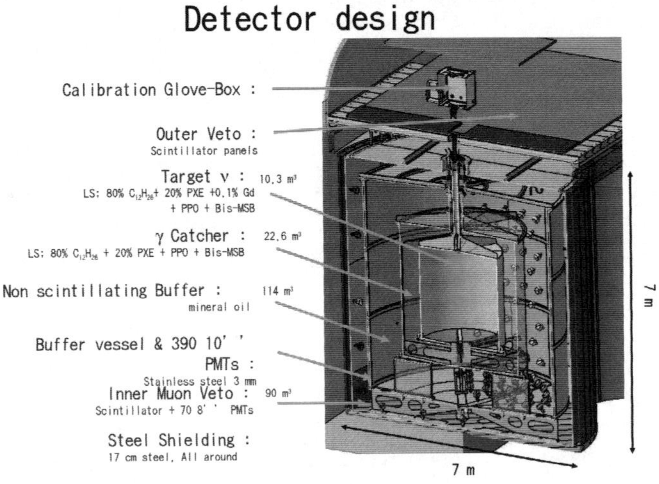

Figure 1. The schematic vew of the Double Chooz Detector. It consists of Target ν (10.3 m^3), γ-Catcher (22.6 m^3), Non scintillating buffer (114 m^3), Inner muon veto (90 m^3), Outer veto.

The neutrino target is surrounded by a gamma catcher. It consists of 22.6 m^3 of unloaded scintillator. This scintillating buffer around the target is necessary for efficiently measuring the gammas from neutron capture on Gd and from positron annihilation.

A non scintillating buffer (mineral oil) surrounds the active scintillating region. It shields the target from the radioactivity of the PMT photocathode. This region is crucial to keep the count rate below 10 Hz in the sensitive region (target+γ-catcher). The 400 10-inch PMTs are installed on the inner wall of the tank. The central detector is surrounded by an inner muon veto. The tank containing the inner veto volume is shielded by 15 cm steel to protect the detector from the external gamma background. In order to improve the coverage for cosmic muons passing near the detector, outer veto counters will be placed on top of detector.

2.3. Expected Experimental Errors and Backgrounds

The expected $\bar{\nu}_e$ rate is ~15,000 events/year at the far site, and ~150,000 events/year at the near site. Then, the statistical error will become ~0.4 % after 5 years of operation.

The uncertainties in the knowledge of the neutrino flux coming from the reactor vanishes by adding a near detector. The detector induced uncertainty has to be reduced. The concentration of Gd will be identical in both near and far detectors since Target liquid will be produced in a single batch. Then, the uncertainty on the scintillator density(0.01%) as well as chemical composition (1%) disappears from the relative normalization error. The uncertainty of the volume determination of each detector is dominant, and it will become ~0.2 %. The total systematic error is estimated ~0.5 %.

There are backgrounds from primordial and man-made radioactivity, as well as backgrounds induced by cosmic ray interactions.

Gamma, beta and neutron signals in the inner detector or in the rock may generate accidental background events which mimic prompt positron-like $\bar{\nu}_e$ interaction signal. The delayed neutron-like background comes mainly from muon spallation neutron captured on Gd. The accidental background rates are 1.6 events/day and 17 events/day at far and near detector, respectively.

Fast neutrons are created by near miss muons interacting around the detector. There large interaction length allows them to cross over the full detector. The recoil proton can mimic the positron signal, and the neutron capture is correlated as a neutrino event. The cosmogenically-produced isotope Lithium 9 can be a serious problem for reactor experiments. Lithium 9 emit neutron after 178 ms. By using the CHOOZ off data, we can estimate this background rate at far detector. The correrated background rates are estimated 1.4 events/day and 10 events/day at far and near detector, respectively.

3. Conclusions

Double Chooz R&D's are in final stages, and detector construction just started. First data taking is expected to start in 2009 with the far detector only, and the sensitivity will become $\sin^2(2\theta_{13}) < 0.06$ in 1.5 year. In 2010 we will take data with both near and far detectors and the sensitivity will become $\sin^2(2\theta_{13}) < 0.03$ in 3 years.

We will know or set a strong limit on θ_{13} within a few years, and the neutrino oscillation studies will go in a new phase.

References

[1] Maki Z, Nakagawa M and Sakata S 1962 *Prog. Part. Nucl. Phys.* **28** 870

[2] Pontecorvo B 1957 *Zh. Eksp. Teor. Fiz.* **33** 549

[3] Pontecorvo B 1958 *Zh. Eksp. Teor. Fiz.* **34** 247

[4] Yao W -M *et al* [Particle Data Group] 2006 *J. Phys. G* **33** 1

[5] Minakata H, Sugiyama H, Yasuda O, Inoue K and Suekane F 2002 (*Preprint* hep-ph/0211111)

[6] Bilenky S M, Nicolo D and Petcov S T 2001 (*Preprint* hep-ph/0112216)

[7] Petcov S T and Piai M 2001 (*Preprint* hep-ph/0112074)

[8] Apollonio M *et al* [CHOOZ Collaboration] 1998 *Phys. Lett. B* **420** 397

[9] Apollonio M *et al* [CHOOZ Collaboration] 2003 *Eur. Phys. J.* **C27** 331 (*Preprint* hep-ex/0301017)

[10] Ardellier F *et al* 2004 Letter of Intent for Double-CHOOZ: a Search for the Mixing Angle θ_{13} (*Preprint* hep-ex/0405032v1)

[11] Berridge S *et al* 2004 Proposal for U.S. participation in Double-CHOOZ: A New θ_{13} Experiment at the Chooz Reactor (*Preprint* hep-ex/0410081v1)

[12] Ardellier F *et al* [Double Chooz Collaboration] 2006 Double Chooz: A Search for the Neutrino Mixing Angle θ_{13} (*Preprint* hep-ex/0606025v4)

[13] Tonazzo A [Double Chooz Collaboration] 2007 *Nucl. Phys. B (Proc. Suppl.)* **172** 41

The Daya Bay Reactor Neutrino Experiment

Shaomin Chen

Department of Engineering Physics, Tsinghua University, Beijing 100084, P.R. China

E-mail: chenshaomin@tsinghua.edu.cn

Abstract. Search for the value of θ_{13} mixing angle is of importance in understanding the lepton flavor mixing matrix, and in motivating future experiments to probe CP violation in the lepton sector. Among the present experimental approaches, reactor experiment can provide a clean laboratory for the θ_{13}-measurement. The Daya Bay experiment will start civil construction this year at Daya Bay, Guangdong, China. The goal of this experiment is to reach a sensitivity in $\sin^2 2\theta_{13}$ of < 0.01 at 90% C.L. by precisely measuring the disappearance and spectral distortion of reactor electron anti-neutrinos with multiple identical detectors at different baselines. The talk will present the current status and prospects of the experiment.

1. Introduction

CP violation has been an important issue of study in both theory and experiment for many years. After neutrino oscillations were discovered, whether or not CP violation can also occur in the lepton sector is now drawing more and more attention [1]. Theoretically, analogous to mixing in the quark sector, CP violation in the lepton sector can be incorporated into the PMNS matrix with one Dirac CP phase and two Majorana CP phases. Experimentally, the last two phases are extremely difficult to determine, while the Dirac CP phase may be accessible through neutrino oscillation as long as the value of $\sin^2 2\theta_{13}$ is not so small. Current knowledge on θ_{13} is $\sin^2 2\theta_{13} < 0.17$ at 90%C.L. [2], and is even smaller as obtained from a three-flavor global fit [3]. Many theoretical models attempt to predict this value but cannot yield a consistent picture [4]. Therefore, θ_{13} is not only a gateway of CP violation in the lepton sector but also a guide in understanding the lepton flavor mixing matrix. Among all the efforts of determining $\sin^2 2\theta_{13}$, a reactor experiment can avoid the cross talk between the θ_{13} and the unknown Dirac CP phase, and thus provide a clean measurement.

2. The Daya Bay experiment

The Daya Bay nuclear power plant, which is located in the Guangdong province, China and 55 km away from the Hong Kong Island, currently has two pairs of reactor cores (Daya Bay and Ling Ao) in operation with a total of 11.6 $\mathrm{GW_{th}}$. One more pair (Ling Ao II) will be added after 2011, giving additional 5.8 $\mathrm{GW_{th}}$. These cores can provides about 2.3×10^{21} (3.5×10^{21} after 2011) electron anti-neutrinos ($\bar{\nu}_e$'s) per second with energies up to 8 MeV. The plant is near a mountainous terrain, which can provide sufficient overburden to suppress cosmic muon induced backgrounds to less than 1% of the antineutrino signal. This greatly reduces the systematic uncertainty due to background.

The principle of the Daya Bay experiment is to look for the disappearance and spectral distortion of $\bar{\nu}_e$'s produced by reactors. The concept of using near and far site detectors to cancel the common systematic uncertainties from both the reactors and the detectors is adopted. Tunnels connecting the near and the far halls will enable the option of interchanging detectors. The target mass for each near site is 40 tons. In addition to the baseline, the far detector is optimized to be 80 tons in weight to achieve the designed sensitivity with minimum cost. The optimization procedure takes into account the $\bar{\nu}_e$ flux, oscillation, energy spectrum, systematic uncertainties from both the reactors and the detectors, overburden, ambient background, spectra of the cosmogenic neutrons and ^9Li, etc. The first near hall is chosen to be 363 m away from the Daya Bay reactor cores with a 98 m overburden; the second near hall is about 500 m from the Ling Ao reactor cores with a 112 m overburden; the far hall is 1,985 m from Daya Bay and 1,615 m from Ling Ao with 350 m overburden.

Eight identical cylindrical $\bar{\nu}_e$ detector modules (AD's) will be built. Each AD, consists of three nested cylindrical zones contained within a stainless steel tank, will be deployed to detect $\bar{\nu}_e$ via the inverse beta-decay reaction: $\bar{\nu}_e + p \rightarrow n + e^+$. Each AD will have 20 tons of 0.1% Gd-doped liquid scintillator (Gd-LS) in the inner-most $\bar{\nu}_e$ target zone (3.1 m in diameter and 3.1 m in height, a detailed design information is given in Fig. 2.3 and 2.4 in Reference [5]). A 45-cm-thick second zone, separated from the target and outer buffer zones by two transparent acrylic vessels, is filled with undoped liquid scintillator (LS) for capturing γ rays that escape from the target thereby improving the detection efficiency. A total of 192 8-inch photomultiplier tubes (PMT's) are arranged along the circumference of the stainless steel tank in the outer-most zone (45 cm in thickness), which contains mineral oil to attenuate γ rays from trace radioactivity in the PMT glass and nearby materials including the outer tank. With reflective surfaces at the top and bottom of the outer acrylic vessel, the energy resolution and the vertex resolution of the detector are estimated to be about 12% at 1 MeV and 12 cm, respectively. The Gd-LS candidate is Linear Alkyl Benzene doped with 0.1% of Gd by weight. It has excellent long-term stability. The current AD design can achieve a neutron detection efficiency of 78% (combining 85% Gd fraction and 93% energy cut) and a 98% e^+ detection efficiency. It is expected that the signal event rates for each AD are 960/day in the Daya Bay near hall, 760/day in the Ling Ao near hall and 90/day in the far hall, respectively. The rate and energy distribution of the $\bar{\nu}_e$'s from the reactors will be measured with two AD's in each near hall. This configuration will enable a cross calibration to verify that the detectors are identical.

To reduce the γ rays and spallation neutrons produced in the surrounding rock from entering the AD's, in each hall, the AD's are separated by 1 m from each other and immersed in a large pool of highly purified water. The pools are octagonal in cross-section. The far hall pool has a cross-section of 16 m^2 with corners cut off at 45° such that the shortest distance from the walls to the steel tank of an AD is 2.5 m. The near hall pools have a cross-section of 16 m $\times$ 10 m rectangles also with corners cut off at 45°. The minimum distance between the outer surface of the detectors and the walls of the pools is 2.5 m. The water shield of each pool is divided into inner and outer sections and instrumented with PMT's to detect Cherenkov photons from muons transverse through the water. The partition is realized with Tyvek film 1070D reflectors stretched over a stainless steel frame. The frame holds PMT's for both the inner and outer sections of the pool. The water Cherenkov detectors are augmented with a muon tracker consisted of four layers of Resistive Plate Chambers (RPC's) placed above the pool. The tracker extends 1 m beyond the edge of the pool in all directions, to allow studies of background caused by muon interactions in the rocks surrounding the pool. The water Cherenkov counters combined with the RPC's give an overall efficiency of cosmic muon detection of better than 99.5%, with an uncertainty $<$0.25%.

The response of the detectors will be calibrated and monitored throughout the experiment. Calibration sources with energy range from about 0.5 MeV to 10 MeV are Am-Be, ^{252}Cf, Pu(C), ^{22}Na, ^{68}Ge, ^{60}Co and ^{238}Pu-^{13}C. These sources will be deployed regularly along a few vertical axes of the detectors to calibrate and monitor the detector response to positrons, and neutron

Table 1. Summary of background, statistical and systematic uncertainties.

Source	Uncertainty
$\bar{\nu}_e$ from reactors	0.087% (4 cores), 0.13% (6 cores)
Detector (per AD module)	0.38% (baseline), 0.18% (goal)
Backgrounds	0.32% (Daya Bay), 0.22% (Ling Ao), 0.22% (far)
Statistics	0.2%

capture. LED's with ns widths are used to check the optical properties of LS, the performance of the PMT's and the timing of the data acquisition systems. The radioactive sources and the LED's are deployed with automatic systems attached to the top lid of the AD. Trigger efficiency and PMT dark noise will be measured by using the random trigger events. Cosmogentic radioactive nuclei and neutrons will also be used to provide continuous calibration of the entire fiducial volume of for the detector.

Each AD and each muon system will use separated front-end electronics, trigger and data acquisition system to minimize possible systematic uncertainty. The systems in all the three halls will share a common clock signal. The $\bar{\nu}_e$ signal will be triggered by the multiplicity of PMT hits with a cross-check from the energy sum trigger. The event rate is estimated to be < 1.5MBytes/s per site. This system is also designed to have capability to readout any detector module based on information from any adjoining module, giving an opportunity to extensively study the cosmogenic neutron background, and independently measure the detection efficiency.

3. Current status and conclusion

The Daya Bay experiment received strong recommendation from the Chinese institutions at the 250th Xiangshan Scientific Meeting held in Beijing in 2005, a standard procedure to initiate a major research project in China. There are currently more than 190 physicists from 34 research institutes in China, the US, Russia and the Czech Republic working on this experiment. The CDR was released in January, 2007 [5]. The TDR is now complete. We have received funding from the Chinese funding agencies. Civil construction began on October 13, 2007. Deployment of the first pair of the AD's to the Daya Bay near hall will take place in 2009. Data taking using the baseline configuration of the two near halls and far hall will begin in 2010. Based on the estimated background, statistical and systematical uncertainties given in Table 1, after three years of running the sensitivity of Daya Bay for $\sin^2 2\theta_{13}$ will reach 0.008, relatively independent of the value of Δm^2_{31} within its allowed range.

Acknowledgments

I am grateful to my colleagues of the Daya Bay Collaboration, in particular to Kam-Biu Luk for many helpful discussions and comments.

References

[1] K. Anderson *et al* 2004 White paper report on using nuclear reactors to search for a value of θ_{13} (*Preprint*:hep-ex/0402041)
[2] F. Boehm *et al* [Palo Verde Collaboration] 2000 *Phys. Rev.* **D**62 072002
[3] G.L. Fogli *et al* 2006 *Prog. Part. Nucl. Phys.* **57** 742
[4] C. Albright and M. Chen, 2006 *Phys. Rev.* **D**74 113006
[5] X. Guo *et al* [Daya Bay Collaboration] 2007 A precision measurement of the neutrino mixing angle θ_{13} using reactor anti-neutrinos at Daya Bay (*Preprint*:hep-ex/0701029)

RENO: Reactor Experiment for Neutrino Oscillation at Yonggwang

Soo-Bong Kim

for the RENO collaboration

Department of Physics and Astronomy, Seoul National University, Seoul 151-742, Korea

E-mail: sbkim@phya.snu.ac.kr

Abstract. The RENO experiment is under construction to measure the value of the smallest and unknown neutrino mixing angle θ_{13}. The experiment will compare the measured fluxes of electron antineutrinos at two detectors located at 290 m and 1.4 km distances from the center of the Yonggwang nuclear reactors in Korea. It is planned to start data-taking in early 2010. An estimated systematic uncertainty associated with the measurement is less than 0.6%. With three years of data, the experiment will search for the mixing angle values of $\sin^2 2\theta_{13}$ down to 0.02.

1. Introduction

There have been great progresses in understanding the neutrino sector of elementary particle physics in the last decade. The discovery of neutrino oscillations is a direct indication of physics beyond the Standard Model and it provides a unique new window to explore physics at the Grand Unification energy scale. The neutrino oscillations are in general described by the PMNS mixing matrix consisting of three mixing angles and a CP violating phase.

The current best limit on the last unknown mixing angle θ_{13} is given by the Chooz reactor neutrino experiment [1]. A high precision measurement of reactor neutrino oscillation can be achieved by multiple detectors because experimental sensitivity would be nearly unaffected by the uncertainties related to antineutrino flux and detection [2]. The RENO experiment will try to measure θ_{13} or at least improve the current constraint using reactor neutrinos.

It is now widely recognized that the possibility exists for a rich program of measuring CP violation and matter effects in future accelerator based neutrino experiments. However, in these long baseline experiments there are degeneracies and parameter correlations among θ_{13}, CP violation phase, neutrino mass hierarchy, and θ_{23}, whereas reactor neutrino experiments provide a clean information on $\sin^2 2\theta_{13}$. The possibility of measuring CP violation can be fulfilled only if the value of θ_{13} is such that $\sin^2 2\theta_{13}$ should be greater than the order of 0.01. A measurement of or stringent limit on θ_{13} would be crucial as a part of a long-term program to measure CP violation parameters at accelerators. The sufficient value θ_{13} measured in this reactor experiment would strongly motivate the investment required for a new round of accelerator neutrino experiments.

2. The RENO Experiment

Consideration of the RENO experiment began in early 2004 based on "White Paper for a New Nuclear Reactor Neutrino Experiment" [3], and it was approved by the Ministry of Science and Technology in

Korea in May 2005. Since the major funding of the experiment is secured, it is expected to start civil engineering and detector construction in 2008 in order to start data-taking in early 2010.

2.1. Yonggwang Nuclear Reactors

The Yonggwang nuclear power plant is located in the west coast of southern part of Korea, about 400 km from Seoul. The power plant has six reactors producing total thermal power of 16.4 GW, the second largest in the world. The six reactors are lined up in roughly equal distances and span ~1.3 km.

2.2. Near and Far Detectors

The experimental setup consists of two identical detectors with one at a near site, roughly 290 m away from the reactor array center, and the other at a far site, roughly 1.4 km away from the reactor array center. Identical arrangement of the near and far detectors will significantly reduce the total systematic uncertainty to less than 0.6%. Each detector will contain 15 tons of liquid scintillator target doped with 0.1% of gadolinium by weight. The near detector is to be under a 70 m ridge with an overburden of ~130 mwe whereas the far detector is to be located under a 200 m mountain with an overburden ~460 mwe.

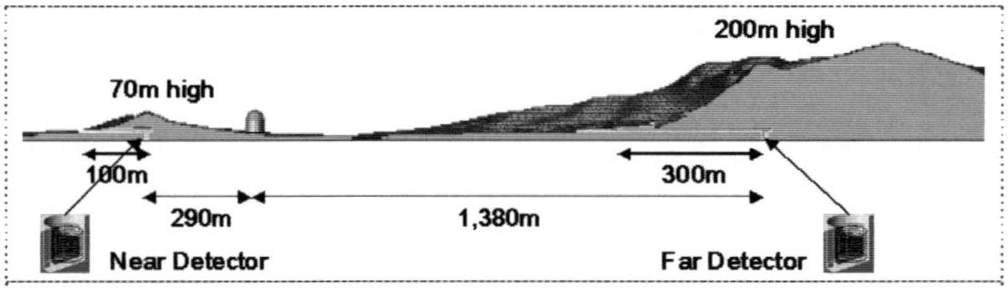

Figure 1. A schematic layout of the RENO experiment.

2.3. Time Scale

An overall schedule of the RENO experiment is given in Fig. 2. The KHNP, the company operating the Yonggwang nuclear power plant, has allowed us to carry out the experiment in the restricted area. The local government and residents have also expressed their best cooperation for RENO. Geological survey and tunnel design are completed in 2007. Civil construction for underground facility, including excavation of two tunnels, is expected to start in early 2008 and will be completed by early 2009. Construction of both near and far detectors will be completed by the end of 2009. Data-taking is expected to start in early 2010. RENO has gone through rapid development stages, to date, of project planning, fund approval, administrative negotiation, detector design, and tunnel design.

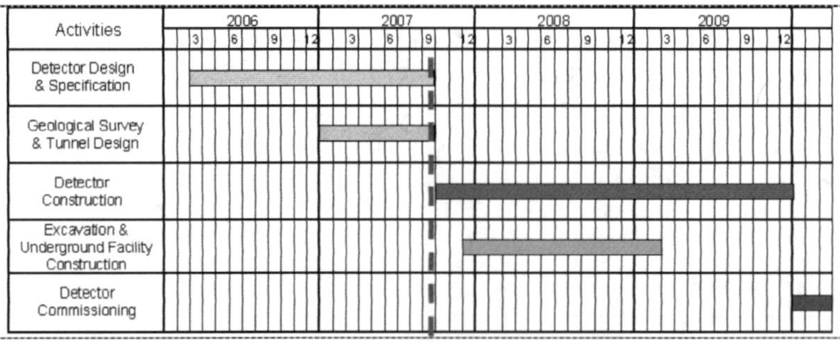

Figure 2. An overall schedule of the RENO experiment. Data-taking is planned to start in early 2010.

3. The RENO Detector

The RENO detector, having a cylindrical shape of 7.8 m in height and 7.4 m in diameter, consists of a neutrino target, a gamma catcher, a buffer and a veto.

3.1. Neutrino Target

The neutrino target consists of 0.1% Gd loaded liquid scintillator in a cylindrical acrylic container of 140 cm in radius, 320 cm in height, and 8 mm in thickness. It has a total volume of 19.7 m^3 and a target mass of 15.4 tons.

3.2. Gamma Catcher

Gamma catcher surrounds the neutrino target with 60 cm thick unloaded liquid scintillator of 35.2 m^3 in volume and 27.5 tons in mass. The gamma catcher is contained in a cylindrical acrylic vessel of 200 cm in radius, 440 cm in height, and 12 mm in thickness.

3.3. Buffer

A 70 cm thick non-scintillating liquid surrounds the gamma catcher to reduce the accidental backgrounds coming from outside, mainly from radioactivity in the photomultiplier tubes, by almost two orders of magnitude. A total 59.2 tons of mineral oil is contained in a stainless steel vessel of 270 cm in radius, 580 cm in height, and 4 mm in thickness. This buffer is necessary for keeping the single rate below 10 Hz in the neutrino target and gamma catcher regions.

3.4. Acrylic Vessels

The acrylic vessels should contain aromatic liquids without leak and changing properties for a long-term period, roughly more than 10 years. They should not develop any chemical reaction with the scintillating liquids of neutrino target and gamma catcher, and with non -scintillating mineral oil of buffer for a long time period.

3.5. Photomultiplier Tube

A total 342 of 10-inch PMTs in a uniform array are mounted from the inner surface of the buffer vessel, providing 12.6% photo-sensitive surface coverage. The cylindrical steel vessel optically isolates the inner detector part from the outer veto system.

3.6. Veto

A 1 m thick water layer of 201.8 tons surrounds the whole inner detector with a volume of 102.1 tons. A total 51 of 8-inch PMTs are mounted in front of Tyvek reflector on a cylindrical steel tank.

3.7. Liquid Scintillator

The neutrino target will be filled with 1g/ℓ Gd loaded liquid scintillator. For Gd loading, two scintillator formulations are investigated, one based on carboxylic acids (CBX) and the other on beta-diketonates (BDK) to guarantee long-term stability. Linear Alkyl Benzene (LAB) will be used as scintillator solvent to replace dodecane and pseudocumene admixture. LAB has a high flash-point of 130 ℃ and a good optical property, and is commercially available in Korea.

4. Sensitivity

RENO is expected to improve the sensitivity on $\sin^2 2\theta_{13}$ by an order magnitude relative to the current best limit. This will be achieved by reducing both statistical and systematic errors, less than 0.6%.

The reactor antineutrinos are detected via the capture on a free proton, emitting a positron and a neutron, with an energy threshold of 1.8 MeV. The signature of a neutrino event is a prompt signal of positron followed by the neutron capture on Gd creating gammas with about 8 MeV.

An expected number of observed antineutrino is roughly 5,000 per day and roughly 100 per day in the near detector and far detector, respectively. An estimated systematic uncertainty associated with the measurement is less than 0.6%.

Based on three years of data, it would be sensitive to measure the neutrino mixing angle in the range of $\sin^2 2\theta_{13} > 0.02$. This sensitivity is five times better than the current limit obtained by Chooz.

5. Summary

The RENO experiment will try to measure θ_{13} or at least improve the current best limit by an order magnitude using two identical detectors. Geological survey and tunnel design are completed. Data-taking is planned to start in early 2010. Based on three years of data, it would be sensitive to measure the neutrino mixing angle in the range of $\sin^2 2\theta_{13} > 0.02$.

References

[1] M. Apollonio *et al.* (Chooz Collaboration), *Eur. Phys. J.* C **27**, 331 (2003).

[2] L. A. Mikaelyan and V. V. Sinev, *Phys. Atomic Nucl.* **63**, 1002 (2000).

[3] K. Anderson *et al.*, "White paper report on using nuclear reactors to search for a value of θ_{13}", hep-ex/0402041 (2004).

Low Energy Neutrino Physics at the Kuo-Sheng Reactor Laboratory in Taiwan

Shin-Ted Lin (*on behalf of TEXONO Collaboration*)
Institute of Physics, Acamedia Sinica, NanKang 11529, Taipei, Taiwan
E-mail: `linst@phys.sinica.edu.tw`

Abstract. The TEXONO Collaboration has been pursuing a research program on low energy neutrino physics at the Kuo Sheng Reactor Neutrino Laboratory in Taiwan. We focus on the final results of our studies on neutrino magnetic moments and radiative decay lifetimes with a high-purity germanium detector. Limits of $\mu_\nu(\bar{\nu_e}) < 7.4 \times 10^{-11}\ \mu_B$ at 90% confidence level were derived at a physics threshold of 12 keV and a background level of 1 $\mathrm{keV^{-1}kg^{-1}day^{-1}}$, comparable to underground dark matter experiments. The other programs at Kuo-Sheng are described.

1. Introduction

The TEXONO Collaboration [1] has been built up since 1997 to pursue an experimental program in Neutrino and Astroparticle Physics. The "flagship" program is on reactor-based low-energy low-background in Neutrino and Astrophysics Physics at the Kuo-Sheng (KS) Power Plant in Taiwan [2].

Results from recent neutrino experiments strongly favor neutrino oscillations which imply neutrino masses and mixings [3]. Their physical origin and experimental consequences are not fully understood. There are strong motivations for further experimental efforts to shed light on these fundamental questions by probing standard and anomalous neutrino properties and interactions. The results can constrain theoretical models necessary to interpret the future precision data or may yield surprises which have been the characteristics of the field. In addition, these studies will also explore new neutrino sources and novel detection channels to provide new tools for future investigations.

2. Kuo-Sheng Neutrino Laboratory

The "Kuo-Sheng Neutrino Laboratory" is located at a distance of 28 m from the 2.9GW reactor core of the Kuo-Sheng Nuclear Power Station at the northern shore of Taiwan. The measureable nuclear and electron recoil spectra due to reactor $\bar{\nu}_e$ are depicted in Figure 1. An ultra low-background high purity germanium (ULB-HPGe) detector started data taking since June 2001, while 200 kg of CsI(Tl) crystal scintillators were added from January 2003.

3. Search of Neutrino Magnetic Moments at Kuo-Sheng

An overview on the particle physics aspects of neutrino magnetic moments can be refered to a recent review [5] and the references therein. The neutrino electromagnetic vertex can be parametrized by terms corresponding to interactions without and with its spin, identified

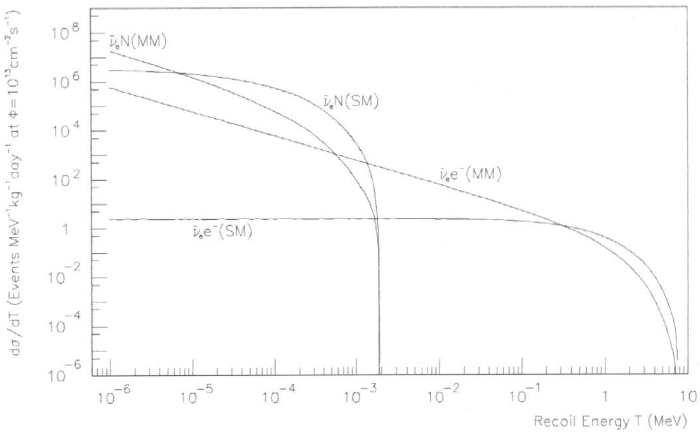

Figure 1. Differential cross section showing the recoil energy spectrum in $\bar{\nu}_e$-e for the Standard Model [$\bar{\nu}_{\rm e}{\rm e}^-$(SM)] and magnetic moment [$\bar{\nu}_{\rm e}{\rm e}^-$(MM)] [4] ,as well as in neutrino coherent scatterings on the nuclei $\bar{\nu}_{\rm e}$N(SM) and $\bar{\nu}_{\rm e}$N(MM), respectively at a reactor neutrino flux of 10^{13} cm^{-2} s^{-1}, due to a neutrino magnetic moment (MM) of 10^{-10} μ_B.

as the "neutrino charge radius" and "'neutrino magenetic moments', respectively. With an ULB-HPGe of mass 1.06kg surrounded by NaI(Tl) and CsI(Tl) crystal scintillators as anti-Compton detectors, the low energy spectrum of Figure 2 is relevant to the studies of neutrino magnetic moments. A detector threshold of 5 keV and a backgound level above 12 keV at the range of 1 keV^{-1}kg^{-1}day^{-1} were achieved. Based on the measured spectra for 570.7/127.8 days of Reactor ON/OFF data [6], at an average Reactor ON electron anti-neutrino flux of 6.4×10^{12} cm^{-2}s^{-1}, shows no excess in Figure 3 and limits of the neutrino magnetic moment $\mu_\nu(\bar{\nu}_{\rm e}) < 7.4 \times 10^{-11}$ $\mu_{\rm B}$ at 90% confidence level (CL) were derived.

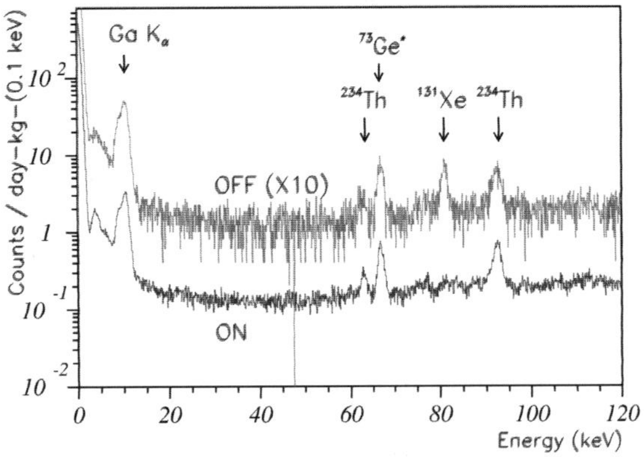

Figure 2. Both Reactor ON and OFF spectra after all the selection criteria were applied are separately displayed for the range below 120 keV. Key γ-lines were identified.

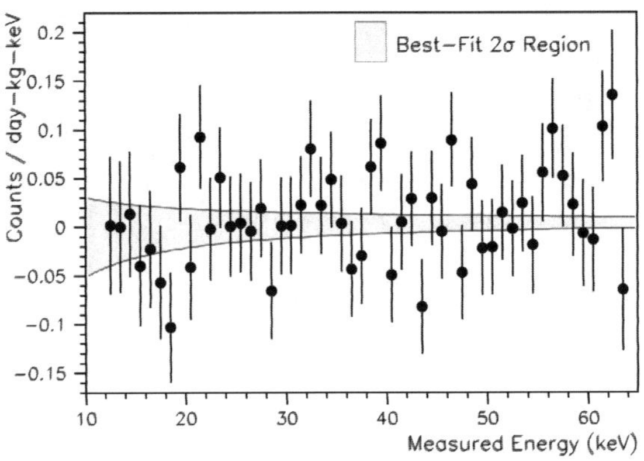

Figure 3. The residual plot on the Reactor ON data of all periods combined over the background spectra.

Depicted in Figure 4 is the summary of the results in $\mu_\nu(\bar{\nu}_{\rm e})$ searches versus the achieved threshold in various reactor experiments [6]. The neutrino-photon couplings probed by μ_ν searches in ν-e scatterings are related to the neutrino radiative decays (γ_ν) [7]. Indirect bounds on γ_ν can be inferred and are displayed in Figure 5.

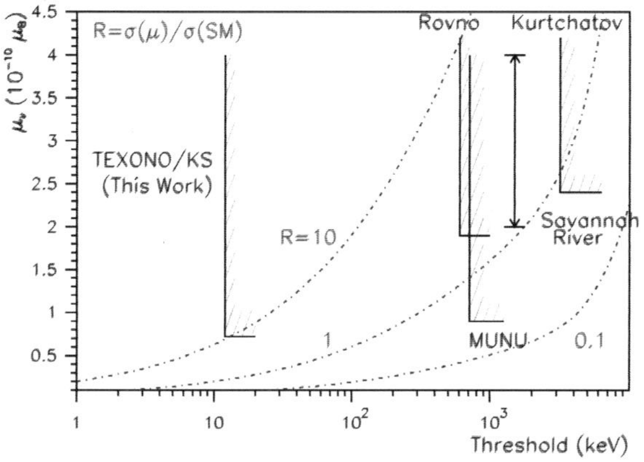

Figure 4. Summary of the KS results in the searches of neutrino magnetic moments with reactor neutrinos.Both the limits and the detection thresholds of the various experiments are shown.

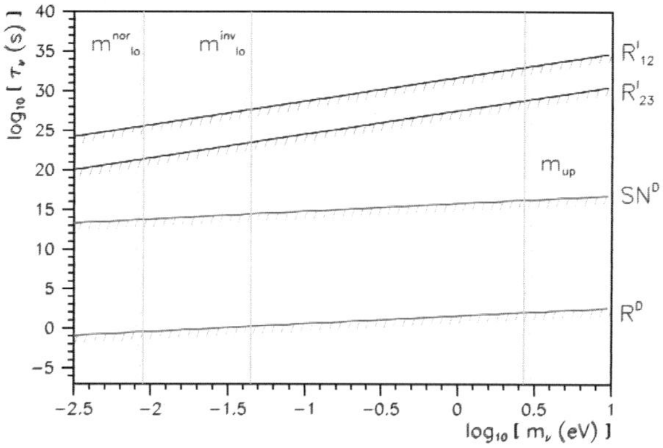

Figure 5. Summary of the KS results in the excluded parameter space on neutrino radiative decay lifetime from $\bar{\nu_e}$ measurements. Bounds on ν_1 and ν_2 from reactor and supernova SN1987a data are denoted by R and SN, respectively.

4. Other Programs at Kuo-Sheng

Nuclear fission at reactor cores also produce electron neutrino (ν_e) through the production of unstable isotopes. A realistic neutron transfer simulation has been performed to estimate the flux and physics analysis on the μ_ν and γ_ν for ν_e will be performed [8]. In addition, studies of neutrino-induced nuclear transitions, as well as searches for possible reactor-produced axions [9], are pursued. The physics goal for the CsI(Tl) scintillating crystal array is to measure the Standard Model neutrino-electron scattering cross sections, and thereby to provide a measurement of $\sin^2\theta_W$ at the untested MeV range.

Our future goal of developing a ultra-low-energy germanium (ULE-HPGe) detector is to reach a 100-eV threshold, a background level of 1 keV^{-1}kg^{-1}day^{-1}, target size of the 1 kg mass range, for neutrino-nucleus coherent scattering and dark matter searches [10].

References

[1] TEXONO home Page http://hepmail.phys.sinica.edu.tw/~texono
[2] H.T. Wong and J. Li 2000 *Mod. Phys. Lett.* A **15**, 2011; H.T. Wong 2004 *Mod. Phys. Lett.* A **19**, 1207
[3] See the respective sections in *Review of Particle Physics* 2006, Particle Data Group, *J. Phys.* G **33**
[4] P.Vogel and J.Engel 1989 *Phys. Rev.* D **39**, 3378.
[5] H.T. Wong 2005 *Nucl. Phys.* B **143**, 205; H.T. Wong, H.B. Li 2005 *Mod. Phys. Lett.* A **20**, 1103
[6] H.B. Li et al.,2003 *Phys. Rev. Lett.* **90**, 131802; H.T. Wong et al., 2007 *Phys. Rev.* D **75**, 012001
[7] G.G. Raffelt 1989 *Phys. Rev.* D**39**, 2066
[8] X Bin et al.,2005 *Phys. Rev.* D **72** , 012006
[9] H.M. Chang et al., 2007 *Phys. Rev.* D **75**, 052004
[10] H.T. Wong 2006 *J. Phys. Conf. Ser.* **39**, 266

Prospects of precision neutrino oscillation studies with Hanohano

M. S. Batygov
The University of Hawaii
Email: batygov@phys.hawaii.edu

1 Indroduction

The recent experimental observation of neutrino oscillations has been a great success of particle physics. Experiments such as SNO [1] and KamLAND [2], managed to not only provide the evidence for the phenomenon but also measure two of the oscillation parameters, namely Δm^2_{12} and θ_{12} with remarkable precision. However other parameters remain unknown or known with a relatively poor precision. These include Δm^2_{13}, Δm^2_{23} and θ_{13} which are, in principle, measurable with a big KamLAND-like reactor $\bar{\nu}_e$ experiment, provided the detector size and the baselines are chosen adequately. Moreover, the measurement of Δm^2_{13} and Δm^2_{23} with sufficient precision will determine the neutrino mass hierarchy.

That said, the currently operating detectors are not suitable for most of those studies because of being too small or inappropriately placed or both. The measurement of θ_{13}, Δm^2_{13} and Δm^2_{23} requires a detector with a baseline finely tuned for respective studies.

Such an ocean-based detector named "Hanohano" has been suggested [3, 4]. Its ocean-based location provides more flexibility in the choice of the baseline. Since the optimal baseline for some of the studies is fairly long, the detector has to be quite big in order not to be statistically limited. Besides the measurement of θ_{13}, Δm^2_{13}, Δm^2_{23} and, possibly, neutrino mass hierarchy, the new experiment will be able to improve the estimation of the "solar" parameters, Δm^2_{12} and θ_{13}, compared with what we know from the current experiments. The concept of a medium-baselined reactor $\bar{\nu}_e$ experiment for such studies has been thoroughly developed in [5].

Once the oscillation study is completed, the detector can be moved to a remote location in the ocean for the dedicated terrestrial neutrino study. Moreover, like any large scintillator detector, Hanohano can be used for a number of other experiments, including but not amounting to nucleon decay studies, solar neutrino observation and supernova monitoring.

The numerical analysis of the prospects of Hanohano and the requirements it must meet are presented below.

2 Oscillation study through precision spectrum measurement

In the vacuum oscillation approximation, we have the following $\bar{\nu}_e$ survival probability:

$$\begin{aligned} P(\bar{\nu}_e \to \bar{\nu}_e) = 1 &- \cos^4(\theta_{13}) \sin^2(2\theta_{12}) \sin^2 \Delta_{12} \\ &+ \sin^2(2\theta_{13}) \cos^2(\theta_{12}) \sin^2 \Delta_{13} \\ &+ \sin^2(2\theta_{13}) \sin^2(\theta_{12}) \sin^2 \Delta_{23}, \end{aligned}$$

For the experiment to be sensitive to Δm_{13}^2 and Δm_{23}^2, the mixing angle θ_{13} must not be zero. If it turns out to be zero or finite but extremely small, the determination of those squared mass differences through reactor $\bar{\nu}_e$ disappearance measurement becomes impossible. However a much tighter limit for the angle itself than the currently known $\theta_{13} < 0.015$ [6] will be set anyway.

This sensitivity analysis was based on the unbinned likelihood approach, where the test statistic was chosen to be the negative logarithm of the likelihood difference, which, if multiplied by two, can be proved to be have a χ^2 distribution for a rather general case. Although the sensitivity estimation can be carried out by the integration of spectrum dissimilarities and does not require the actual experiment simulation, the simulation has been carried out for the purpose of a cross check and was found in a perfect agreement with the "theoretical" prediction.

Only two kinds of systematic uncertainties were considered here: the general detection efficiency error, and the uncertainty in the detector energy resolution.

3 Results

The sensitivity profile for the "solar" oscillation parameters is shown in Figure 1a. The optimum baseline for the Δm_{12}^2 measurement is about 30 km, for θ_{12} – 40 km. Provided the background is kept small and the systematic uncertainties are not extremely large, the 10-kiloton KamLAND-like detector located at 40 km from a 5 GW power plant can measure $\sin^2\theta_{12}$ to the accuracy down to 0.005 and Δm_{12}^2 down to $0.05 \times 10^{-5} eV^2$, which is about four times better than currently known from the combined KamLAND and SNO parameter fit.

For θ_{13}, the shorter baselines seem to be better, based on just statistics alone. However a bigger detector with a longer baseline was found to be much less dependent on the systematic error of the detection efficiency. At 40 km, it will be able to measure $\sin^2 2\theta_{13}$ down to 0.02, in about 2 years, which is quite competitive with the dedicated experiments currently in construction. Still longer baselines may lack in statistics and have an additional disadvantage of having dependence on the detector energy resolution estimation error.

The "atmospheric" Δm_{13}^2 and Δm_{23}^2 determination depends on the actual value of θ_{13}. If $\sin^2 2\theta_{13}$ turns out to be no less than 0.05, these squared mass differences can be determined with the accuracy of about $2.5 \times 10^{-5} eV^2$, which is about 10 times better than currently known from the baselines of up to 40-50 km. This study depends on a good detector resolution.

The hierarchy study is the most challenging one. The sensitivity for baselines shorter than the optimum 50 km deteriorates dramatically (Fig 1b). This study depends on the detector resolution much more than that for the "atmospheric" squared mass difference. Having at worst $0.025\sqrt{E_{vis}[MeV]}$ seems to be absolutely necessary since compromising it to $0.035\sqrt{E_{vis}[MeV]}$ (which would still be excellent by contemporary standards) is enough to require a four-fold increase in livetime to achieve the same accuracy. Even with the best possible resolution, this study relies on a substantially high θ_{13}.

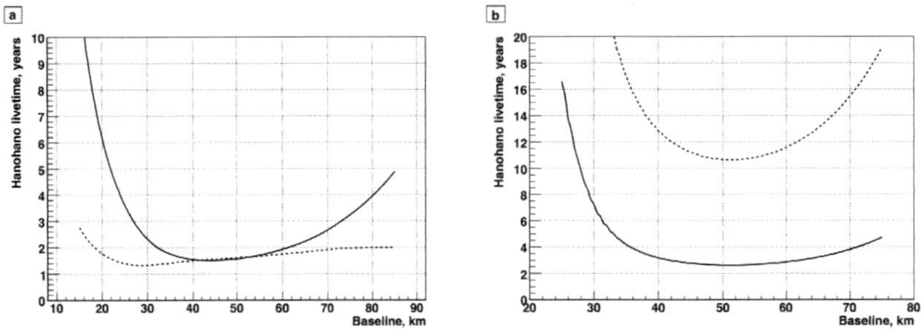

Figure 1: Hanohano livetime necessary to measure $\sin^2 2\theta_{12}$ (solid) to 0.005 and Δm^2_{12} (dashed) to $0.05 \times 10^{-5} eV^2$ (a); Hanohano livetime necessary to discriminate neutrino mass hierarchies to 1σ CL if $\sin^2 2\theta_{13} = 0.05$ (solid) and $\sin^2 2\theta_{13} = 0.025$ (dashed), detector energy resolution being $0.025\sqrt{E_{vis}[MeV]}$ in both cases (b).

4 Conclusion

Hanohano will provide a dramatic improvement in the evaluation of almost all oscillation parameters measurable through $\bar{\nu}_e$. The optimum baseline for such an ocean-based detector is found to be 40 to 50 km but may need further adjustments, once the parameters of the detector itself are measured. The neutrino mass hierarchy study requires certain design choices for such a detector. First of all, it must be big enough. Second, it must have an outstanding energy resolution. Even if the reliable hierarchy turns out to be difficult because θ_{13} proves to be be very small, the limits on all other parameters will be improved.

In addition to the oscillation measurement, the Hanohano detector will be useful for many other experiments, including nucleon decay and supernova monitoring and will provide absolutely unique opportunities for terrestrial neutino studies.

References

[1] B. Aharmim *et al.*, Phys. Rev. **C72** 055502 (2005).

[2] S. Abe *et al.* arXiv:0801.4589v2.

[3] J. G. Learned, S. T. Dye, S. Pakvasa, R. C. Svoboda, "Determination of Neutrino Mass Hierarchy and θ_{13} with a Remote Detector of Reactor Antineutrinos", hep-ex/0612022.

[4] J. G. Learned, S. T. Dye, S. Pakvasa (2007) in *Twelfth International Workshop on Neutrino Telescopes*, ed Baldo Ceolin M (Instituto Veneto di Scienze, Padova), pp 235-269.

[5] S. Choubey, S. T. Petcov, M. Piai, Phys. Rev. D **68**, 113006 (2003) arXiv:hep-ph/0306017v1.

[6] M. Apollonio *et al.*, Eur. Phys. J. C **27**, 331 (2003).

KATRIN: an experiment to determine the neutrino mass

R.G. Hamish Robertson

Center for Experimental Nuclear Physics and Astrophysics, University of Washington, Seattle, WA, 98195, USA

E-mail: rghr@u.washington.edu

KATRIN Collaboration

Abstract. KATRIN is a very large scale tritium-beta-decay experiment to determine the mass of the neutrino. It is presently under construction at the Forschungszentrum Karlsruhe, and makes use of the Tritium Laboratory built there for the ITER project. The combination of a very large retarding-potential electrostatic-magnetic spectrometer and an intense gaseous molecular tritium source makes possible a sensitivity to neutrino mass of 0.2 eV, about an order of magnitude below present laboratory limits. The measurement is kinematic and independent of whether the neutrino is Dirac or Majorana. The status of the project is summarized briefly in this report.

1. Introduction

In beta decay the electron and neutrino share the available energy in a statistical fashion, and in a small fraction of the decays the electron will take almost all the energy unless some must be reserved for the rest mass of the neutrino. The phase-space available to the electron near the endpoint is therefore modified by neutrino mass, a fact realized immediately by Fermi [1] as he formulated the theory of beta decay.

In the intervening seven decades of experimental searches for neutrino mass, tritium has been the beta-active nucleus of choice because it has a low endpoint energy, which makes the modification caused by neutrino mass a larger fraction of the total spectrum:

$$^3\mathrm{H} \rightarrow {}^3\mathrm{He}^+ + \mathrm{e}^- + \overline{\nu}_\mathrm{e} \quad + 18580 \text{ eV}.$$

It is also a very simple atom, and atomic or molecular effects are important at the eV level. The decay is superallowed, with a short half-life, which reduces the amount of source material needed for a given counting rate and sensitivity.

Oscillation experiments now define a lower limit to the mass range [2]. The average mass of the three eigenstates must lie between 20 meV and 2300 meV, the lower limit arising in the 'normal' hierarchy if the lightest mass is also negligibly small, and the upper limit being set by the Mainz experiment on the beta decay of tritium [3]. Above 200 meV the mass spectrum becomes essentially degenerate, and the resulting beta spectrum is indistinguishable from what a single massive 'electron neutrino' would produce if mass eigenstates were also flavor eigenstates.

Figure 1. Layout of the KATRIN Experiment

The dependence of the spectral shape on mass is given by a phase space factor only. Moreover, the mass measured is independent of whether the neutrino is a Majorana or a Dirac particle. A neutrino mass in this quasi-degenerate regime that KATRIN is designed to explore would be of cosmological importance, having a major influence on the formation of large-scale structure in the universe.

2. The KATRIN Experiment

The KArlsruhe TRItium Neutrino experiment [4] consists of seven major subsystems (see Fig. 1), a gaseous tritium source, the tritium processing and recirculation system, a differential pumping section, a cryogenic Ar frost pumping section, a pre-spectrometer, a main spectrometer, and the detector system.

3. Source and Tritium Recirculation

The gaseous molecular source consists of a tube 9 cm in diameter and 10 m in length maintained at the desired temperature (27 K) by circulation of two-phase Ne. An axial magnetic field of 3.6 T guides electrons toward the spectrometers. The source is being constructed by ACCEL Instruments GmbH in two stages, the first being a demonstrator to verify the stringent thermal performance specifications, including temperature regulation to ± 30 mK, and the second the complete system with superconducting magnets, to be delivered in early 2009. Four turbomolecular pumps at the ends of the source tube collect T_2 for recirculation in an 'inner loop' with Pd filters to remove contaminants, particularly ^3He.

4. Differential Pumping Section

Further reduction of tritium pressure is achieved in this section, which consists of a tube with a chicane, superconducting solenoids to guide the electrons, and four 2800 l/s magnetic-bearing turbomolecular pumps to scavenge the tritium. The subsystem is being built by Ansaldo Superconduttori for delivery in 2008.

5. Cryogenic Pumping Section

Any remaining tritium that escapes through the differential pumping section is trapped in Ar frost, which forms a highly efficient, large-area, chemically inert, and radiation-immune surface. The tender for this system is about to be let.

6. Pre-spectrometer

There are two spectrometers in tandem in KATRIN, both of the retarding-potential type. The pre-spectrometer operates at a cutoff potential typically 100 eV below the endpoint, preventing most electrons from reaching the main spectrometer. Electrons can ionize residual gas molecules and create slow electrons that are indistinguishable from the signal. Reducing the electron flux into the main spectrometer is expected to improve the background near the endpoint substantially. The pre-spectrometer was the first KATRIN subassembly to be completed, in order to permit some key concepts to be tested before other design elements were frozen. One important discovery was the existence of a parasitic Penning-trap configuration near the ends, which made it impossible to apply high voltage and a magnetic field simultaneously without breakdown. The addition of appropriate electrodes suppressed this discharge completely, and the spectrometer now runs uneventfully at 35 kV and 4 T. A new feature in this type of spectrometer is a grid inside the shell to suppress the emission of low-energy electrons ejected by cosmic rays and radioactivity from the shell. The performance of the grid will be tested shortly, to verify the design principles of the much larger grid system for the main spectrometer.

7. Main Spectrometer

The main spectrometer is a large stainless-steel vessel 10 m in diameter and 24 m in length. The 200-tonne chamber was fabricated in Deggendorf by MAN-DWE and shipped via the river Danube, the Black Sea, the Mediterranean, the North Sea, and the Rhine to be offloaded at Leopoldshafen near the FZK laboratory. It was placed in its building November 29, 2006. Thermal insulation and heating tubes were installed, and 6 turbopumps. The tank was baked out to 350 C from July 16, 2007 to July 25. The base pressure 2 months later was 0.9×10^{-9} mbar and the outgassing rate (predominantly H_2) 1.2×10^{-12} mbar l s^{-1} cm^{-2}. When the interior grid structure and 1000 m of non-evaporable getter strips are installed later, the predicted base pressure will be 3×10^{-11} mbar.

8. Detector

Electrons surmounting the potential barriers in the spectrometers enter a high axial magnetic field region and are detected in a monolithic 148-pixel Si PIN diode array 10 cm in diameter. Two superconducting solenoids can produce up to 6 T, defining the electron beam diameter and the maximum pitch angle accepted from the source. The magnets are being made by Cryomagnetics, Inc. and the detector by Canberra. Between the solenoids is a region in which various calibration devices – gamma sources and a photoemissive electron gun – can be inserted. Selection of materials, shielding, and an active veto are being adopted to keep non-beam-associated backgrounds below 1 mHz.

9. Summary

The KATRIN experiment is scheduled for initial data-taking in 2010. Subsystems are in an advanced state of construction and commissioning. Funding for the project is being provided by the Helmholtz Gemeinschaft, the Bundesministerium für Bildung und Forschung, and the US Department of Energy.

References

[1] Fermi E 1934 *Nuovo Cim.* **11** 1–19
[2] Yao W M *et al.* (Particle Data Group) 2006 *J. Phys.* **G33** 1–1232
[3] Kraus C *et al.* 2005 *Eur. Phys. J.* **C40** 447–468 (*Preprint* hep-ex/0412056)
[4] Angrik J *et al.* (KATRIN) FZKA-7090, http://www-ik.fzk.de/katrin/

R&D for possible future improvements of KamLAND

A. Terashima*, Y. Takemoto, E. Yonezawa, H. Watanabe, S. Abe, and M. Nakamura
Research Center for Neutrino Science, Tohoku University,
Aramaki Aoba, Aoba, Sendai, Miyagi 980-8578, Japan
* Paper presented at the conference by A. Terashima

E-mail: azusa@awa.tohoku.ac.jp

Abstract. Research and development are being undertaken for possible future plans of KamLAND. Electronics, the liquid scintillator, and the photodetectors are targets of modifications and improvements, aiming at further low-background and high-sensitivity neutrino physics..

1. Introduction

KamLAND, a 1-kt liquid scintillator neutrino detector, succeeded in precisely measuring reactor neutrino oscillation and investigating "geoneutrinos" experimentally for the first time. Currently, the KamLAND scintillator is being re-purified for the detection of solar ^{7}Be neutrinos. For possible plans after that, the detection of solar pep and CNO neutrinos, a search for neutrinoless double beta decay, and directional measurements of antineutrinos are being studied.

2. MoGURA, Dead-Time Free Electronics for KamLAND Solar Phase

2.1. ^{11}C Tagging with KamLAND

One of the interesting KamLAND goals is detection of the CNO and pep neutrinos, which dominate the solar neutrino flux above the ^{7}Be neutrinos' energy region. A contradiction between the recent research results from Standard Solar Model (SSM) and Helioseismology could be investigated by observing these neutrinos. In KamLAND, these neutrinos are overwhelmed by cosmogenic ^{11}C background, which is produced in cosmic muon-induced spallation. We aim to reject those ^{11}C events by tagging almost all cosmogenic neutrons after each muon, since more than 95% of the ^{11}C production modes accompany one or more neutrons. This technique is called "^{11}C Tagging" [1].

To detect all neutron capture signals (mostly 2.2-MeV γ for proton capture) with a multiplicity of up to ~100 within ~210 μs after each muon, a dead-time-free data acquisition (DAQ) is needed. Concerning the amplitude and width of the smallest PMT signals (2 mV and 20 ns), high resolutions of both the voltage and the time are required. On the other hand, the amplitude of huge signals (up to 10 V, corresponding to the showering muon signals) requires a wide dynamic range. This range (up to 10^{5}) requires baseline stabilization, since baseline distortion made by a huge muon signal easily overwhelms the smallest 1 photoelectron (1 p.e.) signals from the following neutron events.

For this purpose, our newly designed front-end electronics (MoGURA: Module for General-Use Rapid-Application) equips Flash ADC (FADC) and FPGA for high-resolution dead-time-free DAQ and a baseline restorer (BLR) for baseline stabilization.

2.2. Dead-Time Free Electronics

The FADC and FPGA used in line are key components for dead-time-free electronics. One 1-GS/s 8-bit FADC and three 200-MS/s 8-bit FADCs are utilized for one MoGURA input channel, each of which accepts waveforms from different gain amplifiers. This combination realizes high resolution (0.1 mV, 1 ns) and a wide dynamic range (0.1 mV~10 V) simultaneously, as well as a dead-time-free ability. We utilize FPGA in three ways: Front-End FPGA (FEF) and System FPGA and User FPGA. FEF works as a transient memory, and also performs Zero Suppress on demand. System FPGA provides versatile functions for in line signal processing: Single Hit Rate for salvaging events from noise, Auto Baseline Scanning for digital baseline determination and Digital Hit Detection, and Independent Hit Time Information Recording for accurate time determination. User FPGA also works as an on-board signal processor, which may perform TQ analysis and data compression.

2.3. Baseline Restorer

KamLAND PMTs have AC coupling on their output circuits. Due to this AC coupling, the overshoot of a huge signal overwhelms the following small signals. This causes fluctuation in effective threshold, as well as the saturation of the highest-gain channel. The time constant of this overshoot (~500 μs) strongly requires BLR, since otherwise we may be blind in the meantime. We call our BLR as Asymmetric LPFed Active Baseline Subtraction BLR. In this BLR, the positive component of the baseline of the input waveform is extracted by ideal diodes then averaged and held by a capacitor, and then subtracted from the input waveform. In this manner, an overshoot, which originated positive bias compared to the original baseline, is restored, but negative pulses, which should be signals, are not distorted. With this circuit, any baseline distortion by overshoot is restored in 1 μs.

2.4. Conclusions

Detection of the CNO and pep neutrinos requires ^{11}C Tagging. This leads us to dead-time free electronics with high resolution and a wide dynamic range. We designed new front-end electronics, MoGURA, by utilizing FADC and in line FPGA. Baseline distortion due to AC coupling in the PMT is restored by 100-times faster with our BLR. MoGURA will be set in KamLAND in 2008.

3. Study of double beta decay in KamLAND

For a possible search for neutrinoless double-beta decay, using the large-volume and low-background KamLAND environment, numerical simulations have been performed for two isotopes, i.e., ^{136}Xe and ^{150}Nd. The data of KamLAND show that the low-energy region (E < 3 MeV) of the single spectrum is dominated by long-lived spallation products (^{10}C and ^{11}C), while, the region E > 3 MeV is dominated by short-lived ones, which can be rejected in an off-line analysis. Then an isotope whose Q-value is more than 3 MeV should be worthy of study. We marked ^{150}Nd because of the advantages of a high Q-value (3.37 MeV) and a high natural abundance (5.6%). On the other hand, it is very attractive that Xe gas can be dissolved [2] in the current liquid scintillator (LS) up to 2 wt%, without further development; we thus also choose ^{136}Xe (Q-value is 2.48 MeV). Possible milestones are: (i) to verify or exclude the KKDC claim [3] on the effective neutrino mass, $\langle m_\nu \rangle$, between 240 and 580 meV with the best-fit 440 meV, and (ii) to explore the case of the inverted mass hierarchy, i.e., $\langle m_\nu \rangle \sim 30$meV.

The results of a simulation show that the expected sensitivity of ^{136}Xe is at the level of 440 meV under the following conditions: 80% enriched ^{136}Xe, 2 wt% of Xe dissolved in 544-ton LS, an energy resolution of $6.5\%/\sqrt{E(MeV)}$, and a 3-year observation. There are much spallation ^{10}C backgrounds around Q-value of ^{136}Xe, so we need to reduce them with dead time free electronics described above. The expectation of ^{150}Nd seems to be promising, reaching not only the KKDC level, but also the inverted mass hierarchy (30 meV) with 5-sigma, assuming 60% enriched ^{150}Nd, 1 wt % of Nd dissolved in 544-ton LS, an energy resolution of $6.5\%/\sqrt{E(MeV)}$, and a 3-year observation. Furthermore, if we develop a liquid scintillator with a 50% higher light yield than the current KamLAND LS, the 5-sigma sensitivity will be 20 meV.

Of course, there are many things to do. The first thing that we have to do is to develop a new 1 wt % Nd-loaded liquid scintillator (good light emission, good transparency, and also achieve stability and safety etc.). We also have to investigate other possible backgrounds (^{208}Tl, gamma-ray, spallation products from target, another radioactive impurity etc), research ways of enrichment (time and cost) and purification, study the energy resolution, and calculate the sensitivity including all backgrounds and systematic uncertainties. Many of these studies will be carried out in the near future.

4. Directional measurement of electron anti neutrinos

The detection of electron antineutrinos ($\overline{\nu}_e$'s) with KamLAND is via the traditional inverse β–decay mode, $\overline{\nu}_e + p \rightarrow e^+ + n$, where the prompt signal from e^+ and the delayed 2.2-MeV γ from neutron capture on a proton make a delayed coincidence. In addition to powerful background rejection, this coincidence can be used to measure the direction of $\overline{\nu}_e$ [4] by measuring the relative position of prompt and delayed vertexes, since, especially for low-energy $\overline{\nu}_e$'s, kinematics and differential cross sections allow only neutrons emitted in forward directions. In the current KamLAND, however, the position resolutions are not enough to detect the expected displacement between the prompt and delayed vertexes, because of the diffusion of thermal neutrons, a long interaction length of 2.2-MeV γ, and a limited time resolution of PMT's. For future geoneutrino research [5], there are plans to improve those issues for directional measurements of $\overline{\nu}_e$'s, which would enable us to 'map' earthly powers of uranium and thorium with a geoneutrino probe. We also aim at it with the following techniques.

4.1. Lithium-loaded liquid scintillator

For effective neutron detection, one can employ liquid scintillator doped with ^{6}Li, ^{10}B, or Gd. For our purpose, ^{6}Li is the best candidate, because, in addition to the large cross section for thermal-neutron capture 940 barns, the reaction $n + {}^6Li \rightarrow \alpha + {}^3T$ yields α and triton (^{3}T), which are confined in a small volume. We estimate that 0.15 wt% of ^{6}Li is required to be dissolved in the liquid scintillator, so that the diffusion of thermal neutrons is sufficiently small and the lifetime of neutrons is about 5 μs. We have already started developing Lithium-loaded sintillator with a high light yield, high transparency, a low quenching effect, long-term stability, and sufficiently low cost per unit volume. We are trying various kinds of solute (e.g., LiCl, Li-carbonate, Li-acetate, etc) and solvents (e.g. Ethylene glycol, Octanol, Toluene, etc).

4.2. Vertex detection by optical imaging

Recently, imaging cameras have become possible to be used for the detection of scintillation track [6]. We are trying to apply an imaging camera to detect prompt and delayed vertexes, although there are many challenging issues. Assuming KamLAND size, the depth of field is very large, e.g., between 2.5 and 15.5 m. Also, high light collection efficiency is required, because the photo-coverage cannot be larger than the current PMT system (34%). A large target size and a small area of possible photodetectors (see the next subsection) require extremely small magnification ($\sim 10^{-3} \sim 10^{-2}$). At the same time, however, a small aberration is needed, because the necessary position resolution is on the order of the displacement between the prompt and delayed vertexes ~10mm. The naïvely estimated focal length is about 10mm.

Considering this small focal-length compared with the large depth of field, one possible design is "pan-focus", where the far boundary of the depth of field is infinity. In the pan-focus design, the number of free parameters is reduced by a strict relation in this framework. Based on such a simplified model, we are performing simulations to find the best optics. Currently, the estimated number of optical modules is unrealistically large, but we expect that it can be reduced by setting less strict requirements for aberration, depth of field, and light collection efficiency etc.

4.3. Photodetector

Candidates for the photodetector systems are (i) an image intensifier (I.I.) and a CCD camera assembly, or (ii) a pixelated photon detector (PPD) with 4 readout channels, just like a position sensitive detector (PSD) assembly. In both cases, the minimal requirements are satisfied, i.e., sufficiently high gain for one photoelectron detection with sufficiently high quantum efficiency, and fast data acquisition that enables two successive events with an interval of 1 μs to be acquired with no data loss.

We started with a test of the I.I. and CCD assembly with cosmic-ray muon events (Fig. 1). Three plastic scintillators ("PS" in the figure) detect muons going through the liquid scintillator (LS). The three-fold coincidence of these PS signals issues a trigger to open the CCD's gate. The I.I. amplifies scintillation track with a gain ~10^6 to display an afterglow at the fluorescence plane at the rear end, which is viewed by the CCD. Fig. 2 shows a resultant image of a cosmic-ray muon. We are now trying the next stage, i.e., detecting low-energy events from a ^{137}Cs source.

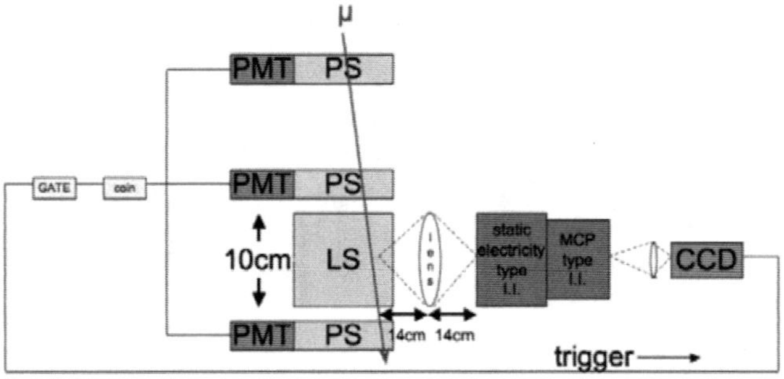

Fig. 1. Schematic view of the test of I.I. and CCD assembly to detect cosmic-ray muons.

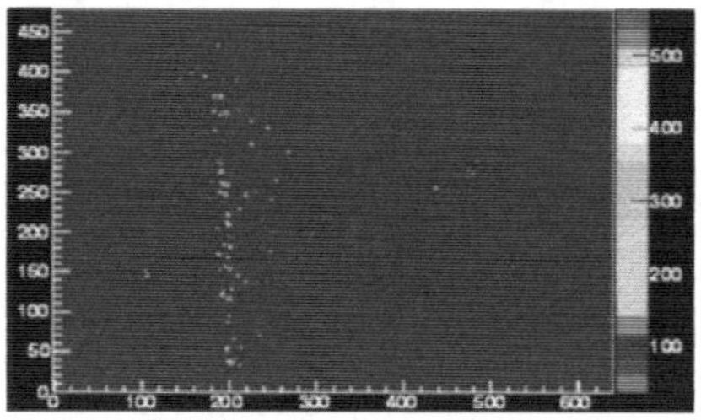

Fig. 2. An acquired image of a cosmic-ray muon.

References

[1] C. Galbiati *et al.*, Phys. Rev. **C71**, 055805 (2005).

[2] R.S. Raghavan, Phys. Rev. Lett. **72**, 1411 (1994).

[3] H.S. Klapdor-Kleingrothaus, I.V. Krivosheina, A. Dietz and O. Chkvorets, Phys. Lett. **B586**, 198 (2004).

[4] M. Apollonio et al., Eur. Phys. J. **C27**, 331 (2003).

[5] See for example F. Mantovani, these proceedings.

[6] A. Sato *et. al.*, "Scintillating Track Image Camera-SCITIC", JJAP **43**, 1593 (2004).

Constraining neutrino oscillation parameters with future observations of supernova neutrinos by various types of detectors

H Suzuki and K Yoshihara

[1] Department of Physics, Faculty of Science and Technology, Tokyo University of Science, 2641 Yamazaki, Noda, Chiba, 278-8510, Japan

[2] National Astronomical Observatory of Japan, 2-21-1 Osawa, Mitaka, Tokyo 181-8588, Japan

E-mail: suzukih@ph.noda.tus.ac.jp

Abstract. We evaluate the potential for constraining neutrino oscillation parameters with future observations of supernova neutrinos. For some selected oscillation parameters, we calculate the expected energy spectra and angular distributions of neutrino events in various detectors (Super-Kamiokande, ICARUS, KamLAND and IceCube) for an assumed galactic core-collapse supernova. We investigate to what extent the oscillation parameters could be constrained using χ^2 analysis when the expected events are really observed.

1. Introduction

It is well known that observations of supernova neutrinos might be one of the useful probes to neutrino oscillation parameters (mass hierarchy and mixing angle θ). Many preceding works[1, 2, 3] discussed possibilities of constraining parameters of neutrino oscillation and/or of supernova neutrino spectra by the future observations of supernova neutrino energy spectra. Especially Skadhauge and Funchal[3] studied rather systematically considering three types of the next generation detectors: water Cherenkov type, liquid scintillator type, and liquid Argon TPC type. In the present study, we have done yet another research on the capabilities of future supernova neutrino observations for determining the oscillation parameters. For some selected oscillation parameter sets, we calculate the numbers or energy spectra and angular distributions of neutrino events which will be observed in four detectors (Super-Kamiokande(SK): water Cherenkov type, ICARUS: liquid Argon TPC type, KamLAND: liquid scintillator type and IceCube: ice Cherenkov type) when the core-collapse supernova occurs near the galactic center. We take into account energy resolution and angular resolution of detectors. We investigate to what extent the oscillation parameters could be constrained using χ analysis when the expected events are really observed.

2. Models and Results

When we calculate the expected event features in detectors, we use the primary spectra of supernova neutrinos motivated from supernova simulations of the Garching group and a typical electron number density profile of supernova envelope (Woosley and Weaver's stellar model). We consider neither shock wave effects nor earth effects in this study. For known

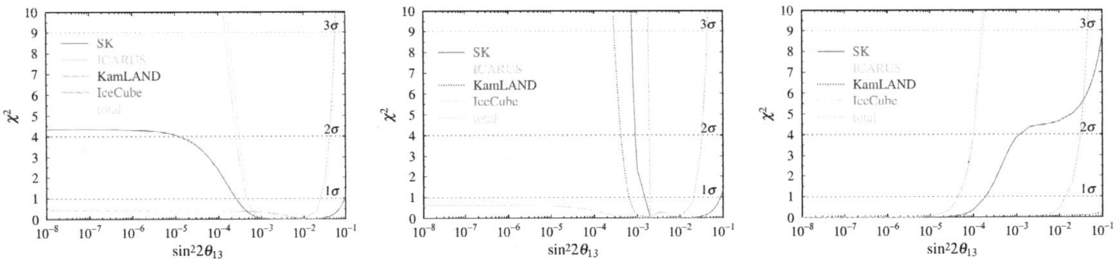

Figure 1. χ value of the fit as function of sin 2θ . Left panel: normal hierarchy and sin 2θ $= 10^{-}$ are assumed as "true" oscillation parameters. Middle panel: inverted hierarchy and sin 2θ $= 10^{-}$. Right panel: normal hierarchy and sin 2θ $= 10^{-}$.

neutrino oscillation parameters, the following values are assumed: sin 2θ $= 0.84$, Δm $=$ 7.9×10^{-5} eV , sin 2θ $= 1.0$, $|\Delta m$ $| = 2.5 \times 10^{-}$ eV . As for unknown parameters, we study three cases: normal mass hierarchy with sin 2θ $= 10^{-}$, inverted hierarchy with sin 2θ $= 10^{-}$, and normal hierarchy with sin 2θ $= 10^{-}$. Since we assume that time-integrated energy spectra of original supernova neutrinos and density profile of the stellar envelope are well known in this study, our results are corresponding to the most optimistic estimation of the potential when the time-integrated event features are analyzed.

We calculate both energy and angular distributions of the supernova neutrino events in SK and ICARUS assuming that charged current interactions between neutrinos and Ar are isotropic. For KamLAND, we calculate only the energy distribution because scintillation lights are emitted isotropically. For IceCube, only the total number of events is calculated after Dighe [2] because IceCube can detect neither energy nor angle of each supernova neutrino event. Details on our calculation of event feature is presented in [4]. For each value of θ selected as "true", an allowed region of θ is obtained by evaluating χ value of the fit for various wrong values of θ . We compile χ of the four detectors as shown in Fig.1. θ with total χ larger than 9 will be excluded (3σ) when the expected events are really observed in the detectors. ICARUS which will be sensitive to ν_e's and IceCube which will have very large statistics play dominant roles in this analysis. For the case of large θ , analysis of time-integrated feature of supernova neutrinos could constrain θ within a width of one order of magnitude (1σ). On the other hand, for the case of small θ , only an upper bound for θ could be obtained. Note that our assumption that we could know the original spectral shape might be crucial to our results. More extended studies are required.

The authors thank T. Sumikama and S. Ando for useful discussion. This work is partly supported by Academic Frontier Project of MEXT "New development of computational science through holistic approach", Large Scale Simulation Program of KEK (No. 07-05) and the Grants in Aid for the Scientific Research No. 19104006 of the MOESSC of Japan.

References

[1] Dighe A S and Smirnov A Y 2000 *Phys. Rev.* **D62** 033007; Lunardini C and Smirnov A Y 2001 *Nucl. Phys.* **B616** 307; Takahashi K, Watanabe M, Sato K and Totani T 2001 *Phys. Rev.* **D64** 093004; Takahashi K and Sato K 2003 *Prog. Theor. Phys.* **109** 919; Takahashi K, Sato K, Burrows A and Thompson T A 2003 *Phys. Rev.* **D38** 113009; Lunardini C and Smirnov A Y 2003 *JCAP* **0306** 009; Gil-Botella I and Rubbia A 2004 *JCAP* **0408** 001; Tomàs R, Kachelrieß M, Raffelt G, Dighe A, Janka H T and Scheck L 2004 *JCAP* **0409** 015

[2] Dighe A S, Keil M T and Raffelt G G 2003 *JCAP* **0306** 005

[3] Skadhauge S and Funchal R Z 2007 *JCAP* **0704** 014

[4] Yoshihara K and Suzuki H 2007 submitted to *Mod. Phys. Let. A*

Measurements with the KATRIN Pre-Spectrometer

F Habermehl for the KATRIN Collaboration

Institute for Experimental Nuclear Physics, University of Karlsruhe, 76133 Karlsruhe, Germany

E-mail: florian.habermehl@kit.edu

Abstract. The KArlsruhe TRItium Neutrino (KATRIN) experiment aims at directly measuring the anti-electron neutrino mass by investigating the kinematics of the tritium-β decay with a sensitivity of $0.2\,\mathrm{eV/c^2}$. The measurement set-up will comprise a windowless gaseous molecular tritium source, a differential and cryogenic pumped electron transport section, a tandem spectrometer system for the energy analysis, followed by a detector-system for counting transmitted β-decay electrons [1][2].

The pre-spectrometer, being the first major component in operation, is used as a prototype for further investigation and validation of the vacuum concept as well as the novel electro-magnetic design of the MAC-E filter. The test measurements presented here verify both vacuum concept and the electro-magnetic design.

1. Measurements with the pre-spectrometer test set-up

The pre-spectrometer test set-up (figure 1) comprises the ultra high vacuum (UHV) recipient, two super-conducting solenoids (max 4.5T), an e-gun system, and a detector chamber either housing a multi-channel plate or a segmented PIN diode array. All-metal gate valves allow to shut off the satellite systems for maintenance if necessary.

The spectrometer tank (volume: $8.5\,\mathrm{m}^3$, surface $25\,\mathrm{m}^2$) is designed to maintain UHV-conditions. Therefore the vacuum vessel is made of stainless steel, specifically 316LN, favourable for its non-magnetic properties.

At the horizontal pump port (not visible in figure 1) a cascaded system of turbo-molecular pumps (TMP) in conjunction with a dry fore-pump are installed, pumping the system with regard to noble gases and non getter-able gases, and entering in that way the $p \leq 10^{-10}\,\mathrm{mbar}$ pressure regime. The 45°-pump port houses a non evaporable getter (NEG) pump. This allows to pump the hydrogen gas load which is, due to outgassing effects, dominant in this regime, thus reaching a final pressure about $10^{-11}\,\mathrm{mbar}$. In the pre-spectrometer vessel the outgassing rate was measured with the pressure rise method to approximately $10^{-13}\,\mathrm{mbar\,l/s\,cm^2}$ [3].

The pre-spectrometer's function as a high pass filter not only requires a magnetic field for the guiding of electrons but also an electro-static retarding potential for the analysis of the electron's energy. Unlike in the predecessor experiments the vacuum recipient itself will be put on high-voltage. An additional inner wire-electrode allows for fine-tuning and stabilization of the retarding potential. This configuration minimizes the occurrence of Penning traps and optimizes the adiabatic transmission properties of the spectrometer. The layout of the electrode offers the possibility to apply different voltages to different parts of the wire-electrode, thus accelerating trapped charged particles into a set direction which effectively empties possible particle traps.

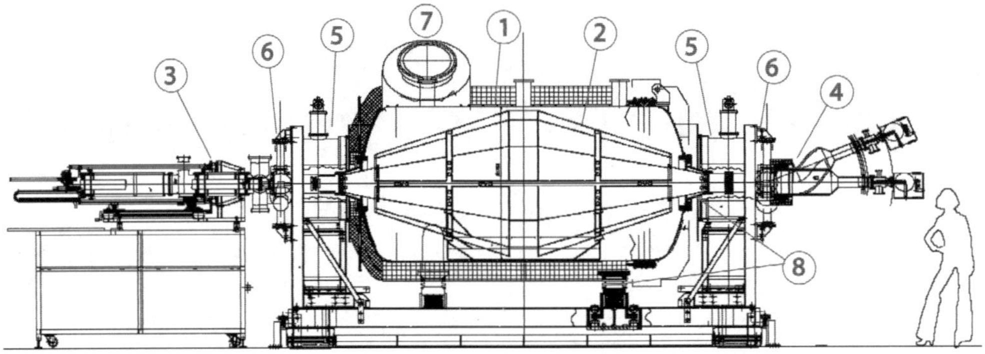

Figure 1. The pre-spectrometer test set-up: UHV recipient (1), inner wire-electrode (2), detector chamber (3), e-gun system (4), super-conducting solenoid (5), all-metal gate valve (6), 45°-pump port (7), ceramic insulators (8).

Instead of using a radioactive isotropically emitting electron source like tritium in the final set-up, the pre-spectrometer test experiment uses a point like source of quasi monoenergetic electrons, also referred to as e-gun. The electrons feature a small energy spread ($< 1\,\mathrm{eV}$ @ $18.6\,\mathrm{kV}$) with a rate that can be changed from about 400 Hz to several kHz.

The electrons guided by the magnetic field lines and passing the analyzing plane, where the retarding potential reaches its maximum, leave the pre-spectrometer tank and hit the detector. For the first commissioning of the high-voltage system and the first measurements with the pre-spectrometer system a robust Hamamatsu F2223-21S multi-channel plate (MCP) was used to detect low energy electrons. After reaching the safe operation mode the segmented PIN diode (SPD) array (64 pixels, $16\,\mathrm{cm}^2$) can be used. The SPD features energy and spatial resolution, necessary for precise investigation of the spectrometer's transmission properties. This detector is an in-house design in collaboration with the University of Washington.

The electro-magnetic tests with the set-up were focused on transmission measurements to deduce the energy resolution of the pre-spectrometer $\Delta E = 10.5 \pm 0.5\,\mathrm{eV}$ at $3\,\mathrm{kV}$ retarding potential, on the commissioning of the SPD, determining its most important properties (average energy resolution $\Delta E = 4.4\,\mathrm{keV}$, overall noise level in the energy interval of 15-22 keV of $(439.8{\pm}2.7) \times 10^{-3}$ events/s, after noise cut $(20.3{\pm}0.6) \times 10^{-3}$ events/s), and on the investigation of Penning traps in the spectrometer volume. The importance of the latter investigation showed after the observation a vacuum breakdown due to a Penning discharge. After analyzing the problem a shielding electrode was designed, manufactured, and built in. Initial measurements with the new set-up indicate that the Penning problem has been solved. Detailed measurements to further validate these first results are under way.

Acknowledgments

This work was supported in parts by the German BMBF (05CK5VKA/5, 05CK5REA/0, 05CK5PMA/0, and 05CK5UMA/3)

References

[1] Angrik J (KATRIN Collaboration) 2004 *FZKA Report* **7090**
[2] Robertson H 2007 *this proceedings*
[3] Bornschein L, Day C, Habermehl F, Luo X and Wolf J 2006 *AIP Conference Proceedings* **837** 200-209

Development of InP detector for *pp*/ 7Be solar neutrino measurement

Y. Fukuda

Faculty of Education, Miyagi University of Education,
149, Aobaku-aza-aoba, Sendai, Miyagi 980-0845, Japan

E-mail: fukuda@staff.miyakyo-u.ac.jp

Abstract. A radiation detectors used a semi-insulating Indium Phosphide (InP) wafer for measurement of solar $pp/^7$Be neutrinos have been developed in last a few years. A volume of the detector has achieved to $7mm^3$. This detector observed γ-rays, and measured a peak for the photoelectric effect at -79 °C. The charge collection is mainly done by a carrier drift along to the electric field at outside of depletion layer, and the efficiency is achieved by 60 to 70%.

1. Introduction

In 2001, Super-Kamiokande [1] and Sudbery Neutrino Observatory [2] using the charged current process have established the ν_e oscillation in their solar neutrino measurement. KamLAND also confirmed the oscillation using a reactor $\overline{\nu_e}$ in sense of Δm^2 [3], however the mixing angel θ_{12} is still not restricted as well as θ_{23} observed in the atmospheric neutrinos. In order to obtain precise θ_{12}, it is necessary to measure not only the flux but also the spectral shape for solar $pp/^7$Be neutrinos.

2. Solar $pp/^7$Be neutrino experiment using ^{115}In

Borexino has reported that the first detection of ^{7}Be neutrinos signal using low background liquid scintillator [4]. In 1976, R.Raghavan proposed the measurement of low energy solar $pp/^7$Be neutrinos [5] via following reaction $^{115}\text{In} + \nu_e \rightarrow ^{115} \text{Sn}^* + e^-$. The prompt electron has an energy with $E_\nu - 125keV$, here E_ν is an energy of incident neutrinos. Therefore the neutrino spectroscopy can be realized. An excited state of ^{115}Sn decays into the ground state with a lifetime of 4.7 μs, and emits two gammas (116keV and 487keV). This signature is also able to use for a triple-coincidence to extract neutrino signal from huge backgrounds. However, ^{115}In itself has natural beta decay into the ground state of ^{115}Sn with a lifetime of 4.4×10^{12} years. The Bremsstrahlung could produce fake coincidence for neutrino signal. Therefore a fine segmented with well energy resolution detector is necessary [6].

3. InP solid state detector

Semi-insulating InP wafer is usually grown by the liquid encapsulated Czochralski (LEC). We have chosen another wafer produced by Sumitomo Electrical Industry Co. LTD with the method of Vapor Czochralski (VCZ). Hamamatsu Co. LTD developed InP detector with $10mm \times 10mm$ in surface, and $0.2mm$ in thickness. The electrodes consist of Cr-Au with $1\mu m$ thickness for top and Au-Ge/Ni/Au with 0.13/0.015/0.5 μm thickness for bottom. The junction between electrode and InP are ohmic contact in room temperature, however actually a Schottky barrier has been formed at -79 °C because of the rectification in the measurement of Hall effect.

4. Performance of InP detector

The performance of InP detector was measured by using gammas emitted by usual radio active sources. Carriers generated by the energy deposit of electrons via photoelectric process or Compton scattering are drifted along to electric field and reached the electrode. The expected charge could be evaluated by $Q[C] = \int_0^R \frac{dE/dx}{\epsilon} e^{-\frac{r(x)}{L_d}} dx \times e$. Here, L_d is carrier drift length, ϵ is an average energy for electron/hole pair production, R is electron range, and $r(x)$ is distance from electrode. Generally speaking, the carrier drift length is expressed by $L_d \equiv \mu\tau\frac{V_0}{d}$, here μ is the carrier mobility, τ is life-time of the carrier trapping. V_0 is the bias voltage, and d is thickness of the wafer. In the depletion layer, carrier life time is long enough to the drift time, so the observed charge should be same as $N_{eh} \times e$. For the measurements of gammas, the detector should be cooled by -79 °C. using Dry-Ice. Figure.1 shows that the observed charge distribution. There found two peaks in each spectra. For instance, the peak for 122keV gamma-ray appears around 0.3×10^{-14} C and 0.55×10^{-14} C. Higher charge peak is produced by the charge collection of depletion layer, and it is consistent with other gammas assuming by 3.5eV of an average energy for electron/hole pair production. This assumption is also confirmed the peak position was not changed as bias voltage increasing. On the other hands, lower charge peak moves to higher position rapidly. The collection efficiency for lower peak is obtained by the 56% in case of 122keV γ, and this peak was formed by charge collection due to carrier drift outside of the depletion layer. According to a simulation, the spectral shape could reproduce assuming both $L_d = 250\mu m$ and $20\mu m$ for the thickness of depletion layer as shown in Figure.1.

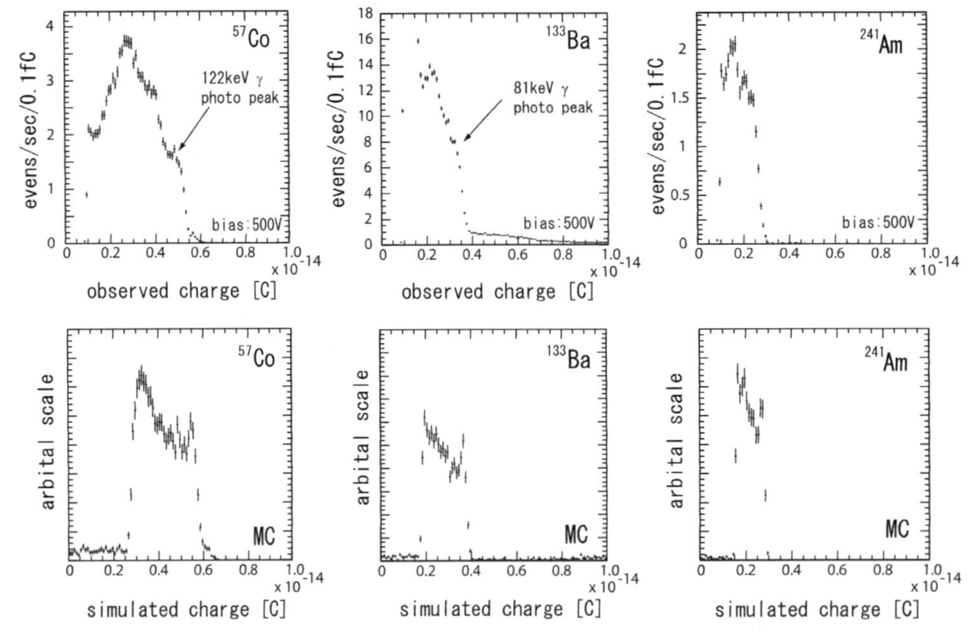

Figure 1. Observed and simulated charge distributions for gamma-rays are shown.

5. Conclusion

An InP detector has been developed, and obtained good performance. Observed charge spectra could be explained by carrier drift with $250\mu m$ and collection efficiency is achieved by 60 to 70 %. This is the first time to demonstrate InP detector with a bulk size crystal, and the detector performance shows us that it is possible to use for solar $pp/^7$Be neutrino measurement.

Acknowledgments

This work was supported by the Grant-in-Aid Scientific Research (B) 17340065 of Japanese Society for the Promotion of Science, Inamori Foundation, and The Asahi Glass Foundation.

References

[1] S.Fukuda *et.al,* Phys.Rev.Lett. **86** (2001) 5651.
[2] SNO Collaboration Phys.Rev.Lett. **87** (2001) 071301.
[3] KamLAND collaboration, Phys.Rev.Lett. **90** (2003) 021802.
[4] Borexino Collaboration submitted to Phys. Lett. B.
[5] R.S.Raghavan Phys. Rev. Lett. **37** (1976) 259.
[6] R.S.Raghavan hep-ex/0106054.

Quality Study of Purified Liquid Scintillator

Y Konno, K Nakajima and Y Kibe
Research Center for Neutrino Science, Tohoku University
Sendai 980-8578, Japan

E-mail: konno@awa.tohoku.ac.jp, kyohei@awa.tohoku.ac.jp, kibe@awa.tohoku.ac.jp

Abstract. We have been distilling the KamLAND liquid scintillator (LS) for the low energy solar neutrino observation. The distillation removes radioactive impurities from LS efficiently. We developed two types of high sensitivity radon detectors to monitor ^{222}Rn contamination which causes a primary background source ^{210}Pb. Their required sensitivity is several mBq/m^3. The features and the measurement results of these detectors are presented. We also report the study of liquid scintillator properties after the distillation : attenuation length, light output and PPO density.

1. MiniLAND : a high sensitivity Rn detector

MiniLAND was developed as a high sensitivity radon detector to monitor ^{222}Rn concentration in LS after the distillation. The detector achieved very low background environment by internal and external shields and optically-isolated cell structure to reduce accidental backgrounds drastically. NaI solution as an internal gamma ray shield has potential applications for some other experiments.

The detector consists of 4 acrylic boxes, 16 cells (100x20x20 cm^3) and contains ~0.36 m^3 LS. Total 32 PMTs (low background type) are set on cells and detect a prompt and delayed signal in a certain time window to tag ^{222}Rn daughter coincidence events (^{214}Bi/^{214}Po). All cells are optically separated from each other with reflectors (sandwiching air and a tyvek sheet between two acrylic boards), which lead to reduction of accidental backgrounds and give better light yield to PMTs by total reflection.

Low background environment is essential for a high sensitivity detector. The miniLAND detector is surrounded by 3 layers of shield which are a copper box, low impurity lead bricks and boron-containing polyethylene bricks. Very low background materials are selected at the construction of these shields based on our studies of contamination and background reduction impacts. For example, ^{210}Po activity from our lead is ~15 % (25 Bq/kg) of one from usual lead and a lead shield reduces gamma rays above 300 keV by 10^{-4} from MC simulation.

External gamma rays are the dominant backgrounds to monitor radon, but gamma rays from a PMT are not negligible. NaI solution between acrylic boxes and PMTs reduces gamma rays from PMTs above 300 keV by 10^{-1} from MC simulation. NaI solution consists of 62 wt% NaI + 0.1 wt% $Na_2S_2O_3$ and both of these are inexpensive. A gamma ray shield for PMT backgrounds must be transparent, high atomic number and high density material. NaI satisfies these requirements : Z=53, 64.1 wt% high solubility (at 20 degrees) and high transmission with $Na_2S_2O_3$. NaI powder of our choice is very low impurity product : ^{238}U < 9.2, ^{232}Th < 17 and ^{40}K < 41 mBq/m^3.

These above features allow a detector to have a high sensitivity to measure several mBq/m^3 radon. All the results of ^{222}Rn measurements in purified LS were less than several mBq/m^3. MiniLAND has been used to qualify the distilled LS before it is supplied to KamLAND.

2. High sensitivity Rn detector by concentration method

A sensitive radon (Rn) detector was newly developed by means of combining conventional concentration in a cold trap with an electrostatic collection of charged decay products. Although various methods had been examined to measure Rn dissolved in liquid scintillator (LS), small and portable types of detectors with the sensitivity of an order of mBq/m^3 have not been realized.

Rn dissolved in LS is extracted by bubbling and then trapped in the copper wool under liquid nitrogen temperature (cold trap). The carrier gas, high purity nitrogen (N_2) produced by newly constructed facility for KamLAND solar phase, is circulated to reduce background from contamination in N_2 itself. Concentrated Rn in the cold trap is released under room temperature, then pushed out of the trap to the the detector vessel to count decay products of Rn. Among the progenies of ^{222}Rn, polonium (^{218}Po and ^{214}Po) is positively charged under less humidity. Charged Po is collected by electrostatic field onto the PIN-photo diode to detect α particles emitted by Po decays. There are two kinds of filters through the path (moderate cold trap and molecular sieves) to reduce vaporized organic components and moisture in processing carrier gas, which make the collection efficiency of Po worse due to neutralization. We evaluated the total detection efficiency using a commercial Rn counter (Model1027, Sun Nuclear Corporation) calibrated in advance. The result obtained was 0.191 ± 0.013.

As a result, the detection sensitivity for Rn dissolved in LS was found to be $7mBq/m^3$ using ~55 litters of measuring LS. Further, the developed method is applicable to measure Rn density in gases whose condensation points are comparable or less than liqued N_2 temperature (sensitivity ~26μ Bq/m^3), and Rn emanation rate from materials (~8 atoms/day). We realized, in the present study, high sensitivity Rn measurement using the proposed method. The developed detector can provide useful information for LS purity during purification period for KamLAND solar phase.

3. Properties of the KamLAND liquid scintillator before and after the purification

We have investigated various methods to find the most effective way to remove the radioactive background sources from the KamLAND LS for the solar neutrino observation. As a result, it turned out that the distillation and nitrogen purge are effective to remove radioactive metals and gases from LS, respectively. There was a concern that properties of LS may change after the distillation because LS is heated in the distillation process. Hence we evaluated attenuation length, light output and PPO density of LS before and after the distillation because these parameters are directly linked to the photon yield.

Attenuation length is measured with a vertically aligned tube monitoring transmission of LED light through various depth of sampled LS. It became slightly shorter after the distillation.

Light output is estimated with a measurement of ^{60}Co Compton. It was almost same as before.

PPO density is measured with a gas chromatography spectroscopy analysis and was consistent with a preset value at the distillation system.

These results showed that properties of distilled LS were not significantly degraded by the distillation.

Background Study for KamLAND Reactor Neutrino Experiment

K. Ichimura and Y. Minekawa

RCNS, Tohoku University, Sendai 980-8578, Japan

E-mail: `ichimura@awa.tohoku.ac.jp`, `yukie@awa.tohoku.ac.jp`

Abstract. One of the goals of the KamLAND experiment is a search for anti-neutrino oscillation via inverse β decay with the characteristic delayed-coincidence method in the liquid scintillator. For a more precise measurement than previous KamLAND result [1], we have improved the background estimations of (α, n) and fast neutrons. We present the estimated number of backgrounds in our data set from Mar. 2002 to May 2007.

In KamLAND, the dominant background is C(α, n) O reaction. This reaction makes prompt signal by proton scattering, 6.05MeV e e^- pair creation or 6.13MeV γ-ray from excited states and delayed signal by neutron capture. Therefore, this reaction can make fake signal of anti-neutrino inverse β decay. The primary α source is the decay of Po, a daughter of the Rn decay chain introduced during the liquid scintillator filling. In Ref.[1], 32% error for the reaction to O grand state was assigned from ambiguities of Po rate, the cross section to O grand state, neutron angular distribution and proton quenching. 100% error was assigned to that of the excited states conservatively since the cross section uses the model calculations[3]. To reduce these large uncertainties, we have done several work. First, the proton quenching factor was measured by n beam at OKTAVIAN, Osaka Univ. Proton quenching affects the prompt energy spectrum of the background. After the measurement, the uncertainty of the quenching factor was reduced from 10% to 2%. Secondly, to know the precise Po decay rate in our fiducial volume, we improved a vertex fitter. The visible energy of 5.304 MeV Po α energy is $\sim$ 0.21MeV in KamLAND. In this energy region, the past vertex fitter has about 15cm bias. Now it is improved to within 3cm. Thus, the uncertainty of Po decay rate is reduced from 10% to 4%. Lastly, to measure the (α,n) prompt spectrum, detector calibration with identical Po C reaction was performed. The expected spectrum was also simulated by the C(α, n) O cross sections from Ref.[2][3] and compared with the data. The newly measurement cross section[2] has 4% accuracy in the total cross section, but it doesn't specify the final state separately. Thus the subtraction of the reaction rate of the excited states estimated from the cross section in Ref.[3] is needed for this new measurement. The first O excited state emits e e^- , but they can't escape from the source capsule . In KamLAND, these events are observed at $\sim$ 1.02MeV due to annihilation gammas. Therefore, the reaction rate of each state can be measured separately by Po C calibration source. The rate of the second excited state was consistent with the calculation[3] without scaling. For the first excited state, the expected spectrum agreed with the calibration data if scaled by 0.6. After the subtraction of the cross sections to excited states, then that to the grand state had been obtained. If the reaction rate of the grand state is scaled by 1.05, the expected spectrum was in good agreement with the data. With these improvement, we assign an uncertainty of 11% for the grand-state and 20% for the excited states. Since the expected spectrum is in good agreement with the Po C data, we don't assign the uncertainty of quenching factor in the analysis. Finally, the number of background event of Po C in our data set from Mar. 2002 to May 2007 is estimated to be 182 $\pm$ 21.7 events.

Cosmogenic neutrons, which are produced by muons going through outer-detector(OD) insensitive regions, sometimes enter the fiducial volume without making OD signals. These neutrons are called "fast neutron backgrounds". They can recoil protons and then make delayed capture signal, just like anti-neutrino's signal. We consider two possibilities of OD inefficiency and rock through muons. In Ref.[1], we estimated the OD inefficiency from muons passing through the detector. However, it is not a definite value for fast neutrons since muons which make fast neutrons may have different topology. To do more precise estimation, we calculated OD inefficiency for fast neutrons by MC including photon information using MUSIC and GEANT4 [4]. As a result, the fast neutron's OD inefficiencies are 0.8 % for muons passing through OD and 37 % for rock through muons. This simulation indicates that the neutron prompt energy spectrum is approximately flat in our energy window from 0.9MeV to 8.5MeV. We also estimate the other backgrounds from atmospheric neutrinos. They are estimated with NUANCE simulation . These simulations indicates that both the neutron prompt energy spectrum and prompt spectrum of the atmospheric neutrinos are approximately flat in our energy window from 0.9MeV to 8.5MeV. Finally, the number of fast neutrons and atmospheric neutrinos are estimated to be less than 9 events in our data set.

Cosmogenic He/ Li events emit beta-ray and neutron, and mimic anti-neutrino. To minimize this background, we apply 2 sec within 3 m cylindrical volume veto along muon tracks for well reconstructed muons and 2 sec whole volume veto for high charge muons. There remains 13.6 $\pm$ 1.0 He/ Li after these veto.

The accidental coincidence background is estimated using off-time window from 10 msec to 20 sec. The expected background is 80.5 $\pm$ 0.1 events above 0.9 MeV prompt energy threshold. The contribution of other backgrounds such as (γ,n) and spontaneous fission are negligible. Finally, the total background in our data set is 276.1 $\pm$ 23.5 events. All backgrounds are summarized in Table 1.

Table 1. Summary of backgrounds in our data-set taking into account of the detection efficiency

Background	contribution
Accidentals	80.5 $\pm$ 0.1
He/ Li	13.6 $\pm$ 1.0
Fast neutrons and Atmospheric ν	< 9.0
C(α, n) O G.S.	157.2 $\pm$ 17.3
C(α, n) O C$(n, n\gamma)$ C (4.4 MeV γ)	6.1 $\pm$ 0.7
C(α, n) O 1st excited state (6.05 MeV e e^-)	15.2 $\pm$ 3.5
C(α, n) O 2nd excited state (6.13 MeV γ)	3.5 $\pm$ 0.2
Total	276.1 $\pm$ 23.5

References

[1] Araki T *et al.* 2005 *Phys.Rev.Lett.* vol 94 081801

[2] Harrissopulos S 2005 *et al.* *Phys.Rev.C* vol 72 062801

[3] JENDL 2005 http://www.ndc.tokai-sc.jaea.go.jp/jendl/jendl.html

[4] GEANT4 Collaboration 2003 *NIM* A vol 506, page 250

Updated results from the 3-ton Gd loaded liquid scintillator target after 2 years of data taking at LNGS

I Barabanov[1], C Cattadori[2], N Danilov[1,3], A di Vacri[4,*], L Ioannucci[4], Yu Krilov[1,3], E Yanovich[1] and M Aglietta[5], G Bari[6], A Bonardi[5], G Bruno[4], W Fulgione[5], E Kemp[7], A Malguine[1] A Porta[5] for the LVD collaboration

[1] Institute of Nuclear Research Moscow Russia
[2] INFN-Milano Italy
[3] Institute of Physical Chemistry Moscow Russia
[4] LNGS INFN Italy
[5] IFSI-Torino INAF and INFN-Torino Italy
[6] INFN-Bologna Italy
[7] Universidade Estadual de Campinas Brasil
[*] Paper presented at the conference by A. di Vacri

E-mail: divacri@lngs.infn.it

Abstract. We performed a 3-m^3 Gd experiment by doping (up to 0.1% in weight) two counters of the LVD experiment, at LNGS, with a Gd organic salt developed and produced as the result of a joint INFN/INR research activity. Feasibility of the experiment and performances of the Gd doped liquid scintillator (Gd-LS) have been presented. The chemical and physical properties of the Gd-LS and its performance as a neutron detector, namely neutron capture efficiency and average capture time are being monitored since the doping time in 2005. From laboratory survey we can state the stability of the trasmittance (T) at the reference wavelength (425 nm) with a C.L. of 81% and 96% respectively for the first and the second doped counter. This is the largest stable Gd-doped organic liquid scintillator target ever produced and continuously operated for a long period.

1. Introduction

The interest of loading the liquid scintillator (LS) used to detect inverse β-decay $\overline{\nu}_e$ interactions with Gd, lies in the effectiveness of this element as neutron capturer. Adding $\sim$ 1 g/l of Gd to the scintillator produces a substantial reduction in the mean n-capture time and the visible energy is significantly above natural radioactivity background (8 MeV instead of 2.2 MeV), increasing the sensitivity to $\overline{\nu}_e$. An example of detector that could benefit for the Gd loading is LVD, devoted to detection of neutrino bursts from Gravitational Stellar Collapses in our galaxy [1]. The main goals of Gd doping 2 LVD detectors are the study of the feasibility of Gd-doping an existing large scale apparatus, the evaluation of the new Gd-LS performances on ton scale and the monitor of their stability when scintillator is kept in contact for few years with detector materials at extreme environmental conditions (large temperature and humidity variability).

2. Properties of the liquid scintillator and the adopted Gd doping technology

LVD uses two kinds of LS: one (LS1) contains $\sim$ 16% of aromatics and the other (LS2) $\sim$ 8%. They show slightly different characteristics, the higher percentage of aromatics causing a higher light yield, measured in laboratory to be about 18% [2]. For both of them the light attenuation length at 425 nm, is $\Lambda \geq 16$ m.

A 15 kg batch of the organic Gd compound (Gd-2Methylvalerate) was prepared at LNGS chemical laboratory in spring 2005, following the recipe developed [3]. The adopted salt formulation was chosen and optimized to be highly soluble in organic solvent even having low aromatic content, minimizing light quenching while keeping good optical properties of final Gd-LS.

Each LVD counter is a 1.5 m^3 stainless steel tank containing 1.2 ton of LS. To dope such a large amount of LS handling a minimum aliquot of the latter, a master solution highly Gd-doped was first prepared and subsequently diluted in the tank; 31 l of LS were extracted from one counter, doped at a Gd concentration of 48 g/l level and then poured back in the counter. We chose to dope one LS1 counter (T40 located in the LNGS external laboratory at [Gd]= 1.05 g/l in May 2005) and one LS2 counter (T3131 in the underground laboratory at [Gd]=0.93 g/l in October 2005) reaching Gd concentrations that differ of 15%. Concentration of stabilizing agent was different as well.

3. Performances of Gd LS and stability survey

The impact of Gd doping on optical characteristics has been determined both through

- *in-tank* measurements: the global counter gain (light yield*transparency) in tank 3131 has been measured before and after Gd introduction by using the cosmic muon signal and the result is a gain decrease of (17±2)%.
- and laboratory measurements on small size LS samples: Gd doping at 0.1 g/l impacts of -15% on the light yield (LY) of the Gd-LS compared to the unloaded and produces a decrease of Λ down to $\sim$ 10 m.

The observed neutron mean capture times are $<\tau_n>= 24.5\pm0.1\mu$s and $<\tau_n>= 27.0\pm0.2\mu$s for T40 and T3131 respectively, corresponding to a decreases of a factor of $\sim$ 10 respect to the value on the mean n-capture time on H ($<\tau_{nH}>\sim 200\mu$s). This impacts of the same value on the S/N ratio. The neutron detection efficiency passes from 70% to $<\eta_n>= 87.5 \pm 0.1\%$ and $<\eta_n>= 86.1 \pm 0.1\%$ respectively for T40 and T3131.

We are monitoring Gd-LS stability

- on large scale: by measuring cosmic muons, neutron detection efficiency and mean n-capture time;
- on small scale: by directly measuring Gd concentration, transmittance and light yield of LS samples periodically extracted from the counters.

4. Results and perspectives

The doped counters have been continuously monitored for a total time of about 700 days. Both of them do not show any indication of variation and degradation in Gd-LS performance, all the parameters monitored turn out to be stable. On the basis of the know-how acquired in metal loading of organic LS and on the satisfactory results obtained on large scale production, the next steps of our R&D phase, ongoing at the LNGS since 2000, will be the development of Neodymium and Cadmium doped scintillators for $0\nu\beta\beta$-decay experiments and the production of Gd doped target based on LAB (Linear Alkyl Benzene) for $\overline{\nu}_e$ detection at reactors.

[1] Aglietta M *et al.* 1992 *Il Nuovo Cimento* A **105** 1793
[2] di Vacri A 2007 *Ph.D. Thesis* L'Aquila University
[3] Danilov N A 2007 *Radiochemistry* **49** 281-9

The Borexino Experiment Liquid Scintillator Purification and Containment System

**F. Dalnoki-Veress, C. Galbiati, A. Goretti, An. Ianni, A. Pocar, F. Calaprice*
on behalf of the Borexino Collaboration**

Princeton University, Physics Dept., Jadwin Hall, 08544, Princeton (NJ), USA
*Paper presented at the conference by F. Calaprice
e-mail: frankc@princeton.edu

Abstract. Borexino is a large volume, real-time, liquid scintillator detector located at the Gran Sasso National Laboratory in Italy. The principal objective of the detector is to measure mono energetic (862 keV) ^{7}Be neutrinos from the sun present with a count rate of several tens of events per day. Measurement at this level requires an extremely low internal background due to natural radioactivity present in the detector components. In this paper the techniques used by Borexino to purify the scintillator and to build the nylon containment vessels are described. The unprecedented high radiopurity reached by Borexino permitted for the first time the real-time detection of ^{7}Be neutrinos from the sun[1].

1. The Borexino Design:

The design of Borexino is based on the principle of graded shielding by which the scintillator is shielded by a series of concentric shells of increasing radiopurity and the entire detector is located at 3500 m.w.e. underground. The scintillator (278 ton), composed of pseudocumene (PC) and 1.5 g/L of the fluor 2,5-diphenyloxazole (PPO), is contained in a thin (125 µm) nylon spherical shell (IV) with a radius of 4.25 m. Concentric with the IV is a second nylon sphere (OV) at a radius of 5.50m and a Stainless Steel Sphere (SSS) at 6.55 m. The medium between both vessels consists of 323+567 tons of PC+5g/L Dimethyl Phthalate (DMP) which quenches the light produced by radioactivity outside the IV. In addition, the OV acts as a radon barrier and the SSS supports the PMT's. Finally the SSS is itself embedded in 21,000 t of ultra-pure water (UPW) providing a shield against external radiation. Within the IV, a fiducial mass of 100 tons is defined by software selection of the events. The scintillator purification and containment systems play a central role in achieving the low backgrounds needed for the detection of the ^{7}Be solar neutrinos.

2. The Borexino Purification Process:

This unprecedented purity (4 decay/day/100 t scintillator) for a large-scale liquid scintillation detector was achieved with a purification system [2] that emphasized cleanliness and leak tightness. Pseudocumene (PC), the scintillator solvent, was delivered by tanker truck to the underground lab at Gran Sasso. It was loaded into storage tanks and purified at a rate of 800 L/h. The PC was fed to a 6 m tall, 0.75 m diameter 6-tray vacuum distillation column to remove low volatility radioactive impurities (e.g. U, Th, K). After distillation the PC was stripped of radioactive noble gas impurities (^{39}Ar and ^{85}Kr) with specialty nitrogen low in Argon and Krypton (LAKN). Gas stripping was done in a 8 m tall, 0.15 m diameter column filled with structured packing. Purified PC was mixed with either a pre-purified concentrated solution of the fluor (PPO) or quencher (DMP) to fill the scintillator and buffer volumes respectively. Five 860 L batches of concentrated fluor master solution (140 g-PPO/L-PC) were purified by a combination of water extraction, filtration, single stage vacuum distillation and nitrogen stripping. The scintillator and buffer regions were filled by periodically switching between volumes to maintain equal levels and hydrostatic head in all three volumes of the Borexino detector. The purification facility is entirely composed of stainless steel, which has been electropolished, and treated to be corrosion resistant. In addition, the equipment has been precision cleaned to a particulate level <30 (MIL-STD-1246C) and assembled in a class 100 clean room. All flanges are double sealed and are continuously purged with N_2. All fittings in the scintillator path have also been surrounded with special aluminum boxes and are continuously purged with N_2. After final assembly, each part of the entire system has been vacuum leak checked to the level of 10^{-8} mbar L/sec.

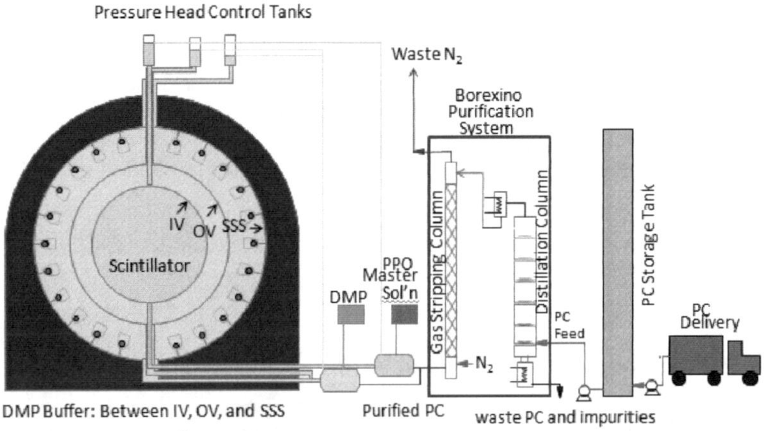

Figure 1: The Borexino Liquid Scintillator Purification System

3. The Liquid Scintillation Containment System (IV,OV):

The nylon chosen for scintillator containment [3] satisfied specific requirements such as chemical compatibility, mechanical strength, low radioactivity, optical transparency, clean fabrication, leak tightness, and low Rn permeability.

The assembly and installation of the nylon vessels was particularly complex, because of the kinds of materials used, and the methodology adopted to cut, assemble and bond the nylon. In addition, all operations were done in a class 100 clean room at Princeton University, in which by using a dedicated pressure-swing adsorption system the air was kept Rn-free. Each of the two nylon spheres is composed of more than 30 separate pieces, carefully stacked to prevent contamination, and bonded utilizing a precisely controlled resorcinol aerosol spray. Insertion of the completed IV within the outer OV was performed prior to OV sealing. A complicated set of winches, pulleys and scaffolding was used to raise the vessels into position within the SSS at Gran Sasso.

After installation, the vessels were purged with several volumes of LAKN N_2. Prior to the final filling with scintillator (completed in May 2007), the vessels were filled with ultra pure water for testing the purification facility and for shedding the nylon surface of nylon monomers and residual radioactive metal ions.

References

[1] Arpesella *et. al.*, accepted for publication Phys. Lett. B., arXiv:physics/0708.2251v2.

[2] J. Benziger *et. al.*, accepted for publication in Nucl. Instr. Meth. A., arXiv:physics/0702162.

[3] J. Benziger *et. al.*, Nucl. Instrum. Meth., A582, 509-534, 2007.

Recent atmospheric neutrino results from Super-Kamiokande

Yoshitaka Itow for the Super-Kamiokande collaboration

Solar-Terrestrial Environment Laboratory, Nagoya University, Furo-cho, Chikusa-ku, Nagoya, Aichi, 464-8601, Japan

E-mail: itow@stelab.nagoya-u.ac.jp

Abstract. The Super-Kamiokande (Super-K) has observed atmospheric neutrinos since 1996. Almost 10 years running of Super-K provides large statistics of atmospheric neutrino sample for precision study of neutrino oscillations such as CC τ-appearance, sub-dominant effect from solar terms and exotic scenarios. Also we present very preliminary results from atmospheric neutrino data from the SK-3 where 40% PMT-coverage is restored by reconstruction works carried out in 2006.

1. Introduction

Super-Kamiokande (Super-K) is a 50 kt water Cherenkov detector 1000 m underground in the Kamioka mine. Operation started with 40% photo-coverage from Apr 1996 until Jul 2001 (SK-1). During this period neutrino oscillations was first discovered in atmospheric neutrinos in 1998. After the PMT accident in Nov 2000 Super-K resumed operation with 20% photo-coverage (SK-2) by remaining PMT's from Oct 2002 until Oct 2005. Re-installation works of rest of PMTs was completed in Jul 2006 and data taking has been resumed with 40% coverage again (SK-3). Here we preset oscillation analyses of atmospheric neutrinos from SK-1 and SK-2. Also we show very preliminary results from SK-3 atmospheric neutrino data.

2. Results in 2-flavor oscillation analysis in SK-1+SK-2

The 2-flavor ν_μ-ν_τ oscillation analysis has been done for a combined data set of SK-1 and SK-2. The SK-1 data set has 1489 live-days and 1646 live-days for a fully-contained/partially-contained sample (FCPC) and an upward going muon sample (UPMU), respectively. Those of SK-2 data are 804 live-days and 828 live-days for FCPC and UPMU, respectively. In this analysis a sample of upward through going muons was divided into "showering" and "non-showering" samples. The upgoing "showering" muons are events where high energy muons created EM showers by radiative processes. Typical values of neutrino energy are ~1TeV and ~100GeV for "showering" and "non-showering" samples, respectively. The best fit parameters for 2-flavor ν_μ-ν_τ oscillations were Δm^2_{23}=2.5$\times$10^{-3}eV^2 and $\sin^2 2\theta$=1.00 with χ^2=839.7/755 dof (18% probability). The 90%C.L. allowed region was obtained as Δm^2_{23}=1.9-3.1$\times$10^{-3}eV^2 and $\sin^2 2\theta > 0.93$ as shown in Fig. 1.

Appearance of charged current (CC) ν_τ events has been studied in the SK-1 FC sample. Two independent methods, maximum-likelihood (LH) and neural net (NN) methods were applied to the sample in order to select CC ν_τ events by checking event shapes such as event sphericity or number of decay electrons. After the selection using the LH method the excess of events in upward going FC events due to CC ν_τ were 59.6±20.7 corresponding to 138±48(stat.)$^{+14.8}_{-31.6}$(sys.) events for total τ excess. This number is consistent to the expected τ excess 78.4±26(sys.). Similar results were derived also from the analysis based on the NN method.

3. Oscillation with a solar term

Since travel lengths of atmospheric neutrinos coming from the other side of earth are large enough, low energy atmospheric ν_e's around 100MeV may oscillate driven by "solar terms", Δm^2_{12}=7.4-9.2×10^{-5}eV^2 and $\tan^2\theta_{12}$=0.28-0.58, which are relevant for solar neutrino oscillations. There is a chance to resolve octant of θ_{23} degeneracy ($\theta_{23} > \pi/4$ or $\theta_{23} < \pi/4$) by checking this sub-dominant oscillation effect driven by a solar term [1]. Such oscillations decrease ν_e rate by P_2, where P_2 is 2 flavor oscillation probability $\nu_e \to \nu_\mu, \nu_\tau$ in matter driven by Δm^2_{12}. On the other hand ν_e rate increases due to ν_μ oscillations into ν_e by $rcos^2\theta_{23}P_2$, where r is ν_μ/ν_e ratio. Since r is ~2 for low energy, two effects are canceled if $\cos^2\theta_{23} = 0.5$ for $\theta = \pi/4$. Therefore deviation of θ_{23} from $\pi/4$ can be observed as small deviation of low energy ν_e rate from standard ν_μ-ν_τ oscillations. The size of effect is expected as ± a few % in sub-GeV e-like sample for $\sin^2\theta_{23}$ =0.5±0.1. Very precise study of atmospheric neutrino data has been done for the SK-1+SK-2 combined data. In this analysis $\sin^2\theta_{13}$=0 is assumed. Without solar terms the best fit gives $\sin^2\theta_{23}$=0.50. With solar terms the best fit of $\sin^2\theta_{23}$=0.52 while it is still consistent to 0.50 within a error. So far no evidence for deviation of $sin^2\theta_{23}$ from $\pi/4$.

4. Exotic scenarios

Advantage of atmospheric neutrinos for studying neutrino oscillations is not only its large statistics but also a large coverage for L/E by which we can check various exotic scenarios. Neutrino decays and neutrino decoherence scenarios have been ruled out in L/E distributions with large χ^2 values, χ^2_{dcy} = 107.1/82dof and χ^2_{decohe} = 112.5/82, respectively. Standard oscillation scenario fits with good χ^2 = 83.9/82/dof. Sterile neutrino admixture in 2+2 mass hierarchy [2] was also tested. The result is consistent with no sterile admixture giving upper limit of 24.5 % admixture at 90% C.L.

Mass Varying Neutrinos (MaVaNs) is originally proposed to explain source of dark energy by relic neutrinos of which masses were varied by ambient neutrino density [3]. Possibly their masses are also varied by matter density or electron density beyond the standard MSW effect. The 2-flavor zenith analysis was made in SK-1 data set to search for additional matter effect. Here the Δm^2 parameter in standard 2-flavor oscillation probability was replaced by $\Delta m^2 \times (\rho_e/\rho_0)^n$, where ρ_0 = 1.0mol/m^3, to check mass varying with ambient electron density. The best fit of n was found to be -0.03 with Δm^2_{23}=1.95×10^{-3}eV^2 (χ^2 = 172.2/178dof). The 90% C.L. allowed region for n and Δm^2 is shown in Fig. 2, resulting no evidence for the MaVaN scenario.

5. Current status and future

Since Aug 2006 SK-3 has been back in online with original 40% photo coverage. Very preliminary analysis of 166 days of SK-3 atmospheric neutrino data (Aug 2006 - Mar 2007) has been made. Event rate and zenith angle dependence are consistent to those in SK-1 and SK-2 as shown in Fig. 3. In summer 2008 DAQ electronics upgrade is scheduled. Super-K will provide atmospheric neutrino data with improved performance such as wider dynamic range of Q or higher decay-e tagging efficiency.

References

[1] O. L. G. Peres and A. Yu. Smirnov, hep-ph/0309312

[2] Fogli et al, PRD 63(2001)053008

[3] R.Fardon, A.E.Nelson and N.Weiner, JCAP 0410:005,2004

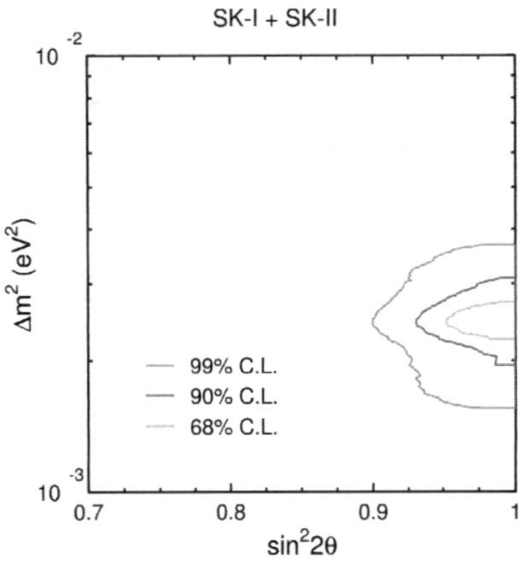

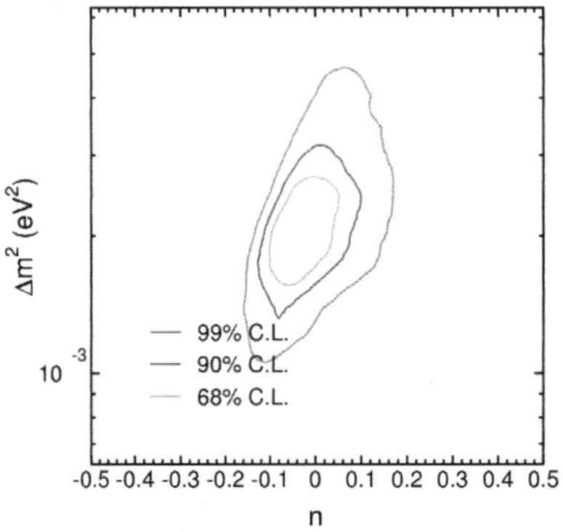

Figure 1. Allowed region of ν_μ-ν_τ oscillations analysis in SK-1+SK-2 combined atmospheric neutrino data. The contour lines corresponds to 68%,90% and 99% C.L. from inside to outside, respectively.

Figure 2. Allowed region for a parameter "n" and Δm^2 in MaVaN scenario (See text). The contour lines corresponds to 68%,90% and 99% C.L. from inside to outside, respectively.

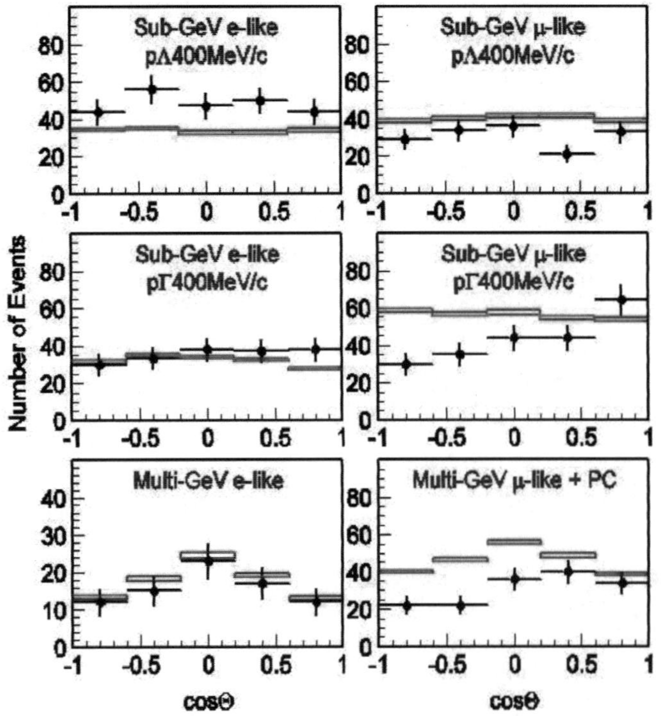

Figure 3. The zenith angle distribution of FC and PC samples from preliminary analysis of 166 days of SK-3 atmospheric neutrino data showing overall consistency to the previous results.

Measurement of the atmospheric neutrino flux with AMANDA-II and IceCube

Paolo Desiati* **for the IceCube Collaboration**[1]

*Department of Physics, University of Wisconsin-Madison, 53706, WI, US

E-mail: `paolo.desiati@icecube.wisc.edu`

Abstract. The IceCube Observatory presently consists of an array of 22 vertical strings, deployed between 1450 m and 2450 m of depth and containing 1320 digital optical sensors, and 26 IceTop surface stations with 104 sensors. The denser AMANDA-II array is integrated into IceCube in order to extend the sensitivity to lower energies. The Observatory is the world's largest neutrino telescope in operation. Recent results on the measurement of the atmospheric neutrino flux with IceCube are discussed, with an emphasis on the acceptance and sensitivity potential of the growing IceCube neutrino telescope.

1. Introduction

Neutrinos are the ideal cosmic messengers that allow us to probe the origin of cosmic rays and high energy mechanisms that shaped our Universe, since they can propagate through huge distances almost undisturbed from their sources. Detection of neutrinos in correlation with possible sources of cosmic rays would be the awaited smoking gun that would prove hadronic acceleration. On the other hand, the very same reasons that make neutrinos interesting cosmic messengers, make them also difficult to detect, requiring large volume detectors deployed in the deep oceans or in the antarctic ice [1].

The IceCube Observatory is a kilometer-cube size neutrino telescope in construction at the geographic South Pole [2], and it presently consists of 22 strings (hence tagged as IC22). The strings are located on an hexagonal array with a spacing of about 125 m, each of which contains 60 digital optical sensors that are 17 m apart. IceCube embeds the predecessor AMANDA-II array, which consists of 677 analog optical sensors compactly arranged in 19 strings [3].

The goal of neutrino telescopes is the detection of high energy neutrinos of extra-terrestrial origin [4]. The major background is the intense flux of downward muons produced by the impact of cosmic rays on the Earth's atmosphere: about a million of them for each neutrino-induced event. The virtually background free upgoing atmospheric neutrinos constitute a foreground of diffuse events among which we search for the high energy extra-terrestrial events. The IceCube Observatory will collect an unprecedented statistics of atmospheric neutrinos, allowing us to measure this foreground and to understand it, and, therefore, to do neutrino astrophysics.

[1] IceCube web site at http://icecube.wisc.edu

2. Selecting Atmospheric Neutrinos

An upward moving event in a neutrino telescope would be guaranteed to be neutrino-induced muon track, if the direction reconstruction was not affected by any ambiguity. In reality, less than 5% of the downward muon bundles are mistakenly reconstructed as upgoing. Moreover, in the IC22 array, about 5% of the recorded events are pairs of uncorrelated muon bundles that happen to pass through the detector in time coincidence. Around 10% of the coincident events mimic upward moving tracks. Both these sources of background are in excess of about $10^3 - 10^4$ with respect to the neutrino-induced events, if no specific selection criteria are applied. The event selection is designed to achieve a good track directional resolution. For instance, requiring an angular resolution of about 2°, we can collect about one atmospheric neutrino per hour in the IC22 array (with a selection efficiency of about 15-20%), of which about 10-15% are fakes. The first analysis of atmospheric neutrinos in IceCube was performed on data collected by the first 9 strings during 2006. 234 events were recorded in 137.4 days livetime, with an expectation of 211±76.1(syst)±14.5(stat) from Monte Carlo simulations [5]. As shown in Fig. 1, most of the selected neutrino events have energy just below 1 TeV. The low energy profile is dominated by the experimental trigger and selection efficiency, while the high energy shape is determined by the steep spectrum and neutrino cross section.

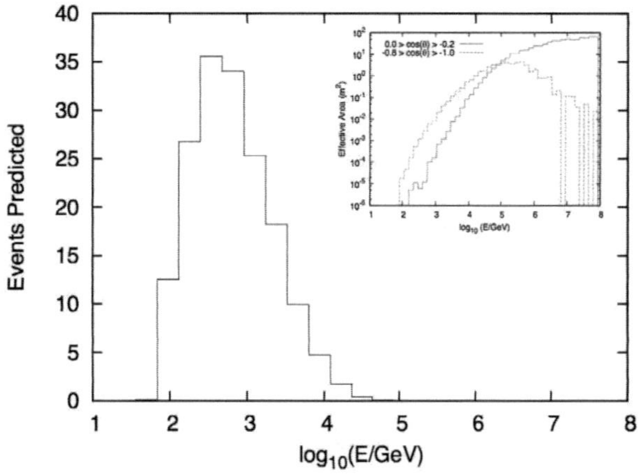

Figure 1. Neutrino energy distribution for the event selected in 2006 with 9 strings of IceCube. Insert: the neutrino effective area as a function of neutrino energy.

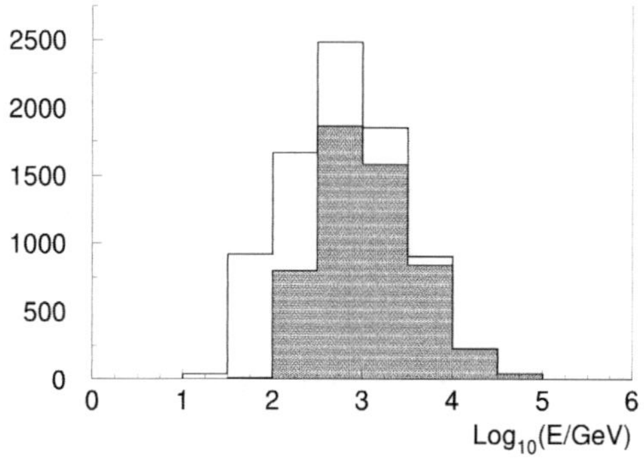

Figure 2. Neutrino energy distribution expected in 200 days with IC22 (shaded area), and with the combined IC22 and AMANDA. In the latter case the energy threshold is about 30 GeV.

The full IceCube array will be able to collect about 5-10 atmospheric neutrinos per hour, depending on the selection criteria and, as a consequence, on the event quality.

The detection of electron neutrino-induced events is more critical than muon events, since their rate is about one order of magnitude lower that the one of muon neutrinos. Such events in IceCube are identified through the hadronic cascades produced by charged current interactions. The main background is due to bremsstrahlung energy loss by muon bundles. A detailed study on background rejection is under way.

2.1. Low Energy Neutrinos

The criteria currently under investigation to achieve a good background rejection, result in a neutrino energy threshold of 100 GeV. At these energies, the uncertainty on the flux of atmospheric neutrinos is estimated to be larger than 30% [6]. Almost one third of the uncertainty is due to the composition of cosmic rays above 1 TeV, and the reminder to the high energy hadronic interaction models. Due to kinematical reasons, neutrinos produced in the decays of kaons (K) dominate the atmospheric flux above about 100 GeV [7], with uncertainties significantly larger than on pion (π) production in the same energy region. Therefore the measurement of atmospheric neutrinos below 100 GeV would probe a region affected by smaller theoretical uncertainties and would help in constraining the absolute normalization. With theoretical uncertainties below 25%, we can test the experimental capabilities of the IceCube experiment and assess its systematics down to that order of magnitude.

The IC22 and AMANDA-II arrays can detect events in time coincidence, lowering the neutrino energy threshold to about 30 GeV (resulting in about 50% more events, see Fig 2), since AMANDA-II is a denser detector and, combined with IceCube, improves the track angular resolution below 100 GeV. At 30 GeV standard oscillations inevitably affect the measurement of neutrino flux, and for vertical tracks, about 70% of muon neutrinos have oscillated into another flavor [8]. On the other hand, experimental energy and angular resolutions dilute the effect of oscillations. A less systematics prone reconstruction of high quality low energy upward moving tracks close to the IceCube strings, is under way. The muon neutrino survival probability has its minimum at 30 GeV for vertical tracks. Since the optical sensors are only 17 m apart, it is possible to statistically estimate the muon track length L_μ and its energy E_μ, and deconvolve them in a $\mathrm{L}_\nu/\mathrm{E}_\nu$ measurement with minimized systematics due to the $\nu_\mu - \mu$ kinematics, where L_ν is the neutrino path length and E_ν the neutrino energy.

2.2. High Energy Neutrinos

Due to the steep atmospheric neutrino spectrum most of the events are around a few TeV. Above this energy the detailed measurement of the spectrum slope depends on the primary cosmic ray spectrum and composition, on the K production in the atmosphere and, mainly above 100 TeV and on the yet unknown contribution of charm production. The atmospheric muon flux is also used to probe interaction models. The μ^+/μ^- ratio measured by the MINOS experiment at about 1 TeV has provided an important constraint on the $\mathrm{K}^+/\mathrm{K}^-$ ratio in the atmosphere [9]. Neutrinos are even more heavily affected by the K ratio above 1 TeV, therefore this constraint is more relevant for neutrino production. With IceCube it will be possible to probe the K/π relative contribution to the atmospheric neutrino flux, with a combined analysis of energy and zenith angular shapes.

Atmospheric neutrinos above 10 TeV can be used to probe the internal structure of the Earth, that starts to be opaque to neutrino propagation at about 25 TeV, where the Earth's diameter equals the neutrino absorption length [10]. Due to the steepness of the atmospheric neutrino energy spectrum, IceCube will take 10 years to become sensitive to the Earth's matter profile, providing direct geophysical information. Above 100 TeV neutrinos starts to be dominated by the decay of heavy mesons and baryons containing the charm quarks [11]. This component of the atmospheric neutrinos is still very uncertain, mainly because of the unknown baryon contribution. A better assessment of the charm background is critical for the discovery of a high energy extra-terrestrial signal.

References

[1] Desiati P for the IceCube Collaboration 2006 *in Proc of 20th ECRS, Lisbon, Portugal*, astro-ph/0611603.
[2] Achterberg A, et al. (IceCube Collaboration) 2006 *Astrop. Phys.* **26** 3.
[3] Andrés E, et al. (AMANDA Collaboration) 2000 *Astrop. Phys.* **13** 1.
[4] Hoshina K and Montaruli T for the IceCube Collaboration *these proceedings*.
[5] Achterberg A, et al. (IceCube Collaboration) 2000 *Phys. Rev.* D **76** 027101.
[6] Barr GD et al. 2006 *Phys. Rev.* D **74** 094009.
[7] Gaisser TK, hep-ph/0209195.
[8] Akhmedov ET, et al., hep-ph/0612285.
[9] Adamson P et al. (MINOS Collaboration) 2007 *Phys. Rev.* D **76** 052003.
[10] Golzalez-Garcia MC et al., hep-ph/0711.0745.
[11] Battistoni G et al. 2006 *in Proc of 2nd Workshop on TeV Particle Astrophysics, Madison, WI.*

Solar atmosphere neutrino oscillations

G. L. Fogli[1], E. Lisi[1], A. Mirizzi[2] *, D. Montanino[3] and P. D. Serpico[4]

[1]Dipartimento di Fisica and Sezione INFN di Bari, Via Amendola 173, 70126 Bari, Italy
[2]Max-Planck-Institut für Physik (Werner-Heisenberg-Institut), Föhringer Ring 6, 80805 Munich, Germany
[3]Dipartimento di Fisica and Sezione INFN di Lecce, Via Arnesano, 73100 Lecce, Italy
[4]Center for Particle Astrophysics, Fermi National Accelerator Laboratory, Batavia, IL 60510-0500 USA
* speaker. e-mail: amirizzi@mppmu.mpg.de

Abstract. The Sun is a source of high energy neutrinos ($E > 10$ GeV) produced by cosmic ray interactions in the solar atmosphere. We study the impact of three-flavor oscillations on the solar atmosphere neutrino fluxes observable at Earth. We find that peculiar matter oscillation effects in the Sun do exist, but are significantly suppressed by averaging over the production region and over the neutrino and antineutrino components. In particular, the relation between the neutrino fluxes at the Sun and at the Earth can be approximately expressed in terms of phase-averaged "vacuum" oscillations, dominated by a single mixing parameter (the angle θ_{23}).

1. Introduction

The Sun is a source of high energy neutrinos ($E > 10$ GeV) generated when cosmic rays interact in the solar atmosphere [1, 2], producing secondary mesons which subsequently decay into ν_e, $\overline{\nu}_e$, ν_μ, and $\overline{\nu}_\mu$, collectively referred to as "solar atmosphere neutrinos" (SAν) hereafter. At the Earth, within the solid angle subtended by the Sun, the SAν flux significantly exceeds the familiar "atmospheric neutrino" flux generated by cosmic rays interacting with the Earth atmosphere: the density at the first interaction point is lower in the solar atmosphere than in the terrestrial one, therefore a larger fraction of mesons decay into neutrinos instead of being absorbed.

The characterization of the SAν signal is an important goal, both in itself and because it might represent a "background" in searches for signatures of possible weakly interacting massive particle (WIMP) annihilation in the Sun. Here we characterize the impact of flavor transitions (in vacuum and in matter) on SAν. In Sec. 2 we describe the input SAν fluxes at the Sun. In Sec. 3 we study SAν oscillations in vacuum and in matter within the standard 3ν framework. In Sec. 4 we evaluate the impact of flavor oscillations on the observable events rates in a km^3 neutrino telescope. We conclude in Sec. 5. This talk in based on the result of [3] to which we address the interested reader for further details.

2. Solar atmosphere neutrino fluxes

The most recent calculation of the SAν flux has been performed in [1] by using Monte Carlo codes for high-energy particle interactions. From pion production dominance, at the source the SAν flavor ratio is expected to be $\phi_e : \phi_\mu : \phi_\tau \simeq 1 : 2 : 0$, with approximately equal fluxes of ν and $\overline{\nu}$. The simulations in [1] find a power-law decrease $\sim E^{-3}$ for the SAν energy spectrum at $E > 100$ GeV. At lower energies, heliomagnetic effects are expected to harden the spectrum (as $\sim E^{-2}$) [2]. After production, SAν's are affected by the solar medium through both absorption on nucleons and oscillations in matter. In the energy range of interest for our work, absorption can be accounted for by an overall attenuation function multiplying the original ν fluxes [1], while oscillation effects can be applied afterwards.

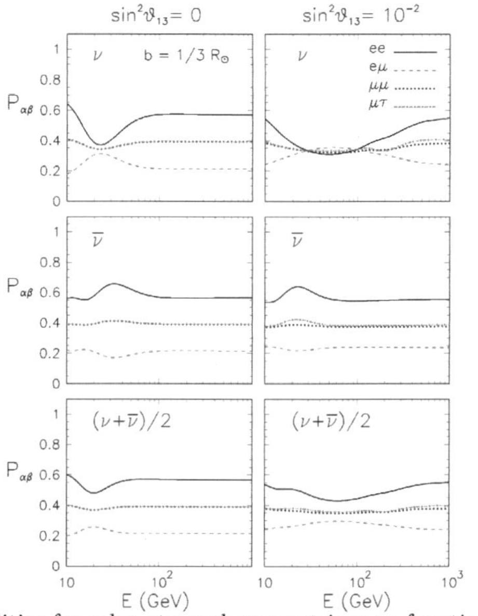

Figure 1. Oscillation probabilities for solar atmosphere neutrinos as functions of energy, for $\sin^2\theta_{13} = 0$ (left panels) and $\sin^2\theta_{13} = 10^{-2}$ (right panels) at fixed impact parameter $b = 1/3R_\odot$.

3. Solar atmosphere neutrino oscillations

Solar atmosphere neutrinos are affected by flavor oscillations (both in matter and in vacuum) during their propagation. Within the standard 3ν framework, the oscillation parameters include two squared mass differences (δm^2, $\pm\Delta m^2$) and three mixing angles ($\theta_{12}, \theta_{23}, \theta_{13}$), whose ordered rotations form a unitary mixing matrix U. We take their best-fit values and 2σ ranges from [4]. In the following, we assume the normal mass hierarchy ($+\Delta m^2$). In matter, the ν oscillation dynamics also depends on the $\nu_e - \nu_{\mu,\tau}$ interaction energy difference $\pm V$ (where $+/-$ for $\nu/\overline{\nu}$), which is proportional to the electron density along the path. Potentially large matter effects are expected when one of the "resonance conditions" is realized, roughly when $V \sim \delta m^2/2E$ or $V \sim \Delta m^2/2E$.

To evaluate the SAν oscillation probabilities $P_{\alpha\beta}$, we evolve numerically the ν and $\overline{\nu}$ flavor evolution equations for many different values of the impact parameter b of the neutrino trajectory ($0 \le b \le R_\odot$). From the unitarity property of the probabilities, one can express all the $P_{\alpha\beta}$'s in terms of only four independent quantities. In the following, we choose P_{ee}, $P_{e\mu}$, $P_{\mu\mu}$ and $P_{\mu\tau}$ as independent set.

In Figure 1 we plot these probabilities as functions of neutrino energy E after gaussian energy resolution smearing (with $\Delta E/E = 10\%$ for definiteness), for a representative impact parameter $b = 1/3\ R_\odot$, and for both $\sin^2\theta_{13} = 0$ (left panels) and $\sin^2\theta_{13} = 10^{-2}$ (right panels), assuming maximal 2-3 mixing and normal mass hierarchy in both cases. The upper and middle panels refer to neutrinos and antineutrinos, respectively. Since solar atmosphere ν's and $\overline{\nu}$'s have approximately equal fluxes, and are basically indistinguishable in neutrino telescopes, we find it useful to plot the arithmetic average of the ν and $\overline{\nu}$ oscillation probabilities in the lower panels of Fig. 1. For $\sin^2\theta_{13} = 0$ (left panels of Fig. 1), the energy-flat vacuum limit ($V = 0$) is reached for $E > 50$ GeV. Matter effects emerge around $E \sim 20$ GeV, and are more pronounced in the neutrino channel (where the δm^2-induced resonance can be realized). For $\sin^2\theta_{13} = 10^{-2}$ (right panels of Fig. 1), additional three-flavor mixing effects, due to a Δm^2-induced resonance, are seen to emerge in the neutrino channel around $E \sim 100$ GeV, with broad features entangled with the previous, lower resonance range. As compared with the left panels, the vacuum limit is reached at higher energies, but with almost the same limiting values for $P_{\alpha\beta}$. The two lowest panels in Fig. 1 show that matter effects, averaged over ν and $\overline{\nu}$ components, produce only mild variations with respect to the energy-flat vacuum limit, independently of the value of $\sin^2\theta_{13}$. Integration over the impact parameter b (see Fig. 5 in [3]) strengthens these conclusions, since trajectories with relatively high impact parameter b (and thus with short path in the Sun and small matter effects) have large geometrical weight. The energy dependence of all probabilities is rather flat and roughly insensitive to $\sin^2\theta_{13}$, with only minor deviations from the vacuum limit governed by the parameter $\sin^2\theta_{23}$. For the specific case of 2-3 maximal mixing ($\sin^2\theta_{23} = 1/2$), one would recover the well-known result for the phase-averaged, vacuum neutrino flavor ratio at detection $\phi_e^D : \phi_\mu^D : \phi_\tau^D \simeq 1 : 1 : 1$. The current error on $\sin^2\theta_{23}$ represents the main uncertainty associated to oscillation effects in calculating SAν fluxes. To give an example, the $\pm 2\sigma$ spread on $\sin^2\theta_{23}$ produces a $\pm 15\%$ variation on the flavor ratio ϕ_e^D/ϕ_μ^D.

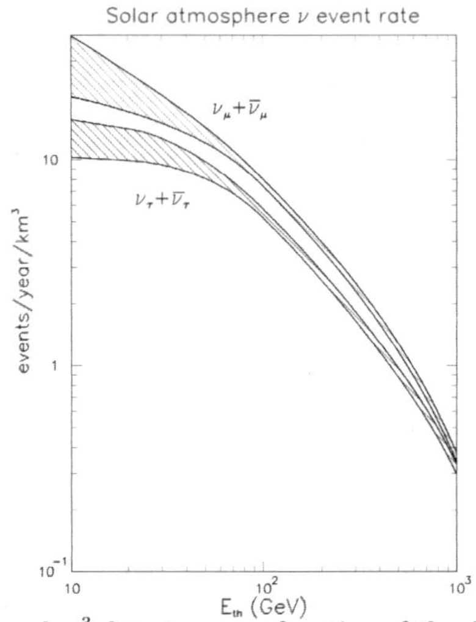

Figure 2. ν_μ and ν_τ event rate in a km^3 detector, as a function of the detector threshold energy $E_{\rm th}$. The ν_e event rate (not shown) is very similar to that of muon events. The width of the bands includes the uncertainties associated to $\sin^2\theta_{23}$ and to the initial fluxes [3].

4. Event rates at Earth

Solar atmosphere neutrinos may be observable in large-volume detectors. Figure 2 shows the results of our calculation for the muon and tau event rates (summing over ν and $\overline{\nu}$) in a km^3 detector like IceCube, as functions of the threshold energy $E_{\rm th}$, including oscillations in matter. A low energy threshold $E_{\rm th} < 100$ GeV is crucial to get a handful of events per year in km-sized detectors. However, even a few SAν-induced muon events per year at $E > 100$ GeV (vs. a prediction twice as large in the absence of θ_{23}-driven oscillations) might allow an independent check in a decade of the large 2-3 mixing established by terrestrial atmospheric neutrinos. This is particularly viable since the theoretically calculated normalization of the SAν flux can be checked indirectly by the GLAST satellite via the measurements of γ's coming from π_0 decays, produced in the very interactions generating the SAν flux.

5. Conclusions

We have recently studied solar atmosphere neutrino oscillations in detail, in the light of the most recent determinations of the neutrino mass mixing parameters, and paying particular attention to matter effects (generally neglected a priori in this context) [3]. We found that, in the energy range $E \in [10, 1000]$ GeV relevant for detection, SAν oscillations in matter have in principle a rich phenomenology. Unfortunately, these effects are smeared out to a large extent by summing over the neutrino and antineutrino channel, and by integrating over the solar disk angle (as well as over energy). In practice, oscillations are dominated by the "vacuum" effect of the θ_{23} mixing angle. Therefore, even if matter effects can and do occur, the original flavor ratio $\phi_e : \phi_\mu : \phi_\tau \simeq 1:2:0$ at the source approximately yields the well-known ratio $\phi_e^D : \phi_\mu^D : \phi_\tau^D \simeq 1:1:1$ at the detector (for maximal 2-3 mixing), just as in the case of phase-averaged vacuum oscillations.

Even if only a handful of SAν events per year may be detectable in a km^3 telescope, their theoretical fluxes are relatively under control, and thus a measurement might provide a high-energy, astrophysical probe of neutrino flavor mixing.

References

[1] G. Ingelman and M. Thunman, Phys. Rev. D **54**, 4385 (1996).
[2] C. Hettlage, K. Mannheim and J. G. Learned, Astropart. Phys. **13**, 45 (2000).
[3] G. L. Fogli *et al.*, Phys. Rev. D **74**, 093004 (2006).
[4] G. L. Fogli *et al.*, Phys. Rev. D **75**, 053001 (2007).

Results from the HARP Experiment

Emilio Radicioni

Istituto Nazionale di Fisica Nucleare - Bari - Italy

E-mail: emilio.radicioni@cern.ch

Abstract. Hadron production is a key ingredient for precise prediction of atmospheric ν fluxes, characterization of accelerator ν beams, and quantification of π production and capture for ν-factory designs.
HARP at the CERN PS was the first hadron production experiment designed on purpose to match all these requirements. We briefly describe here its most recent results.

1. Introduction and motivations

Hadron production data are relevant in several branches of ν physics. The study of atmospheric νs had shown strong evidence for ν oscillations[1, 2]; at the same time, to quantitatively understand this phenomenon, several accelerator-based ν experiments are being built, and new types of ν beams are being designed. In either cases, detailed knowledge of the hadron cross sections at the relevant energies is now considered a must.

2. HARP

HARP at the CERN PS [3][4] was the first hadron production experiment designed on purpose, combining a full phase space acceptance with low systematic errors and large statistics.
HARP performed extensive measurements of hadron production cross sections and secondary particle yields in the energy range 3-15 GeV over the full solid angle, using a large set of cryogenic and solid (thin and thick) targets. A forward spectrometer covers polar angles up to 250 mrad, well matched to the angular range of interest to calculate the properties of conventional neutrino beams, like those of K2K and MiniBooNE; a dipole magnet and large planar drift chambers for particle tracking, and the particle identification is performed with a combination of time-of-flight, Cherenkov, and calorimeter information. . In the large-angle region a TPC positioned in a solenoidal magnet is used for tracking; particle identification is performed by dE/dx in the TPC and by time-of-flight measurements with RPCs. The large-angle spectrometer acceptance is optimized for the measurement of pions relevant to the production of muons beams in future neutrino factories.

3. π production relevant for K2K and MiniBoone

The K2K experiment is most sensitive to uncertainties in the predicted ν spectrum in the energy range between 0.5 and 1 GeV. The distortion of the spectrum measured with the far detector is predicted to be maximal in this range according to the ν oscillation parameters measured in atmospheric ν experiments. The transmission properties of the beam line in the K2K experiment match very well the HARP forward spectrometer acceptance. The measurement of π production

for the K2K experiment using a 5%λ Al target and incident protons of 12.9 GeV/c momentum was published [5] . In Fig.1 a ratio is shown between the ν flux measured at the near and at the far detector as a function of the ν energy. A comparison of the errors on the near/far ratio before and after the inclusion of the HARP results shows a reduction by a factor of two. The neutrino flux calculation for final disappearance analysis of K2K [6] is based on the HARP pion production data.
The case for the MiniBooNE measurement is equally compelling. Here too the ν flux comes predominantly from the π decay into $\mu + \nu$. Again in this case the HARP experiment covers the relevant range, 0.75-6.5 GeV/c in energy and 30-210 mrad in angle (more than 80% of the phase space). The cross section results published in [7] have a direct impact on the prediction of ν fluxes at Miniboone. Results recently presented by Miniboone [8] are based on a parametrization of the HARP π^+ cross sections.

4. Atmospheric neutrino fluxes and extended air showers

The first relevant HARP measurement in this field is the double-differential production cross-section of $\pi^\pm$ in the collision of 12 GeV/c protons with a thin 5% λ_I Carbon target. A systematic error analysis has also been done yielding integral errors (statistical + systematic) of 6.1% and 7.0% for π+ and π- in p–C interactions, 10.2% and 8.5% for π+ and π- in π+–C interactions, 6.5% and 8.2% for π+ and π- in π–C interactions, and an overall normalization error of 2% for the proton beam and 3% for the pion beams. The results are final and a paper is in preparation. In addition, preliminary results using N_2 and O_2 cryogenic targets are now available.

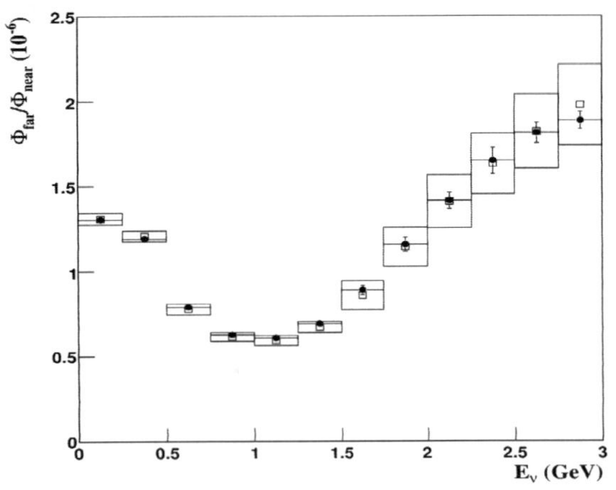

Figure 1. Near to far ratio in the K2K experiments measured by the K2K Monte-Carlo (boxes) and results from the Harp experiment (errored points).

5. Results obtained with the HARP Large Angle spectrometer

One of the main motivations of the HARP experiment is the measurement of the yields of positive and negative πs for a quantitative design of a proton driver and a target station in a future ν-factory. The variables affecting the π production are incident proton beam energy, target material and target geometry (diameter and length). In order to achieve the highest number of potentially collected πs of both charge signs per unit of energy a π production measurement should give the information necessary to optimize both proton beam energy and target material.
In most of the cases high-Z materials are proposed as targets; for this reason it was decided to first analyze a series of settings taken with a range of different beam momenta (from 3 to 12 GeV/c) incident on a Tantalum ((5% λ_{int}) target [9]. Produced πs are best measured in the

large-angle spectrometer, covering the full momentum range of interest for production angles above 0.35 rad.

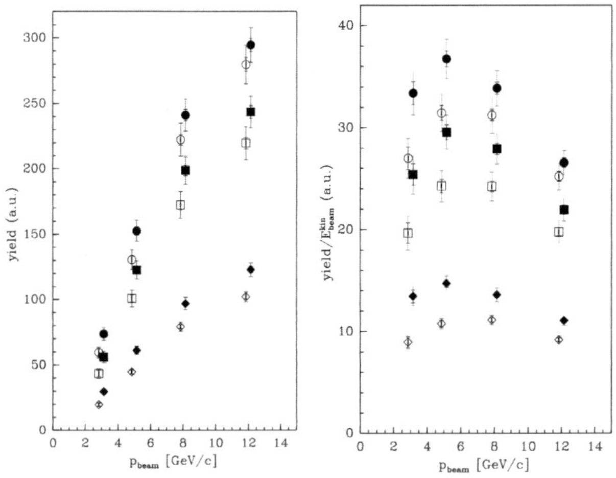

Figure 2. Prediction of the $\pi+$ (closed symbols) and $\pi-$ (open symbols) yield from a tantalum target as a function of incident proton beam momentum for different designs of the neutrino factory focusing stage. Shown are the integrated yields (left) and the integrated yields normalized to the kinetic energy (right). The circles indicate the integral over the full HARP acceptance, the squares are integrated over $0.35rad < \theta < 0.95rad$, while the diamonds are calculated for the restricted angular range and $250MeV/c < p < 500MeV/c$.

The results are shown in Fig. 2. It is shown that the pion yield increases nearly linearly with momentum and that in our kinematic coverage the optimum yield is between 5 GeV/c and 8 GeV/c.

6. Systematic studies

Systematic studies over global characteristics of hadron production cross sections can now be performed. As an example, production off Be, C, Al, Cu, Sn and Pb targets using the same detector [10, 11] have been analysed. They allow, for example, to study the integrated pion yield as a function of beam momentum, its A-dependence, the π-/π+ ratio and all other relevant behaviours with a reduced systematic uncertainty. The fact that these data are taken with the same apparatus and analysed using the same methods improves the precision of the relative comparisons. Papers listing cross section data for all target materials at all available energies are in course of preparation.

7. Conclusions

The HARP experiment, combining almost full phase space acceptance with low systematic errors and high statistics, provide the required reference samples for simulation, and the needed information to characterize the present and future ν beams, and to help a quantitative description of the air showers development.

8. References

[1] Fukuda Y et al. *Phys. Rev. Lett.* **81** (1998) 1562.
[2] Fukuda Y et al. *Phys. Rev. Lett.* **82** (1999) 2644.
[3] Catanesi M G et al. CERN-SPSC-99-35.
[4] Catanesi M G et al. *Nucl. Instr. Meth.* **A571** (2007) 527.
[5] Catanesi M G et al. *Nucl.Phys.* **B732** (2006) 1-45, .
[6] Aliu E et al. *Phys. Rev. Lett.* **94** (2005) 081802
[7] Catanesi M G et al. *E.P.J.* **C52**(2007) 29.
[8] Aguilar A A et al. *Phys. Rev. Lett.* **98** (2007) 231801.
[9] Catanesi M G et al. *E.P.J.* **C51** (2007) 787.
[10] Catanesi M G et al. in print on *E.P.J.* C
[11] Catanesi M G et al. sub. to *E.P.J.* C

Neutrino Oscillation Results from MINOS.

Andrew Blake[1] **(for the MINOS collaboration).**

[1] Cavendish Laboratory, University of Cambridge, J. J. Thomson Avenue, CB3 0HE, UK.

E-mail: blake@hep.phy.cam.ac.uk

Abstract. We report an updated measurement of muon neutrino disappearance in the NuMI neutrino beam performed using the MINOS detectors. This preliminary result is based on an increased data set corresponding to 2.50×10^{20} protons on the NuMI target, and incorporates a number of improvements to our analysis. We observe 563 candidate ν_μ CC interactions in the far detector data, compared with an expectation of 738 ± 30 in the absence of oscillations. A maximum likelihood fit to the observed far detector energy spectrum assuming two-flavour $\nu_\mu \rightarrow \nu_\tau$ oscillations yields best fit oscillation parameters of $\Delta m^2_{23} = 2.38^{+0.20}_{-0.16} \times 10^{-3}\mathrm{eV}^2$ and $\sin^2 2\theta_{23} = 1.00_{-0.08}$ with errors quoted at the 68% confidence level.

1. Introduction.

MINOS is a long baseline neutrino experiment performing a precision study of neutrino oscillations using an accelerator beam of muon neutrinos. The primary aim of the experiment is to precisely determine the values of the oscillation parameters Δm^2_{23} and $\sin^2 2\theta_{23}$ by measuring the disappearance of muon neutrinos in the beam due to the dominant $\nu_\mu \rightarrow \nu_\tau$ oscillation mode; an additional aim is to discover or improve the limit on the mixing angle θ_{13} by searching for the appearance of electron neutrinos due to the sub-dominant $\nu_\mu \rightarrow \nu_e$ oscillation mode.

The neutrino beam is generated by the NuMI facility at Fermilab by directing 120 GeV protons extracted from the Main Injector ring onto a graphite target. The protons are delivered to the target in 10 μs spills every 2.4 seconds, with each spill containing up to 4.0×10^{13} protons. The positively charged secondaries emitted from the target (mainly π^+ and κ^+) are focused by two pulsed parabolic horns into a 675 m evacuated pipe where they decay to produce neutrinos. The relative position of the target and first horn, and the current through the horns, are variable and are used to configure the beam spectrum. The majority of running has been carried out in the "low energy" configuration, which optimizes neutrino production in the 1-3 GeV region, where the maximum oscillations are expected. The beam is described in detail in [1].

The NuMI beam is sampled by the two MINOS detectors. A 1 kT near detector located at Fermilab is used to measure the beam spectrum prior to oscillations; a 5.4 kT far detector constructed 735 km away at the Soudan mine is then used to measure the oscillation signal. The MINOS detectors are designed to be as functionally similar as possible in order to minimize systematic uncertainties. Both detectors are fine-grained steel-scintillator sampling calorimeters with toroidal magnetic fields. The detectors are synchronized to the beam by GPS receivers, enabling the identification of beam spills. The detectors are described in detail in [2].

The MINOS collaboration has previously reported oscillation results based on the data collected during the first period of NuMI operations between May 2005 and February 2006, which corresponded to an overall exposure of 1.27×10^{20} protons on the NuMI target [3] [4].

Here, we report updated results on muon neutrino disappearance obtained from a re-analysis of this data set along with an additional data set collected between June 2006 and March 2007. This has increased the overall exposure to 2.50×10^{20} protons on target.

2. Analysis Procedure.

Most aspects of the updated analysis follow on from the previous analysis. In each detector, candidate neutrino events are reconstructed from clusters of tracks and showers coincident with beam spills. A ν_μ CC event is characterized by the presence of a muon track, with the vertex reconstructed inside the fiducial volume of the detector and the propagation direction correlated with the incoming neutrino beam. A multivariate algorithm is used to separate the ν_μ CC events from the background of NC events using a number of event topology and kinematic variables. The neutrino energy is then reconstructed by summing the energy of the muon track with the visible energy of the hadronic system. For the preliminary results presented here, only neutrinos (tagged by a negatively charged muon) are considered.

A number of improvements have been incorporated in this updated analysis. A new track reconstruction algorithm is employed which identifies and fits an additional 4% of the muons produced by ν_μ CC interactions. The far detector fiducial volume has also been expanded, generating a 3% increase in the event yield. The separation of the ν_μ CC signal from the NC background is now based on an extended multivariate likelihood procedure which combines an increased number of event variables and also accounts for their correlation with event length. This new selection procedure generates a modest increase in the ν_μ CC selection efficiency, and reduces the NC background to approximately half its previous value. The Monte Carlo simulation uses an upgraded neutrino interaction simulation featuring more accurate models of hadronization, intra-nuclear re-scattering and deep inelastic scattering. To further improve the Monte Carlo simulation, the hadron production model and other elements of the beam and detector simulation are tuned by fitting the neutrino and anti-neutrino energy spectra obtained at the near detector in different beam configurations.

An oscillation analysis is performed based on the neutrino energy spectrum in each detector. The observed far detector energy spectrum is compared with a set of oscillated predictions derived from the observed near detector energy spectrum, and a maximum likelihood procedure is used to fit the oscillation parameters. The extrapolation from the near to the far detector is performed using a beam transfer matrix derived from the tuned Monte Carlo simulation. This matrix relates neutrino interactions in the near and far detectors via their parent hadrons. The beam matrix algorithm has been cross-checked by a number of alternative extrapolation methods, and the fluctuations between different methods are found to be much smaller than the expected statistical error bands.

3. Oscillation Results.

For an exposure of 2.5×10^{20} protons on target, a total of 563 candidate muon neutrinos are observed in the far detector data, compared with an expectation of 738 ± 30 for no oscillations. In the low energy region of the spectrum, the observed deficit of events is highly significant. Below 5 GeV, a total of 198 events are selected, compared with an expected 350 ± 14 events. The Monte Carlo expectation for the NC contamination in this sample is 5.6 ± 2.8 events.

The systematic errors in the oscillation parameters resulting from uncertainties in the predicted far detector spectrum are evaluated by performing oscillation fits to Monte Carlo data sets which have the assigned uncertainties applied to them. The three largest systematic errors are found to arise from the uncertainty in the relative near to far normalization (4%), the absolute hadronic shower energy scale (10%), and the neutral current normalization (50%). These systematic uncertainties are incorporated as nuisance parameters in the oscillation fit. Our systematic studies show that this analysis is dominated by statistical errors.

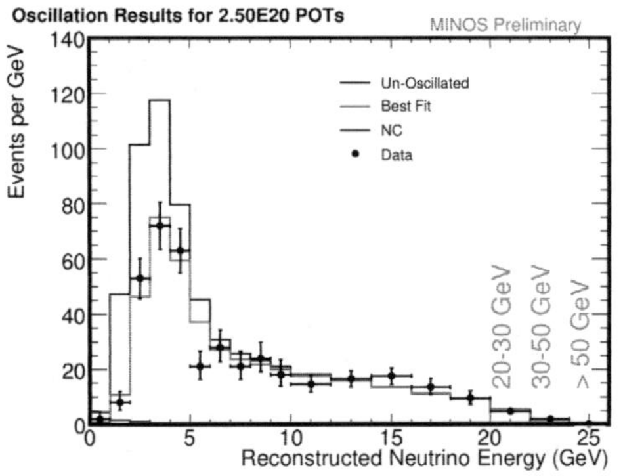

Figure 1. The reconstructed charged current ν_μ energy spectrum at the far detector, showing the prediction for the best fit oscillation parameters (red), null oscillations (black), and the observed data (points).

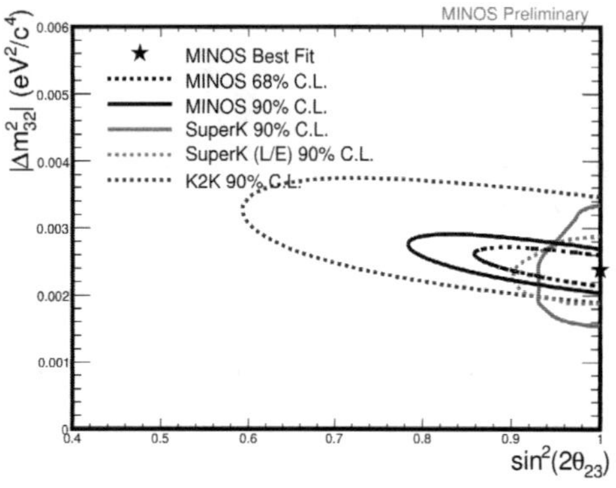

Figure 2. The new MINOS best fit point, along with the 68% and 90% contours. Overlaid are the 90% contours from the Super-Kamiokande zenith angle [5] and L/E [6] analyses, along with the K2K contour [7].

A maximum likelihood fit to the observed far detector energy spectrum is used to extract the oscillation parameters Δm^2_{23} and $\sin^2 2\theta_{23}$ approximating two-flavour $\nu_\mu \to \nu_\tau$ oscillations. The best fit oscillations occur at: $\Delta m^2_{23} = 2.38^{+0.20}_{-0.16} \times 10^{-3}\mathrm{eV}^2$ and $\sin^2 2\theta_{23} = 1.00_{-0.08}$, corresponding to $\chi^2 = 41.2$ for 34 degrees of freedom. The quoted uncertainties represent the 68% confidence intervals and are estimated from the parameter values that increase the χ^2 by one unit relative to the best fit value when minimized with respect to all the other parameters. Figure 1 shows the predicted far detector spectrum for the best fit oscillation parameters (red), and without any oscillations (black), compared to the spectrum observed for the data (points). The characteristic dip expected for neutrino oscillations is evident in the data.

Figure 2 shows the best fit point and 68% and 90% confidence intervals in oscillation parameter space, calculated in the approximation of Gaussian errors. We have confirmed the coverage of these confidence intervals using the unified approach of Feldman and Cousins [8]. A more detailed description of the preliminary results reported here can also be found in [9]. We are currently analysing the full data set collected in the first two years of beam running, corresponding to an exposure of 3.5×10^{20} protons on the NuMI target.

Acknowledgements

This work is supported by: the US DOE; the UK STFC; the US NSF; the State and University of Minnesota; the University of Athens; and the FAPESP and CNPq in Brazil. We gratefully acknowledge the Minnesota Department of Natural Resources; the crew of the Soudan Underground Laboratory; and the staff of Fermilab for their contributions to this effort.

References

[1] Kopp S 2005 *Proc. of IEEE Part. Accel. Conf. (Knoxville, TN)* (*Preprint* ArXiv:physics/0508001)
[2] *to be submitted to Nucl. Instrum. Meth.*
[3] Michael D et al (The MINOS collaboration) 2006 *Phys. Rev. Lett.* 97 191801
[4] Adamson P et al (The MINOS collaboration) 2007 *Preprint* arXiv:0711.0769 [hep-ex]
[5] Ashie Y et al (The Super-Kamiokande collaboration) 2005 *Phys. Rev.* D71 112005
[6] Ashie Y et al (The Super-Kamiokande collaboration) 2004 *Phys. Rev. Lett.* 93 101801
[7] Ahn M et al (The K2K collaboration) 2006 *Phys. Rev.* D74 072003
[8] Feldman G and Cousins R 1998 *Phys. Rev.* D57 3873
[9] The MINOS collaboration 2007 *Preprint* arXiv:0708.1495 [hep-ex]

The OPERA experiment and the first events from the CNGS neutrino beam

Cécile Jollet for the OPERA collaboration
Institut Pluridisciplinaire Hubert Curien - ULP-IN2P3, Strasbourg, France
E-mail: cecile.jollet@ires.in2p3.fr

Abstract

The aim of the OPERA experiment is to provide unambiguous evidence for the $\nu_\mu \leftrightarrow \nu_\tau$ oscillation by looking at the appearance of ν_τ in a pure ν_μ beam. The detector is located on the high-energy, long-baseline CERN to LNGS beam (CNGS) 730 km away from the neutrino source. The apparatus consists of a target made of lead/emulsion-films bricks complemented by electronic detectors. The neutrino interaction inside the bricks is tagged using the electronic trackers. In August 2006, a short CNGS run was successfully performed and a first sample of neutrino events was collected. Experiment description and results from the first CNGS run are reported in details.

1 Introduction

The "appearance" long baseline neutrino experiment OPERA [1] has been motivated by the atmospheric neutrino disappearance. Given the distance of 730 km between the neutrino source (at CERN) and the detector (in the Gran Sasso underground laboratory), the CNGS beam was designed in order to maximise the number of ν_τ charged current interactions detectable at Gran Sasso. To be sensitive in the oscillation parameter region delimited mainly by the Super-Kamiokande results [2], the average energy of the CNGS beam is about 17 GeV. With the CERN SPS accelerator operating in a shared mode, 4.5×10^{19} protons on target (pot) will be delivered per year, and 2900/kton/year ν_μ charged current interactions are expected at Gran Sasso. If the $\nu_\mu \leftrightarrow \nu_\tau$ oscillation hypothesis is confirmed, the number of τ's produced via charged current interaction at Gran Sasso should be of the order of 16/kton/year for $\Delta m^2 = 2.4\times10^{-3}$ eV^2 at full mixing.

2 The OPERA detector and experimental strategy

The principle is to detect the τ leptons produced by ν_τ interactions. According to the mean life time of the τ lepton, an unambiguous signature of its presence is the detection of its decay topology in one prong (electron, muon, hadron) or in three-prongs. Consequently, the photographic emulsions having an accuracy better than 1 μm are used for the charged track detection. Moreover, in order to increase the number of ν_τ interactions, lead is used as target. The detection principle is depicted in Figure 1. Each track segment is reconstructed using 15 to 20 visible grains produced by the charged particles. The spatial resolution is 0.21 μm while the angular resolution is 2.1 mrad. On the reconstructed events, the decay *kink* search is performed as well as the energy reconstruction of electromagnetic showers and the determination of momenta of charged particles by multiple scattering.
The detector is made of two identical supermodules each one consisting of a target followed by a muon spectrometer. The schematic view is given in Figure 2. Each target combines "passive" elements such as emulsions with electronic detectors.

2.1 Target Bricks

The base component of the detector is the target brick. Each brick is made of a sandwich of 56 1 mm thick lead sheets interleaved with 57 emulsion layers. The dimensions of each brick are $12.8\times7.5\times10.3$ cm^3 and the weight is 8.3 kg. The bricks are contained in a 6.7 m high wall made of steel. The target mass will reach 1.35 kton, consequently 154700 bricks will be installed into 62 walls. The production is carried out by a dedicated apparatus called Brick Assembly Machine installed in Gran Sasso. It is a chain of different stations (for piling, pressing, wrapping) using robots operating in dark rooms with controlled environment. The goal is to construct 936 bricks/day in order to finish the filling by April 2008.

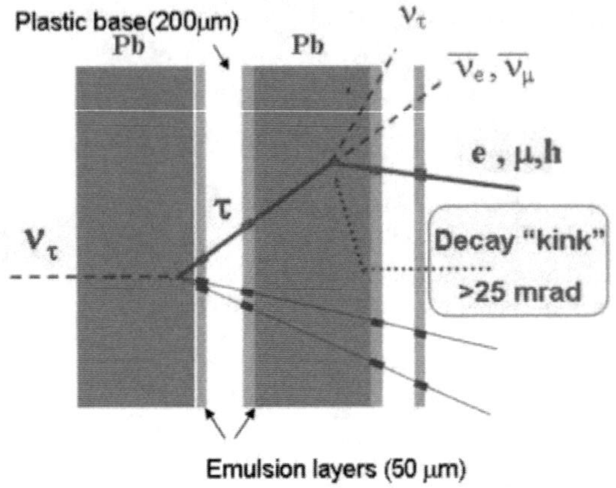

Figure 1: ν_τ and τ detection principle used by OPERA.

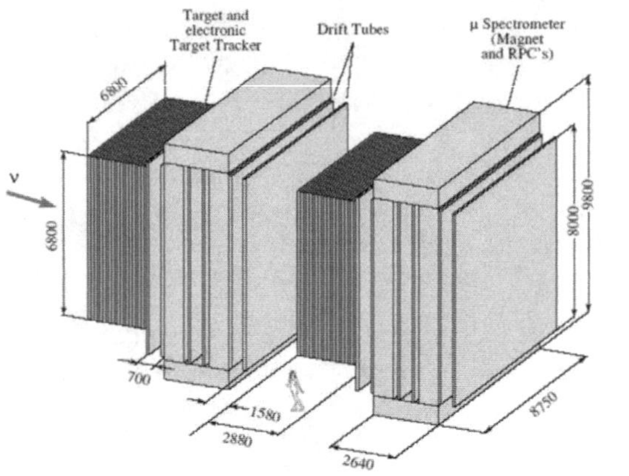

Figure 2: Schematic view of the OPERA detector.

2.2 Electronic Detectors

The main role of the Target Tracker (TT) [3] is to localize the right brick to extract. Each wall of the TT will provide in 2D the position of the brick. It is composed of X and Y planes of plastic scintillator strips (6.8 m×2.6 cm×1 cm). A particle crossing the strips will create blue scintillation light. A wavelength shifting fiber glued in a groove made in the center of the strip will absorb the blue light, reemit it in green wavelength and then propagate the light up to the extremities of the fiber. Each TT wall is divided in 4 horizontal modules and 4 vertical modules. Each module is made of 64 scintillator strips, and the 64 fibers are connected to a multianode Hamamatsu photomultiplier tube at both strip ends. For a particle at the minimum of ionization, at least 5 photoelectrons are detected permitting to reach a detection efficiency of 99% which will provide the trigger of the experiment.
When the brick is selected, a robot called BMS (Brick Manipulator System) will extract it. The emulsions are developed and scanned with automatic microscopes [4] [5] having a scanning speed of 20 cm^2/h.
The aim of the muon spectrometer is to measure precisely the charge and the momentum of the muon and provide an efficient muon tagging. It is composed of an inner tracker and a precision tracker [6]. The magnet of the inner tracker delivers a magnetic field of 1.55 T. Each wall of the magnet is made of 12 iron slabs interleaved with RPC's (Resistive Plates Counters). While the RPC's give the range of the muons, the precision tracker measure precisely their momentum. It is made of drift tubes covering an area of 8×8 m^2. Drift tubes stations are placed in front, in the middle, and behind each dipole magnet. The precision on the momentum measurement is lower than 20% for momentum lower than 50 GeV. Combining the muon spectrometer data with the TT information, the muon identification probability is around 95%.

2.3 Physics performance

The expected results are compiled in Table 1. They are given for a full mixing, after 5 years of data taking and for an intensity of 4.5×10^{19} pot per year. The detection efficiency is given for the studied τ decay channels and the number of expected events is given for two values of Δm^2. At 90% Confidence Level, the OPERA experiment will test all the region of the parameters given by Super-Kamiokande.

3 First CNGS neutrino events

The first CNGS run took place in August 2006 [7]. The beam intensity was 70% of the nominal one, the total integrated intensity was 7.6×10^{17} pot in 121 hours of run, and no bricks were yet inserted into the target. The chosen SPS cylce was 16.8 seconds, and the CNGS proton beam was extracted in two 10.5

channel	ϵ(%)	BR(%)	Signal , Δm^2 (eV^2)		Background
			$2.5\ 10^{-3}$	$3.0\ 10^{-3}$	
$\tau \to \mu$	17.5	17.7	2.9	4.2	0.17
$\tau \to$ e	20.8	17.8	3.5	5.0	0.17
$\tau \to$h	5.8	50	3.1	4.4	0.24
$\tau \to$3h	6.3	15	0.9	1.3	0.17
total	$\epsilon\times$BR	=10.6%	10.4	15.0	0.76

Table 1: Number of expected events and background for the $\nu_\mu \leftrightarrow \nu_\tau$ oscillation.

μs short pulses separated by 50 ms.
The events were selected by a comparison of their absolute timestamps with respect to the beam time information with a measured resolution of about 100 ns. Doing this time selection, 319 beam events were collected and examples of two reconstructed events are shown in Figure 3.

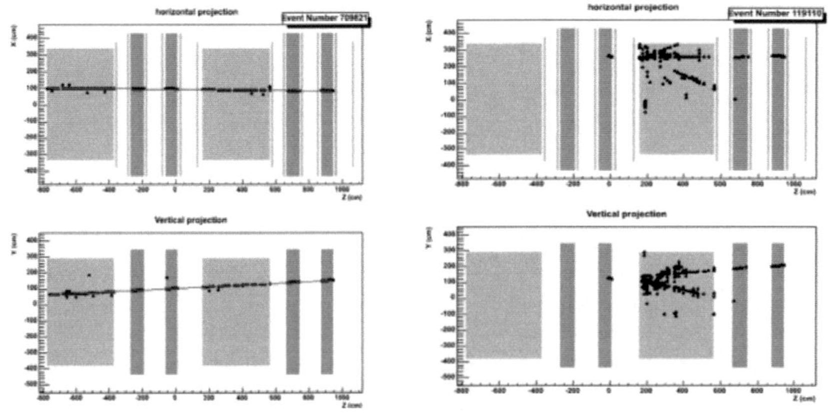

Figure 3: Event display (top and side views) of reconstructed CNGS ν beam events in the OPERA detector: muon produced in a CC interaction in the upstream material (left); ν interaction in the iron of the magnet (right).

The angular distribution with respect to the horizontal axis observed for the sample of single-track events is shown in Figure 4 and compared with simulated cosmic ray muons. Beam-induced events appear as a peak around the horizontal direction. A gaussian fit yields an inclination of 3.4±0.3° in agreement with the expected value of 3.3°.
Another CNGS run with real bricks was foreseen in October 2006, but a serious leak problem in the water cooling circuit of the CNGS reflector caused the early stop of the run. An integrated intensity of 0.6×10^{17} pot was collected and only 29 neutrino-induced events were registered. Anyway, the TT to brick connection has been studied using cosmic rays and a good agreement has been found.

4 Conclusions and perspectives

The first detection of neutrino events from the long baseline CNGS neutrino beam has been performed by OPERA. 319 neutrino-induced events were collected for an integrated intensity of 7.6×10^{17} pot in agreement with the expectations. The reconstructed zenith-angle distributions and the time structure of the events demonstrate the capability of the electronic detectors, build up during the last three years, to reach the experiment goals. The next CNGS run begins the 18th of September and it will last 6 weeks, 70% of the nominal intensity will be delivered and the target mass will be around 600 tons. The OPERA detector faces the last effort of brick production and insertion and the detector is ready for the next phase which is the observation of neutrino interactions inside bricks.

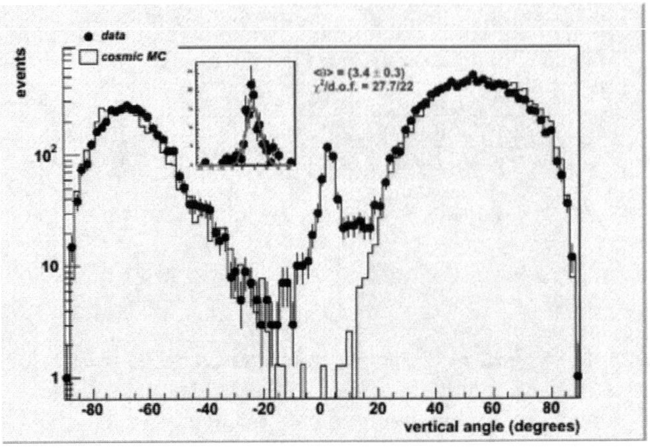

Figure 4: Angular distribution of beam-induced and cosmic-muon events taken with the electronic detectors. The histogram indicates the predictions from cosmic-ray simulations. The inset shows the angular distribution of on-time beam events.

References

[1] M. Guler *et al.* [OPERA and collaboration], CERN-SPSC-2000-028.

[2] Y. Fukuda *et al.*, Phys. Rev. Lett. **81** (1998) 1562.

[3] A. Adam *et al.*, Nucl. Instr. Meth **A577** (2007) 523.

[4] L. Arrabito *et al.*, Nucl. Instr. Meth. **A568** (2005) 261.

[5] S. Aoki *et al.*, Nucl. Instr. Meth **B51** (1990) 466.

[6] R. Zimmermann *et al.*, Nucl. Instr. Meth. **A555** (2005) 435.

[7] R. Acquafredda *et al.*, New Journal of Physics 8 2006 303.

Status of FNAL SciBooNE experiment

Yasuhiro Nakajima, for the SciBooNE collaboration
Department of Physics, Kyoto University, Kyoto 606-8502, Japan
E-mail: nakajima@scphys.kyoto-u.ac.jp

Abstract. SciBooNE is a new experiment at FNAL which will make precision neutrino-nucleus cross section measurements in the one GeV region. These measurements are essential for the future neutrino oscillation experiments. We started data taking in the antineutrino mode on June 8, 2007, and collected 5.19×10^{19} protons on target (POT) before the accelerator shutdown in August. The first data from SciBooNE are reported in this article.

1. Introduction

Neutrino oscillation is a tool to study the nature of neutrino mass and mixing, and is a probe of physics at high energy scales beyond the standard model. In future accelerator neutrino experiments, precise knowledge of neutrino cross sections with nuclei are important to achieve the best sensitivity. On the other hand, the typical accuracy of neutrino-nucleus cross sections in the one GeV region, which is relevant to many future experiments, is ~20% with our current knowledge.

The SciBooNE experiment (FNAL E954) [1] is designed to measure the neutrino cross sections on carbon in the one GeV region. SciBooNE's neutrino beam energy is well matched to T2K[2]. Additionally, SciBooNE serves as a near detector for MiniBooNE[3] by constraining the neutrino fluxes.

2. Experimental setup

2.1. Booster neutrino beam at FNAL

The primary proton beam, with kinetic energy 8 GeV, is extracted from the Fermilab Booster and steered to strike a 71 cm long, 1 cm diameter beryllium target. The proton beam typically has 4×10^{12} protons per ~1.6 μsec beam spill. The mesons, primarily π^+, generated by the p–Be interactions are focused with a magnetic horn and decay in the following 50m decay volume, producing an intense neutrino beam with the peak energy of 0.7 GeV. When the horn polarity is reversed, π^- are focused and hence a predominantly antineutrino beam is created.

We plan to run 10^{20} POT each in neutrino mode and in antineutrino mode. Since the horn was operating in antineutrino mode for MiniBooNE, we started running with antineutrino beam.

2.2. SciBooNE detector

The SciBooNE detector is located 100 m downstream from the beryllium target, on the axis of the neutrino beam direction. The detector complex consists of three sub-detectors: a fully active fine grained scintillator tracking detector (SciBar), an electromagnetic calorimeter (EC) and a muon range detector (MRD).

The SciBar detector was originally built and operated in the K2K experiment[4]. After the K2K was completed, the SciBar was shipped and reassembled at FNAL together with the EC which was used in the CHORUS[5] and K2K experiments. The MRD is a newly constructed hodoscope made with recycled iron plates and scintillators from past FNAL experiments.

The SciBar consists of 14,336 extruded plastic scintillator strips, each with dimension of $1.3 \times 2.5 \times 300$ cm^3. The scintillators are arranged vertically and horizontally to construct a $3 \times 3 \times 1.7$ m^3 detector. The detector itself is the neutrino target and can reconstruct all the charged particles from a neutrino interaction together with particle identification capability using dE/dx.

The EC is installed downstream of the SciBar, and is made of scintillating fibers embedded in lead foil. It has a thickness of 11 radiation length along the beam direction to measure π^0 produced in neutrino interactions and the intrinsic ν_e contamination. The energy resolution is $14\%/\sqrt{E[\mathrm{GeV}]}$. The MRD is located downstream of the EC in order to measure the momentum of muons up to 1.2 GeV/c using the muon range. It consists of 12 layers of 2"-thick iron plates sandwiched between layers of 6 mm-thick plastic scintillator planes.

3. Detector commissioning

The detector construction was completed in March 2007. Immediately after construction, the detectors were commissioned with cosmic ray muons. In April 2007, the detectors were moved and installed in the experimental hall on the neutrino beam line. The second commissioning using the antineutrino beam started in May 2007, and we successfully detected events on May 29th, 2007.

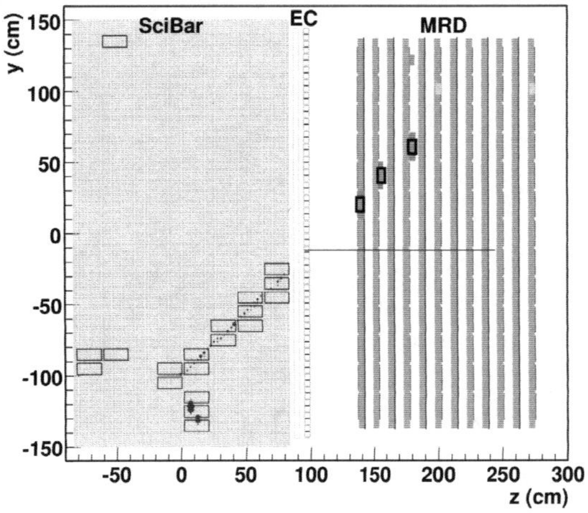

Figure 1. A candidate of an antineutrino charged current quasi-elastic event ($\bar{\nu}_\mu + p \rightarrow \mu^+ + n$). Circles and boxes on the SciBar indicate hits of ADC and TDC respectively. The area of the circle is proportional to the ADC value. For the EC, the size of a bar corresponds to the energy deposit. Finally, framed boxes on the MRD indicate TDC hits within the beam-timing.

Figure 1 shows a display of an antineutrino charged current quasi-elastic event ($\bar{\nu}_\mu + p \rightarrow \mu^+ + n$) candidate from the commissioning period. The longer track penetrating through the SciBar and the EC to the MRD is a muon, and the short track with large energy deposit in the SciBar is presumed to be a proton recoiled by a neutron from the vertex of neutrino interaction.

4. First antineutrino data taking

After commissioning, we took data in antineutrino mode from June 8, 2007 to the accelerator shutdown in August 2007. We collected 5.19×10^{19} POT in two months, which is about half of our goal for antineutrino data.

In order to confirm the stability of the detector and the beam, we select charged current (CC) candidate events by requiring a track exiting from the SciBar whose vertex is inside the fiducial volume of the SciBar, corresponding to a mass of 10.6 tons.

Figure 2 shows the timing distribution of the CC candidate events. One can clearly see events clustered in the beam-timing window. Figure 3 shows the number of events after cosmic ray background subtraction per week per 4×10^{16} POT in the beam-timing window without any acceptance corrections. The number of cosmic ray background events is estimated from the events outside the beam-timing window. The rate is stable within the statistical error for the whole period of the running.

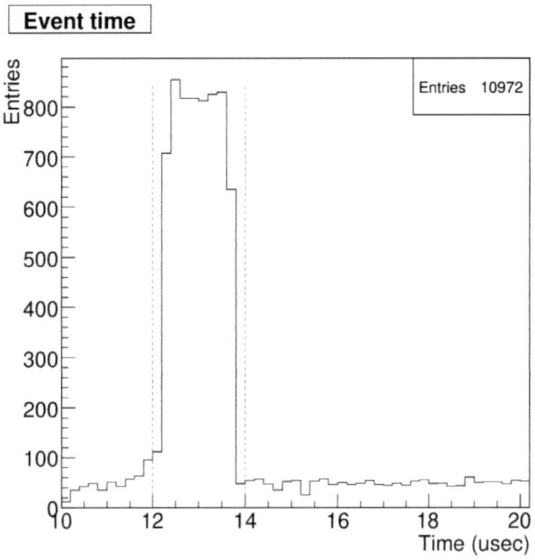

Figure 2. Timing distribution of charged current candidate events with the vertices inside the SciBar. Dashed lines indicate the 2 μsec beam-timing window.

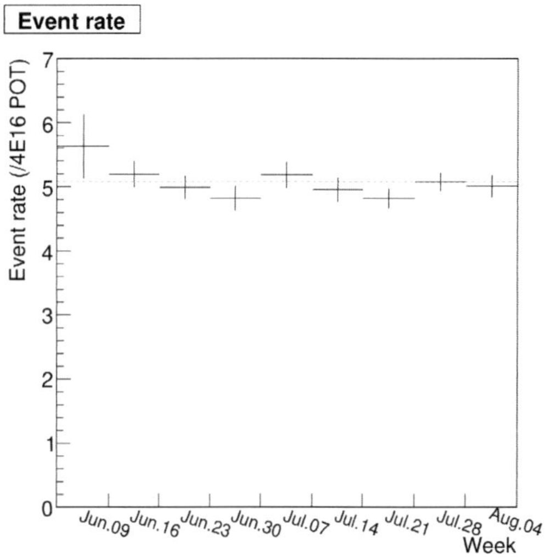

Figure 3. Event rate of the CC candidate events in the beam timing. The rate is normalized by the number of protons on target. The error bars show the statistical errors. Dashed lines indicate the average rate.

5. Summary and Future prospects

The SciBooNE data taking successfully started in antineutrino mode running in June 2007. We confirm that the detector and beam ran stably for more than 2 months.

After the 2007 summer shutdown, we started taking data in neutrino mode and plan to collect 10^{20} POT by spring of 2008. After the neutrino mode run, we will switch back to antineutrino mode and collect the additional 5×10^{19} POT data before summer 2008.

Acknowledgments

SciBooNE collaboration gratefully acknowledges the support from various grants and contracts from the Department of Energy (U.S.), the National Science Foundation (U.S.), the MEXT (Japan), the INFN (Italy) and the Spanish Ministry of Education and Science. The author was supported by Japan Society for the Promotion of Science.

References

[1] Aguilar-Arevalo A A *et al.* 2006 (*Preprint* hep-ex/060122)
[2] Itow Y *et al.* 2001 (*Preprint* hep-ex/0106019)
[3] Aguilar-Arevalo A A *et al.* 2007 *Phys. Rev. Lett.* **98** 231801 (*Preprint* arXive:0704.1500 [hep-ex])
[4] Nitta K *et al.* 2004 *Nucl. Instrument. Meth.* A **535** 147
[5] Buontempo S *et al.* 1995 *Nucl. Phys. Proc. Suppl.* **44** 45

Status of the NOvA Experiment

Robert Plunkett, for the NOvA Collaboration

Fermilab, P.O. Box 500, Batavia, IL 60510, USA

plunk@fnal.gov

Abstract. The NOvA experiment, using the existing NuMI beamline, is planned for construction at Ash River, Minnesota. The experiment will provide a measurement of, or strong limit on the neutrino mixing angle θ_{13}, and for sufficiently strong mixing, establish the hierarchy of the neutrino masses.

1. Introduction

The discovery of neutrino oscillations in atmospheric and terrestrial neutrino experiments [1], has focused the attention of the physics community on the unusual pattern of the mixing among the neutrino mass eigenstates. In particular, the mixing of the weak eigenstate ν_e with the mass eigenstate ν_3 has been limited to relatively small values [2], in sharp contrast with the other mixing angles, which are near-maximal. A number of planned experiments will measure or further constrain this angle. In addition, there are a number of fundamental parameters of the neutrino system which are not currently known – one of the most important of which is the ordering, or hierarchy, of the neutrino mass eigenstates ν_1, ν_2, and ν_3 (since the oscillation phenomenon in vacuum measures only mass differences). The NOvA experiment is proposed for construction at a site (Ash River) in northern Minnesota, with a baseline of 810 km with respect to the existing NuMI beam facility at Fermilab. Together with upgrades to the intensity of the NuMI beamline, NOvA will provide a powerful tool for addressing the most pressing questions in the area of neutrino physics.

2. Construction of Detector and Beamline

The NuMI beamiline at Fermilab will be significantly upgraded as part of the NOvA project. Since production of neutrinos is proportional to beam power on the target, the flux into an experiment depends both on the beam intensity and on the frequency with which it is sent to a production target. The expected end of the Collider program at the Fermilab Tevatron will the make the Recylcler, currently used to cache antiprotons, available as a storage ring for protons. This allows the Fermilab Main Injector magnets to complete a full cycle of acceleration and extraction every 1.3s, an increase over the current 2.2 s. Together with modest increases in pulse intensity, this should provide 700 kW of beam power to NOvA. Further accelerator projects may increase the power to greater than 1 MW. The detector location is 14.6 mradians off-axis, so the Main Injector proton beam will create a narrow-band neutrino beam with a spectrum sharply peaked around 2 GeV.

The NOvA detector design is fully active. It will be constructed of 15.6 m PVC extrusions filled with a mixture of mineral oil and liquid scintillator. Readout of each extrusion tube is via an imbedded wave-length shifting fiber to an HPD. There are 384 tubes per plane and a design with 18kT mass includes approximately 1200 layers.

The fine segmentation (approx. 4 cm transversely and 6 cm longitudinally) and transparency of the detector medium allows good (~30%) efficiency for identification of electron showers induced by ν_e CC events, with acceptable background rates. Figure 2 shows a Monte Carlo representation of an electron-induced shower in the NOvA detector.

Backgrounds to the appearance signal arise from mis-identified ν_μ NC and CC events and intrinsic ν_e's in the NuMI beam. The narrow-band off-axis beam causes NC events to migrate to low energies not underneath the signal region. Similarly, intrinsic beam ν_e's are spread out in energy, reducing the background level from this source to manageable levels. Mis-identified ν_μ CC events are rejected by pattern recognition using an ANN or other similar algorithm.

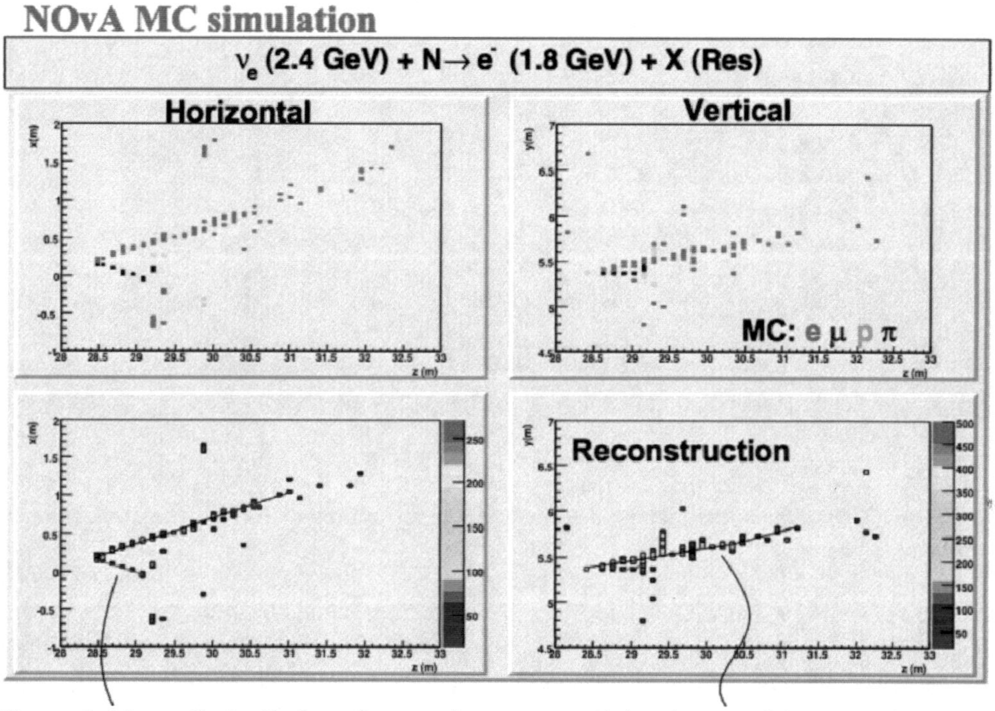

Fig;ure 1: Monte Carlo display of raw and reconstructed data in a candidate event in NOvA.

3. Physics Sensitivities

The combination of active detector and intense neutrino beam discussed above gives NOvA sensitivity to a large range of values of $\sin^2 2\theta_{13}$ (appearing in the combination $\sin\theta_{23}\sin^2 2\theta_{13}$). At the NOvA baseline, matter effects on the ν_e component of the oscillating neutrino state cannot be neglected. These depend on the mass hierarchy, increasing the appearance probability for neutrinos for normal hierarchy over the inverted case (and conversely for anti-neutrinos). A further complication for all NOvA measurements is the effect of a possible CP-violating phase δ in the mixing matrix. Optimized running for NOvA with a mixture of both neutrino and anti-neutrino beams is therefore to be expected. Figure 2 shows the sensitivity for detection of non-zero θ_{13} for a typical NOvA running scenario, presented as a function of δ.

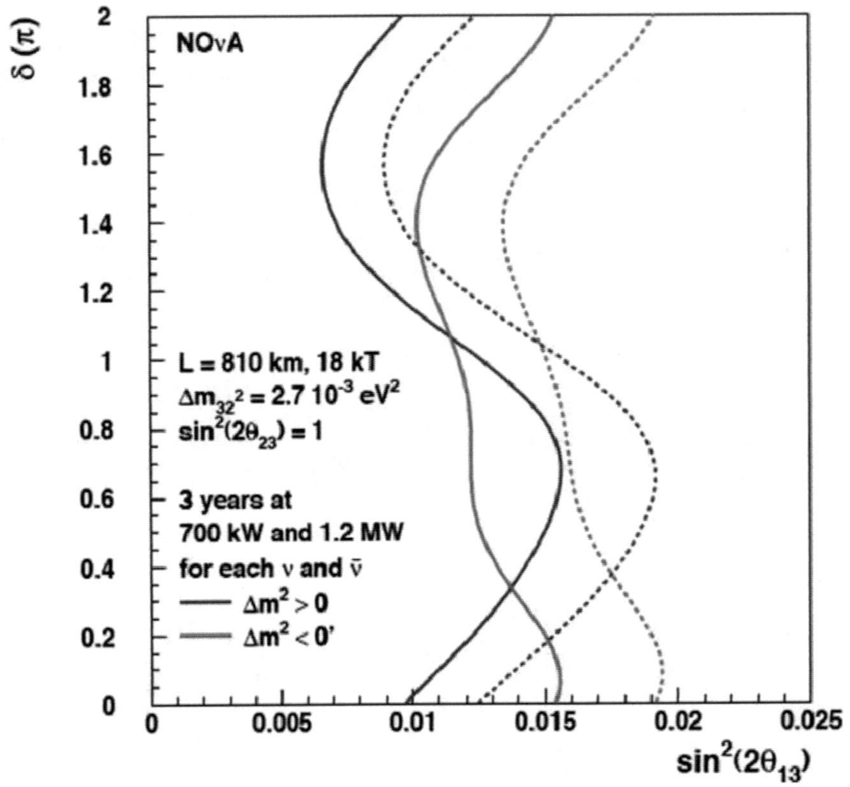

Figure 2: Three-σ sensitivity to establish non-zero θ_{13} for an 18 kT NOvA detector with two running scenarios. Dotted curves show results for NuMI beam running at 700 kW, and solid curves for 1.2 MW of beam power.

Matter effects due to the long NOvA baseline provide the mechanism by which the experiment will address the question of whether the neutrino mass eigenstate differences are grouped as Δm(solar): Δm(atmospheric) or in the inverted hierarchy Δm(atmospheric):Δm(solar) (in increasing order of mass). The matter effects are exploited by comparison of measurements using neutrino and anti-neutrino beams. These measurements are also affected by the possible CP phase δ. In Figure 3, one sees the 95% confidence limit curves for NOvA to distinguish the two situations by running with different beam powers. The contours are closed below the CHOOZ limit for approximately ½ of the allowed CP phases in each of the two cases of hierarchy. It is important to notice that addition of expected information from the T2K experiment improves the experiment significantly for all values of δ. Here NOvA provides the vital information on neutrino-antineutrino differences, which can be combined with the appearance probabilities which will be measured by T2K at a different baseline.

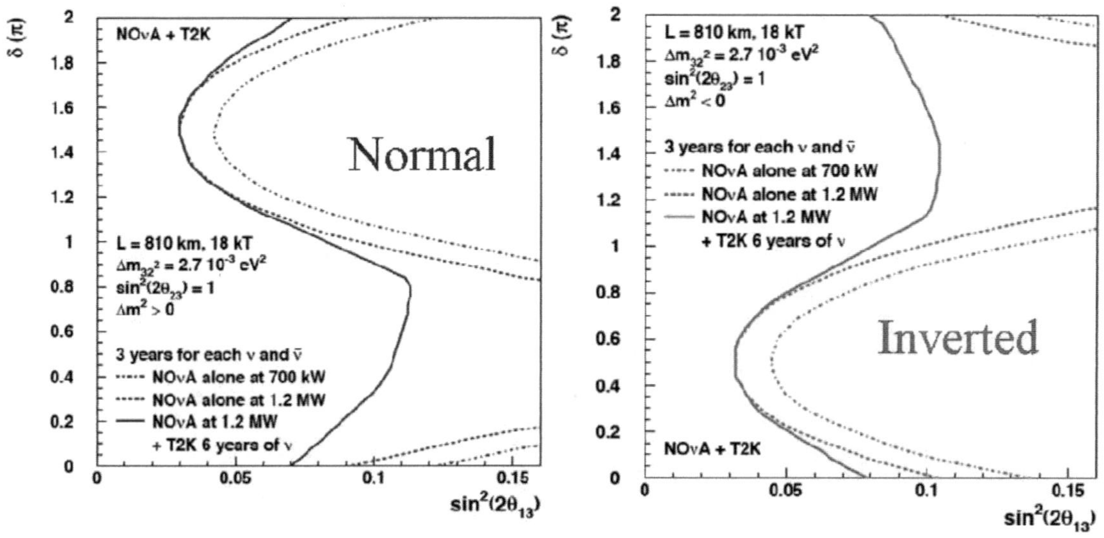

Figure 3: Sensitivity of NOvA and NOvA/T2K experiments to the neutrino mass hierarchy. Contours represent values of the parameters for which a 95% confidence level separation is possible.

4. Conclusions

The NOvA experiment represents the next step in the long-baseline neutrino program using the Fermilab NuMI facility. In addition to sensitivity to a wide range of θ_{13} and the ability to enable discrimination between the mass hierarchy scenarios, it will provide a precision measurement of the the parameters of nm mixing, currently being addressed by MINOS. Together with data from reactor experiments, which measure θ_{13} directly, NOvA may allow the determination of the octant of the mixing angle θ_{23}. NOvA will also provide clues to future directions of neutrino physics. For example, the results of the comparison between neutrino and anti-neutrino measurements will likely indicate if extremely high-precision neutrino sources such as a neutrino factory will be needed to unravel the puzzle of CP- violation in the neutrino sector.

References

[1] Y. Ashie et al. (Super-Kamiokande), Phys. Rev. **D71**, 112005 (2005);
M.H. Ahn et al. (K2K), Phys. Rev. **D74**, 072003 (2006);
D.G. Michael et al. (MINOS), Phys. Rev. Lett. **97**, 191801 (2006).

[2] M. Apollonio et al. (CHOOZ), Phys. Lett. **B466**, 415 (1999)

Future neutrino oscillation experiment with 2 detectors at Kamioka and Korea

Takaaki Kajita

Research Center for Cosmic Neutrinos, Institute for Cosmic Ray Research (ICRR) and Institute for the Physics and Mathematics of the Universe (IPMU), Univ. of Tokyo, Kashiwa-no-ha 5-1-5, Kashiwa, Chiba 277-8582, Japan

E-mail: kajita@icrr.u-tokyo.ac.jp

Abstract.

The J-PARC neutrino facility will produce a very intense neutrino beam. The neutrinos can be detected at Korea as well as at Kamioka. We explore a possibility of two identical detectors at Kamioka and Korea exposed to the same neutrino beam. We show that this setup can be a promising experiment to measure the yet unknown neutrino parameters.

1. Introduction: Initial studies of T2KK

T2K is the next generation neutrino oscillation experiment [1]. The neutrino beam is produced by the J-PARC facility. The neutrinos are detected in Super-Kamiokande, which is located 295 km away from J-PARC. T2K plans to proceed to Phase-2 of the experiment once θ_{13} is found to be non-zero. In Phase-2, it is planned to upgrade the beam power to about 4 MW. In addition, a much larger detector than Super-Kamiokande is planned. It is called Hyper-Kamiokande, and will have the fiducial mass of 0.54 Mton. We refer this setup as T2K-II.

The T2K experiment will use an off-axis beam. The 2.5 degree off-axis beam in Kamioka and Korea (which is more than 1,000 km away from the target) is simultaneously available with the present T2K beam-line configuration. Therefore, we consider a setup with two identical detectors in Kamioka and Korea, both of which will have the fiducial mass of 0.27 Mton, i.e., half of the full Hyper-Kamiokande fiducial volume. The peak energy is about 0.7 GeV for the 2.5 degree off-axis beam. We refer this setup as T2KK in this paper.

Throughout this paper, we assume $\Delta m^2_{21} = 7.9 \times 10^{-5}$ eV^2, $\sin^2\theta_{12} = 0.29$, and $\Delta m^2_{31} = \pm 2.5 \times 10^{-3}$ eV^2. We take the baseline length of 1050 km as a typical distance to the Korean detector. We also assume that both the neutrino and anti-neutrino runs will be for 4 years.

Since both the detectors will be exposed to the same 2.5 degree off-axis beam, the background must be almost identical if the trivial flight-length correction is made. On the other hand, the neutrino and anti-neutrino oscillation probabilities, which depend on the CP phase and mass hierarchy, must be different due to the different baseline length and the different amount of matter traversed. When we move from the first to the second oscillation maxima, the size of CP phase effects become larger by about a factor of 3. In order to compare the sensitivities for T2K-II and T2KK, a detailed χ^2 analysis was carried out [2, 5] taking into account various detector effects, such as background contamination, detection efficiency, and their systematic errors. Figure 1 shows the contours for the sensitivity to the mass hierarchy at 2 and 3 standard deviations on the δ-$\sin^2 2\theta_{13}$ plane. The 2 and 3 standard deviations are defined to be χ^2_{min}(wrong hierarchy)$-\chi^2_{min}$(true hierarchy) > 4 and 9, respectively. As expected, better sensitivity to the mass hierarchy is obtained for T2KK. Also shown in Fig. 1 is the sensitivity to the leptonic CP violation. For large $\sin^2 2\theta_{13}$, T2KK gives the better sensitivity. It is due to the fact that in T2KK, the mass hierarchy can be resolved for large $\sin^2 2\theta_{13}$, and therefore the value of δ is uniquely measured in the case of T2KK [2].

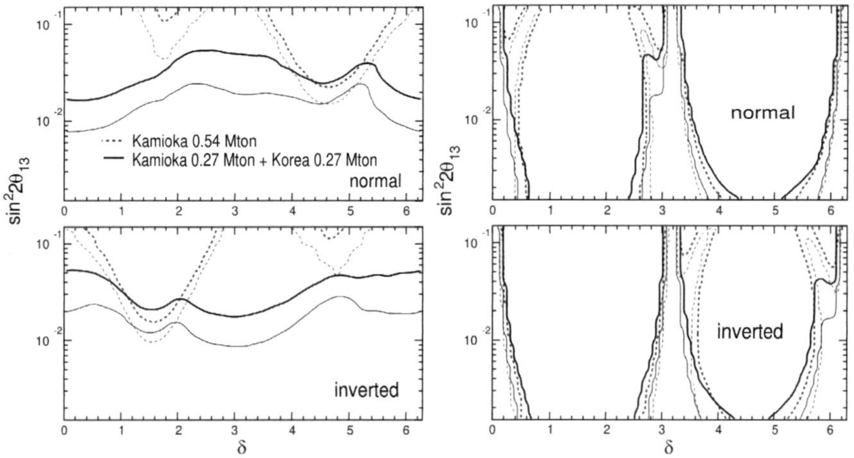

Figure 1. 2(thin lines) and 3(thick lines) σ sensitivities to the mass hierarchy (left) and the CP violation (right) for T2K-II and T2KK.

2. Further studies of the T2KK sensitivities

2.1. Baseline Length dependence of the Sensitivity

The baseline length to the Korean detector has been assumed to be 1050 km. However, we notice that the 2.5 degree off-axis beam can be available across the Korean peninsula. Therefore the possible baseline length can be between 1000 and 1250 km. A study was carried out to understand the baseline length dependence of the sensitivities to the mass hierarchy and CP violation [3]. It was found that the sensitivities do not depend on the baseline length strongly. This is good, since the selection of the Korean detector site could have many possibilities.

2.2. Higher Energy Option for the Korean Detector

Since the matter effect is proportional to the neutrino energy, higher energy beam must be useful to determine the sign of Δm^2. The higher energy beam is available near the on-axis in Korea. (The Kamioka detector is always exposed to the 2.5 degree beam.) However, higher energy beam might have more background for the electron appearance search, especially in the lower energy part of the spectrum. Therefore, a careful analysis should be carried out to understand the relative merit and demerit of the higher energy beam for the Korean detector. A maximum likelihood analysis was carried out to search for electron appearance signal in the higher energy beam [4]. Figure 2 shows the expected signal and background for 1.0 and 2.5 degree off-axis beams. As expected there are much more background in the sub-GeV region for the 1.0 degree off-axis beam, while the background level from this analysis for the 2.5 degree off-axis beam is similar to the standard T2K analysis. We note that the background level for the 1.0 degree off-axis beam in the multi-GeV region, where the electron appearance signal from the first oscillation maximum is expected, is not very high compared with that of the 2.5 degree off-axis beam. This suggests that this configuration might be useful for the mass hierarchy determination. Figure 2 shows the sensitivities to mass hierarchy at 3 σ for various off-axis beams. As expected, the sensitivity of the T2KK experiment to the mass hierarchy improves with the higher energy neutrino beam. The sensitivity to the CP violation is almost independent of the off-axis angle for the Korean detector [4].

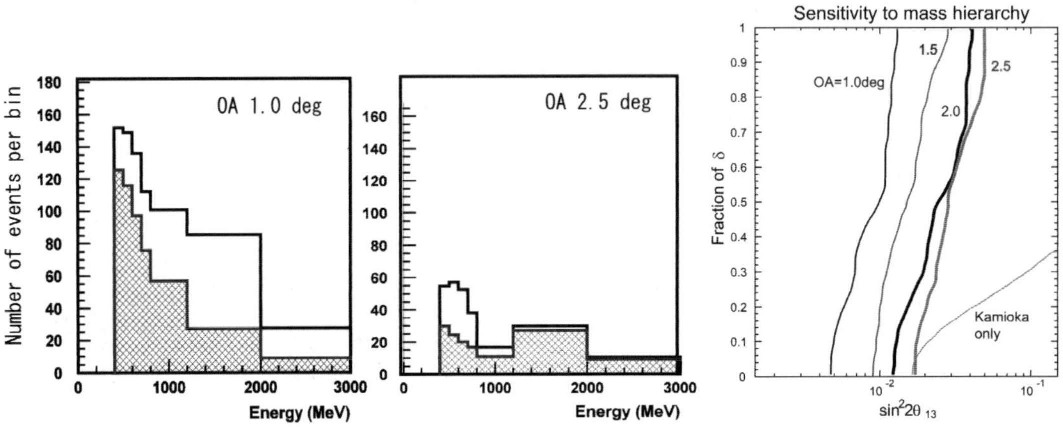

Figure 2. (Left) Expected background for the electron appearance search for the Korean detector with off-axis angles 1.0 and 2.5 degrees. Also shown are the expected signals for $\sin^2 2\theta_{13} = 0.1$ and $\delta = \pi/2$. (Right) 3 σ sensitivities for the mass hierarchy determination in T2KK with off-axis angles for the Korean detector between 1.0 and 2.5 degrees (preliminary). The true mass hierarchy is assumed to be normal.

2.3. Resolving the octant ambiguity of θ_{23}

So far, we assumed that $\sin^2 2\theta_{23}$ is maximal. However, if it is not the case, there is a parameter degeneracy due to the octant ambiguity of θ_{23}. The ν_e appearance probability driven by Δm^2_{23} are proportional to $\sin^2 2\theta_{13} \cdot \sin^2\theta_{23}$. This suggests that if $\sin^2\theta_{23}^{1st}$ and $\sin^2 2\theta_{13}^{1st}$ give a good fit to the data, $\sin^2(\pi/2 - \theta_{23}^{1st})$ and $\sin^2 2\theta_{13}^{2nd}$ also give a good fit, as far as $\sin^2 2\theta_{13}^{1st} \cdot \sin^2\theta_{23}^{1st} = \sin^2 2\theta_{13}^{2nd} \cdot \sin^2(\pi/2 - \theta_{23}^{1st})$. On the other hand, the appearance probability driven by the solar Δm^2_{21}, is proportional to $\cos^2\theta_{23}$. Therefore, the θ_{23} octant degeneracy can be resolved by measuring the solar-Δm^2 effect at the Korean detector, which is located at a very large L/E site. From the detailed χ^2 analysis, it was found that T2KK is able to solve the octant ambiguity at 2σ CL, if $\sin^2 2\theta_{23} < 0.97$. This conclusion depends weakly on the value of $\sin^2 2\theta_{13}$, as well as the value of the CP phase and the mass hierarchy [5].

3. Summary

In this paper, we discussed the sensitivity of the oscillation experiment with the detectors at Kamioka and Korea (T2KK) to the yet unknown neutrino parameters. The possible measurements include the CP violation in the neutrino sector, the neutrino mass hierarchy and the octant of θ_{23}. We have found that T2KK has high sensitivities to these measurements as far as $\sin^2 2\theta_{13}$ is not too small. Finally, we point out that there are possibilities to optimize the beam energy and the baseline length by selecting the proper location of the Korean detector.

[1] Y. Itow *et al.* hep-ex/0106019, see also H. Kakuno, for the T2K collaboration, in these proceedings.
[2] M. Ishitsuka, T. Kajita, H. Minakata and H. Nunokawa, *Phys. Rev. D* **72** (2005) 033003.
[3] K. Okumura, talk at the 2nd International Workshop on a Far Detector in Korea for the J-PARC Neutrino Beam, Seoul, Korea, July 2006.
[4] F. Dufour, talk at the 2nd International Workshop on a Far Detector in Korea for the J-PARC Neutrino Beam, Seoul, Korea, July 2006. (The results were updated after the workshop.)
[5] T. Kajita, H. Minakata, H. Nunokawa and S. Nakayama, *Phys. Rev. D* **75** (2007) 013006.

Comprehensive Analysis of Neutrino Events in Super-Kamiokande Type Detector from the View Point of Computer Numerical Experiment

E.Konoshi(1), N.Takahashi(1), V.I.Galkin(2) and A.Misaki(3)

(1) Graduated School of Science and Engineering, Hirosaki University, Hirosaki, Japan
(2) Department of Physics, Moscow State University, Moscow, Russia
(3) Research Institute for Science and Engineering, Waseda University, Tokyo, Japan

E-mail: `misaki@kurenai.waseda.jp`

Abstract. Comparisons are made on the direction of the incident neutrino, zenith angle distribution of neutrino events produced both inside and outside the detector between the Super-Kamiokande Experiment and our Computer Numerical Experiment.

1. Introduction The Computer Numerical Experiments (CNE) obtains virtual neutrino events and analyzes them constructing the virtual Super-Kamiokande detector in the computer. We pursue every physical process of the neutrino events in a stochastic manner as exactly as possible. In the CNE, we could obtain any physical results which Super-Kamiokande(SK) could obtain for any given neutrino oscillation parameter. Namely, we could compare our results with SK ones directly, expecting a productive discussion around the points in dispute.

2. Points in dispute between SK and CNE

The SK assumption: In Super-Kamiokande, they assume that the direction of an observed lepton is in the direction of the incident neutrino (SK assumption)([1],[2]). We obtain $\cos\theta_\mu$, the zenith angles of observed leptons starting from $\cos\theta_\nu$, the zenith angles of neutrinos, in a stochastic manner. The correlations between $\cos\theta_\nu$ and $\cos\theta_\mu$ are shown in Figure 1 (without oscillation) and Figure 2 (with oscillation) for 1489.2 live days. It is clear from the figures that the SK assumption does not hold even statistically.

Zenith Angle Distributions for Single Ring Muon-like Events: The SK data[3] is compared directly in Figure 3 with our data for 1489.2 live days. It is clear from the figure that our results, both with and without oscillation disagree with the SK results.

Zenith Angle Distribution for Upward Through Going Muon Events and Stopping Muon Events: Our results with and without oscillation are compared in Figure 4 (Stopping Muon events) and in Figure 5 (Upward Through Going Muon Events) with the SK result[3] for 1645.9 live days. It is reasonable from figures 4 and 5 that we could not say that the SK data agrees with either our data with oscillation, or with our data without oscillation in the cases for both Upward Through Going Muon Events and for Stopping Muon Events.

It is further noticed that the frequency of events with oscillation is larger than that without oscillation. This tells us that 1645.9 live days are statistically insufficient to extract a definite conclusion about neutrino oscillations.

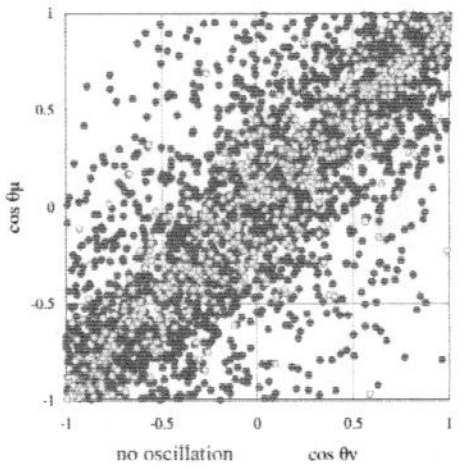

Figure 1. Correlation diagram between zenith angles of neutrinos and those of muons for null oscillation. Green circles denote $1 < E_\nu < 2$GeV, yellow ones $2 < E_\nu < 5$GeV and red ones $E_\nu > 5$GeV.

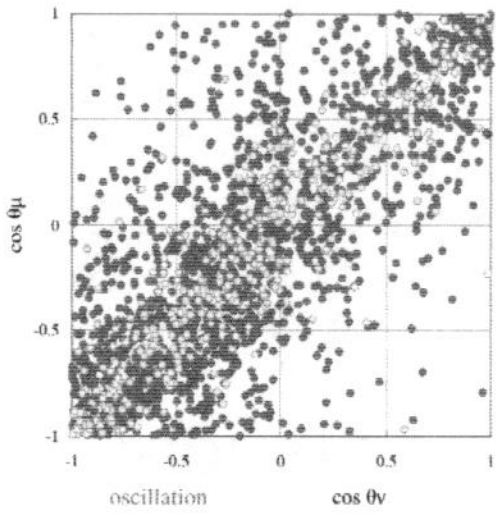

Figure 2. Correlation diagram between zenith angles of neutrinos and those of muons for oscillation. The color symbols have same meanings as in Figure 1.

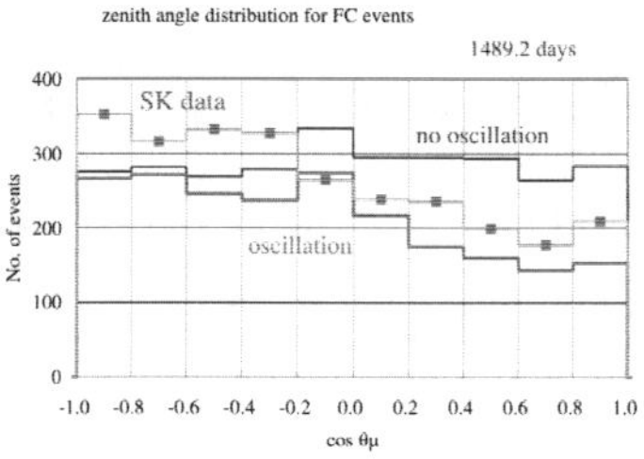

Figure 3. Zenith angle distribution for Single-Ring Muon-like Fully Contained Events for 1489.2 live days.

The approaches to the experimental data or "experimental data" between us and Super-Kamiokande are quite different. Therefore, a careful examination of both methods is highly desirable to clarify the disagreements between real experiments and computer numerical experiments.

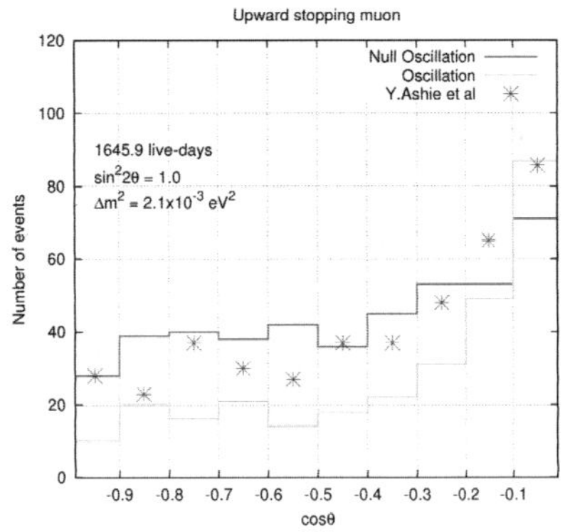

Figure 4. Zenith angle distribution for Upward Stopping Muons.

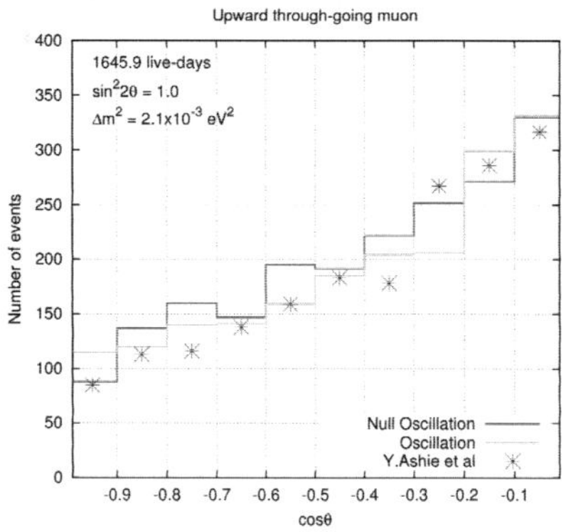

Figure 5. Zenith angle distribution for Upward Through Going Muons.

3. Conclusion

It is evident that our data disagrees with the SK data about neutrino oscillation. Therefore, an open discussion between SK and CNE is desired for the common view points.

References

1 Kajita,T.and Totsuka,Y. Rev.Mod.Phys., **73**(2001) p.101

2 Ishitsuka,M., Ph.D Thesis, University of Tokyo(2004) p138

3 Ashie Y., Phys.Rev.**D71**,112005(2005) pp 112005-8.

NEMO 3 double beta decay experiment

Hideaki Ohsumi* **on behalf of the NEMO Collaboration**

*Saga Univesity, Saga, 840-8502, Japan

E-mail: ohsumi@cc.saga-u.ac.jp

Abstract.

The NEMO 3 detector is devoted to the search for neutrinoless double-beta decay ($\beta\beta0\nu$). The expected performance of the detector has successfully achieved after the operation of a radon free air facility. Half-lives of double beta decay with two neutrinos ($\beta\beta2\nu$) have been measured systematically for the 7 major nuclei (^{100}Mo ^{82}Se, ^{116}Cd ^{150}Nd. ^{96}Zr, ^{48}Ca, ^{130}Te). There is no evidence of $\beta\beta0\nu$, and the corresponding lower limits for the half-lives are $T^{0\nu}_{1/2}(^{100}\text{Mo}) > 5.8 \times 10^{23}$y and $T^{0\nu}_{1/2}(^{82}\text{Se}) > 2.1 \times 10^{23}$y at 90% C.L..

1. Introduction

Neutrino oscillations discovery showed that neutrinos are massive particles. However the neutrino oscillations are only sensitive to the squared mass difference between two neutrino flavor mass eigenstates, and the absolute value of the neutrino mass remains a fundamental problem to be solved. Moreover, the nature of the neutrino mass (either Dirac or Majorana) is unknown. The best way to probe these questions is to look for neutrinoless double decays ($\beta\beta0\nu$). This decay means the non-conservation of lepton number, which is a signature of physics beyond Standard Model. Observation of this process would be a proof for Majorana neutrinos and it would give us a value for the absolute neutrino mass scale.

The main goal of the NEMO 3 experiment is to study $\beta\beta0\nu$ and to probe the half-life up to 10 $^{\sim 5}$y, which corresponds to the expected sensitivity of the effective neutrino mass ($< m_\nu >$) down to Sub-eV region, which depends on nuclear matrix element (NME). This sensitivity is the highest one among the present direct beta-ray tracking experiments, and very similar to the present best inclusive measurements of the past enriched Ge experiments (HM, IGEX) and the recent low temperature Te borometer experiment (CUORICINO).

In addition to this, the systematic study of two neutrino double beta decays ($\beta\beta0\nu$) of various several isotopes in NEMO 3 will be an important contribution to nuclear physics and NME calculations.

2. The NEMO 3 detector

The NEMO 3 detector (see fig. 1) has been operating in the Fréjus Underground Laboratory (LSM, France) since February 2003. The detector is cylindrical in design and divided into 20 equal sectors. Thin foils (about 30 - 60 mg/cm) of various kinds of $\beta\beta$ sources, which are Mo (6.9 kg), Se (0.93 kg), Cd (0.4kg), ^{5} Nd (37g), Zr (9.4g), Ca (7g), Te (0.45 kg), are surrounded by a tracking chamber with 6180 open dirft Geiger cells and a carorimeter with 1940 plastic scintillators coupled to low radioactive phototubes. A solenoid coils surrounds the detector and produces a weak magnetic field (25G) in the tracking volume. The magnetic field is used for the electron/positron charge recognition by measureing the curvature of the tracks. The system provides tracking vertex resolution of about 1 cm for 1 MeV electron, an energy resolution at FWHM of 14-17 % $/\sqrt{E}$, and time resolution of 250 ps. Thus the combination of a tracking detector, a calrimeter and a magnetic field allow the identification of electros, positrons, γ and delayed α particles.

Finally an external shield of 18 cm of low radioactive ion covers the detector to reduce external γ and thermal neutrons. Outside the iron, a water shield and a wood shield thermalize neutrons.

Detailed description of the detector and the performance is in the article [1].

In the beginning of the NEMO 3 experiment, it was found that the Rn level inside the tracking gas is higher than the initial requirements. A radon-free air facility was built to purify the air surrounding the detector prevent from the laboratory air penetration through unknown small leaks.

The radon free air facility started at October 2004. Later we will refer to the experiment before the date as Phase I, and the experiment after the date as Phase II.

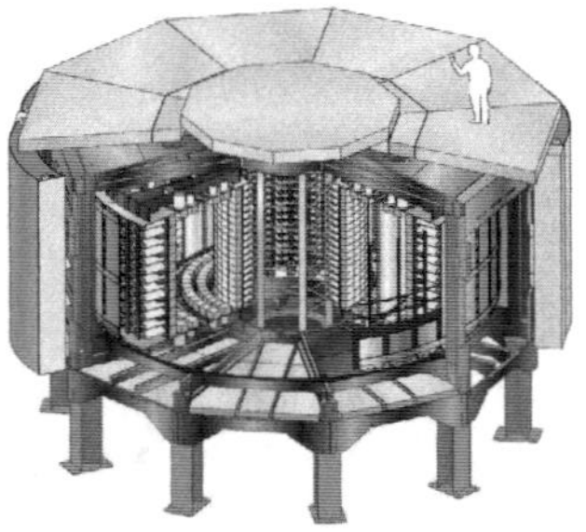

Figure 1. NEMO 3 detector.

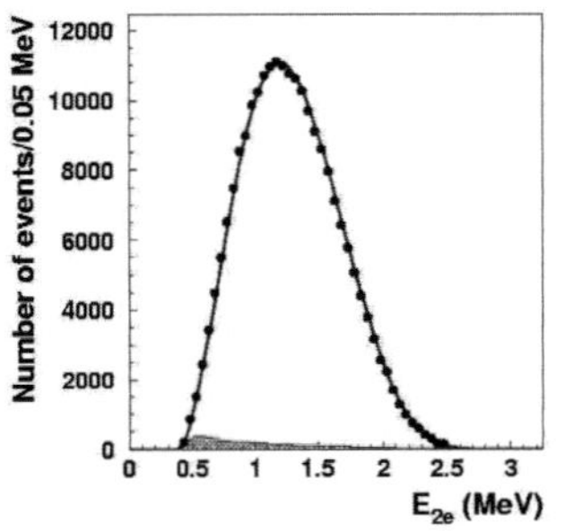

Figure 2. Sum energy spectrum of two electrons for $\beta\beta$ decay of Mo (7.369 kg·y). Background is extremely low (S/N~40). Solid line is MC simulation.

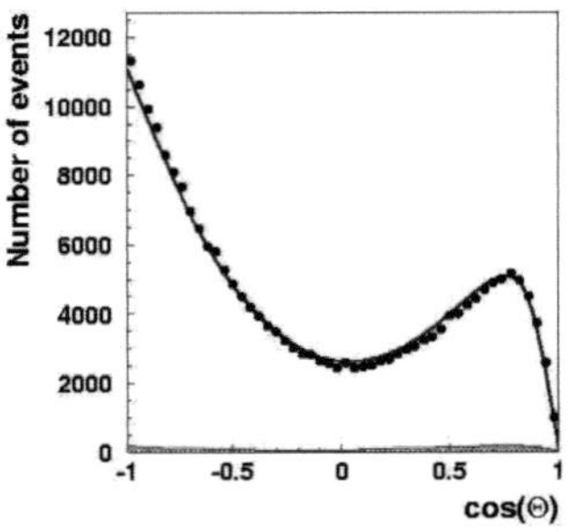

Figure 3. Angular distribution between two electrons for $\beta\beta$ decay of Mo. Solid line is MC simulation using the acceptance of NEMO 3 detector.

3. NEMO 3 results

Measured $\beta\beta 2\nu$ decay of Mo in Phase I experiment is shown in the fig. 2 and fig. 3 and the half life is $T^{\nu}_{/}$ =7.11±0.02(stat)±0.54(syst)×10 y [2]. It is noted that the biggest statistics in the world of double beta decay measurements (219000 events) and a good signal to background (S/B) ratio of 40. No significant discrepancy to the Monte Carlo simulations with the standard model parameters, which means the $\beta\beta 2\nu$ decay is really standard process.

No evidence for $\beta\beta 0\nu$ signal was found in Phase I and Phase II(part) experiment (see fig. 4). A lower limit on the half life of $T^{\nu}_{/} > 5.8 \times 10$ y (90%CL). This corresponds to the effective Majorana neutrino mass $< m_\nu > < 0.8$ - 1.3 eV, which depends on the recent QRPA NME calculations [3].

The $\beta\beta$ decays have a possibility to proceed through transitions to the various excited states of daughter nucleus as well as to the normal ground state. Studies of the transitions to excited states are also possible using the γ-ray tagging by NEMO 3 calorimeters. After the analysis of 8024 h of data, the half life for $\beta\beta 2\nu$ decay of Mo to the excited 0 state is measured $T^{\nu}_{/}$ =5.7$_{-}$: (stat)±0.8(syst)×10 y [4].

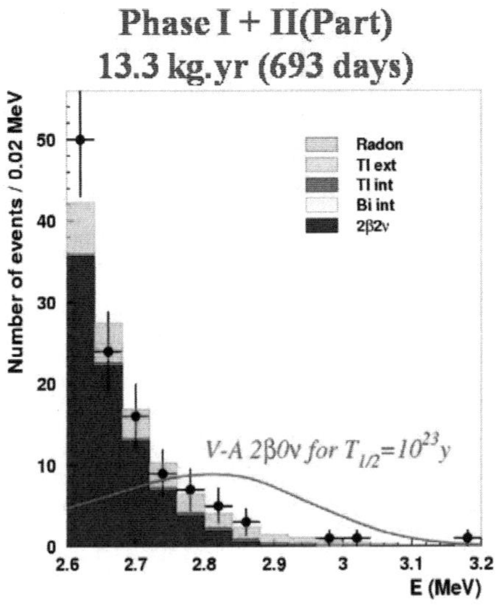

Figure 4. Spectrum of the energy sum of the two electrons for Mo $\beta\beta$ decay, the energy region above 2.6MeV is shown here (Phase I and Phase II(part), 13.3 kg·y).

Table 1. Main $\beta\beta2\nu$ results of NEMO 3 experiment. S/B is a signal to background ratio

Nuclei	S/B	$T_{/}$, y
Mo	40	7.11±0.02(stat)±0.54(syst)×10
Se	4	9.6±0.3(stat)±1.0(syst)×10
Cd	7.5	2.8±0.1(stat)±0.3(syst)×10
^{5} Nd	2.8	9.7±0.7(stat)±1.0(syst)×10
Zr	1	2.0±0.3(stat)±0.2(syst)×10
Ca	~10	3.9±0.7(stat)±0.6(syst)×10
Te	0.25	7.6±1.5(stat)±0.8(syst)×10

$\beta\beta2\nu$ half life of Se is T $^{\nu}_{/}$ =7.11±0.02(stat)±0.54(syst)×10 y (Phase I, S/B ratio is about 4) and $\beta\beta0\nu$ half life limit is $T^{\nu}_{/} > 2.1 \times 10$ y (90%CL) (Phase I and Phase II(part), 1.76 kg·y), which corresponds to an upper neutrino mass limit of $< m_\nu > < 1.4$ -2.2 eV.

$\beta\beta$ ν

$\beta\beta2\nu$ results of various isotopes measured in NEMO 3 are summarized as Table 1. Besides Te, data are taken from Phase I. Here recent preliminary analysis for $\beta\beta2\nu$ half life of Te is reported. Phase II data with extremely low Rn background are essential to the analysis for $\beta\beta2\nu$ decay of Te because of the extremely low decay rate.

References

[1] Arnold R et al., 2005 *Nucl. Instr. Meth.* A **536** 79

[2] Arnold R et al., 2005 *Phys. Rev. Lett.* **95** 182302

[3] Rodin V A et al., 2007 *Nucl. Phys.* A **793** 213; Kortelainen M and Suhonen J 2007 *Phys. Rev.* C **75** 051303(R); Kortelainen M and Suhonen J 2007 *Phys. Rev.* C **76** 024315; Aunola M and Suhonen J 1998 *Nucl. Phys.* A **463** 207

[4] Arnold R et al., 2007 *Nucl. Phys.* A **781** 209

The COBRA Experiment

[1]J.R. Wilson[2]

Department of Physics and Astronomy, University of Sussex, Brighton, BN1 9QH, UK

E-mail: drjeannewilson@gmail.com

Abstract. The COBRA experiment aims to use a large array of CdZnTe semiconductor detectors to search for neutrinoless double beta decay. The COBRA collaboration are currently operating a small proto-type array of crystals in a low-background environment. This paper presents the current status of the experiment, results from current and previous proto-types and future prospects for COBRA.

1. Concept

COBRA aims to search for neutrinoless double beta decay ($0\nu\beta\beta$) in CdZnTe semiconductor crystals[1]. There are actually 9 candidate $0\nu\beta\beta$ isotopes, 5 in the form of $\beta^-\beta^-$ emitters, intrinsic to the detector material, but measurements will focus on ^{116}Cd. This has the highest Q-value at 2.8 MeV, beyond all the single gamma-lines from the natural decay chains which would be problematic in contaminating signal regions at lower energies. CdZnTe, in common with other semiconductors, can be produced cleanly with low levels of intrinsic radioactive background and offers good energy resolution. Unlike many semiconductors though, CdZnTe can be operated at room temperature without the need for expensive cryogenics close to the detectors.

For a neutrino mass of about 50 meV, as suggested by neutrino oscillation data, the half-life of ^{116}Cd is of order 10^{26} years (exact values depend on the nuclear matrix elements). As with all neutrinoless double beta decay searches, the COBRA experiment will require low backgrounds, good energy resolution and a large detector mass to measure this. To obtain sensitivity to $T_{1/2} = 10^{26}$ y in 5 years of operation, a resolution of $\Delta E < 2\%$ (FWHM), and a background rate in the signal region of <0.001 counts/keV/kg/year are required. Resolutions of <2% at 2805 keV have already been achieved, and improvements are anticipated with better quality detectors and by cooling the detectors and electronics by a few degrees.

2. Status

Small proto-types for the COBRA experiment have been operating in the LNGS laboratory in Italy, which provides a shielding equivalent to ≈3500 m of water. 4.34 kg·days of data were collected with an initial proto-type consisting of four 1 cm^3 crystals. This was the first operation of CdZnTe crystals in a low background, underground environment and the data were used to study the detector properties and the major sources of background in the crystals. Despite the small detector mass, less than 25 g, the data revealed a number of interesting results including a measurement of the 4-fold forbidden beta decay of ^{113}Cd[2] which yields

[2] For the COBRA collaboration

Isotope and Decay		T_{half} limit (years, 90% C.L.)	
		Current Data	Previous
		$\beta^-\beta^-$ Decays	
^{116}Cd	to g.s	6.05×10^{19}	3.14×10^{19}
^{130}Te	to g.s	3.44×10^{20}	9.92×10^{19}
^{130}Te	to 536 keV	2.49×10^{20}	3.73×10^{19}
^{116}Cd	to 1294 keV	2.80×10^{19}	4.92×10^{18}
^{116}Cd	to 1757 keV	3.03×10^{19}	9.13×10^{18}
^{116}Cd	to 2027 keV	3.14×10^{19}	1.37×10^{19}
^{116}Cd	to 2112 keV	4.16×10^{19}	1.08×10^{19}
^{116}Cd	to 2225 keV	2.67×10^{19}	9.46×10^{18}
^{130}Te	to 1794 keV	1.45×10^{20}	
^{130}Te	to 1122 keV	9.48×10^{19}	
^{114}Cd	to g.s.	4.71×10^{20}	
		$\beta^+\beta^+$ Decays	
^{64}Zn	$0\nu\beta^+$EC to g.s.	1.18×10^{18}	2.78×10^{17}
^{64}Zn	0ν2EC to g.s.	$\mathbf{7.43\times10^{18}}$	1.19×10^{17}
^{120}Te	0ν2EC to g.s.	$\mathbf{1.13\times10^{17}}$	2.68×10^{15}
^{120}Te	0ν2EC to 1171keV	$\mathbf{3.43\times10^{16}}$	9.72×10^{15}
^{106}Cd	$0\nu\beta^+\beta^+$ to g.s.	5.12×10^{18}	4.50×10^{17}
^{106}Cd	0ν2EC to g.s.	$\mathbf{5.48\times10^{18}}$	5.70×10^{16}
^{106}Cd	$0\nu\beta^+\beta^+$ to 512keV	7.17×10^{17}	1.81×10^{17}

Table 1. Preliminary half-life limits derived from data obtained with the first 16 crystals of the current COBRA proto-type, compared to results previously published by the collaboration in Ref. [3]. Values in bold exceed current World-best limits.

$T_{1/2} = (8.2 \pm 0.2(\text{stat.})^{+0.2}_{-1.0}(\text{sys.})) \cdot 10^{15}$ years. Half-life limits were calculated for a variety of possible $0\nu\beta\beta$-decays and are presented in Ref. [3].

In 2006, installation of a larger proto-type consisting of a 4×4×4 array of 1 cm^3 crystals commenced. So far, 7.85 kg·days of data have been collected with the first layer of 16 crystals, which is shown in figure 1. A refined likelihood analysis, which allows for different background levels, energy thresholds, resolutions and livetimes for each crystal was developed and applied to this data resulting in the new half-life limits presented in table 1.

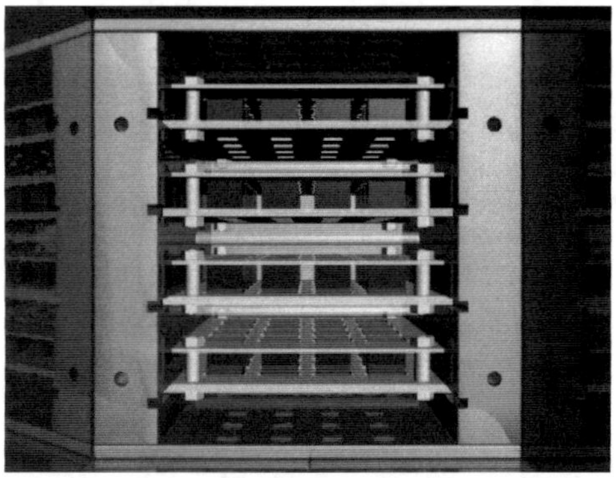

Figure 1. A photograph of the first layer of the 64-crystal proto-type (left) and an artistic impression of the whole array in the copper mounting structure (right).

The other 48 crystals will be installed at LNGS in October 2007 with a number of modifications to the setup, including revised contacting methods to reduce cross-talk and pick-up. One major background was identified in the red-coloured passivation coating on the crystal surfaces. Alternative passivation coatings have been tested, resulting in a factor $\approx$5 reduction in the background in the region of interest between 2–3 MeV on the current background level of $\approx$150 counts/keV/kg/year. Simulations of the measured contamination levels indicate greater improvements are expected, suggesting other background components are now being observed.

Radon in the air surrounding the detectors could be an additional source of background. Early tests of flushing the system with nitrogen gas indicate a further factor of at least 3–4 reduction in background levels. New crystal production techniques are being investigated in collaboration with the Freiburg materials research institute, with the aim of producing radioactively pure crystals with enriched ^{116}Cd in the near future. A sophisticated shield has been designed for future phases of COBRA, which incorporates an active scintillator veto component for background rejection. Details of the design can be found in Ref. [4].

Techniques are also being investigated to actively reject backgrounds through the event topology. The 1 cm^3 cube size ensures that for $0\nu\beta\beta$-events, the emitted electrons are unlikely to escape the crystal, but a large proportion of high energy gamma ray events will deposit energy in more than one crystal through Compton scattering. The COBRA design is modular, allowing easy upgrade and offering the ability to discriminate such backgrounds by rejecting events where more than one crystal is triggered. Simulations for a 64,000 crystal array indicate that ^{232}Th chain backgrounds in the 2–3 MeV region can be reduced by more than 50% in this way.

The COBRA collaboration is also investigating the exciting possibility of pixellising the detectors to obtain tracking information that provides significantly more information for background rejection. For example, a 2.8 MeV α is only expected to travel a few μm in CdZnTe but βs from ^{116}Cd $0\nu\beta\beta$-decay are expected to travel 1–1.5 mm. Crystals with a pixel-pitch of $\approx$200 μm will be able to veto decays involving α-emission with nearly 100% efficiency (see Ref. [5] for more details). Tests of detectors with 1.6 mm pixels have produced promising results and recently, a detector with 50 μm sized pixels has been successfully created.

3. Summary

The COBRA experiment aims to search for neutrinoless double beta decay of ^{116}Cd with a sensitivity to half-lives greater than 10^{26} years with a large array of CdZnTe crystals. Resolutions of FWHM=2% at the peak energy (2.8 MeV) have already been obtained and improvements are expected. The background in the signal region will be minimised through careful selection of materials, a comprehensive shielding with active veto, and the use of timing and spatial coincidences to reject radioactive decay events. A new proto-type experiment is being deployed to investigate these issues.

Acknowledgments

This research was supported by PPARC and the Deutsche Forschungsgemeinschaft (DFG). We thank V. Tretyak, eV-PRODUCTS, K. Eitel and the mechanical workshop of the University Dortmund for their support and the Laboratori Nazionali del Gran Sasso (LNGS) for offering the possibility to perform measurements underground.

References

[1] Zuber K., *Phys. Lett.* B **519** 262 (2001)
[2] Goeßling G. et al, *Phys. Rev.* C **72** 064328 (2005)
[3] Bloxham T. et al, *Phys. Rev.* C **76** 025501 (2007)
[4] Stewart D. et al, *NIM* A **571** 651–662 (2007)
[5] Bloxham, T. and Freer, M., *NIM* A **572**, 2, 722–728 (2007)

DCBA experiment for searching for neutrinoless double beta decay (I)

N Ishihara[1], Y Kato[1], T Inagaki[1], G Iwai[1], T Ohama[1], S Takeda[1], Y Yamada[1], T Haruyama[1], Y Makida[1], M Kawai[1], S Kitamura[2], T Ishikawa[3], T Sakuma[3], T Sumiyoshi[3], Y Teramoto[4], I Nakano[5], Y Sakamoto[6], Y Nagasaka[7], N Tamura[8], K Tanaka[9] and R Ito[10]

[1] High Energy Accel. Research Organization (KEK), Tsukuba, Ibaraki 305-0801 Japan
[2] Tokyo Metropolitan University, Arakawa, Tokyo 116-8551 Japan
[3] Tokyo Metropolitan University, Hachioji, Tokyo 192-0398 Japan
[4] Osaka City University, Sumiyoshi, Osaka 558-8585 Japan
[5] Okayama University, Okayama 700-8530 Japan
[6] Tohoku Gakuin University, Sendai, Miyagi 981-3193 Japan
[7] Hiroshima Institute of Technology, Saeki, Hiroshima 731-5193 Japan
[8] Niigata University, Niigata 950-2181 Japan
[9] Saitama Science Institute, Inarimachi-kita, Fukaya, Saitama 366-0028 Japan
[10] Zentek Technology Japan Inc., Uchikanda, Chiyoda, Tokyo 101-0047 Japan

nobuhiro.ishihara@kek.jp

Abstract. A magnetic tracking detector called DCBA (Drift Chamber Beta-ray Analyzer) is being developed at KEK in order to search for neutrinoless double beta decay. Kinetic energy of each beta ray is obtained from its momentum, which is measured with a drift chamber installed in a uniform magnetic field. Energy resolution has been studied using a test prototype DCBA-T2. The study results are briefly described together with the structure and the principle of DCBA detector.

1. Introduction

The existence of neutrino mass was confirmed by the neutrino oscillation experiments. Next important issues are both the absolute mass scale and the Majorana nature of neutrinos. Neutrinoless double beta decay ($0\nu\beta\beta$) experiment is only the realistic method to investigate the issues, because $0\nu\beta\beta$ takes place when neutrinos are Majorana particles, and the effective neutrino mass ($<m_\nu>$) is obtained from the decay life. Though Klapdor et al. insist that they measured the $0\nu\beta\beta$ half-life of ^{76}Ge [1], their results elicited a number of critical replies [2]. Therefore other experiments are now strongly desired to decide that their results are valid or not.

DCBA (Drift Chamber Beta-ray Analyzer) consists of gas drift chambers and a solenoid magnet [3]. Since the drift chamber is sensitive to only charged particles, gamma ray backgrounds are automatically rejected. DCBA also can easily separate the $0\nu\beta\beta$ events from background events such as single electrons, alpha particles and electron-positron pair creations. However, the energy resolution of DCBA was not well known so far. The results of energy resolution study using a test prototype DCBA-T2 are discussed below together with the detector structure and its operational principle.

2. DCBA-T2

A test prototype DCBA-T2 shown in figure 1 has been constructed and operated at KEK.

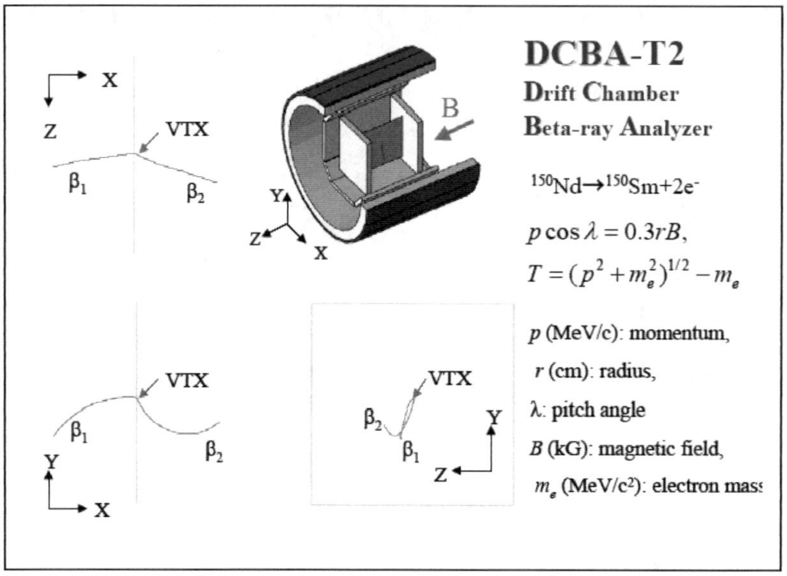

Figure 1. Illustration of DCBA-T2.

DCBA-T2 consists of a source plate, two drift chambers, a gas container and a solenoid magnet. Source plate is made of natural Nd_2O_3 including 5.6% ^{150}Nd. Two drift chambers are located at both the sides of the source plate. Each chamber has wire planes of anode, pickup and cathode. A wire pitch is 6 mm for every wire. An anode wire is gold -plated tungsten of 20 μm diameter, and other wires are aluminum of 80 μm diameter. The anode wires are horizontally strung with 4 mm apart from the source plate. The pickup wires are vertical with a gap of 2 mm from anode wires. The volume of 90 mm between the pickup wire plane and the cathode one is sensitive to the trajectory of charged particle. The chambers are filled with the premixed gas of 85% He and 15% CO_2.

2.1. Operational principle

In figure 1, X, Y and Z-coordinates are defined as well as magnetic field direction B. Suppose 0νββ takes place at the vertex point (VTX) back-to-back, each beta ray makes ionization along its track in the camber gas. Produced ionization electrons drift to anode wires along lines of electric force, and reach to anode wires. The coming electron is accelerated by strong electric field near the anode wire, and makes the so-called electron avalanche on the wire surface. While huge number of electrons produced in the avalanche is immediately absorbed in the anode wire, ions leave the anode wire for other electrodes such as pickup wires and the source plate of ground potential. An anode signal is induced by the leaving ions, and a pickup signal is done by the coming ions. Completely same timing of both the anode and pickup signals is useful to determine X, Y and Z-coordinates of a track point. The X-coordinate is the product for the velocity and the drift time of the drift electron. The Y and Z-coordinates corresponding X are the positions of anode and pickup wires, respectively. In figure 1, tracks named β1 and β2 are shown in three projected planes after making the three dimensional reconstruction of positions. The measurement of helical radius and pitch angle of each track enables us to obtain its momentum (p) and kinetic energy (T) from equations shown in figure 1.

2.2. Actual tracks of charged particles

Figure 2, 3 and 4 show actual tracks of an electron, alpha and electron-positron pair creation, respectively. In the figures, X-Y planes show anode wire signals and X-Z planes show pickup wire signals. We can make three-dimensional reconstruction from these tracks.

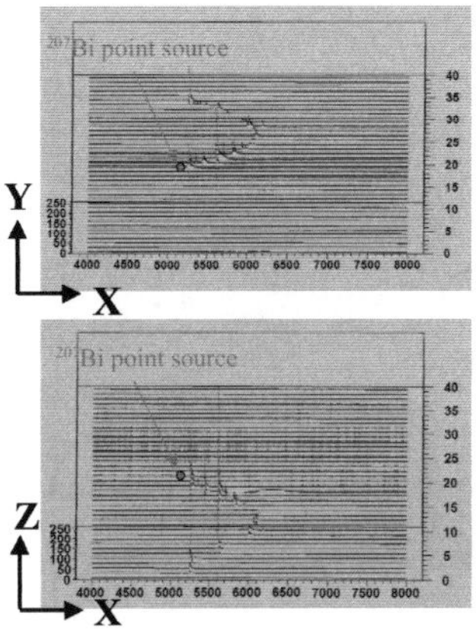

Figure 2. Electron track

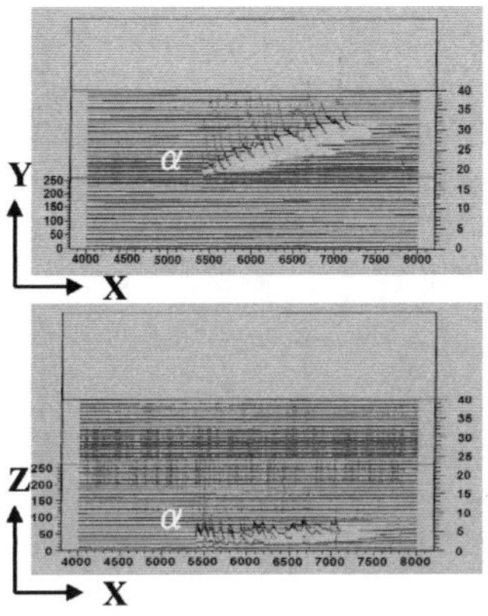

Figure 3. Alpha track

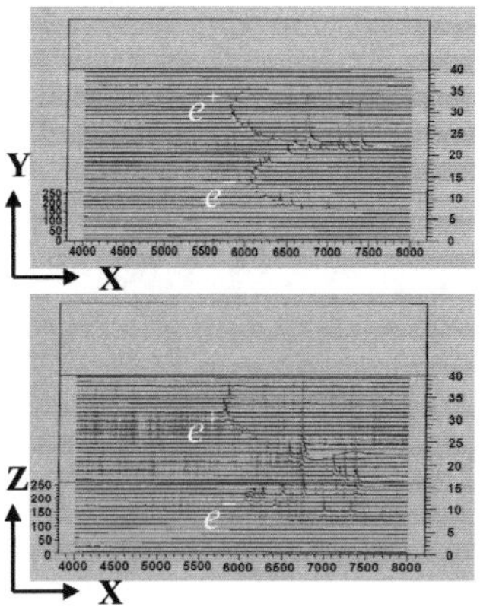

Figure 4. Pair creation

2.3. Energy resolution

In order to study the energy resolution of DCBA-T2, a point source of ^{207}Bi has been installed at the central part of the source plate. Figure 5 shows the energy spectra of ^{207}Bi including background events. Most intensive electron energy is 0.98 MeV. The energy resolution at 0.98 MeV is about 150 keV (FWHM). If same resolution is obtained for every beta ray, the resolution of sum-energy of two betas becomes about 210 keV (FWHM). This means that the relative energy resolution is about 6.2% (FWHM) at the Q-value of ^{150}Nd. Since the energy resolution depends on the multiple scattering and the energy loss of a beta ray in the chamber gas, it is very useful to make the track length short in a stronger magnetic field in order to improve the energy resolution.

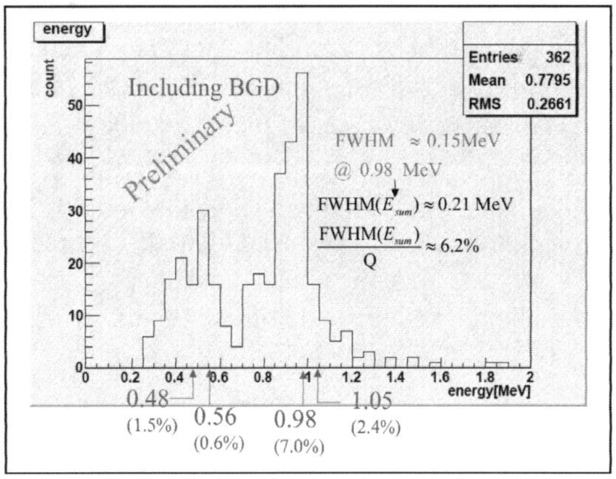

Figure 5. Energy spectra of ^{207}Bi .

3. Conclusions

DCBA-T2 has a possibility of 6.2% FWHM energy resolution at Q value of ^{150}Nd. More improvement is expected by using a stronger magnetic field, in order to investigate the effective neutrino mass down to 30 meV [4].

References

[1] H. V. Klapdor-Kleingrothaus, I. V. Krivosheina, A. Dietz, O. Chkvorets, Phys. Lett. **B 586** (2004) 198-212.

[2] C. E. Aalseth *et al.*, Mod. Phys. Lett. **A 17** (2002) 1475; F. Feruglio *et al.*, Nucl. Phys. **B 637** (2002) 345; Zdesenco *et al.*, Phys. Lett. **B 546** (2002) 206.

[3] N. Ishihara et al., Nucl. Instr. Meth. Phys. Res. **A 373** (1996) 325; N. Ishihara et al., Nucl. Instr. Meth. Phys. Res. **A 443** (2000) 101; Y. Kato et al., Nucl. Instr. Meth. Phys. Res. **A 498** (2003) 430.

[4] N. Ishihara et al., DCBA experiment for searching for neutrinoless double beta decay (II), this conference.

Double beta decay experiments and ν-mass sensitivities

H. Ejiri[1,2,3], K. Fushimi[4], R. Hazama[5], H. Nakamura[2] and T. Shima[1]

[1]RCNP, Osaka University, Ibaraki, Osaka 567-0047, Japan,
[2]NIRS, National Institute of Radiological Sciences, Chiba, 263-8555, Japan,
[3] Nuclear Science, Czech Technical University, Prague, Czech Republic,
[4]IAS, University of Tokushima, Tokushima 770-8592, Japan,
[5]Hiroshima University, Higashi Hiroshima, Hiroshima 739-8527, Japan,

E-mail: ejiri@rcnp.osaka-u.ac.jp

Abstract. This is a brief report on Majorana neutrino(ν) mass to be studied by neutrino-less double beta decays (0νββ). The ν-mass depends on the nuclear structure of the ββ isotope, the effective detector volume and the background rate. Subjects discussed include 1 0νββ and ν-mass, 2 0νββ nuclear matrix elements, and 3 0νββ experiments and ν-mass sensitivities.

1. Neutrino-less double beta decays and Majorana neutrino masses

Neutrino-less double beta decays (0νββ) is quite realistic and sensitive probe for the Majorana nature of the neutrino, the absolute mass scale, and the CP phase, which are beyond the standard mode. Recent experimental and theoretical studies are given in review articles and references therein [1,2,3].

The (0νββ) process may include the left- and right-handed weak currents. The left-handed current may be caused by the light Majorana-ν exchange and the SUSY particle exchange. In case of the light Majorana-ν exchange, one can derive the effective Majorana-ν mass.

There are two types of ββ experiments, calorimetric experiments with ββ sources integrated into ββ detectors and spectroscopic experiments with ββ sources separated from ββ detectors. Spectroscopic experiments identify the left-handed current process by measuring individual ββ energy and angular correlations. The experiments with ββ sources ≠ ββ detector can measure two ββ isotopes and also ββ decays to the ground and excited states. These measurements are useful to identify the light Majorana-ν exchange and to derive the Majorana –ν mass [1,3].

The 0νββ transition rate is expressed in terms of the effective mass $<m_\nu>$ as

$$T^{0\nu} = G^{0\nu}\ |M^{0\nu} <m_\nu>|^2, \tag{1}$$

where $G^{0\nu}$ is the kinematical factor for 0νββ and $M^{0\nu}$ is the nuclear matrix element. The effective neutrino mass to be studied by 0νββ experiments with 90 % confidential level is expressed as

$$<m_\nu> = (S_N)^{-1/2}\ (N_{\beta\beta})^{-1/2}\ \delta^{1/2}, \tag{2}$$

where S_N, $N_{\beta\beta}$, and δ are the nuclear sensitivity, the effective number of ββ isotopes and the effective number of 0νββ signals required for the 90 % C.L. They are written as

$$S_N = 6\ 10^{15}\ G^{0\nu}\ |M^{0\nu}|^2\ (A)^{-1} m_e^{-2},\quad N_{\beta\beta},= \varepsilon^{0\nu}\ N\ T,\quad \delta \sim 1.6 + 1.7\ (BT)^{1/2} \qquad (3)$$

where $G^{0\nu}$ is the phase space volume in unit of 10^{-24} y^{-1} , A is the mass number, $\varepsilon^{0\nu}$ is the efficiency, N is the total ββ isotopes in unit of ton, T is the run time in unit of year and B is the background rate per ton (N=1) per year.

2. Nuclear matrix elements for neutrino-less double beta decays

The nuclear matrix element $M^{0\nu}$ is crucial for the mass sensitivity and thus for designing ββ experiments as well as for extracting the neutrino mass from the 0νββ rate. It is found that nuclear matrix elements by recent QRPA calculations [4,5] are well expressed as

$$M^{0\nu} \sim 18\ A^{-1/3} \qquad (4)$$

Then the nuclear matrix element and nuclear sensitivity for ββ nuclides are shown in Table 1. The mass sensitivities required for QD (Quasi-degenerate mass spectrum) and IH (Inverted Hierarchy spectrum) are around 100 and 30 meV.

Table 1. Nuclear sensitivity $(S_N)^{-1/2}$ in unit of meV. $G^{0\nu}$ is the phase space volume, and $M^{o\nu}$ is the muclear matrix element. evaluated by using equ. (4). ^{100}Mo* stands for the ββ decay to the excited state. All others are for the ground state transitions.

Isotope	$G^{0\nu}$ 10^{-14} y^{-1}	$M^{o\nu}$	$(S_N)^{-1/2}$ meV
^{76}Ge	0.44	4.3	19.5
^{82}Se	1.89	4.1	10.5
^{100}Mo	3.17	3.9	9.6
^{100}Mo *	0.77	3.9	19.5
^{116}Cd	3.24	3.7	7.55
^{130}Te	2.86	3.5	12.5
^{136}Xe	3.03	3.45	9.5
^{150}Nd	13.4	3.4/3	19.5

Calorimetric experiments with ββ source= ββ detectors are made for ^{76}Ge, ^{130}Te, and ^{136}Xe by using ^{76}Ge semi-conductive detectors [6], TeO_2 bolometers [7] and ^{136}Xe scintillation detectors [8], respectively. The nuclear sensitivities for ^{130}Te, and ^{136}Xe are around 10 meV, while the sensitivity for ^{76}Ge is around 20 meV. They are used for studying QD and IH neutrino masses by using low BG detectors (B ~ 1/t y) with $N_{\beta\beta}$ ~ 0.1 t and ~1t, respectively.

Spectroscopic experiments such as MOON [9-11] and NEMO [12] with detector ≠ ββ source can select ββ isotopes to be studied. The ^{82}Se and ^{100}Mo are considered in MOON. The sensitivity for their ground state transitions are S_N~10 meV, and that for the excited transition of ^{100}Mo is S_N ~20 meV. The ground state transition of ^{150}Nd, which is studied by SuperNEMO [12], may be intersting because of the large phase space volume, but the matrix element may be reduced because of the large difference of the deformations between the initial and final nuclei. The nuclear sensitivity is around 20 meV in case of the nuclear matrix element reduced by a factor 3, as shown in table 1.

3. Neutrino mass sensitivity of neutrino-less double beta decay experiment.

Neutrino mass sensitivities are evaluated for several ββ isotopes by using equ.(3) as given in table 2. We assume the efficiency $\varepsilon^{0\nu} = 0.6$ and B=1 / t y for ^{76}Ge [6], the E-resolutions σ=2.2 % and 1.7% and 2νββ BG's for ^{82}Se ^{100}Mo and ^{82}Se* ^{100}Mo* [9-11], and BG B=200 / t y and 20 / t y for ^{130}Te and ^{130}Te*. The BG's for ^{130}Te correspond to 0.01 and 0.001 / keV, kg, y for TeO_2. The QD and IH mass sensitivities are achieved by the ^{76}Ge, ^{82}Se ^{100}Mo and ^{130}Te experiments with $NT \sim 0.1$ t y and $N \sim 2$ t y experiments. Improvement of the E-resolution from σ=2.2 % to 1.7% for ^{82}Se - ^{100}Mo in MOON/SuperNEMO [9-12] improve the mass sensitivity by a factor 1.5. The IH mass of $<m_\nu> \sim 30$ meV can be achieved by NT~5 t y experiments of ^{76}Ge with B~1, ^{82}Se -^{100}Mo with σ=2.2 % and 1.7% and ^{130}Te with B=20 / t y. The ^{82}Se experiment with $\sigma \sim 1.7\%$ is promising to reach the lower bound IH mass of 23 meV.

Table 2. Neutrino mass sensitivities in meV for NT tonne year runs. The E-resolutions for ^{82}Se ^{100}Mo are σ=2.2 % and 1.7% (*). BG 's for ^{130}Te are B=200 and 20(*)

NT t y	0.06	0.12	0.48	2	5
^{76}Ge	162	112	60	35	26
^{82}Se	134	98	60	38	29
^{82}Se*	130	92	52	31	23
^{100}Mo	161	124	82	56	44
^{100}Mo*	137	103	66	44	34
^{130}Te	167	131	92	63	50
^{130}Te*	123	88	57	37	29

Accordingly, Majorana [6], MOON/SuperNEMO [9-12] and CUORE with $N \sim 0.1$ and 1 t ββ sources are of great interest for QD and IH mass studies. There are other promising experiments like EXO [8] with high efficiency laser tagging for QD-IH. It is noted that the MOON experiment with multi-layer structure [9-11] can be expanded to challenge to NH mass of a few meV.

1. References

[1]. Ejiri H 2005 *J. Phys. Soc. Jap.* **74,** 2101
[2]. Avignone F, Elliott S, and Engel J 2007 arXiev 0708 1033v1 [nucl-exp]
[3]. Vergados J 2002 *Phys. Rept.* **361** 1 ; *Private communication* 2007
[4]. Kotelainen N and Suhonen J 2007 arXiv ;0708.0115v1[nucl-th], 0705.0469 [nucl-th]
[5]. Rodin V et al. 2007 arXiv;0706.4304 [nucl-th].
[6]. Aalseth C et al. 2004 *Yad. Fiz.* 67, No. 11
[7]. Fiorini E 2007 *Proc. ICNP 2007, Tokyo, June*
[8]. Piepke A 2006 *Proc. Neutrion 2006, Santa Fe*
[9]. Ejiri H, et al., 2000 *Phys. Rev. Lett.* **85**, 2917
[10]. Ejiri H, 2007 *Mod. Phys. Lett. A*, **22**, 1277
[11]. Nakamura H et al. 2007 J. Phys. Soc. Japan 114201
[12]. Piquemal F 2007 Proc. TAUP 2007, Sendai Japan

From Cuoricino to CUORE: investigating the inverted hierarchy region of neutrino mass

Andrea Giuliani on behalf of the CUORE collaboration
INFN - Sez. Milano-Bicocca, P.zza della Scienza 3, 20126 Milano, Italy
E-mail: andrea.giuliani@mib.infn.it

Abstract. Cuoricino is a Double Beta Decay experiment operating deep underground, in the Laboratori Nazionali del Gran Sasso (Italy) at a depth of about 3500 m.w.e. The search for the $0\nu\beta\beta$ of ^{130}Te is carried out with the bolometric technique and an upper limit of 3×10^{24} y (@ 90%C.L.) is set for this process. Cuoricino represents not only the most sensitive DBD Experiment presently operating but also a prototype for a next generation experiment, CUORE (Cryogenic Underground Observatory for Rare Events). The expected performance and sensitivity, based on Monte Carlo simulations and extrapolations of the present Cuoricino results, indicate that CUORE will be able to test the 0.02-0.05 eV region for the effective neutrino mass, having a high discovery potential in the inverted hierarchy region of the neutrino mass pattern.

1. Introduction

The search for the neutrino mass is one of the most relevant and exciting field in particle physics and cosmology, the neutrino mass scale being one of the key quantities in the new theories beyond the Standard Model.

A physical tool able to help in answering crucial questions about neutrinos is the nuclear process called neutrinoless double beta decay ($0\nu\beta\beta$) [1]. It is a rare nuclear transition that, if observed, would imply that the neutrino mass is different from zero and that neutrinos are Majorana particles.

To detect this process, the bolometric technique is very powerful. It requires very low temperatures to read appreciable thermal signals induced by the investigated decays.

Cuoricino, described in the following sections, is a $0\nu\beta\beta$ experiment looking for the decay of ^{130}Te and using the bolometric technique.

2. The Cuoricino detector

Cuoricino is an array of 62 bolometers of TeO_2 with an active mass of 40.7 kg. There are 44 bolometers with $5\times5\times5$ cm^3 crystals and 18 bolometers with $3\times3\times6$ cm^3 crystals. Of the small crystals, 2 are enriched in ^{128}Te and 2 in ^{130}Te. The array is cooled down to ~8 mK in a dilution refrigerator, shielded from environmental radioactivity and energetic neutrons. It is running in the Laboratori Nazionali del Gran Sasso (LNGS), Italy, a location that guarantees a high degree of suppression of cosmic ray flux thanks to the ~3500 m.w.e. depth.

To fulfill the background requirements typical of rare events physics, a particular care was dedicated to the selection and treatment of the materials used for the construction of the

Cuoricino array.

The mechanical structure of the array was made exclusively in OFHC copper and PTFE, due to their extremely low radioactive content. All the copper and PTFE parts of the mounting structure were separetely treated with acids to remove surface contamination. Finally, the array was assembled in an underground clean room in a nitrogen atmosphere to avoid radon contamination, closed in a copper structure and hung in vacuum inside the Inner Vacuum Chamber of the refrigerator; it is surrounded by a ~ 1 cm thick roman lead cylindrical shield closed with bottom and top lead discs of thicknesses of 7.5 cm and 10 cm respectively. The refrigerator itself is shielded with low activity lead with 20 cm thickness and with borated PET with 10 cm thickness. Nitrogen is fluxed between the external lead shield and the cryostat to avoid any Rn contribution to the detector background; in addition the cryostat is placed inside a Faraday cage to reduce electromagnetic interferences. In order to prevent vibrations induced by the overall facility from reaching the detectors, the tower is mechanically decoupled from the cryostat through a stainless steel spring.

Thermal pulses are recorded by means of Neutron Transmutation Doped (NTD) Ge thermistors operated in the Variable Range Hopping (VRH) conduction regime. Electrical connections are made by two gold wires, ball bonded to metalized surfaces on the thermistor, which are then attached to each bolometer by 9 spots of two-component epoxy. The front-end electronics is at room temperature for most of the channels; some channels are read-out by a Si JFET in emitter-follower configuration operating in a 150 K environment placed at the 4 K plate of the dilutione fridge.

3. Cuoricino results

Cuoricino is the most massive bolometric experiment taking data in the world. The present limit on the half life for the $0\nu\beta\beta$ decay of ^{130}Te is

$$T^{0\nu}_{1/2} > 3 \cdot 10^{24}\ y \quad (90\% C.L.) \tag{1}$$

with an exposure of 11.83 kg·y of ^{130}Te. This translates into a limit on the effective neutrino mass of

$$m_\nu < 0.20 - 0.98\ eV \tag{2}$$

where the range for the limit is spanned by the most and the less favourauble nuclear matrix elements. In the context of a number of nuclear models, these values fall within the range corresponding to the claim of evidence of decay by H.V. Klapdor-Kleingrothaus and his co-workers [2]. Cuoricino is still collecting data and the sensitivity to half life should be twice the present limit when it will be stopped in coincidence with the start-up of CUORE, its natural continuation.

Looking at the background energy spectrum in the region of interest for DBD summed over all the channels, no peak appears at the $0\nu\beta\beta$ Q value; the upper limit, obtained with a Maximum Likelihood procedure, is evaluated using anticoincidence sum spectra and considering separately the sum spectra of big, small and ^{130}Te enriched crystals and distinguishing two runs; to account for the spread in the detector energy resolutions, the response function used in the fit is the sum of N gaussian (N being the number of detectors summed in the considered spectrum) centered at the same energy but with different sigmas.

4. From Cuoricino to CUORE

CUORE is the natural extension of the Cuoricino experiment [4]. It will be also located in the Laboratori Nazionali del Gran Sasso. It is a proposed tightly packed array of 988 TeO_2 bolometers, each being a cube of 5 cm on a side with a mass of 760 g. The array will consist of

19 vertical towers, arranged in a cylindrical structure. CUORE is a funded project expected to take data during 2011.

The temperature sensors will be NTD Ge thermistors and proper resistive heaters will be used to calibrate and stabilize the gain of the bolometers over long running periods. CUORE crystals will be grouped in elementary modules of 4 elements held between two copper frames joined by copper columns. PTFE pieces are inserted between the copper and TeO_2, as a heat impedance and to clamp the crystals. The bolometers will require an operational working point of ~7-10 mK that will imply an extremely powerful dilution refrigerator.

The background goal of CUORE is to achieve a counting rate in the range from 0.001 to 0.01 counts/keV/kg/y at the $0\nu\beta\beta$ transition energy of ^{130}Te (2530 keV). A low counting rate near threshold (that will be of the order of ~5 keV) is also foreseen and will allow to have results in the WIMP and solar axions research fields. Radioactive contaminations of individual construction materials, as well as the laboratory environment, were measured and the impact on detector performance determined by Monte Carlo computations.

Although background levels of the order of 0.001 counts/(keV kg y) seem to be viable thanks to an accurate selection of construction material and to the optimization of the surface treatment, a more conservative level of 0.01 counts/(keV kg y) can be also considered for CUORE sensitivity. A full Monte Carlo simulation of the CUORE background shows that all its sources are controlled at the level of 0.001 counts/(keV kg y), with the exception of *surface contamination*. The presently achieved level in the R&D tests for CUORE allows to predict a background level from 2 to 4 times higher than the conservative scenario of 0.01 counts/(keV kg y). Therefore some work is still to be done to reach even the *conservative* goal of CUORE, and this work regards solely the issue of surface contamination. The radical solution of this problem would have a dramatic impact on the physical reach of CUORE. In fact, a background level of 0.01 counts/(keV kg y) would allow just to touch the inverted hierarchy region of the neutrino mass pattern, while a factor 10 lower would allow to explore deeply this region, with an obvious increase of the CUORE discovery potential. This explains why the struggle against surface background is presently the central issue in the CUORE preparation work.

5. Conclusions

Cuoricino is the most sensitive running experiment on $0\nu\beta\beta$; while being a self consistent experiment, it is also a fundamental test for the next generation CUORE experiment for what concerns detector performances and background value in the $0\nu\beta\beta$ region. For a background value in the $0\nu\beta\beta$ region of the order of 0.001 c/keV/kg/y CUORE could reach a sensitivity for the effective neutrino mass in the range 15-50 meV in 5 years, just in the region favoured by current oscillation experiments for the inverted hierarchy.

Acknowledgments

The Cuoricino Collaboration owes many thanks to the Director and Staff of the Laboratori Nazionali del Gran Sasso. The experiment was supported by the Istituto Nazionale di Fisica Nucleare (INFN), the European Commission under Contract No. HPRN-CT-2002-00322 and Contract No. RII3-CT-2004-506222 (ILIAS Integrating Activity), by the U.S. Department of Energy under Contract No. DE-AC03-76-SF00098, and DOE W-7405-Eng-48, and by the National Science Foundation Grants Nos. PHY-0139294 and PHY-0500337.

References

[1] S. R. Elliott and P. Vogel, *Ann. Rev. Nucl. Part. Sci.* 52, 115 (2002).
[2] H. V. Klapdor-Kleingrothaus, I. V. Krivosheina, A. Dietz, O. Chkvorets, *Phys. Lett.* **B** 586, 198 (2004).
[3] C. Arnaboldi et al., *Phys. Rev. Lett.* 95, 142501 (2005).
[4] R. Ardito et al., *hep/ex 0501010*.

Status of the GERDA experiment

Hardy Simgen
on behalf of the GERDA collaboration
Max-Planck-Institut für Kernphysik, Saupfercheckweg 1, D-69117 Heidelberg / Germany
E-mail: H.Simgen@mpi-hd.mpg.de

Abstract.
The GERmanium Detector Array GERDA is designed to search for neutrinoless double beta decay of ^{76}Ge. This search is necessary to establish the nature of the neutrino (Dirac or Majorana) and is emphasized by the evidence of a non-zero neutrino mass from flavour oscillations and by the claim for a positive signal based on data of the Heidelberg-Moscow experiment. GERDA will be installed in the Gran Sasso underground laboratory of INFN/Italy. The experiment is designed to collect at the end of phase II an exposure of about 100 kg·y quasi background free. This leads to a requirement of a background index of the order of 10^{-3} counts/(kg·keV·y) at the $Q_{\beta\beta}$–value of 2039 keV.

1. Introduction

The GERmanium Detector Array GERDA [1] is a new double beta decay experiment using ^{76}Ge. To achieve better sensitivity a significant reduction of background compared to previous experiments is required. The main concept for background suppression followed by GERDA is the operation of bare Ge-diodes enriched in ^{76}Ge in ultrapure liquid argon. Liquid argon acts as passive shield and cooling medium for the diodes. It is contained in a stainless steel vacuum cryostat. Additional shield against external neutrons and gamma-rays is provided by pure water surrounding the cryostat. The water tank is equipped with photo multiplier tubes and operated as a Cherenkov detector to reject residual cosmic muons. Enriched ^{76}Ge-diodes will be prepared in a cleanroom and inserted into a lock system from where they are lowered inside the cryostat.

2. Sensitivity

The GERDA experiment is performed in different phases: In phase I 17.9 kg of existing ^{76}Ge diodes from the previous Heidelberg-Moscow and IGEX experiments will be re-used. At the expected background rate of 10^{-2} counts/(kg·keV·y) around the Q-value of the ^{76}Ge decay (2039 keV) the resulting sensitivity for the half-life of the neutrinoless double beta decay is $2 \cdot 10^{25}$ y after 1 year of exposure. This is sufficient to scrutinize the existing claim for a positive signal [2]. In phase II new diodes will be added to increase the active mass up to 40 kg. A major progress in sensitivity can only be obtained if the background can be suppressed to 10^{-3} counts/(kg·keV·y). This will be achieved by segmented detector readout, minimization of cosmic ray exposure, pulse shape discrimination and careful selection of construction materials. The expected sensitivity lies around 130 meV for the effective neutrino mass at an exposure of 100 kg·y. A third phase aiming to the inverted neutrino mass hierarchy regime would require about 1 ton of ^{76}Ge target material as well as a further reduction of background.

3. Status of selected sub-projects

3.1. Liquid argon purity

Liquid argon will be in direct contact to the Germanium diodes. Thus, it has to fulfill stringent purity requirements. The most worrisome radio-active contamination in argon is ^{222}Rn and in particular its progeny ^{214}Bi. Radon can be removed from argon by cryogenic adsorption on activated carbons. In a series of measurements it was found that a concentration reduction of more than a factor 2500 can be achieved with 1 kg of activated carbon if the purification happens in gas phase. For liquid phase purification which is relevant for GERDA the efficiency is about one order of magnitude worse. However, due to the relatively short half-life of ^{222}Rn a reduction can also be obtained by storing liquid argon in a clean storage tank. For this reason the ^{222}Rn emanation rate of cryogenic storage tanks was investigated. Although many tanks had rather high ^{222}Rn emanation rates a few sufficiently clean tanks could be identified. One of them with a volume of 6.3 m^3 and a ^{222}Rn activity of (3.5 $\pm$ 0.2) mBq in saturation might be available for the GERDA experiment.

3.2. Cryostat

To achieve the ambitious background goals also the GERDA stainless steel cryostat must fulfill stringent radio-purity requirements. In particular, the high energy gamma lines from ^{208}Tl and ^{214}Bi from U/Th contaminations must be suppressed sufficiently. This is achieved by a careful selection of radio-pure stainless steel and the installation of copper plates inside the inner wall of the cryostat. More than 10 batches of stainless steel type 1.4571 were screened with low background Germanium spectrometers. The activities found for ^{226}Ra/^{228}Th lie in most cases below 1 mBq/kg [3]. This is significantly lower than what was expected from results of previous stainless steel screening campaigns [4]. As a consequence the required amount of copper could be reduced by more than a factor 2 which relaxed mechanical constraints and costs significantly. The cryostat is under construction now and is planned to be delivered to Gran Sasso beginning of 2008.

3.3. GERDA detector laboratory

The GERDA detector laboratory is located in close neighborhood to the main GERDA site in hall A of the Gran Sasso underground laboratory. It is a facility offering all equipment for handling and manipulations of bare Germanium diodes like a chemical hood and clean benches. One of the clean benches is air tight so that a ^{222}Rn-free atmosphere can be provided. Diodes mounted there can directly be tested in an attached liquid argon test stand without being exposed to air. The GERDA detector laboratory also hosts a larger (1 m^3) liquid argon test stand called LArGe. Besides detector testing it will be used for liquid argon scintillation light studies. LArGe is under construction and will be operational in 2008.

3.4. Preparation of existing diodes

For phase I of GERDA existing Germanium diodes enriched in ^{76}Ge from previous experiments are used. Altogether 8 diodes with a mass of 17.9 kg are available corresponding to $\sim$ 15 kg of pure ^{76}Ge. After a careful characterization of the diodes they have been removed from their cryostats. Then their dimensions were measured and they were stored in an underground location (HADES) in Belgium. Together with Canberra / Olen in Belgium they were prepared for the operation without vacuum cryostat. A refurbishment including a loss of a few grams of ^{76}Ge was tolerated for some of the diodes which had a different contact scheme. Now all diodes have the same well tested and well understood contact scheme which increases the confidence of save operations in liquid argon.

3.5. Prototype testing for phase I

For the last 2 years intensive tests with a phase I prototype crystal have been performed. It is mounted in a low mass ($\sim$ 80 g) support structure made out of copper and PTFE. Both materials were screened by gamma ray spectroscopy and only upper limits for U/Th contaminations were found (in the range of 20 μBq/kg for copper and 100 μBq/kg for PTFE). Monte Carlo-simulations have shown that the background from the material of the detector holder fulfills the requirements for phase I. Handling of bare diodes is very delicate, because the passivation layer is very sensitive. In particular, it is important to be able to warm up and cool down the diodes without damaging them. Meanwhile, more than 40 cooling and warming cycles have been performed and the passivation layer had to be renewed only twice. This shows that the detector handling process is well under control. Also the response of the diode's leakage current to irradiation with gamma ray sources was investigated. A reversible increase of the leakage current was observed which was strongest when the passivation layer was irradiated directly. Covering this side of the diode with a PTFE/Cu/PTFE sandwich disk solved the problem. A several months lasting long-term test showed only a negligible increase of leakage current. Further improvements are under development.

3.6. Preparations for phase II

In 2005 37.5 kg of enriched ^{76}Ge was produced for the phase II diodes. Currently, techniques are being optimized to reach the chemical purity necessary for crystal pulling and to improve the yield and thus to minimize the loss of enriched Germanium. In parallel different options for crystal pulling are investigated. Until the preparation of the next steps is finished the material is stored in an underground location to avoid cosmogenic activation. The phase II detectors will be n-type detectors with segmented readout electrodes. The background suppression by segmented detectors is currently tested with several test setups. Promising results were obtained with the chosen segmentation scheme of 3x6 segments (3 along z-coordinate and 6 azimuthal segments).

4. Outlook

The GERDA experiment will become operational in 2009. After 1 year of data taking with the existing diodes the claim by [2] for a $0\nu\beta\beta$-decay will be checked unambiguously. If it would not be confirmed even after phase II the goal must be to do a $\sim$ 1 ton experiment to access the range of inverted neutrino mass hierarchy. Since this can only be afforded in a world-wide collaboration a close contact with the Majorana project has already been established aiming to a common large-scale experiment in the future.

[1] I. Abt et al., Proposal to LNGS (2004) http://www.mpi-hd.mpg.de/ge76/proposal.pdf.

[2] H.V. Klapdor-Kleingrothaus et al., *Data acquisition and analysis of the ^{76}Ge double beta experiment in Gran Sasso 1990–2003*, Nucl. Inst. and Meth. A 522 (2004) 371406.

[3] W. Maneschg et al., *Measurements of extremely low radioactivity levels in stainless steel for* GERDA, in preparation.

[4] BOREXINO collaboration, C. Arpesella et al., *Measurements of extremely low radioactivity levels in* BOREXINO, Astrop. Phys., 18 (2002) 1-25.

Study of Double Beta Decay of ^{48}Ca with CANDLES

Y.Hirano[a], T.Kishimoto[a], I.Ogawa[a], R.Hazama[b], S.Umehara[a], K.Matsuoka[a], G.Ito[a], Y.Tsubota[a] For the CANDLES Collaboration

[a]Graduate School of Science, Osaka University, Toyonaka, Osaka 560-0043, Japan
[b]Graduate School of Engineering, Hiroshima University, Higashi-Hiroshima, Hiroshima, 739-8527, Japan

E-mail: hirano@km.phys.sci.osaka-u.ac.jp

Abstract. CANDLES system had proposed for the study of double beta decay of ^{48}Ca by using of undoped CaF_2 scintillator. We have constructed CANDLES III detector at sea level. The result of first performance check is described.

1. Introduction

The study of neutrinoless double beta decay is of most importance for neutrino physics. We have been studying the double beta decay of ^{48}Ca. The most advantage of using ^{48}Ca isotope is background of natural radioactivities are strongly limited at the Q-value region. Because the Q-value which is the highest (4.27MeV) among double beta isotopes, is higher than natural radioactivities. we have developed the CaF_2(Eu) scintillation detector system (ELEGANT VI) which have being operated at the underground laboratory (Oto Cosmo Observatory). the most stringent limit on the half-life of $0\nu\beta\beta$ of ^{48}Ca has been obtained [1].

In order to enhance the sensitivity for the neutrino effective mass, we must scale up the detector. However, CaF_2(Eu) is not suitable for large detector because of the short attenuation length of scintillation lights, Then we proposed CANDLES (Calcium fluoride for the study of Neutrinos and Dark matters by Low Energy Spectrometer)

2. CANDLES system

CANDLES system is suitable for a large detector and powerful 4π active shielding system with use of updoped CaF_2 scintillators and liquid scintillator [2]. The large number of CaF_2s with a dimension of 10 cm cube are immersed in liquid scintillator. Scintillation lights from both CaF_2 and liquid scintillator are viewed by the large PMTs. The outside of the liquid scintillator vessel filled with pure water for the passive shield. The CaF_2 scintillator has a decay time of 1μ sec, on the other hand that of liquid scintillator is a few tens nsec. Thus by observing pulse shape the signals from liquid scintillator (external background) can be discriminated.

We constructed prototype detector CANDLES III in our laboratory at sea level, which consist of 40 PMTs (15inch$\times$8, 13inch$\times$32) and 60 CaF_2s with the total mass of 191kg. Twelve CaF_2s are located with the 20 cm distance from the center in a layer. Total five layers are supported by the wires inside a liquid scintillator vessel of 1 m in diameter and 1 m in height. These are contained in a water tank, which is 2.8 m in diameter and 2.6 m in height. The schematic view of the CANDLES III is shown by Figure 1.

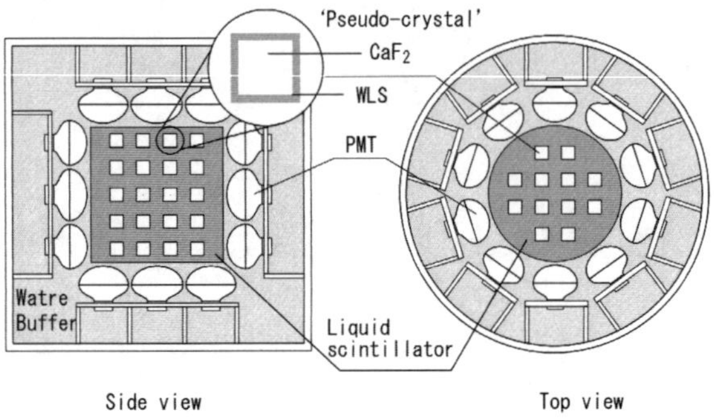

Figure 1. Schematic cross sectional view of the CANDLES-III detector. we constructed at sea level.

3. Performance test

The Data taking for the performance test started, and present result from the measurement of total live time of 30 hours. The total number of CaF_2s is 59 and the one of the crystals has much radioactive contamination. The contamination of the dirty one is equivalent to 46mBq/kg. The average of the other crystals contamination is about 42μBq/kg. 35 PMTs are used ,which correspond to 30% photo-coverage. In order to discriminate liquid scintillator pulse shape from CaF_2 scintillator pulse shape, different length of charge sensitive ADC gate (*total gate* and *prompt short gate*are introduced to each PMT. The demonstration of the pulse shape discrimination is shown in Figure 2. The ratio defined by ADC value of (*prompt short gate*)/(*total gate*),

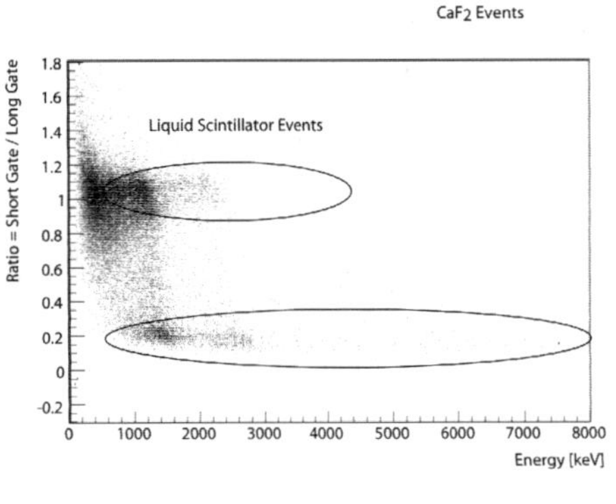

Figure 2. The pulse shape discrimination. the events distributed around 0.15 on vertical axis from CaF_2 scintillator, 1.0, from liquid scitillator, respectively. The events are selected around the CaF_2No.59

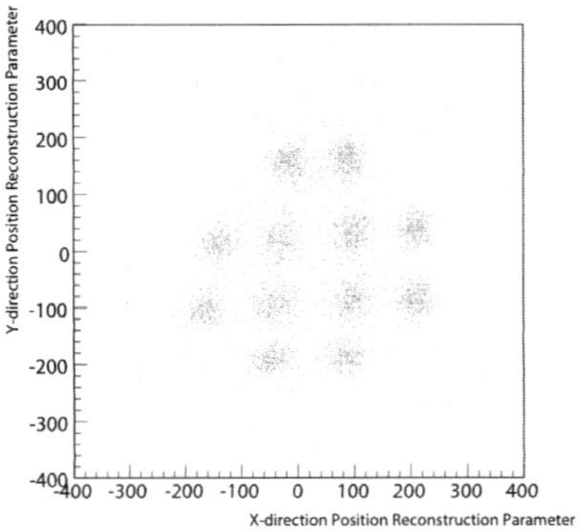

Figure 3. Two dimensional plots of position reconstruction parameters at the third layer. The events are higher than 2MeV and CaF_2-like events from the value of the ratio.

One of the important subject of the detector performance is to identify the crystals where scintillation lights are emitted. This performance useful to the rejection of background origin of decay of ^{208}Tl, which can be rejected by identifying preceding α decay in the same crystal. For the preliminarily analysis, We use the position reconstruction parameter defined by as follows; R=$\sum_{i=1}^{npmt}$ADC(i)$\times$PMT(i)/$\sum_{i=1}^{nmpt}$ADC(i), where *npmt* is ID of PMT, ADC(i) is observed pulse hight in i-th PMT, and PMT(i) is the coordinates of i-th PMT. The Result of position reconstruction is showd by Figure 3. The events are higher than 2MeV and CaF_2-like events from the result of the pulse shape discrimination. The identification of each crystal was found to be possible, though this algorithm for the reconstruction will be improved.

The Energy resolution influences the rejection of $2\nu\beta\beta$ decay events. In order to estimate the energy resolution of the detector system, we used α events cased by decay of ^{214}Po (U-Chain). Considering the quenching factor, the peak energy of the α events is 2.3MeV. The obtained resolution is 9%, which is correspond to 6.6% at Q-value region.

The obtained energy spectra of crystal number 59 and 60 are shown in Figure 4. No.60 is a dirty crystal. These are located at bottom layer and next to each other. In the Q-Value region (4.0~4.5MeV) of the dirty crystal spectrum, 8.6 events/hour was obtained. Since the contamination of clean crystals are three order of magnitude less than the dirty one, $\sim 8\times10^{-3}$ events/hour is expected for each clean crystal. However 3 events were remained in the Q-value region of the clean crystal No.59. At the current understanding, these events attribute to cosmic rays. muons which are most part of cosmic rays are almost rejected by active shield. However μ^{-}s lost the kinetic energy and captured by a nucleus, there can be emitting high energy gammas. In addition, an inelastic scattering process by environmental fast neutrons or a bremsstrahlung process by muons, these process have possibility of the background. we are investigating the influences of cosmic rays by using of the simulation or same experiments. In case that these remaining several events in the clear crystal are cased by the cosmic rays, these events will be negligible moving to the underground laboratory.

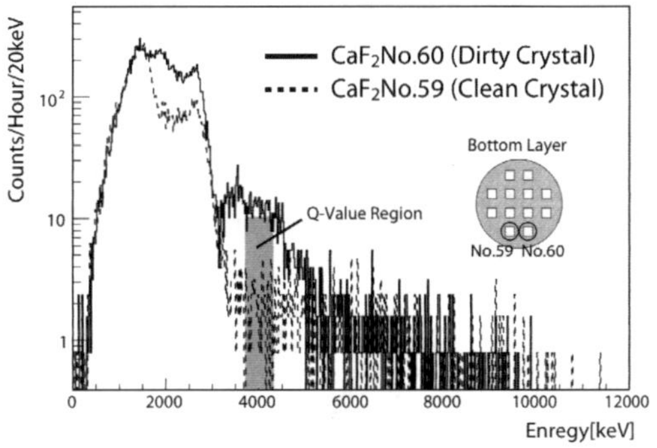

Figure 4. Energy Spectra of CaF_2No.59 and No.60. These are located at the bottom layer.

4. Discussion

We evaluated the performance of the CANDLES III at sea level. We obtained clear separation of generation source of the CaF_2 scintillation lights. Comparing the spectrum of the dirty crystal to that of clean crystal, $\sim 10^{-3}$ events/hour to each crystal is expected. Moreover pulse shape discrimination by use of the FADC, internal background can be rejected. The rejection efficiency is expected about 99%. The performance check is under the investigation using the 500MHz FADC. We have plan to move CANDLES III, which will be somewhat improved to the underground laboratory next summer (2008). Introducing the 500MHz FADC and moving to the underground laboratory, almost all zero background measurement is expected.

References

[1] I.Ogawa*et al*, 2003 *Nucl. Phys.* A **721** 525c-528c
[2] S.Yoshida *et al*, 2005 *Nucl. Phys. (Proc.Suppl.)* B **138** 214-216

SuperNEMO project

Hideaki Ohsumi* **on behalf of NEMO and SuperNEMO Collaboration**
*Saga Univesity, Saga, 840-8502, Japan
E-mail: `ohsumi@cc.saga-u.ac.jp`

Abstract.
The neutrinoless double beta decay ($\beta\beta0\nu$) is the most sensitive process for the search of lepton number violation and its discovery would prove the neutrino is a massive Majorana particle. Based on the experience of the NEMO 3 detector construction and its data analysis, the SuperNEMO collaboration has been formed and started the feasibility study to construct next stage detector for the $\beta\beta$ experiment with a mass of at least 100 kg enriched $\beta\beta$ isotpes. The goal of the sensitivity is to reach $\beta\beta0\nu$ half-life limit of a few 10^{26} years and 50 meV of the effective Majorana neutrino mass.

1. Introduction

The absolute neutrino mass scale is recognized as one of the most fundamental questions to be solved experimentally. Recent experimental results of neutrino oscillations suggest important information related to the small absolute neutrino mass scale. However neutrino oscillations are only sensitive to the squared mass difference between two flavor mass eigenstates. Therefore neutrinoless double beta decay ($\beta\beta0\nu$), which process is beyond the standard model with lepton number violation, is the only practical way to do the most precise laboratory measurement of absolute Majorana neutrino mass.

The present experiments, for example NEMO 3 [1] and CUORICINO, are searching for neutrino mass in Sub-eV region. On the other hand, recent analysis of neutrino oscillations combined with the inverted neutrino mass hierarchy hypothesis strongly suggests the ~50 meV effective $< m_{\beta\beta} >$ region. So the sensitivity of the next stage experiment should be improved one order of magnitude.

Experimental techniques of $\beta\beta$ decay measurements are mainly divided by two methods. One is experiment using pure calorimeter, as Ge semi-conductor diodes or bolometer detector. Here the $\beta\beta$ source is used as the detector and the measured quantity is the total deposited energy. Second technique is "tracko-calo" detector. It is not only the summed energy measurement but also available tracking reconstruction technique to identify the two emitted electrons.

The main advantage of pure calorimeter experiment is the good energy resolution and detection efficiency. The disadvantage is that there is no direct signature of the two electrons and the origins of the backgroud signals are more dificult to identify. On the contrary, "tracko-calo" experiment has less energy resolution but the capability to identify the two electron emissions and to reject background signals with higher efficiency than calorimeter experiment. It also measure individual energy for each of the two electrons and their angular distribution, which allows distinguishing $\beta\beta0\nu$ decay mechanisms in detail. Another advantage is that the detector

Table 1. From NEMO 3 to SuperNEMO

NEMO 3		SuperNEMO
Mo $T_{/}(\beta\beta2\nu) \sim 7{\cdot}10$ y	Choice of isotope	Se (or ^{5}Nd) $T_{/}(\beta\beta2\nu) \sim 10$ y
7 kg	Isotope Mass	100 - 200 kg
$\epsilon(\beta\beta0\nu) \sim 8\%$	Efficiency	$\epsilon(\beta\beta0\nu) \sim 30\%$
Bi $<$ 300 μBq/kg Tl $<$ 20 μBq/kg (Tl, Bi) $\sim$ 1 evt/7kg/y $\beta\beta2\nu \sim$ 2 evt/7kg/y	Background	Bi $<$ 10 μBq/kg Tl $<$ 2 μBq/kg (Tl, Bi) $\sim$ 1 evt/100 kg/y $\beta\beta2\nu \sim$ 1 evt/100 kg/y
$\sim$ 8% @ 3 MeV (FWHM)	Energy Resolution	$\geq$ 4% @ 3 MeV (FWHM)
$T_{/}(\beta\beta0\nu) > 2{\cdot}10$ y $< m_\nu > <$ 0.3 - 1.3 eV	Sensitivity	$T_{/}(\beta\beta0\nu) > 10$ y $< m_\nu > <$ 50 meV

is independent from $\beta\beta$ source foils, which allows to select best $\beta\beta$ isotope or to measure different $\beta\beta$ isotopes at the same time.

Thus, based on the good experience of the typical "tracko-calo" detector of NEMO 3 detector, the SuperNEMO collabollation has been formed and started the feasibility study to construct next stage detector with a mass of at least 100kg of enriched $\beta\beta$ isotope in order to reach the sensitivity of $\beta\beta0\nu$ half-life limit a few 10 years and 50 meV of the effective Majorana neutrino mass.

2. From NEMO 3 to SuperNEMO

The performance characteristics to improve from NEMO 3 to SuperNEMO are summarized in Table 1.

In spite of the recent progress in the calculation methods for the nuclear matrix elements, these calculations are always too uncertain and nuclear theory is not yet able to predict the best candidate from the point of view of nuclear matrix element. In order to reduce the background coming from the $\beta\beta2\nu$ decay, a good criterion for isotope selection would be to choose an isotope with a half-life $T_{/}(2\nu)$ longer than 10 years and higher $Q_{\beta\beta}$ value. Higher $Q_{\beta\beta}$ value gives larger phase space factor and exceeds crucial energies of β and γ decays of U and Th daugter nuclei, which are major background sources to the $\beta\beta0\nu$ experiment.

Se is one of the best candidate with $T_{/}(2\nu) \sim 10$ y and $Q_{\beta\beta} = 2.992$ MeV. Using the actual enrichment factories, the production of an order of 100 kg is feasible in a few years with a reasonable cost. Already R&D program of the production of 2 kg source for the study of radiopurity, chmical purification, souce foil making process etc. is in progress.

There is an interesting isotope candidate ^{5}Nd, which has 8 times larger phase space factor than Se and enough $Q_{\beta\beta}$ value (3.368 MeV). However, it is very difficult to enrich an order of 100 kg quantities by known methods including a conventional centrifugal method. Recently an alternative method of atomic vapor laser ion separation (AVLIS) technology is discussed to be effective for ^{5}Nd isotope enrichment and the possibilities of the reuse of the MENPHIS facility in France is considered. If the facility will be available, ^{5}Nd will be another good candidate of the SuperNEMO $\beta\beta$ source.

The contribution from $\beta\beta2\nu$ high energy tail to the concerned peak region of $\beta\beta0\nu$ is directly related to the detector energy resolution. The SuperNEMO detector is required to the calorimeter resolution better than 4% (FWHM) at 3 MeV ($\beta\beta0\nu$ peak region) to achieve the sensitivity goal. The collaboration has been studying various possibilities to keep the NEMO 3 techniques but improving plastic scintillator (including the use of liquid scintillator option etc.), PMT properties and so on.

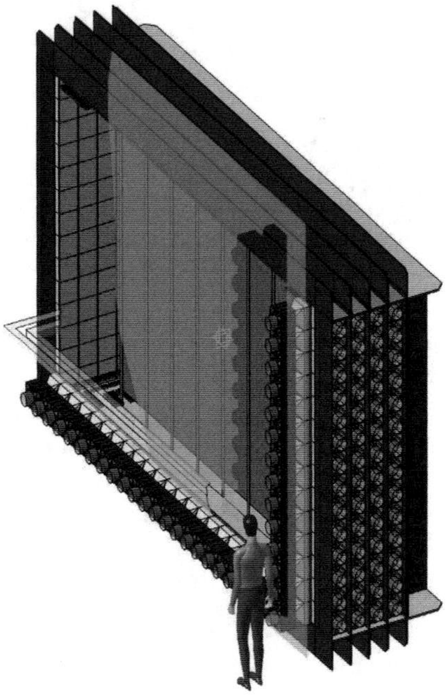

Figure 1. View of a possible one SuperNEMO module

To reach sensitivity goal, radiopurity levels of the source foils of Bi and Tl have to be lower than 10μ Bq/kg and 2μ Bq/kg, which means reduction factor 10 to the present NEMO 3 experiment. It is necessary to develope new methods to measure radiopurity levels for Bi and Tl as low as 10μ Bq/kg and 2μ Bq/kg, respectively. "Bi-Po" detector project has been developing to the detection before source installations using delayed coincidence of Bi-Po sequential decays related series of the radioactive decay.

The R&D program of the tracking detector is also important to improve its transparency by optimizing the length, diameter, wire materials etc. to keep the better efficiency.

The first start point of the design of the SuperNEMO detector would be composed of several identical modules for example 20 modules with 5 kg $\beta\beta$ source. The size of the source plane would be, for example, 4 m long, 3 m high and 40 mg/cm (see fig. 1).

After three years various R&D programs to achieve the requirements from NEMO 3 to SuperNEMO, a detailed technical design proposal for SuperNEMO will be written by the end of 2008.

References

[1] Arnold R et al., 2005 *Nucl. Instr. Meth.* A **536** (2005) 79, Arnold R et al., 2005 *Phys. Rev. Lett.* **95** 182302

MOON for next-generation neutrino-less double-beta decay experiment; present status and perspective

T. Shima[1], P.J. Doe[2], H. Ejiri[1,3,4], S.R. Elliot[2,5], J. Engel[6], M. Finger[7], M. Finger, Jr.[7], K. Fushimi[8], V.M. Gehman[2,5], M.B. Greenfield[9], R. Hazama[10], H. Imaseki[3], P. Kavitov[11], V.D. Kekelidze[12], H. Kitamura[3], K. Matsuoka[13], H. Nakamura[3], M. Nomachi[13], A. Para[14], R.G.H. Robertson[2], M. Slunecka[6], G.D. Shirkov[12], A.N. Sissakian[12], A.I. Titov[12], Y. Uchihori[3], S. Umehara[15], V. Vaturin[11], V.V. Voronov[12], J.F. Wilkerson[2], D.I. Will[2], K. Yasuda[8], and S. Yoshida[16]

[1]Research Center for Nuclear Physics, Osaka University, Ibaraki, Osaka 567-0047
[2]CENPA, University Washington, Seattle, WA 98195, USA
[3]NIRS, National Institute of Radiological Sciences, Chiba 263-8555
[4]Nuclear Science, Czech Technical University, Prague, Czech Republic
[5]4LANL, P.O.Box 1663, MSH 803, Las Alamos, NM 87545, USA
[6]Physics and Astronomy, University of North Carolina, Chapel Hill, NC 27599, USA
[7]Charles University, FMP, CZ-18000 Prague 8, Czech Republic
[8]IAS, University of Tokushima, Tokushima 770-8592
[9]Physics, International Christian University, Tokyo 181-8585
[10]Hiroshima University, Higashi Hiroshima, Hiroshima 739-8527
[11]VNIIEF, 607188, Sarov, Nizhny Novgorod Region, Mira Ave, 37, Russia
[12]Joint Institute for Nuclear Research, 141980 Dubna, Russia
[13]OULNS, Osaka University, Toyonaka, Osaka 560-0043
[14]FNAL, P.O.Box, 500, Batavia, IL 60510-0500, USA
[15]Dept. Physics, Osaka University, Toyonaka, Osaka 560-0043
[16]RCNS, Tohoku University, Sendai 980-8578

E-mail: `shima@rcnp.osaka-u.ac.jp`

Abstract. The performance of the MOON detector for a next-generation neutrino-less double-beta decay experiment was evaluated by means of the Monte Carlo method. The MOON detector was found to be a feasible solution for the future experiment to search for the Majorana neutrino mass in the range of 100−30 meV.

1. Introduction

Neutrino-less double-beta decays ($0\nu\beta\beta$) provide a unique probe for lepton-number violating weak processes by ΔL =2, the Majorana nature of the neutrino, the absolute scale and the ordering of the neutrino masses, possible right-handed weak currents, and so on. The effective neutrino masses to be studied by $0\nu\beta\beta$ are around 100 meV and 30 meV in cases of the quasi-degenerate (QD) and the inverted-hierarchy (IH) mass spectra, respectively[1]. To search for $0\nu\beta\beta$ in QD$\sim$IH mass ranges, one has to use $\beta\beta$ isotopes of the order of 0.1$\sim$1 t (tonne) even in cases of such favorable $\beta\beta$ nuclides as ^{82}Se, ^{100}Mo and ^{150}Nd with large phase volumes.

MOON (Mo/Majorana Observatory Of Neutrinos) has been proposed for a next-generation spectroscopic $0\nu\beta\beta$ experiment[2-6]. This is a brief report on the expected background and the neutrino mass sensitivity of MOON.

2. Outline of MOON Detector

The MOON detector is a stack of multi-layer detector modules, each consists of a scintillator plate for measuring energy and time, two thin detector layers for position and particle identifications, and a thin $\beta\beta$ source film interleaved between them. At present, plastic scintillators and NaI(Tl) scintillators are considered as the candidates for the scintillator plates. For position-sensitive detectors, possible candidates are multi-wire proportional chambers (MWPCs) and Si-strip detectors. All detector components are surrounded by passive shields against external γ-rays and neutrons. Since the $\beta\beta$ source is separated from the detector components, one can measure the decay rate as well as the energy and angular correlations of the two β-rays. This spectroscopic method also enables one to obtain extremely good signal-to-noise ratio by requiring the following conditions for the $0\nu\beta\beta$ event identification;
(a) two adjacent scintillators provide signals with energy losses greater than 500 keV,
(b) no γ-ray or a γ-ray from the excited states of a daughter nucleus is observed in coincidence,
(c) two β-rays are emitted from the same position on a $\beta\beta$ source film, and the tracks of those β-rays are detected by the position-sensitive detector layers on both sides of the source film, and
(d) sum energy of the two β-rays is in the $0\nu\beta\beta$ energy window.

The performance of the MOON detector with plastic scintillators and MWPCs was estimated using the Monte Carlo simulations for various background sources as described below.

3. Backgrounds

The backgrounds due to radioactive isotopes are expected to be dominated by ^{208}Tl and ^{214}Bi from the impurities of Th and U in the $\beta\beta$ source films because of their large Q-values. By assuming realistic contamination levels of 20 μBq/kg and 300 μBq/kg for ^{208}Tl and ^{214}Bi, respectively[9], the background rates in the optimized $0\nu\beta\beta$ window of 2900–3200 keV are estimated as 0.6 counts/t/y (counts per 1 tonne of isotopes for 1 year exposure) and 1.5 counts/t/y, respectively. Fig. 1 shows a calculated sum energy spectra for the β-γ(2.615MeV)-γ(0.583MeV) decay of ^{208}Tl in the source film.

Another main source of the external background is the γ-rays produced by the neutrons due to weak capture reactions of cosmic muons. The energy spectrum of the μ-induced neutrons in an underground laboratory was calculated at a 2000 m water-equivalent location[7]. It is dominated by fast neutrons with the energy up to ~20 MeV. In case of the MOON detector with multi-layer plastic scintillaors, the neutron-induced background is dominated by the inelastic scattering of fast neutrons from ^{12}C in the plastic scintillators, which can be reduced by a factor of $\sim 10^{-4}$ with a 2 m thick water shield. The production rate of the inelastic γ-rays from ^{12}C is about 10^{-6} photons/t/s, and its contribution to the $0\nu\beta\beta$ window amounts to ~0.3 counts/t/y.

As for the intrinsic background due to the high energy tail of the $2\nu\beta\beta$ spectrum in the $0\nu\beta\beta$ window, their counting rates are estimated to be 6.8 counts/t/y, 3.4 counts/t/y, 1.9 counts/t/y for ^{82}Se and 19 counts/t/y, 5.4 counts/ty, 4.6 counts/t/y for ^{100}Mo, depending on the energy resolutions of σ = 2.9%, 2.2%, and 1.7% at 3 MeV. They are much reduced if the energy resolution of the β-ray detectors is improved from the presently achieved value of σ = 2.9%[8].

4. Sensitivity for $0\nu\beta\beta$

The sensitivity for the effective Majorana neutrino mass was evaluated for ^{82}Se and ^{100}Mo by using the backgrounds discussed above. In this estimation a typical thickness of the $\beta\beta$ source film of 30 mg/cm^2 and the energy resolutions of the plastic scintillator of σ = 2.9%, 2.2%, and

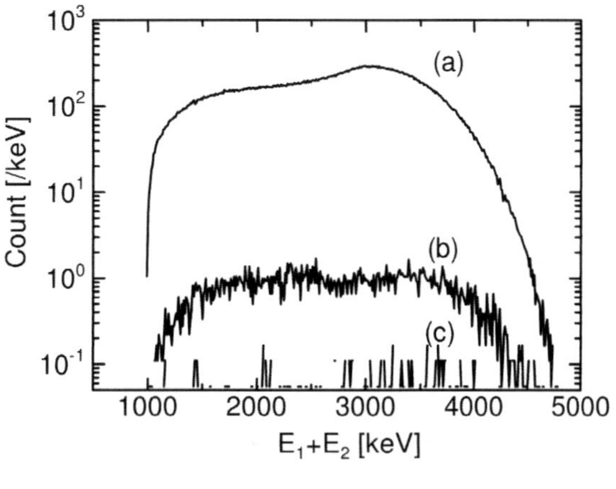

Figure 1. Sum-energy spectra of a pair of adjacent plastic scintillators for β-γ(2.615MeV)-γ(0.583MeV) decay of ^{208}Tl contained in the $\beta\beta$ source film. The curves (a)-(c) are obtained for 10^7 decays with the conditions (a)-(c) described in the text for event selections.

1.7% were employed. Adopting the nuclear matrix elements $M^{0\nu} = 3$ for both ^{82}Se and ^{100}Mo, the mass sensitivities with 90% confidential level for 2 t·y exposure are estimated as 70 meV, 57 meV, and 52 meV for ^{82}Se, and 86 meV, 64 meV, and 56 meV for ^{100}Mo, depending on the energy resolutions of $\sigma = 2.9\%$, 2.2%, and 1.7% at 3 MeV, respectively.

5. Conclusion

MOON is expected to be a realistic solution for exploring $0\nu\beta\beta$ in the effective neutrino mass regions from QD to IH, because of its high sensitivity due to the good resolutions for energies and positions of the β-rays as well as its ability to be expanded from a few ten kilograms to multi tons in the scale of the $\beta\beta$ source quantity.

The background at the $0\nu\beta\beta$ window is expected to be dominated by $2\nu\beta\beta$, and therefore it is crucial for high sensitivity experiment to use β-ray detectors with good energy resolutions. The resolution of $\sigma = 2.9\%$ is fine for the QD mass studies, but $\sigma = 2.2\%$ and 1.7% are required for the IH mass studies. Improvement of the energy resolution is now in progress.

MOON has capabilities to measure various source nuclei, the energy and angular correlations of β- and γ-rays, and the transitions to both the ground and excited states of the daughter nucleus. These unique features of the MOON detector are useful for identifying the light Majorana neutrino process and also for measuring low-energy solar neutrinos in real-time[10].

6. References

[1] S. Pascoli, S.T. Petcov, and T. Schwetz, Nucl. Phys. B734 (2006) 24-49.
[2] H. Ejiri, Mod. Phys. Lett. A22 (2007) 1277-1291.
[3] H. Ejiri, Czech. J. Phys. 56 (2006) 459-466.
[4] H. Ejiri, J. Engel, R. Hazama, P. Krastev, N. Kudomi, and R.G.H. Robertson, Phys. Rev. Lett. 85 (2000) 2917-2920.
[5] M. Nomachi *et al.*, Nucl. Phys. B, Proc.Suppl.138 (2005) 221-223.
[6] P. Doe *et al.*, Nucl. Phys. A721 (2003) 517-520.
[7] R.J. Gaitskell, 3rd Int. Workshop on Identification of Dark Matter (World Scientific, September 2001), Eds. N.J.C. Spooner and V. Kudryavtsev. arXiv: astro-ph/0106200
[8] H. Nakamura *et al.*, J. Phys. Soc. Japan 76 (2007) 114201.
[9] R. Arnold *et al.*, Nucl. Instr. Meth. in Phys. Res. A536 (2005) 79-122.
[10] R. Hazama *et al.*, Nucl. Phys. B138, Proc. Suppl. (2005) 102-105.

The Enriched Xenon Observatory

K. O'Sullivan on behalf of the EXO Collaboration
Physics Department, Stanford University, Stanford CA,USA
E-mail: kosok@stanford.edu

Abstract. The Enriched Xenon Observatory (EXO) is a program designed to search for the $0\nu\beta\beta$-decay of ^{136}Xe. EXO-200, a 200 kg enriched Xe TPC, is nearing completion, with data taking expected to begin by the end of 2008. The cryogenics and fluid handling systems are being installed in its underground location at WIPP. The EXO collaboration is also pursuing in parallel the development of a technique to identify the production of the Ba daughter ion of the decay, which would allow for an extreme background reduction.

1. Introduction

LXe offers the possibility of a novel background rejection technique in a $0\nu\beta\beta$-decay search by identifying the $^{136}Ba^{++}$ daughter particle of the decay. In the LXe this ion will capture an electron and become $^{136}Ba^{+}$, which can be identified using laser induced fluorescence spectroscopy, as first suggested by Moe [1]. If one can transfer the $^{136}Ba^{+}$ daughter to an ion trap, where spectroscopy can take place, one can positively identify the event as a $\beta\beta$-decay; the goal of EXO is to integrate this rejection technique into a ton-scale Xe TPC. This requires achieving two goals: determining how to build a large, low-background LXe TPC with good energy resolution, and how to identify Ba ions from the LXe. Two paths will proceed in parallel: a $0\nu\beta\beta$ search with LXe enriched to 80% in ^{136}Xe, which will not use this Ba tagging event rejection, called EXO-200 is currently under construction while long term R&D on ion tagging proceeds in the laboratory.

2. EXO-200

EXO-200 will search for the the $0\nu\beta\beta$ of ^{136}Xe with 200 kg of Xe enriched to 80% in ^{136}Xe; the TPC sits inside a thin-walled (and thus low mass) Cu vessel. The experiment takes place underground in the Waste Isolation Pilot Plant (WIPP). The energy resolution of LXe TPCs is being improved by measuring both the scintillation and ionization signals, which reduces the effects of fluctuations in recombination as demonstrated by this collaboration and later improved upon by the XENON collaboration [2,3]. The Cu vessel will has a high voltage cathode plane in the middle and an anode of photo-etched phosphor-bronze wires and a plane of large area avalanche photo-diodes (LAAPDs) on each side. The detector sits in a bath of HFE-7000, a heat transfer fluid, inside a two-walled copper cryostat cooled down by three closed circuit refrigerators. The cryostat, along with the Xe and HFE plumbing and some electronics sit in six modular clean rooms. An extensive screening of materials for the, lead, cryostat, and detector has been carried out, building a database of over 300 materials, using several different radiopurity qualification methods [4]. The Xe handling and cooling system was commissioned

Figure 1. The EXO-200 modular clean rooms reassembled underground at WIPP from October 2007, with an sheltered entrance way added.

in March and April of 2007 and 30 kg of Xe was liquefied in a stainless steel vessel. In June 2007 the EXO clean rooms were disassembled for relocation to WIPP with the last module arriving in August; by mid-September the modules were re-assembled and are shown in figure 1.

3. Barium Ion Transfer and Identification

The collaboration is following several possible paths for identifying Ba ions: identification in situ in LXe, in situ in GXe, and transfer from LXe to an external linear ion trap. In the the case of the external ion trap, Ba ions are trapped in a quadrapole radio frequency (RF) Paul trap; four electrodes extend over the length of the trap and are broken up into individually contacted segments which apply a DC potential in the longitudinal direction coming to a trough at the trapping region. Also, two opposite electrodes apply an RF potential with the other two at RF ground which creates a quadrupole saddle potential to confine the ion in the transverse plane. The ions are collisionally cooled by a buffer gas: 10^{-4}-10^{-2} Torr of either He or Ar. A full technical description of the trap can be found in [5]. Once trapped, the spectroscopy of the ion works based on cycling through its states; a laser of the appropriate frequency pumps the electron from the $6S_{1/2}$ to the $6P_{1/2}$ excited state, with the de-excitation observed as fluorescence. In the case of Ba, there is also a $5D_{3/2}$ metastable state whose transition to the ground state is forbidden; this state must also be pumped to prevent loss of fluorescence. If the correct frequencies for $^{136}Ba^{+}$ are used, only $^{136}Ba^{+}$ will display this fluorescence signature. The fluorescence is measured using an EMCCD with the signal integrated over a region of interest and a 5 s exposure time; this allows ions to be detected with a high signal to noise ratio and sensitivity to single ions as can be seen in figure 2 [6]. As for actually transferring ions from LXe to the trap, there are two main methods in development: a capacitive cryo-tip and a optical fiber for Resonant Ionization Spectroscopy (RIS). The cryo-tip method works as such: a thin-layer (100 nm) of Xe ice is frozen on the end of the tip and the ion is subsequently attracted to the tip and embedded in the surface of the ice; the tip then moves to the trap where the ice is sublimed and the ion ejected. The issue with this concept is that the gas purity requirements for the trap are very stringent, implying the need for thin ice layers. In order to control the ice thickness, the thickness is measured using an interdigitated coplanar capacitor, which is a set of electrically isolated fingers. Because SXe has a different dielectric constant than LXe, the ice build up changes the capacitance; this effect has been measured in the laboratory and the ice thickness can be controlled to within a few mono-layers [7]. However, such a thin layer will sublime in 10^{-6} s in 10^{-4} Torr of vacuum, and in order to avoid this problem the cryostat must be cooled down to 40 K with a LHe cryostat to reduced the vapor SXe pressure. The optical fiber tip works differently in that the Ba ion actually does attach itself to the fiber and

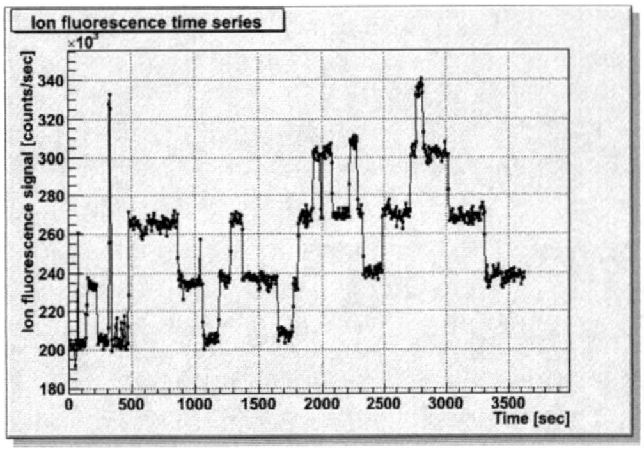

Figure 2. Fluorescence trace over time from the ion trap showing discrete jumps corresponding to individual ions coming in and out of the ion cloud indicating high signal to noise ratio and single ion sensitivity.

is desorbed by thermal emission (coming off as neutral Ba) using a 1064 nm Nd:Yag laser pulse. The ion is then hit with two simultaneous 10 ns pulses of 553 nm, which excites the valence electron, and 390 nm, which ionizes the atom. The reason for doing with in two steps is that the cross section for ionization Ba with one pulse is too low to achieve high efficiency. The lasers have just recently been acquired and initial R&D is underway.

4. Future Prospects

The EXO-200 detector is being installed at WIPP in early 2008 with data collection beginning before the end of the year. With two years of data and an energy resolution of 1.6%, EXO-200 can set a limit of $T_{1/2}^{0\nu\beta\beta} > 6.4 \times 10^{25}$ yr. implying a sensitivity for $< m_\nu >$ of 95-224 meV [8,9,10]. Also, a device has been designed for testing ion transfer from a LXe cell to the ion trap with various tips, with a a taggable Ba^+ source in development. A full EXO will be a one to ten ton detector running for five to ten years with a sensitivity to $< m_\nu >$ of 4-40 meV [8,9,10].

5. References

[1] M.K. Moe *Phys. Rev. C* **44** (1991)
[2] E. Conti *et al.*, *Phys. Rev. B* **68** (2003)
[3] E. Aprile *et al.*, *Phys. Rev. Lett.* **97**, 081302 (2006)
[4] D.S. Leonard *et al.*, *Preprint* arXiv:0709.4524v1
[5] B. Flatt *et al.*, *Nucl. Inst. Meth. A* 578, 399 (2007)
[6] M. Green *et al.*, *Phys. Rev. A* **76** 023404 (2007)
[7] P. Fierlinger *et al.*, *Preprint* arXiv:0706.0540v1
[8] V.A. Rodin *et al.*, *Nucl. Phys. A* **766** (2006) 107131
[9] V.A. Rodin *et al.*, *Preprint* arXiv:0706.4304v1
[10] E. Caurier *et al.*, *Preprint* arXiv:0709.2137v1

New limits on the β^+EC and ECEC processes in ^{74}Se and ^{120}Te

A S Barabash[1], F Hubert[2], Ph Hubert[2], A Nachab[2] and V Umatov[1]

[1] Institute of Theoretical and Experimental Physics, B. Cheremushkinskaya 25, 117218 Moscow, Russia

[2] Centre d'Etudes Nucléaires, IN2P3-CNRS et Université de Bordeaux, 33170 Gradignan, France

E-mail: barabash@itep.ru

Abstract. Limits on double-beta processes in ^{74}Se and ^{120}Te have been obtained using a 400 cm^3 HPGe detector and an external source consisting of natural Se or TeO_2 powder. In particular, limit with a 90% C.L. on the ^{74}Se half-life of 5.5×10^{18} y for ECEC(0ν) transitions to the 2_2^+ excited state in ^{74}Ge (1204.2 keV) has been obtained. This transition is discussed in association with a possible enhancement of the decay rate, in this case by several orders of magnitude, because the ECEC(0ν) process is nearly degenerate with an excited state in the daughter nuclide. For other β^+EC and ECEC transitions (to the ground and excited states) obtained limits are on the level $\sim 10^{18} - 10^{19}$ y.

1. Introduction

Most double beta decay investigations have concentrated on the $\beta^-\beta^-$ decay (see recent reviews [1, 2]). Much less attention has been given to the investigation of $\beta^+\beta^+$, β^+EC and ECEC processes (see review [3]). The $\beta^+\beta^+$ and β^+EC processes are less favorable due to smaller kinetic energy available for the emitted particles and Coulomb barrier for the positrons. However, an attractive feature of these processes from the experimental point of view is the possibility of detecting either the coincidence signals from four (two) annihilation γ-rays and two (one) positrons, or the annihilation γ-rays. It is difficult to investigate the ECEC process because one only detects the low energy X-rays. It is interesting to search for transitions to the excited states of daughter nuclei, which are easier to detect given the cascade of higher energy gammas [4]. One can foresee that with improvement in the experimental techniques and increasing the mass of the detectors, the search for $\beta^+\beta^+$, β^+EC and ECEC processes will be considerably extended.

In this paper the results of an experimental investigation of the β^+EC and ECEC processes in ^{74}Se and ^{120}Te (transitions to the ground and excited states of daughter nuclei) are presented.

2. Experimental details

The experiment has been performed in the Modane Underground Laboratory (depth of 4800 m w.e.). The natural selenium and TeO_2 powder samples were measured using a 400 cm^3 low-background HPGe detector with energy resolution of 2.0 keV for the 1332-keV line of ^{60}Co.

The passive shielding consisted of 4 cm of Roman-era lead and 3-10 cm of OFHC copper inside 15 cm of ordinary lead. To remove ^{222}Rn gas, one of the main sources of the background, a special effort was made to minimize the free space near the detector. In addition, the passive shielding was enclosed in an aluminum box flushed with high-purity nitrogen.

The sample of natural selenium powder was placed in a circular plastic box and put on the endcap of the HPGe detector. The mass of the powder was 563 g, 4.69 g of which was ^{74}Se (as the natural abundance is 0.89%). The duration of measurement was 436.56 hours.

The sample of natural TeO_2 powder was placed in a delrin Marinelli box surrounding the HPGe detector. The mass of the powder was 1004.2 g, 0.72 g of which was ^{120}Te. The duration of the measurement was 475.4 hours.

Table 1. The limits on double-beta processes in ^{74}Se. The second column presents γ-ray energies in keV and their efficiencies used to estimate half-lives. Limits on half-lives $T_{1/2}$ are given at a confidence level of 90%. $^{*)}$ For transition with irradiation of e^+e^- pair. $^{**)}$ This transition is suppressed (forbidden).

Transitions	γ-ray(efficiency)	$T_{1/2}$, 10^{19} y
$L^1L^2, Q' = 1206.9$ keV, ECEC(0ν) to the 2_2^+(1204.2-keV) level of ^{74}Ge	595.8 keV (1.88%) 608.4 keV (1.84%) 1204.2 keV (0.757%)	0.55
$L^1L^2, Q' = 1206.9$ keV, ECEC(0ν) to the 2_1^+(595.8-keV) level of ^{74}Ge	595.8 keV (2.76%) 611.1 keV (2.70%)	1.30
$L^1L^2, Q' = 1206.9$ keV, ECEC(0ν) to the ground state of ^{74}Ge	1206.9 keV (2.13%)	0.41
$K^1L^2, Q' = 1197.2$ keV, ECEC(0ν) to the 2_1^+(595.8-keV) level of ^{74}Ge	595.8 keV (2.75%) 601.4 keV (2.71%)	1.12
$K^1L^2, Q' = 1197.2$ keV, ECEC(0ν) to the ground state of ^{74}Ge	1197.2 keV (2.13%)	0.64
$K^1K^2, Q' = 1185.9$ keV, ECEC(0ν) to the 2_1^+(595.8-keV) level of ^{74}Ge	595.8 keV (2.75%) 590.1 keV (2.76%)	1.57
$K^1K^2, Q' = 1185.9$ keV, ECEC(0ν) to the ground state of ^{74}Ge	511 keV (6.74%) 1185.9 keV (2.15%)	$0.19^{*)}$ $0.62^{**)}$
β^+EC($0\nu + 2\nu$) transition to the ground state of ^{74}Ge	511 keV (6.74%)	0.19
ECEC(2ν) to the (2_2^+)(1204.2-keV) level of ^{74}Ge	595.8 keV (1.88%) 608.4 keV (1.84%) 1204.2keV (0.757%)	0.55
ECEC(2ν) to the 2_1^+(595.8-keV) level of ^{74}Ge	595.8 keV (3.13%)	0.77

3. Results

3.1. Results with ^{74}Se [5]

A search for different β^+EC and ECEC processes in ^{74}Se was carried out using the germanium detector to look for γ-ray lines corresponding to these processes. Hereinafter, Q' is the effective Q-value defined as $Q' = \Delta M - \epsilon_1 - \epsilon_2$ where ΔM is the difference of parent and daughter atomic masses, ϵ_i is an electron binding energy in a daughter nuclide. The photon detection efficiency for each investigated process has been computed with the CERN Monte Carlo code GEANT 3.21.

The obtained results are presented in Table 1. It is necessary to stress that ^{74}Se has never been investigated before and all results presented here are obtained for the first time.

Table 2. The limits on β^+EC and ECEC processes in ^{120}Te.

Transition	Energy of γ-rays (efficiency)	$T_{1/2}$, 10^{18} y (C.L. 90%)	
		This work	other works
β^+EC($0\nu + 2\nu$); g.s.	511.0 (7.38 %)	0.19	0.121 (0ν) [8]
ECEC(0ν) L^1L^2; g.s.	1691.2 (2.05 %)	0.29	0.00268 (0ν) [8]
ECEC(0ν) K^1L^2; g.s.	1666.4 (2.08 %)	0.39	0.00268 (0ν) [8]
ECEC(0ν) K^1K^2; g.s.	1641.7 (2.08 %)	0.60	0.00268 (0ν) [8]
	511.0 (7.38 %)	0.19	-
ECEC(2ν); g.s.	-	-	0.0094 (2ν) [7]
ECEC($0\nu + 2\nu$); 2^+	1171.26 (2.60 %)	0.75	0.0084 (2ν) [7]

3.2. Results with ^{120}Te [6]

The search for different β^+EC and ECEC processes in ^{120}Te looked for γ-ray lines corresponding to these processes. For ^{120}Sn ϵ is equal to 29.2 keV for the K shell and 4.46 keV, 4.15 keV and 3.93 keV for the L shell (2s, $2p_{1/2}$ and $2p_{3/2}$ levels) . In the case of the L shell the resolution of the detector prohibits separation of the lines so we center the study on the 4.15 keV line.

In the case of ECEC(0ν) transition to the ground state of ^{120}Sn there is a large uncertainty in the energy of the bremsstrahlung γ-quantum because of poor accuracy for ΔM ($\pm$10.3 keV). Thus the position of the peak was varied in the region of the uncertainty and the most conservative value of the limit for the half-life was selected.

The photon detection efficiency for each investigated process has been computed with the CERN Monte Carlo code GEANT 3.21.

The final results are presented in Table 2. In the last column the best previous results are presented for comparison. One can see that the present work is 1.5 to 200 times better than previous measurements [7, 8].

4. Conclusion

For the first time, limits on double-beta processes in ^{74}Se have been obtained. In addition, in this paper we have presented new, more stringent, experimental limits for the different β^+EC and ECEC processes in ^{120}Te.

Acknowledgments

The authors would like to thank the Modane Underground Laboratory staff for their technical assistance in running the experiment. Portions of this work were supported by a grant from RFBR 06-02-72553 and a State contract N 02.516.11.6099. This work was supported by the Russian Federal Agency for Atomic Energy.

References

[1] Avignone F T, Elliott S R and Engel J 2007 *Preprint* nucl-ex/0708.1033
[2] Barabash A S 2006 *JINST* **1** P07002 (*Preprint* hep-ex/0602037)
[3] Barabash A S 2004 *Phys. At. Nucl.* **67** 438
[4] Barabash A S 1994 *JETP Lett.* **59** 644
[5] Barabash A S et al 2007 *Nucl. Phys.* A **785** 371
[6] Barabash A S et al 2007 *J. Phys.* G **34** 1721
[7] Kiel H, Munstermann D and Zuber K 2003 *Nucl. Phys.* A **723** 499
[8] Bloxham T et al 2007 *Phys. Rev.* C **76** 025501

Double beta decay of ^{48}Ca studied by CaF_2(Eu) scintillators

S Umehara[1], T Kishimoto[1], I Ogawa[1], R Hazama[2], H Miyawaki[1], S Yoshida[1], K Matsuoka[1], K Kishimoto[1], A Katsuki[1], H Sakai[1], D Yokoyama[1], K Mukaida[1], S Tomii[1], Y Tatewaki[1], T Kobayashi[1] and A Yanagisawa[1]

[1]Graduate School of Science, Osaka University, Toyonaka, 560-0043 Osaka, JAPAN
[2]Graduate School of Engineering, Hiroshima University, Higashi-Hiroshima, 739-8527 Hiroshima, JAPAN

E-mail: umehara@km.phys.sci.osaka-u.ac.jp

Abstract. We report the new limit on the half-life of neutrino-less double beta decay($0\nu\beta\beta$) of ^{48}Ca. In order to achieve the background free measurement, measurement of pulse shape was essential. We have observed no event in the $Q_{\beta\beta}$-value region. We set the new limit on $0\nu\beta\beta$ half-life as $T_{1/2}^{0\nu\beta\beta} > 4.0 \times 10^{22}$ year.

1. Double beta decay of ^{48}Ca

Neutrino-less double beta decay($0\nu\beta\beta$) is one of great interest in the view point of neutrino physics. $0\nu\beta\beta$, which violates the lepton number conservation, provides an unique test for a property of neutrinos (Majorana or Dirac particle). The decay rate is directly related to the Majorana neutrino mass.

^{48}Ca has an advantage among all $\beta\beta$ isotopes since it has the highest $Q_{\beta\beta}$-value (4.27 MeV) which ensures a good signal to background ratio. The large $Q_{\beta\beta}$-value gives a large phase-space factor to enhance the $0\nu\beta\beta$ rate and a minimum background contribution of radiations from natural radioactivities of which maximum energy is effectively 3.3 MeV from ^{214}Bi β decay. We have studied $0\nu\beta\beta$ of ^{48}Ca by CaF_2(Eu) scintillators and set the best limit of its half-life in the last publication[1].

We have been running the experiment after the last publication. In the previous measurement we observed no event in the $Q_{\beta\beta}$-value region although expected background exceeds 1 event. In the present measurement we employed pulse shape measurement which is a key to realize further background reduction.

2. The ELEGANT VI system

The ELEGANT VI system have been developed to achieve background free measurements by a 4π active shield. The design of the system is schematically shown in figure 1. The system consists of CaF_2(Eu), CaF_2(pure) and CsI(Tl) scintillators. The 23 CaF_2(Eu) scintillators with total mass of 6.7 kg act as a $0\nu\beta\beta$ source and a detector. The CaF_2(Eu) scintillator is sandwiched by two CaF_2(pure) scintillators and viewed by two photomultiplier tubes mounted both sides of scintillators. The CaF_2(pure) and CsI(Tl) scintillators act as veto counters for the CaF_2(Eu) scintillators[1].

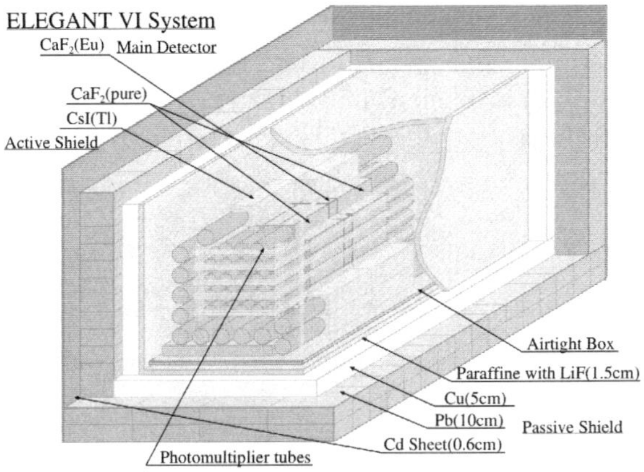

Figure 1. Inside view of the ELEGANT VI system. The CaF_2(Eu) scintillators are shielded by CaF_2(pure) and CsI(Tl) scintillators from external background radiations.

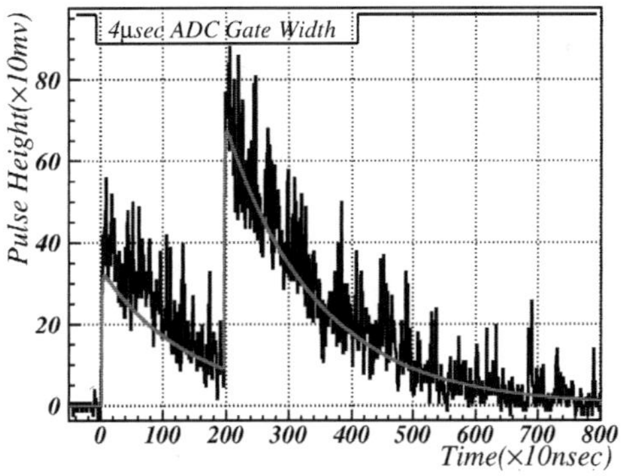

Figure 2. Typical β-α sequential event and ADC gate width for the CaF_2(Eu) scintillators. Thin line represents fitted sequential shape.

The energy resolution of each CaF_2(Eu) scintillator was obtained by measuring peak width of α-rays from ^{216}Po contained in the scintillator. The α-ray energy is 1.3 MeV in electron equivalent energy. A peak from α-rays was clearly observed by off-line analysis of which procedure is described in ref.[1]. The energy resolution was assumed to be inversely proportional to square root of deposited energy in an energy region up to the $Q_{\beta\beta}$-value. This assumption is confirmed by neutron scatting up to 2.33 MeV by exciting the first excited state (3.35 MeV) of ^{40}Ca[1]. The extrapolated resolutions at the $Q_{\beta\beta}$-value are from 4% to 6% in FWHM depending on the CaF_2(Eu) scintillators.

3. Backgrounds

The serious background process is pile-up events by β- and α-rays in U- and Th-chain which are contained within the CaF_2(Eu) scintillators. Other decays from outside the system are almost negligible by the 4π active shield and the large $Q_{\beta\beta}$-value.

The pile-up events come from the following sequential decays.

(a) $^{212}\text{Bi}\ (Q_\beta = 2.25\ \text{MeV}) \xrightarrow[\beta]{} {}^{212}\text{Po}\ (Q_\alpha = 8.95\ \text{MeV}, T_{1/2} = 0.299\ \mu\text{sec}) \xrightarrow[\alpha]{} {}^{208}\text{Pb}$ (Th-chain),

(b) $^{214}\text{Bi}\ (Q_\beta = 3.27\ \text{MeV}) \xrightarrow[\beta]{} {}^{214}\text{Po}\ (Q_\alpha = 7.83\ \text{MeV}, T_{1/2} = 164\ \mu\text{sec}) \xrightarrow[\alpha]{} {}^{210}\text{Pb}$ (U-chain).

Typical sequential event is shown in figure 2. In particular, the sequential event by (a) decays is serious because ^{212}Po has a short half-life 0.3 μs against ADC gate width of 4 μs due to 1μs decay constant of CaF_2(Eu). The ADC gives information of sum energy of β- and α-rays, which may exceed the $Q_{\beta\beta}$-value. The measured radioactivities in the scintillators are 0.29-4.53 mBq/kg and 0.042-0.276 mBq/kg for U- and Th-chain, respectively [1].

4. Measurement for neutrino-less double beta decay of ^{48}Ca

In order to reject the β-α sequential events, we included pulse shape information to a data acquisition system by applying 100MHz Flash ADC (FADC). Data with FADC were obtained in a period from January in 2003 to September in 2006 at Oto cosmo observatory. Total live time after FADC installation is 3394 kg·day.

In order to estimate the $0\nu\beta\beta$ rate, the energy window was set at 4.17 - 4.37 MeV as shown in figure 3. Background contributions from the natural radioactive contaminations were estimated by Monte Carlo simulation based on the measured radioactivities in the scintillators and background rejection efficiency of analyses. The estimated background yield, which is including a small contribution from $2\nu\beta\beta$ decay, are given in table 1.

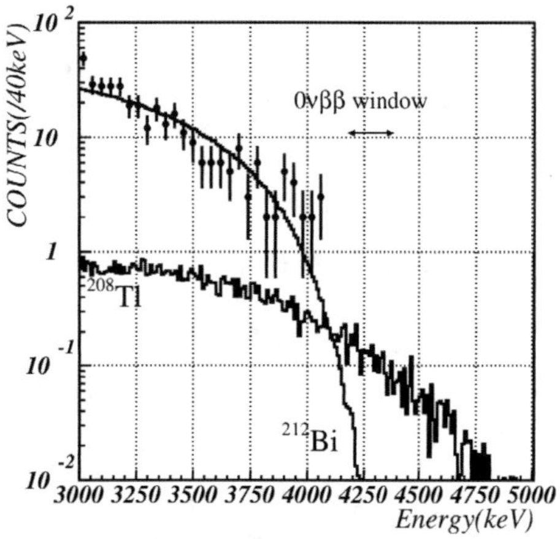

Figure 3. The obtained energy spectrum and the expected background spectra. Solid circles represent the obtained experimental energy spectrum after the pulse shape analysis. There is no events in the $Q_{\beta\beta}$-value window. Upper and lower solid lines in low energy region correspond to the expected background yield from ^{212}Bi and ^{208}Tl. Effect from ^{214}Bi was negligible in this measurement because the β-α sequential events from ^{214}Bi were effectively rejected by the pulse shape analysis.

Table 1. Experimental run summary with the applied analysis. The obtained yield, the expected background rate and live time also are given.

Date	Jun1998-Nov1999[1]	Jan2003-Sep2006
Pulse Shape Analysis(PSA)	without PSA	with PSA
Observed events (counts)	0	0
Expected background (counts)	1.30	0.70
Live time(kg·day)	1553	3394

There are no events in the energy window during live time 4947 kg·day, which includes data without the pulse shape analysis. The new limit with the 90% C.L. on the half-life is obtained $T_{1/2}^{0\nu\beta\beta} > 4.0 \times 10^{22}$ year by a procedure in ref.[2]. Using a nuclear matrix element given in ref.[3], the half-life limit leads to upper limit of $\langle m_\nu \rangle <$ (4.2-26)eV (90%C.L.).

5. Conclusion and outlook

Sensitivity of our measurement is not limited by the backgrounds for the small detector volume of 6.7kg. We observed no background events in the $0\nu\beta\beta$ energy window.

We have developed the large scale detector system CANDLES which features CaF_2(pure) scintillators[4]. The CANDLES system is currently under development at sea level laboratory[5]. Backgrounds in the system is expected to be similar to that of the ELEGANT VI system. We studied purification of the CaF_2(pure) scintilltors and made improvements in the pulse shape analysis. As a consequence, the backgrounds from the radioactive contaminations in the scintillators will be reduced by 2 orders of magnitude than that of the ELEGANT VI system[6]. The CANDLES III system, which consists of the 96 CaF_2(pure) scintillators with total mass of 305 kg, will be installed at Kamioka underground laboratory next year. The expected sensitivity is 0.5 eV for effective neutrino mass.

References

[1] Ogawa I *et al* 2004 *Nucl. Phys.* A **730** 215-223
[2] Feldman G J and Cousins R D 1998 *Phys. Rev.* D**57** 3873-89
[3] Suhonen J and Civitarese O 1998 *Phys. Rep.* **300** 123-214
[4] Kishimoto T *et al* 2003 *Proc. of 4th Int. Workshop on Neutrino Oscillations and their Origin (Kanazawa)* (Singapore : World Scientific) pp 338-349
[5] Hirano Y *et al* in this volume
[6] Umehara S *et al* 2006 *J. Phys.: Conf. Series* **39** 356-358; Umehara S *et al* 2003 *Proc. of Symp. on Neutrino and Dark Matters in Nuclear Physics (Nara)* (Osaka : Yamada Science Foundation) pp XII23-33

Study of light guide system for CANDLES

Yushi Tsubota

Kishimoto lab., Department of Physics, Graduate School of Science, Osaka University, Machikaneyama-cho 1-1, Toyonaka, Osaka, 560-0043 JAPAN

E-mail: tsubota@km.phys.sci.osaka-u.ac.jp

Abstract. We are trying to observe the neutrinoless double beta decay at Osaka University with the detector system named CANDLES III that uses CaF_2(pure) crystals and liquid scintillator. CANDLES III has the photo-caverage of about 30% now. We will install the mirror light guide system to improve this photo-coverage. This means the improvement of the energy resolution. We will report on the result of the experiment to evaluate the performance of theis light guide system.

1. Introduction

If the neutrino is Majorana particle, the neutrinoless double beta decay is predicted to happen. Then this rate is related to the mass of the neutino.

We have tried to observe the neutrinoless double beta decay with the detector system named ELEGANT VI that has used CaF (Eu) scintillators up to now [1]. The CANDLES system that uses CaF (pure) as a new detector that replaces this is designed, and the test experiment at the sea level by the third machine tries to be conducted in Osaka University now [2].

CaF (pure) has some advantages. One is the decreasing of the background in the crystal because their densities of impurities is about one tenth lower than CaF (Eu). Another is suitable for the enlargement of the detector because CaF (pure) has the longer attenuation length than CaF (Eu). However, CaF (pure) has only the 40% amount of luminescence of CaF (Eu). Therefore, to maintain energy resolution, improvement of light collection mechanism becomes important.

It seems that we only have to bring the distance of a crystal and a photomultiplier tube (PMT) close to improve the light collection efficiency. By analyzing data taken in ELEGANT VI, however, it is understood that the gamma-rays emitted from radioactive impurities (mainly K) in PMT windows becomes a serious background. To veto external radiations efficiently, we want to install the liquid scintillator layer between the crystal and the PMT and to achieve 4π active shield.

The request on both sides of such the light collection efficiency improvement and the background reduction makes us need a light guide system. A light guide is able to take the distance from a PMT that is a gamma-rays source, and to lead an optical photon from a crystal to a PMT efficiently.

The loss on a reflector means that the reduction of the number of the reflection brings us to improvement of the efficiency. That is, we need decreasing the number of acute angle incidences to the light guide surface of a photon. Therefore it is necessary to make the multi-reading light guide system from a cubic crystal to shoal the angle of incidence degree [3].

2. PERFORMANCE EVALUATION EXPERIMENT OF LIGHT GUIDE

A small prototype detector with the six-reading light guide system that uses CaF (pure) is being produced this time (Fig.1). As a mirror-type reflector, I selected Enhanced Supecular Reflector(ESR)(Sumitomo 3M Co., Tokyo, Japan) that has the hightest reflectivity of 98.4% in the visible light range. The light guide was composed of a square acrylic cylinder and this ESR.

With this system, I measured the light collection efficiency by using Cs gamma-ray source, and compared it with the Monte Carlo simulation result (Fig.2).

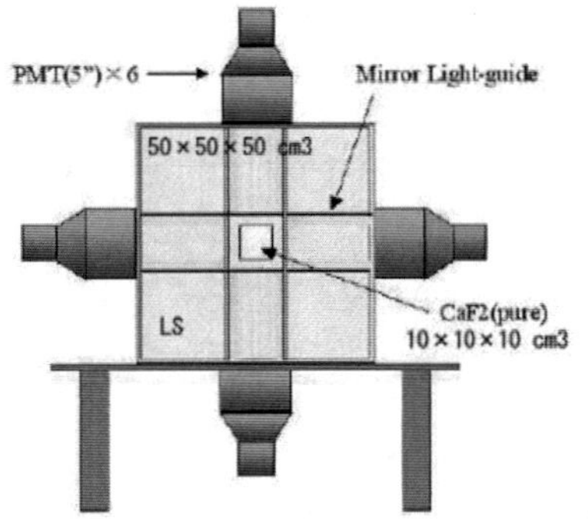

Figure 1. The prototype six-reading mirror light guide system.

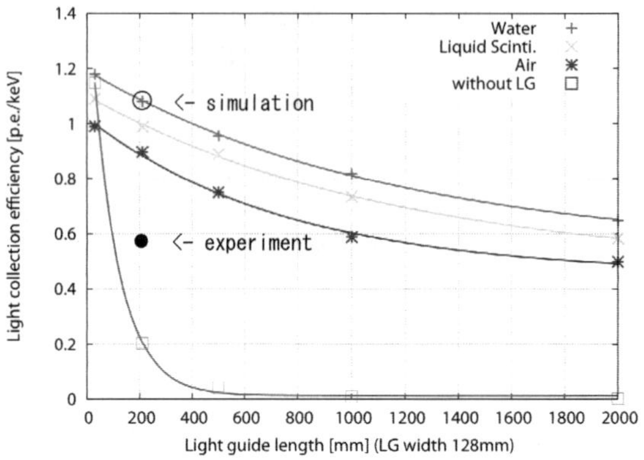

Figure 2. The dependence of the efficiency on the light guide length in the six-reading system with ESR.

3. CONCLUSION AND FUTURE PROSPECTS

The experimental collection efficiency was about one half of the simulation result, though this is naturally better than the case without the light guide. The cause of this poor efficiency is possibly the coarse of how to paste ESR. The coarse might give ESR a bigger diffuse component. When the effective reflectivity of ESR was requested by the simulation to reproduce the experimental result, it became about 65%.

In the future, I will devise how to paste ESR, and I want to apply the effective reflectivity to the simulation of CANDLES III. And I plan to produce the light guide system for CANDLES III dectector after optimization of its configuration by the simulation. Moreover, I will adjust the design of the light guide system for CANDLES III(underground) to achieve aimed energy resolution of 4% at Q value(4.27 MeV).

Acknowledgments

I am grateful to T Kishimoto, I Ogawa and all CANDLES collaborators for enlightening discussion.

References

[1] Ogawa I *et al.* 2003 *Nucl. Phys. A* **721** C525-C528

[2] Ogawa I *et al.* 2007 *Proc. of International Nuclear Physics Conference(INPC2007)*

[3] Yanagisawa A 2005 *Master thesis, Osaka University*

Study of Design for CANDLES Trigger System

G.Ito, T.Kishimoto, I.Ogawa, S.Umehara, Y.Hirano, Y.Tsubota, K.Matsuoka, R.Hazama* for the CANDLES collaboration

Department of Physics, Oska University, * Graduate School of Engineering, Hiroshima University

E-mail: ito@km.phys.sci.osaka-u.ac.jp

Abstract. We are searching for 0ν double beta decay of ^{48}Ca with CANDLES detector. We have a plan to construct a more larger detector underground from next year and considering the deadtime, we have to reduce background events at online trigger. Because FPGA is useful high integrated module, we are experimenting for the performance evaluation with prototype detector now.

1. Introduction

Observation of the neutrinoless double beta decay is most sensitive to Majorana effective mass and test for Majorana nature of neutrinos.[1] The CANDLES(Calcium fluoride for the study of Neutrinos and Dark matters by Low Enery Spectrometer) is a detector to search for the double beta decay ^{48}Ca with undoped calcium fluoride crystal (CaF_2) scintillator.[2] The light from calcium fluoride crystal is converted into visible light range with the wavelength shifter and detected by 40 PMTs. Background is rejected by 4π active shield of liquid schintillator(LS). Though experimental trigger is made by using many NIM modules now, we will make the trigger by using FPGA in the future. The advantage of using FPGA is high integration circuit and the easiness of the circuit change. I did test experiment with ULM(Universal Logic Module) for performance evaluation.

Figure 1. CANDLES detector

2. Evaluation of Trigger on FPGA

In a present experimental trigger, we use the sum of all PMT's signal through lower and upper threshold which cuts noise and high energy liquid scintillator signal respectively. Trigger rate of LS, however, is about 70% and that contributes to the dead time($\sim$ 40%). We can make more efficient trigger if we use each PMT signal and signal characteristic of CaF_2($\sim 1\mu s$) and LS($\sim 10ns$). This means that lower multiplicity will reject noise signal and higher multiplicity can be used to veto LS events. I searched for the good parameters that was able to distinguish two events efficiently.

3. Experiment for Performance Evaluation

For the performance evaluation of FPGA trigger, I made several multiplicity condition and examined the influence on the events of CaF_2 and liquid scintillator. The output from FPGA was input to scaler and count the signals during long gate. Coincidence width is changed from 8ns to 48ns by 8ns and discrim. level is 10mV. (one photoelectron efficiency is $\gg$ 99%.) In multiplicity 2, for example, there are 3 or 4 signals in a low energy event of CaF_2 while almost only one signal in a liquid scintillator event. Fig.2 shows that S/N of CaF_2 events(3σ) has been improved from 3.37 to 4.49 compared with present experimental trigger condition. This means that ratio of CaF_2 to liquid scintillator rise from 4% to 9% by multiplicity cut.

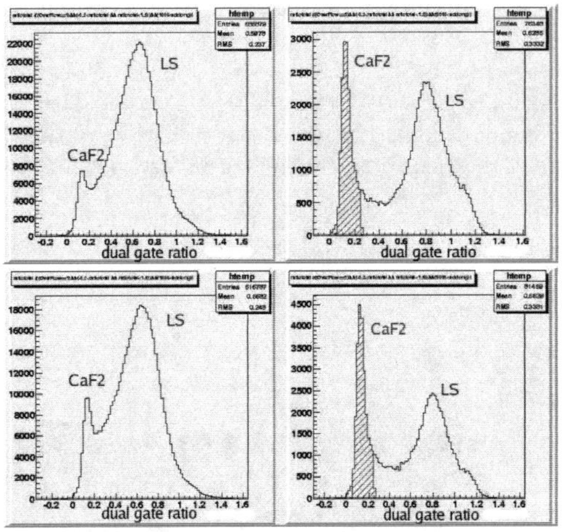

Figure 2. above:only 1st level trigger, below:include 2nd level trigger, left:all data, right:energy cut($\geq 1MeV$). Shaded part means CaF_2 ratio cut ≤ 0.25.

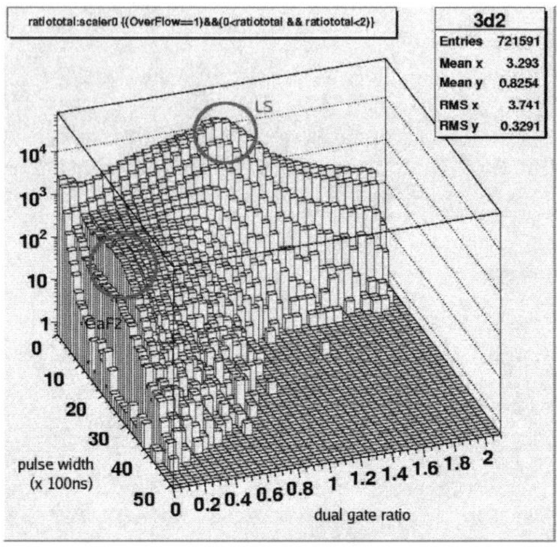

Figure 3. Dual gate ratio(Short gate charge/All charge) vs pulse width lego plot.

Table 1. evaluation performance based on multiplicity

multiplicity	2	3	4	6	8
trigger rate(cps)	135.9	54.0	22.4	6.4	1.7
CaF_2(%)	23.68	8.33	5.02	1.87	0.33
LS(%)	56.88	23.96	10.69	2.52	0.64
relative ratio	2.40	2.88	2.13	1.35	1.64

4. Conclusion

Ratio of CaF_2 to liquid scintillator was confirmed to rise from by the twice or more compared with present trigger condition. This means there is a possibility to be able to separate them by FPGA efficiently. Coincidence width is so severe that both CaF_2 and LS events rate are suppressed to low this time. I'll search more effective trigger condition for background rejection.

5. References

[1] S.T. Petcov, New J. of Phys. 6(2004) 109.
[2] I. Ogawa et al., OULNS Annual Report (2004), p1.

DCBA experiment for searching for neutrinoless double beta decay (II)

N Ishihara[1], Y Kato[1], T Inagaki[1], G Iwai[1], T Ohama[1], S Takeda[1], Y Yamada[1], T Haruyama[1], Y Makida[1], M Kawai[1], S Kitamura[2], T Ishikawa[3], T Sakuma[3], T Sumiyoshi[3], Y Teramoto[4], I Nakano[5], Y Sakamoto[6], Y Nagasaka[7], N Tamura[8], K Tanaka[9] and R Ito[10]

[1] High Energy Accel. Research Organization (KEK), Tsukuba, Ibaraki 305-0801 Japan
[2] Tokyo Metropolitan University, Arakawa, Tokyo 116-8551 Japan
[3] Tokyo Metropolitan University, Hachioji, Tokyo 192-0397 Japan
[4] Osaka City University, Sumiyoshi, Osaka 558-8585 Japan
[5] Okayama University, Okayama 700-8530 Japan
[6] Tohoku Gakuin University, Izumi, Sendai, Miyagi, 981-3193 Japan
[7] Hiroshima Institute of Technology, Saeki, Hiroshima 731-5193 Japan
[8] Niigata University, Niigata 950-2181 Japan
[9] Saitama Science Institute, Inarimachi-kita, Fukaya, Saitama 366-0028 Japan
[10] Zentek Technology Japan Inc., Uchikanda, Chiyoda, Tokyo 101-0047 Japan.

nobuhiro.ishihara@kek.jp

Abstract. R&D studies using DCBA (Drift Chamber Beta-ray Analyzer) are in progress at KEK in order to search for neutrinoless double beta decay. A test prototype DCBA-T3 has provided the information of new technologies for constructing a big size Magnetic Tracking Detector (temporarily called MTD).

1. Introduction

Next generation experiments of neutrinoless double beta decay ($0\nu\beta\beta$) require the sensitivity of effective neutrino mass ($<m_\nu>$) around 30 meV, which is predicted by the so-called inverted hierarchy mass spectrum. R&D studies using DCBA (Drift Chamber Beta-ray Analyzer) have shown that the track detection using gas chambers in a uniform magnetic field can identify single electrons, alpha particles and electron-positron pair creations [1]. This is very useful for rejecting background events. On the other hand, the energy resolution is the key issue to separate the event of $0\nu\beta\beta$ from that of $2\nu\beta\beta$. Suppose $<m_\nu>$ is 30 meV, it is estimated by the simulation studies that the relative energy resolution should be less 5% (FWHM) at the Q-value of decay source. In order to study the energy resolution of DCBA, a test prototype called DCBA-T2 has been constructed, and FWHM resolution of about 6% at Q-value of ^{150}Nd has been expected from the experimental results [2], but not enough. Further R&D studies using DCBA-T3 are described in this report.

2. DCBA-T3

A new prototype DCBA-T3 shown in figure 1 is now under construction in order to obtain better energy resolution.

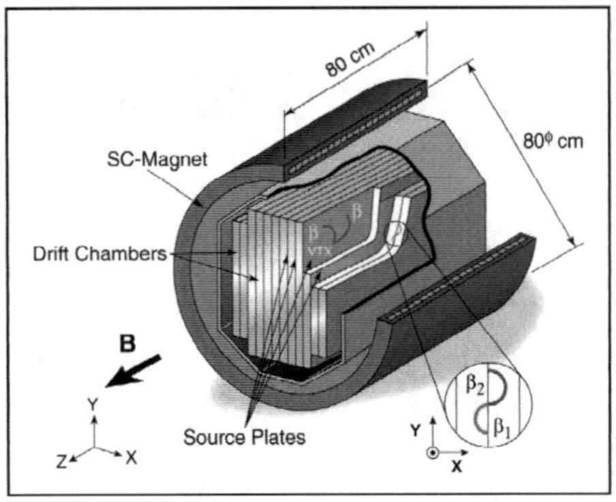

Figure 1. Schematic view of DCBA-T3.

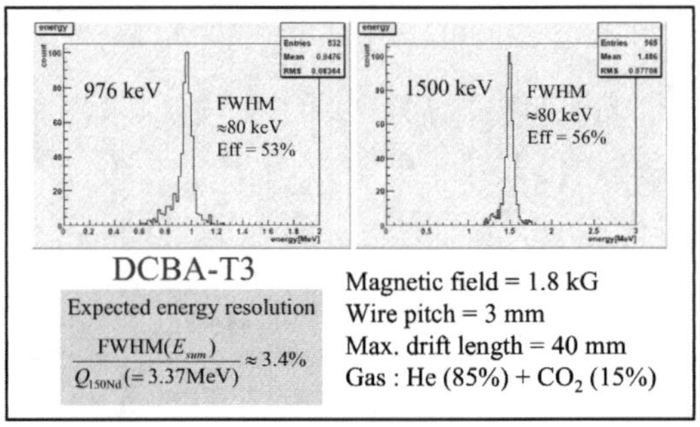

Figure 2. Simulation studies on the energy resolution of DCBA-T3.

Main improved items in DCBA-T3 are 2 kG (0.8 kG in T2 case) magnetic field and 3 mm (6 mm in T2 case) pitch of sampling points in both X and Y coordinates. Since the energy resolution in DCBA is strongly affected by multiple scatterings and energy losses in the chamber gas, the shorter track length in a stronger magnetic field can improve the energy resolution with enough number of sampling points. The energy resolution is expected to be 3.4% at Q-value of ^{150}Nd as shown in figure2.

3. Magnetic Tracking Detector (MTD)

On the basis of DCBA studies, a future project temporarily called MTD has been planed as shown in figure 3. An MTD module consists of the same components as DCBA-T3, though the volume of MTD is much larger than that of DCBA-T3. Since the total area of source plates is about 80 m^2, the source amount of about 32 kg is installed in one module, if the source thickness is 40 mg/cm^2. The sensitivity of <m$_\nu$> is estimated to be 0.1 eV for 1 year measurement, when 60% enriched ^{150}Nd is used. In the viewpoint of energy resolution, the source thickness comparable to the chamber resolution is 15 mg/cm^2. Therefore, 50 modules of MDT with 15 mg/cm^2 of 60% enriched ^{150}Nd can search for <m$_\nu$> down to 30 meV.

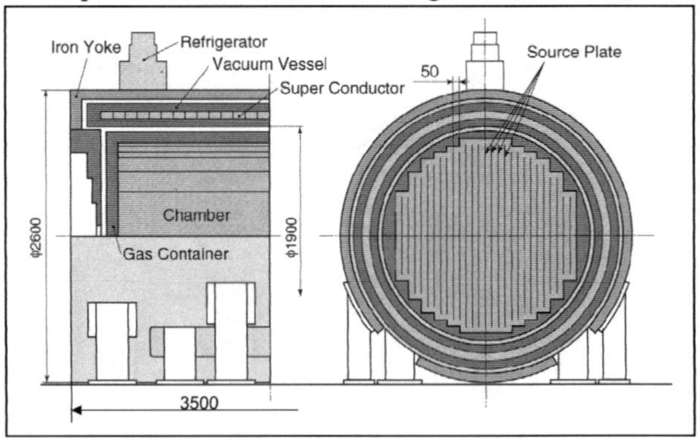

Figure 3. Conceptual drawings of MTD.

4. Conclusions

A test prototype DCBA-T3 has a possibility of 3.4% FWHM energy resolution at Q value of ^{150}Nd. A future project temporarily called MDT has been designed on the basis of DCBA-T3 in order to investigate <m$_\nu$> down to 30 meV.

Acknowledgments

The work was supported by Grants-in-Aid for Scientific Research from Japan Society for the Promotion of Science and Yamada Science Foundation.

References

[1] Ishihara N *et al.*, *Nucl. Instr. Meth. Phys. Res.* A **373** (1996) 325; Ishihara N *et al.*, *Nucl. Instr. Meth. Phys. Res.* A **443** (2000) 101; Kato Y *et al.*, *Nucl. Instr. Meth. Phys. Res.* A **498** (2003) 430.

[2] Ishihara N *et al.*, DCBA experiment for searching for neutrinoless double beta decay (I), this conference.

Origin of ultra-high energy cosmic rays in the era of Auger and Telescope Array

Susumu Inoue

National Astronomical Observatory of Japan, 2-21-1 Osawa, Mitaka, Tokyo 181-8588, Japan [1]

E-mail: inoue@th.nao.ac.jp

Abstract. We discuss the origin of ultra-high energy cosmic rays in light of the latest observational results from the Pierre Auger Observatory, highlighting potential astrophysical sources such as active galactic nuclei, gamma-ray bursts, and clusters of galaxies. Key issues include their energy budget, the acceleration and escape of protons and nuclei, and their propagation in extragalactic radiation and magnetic fields. We briefly address the prospects for Telescope Array and future facilities such as JEM-EUSO, and also emphasize the importance of multi-messenger X-ray and gamma-ray signatures in addition to neutrinos as diagnostic tools for source identification.

1. Introduction

Several decades after their discovery, the origin of ultra-high energy cosmic rays (UHECRs), cosmic particles with energies 10^{18}-10^{20} eV and above, continue to be one of the biggest mysteries in physics and astrophysics [1]. During the last few years, the field has witnessed great advances, thanks to the advent of new generation, hybrid detector facilities spearheaded by the Pierre Auger Observatory in the southern hemisphere [2]. The Telescope Array [3] has begun operation in the northern hemisphere, and we eagerly await their data. Expectations are already high for future projects such as Auger North [4] or the Extreme Universe Space Observatory on the Japanese Experiment Module (JEM-EUSO), to be deployed on the International Space Station [5]. This article, by no means a thorough review, discusses selected topics in the rapidly developing field of UHECRs, focusing on the theory of potential astrophysical sources confronted with the latest observational results. Some of the issues below are covered in more detail in earlier articles [6].

2. General issues on propagation and recent observational results

We first touch upon some general issues concerning the propagation of UHECRs (see e.g. [7, 8] for more details). If UHECRs are protons of extragalactic origin, photopion interactions with cosmic microwave background (CMB) photons must induce severe energy losses at $\gtrsim 6 \times 10^{19}$ eV for propagation lengths $\gtrsim$ 100 Mpc. Unless the sources lie much nearer, a spectral steepening (Greisen-Zatsepin-Kuzmin or GZK cutoff [9]) is expected above these energies. The Pierre Auger Observatory [10] as well as the HiRes experiment [11] have recently demonstrated the existence

[1] Present address: Department of Physics, Kyoto University, Oiwake-cho, Kitashirakawa, Kyoto 606-8405, Japan; E-mail: inoue@tap.scphys.kyoto-u.ac.jp

of a strong spectral steepening at just this energy with high statistical significance, although at present it cannot be excluded that this represents instead some limit to acceleration at the sources.

The absence of super-GZK events reduces the motivation for the so-called top-down scenarios, in which UHECRs arise as decay products of higher energy entities envisaged in non-standard particle physics models, such as topological defects, superheavy dark matter, etc. [7]. Moreover, they are seriously constrained by the tight limits on the photon fraction from the latest Auger measurements [12]. In the following, we concentrate on the bottom-up, astrophysical scenarios.

A spectral steepening analogous to that for protons can also result if the highest energy UHECRs are composed mainly of heavy nuclei such as iron. For nuclei, the dominant energy loss process above 10^{19} eV during intergalactic propagation is photodisintegration and pair production interactions with photons of the far-infrared background (FIRB) and the CMB [13]. Based on recent determinations of the FIRB, the energy loss length for iron nuclei at 10^{20} eV is $\gtrsim 300$ Mpc, somewhat larger than that for protons [14]. Observationally, the nuclear composition of UHECRs, particularly at the highest energies, is quite uncertain. Fluorescence measurements of shower elongation rates by the HiRes experiment have indicated that the composition changes from heavy-dominated below 10^{18} eV (presumably of Galactic origin) to light-dominated above this energy [15]. However, the latest results from Auger, based on hybrid events with higher precision and considerably higher statistics, instead reveal an intermediate composition at all energies up to 4×10^{19} eV, with hints of a trend toward a heavier composition at the highest energies [16]. A caveat is that systematic uncertainties in interaction models and atmospheric optical attenuation are still quite large [17]. These issues are also critical for the interpretation of the ankle feature in the spectrum at $\sim 3 \times 10^{18}$ eV [18, 19].

Another unavoidable aspect is deflection by Galactic and extragalactic magnetic fields (EGMF), which affect the arrival directions of UHECRs and can also act to lengthen their effective propagation distance. The strength and distribution of magnetic fields in the intergalactic medium as well as at high latitudes in the Galaxy are very poorly known, both observationally and theoretically. Faraday rotation measurements of distant radio sources give only upper limits in the nanogauss range for intergalactic fields on average, subject to assumptions on the field coherence scale and ionized gas distribution [20]. Realistically, whatever their origin, EGMF can be expected to have some correlation with the distribution of large scale structure, and various EGMF models that account for this have been introduced to assess their effects on UHECR propagation [21]. The effect of Galactic magnetic fields (GMF) is also model-dependent [22, 23].

The Auger collaboration has recently announced statistically significant evidence for UHECR anisotropy [24]. They find a $\sim 3\sigma$ correlation between the arrival directions of 20 UHECR events above 5.6×10^{19} eV, and the positions of active galactic nuclei (AGNs) in the Veron-Cetty and Veron catalog with distances <75 Mpc, to within 3.1 degrees. Some robust implications of this important result are that: 1) the UHECR sky is not isotropic, 2) the sources of UHECRs are extragalactic and trace large-scale structure on scales of order 100 Mpc, ruling out Galactic sources, and 3) deflections of UHECRs in the GMF and EMGF are not so severe as to isotropize the arrival directions, so that charged particle astronomy is possible.

Not much more can be said with confidence, however. A correlation with the global distribution of AGNs does not necessarily mean that they are the true sources; any other source that follows the distribution of large-scale structure, such as gamma-ray bursts (GRBs) and clusters of galaxies, may also be feasible. Inferences on the source density or UHECR composition from the current anisotropy results are dependent on assumptions regarding the EGMF and GMF (see below). In addition, the Veron-Cetty catalog is known to be highly non-uniform and incomplete, and further studies with higher statistics utilizing more suitable catalogs are warranted. Since the northern hemisphere should be less affected by the GMF, data from the Telescope Array will be crucial, and all-sky observations by JEM-EUSO even better. We note that the AGN correlation results have not been confirmed by the northern HiRes, although their statistics is smaller [25]. At present, the true identity of the UHECR sources is still wide open.

3. Acceleration and energetics in astrophysical sources of UHECRs

A minimum requirement for astrophysical sources of UHECRs is the ability to magnetically confine particles of the requisite energies. For particles with energy E and charge Z, this implies the Hillas condition $(R/\mathrm{pc})(B/1\mathrm{G}) \gtrsim (E/10^{20}\mathrm{eV})/Z$ between the system's size R and magnetic field B. Only a few types of objects are known to meet this criterion, among them the jets of radio-loud AGNs, GRBs, and clusters of galaxies [26]. Below we focus on these three as representative types of potential UHECR sources (see [26] for other possibilities). The actual maximum energy attainable under different circumstances must be evaluated by comparing the timescales for particle acceleration, often that for the first order Fermi mechanism in shocks, against the timescales for limiting processes such as source lifetime, particle escape, adiabatic or radiative energy loss, etc.

Equally important is the available energy budget. Fig.3 shows estimates of the kinetic energy output averaged over the universe as a function of redshift z due to AGN jets, GRB explosions and accretion onto clusters, which should be proportional to their cosmic ray output. The plotted quantity is differential per unit z, $dE_{\mathrm{kin}}/dz = (dt/dz) \int L(dn/dL)dL$, where L is the kinetic luminosity per object and dn/dL is the z-dependent luminosity function, with cosmological parameters h=0.7, Ω_m=0.3 and Ω_Λ=0.7. For AGN jets, we have made use of the observed radio luminosity function along with the observed correlation between the radio and jet kinetic luminosities of radio galaxies [27]. GRBs were assumed to occur each with kinetic energy $E_{\mathrm{GRB}} = 10^{53}$ erg at a rate that follows the star formation history and matches the $\log N$-$\log S$ distribution observed by BATSE [28], the estimate being roughly independent of the beaming factor (see below for modifications required by the latest Swift observations). The three curves each for AGNs and GRBs correspond to different evolutionary assumptions at the highest z, with only small differences at low z. For clusters with mass M, the rate of gas kinetic energy dissipation through accretion shocks can be estimated as $L_{\mathrm{acc}} \simeq 9 \times 10^{45}(M/10^{15}M_\odot)^{5/3}$ erg/s [29, 30]. This can be combined with the Press-Schechter mass function to evaluate dE_{kin}/dz for clusters of different M. Note that due to the hierarchical nature of structure formation together with the nonlinear nature of gravity, the maximum is reached at $z = 0$.

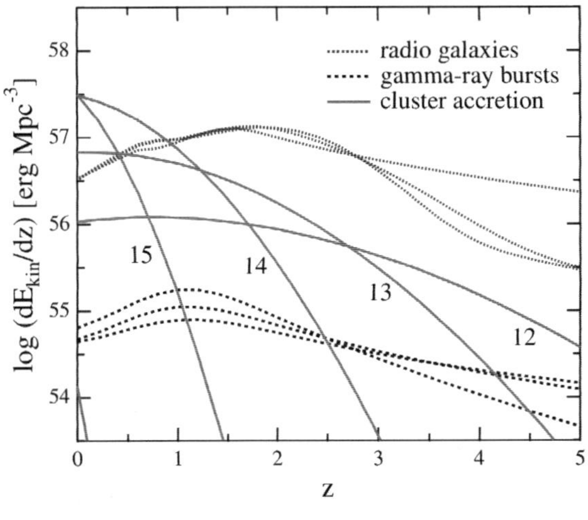

Figure 1. Energy budget of candidate UHECR sources: AGN jets (dotted), GRBs (dashed) and cluster accretion shocks (solid), the latter separately for each $\log M$ as labelled. See text for more details.

The results at low z can be compared with the observed energy density of UHECRs, $\simeq 10^{-19}$erg cm^{-3} $\simeq 3 \times 10^{54}$erg Mpc^{-3} above 10^{19} eV. Considering further factors for energy loss during propagation, it is apparent that whereas AGN jets and cluster accretion shocks have reasonable margins to accommodate the energetics of UHECRs, GRBs, with a substantially smaller energy budget, require a high efficiency of energy conversion into UHECRs.

4. Active galactic nuclei

AGNs are luminous active objects in the centers of some galaxies that are thought to be powered by accretion of matter onto supermassive black holes [31]. Roughly 10 % of them are known to be radio-loud, ejecting powerful, relativistic jets away from the nucleus, but the majority of AGNs are of the radio-quiet type with no strong jets. In contrast to the former, the latter are considered unlikely to be UHECR sources due to the severe radiative losses expected near the nucleus [32], and the absence of nonthermal emission components.

In the jets of radio-loud AGNs, different locations along the outflow can be UHECR production sites, since one expects $B \propto R^{-1}$ under the naive assumption that the ratio of magnetic to kinetic energy is constant. One candidate is the inner jet region with $R \sim 10^{16}$-10^{17} cm and $B \sim 0.1$-1 G, known through observations of blazars to be a site of particle acceleration, perhaps due to internal shocks. Estimates assuming Bohm-type acceleration (i.e. acceleration time comparable to gyration time in fully turbulent magnetic fields) show that the maximum proton energy should be limited by photopion interactions with low frequency internal radiation to somewhat below 10^{20} eV [33]. However, plausible models of broadband blazar emission based on electron acceleration generally indicate that the acceleration conditions in inner jets are far from the Bohm limit [34]. Furthermore, efficient conversion to neutrons may be necessary for the particles to escape the jet without suffering adiabatic expansion losses and contribute to UHECRs by decaying back to protons outside. A less problematic site may be the hot spots of powerful radio galaxies, termination shocks where large-scale jets are decelerated by the external medium, with $R \sim 10^{21}$ cm and $B \sim 1$ mG. Here the maximum energy may exceed 10^{20} eV, limited by escape [35]. However, note that these particles must further traverse the extensive cocoon of shocked, magnetized jet material in order to completely escape the system, an issue that has not been examined in detail. A further possibility is acceleration by the bow shocks being driven into the ambient gas by the expansion of the cocoon [36]. In view of the Auger composition results [16], another interesting question is whether heavy nuclei such as iron can be accelerated and survive photodisintegration in these environments [37].

In order to verify an AGN origin, detailed analysis of observed UHECR arrival directions and cross correlations with known source positions is undoubtedly essential, and the recent Auger results are an important step in this direction. However, it is not a trivial task, given the uncertainties in the intervening GMF and EGMF and the UHECR composition (Sec. 2). In addition to deviations in the arrival directions, deflections in the EGMF during propagation also entail significant delays in the arrival times of UHECRs, which can be of order 10^7 yr or more for distances 100 Mpc, depending on the properties of the EGMF [38]. This is in fact comparable to or longer than the typical lifetime or duration of AGN activity, particularly for radio-loud AGNs [39], so that they can effectively behave as transient sources, in much the same way as GRBs. Before the UHECRs emitted by an AGN reaches the observer, the object may significantly evolve in its radio power, or possibly shut off its activity altogether. Another implication is that an effectively larger number of sources contribute to small scale clustering, underscoring the need for high event statistics as might be provided by JEM-EUSO or Auger North.

Because of these concerns, it is also highly desirable to have some means to pinpoint individual sources through characteristic, UHECR-induced signatures of secondary neutral radiation. Neutrinos are ideal in providing an unambiguous earmark of high energy hadrons

[40]. However, even with km^3 detector facilities such as IceCube or KM3NeT, it may not be easy to resolve individual AGNs. Thus, distinctive electromagnetic signals will also be extremely valuable. For radio galaxy hot spots, synchrotron emission from UHE protons can produce nonthermal X-rays that could be distinguished from other processes such as electron inverse Compton through multiwavelength observations [41]. If the medium surrounding the source is sufficiently magnetized, UHE protons propagating in the source's vicinity can lead to diffuse gamma-ray emission from photomeson-triggered cascades that may be detectable by current and upcoming instruments [42].

5. Gamma-ray bursts

GRBs are explosive phenomena that are believed to arise in a fraction of stellar collapse events, characterized by highly variable, ultrarelativistic outflows lasting for $\sim 10 - 100$ s [43]. As with AGN jets, these outflows can be host to different sites for UHECR acceleration. Potential locales include internal shocks, external reverse shocks, and external forward shocks, believed to be the emission sites of the prompt X-rays and gamma-rays, optical flash and radio flare, and the radio to X-ray afterglow, respectively [44]. The external forward shock may possibly be disfavored due to the ultrarelativistic velocity and weak upstream magnetic field [45], but this is controversial [46]. For the mildly relativistic internal and external reverse shocks, a different problem is that for the particles to escape the acceleration site without significant losses, neutron conversion may be required, as with AGN inner jet regions. This entails some reduction in efficiency, which cannot be too severe in view of the tight energy budget, as discussed above.

A crucial distinguishing feature of UHECRs from GRBs is the narrow dispersion in CR energy expected for individual sources, due to the time delay during propagation in EGMF [38]. Clear detections of this effect demand substantial event statistics that may only be achievable with future facilities. Even then, it would only demonstrate that UHECRs come from bursting sources, and their GRB origin will remain ambiguous. Thus, multi-messenger photon and neutrino signatures are particularly essential in the case of GRBs.

Identification of high energy neutrino signals from GRBs are facilitated through time coincidence, but studying individual bursts in detail may be difficult. Thus, photon signatures of UHECR production will also be crucial [47]. Detailed studies of this issue were conducted for the GeV-TeV emission from internal shocks, utilizing a comprehensive Monte Carlo code that includes a wide variety of physical processes related to high energy protons and electrons [48, 49]. Besides inverse Compton emission from primary electrons, interesting proton-induced components can become clearly visible, such as synchrotron emission from high energy protons and muons, as well as emissin from secondary pair cascades injected by photomeson interactions, serving as unique signatures of ultra-high-energy protons.

Particularly interesting are proton-dominated GRBs, which contain a significantly larger amount of energy in accelerated protons compared to accelerated electrons [49]. Such conditions are motivated not only by the physics of particle acceleration in collisionless shocks, but also by the latest Swift measurements of the GRB z-distribution. Post-Swift estimates of the local GRB rate [50] may imply that the proton energy content per burst may need to be substantially higher than previously thought in order for them to remain viable as UHECR sources. Unique UHECR-induced GeV-TeV components are expected, n some cases leading to GeV-dominated bursts or luminous TeV bumps (Fig.2).

Note that the conditions favorable for GeV-TeV emission may also imply efficient UHECR acceleration and escape, but not necessarily strong neutrino emission [48]. The observational prospects are promising for GLAST, atmospheric Cerenkov telescopes such as MAGIC (II), H.E.S.S. (II), VERITAS and CANGAROO III, as well as wide-field surface detector facilities, which should test these expectations and provide important information on the physical conditions in GRB outflows.

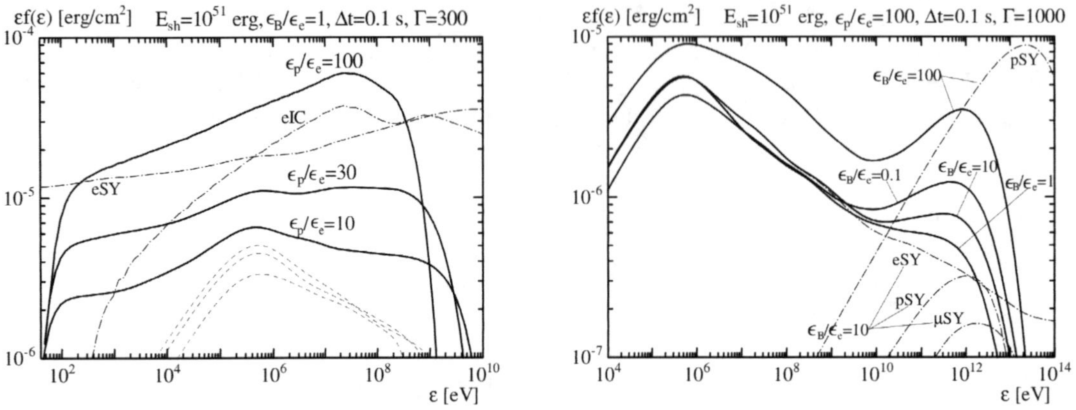

Figure 2. Prompt emission spectra from internal shocks in proton-dominated GRBs. See [49] for more details. (Left) For for $\Gamma = 300$ and different proton/electron ratios. Solid curves are the total emission, and dot-dashed curves denote separately electron synchrotron (eSY) and electron inverse Compton (eIC) components without $\gamma\gamma$ absorption effects. Dashed curves are primary electron components only. (Right) Similar to left panel, except for $\Gamma = 1000$ and different magnetic fields. Dot-dashed curves denote separately proton (pSY) and muon synchrotron (μSY) components.

A different class of low luminosity GRBs have been proposed to be more important than high luminosity GRBs for the integrated energy budget, and hence potentially more relevant as sources of UHECRs [51]. As with AGN jets, the acceleration and survival of heavy nuclei in in the environments of both high and luminosity GRBs is an important question [52].

6. Cluster accretion shocks

Clusters of galaxies are the largest gravitationally bound systems in the universe, composed of stars, hot gas and dark matter. According to the currently favored picture of hierarchical structure formation in the CDM cosmology, they are the latest objects to form, and should be surrounded by strong accretion shocks as a consequence of continuing infall of dark matter and baryonic gas [54]. Such shocks should be interesting sites of particle acceleration and have been proposed as sources of UHECRs [36]. However, estimates of the maximum energy $E_{\rm max}$ for protons seem to fall short of 10^{20} eV by 1-2 orders of magnitude [55, 30].

We have shown that invoking UHECR nuclei in cluster accretion shocks may offer a possible solution [53]. Heavy nuclei with higher Z have correspondingly shorter acceleration time so that Fe may be accelerated up to 10^{20} eV in the same shock conditions, notwithstanding energy losses by photodisintegration with the FIRB and CMB. With reasonable assumptions regarding the number density and CR power of the sources, detailed propagation calculations demonstrate that the model compares favorably with the existing observations, as long as the source composition of Fe is similar to that of Galactic CRs (Fig.3). See also [56] for related studies.

The mass composition at $\lesssim 3 \times 10^{19}$ eV should be predominantly light and consistent with HiRes reports [15]. The rapid increase of the average mass at higher energies is a clear prediction of the scenario, in line with the latest Auger elongation measurements [16], and to be tested further by new generation facilities including Telescope Array and JEM-EUSO.

Despite the relative rarity of massive clusters in the local universe, strong deflections of the highly charged nuclei in EGMF allow consistency with the global isotropy. On the other hand, with a sufficient number of accumulated events, anisotropies toward a small number of

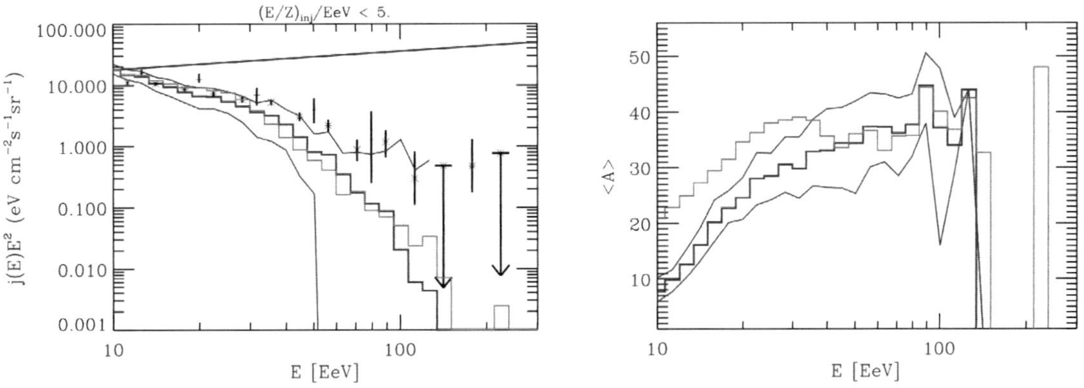

Figure 3. Observed UHECR spectrum (left) and mean mass composition (right) versus energy E (1 EeV $\equiv 10^{18}$ eV) from cluster accretion shocks for $\alpha = 1.7$ and $\beta = 0.5$ [53]. Compared are current data for HiRes [57] (bars) and Auger [10] (asterisks). The histograms are the average over different model realizations for the cases with (thick) and without (thin) EGMF, and the thin curves outline the median deviations due to cosmic variance for the former case only. The straight line in the top panel denotes the injection spectrum.

individual sources should appear, although this expectation is subject to uncertainties in the EGMF and GMF. Whether it is consistent with the latest Auger anisotropy results calls for a deeper investigation. It could possibly be the case, including the observed deficit of events from the Virgo direction, if the GMF is of the plausible, bisymmetric spiral type, for which deflection angles can be minimal in selected regions of the sky around the supergalactic plane, particularly around Centaurus [23] (see also [58] for related discussions). Note that the Centaurus cluster at distance ~50 Mpc is one of most prominent clusters in the local universe, somewhat more massive than Virgo [59].

We mention that the escape of CRs from clusters can be mediated by diffusion in directions away from the filaments, or perhaps by advection in partial outflows during merging events. Energy-dependent escape upstream of the shock may also be possible. An aspect of this scenario that warrants further study is the spectral domain $< 10^{19}$ eV and the implications for the Galactic-extragalactic transition region. Alternatively, provided that EGMF horizon effects are effective $\lesssim 10^{17}$ eV [60], cluster accretion shocks may give an appreciable contribution to CRs between the second knee and the ankle, in which case they may also be interesting as high energy neutrino sources [61] (see also [62]).

Finally, we may look forward to very unique signatures in X-rays and gamma-rays. Protons accelerated to 10^{18}-10^{19} eV in cluster accretion shocks should efficiently channel energy into pairs of energy 10^{15}-10^{16} eV through interactions with the CMB, which then emit synchrotron radiation peaking in hard X-rays and inverse Compton radiation in TeV gamma-rays [30]. The detection prospects are promising for Cerenkov telescopes such as HESS, and hard X-ray observatories such as Suzaku and the future NeXT mission. Photopair production by nuclei may also be efficient and induce further interesting signals that are worth investigating.

7. Conclusion and outlook

Besides AGNs, GRBs and clusters, other proposals for the astrophysical origin of UHECRs not covered in this brief review include starburst galaxies, merging galaxies, extragalactic magnetars, dormant black holes, etc. (see [26]). The true source of UHECRs could be one or more of these; alternatively, we should not discount the possibility that it is actually none of them, and the truth something totally unexpected.

In this regard, we may recall the history of GRB research. Before the launch of BATSE in 1991, the sample of GRBs numbered $\lesssim 200$, with no clear anisotropy observed in the sky distribution. However, most believed, based on theoretical plausibility, that they were generated by Galactic neutron stars, while a very small minority advocated neutron star mergers at cosmological distances. After several years and a sample of more than 3000 bursts, the currently most favored scenario for the progenitors of long GRBs is related to the collapse of massive stars, a concept that didn't exist before BATSE. The lesson is that we should be cautious about jumping to conclusions on the origin of UHECRs based on just 20 observed events and our limited theoretical understanding of cosmic objects.

Regardless of theoretical prejudice, the quest for the solution of the UHECR mystery is progressing in earnest with ongoing measurements by Auger and Telescope Array. An order of magnitude greater statistics for the highest energy events expected from JEM-EUSO or Auger North+South, and the combined effort of upcoming neutrino, X-ray and gamma-ray observations is expected to lead us toward the true answer in the near future.

Acknowledgments
The author is thankful for past and ongoing collaborations with F. Aharonian, E. Armengaud, K. Asano, P. Meszaros, F. Miniati, K. Murase, K. Sato, G. Sigl, N. Sugiyama, H. Takami and T. Yamamoto.

References

[1] Nagano M and Watson A A 2000 *Rev. Mod. Phys.* **72** 689
Stanev T 2004 *High Energy Cosmic Rays* (Chichester: Springer)
Cronin J W 2005 *Nucl. Phys. B Proc. Suppl.* **138** 465
[2] Watson A A *Preprint* arXiv:0801.2321
Kampert K-H *Preprint* arXiv:0801.1986
[3] Fukushima M 2007 *Energy Budget in the High Energy Universe* ed K Sato and J Hisano (Singapore: World Scientific) p1
[4] Nitz D [Pierre Auger collaboration] *Preprint* arXiv:0706.3940
[5] Teshima M 2007 *Astron. Nach.* **328** 621
[6] Inoue S 2007 *Energy Budget in the High Energy Universe* ed K Sato and J Hisano (Singapore: World Scientific) p17 (*Preprint* arXiv:astro-ph/0701835)
Inoue S 2008 *Astronomy and Astrophysics of the Extreme Universe* ed Y Takahashi et al (Tokyo: Universal Academy Press), in press
[7] Bhattacharjee P and Sigl G 2000 *Phys. Rep* **327** 109
[8] Protheroe R J and Clay R W 2004 *Pub. Astron. Soc. Aus.* **21** 1
[9] Greisen K 1966 *Phys. Rev. Lett.* **16** 748
Zatsepin G T and Kuzmin V A 1966 *JETP Lett.* **4** 78
[10] Yamamoto T [Pierre Auger collaboration] *Preprint* arXiv:0707.2638
Pierre Auger Collaboration *Preprint* arXiv:0806.4302
[11] Abbasi R U et al. 2008 *Phys. Rev. Lett.* **100** 101101
[12] Pierre Auger Collaboration 2008 *Astropart. Phys.* **29** 243
[13] Puget J L, Stecker F W and Bredekamp J H 1976 *Astrophys. J.* **205** 638
[14] Stecker F W and Salamon M H 1999 *Astrophys. J.* **512** 521
Hooper D, Sarkar S and Taylor A M 2007 *Astropart. Phys.* **27** 199
[15] Abbasi R U et al. 2005 *Astrophys. J.* **622** 910
[16] Unger M [Pierre Auger collaboration] *Preprint* arXiv:0706.1495
[17] Watson A A 2004 *Nuc. Phys. B Proc. Suppl.* **136** 290.
[18] Berezinsky V, Gazizov A and Grigorieva S 2006 *Phys. Rev.* D **74** 43005 (*Preprint* hep-ph/0204357)
Berezinsky V *Preprint* arXiv:0801.3028
[19] Allard D, Parizot E and Olinto A V 2007 *Astropart. Phys.* **27** 61
[20] Kronberg P P 1994 *Rep. Prog. Phys.* **57** 325
[21] Sigl G, Miniati F and Ensslin T A 2003 *Phys. Rev.* D **68** 43002
Dolag K et al. 2004 *JETP* **79** 583
Takami H, Yoshiguchi H and Sato K 2006 *Astrophys. J.* **639** 803

[22] Kachelrieß M, Serpico P D and Teshima M 2007 *Astropart. Phys.* **26** 378
[23] Takami H and Sato K *Astrophys. J.* submitted (*Preprint* arXiv:0711.2386)
[24] Pierre Auger collaboration 2007 *Science* **318** 938
Pierre Auger collaboration *Preprint* arXiv:0712.2843
[25] Abbasi R U et al. *Preprint* arXiv:0804.0382
[26] A. M. Hillas 1984 *Ann. Rev. Astron. Astrophys.* **22** 425
Torres D and Anchordoqui L A 2004 *Rep. Prog. Phys.* **67** 1663
[27] Inoue S and Sasaki S 2001 *Astrophys. J.* **562** 618
[28] Porciani C and Madau P 2001 *Astrophys. J.* **548** 522
[29] Keshet U, Waxman E and Loeb A 2004 *Astrophys. J.* **617** 281
Pavlidou V and Fields B D 2006 *Astrophys. J.* **642** 734
[30] Inoue S, Aharonian F A and Sugiyama N 2005 *Astrophys. J.* **628** L9
[31] Krolik J H 1998 *Active Galactic Nuclei* (Princeton: Princeton Univ. Press)
[32] Szabo A P and Protheroe R J 1994 *Astropart. Phys.* **2** 375
[33] Mannheim K, Protheroe R J and Rachen J P 2001 *Phys. Rev.* D **63** 3003
Atoyan A M and Dermer C D 2003 *Astrophys. J.* **586** 79
[34] Inoue S and Takahara F 1996 *Astrophys. J.* **463** 555
[35] Rachen J P and Biermann P L 1993 *Astron. Astrophys.* **272** 161
[36] Norman C A, Melrose D B and Achterberg A 1995 *Astrophys. J.* **454** 60
[37] Dermer C D *Preprint* arXiv:0711.2804
[38] Waxman E and Miralda-Escudé J 1996 *Astrophys. J.* **472** L89
[39] Martini P, *Preprint* arXiv:astro-ph/0304009
[40] Halzen F and Hooper D 2002 *Rep. Prog. Phys.* **65** 1025
[41] Aharonian F A 2002 *Mon. Not. R. A. S.* **332** 215
[42] Gabici S and Aharonian F A 2005 *Phys. Rev. Lett.* **95** 251102
Armengaud E, Sigl G and Miniati F 2006 *Phys. Rev.* D **73** 083008
[43] Meszaros P 2006 *Rep. Prog. Phys.* **64** 429
[44] Waxman E 1995 *Phys. Rev. Lett.* **75** 386
Vietri M 1995 *Astrophys. J.* **453** 883
Waxman E and Bahcall J N 2000 *Astrophys. J.* **541** 707
[45] Gallant Y A and Achterberg A 1999 *Mon. Not. R. A. S.* **305** L6
[46] Vietri M, De Marco D and Guetta D 2003 *Astrophys. J.* **592** 378
Waxman E 2004 *Astrophys. J.* **606** 988
[47] Dermer C D and Atoyan A 2006 *New J. of Phys.* **8** 122
[48] Asano K and Inoue S 2007 *Astrophys. J.* **671** 645
[49] Asano K, Inoue S and Meszaros P 2008 *Astrophys. J.* submitted (*Preprint* arXiv:0807.0951)
[50] Le T and Dermer C D 2007 *Astrophys. J.* **661** 394
Guetta D and Piran T 2007 *JCAP* **07** 003
[51] Murase K et al. 2006 *Astrophys. J.* **651** L05
[52] Anchordoqui L A et al. 2007 *Astropart. Phys.* **29** 1
Wang X-Y, Razzaque S and Meszaros P *Astrophys. J.* **677** 432
Murase K et al. *Phys. Rev.* D in press (*Preprint* arXiv:0801.2861)
[53] Inoue S, Sigl G, Miniati F and Armengaud E *Phys. Rev.* D submitted (*Preprint* arXiv:astro-ph/0701167)
Inoue S, Sigl G, Miniati F and Armengaud E *Preprint* arXiv:0711.1027
[54] Ryu D et al., 2003 *Astrophys. J.* **593** 599
[55] Kang H, Ryu D and Jones T W 1996 *Astrophys. J.* **456** 422
Kang H, Rachen J P and Biermann P L 1997 *Mon. Not. R. A. S.* **286** 257
Ostrowski M and Siemieniec-Ozięblo G 2002 *Astron. Astrophys.* **386** 829
[56] Arisaka K et al. 2007 *JCAP* **12** 002
Allard D et al. 2008 *Preprint* arXiv:0805.4779
[57] Abbasi R U et al. 2004 *Phys. Rev. Lett.* **92** 151101
[58] Wibig T and Wolfendale A W *Preprint* arXiv:0712.3403
Gorbunov D S et al. *Preprint* arXiv:0804.1088
[59] Reiprich T H and Böhringer H 2002 Astrophys. J. **567** 716
[60] Kotera K and Lemoine M 2008 *Phys. Rev.* D **77** 023005
[61] Murase K, Inoue S and Nagataki S *Preprint* arXiv:0805.0104
[62] D. De Marco et al., Phys. Rev. D 73 (2006) 043004.

Ultra High-Energy Cosmic Ray Observations

Karl-Heinz Kampert

Bergische Universität Wuppertal, Fachbereich C - Physik, Gaußstr. 20, D-42119 Wuppertal

E-mail: `kampert@uni-wuppertal.de`

Abstract. The year 2007 has furnished us with outstanding results about the origin of the most energetic cosmic rays: a flux suppression as expected from the GZK-effect has been observed in the data of the HiRes and Auger experiments and correlations between the positions of nearby AGN and the arrival directions of trans-GZK events have been observed by the Pierre Auger Observatory. The latter finding marks the beginning of ultra high-energy cosmic ray astronomy and is considered a major breakthrough starting to shed first light onto the sources of the most extreme particles in nature. This report summarizes those observations and includes other major advances of the field, mostly presented at the 30th International Cosmic Ray Conference held in Mérida, Mexico, in July 2007. With increasing statistics becoming available from current and even terminated experiments, systematic differences amongst different experiments and techniques can be studied in detail which is hoped to improve our understanding of experimental techniques and their limitations.

1. Introduction

Understanding the origin of the highest energy cosmic rays is one of the most pressing questions of astroparticle physics. Cosmic rays with energies exceeding 10^{20} eV have been observed for more than 40 years (see e.g. [1]) but due to their low flux only some ten events of such high energies could be detected up to recently. There are no generally accepted source candidates known to be able to produce particles of such extreme energies. Moreover, there should be a steeping in the energy spectrum near 10^{20} eV due to the interaction of cosmic rays with the microwave background radiation (CMB). This Greisen-Zatsepin-Kuzmin (GZK) effect [2] severely limits the horizon from which particles in excess of $\sim 6 \cdot 10^{19}$ eV can be observed. For example, the sources of protons observed with $E \geq 10^{20}$ eV need to be within a distance of less than 50 Mpc [3]. The non-observation of the GZK-effect in the data of the AGASA experiment [4] has motivated an enormous number of theoretical and phenomenological models trying to explain the absence of the GZK-effect and has stimulated the field as a whole. Only this year, with the final analysis of the HiRes-data [5] and the advent of high-statistics and high quality hybrid data from the Pierre Auger Observatory (PAO) [6], the situation has changed considerably: a suppression such as expected from the GZK-effect is now observed with high statistical significance. The very recent breaking news about the observation of directional correlations of the most energetic Pierre Auger events with the positions of nearby AGN [7] complements the observation of the GZK effect very nicely and provides evidence for an astrophysical origin of the most energetic cosmic rays. Another key observable allowing to discriminate different models about the origin of high-energy cosmic rays is given by the mass composition of cosmic rays. Unfortunately, such measurements are much more difficult due to their strong dependence on hadronic interaction models. Only primary photons can be discriminated safely from protons and nuclei and recent upper limits to their flux largely rule out top-down models, originally invented to explain the apparent absence of the GZK-effect in AGASA data.

In this article, prepared for the TAUP conference in Sendai (Japan), we describe the status of each of these topics, as reported during the recent International Cosmic Ray Conference held in Mérida, Mexico, in July 2007 (ICRC2007) and in publications becoming available since.

2. Experiments and their Exposures

Most of the data available today at energies above $\sim 10^{18}$ eV are provided by the 100 km^2 AGASA array [4], the HiRes fluorescence telescopes [5] and the 3000 km^2 PAO [8]. Both, the AGASA and HiRes instruments have now closed while the PAO started operation during its construction phase in 2004. Even though its last surface detector stations are planned to be deployed in March 2008 only, it has already accumulated the largest exposure available today and it will continue to deliver more than about 7000 km^2 sr for each year of operation. Table 1 lists for various experiments the approximate accumulated exposures and currently observed numbers of events claimed to be above 10 and 50 EeV, respectively. The table (based on [9]) includes for comparison smaller air shower arrays, such as the phased out Haverah Park [10] and AKENO [11] arrays, the Yakutsk [12], and the KASCADE-Grande array [13]. The Telescope Array (TA), led by a Japanese consortium in Millard County, Utah, USA, has just begun operation this year [14]. Like the PAO, the TA is a hybrid detector. It covers an area of 860 km^2 and comprises 576 scintillator stations and three FD sites on a triangle with about 35 km separation each equipped with 12 fluorescence telescopes.

The aperture assumed in calculating each exposure is appropriate to about 50 EeV. Please note that the apertures of fluorescence telescopes such as employed by HiRes are a growing function of energy while those of ground arrays are flat above their respective threshold energies. A cursory comparison of the rate of events above 50 EeV (the energy at which the exposures in Table 1 are calculated) makes clear that the differences between the integral rates are much larger (by more than a factor of 2) than can be accounted for by Poissonian variations. Possible reasons for the discrepancies will be discussed below.

Table 1. Exposure and approximate event numbers from various instruments.

Experiment	**status**	**km^2 sr yr @ 50 EeV**	**# events > 10 EeV**	**# events > 50 EeV**	**based on Ref.**
Haverah Park	1962-1987	$\sim$ 245	106	10	[10]
Yakutsk	1974-present	$\sim$ 900	171	6	[9, 12]
AGASA	1993-2005	1620	886	46	[15]
HiRes-I mono	1997-2006	$\sim$ 4500	561	31	[16, 17, 18]
HiRes-II mono	1999-2006	$\sim$ 1500	179	12	[16, 17]
HiRes stereo	1999-2006	$\sim$ 2400	270	11	[19, 20]
Auger	2004-present	$\sim$ 7000	1644	38	[21, 22]
TA	2007-present	860$\times$yrs			[14]

In case of ground arrays, the aperture is calculated in a straight forward and model independent way, once the energy threshold for CR detection and reconstruction is exceeded. The only uncertainty arises from the reconstruction of the landing point of the shower, which again is safely reconstructed if only showers within the geometry covered by the array are considered. The situation is quite different for fluorescence telescopes. Here, the maximum distance out to which showers can be observed increases with increasing fluorescence light and thereby increasing energy. On the one hand, the growing aperture is very attractive and cost effective, as it allows to observe more showers at high energies. On the other hand, the maximum distance out to which EAS can be seen is directly related to the signal-to-noise ratio in the light sensors of the cameras, the varying atmospheric conditions, etc. Moreover, an accurate reconstruction of the CR energy requires the observation of the position of the shower maximum, $X_{\max}$, in the field-of-view of the telescopes. This condition imposes a sensitivity also to the primary mass and consequently also to the hadronic interaction models employed in the aperture calculation.

The effect of growing apertures in case of fluorescence telescopes and constant apertures in case of ground arrays can be seen in Fig. 1. Note also the difference in the detection-/reconstruction thresholds of HiRes-I and -II. This is mostly because of the different ranges of elevation angles viewed by HiRes-I (3°-17°) and -II (3°-31°) cameras: low energy showers, which can - due to their low light level - only be observed near the telescopes, reach their shower maximum above the field-of-view of the cameras and thus cannot be reconstructed. Fig. 2 compares in more details the aperture of HiRes-II for p- and Fe-induced showers [23]. Note that the p- and Fe-apertures differ by more than a factor of 20 at $E \simeq 3 \cdot 10^{17}$ eV! Thus, without knowledge of the primary mass, the flux in this energy range is uncertain by a factor of 20 (or more, dependent on the hadronic interaction model used in the aperture calculation) and composition measurements will be very biased. We also illustrate the width of the experimental energy resolution which indicates that the rapid fall of the aperture distribution almost equals the energy resolution function. Fluctuations in the energy reconstruction by only one standard deviation or a shift of the overall energy scale within the uncertainty of the experiment causes changes in the aperture (and thereby the CR-flux) by a factor of about 6! Clearly, controlling all these uncertainties particularly at energies below 10^{18} eV appears very difficult.

3. The Energy Spectrum

A very important step towards unveiling the origin of the sources of UHECR is provided by measurements of the CR energy spectrum. Four features are known to provide information about the CR origin: the prominent *knee* at $E \simeq 4 \cdot 10^{15}$ eV may signal the limiting energy of galactic CR accelerators and/or the onset of diffusion losses from the galaxy, the *second knee* at $E \simeq 10^{17}$ eV, still to be confirmed [24], is considered to be caused by the fading of

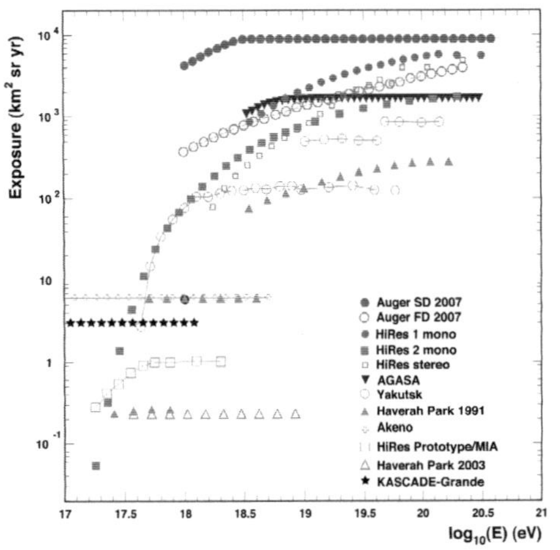

Figure 1. Accumulated exposures of various experiments. The data and exposures are based on Refs. [16, 19, 25, 13, 22].

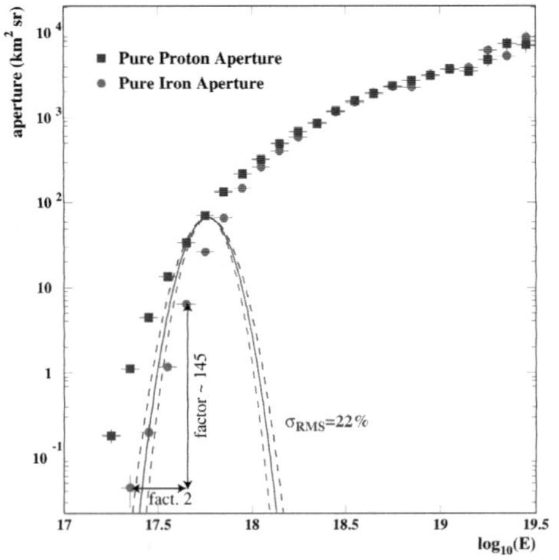

Figure 2. HiRes apertures (from [23]) for pure proton and iron simulations compared to the energy resolution.

the heavy galactic CR component, the *ankle* at $E \simeq 4 \cdot 10^{18}$ eV is either due to the onset of the extragalactic CR component or due to energy losses of extragalactic protons by e^+e^- pair production in the CMB [26], and the *GZK cut-off* at $E \simeq 6 \cdot 10^{19}$ eV [2] is due to photo-pion production of extragalactic protons in the CMB.

Recent measurements of the CR energy spectrum by AGASA and HiRes have yielded results which differ in their shape and overall flux. A comparison including data from the PAO as presented at the ICRC 2007 is shown in Fig. 3. Generally, the error bars in such plots are of statistical nature only and neglect systematic uncertainties in the determination of the energy scale and exposure. Typical uncertainties of the energy scale are on the order of 20-25 %. Ground arrays like AGASA rely entirely on EAS simulations with their uncertainties originating from the limiting knowledge of hadronic interactions (total inelastic cross sections, particle multiplicities, inelasticities, etc.). CORSIKA simulations [27] have shown that the muon density at ground predicted by different hadronic interaction models differ by up to 30 %. Fluorescence telescopes, such as operated by HiRes and the PAO, observe the (almost) full longitudinal shower development in the atmosphere. In this way, the atmosphere is employed as a homogenous calorimeter with an absorber thickness of 30 radiation lengths or 11 hadronic interaction lengths. Corrections for (model dependent) energy 'leakage' into ground -mostly by muons and neutrinos - are below 10 % and their uncertainties are only a few percent. As a consequence, fluorescence detectors provide an energy measurement which is basically independent from hadronic interaction models. Uncertainties in the energy scale arise most dominantly from the fluorescence yield in the atmosphere. Several measurements have been made in the past, e.g. the Auger Collaboration uses the fluorescence yield by Nagano et al. [28] and HiRes by Kakimoto et al. [29]. This is an unpleasant situation, which by itself may account for a ~ 10 % shift between the energy scales of Auger and HiRes. For this reason, major efforts have been started to remeasure the fluorescence yield as a function of temperature, pressure and humidity with high precision [30] in order to reduce this source of uncertainty.

Taking benefit of the Auger hybrid detector, the Auger Collaboration has used a clean set of hybrid data, in which EAS have been detected simultaneously by at least one florescence eye and the ground array, to calibrate their observatory [21]. To evaluate the observed differences in the energy scale of AGASA, HiRes and Auger in Fig. 3, an overall shift of +17 % and −25 % has been applied to the Auger and AGASA energy scale, respectively (c.f. Fig. 4). Such a shift remains well within the quoted uncertainties of the experiments (particularly when accounting for the different fluorescence yields used by HiRes and Auger) and yields a fairly good agreement

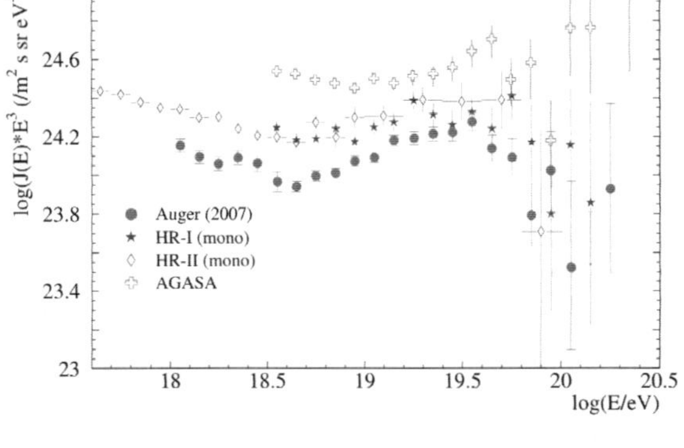

Figure 3. Cosmic ray flux measurements (multiplied by E^3) from AGASA [4], HiRes [17], and the PAO [21].

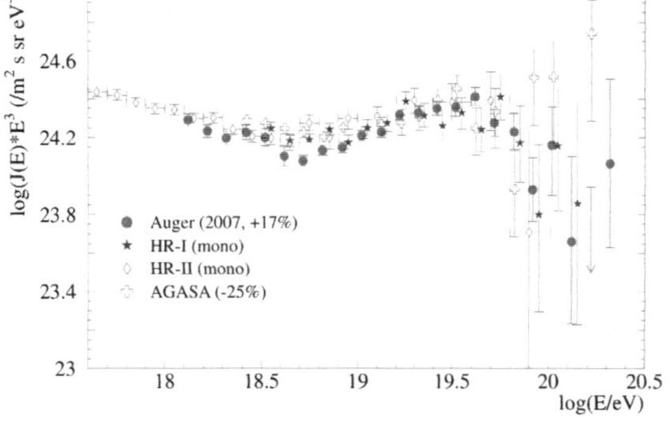

Figure 4. Same as Fig. 3 but with the energy scale of Auger and AGASA shifted by +17 % and −25 %, respectively.

of the data points, except perhaps at $E \gtrsim 10^{20}$ eV for AGASA and in the ankle region which appears sharper in the PAO data.

The GZK-like suppression is clearly visible in both the HiRes and Auger data. Using different statistical approaches, HiRes quotes a significance of about 4.5 standard deviations and the PAO of more than 6σ. For example, fitting a power law to the Auger spectrum between $4 \cdot 10^{18} < E < 4 \cdot 10^{19}$ eV using a binned likelihood method yields $\gamma = -2.69 \pm 0.02(\mathrm{stat}) \pm 0.06(\mathrm{sys})$. Extrapolating this slope to higher energies one expects 167 ± 3 and 35 ± 1 events at $E > 4 \cdot 10^{19}$ and 10^{20} eV, respectively, whereas 66 and 1 event are observed. The observation of the GZK-effect 40 years after its prediction provides for the first time clear evidence for an extragalactic origin of EHECRs. Of course, this interpretation is challenged if the sources would happen to run out of acceleration power just at the value of the GZK threshold. However, this would be a strange coincidence and in fact is not supported by Pierre Auger data (see Sect. 5).

The shape of the energy spectrum around 10^{20} eV carries information about the distance distribution of CR sources and their injection spectrum [6]. However, more statistics is required before firm conclusions can be drawn. Answering the question about the origin of the ankle and discriminating the e^+e^- dip-model [26] from the traditional ankle model cannot be done based on the energy spectrum alone but requires measurements of the CR composition.

4. Chemical Composition, Photon and Neutrino Limits

As noted above, the mass composition of CRs allows to discriminate models of UHECR origin and may be the only measurement allowing to answer the question about the transition from galactic to extragalactic CRs. The Berezinsky dip-model [26] predicts the transition taking place at energies significantly below the ankle but requires a proton dominant composition in the ankle region to make the Bethe-Heitler pair production process work. In the classical picture, on the other hand, the ankle itself marks the transition region and one expects a change from a heavy to light composition at the ankle. Unfortunately, the mass composition can be inferred only indirectly from EAS experiments by making assumptions about the hadronic interactions at the highest energies. In contrast to energy measurements, this model dependence is true also for fluorescence detectors. The key observable here is the position of the shower maximum, X_{max}, which is directly observed by fluorescence telescopes and can be inferred also from surface detector data. New results based on HiRes-Stereo and PAO hybrid data were reported at the ICRC [31, 32]. As can be seen from Fig. 5, both data sets agree very well up to $\sim 3 \cdot 10^{18}$ eV but differ slightly at higher energies. The differences between the two experiments is of the same order as the differences observed between p- and Fe-predictions for different hadronic interaction models. With these caveat kept in mind, both experiments observe an increasingly lighter composition towards the ankle. At higher energies, the HiRes measurement yields a lighter composition than Auger. Also shown are predictions of X_{max} based on the QGSJET01 model for the traditional G-EG transition [33] (labelled “A”) and the Berezinsky dip-model [26] (labelled “B”). None of the two models appears to describe the preliminary data well, but they demonstrate the power of such measurements which will be particularly important in the energy range 10^{17}-10^{18} eV. The low energy upgrade HEAT with additional high elevation telescopes, infill stations and muon detectors in Auger [34], the low energy extension TALE of the TA [35], KASCADE-Grande [13, 36], and IceTop/IceCube [37] will provide powerful data in this energy range to provide a definite answer about the G-EG-Transition already in the very near future.

Primary photons are easier to separate experimentally as they penetrate deeper into the atmosphere, particularly at energies above 10^{18} eV. Their EAS development is also much less affected by uncertainties of hadronic interaction models due to the dominant electromagnetic shower component. They are of interest for several reasons: top-down models, invented to explain the apparent absence of the GZK-effect in AGASA data, predict a substantial photon flux at high energies [38]. In the presence of a GZK effect, UHE photons can act as tracers of the

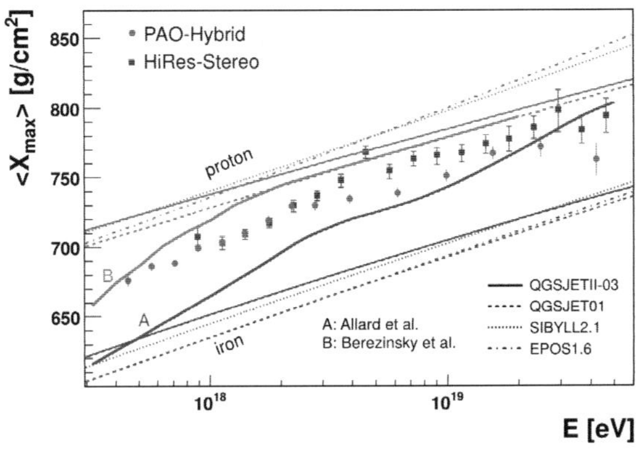

Figure 5. $\langle X_{\max} \rangle$ as a function of energy for HiRes Stereo [32] and PAO hybrid data [31] compared to proton and Iron predictions using different hadronic interaction models and different models of EHECR origin.

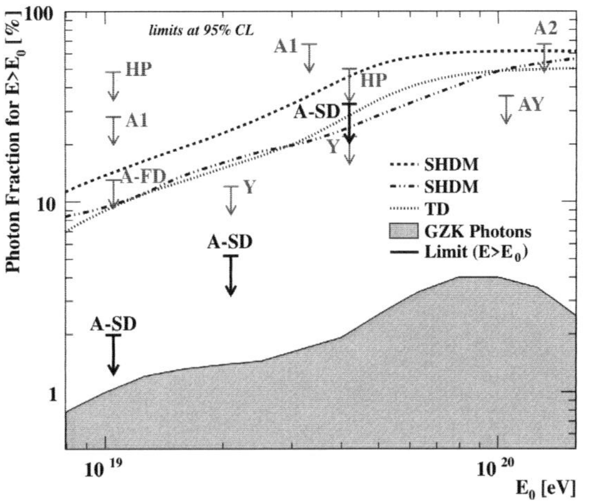

Figure 6. Upper limits on the fraction of photons in the integral CR flux compared to predictions. For references see [39].

GZK process and provide relevant information about the sources and propagation. Moreover, they can be used to obtain input to fundamental physics and UHE photons could be used to perform EHE astronomy.

Experimentally, photon showers can be identified by their longitudinal shower profile, most importantly by their deep $X_{\max}$ position and low muon numbers. Up to now, only upper limits could be derived from various experiments, either expressed in terms of the photon fraction or the photon flux. Figure 6 presents a compilation of present results on the photon fraction. The most stringent limits are provided by the Auger surface detector [39]. Current top-down models appear to be ruled out by the current bounds. This result can be considered an independent confirmation of the GZK-effect seen in the energy spectrum. It will be very exciting to possibly even touch the flux levels expected for GZK-photons ($p+\gamma_{CMB} \to p+\pi^0 \to p+\gamma\gamma$) after several years of data taking.

The detection of UHE cosmic neutrinos is another long standing experimental challenge. All models of UHECR origin predict neutrinos from the decay of pions and kaons produced in hadronic interactions either at the sources or during propagation in background fields. Similarly to GZK-photons one also expects GZK-neutrinos, generally called 'cosmogenic neutrinos'. Moreover, top-down models predict dominantly neutrinos at UHE energies. Even though neutrino flavors are produced at different abundances, e.g. a 1:2 ratio of ν_e:ν_μ results from pion decay, neutrino oscillations during propagation will lead to equal numbers of ν_e, ν_μ, and ν_τ at Earth. At energies above 10^{15} eV, neutrinos are absorbed within the Earth so that upgoing neutrino induced showers cannot be detected anymore. Only tau neutrinos entering the Earth just below the horizon (Earth-skimming) can undergo charged-current interactions to produce τ leptons which then can travel several tens of kilometers in the Earth and emerge into the atmosphere to eventually decay in flight producing a nearly horizontal air shower above the detector. Such showers can be searched for in ground arrays and fluorescence detectors. The absence of any candidates observed in the detectors has been used to place upper limits on diffuse neutrino fluxes. As can be seen from Fig. 7, AMANDA and the PAO provide at present the best upper limits up to energies of about 10^{19} eV and, similarly to the photons discussed above, they already constrain top-down models and are expected to reach the level of cosmogenic neutrinos after several years of data taking.

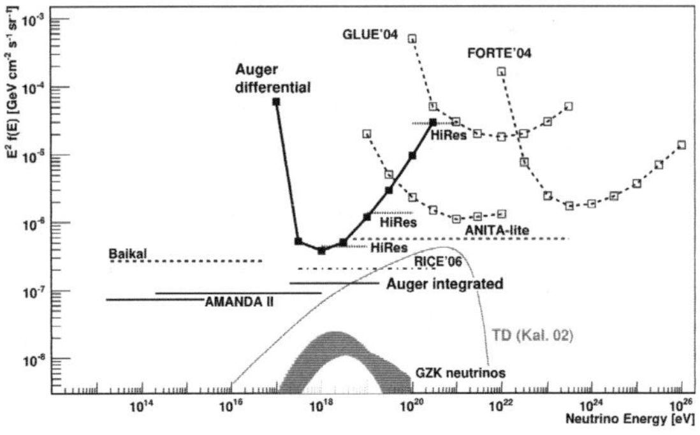

Figure 7. Limits at the 90 % C.L. for a diffuse flux of ν_τ assuming a 1:1:1 ratio of the 3 neutrino flavors ([40] and references therein) and predictions for a top-down model [41].

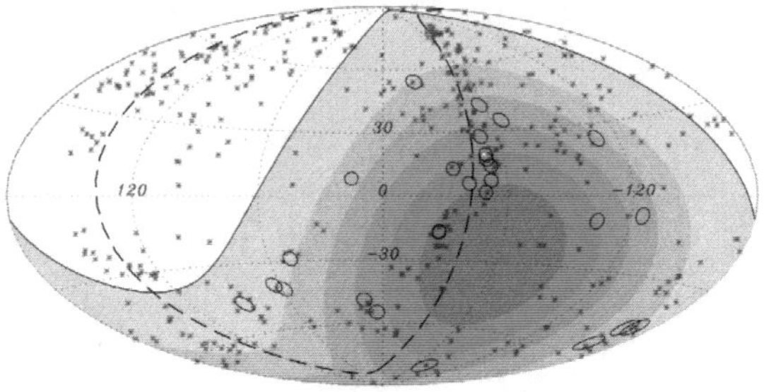

Figure 8. Aitoff projection of the celestial sphere in galactic coordinates. The positions of the AGN within $D < 71$ Mpc (stars) and of the events with $E > 57$ EeV (circles) are marked The colors indicate equal exposure [22].

5. Arrival Directions and Correlations with AGN

Recently, the Pierre Auger Collaboration reported the observation of a correlation between the arrival directions of the highest energy CRs and the positions of nearby AGN from the Véron-Cetty catalogue at a confidence level of more than 99 % [7, 22]. Since several claims about seeing clustering of EHECRs were already made in the past with none of them being confirmed by independent data sets, the Auger group has performed an 'exploratory' scan of parameters using an initial data-set and applied these parameters to a new independent data-set for confirmation. With the parameters specified *a priori* the analysis avoids the application of penalty factors which otherwise would need to be applied for in *a posteriori* searches. The correlation has maximum significance for CRs with energies greater than $5.7 \cdot 10^{19}$ eV and AGN at a distance less than ~ 71 Mpc. At this energy threshold, 20 of the 27 events correlate within $3.2°$ with positions of a nearby AGN.

Observing such kind of anisotropy can be considered the first evidence for an extragalactic origin of the most energetic CRs because none of any models of galactic origin even including a very large halo would result in an anisotropy such as observed in the data. Besides this, the correlation parameters itself are highly interesting as the energy threshold at which the correlation becomes maximized matches the energy at which the energy spectrum shows the GZK feature (~ 50 % flux suppression), i.e. CRs observed above this threshold need to originate from within the GZK-horizon of ~ 100-200 Mpc. This number again matches (within a factor of two) the maximum distance for which the correlation is observed! Thus, the set of the two parameters suggests that the suppression in the energy spectrum is indeed due to the GZK-effect, rather than to a limited energy of the accelerators. Thereby, the GZK-effect acts as an effective filter to nearby sources and minimizes effects from extragalactic magnetic field deflections. On top of this, it is also the large magnetic rigidity which helps to open up the window for performing charged particle astronomy.

The correlation may tell us also about the strength of galactic and extragalactic magnetic fields. The galactic fields are reasonably well known and one expects strong deflections for particles arriving from nearby the galactic plane even at energies of 60 EeV. And in fact, 5 of the 7 events that do not correlate with positions of nearby AGN arrive with galactic latitudes $|b| < 12°$. The angular scale of the observed correlation also implies that the intergalactic magnetic fields do not deflect the CRs by more than a few degrees and one can constrain models of turbulent magnetic fields to $B_{\rm rms}\sqrt{L_c} \leq 10^{-9}$ G$\sqrt{\rm Mpc}$ within the GZK horizon assuming protons as primary particles [22].

6. Concluding Remarks

Remarkable progress has been made in cosmic ray physics at the highest energies, particularly by the start-up of the (still incomplete) Pierre Auger Observatory. The event statistics above 10^{19} eV available by now allows detailed comparisons between experiments and indicates relative shifts of their energy scales by $\pm 25\,\%$. Given the experimental and theoretical difficulties in measuring and simulating extensive air showers at these extreme energies, this may be considered a great success. On the other hand, knowing about overall mismatches of the energy scales between experiments may tell us something. Clearly, in case of fluorescence detectors better measurements of the spectral and absolute fluorescence yields and their dependence on atmospheric parameters are needed and will hopefully become available in the very near future [30]. This should furnish all fluorescence experiments with a common set of data. Differences in the calibration between surface detectors and fluorescence telescopes, best probed by hybrid experiments like Auger and TA, may then be used to test the modelling of EAS. The muon component at ground, known to be very sensitive to hadronic interactions at high energies [27], could in this way serve to improve hadronic interaction models in an energy range not accessible at man-made accelerators. In fact, several studies (e.g. [42]) indicate a deficit of muons by 30 % or more in interaction models like QGSJET.

The energy scale is of great importance also for the AGN correlation discussed in the previous section. As shown in [22], the correlation sets in abruptly at an (Auger) threshold energy of about 57 EeV. Already a downshift in energy by 17 % (the mismatch between Auger and HiRes) would weaken the signal by more than 3 orders of magnitude to make it basically disappear. Thus, verification of the correlation signal by HiRes or AGASA would need to be done for a threshold energy (on their scale) of 67 EeV and 85 EeV, respectively. In this energy range, HiRes observes a spectral slope of $\gamma = 5.1 \pm 0.7$ [17], i.e. the number of events available for a correlation analysis would, according to Table 1, drop to about 12 (HR-I) and 4 (HR-II) when taking the rise of the apertures into account. This would amount to about half the statistics of Auger, well in agreement with the quoted exposures. Unfortunately, the angular resolution of monocular reconstruction is by far too poor for such a test. Only stereo data could provide the required angular resolution. However, in this case the expected statistics of about 5 events above threshold (based upon the numbers and exposures given above) appears too small for any verification.

In fact, the distance parameter of the correlation of 71 Mpc may indicate a mismatch of the energy scale: For protons above 57 EeV the GZK horizon would be 200 Mpc [3] but already for 20 % higher energy it would shrink by more than a factor of two to become consistent to the correlation parameter. Another puzzling feature is the observed small deflection of particles which suggests dominantly protons as primaries. Note that 90 % of the events (20/22) off the galactic plane are correlated to within $\sim 3^\circ$ which AGN positions which is very unlikely for heavy nuclei. On the other hand, the elongation curves in Fig. 5 suggests an admixture of heavy nuclei by more than 10 %. This may be related again to imperfections of the hadronic interaction models used for comparison in Fig. 5.

All of this tells us that the near future will be highly exciting: The question of the energy scales will soon be settled and more detailed comparisons between experiments will become possible. The shape of the energy spectrum in the GZK region will tell us about the source evolution, the composition in the ankle region will answer the question about the G-EG transition, observations of cosmogenic photons and neutrinos are in reach and in case of neutrinos will probe the GZK effect over larger volumes, the correlations will be done with better statistics, with improved search techniques and with more appropriate source catalogues and source selection parameters to tell us about source densities, and the true sources of EHECRs. Very important to note is that different pieces of information start to mesh and are being accessed from different observational techniques and can be cross-checked: *The big picture is being painted!*

Given the scientific importance of this, it would be a mistake to have only one observatory - even when operated as a hybrid detector - taking data. The TA project and its extensions will be very important particularly in the sub-GZK range but, unfortunately, will be too small to collect sufficient statistics at the highest energies. Auger-North will be imperative here and needs immediate vigorous support. The next generation experiment JEM EUSO to be mounted at the Exposed Facility of Japanese Experiment Module JEM EF will potentially reach much larger exposures but still faces many experimental challenges to be addressed.

Acknowledgement

I would like to thank the organizers of the excellent TAUP meeting in Sendai for the invitation to give a review talk at this exciting time. Also, its a pleasure to thank many of my colleagues for stimulating discussions. The German Ministry for Research and Education (BMBF) and the Deutsche Forschungsgemeinschaft (DFG) are gratefully acknowledged for financial support.

References

[1] M. Nagano and A.A. Watson, *Rev. Mod. Phys.* **72** (2000) 689.
[2] K. Greisen, *Phys. Rev. Lett.* **16** (1966) 748, and G.T. Zatsepin and V.A. Kuz'min, *Sov. Phys. JETP Lett.* (Engl. Transl.), **4** (1966) 78.
[3] D. Harari, S. Mollerach, E. Roulet, *JCAP* **012** (2006) 611; (arXiv:astro-ph/0609294).
[4] M. Takeda et al. [AGASA Collaboration] *Astropart. Phys.* **19** (2003) 447.
[5] R. Abbasi et al. [HiRes Collaboration], submitted to *Phys. Rev. Lett.*; [arXiv:astro-ph/0703099].
[6] T. Yamamoto et al. [Pierre Auger Collaboration], *Proc. 30th ICRC, Mérida* (2007) p.0318, [arXiv:0707.2638].
[7] J. Abraham et al. [Pierre Auger Collaboration], *Science* **318** (2007) 938; [arXiv:0711.2256].
[8] J. Abraham et al. [Pierre Auger Collaboration], *Nuc. Instr. Meth.* **A 523** (2004) 50-95.
[9] A.A. Watson, *J. Phys. Conf. Ser.* **39** (2006) 365-371; [arXiv:astro-ph/0511800].
[10] M.A. Lawrence, R.J.O. Reid, A.A. Watson *J. Phys.* **G 17** (1991) 733.
[11] M. Nagano et al., *J. Phys.* **G 18** (1992) 423.
[12] A. V. Glushkov and M. I. Pravdin, *JETP* **101** (2005) 88.
[13] A. Haungs et al. [KASCADE Collaboration], *Proc. 30th ICRC, Mérida* (2007) p.828
[14] M. Fukushima et al. [Telescope Array Collaboration], *Proc. 30th ICRC, Mérida* (2007) p.955
[15] Homepage of the AGASA Experiment: `http://www-akeno.icrr.u-tokyo.ac.jp/AGASA/results.html#100EeV`
[16] D.R. Bergman et al. [HiRes Collaboration], *Nucl. Phys. B (Proc. Suppl.)* **165** (2007) 19
[17] D.R. Bergman et al. [HiRes Collaboration], *Proc. 30th ICRC, Mérida* (2007) p.1128
[18] B.T. Stokes et al. [HiRes Collaboration], *Proc. 30th ICRC, Mérida* (2007) p.1146
[19] W. Hanlon et al. [HiRes Collaboration], *Proc. 30th ICRC, Mérida* (2007) p.1247
[20] P. Sokolsky, G. Thomson, arXiv:0706.1248
[21] M. Roth et al. [Pierre Auger Collaboration], *Proc. 30th ICRC, Mérida* (2007) p.313
[22] J. Abraham et al. [Pierre Auger Collaboration], arXiv:0712.2843, Astropart. Phys. (2008) in press
[23] R. Abbasi et al. [HiRes Collaboration], *Astropart. Phys.* **27** (2007) 370.
[24] K.-H. Kampert, *Nucl. Phys. (Proc. Suppl)* **B165** (2007) 294; [astro-ph/0611884]
[25] D. R. Bergman and J. W. Belz, arXiv:0704.3721
[26] V. Berezinsky, A.Z. Gazizov, S.I.Grigorieva, *Phys. Lett.* **B612** (2005) 147.
[27] H.J. Drescher et al., *Astropart. Phys.* **21** (2004) 87.
[28] M. Nagano et al., *Astropart. Phys.* **22** (2004) 235.
[29] F. Kakimoto et al., *Nucl. Instr. Meth.* **A372** (1996) 527.
[30] see 5th Fluorescence Workshop, El Escorial, Spain, 16.-20.9. 2007, to appear in *Nucl. Instr. Meth.* **A**
[31] M. Unger et al. [Pierre Auger Collaboration], *Proc. 30th ICRC, Mérida* (2007) p.594
[32] Y. Fedorova et al. [HiRes Collaboration], *Proc. 30th ICRC, Mérida* (2007) p.1236; and talk by P. Sokolsky
[33] D. Allard, E. Parizot, A.V. Olinto, *Astropart. Phys.* **27** (2007) 61.
[34] H.O. Klages et al. [Pierre Auger Coll.], *Proc. 30th ICRC, Mérida* (2007) p.65; A. Etchegoyan, p.1307 ibid.
[35] D. Bergman et al. [TA Collaboration], *Proc. 30th ICRC, Mérida* (2007) p.1130
[36] J. Blümer et al. [KASCADE-Grande Collaboration], *Proc. to this conference*
[37] T. Gaisser et al. [IceCube Collaboration], *Proc. 30th ICRC, Mérida* (2007) p.758
[38] M. Risse, P. Homola, *Mod. Phys. Lett.* **A22** (2007) 749 - 766; [astro-ph/0702632]
[39] J. Abraham et al. [Pierre Auger Collaboration], subm. to *Astropart. Phys.* (2008); arXiv:0712.1147
[40] J. Abraham et al. [Pierre Auger Collaboration], subm. to *Phys. Rev. Lett.* (2008); arXiv:0712.1909
[41] O.E. Kalashev et al., *Phys. Rev.* **D66** (2002) 063004
[42] R. Engel et al. [Pierre Auger Collaboration], *Proc. 30th ICRC, Mérida* (2007) p.605

Acceleration of cosmic rays

Evgeny Berezhko
Yu.G. Shafer Institute of Cosmophysical Research and Aeronomy,
31 Lenin Ave., 677980 Yakutsk, Russia
E-mail: `berezhko@ikfia.ysn.ru`

Abstract. Cosmic ray (CR) origin problem is briefly discussed. It is argued that CRs with energies up to 10^{17} eV are produced in galactic supernova remnants, whereas ultra high energy CRs are extragalactic. CR composition strongly changes within the transition from galactic to extragalactic CR component, therefore precise measurements of CR composition at energies $10^{17} - 10^{19}$ eV are needed for the reliable determination of this transition. The possible sources of extragalactic CRs are briefly discussed. It is argued that CR acceleration at the shock created by the expanding cocoons around active galactic nuclei has to be considered as a prime candidate for the sources of extragalactic CRs.

1. Introduction

The origin of cosmic rays (CR) is steel unresolved problem in astrophysics. The understanding of CR origin requires determination of astrophysical objects, which are the CR sources, and appropriate acceleration process, which form CR spectrum in these objects. In this regard, during last several years considerable progress has been achieved experimentally and theoretically as well. Recently the steepening of CR spectrum above 3×10^{19} eV was established in HiRes [1] and Auger [2] experiments. It is presumably Greizen-Zatsepin-Kuzmin (GZK) cutoff, caused by CR energy losses in their interactions with cosmic microwave background radiation. Therefore it becomes evident that the highest energy part of CR spectrum is of extragalactic origin. At the same time it was also recently demonstrated [3] that CRs with energies up to $\epsilon \sim 10^{17}$ eV are presumably produced in supernova remnants (SNRs).

Below we briefly discuss the most essential aspects of CR spectrum formation.

2. CR acceleration in SNRs

The main reason why SNRs are usually considered as the CR source is a simple argument about the energy required to sustain the galactic CR population against loss by escape, nuclear interactions and ionization energy loss. Supernovae have enough power to drive the CR acceleration if there exists a mechanism for channeling about 10% of the mechanical energy release into relativistic particles. The high velocity ejecta produced in the supernova explosion interacts with the ambient medium to produce a strong blast wave. This outer shock may accelerate a small suprathermal fraction of the ambient plasma to high energies (e.g. [4, 5, 6]).

The only theory of particle acceleration which at present is sufficiently well developed and specific to allow quantitative model calculations is diffusive acceleration [7, 8] applied to the strong outer shock associated with SNRs (e.g. see [4, 5, 6] for review). Considerable efforts have been made during the last years to empirically confirm the theoretical expectation that the main part of CRs indeed originates in SNRs. Theoretically, progress in the solution of this problem has been due to the development of a kinetic nonlinear theory of diffusive shock acceleration (e.g. [9]). This theory includes all the most relevant physical factors, essential for the SNR evolution and CR acceleration in a SNR, at least in its early very energetic stages, and it is able to make quantitative predictions of the expected properties of CRs produced in young SNRs and their nonthermal radiation. The application of the theory to individual SNRs and their known synchrotron emission [5, 10, 6] has demonstrated its capability of explaining the observed SNR properties and in calculating new effects like the extent of magnetic field amplification which leads to the concentration of the highest-energy electrons in a very thin shell just behind the shock.

Nonlinear kinetic theory of CR acceleration in SNRs explains the existing measurements of CRs with energies up to $\epsilon \sim 10^{17}$ eV [3]. As it is seen in Fig.1, where CR differential intensity $J(\epsilon)$ as a function of total energy ϵ of CR particle is shown, the observed CR spectrum can be well represented by two components [3]. The first one $J^g(\epsilon)$, dominated up to 10^{17} eV, consists of CRs, produced in galactic SNRs, whereas the second, $J^{eg}(\epsilon)$,is produced in extragalactic sources. This so called dip scenario [11] requires relatively steep CR spectrum produced in extragalactic sources, $J_s^{eg} \propto \epsilon^{-\gamma}$ with $\gamma \approx 2.7$ at energies $\epsilon > 10^{18}$ eV. Note that compared with the source spectrum $J_s^{eg}(\epsilon)$ the extragalactic component $J^{eg}(\epsilon)$, observed in the Galaxy, is modified by two factors. At energies $\epsilon > 10^{18}$ eV the shape of $J^{eg}(\epsilon)$ is influenced by the energy losses of CRs in intergalactic space as a result of their interaction with the cosmic microwave background that leads to the formation of a "dip" structure at $\epsilon \sim 10^{19}$ eV, and to a GZK-cutoff for $\epsilon > 3 \times 10^{19}$ eV [11]. At $\epsilon < 10^{18}$ eV the spectrum $J^{eg}(\epsilon)$ is determined by the character of CR propagation in intergalactic space, followed by adiabatic cooling [11]. Since a galactic wind is expected to exist, CRs penetrating into the Galaxy from outside are in addition subject to modulation by the wind [3].

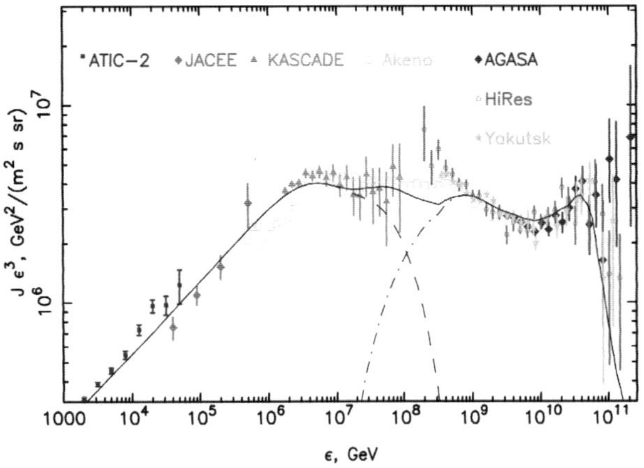

Figure 1. CR intensity as a function of energy [3]. The dashed and dash-dotted lines represent galactic and extragalactic components respectively. Data obtained in the ATIC-2 [12], JACEE [13], KASCADE [14], Akeno - AGASA [15], HiRes [16] and Yakutsk [17] experiments are shown as well.

3. Reacceleration of galactic CRs

If the spectrum of extragalactic CRs is as hard as $J^{eg} \propto \epsilon^{-2}$, then in this so called ankle scenario extragalactic CRs dominate in the observed CR spectrum only above 10^{19} eV [11, 18]. In this case except SNRs there is another galactic CR source (component B according to Hillas's [19] definition), which significantly contributes in the energy interval $10^{17} - 10^{19}$ eV. The most natural way to produce a smooth extension of the spectrum of CRs created in SNRs towards the higher energies is some kind of reacceleration process which picks up the most energetic CRs from SNRs and increases their energy up to a factor of 100. Such a process can be modeled in a following

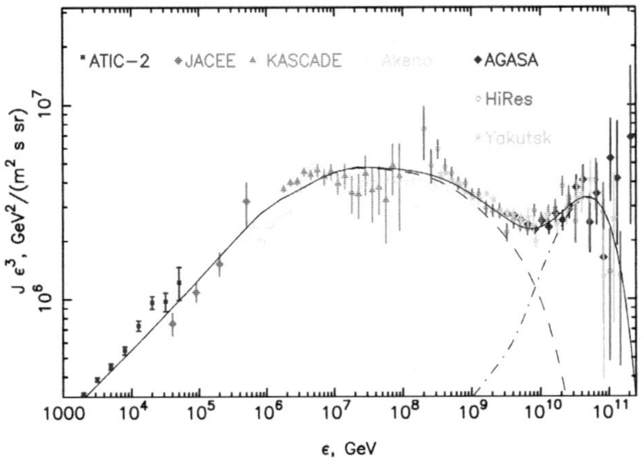

Figure 2. The same as in Fig.1 but for the ankle scenario. Dashed line represents galactic component, which includes CRs produced in SNRs and reaccelerated CRs. Dash-dotted line represents extragalactic component, which corresponds to the source spectrum $J_s^{eg} \propto \epsilon^{-2}$ [11].

way. Instead of $J_i^g(\epsilon)$ as in the dip scenario we use for every CR element i the spectrum $J_i'^g(\epsilon)$, which coincides with $J_i^g(\epsilon)$ at $\epsilon < \epsilon^i_{max1}$ and has a power-law form with exponential cutoff

$$J_i'^g(\epsilon) = J_i^g(\epsilon^i_{max1})(\epsilon/\epsilon^i_{max1})^{-\gamma}\exp(-\epsilon/\epsilon^i_{max2}) \quad (1)$$

at $\epsilon > \epsilon^i_{max1}$. Here ϵ^i_{max1} is minimal energy of particles involved into reacceleration and ϵ^i_{max2} is maximal particle energy achieved during reacceleration. It is natural to assume that these energies are proportional to the particle rigidity $\epsilon^i_{max} = Z_i\epsilon^p_{max}$. Here Z_i is charge number of CR element i and subscript p denotes protons. Quantities ϵ^p_{max1}, ϵ^p_{max2} and γ are treated as a free parameters, which values are determined as a result of best fit. CR acceleration by spiral shocks in the galactic wind [20] or CR acceleration in the pulsar vicinity [21, 22] could play a role of reacceleration mechanism.

According to Fig.2 the CR spectrum calculated within the ankle scenario with $\epsilon^p_{max1} = 5\times10^{15}$ eV, $\epsilon^p_{max2} = 1.5\times10^{17}$ eV and $\gamma = 3$ is also in reasonable agreement with the experiment. Note, however that such a well-known peculiarity in CR spectrum as a knee at $\epsilon \approx 3\times10^{15}$ eV is much less pronounced in the theoretical CR spectrum. This can be considered as indication against the ankle scenario.

4. Transition from galactic to extragalactic CRs

It was noted [3, 23] that CR chemical composition is expected to be very different in these two cases - dip and ankle scenario - at energies $10^{17} - 10^{19}$ eV. It is indeed very clearly seen in Fig.3, where the mean logarithm of CR atomic number as a function of CR kinetic energy ϵ_k is presented, for these two scenarios [23]. Since maximal energy of CRs produced in SNRs scales as $\epsilon^i_{max} = 3Z_i\times10^{15}$ eV CR composition progressively changes from proton-dominated to iron-dominated as the energy increases from $\epsilon = 3\times10^{15}$ eV to $\epsilon = 3\times10^{17}$ eV. Within the dip scenario after reaching the peak value at $\epsilon \approx 3\times10^{17}$ eV the quantity $< \ln A >$ goes down at higher energies $\epsilon > 3\times10^{17}$ eV towards the value $< \ln A > \approx 0.3$, which corresponds to extragalactic CRs (see Fig.3).

Since reacceleration produces a power law tail of CR iron spectrum originally produced in SNRs up to the energy $\epsilon \approx 4\times10^{18}$ eV, within the ankle scenario CR spectrum is expected to be dominated by the iron contribution at energies $10^{17} < \epsilon < 10^{19}$ eV. As it is seen in Fig.3 CR atomic number is indeed almost constant $< \ln A > \approx 2.5$ at these energies, that is very much different from what is expected within the dip model.

Yakutsk data as it is seen in Fig.3 better agree with the ankle scenario. Note however that CR composition determined on the basis of measurements accomplished by different instruments are not in agreement with each other. Contrary to the Yakutsk data, HiRes [16]

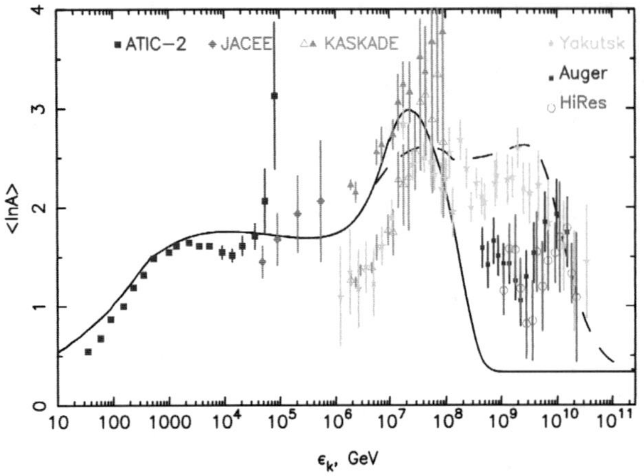

Figure 3. Mean logarithm of the CR nucleus atomic number as a function of kinetic energy represented for dip and ankle scenarios by solid and dashed lines respectively [23]. Experimental data obtained in the ATIC-2, JACEE, KASCADE (QGSJET and SYBYLL [24]), HiRes [16], Auger[30] and Yakutsk experiments are shown.

and Auger [30] data reveals considerably lighter CR composition with $< \ln A > \approx 1.5$ at energies $3 \times 10^{17} < \epsilon < 2 \times 10^{19}$ eV.

5. Extragalactic CR sources

Astrophysical objects, which can be considered as potential extragalactic sources of ultra high energy CRs, should fulfill a number of conditions (e.g. [18]). First, they should have sufficiently high energy output, not less than $L_p \sim 2 \times 10^{45} - 3.5 \times 10^{46}$ erg Mpc^{-3}yr^{-1} depending on the form of CR spectrum, which they produce. According to this energetic requirements active galactic nuclei (AGN) (e.g. [18]) and gamma-ray bursts (GRBs) [25, 27, 26] are considered as a potential extragalactic sources of ultra high energy CRs. However, as it was argued [18], the energy output of GRBs has a serious problem to be considered as a main source of extragalactic CRs.

The second requirement to extragalactic CR sources is their ability to produce power-law CR spectrum up to the maximal energy, not less than $\epsilon_{max} = 10^{20}$ eV, that is well above GZK-cutoff. Particle acceleration at relativistic shocks in AGNs [31] and GRBs [25, 27, 26] and frictional acceleration at the sides of AGN jets [32] are discussed as an appropriate acceleration processes. Particles are effectively accelerated at the relativistic shock under the condition of their nearly isotropic scattering in the downstream region [33]. However, simulations of CR shock acceleration performed for realistic parameters of the upstream medium [34] demonstrated insufficient CR scattering in downstream region leading to low efficiency of CR acceleration.

Since many aspects of CR acceleration by nonrelativistic shocks are very well studied one can obtain the most reliable estimate of the expected spectrum of CRs produced by nonrelativistic shocks in extragalactic objects [35]. In this case the maximal energy of CRs, accelerated at the expanded shock of size $R(t)$ and speed $V = dR/dt$, is determined by the expression [36]

$$\epsilon_{max} \approx ZeBRV/c, \tag{2}$$

where B is the upstream magnetic field, c is the speed of light, e is the proton charge. Magnetic field near the shock front, as it was established for all young SNRs [5, 6], is strongly amplified up to the level

$$B^2/(8\pi) \sim 10^{-3}\rho V^2, \tag{3}$$

which is presumably also appropriate for extragalactic shocks. Here ρ is the external gas density. If the shock is produced due to the prompt energy release E, like the supernova shock, then the CRs with maximal energy are produced at the beginning of the Sedov evolutionary phase $t = t_0$, where $t_0 = R_0/V_0$ $R_0 = [3M_{ej}/(4\pi\rho]^{1/3}$, $V_0 = \sqrt{2E/M_{ej}}$, M_{ej} is the ejected mass.

In the case of GRBs we have $M_{ej} = E/(\Gamma c^2)$, where $\Gamma \approx 100$ is fireball Lorenz factor. Then Eq.(2) gives

$$\epsilon_{max} = 10^{18} Z (E/10^{51} \text{ erg})^{1/3} (N_g/1 \text{ cm}^{-3})^{1/6}, \tag{4}$$

where $N_g = \rho/m_p$ is the ambient gas number density, m_p is the proton mass. This is too small value to consider GRBs as a main sources of extragalactic CRs, if energy release in the GRBs is not essentially higher than 10^{51} erg.

Relativistic jet in AGN is surrounded by the hot cocoon, which expands with the speed (e.g. [37])

$$V_c \approx [L_j/(\rho V_h)]^{1/4} t^{-1/2}, \tag{5}$$

where L_j is the mechanical luminosity of the jet, V_h is the hot spot speed. In this case Eq.(2) gives

$$\epsilon_{max} \approx 10^{20} Z \left(\frac{L_j}{10^{47} \text{ erg/s}}\right)^{3/4} \left(\frac{N_g}{10^{-4} \text{ cm}^{-3}}\right)^{-1/4} \left(\frac{V_h}{10^{10} \text{ cm/c}}\right)^{-1/4} \left(\frac{l_j}{1 \text{ kpc}}\right)^{-1/2} \text{eV}, \tag{6}$$

where $l_j = V_h t$ is the jet length. According to this expression proton maximal energy decreases from $\epsilon_{max} = 10^{20}$ eV to $\epsilon_{max} = 10^{18}$ eV when jet length increases from $l_j = 1$ kpc to $l_j = 10$ Mpc. Thus we conclude, that AGNs fulfill also the second requirement for extragalactic CR sources.

Note, that the above expression is based on the amplified magnetic field value (3) and therefore does not depends on the assumed value of intergalactic magnetic field opposite to earlier estimate for ϵ_{max} [29].

The third requirement to CR sources is related with the form of CR spectrum. Thus for two considered cases one needs power law CR spectrum $J_s^{eg} \propto \epsilon^{-\gamma}$ with $\gamma \approx 2.7$ and $\gamma \approx 2$ for dip and ankle scenarios respectively. The form of resultant CR spectrum produced during the whole evolution of the expanding shock is determined by three physical factors: i) nonlinear shock modification due to the CR backreaction; ii) adiabatic energy losses in downstream region; iii) diffusive CR escape from the shock vicinity in the outer space. The existence of CR escape, makes it possible to estimate the shape of spectrum of the most energetic CRs, produced during the source evolution. As it was demonstrated analytically [4, 36] and confirmed numerically [9], particles accelerated by the expanding shock which produces decreasing maximal energy $\epsilon_{max}(t)$ undergo diffusive escape from the shock vicinity into the surrounding space. If such a behavior happens during the period from t_1 to t_2, then at any given epoch $t_1 < t < t_2$ all accelerated particles with energies $\epsilon > \epsilon_{max}(t)$ are already escaped and inside the source there are only particles with $\epsilon \leq \epsilon_{max}(t)$.

Note, that this phenomenon, particle escape, has two important aspects. First, escaped CRs reach the outer observer with the same energy without any influence of adiabatic cooling inside the parent expanding source. Second, the spectrum of the escaped CRs can be essentially different from the canonical spectrum $N \propto \epsilon^{-2}$, produced by strong nonrelativistic shock. To estimate the spectrum of escaped CRs one can use the simple relation [4, 35]

$$N(\epsilon)\epsilon d\epsilon \propto \rho V^2 R^{2-\beta} dR, \tag{7}$$

which determines at any given phase t the overall number of CRs $N(\epsilon)$ with highest energy $\epsilon \sim \epsilon_{max}(t)$. Due to hard selfconsistent spectrum of CRs, produced by the strong shock, the main contribution to CR energy content is given by these particles with highest energy $\epsilon \sim \epsilon_{max}(t)$. Therefore their energy content scales as the shock ram pressure ρV^2 times the shock volume, as it is given by Eq.(7). Factor $R^{-\beta}$ describes the progressive decrease of the acceleration efficiency due to the shock weakening. For the expansion law $R \propto t^{-\nu}$ relation (7) gives

$$N(\epsilon) \propto \epsilon^{-\gamma} \quad \text{with} \quad \gamma = 1 + (2-\beta)/(2/\nu - 3). \tag{8}$$

In our case $\nu = 1/2$ that gives $\gamma = 3 - \beta$. For the constant acceleration efficiency $\beta = 0$ and we have $\gamma = 3$. However, the shock deceleration accompanies by the decrease of the acceleration efficiency. This effect is described by the amount of shock modification, which is characterized by the shock compression ratio, which for the case of strong shock depends on the shock speed as $\sigma \propto V^{3/8}$ [9]. Since in our case $V \propto 1/R$ we have $\beta = 3/8$ that gives $\gamma \approx 2.6$. Such a spectrum of extragalactic CRs at energies $\epsilon > 10^{18}$ eV very well corresponds to observations [18].

Particles with energies $\epsilon \leq \epsilon_{max}(t)$ at any given epoch are contained inside the source and have the spectrum close to $N \propto \epsilon^{-2}$. At very late epoch when the outer shock becomes weak these particles leave the source. Note however, that this is already not important for the Galaxy, because the contribution of extragalactic CRs expected to be low at energies $\epsilon < 10^{18}$ eV [18].

Note, that based on the data collected at the Auger experiment, correlation between the arrival directions of CRs with energy above 6×10^{19} eV and the position of nearby AGNs has been find [38], that strongly supports AGNs as a prime candidate for the source of ultra high energy CRs.

6. Summary

Magnetic field amplification due to shock accelerated CRs leads to a considerable increase of the maximal energy of CRs produced in SNRs. Calculations performed within nonlinear kinetic theory demonstrate that the expected CR spectrum, produced in SNRs, fits the existing CR data in a satisfactory way up to the energy 10^{17} eV [3]. The first knee in the observed all-particle CR spectrum is attributed to the maximum energy of protons that are produced in SNRs. The steepening of the all-particle CR spectrum above the knee energy 3×10^{15} eV is then the result of a progressive depression of the contribution of light CR nuclei with increasing energy. Such a scenario is confirmed by the KASCADE experiment which shows relatively sharp cutoff spectra of CR species at energies $\epsilon_{max} \approx 3Z \times 10^{15}$ eV [14].

If the extragalactic CR source spectrum is rather steep, $J_s^{eg} \propto \epsilon^{-\gamma}$ with $\gamma = 2.5 - 2.7$, so that it dominates the observed CR spectrum already at $\epsilon \geq 10^{18}$ eV, then CRs at lower energies are produced in SNRs without significant contribution from other Galactic CR sources. The CR spectrum up to the energy $\epsilon = 10^{17}$ eV is dominated by the contribution of SNRs, whereas at $\epsilon > 10^{18}$ eV CRs are predominantly extragalactic. If the extragalactic source spectrum is much harder $J_s^{eg} \propto \epsilon^{-2}$, then the transition from the galactic to the extragalactic component is expected at higher energy $\epsilon \sim 10^{19}$ eV. Since the expected CR chemical composition at $10^{17} - 10^{18}$ eV is very different in these two cases, an experimental study of CR composition could discriminate them.

Chemical CR composition determined at Yakutsk experiment is characterized by CR mean atomic number which increases with energy at $\epsilon < 3 \times 10^{16}$ eV, it remains nearly constant $< \ln A > \approx 2.3$ at $3 \times 10^{16} < \epsilon < 4 \times 10^{18}$ eV and progressively decreases with energy at $\epsilon > 4 \times 10^{18}$ eV. Such a behavior is well consistent with ankle scenario in which CR spectrum consists of three components: CRs produced in SNRs; reaccelerated galactic CRs and extragalactic CRs with hard spectrum. However CR composition deduced from HiRes and Auger experiments is significantly different: at energies $\epsilon > 3 \times 10^{17}$ eV it reveals almost constant relatively light CR composition with $< \ln A > \approx 1.5$. Such a rather controversial situation does not allow to make any reliable conclusion about the transition from galactic to extragalactic CR component. More precise determination of CR composition at $10^{16} < \epsilon < 10^{19}$ eV is needed to determine this transition.

The outer shock created by the expanding cocoon around AGNs is expected to produce CR spectrum $N \propto \epsilon^{-\gamma}$ with $\gamma \approx 2.6$ at energies $\epsilon > 10^{18}$ eV up to $\epsilon_{max} \sim 10^{20}$ eV [35], that very well fulfill the requirements to extragalactic CR sources. Therefore AGNs have to be considered as a prime candidate for the sources of extragalactic CRs.

Acknowledgments

The work was supported by the Presidium of RAS (program No.16) and by the Russian Foundation for Basic Research (grants 06-02-96008, 07-02-00221) The author thanks the organizers for their kind invitation and for some financial support.

References

[1] Bergman D R et al. 2007 *Nuclear Phys. B (Proc. Supl.)* **165** 19
[2] Yamamoto T et al. 2007 *Preprint* arXiv:0707.2638v3[astro-ph]
[3] Berezhko E G and Völk H J 2007 *Astrophys. J.* **661** L175
[4] Berezhko E G and Krymsky G F 1988 *Soviet Phys.-Uspekhi* **12** 155
[5] Berezhko E G 2005 *Adv. Space Res.* **35** 1031
[6] Berezhko E G 2008 *Adv. Space Res.* **41** 429
[7] Krymsky G F 1977 *Soviet Phys. Dokl.* **23** 327
[8] Bell A R 1978 *Mon. Not. R. Astron. Soc.* **182** 147
[9] Berezhko E G, Elshin V K and Ksenofontov L T 1996 *JETPh* **82** 1
[10] Berezhko E G and Völk H J 2006 *A&A* **451** 981
[11] Aloisio R et al. 2007 *Astropart. Phys.* **27** 76
[12] Panov A D et al. 2006 *Preprint* astro-ph/0612377
[13] Asakimori K et al. 2003 *Astrophys. J.* **502** 278
[14] Antoni T et al. 2005 *Astropart. Phys.* **24** 1
[15] Takeda M et al. 2003 *Astropart. Phys.* **19** 447
[16] Abbasi R U et al. 2005 *Astrophys. J.* **622** 910
[17] Egorova V P et al. 2004 *Nucl. Phys. B (Proc.Suppl.)* **136** 3
[18] Berezinsky V S et al. 2006 *Phys. Rev. D* **74** 043005
[19] Hillas A M 2006 *J. Phys.: Conf. Ser.* **47** 168
[20] Völk H J and Zirakashvili V N 2004 *Astron. Astrophys.* **417** 807
[21] Bell A R 1992, *Mon. Not. R. Astron. Soc.* **257** 493
[22] Berezhko E G 1994 *Astronom. Let.* **20** 75
[23] Berezhko E G, Ivanov A A, Knurenko S P and Saburov A V 2008 *Astrophys. J.* submitted
[24] Hörandel J R 2005 *Preprint* astro-ph/0508014
[25] Milgrom M and Usov V 1995 *Astrophys. J.* **449** L37
[26] Vietri M 1995 *Astrophys. J.* **453** 883
[27] Waxman E 1995 *Phys. Rev. Let.* **75** 386
[28] Abraham J (Pier Auger Collaboration) 2007 *Science* **318** 938
[29] Norman C A, Melrose D B and Achterberg A 1995 *Astrophys. J.* **454** 60
[30] Unger M et al. 2007 *Preprint* arXiv:0706.1495v1[astro-ph]
[31] Rachen J P and Biermann P 1993 *Astron. Astrophys.* **272** 161
[32] Ostrowski M 2000 *Mon. Not. R. Astron. Soc.* **312** 579
[33] Achterberg A, Gallant Y A, Kirk J G and Guthmann A W 2000 *Mon. Not. R. Astron. Soc.* **328** 393
[34] Niemiec J, Ostrowski M and Pohl M 2006 *Astrophys. J.* **650** 1020
[35] Berezhko E G 2008 *Astrophys. J.* submitted
[36] Berezhko E G 1996 *Astropart. Phys.* **5** 367
[37] Ferrari A 1998 *Annu. Rev. Astron. Astrophys.* **36** 539
[38] Abraham J (Pier Auger Collaboration) 2007 *Science* **318** 938

The Highest Energy Neutrinos

Francis Halzen

Department of Physics, University of Wisconsin, Madison, WI 53706, USA

Abstract. Measurements of the arrival directions of cosmic rays have not revealed their sources. High energy neutrino telescopes attempt to resolve the problem by detecting neutrinos whose directions have not been scrambled by magnetic fields. The key issue is whether the neutrino flux produced in cosmic ray accelerators is detectable. It is believed that the answer is affirmative, both for the galactic and extragalactic sources, provided the detector has kilometer-scale dimensions. We revisit the case for kilometer-scale neutrino detectors in a model-independent way by focussing on the energetics of the sources. The real breakthrough though has not been on the theory but on the technology front: the considerable technical hurdles to build such detectors have been overcome.

1. Cosmic Rays and Neutrinos

An illustration of the neutrino sky is shown in Fig. 1 displaying a spectrum ranging from microwave neutrinos produced in the big bang to GZK neutrinos associated with the highest energy cosmic rays[1]. The GZK neutrinos are the decay products of pions produced in the interaction of cosmic rays with microwave photons. These are the same interactions that shape the Greissen-Zatsepin-Kuzmin absorption feature in the spectrum, hence their name. Prominently displayed in the figure is the flux of the highest energy atmospheric neutrinos observed up to $\sim 100\,\mathrm{TeV}$ by the AMANDA experiment[2]. This beam, very successfully mined for particle physics by the Superkamiokande-generation of experiments, will be exploited at yet higher energies[3]. Because of its steep spectrum, events above several hundreds of TeV become very rare, leaving a clear neutrino sky to be explored for sources of cosmic neutrinos beyond the sun. Neutrino telescopes will open some ten orders of magnitude in neutrino wavelength, from their tens of GeV threshold to the EeV energy of GZK neutrinos.[1] The existence of cosmic neutrinos with yet higher energy are a matter of speculation; they could be the decay products of cosmic remnants or topological defects associated with phase transitions in the early universe[4].

The neutrino fluxes anticipated[4] from non-thermal astronomical sources and, therefore, candidate cosmic ray accelerators such as active galaxies (AGN) and gamma ray bursts (GRB), dominate the atmospheric flux above $\sim 1\,\mathrm{PeV}$; see Fig. 1. The energy fluxes are at the level of the flux, as far as we know unobservable, of big bang neutrinos. The high energy fluxes may nevertheless be observed by exploiting the relatively large neutrino interaction cross section in combination with detectors of gigaton size. Standard model physics is sufficient to establish that the cosmic fluxes shown are observable in a volume of 1 kilometer cubed instrumented with photomultipliers[4]. Neutrino telescopes detect the Cherenkov radiation from secondary particles produced in the interactions of high energy neutrinos in highly transparent and well shielded

[1] We use units GeV, TeV, PeV and EeV, increasing energy in steps of one thousand.

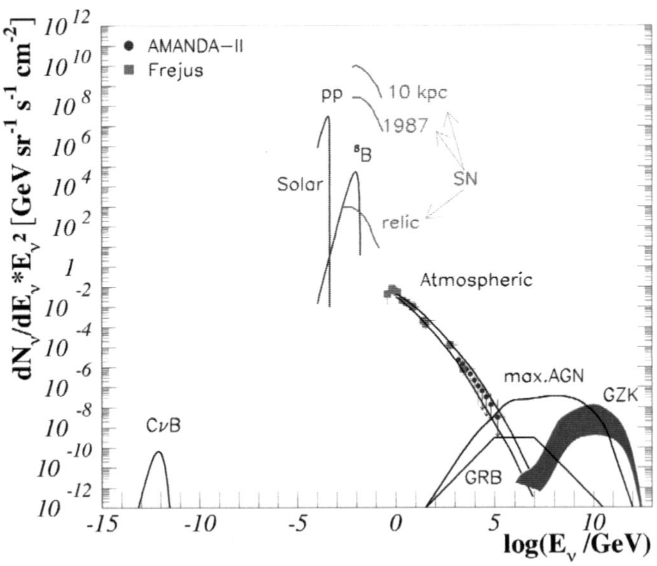

Figure 1. The neutrino sky from the lowest energy neutrinos produced in the big bang to the highest energies associated with the sources of the cosmic rays, here assumed to be gamma ray bursts or, alternatively, active galaxies. These will be the target of kilometer-scale neutrino detectors such as IceCube and KM3NeT. Neutrinos at intermediate energies, produced in the sun, supernovae and in collisions of cosmic rays in the atmosphere, have been studied by SuperK and similar detectors[8]

deep water or ice. At the higher energies the neutrino cross section grows and secondary muons travel up to tens of kilometers to reach the detector from interactions outside the instrumented volume.

The construction of kilometer-scale instruments such as IceCube at the South Pole and the future KM3NeT detector in the Mediterranean, have been made possible by development efforts that resulted in the commissioning of prototypes that are two orders of magnitude smaller, AMANDA and ANTARES[5]. Their successful technologies have, in turn, relied on pioneering efforts by the DUMAND[6] and Baikal[7], as well as the Macro and SuperK collaborations[8]. While much larger than the latter, kilometer scale neutrino telescopes are insensitive to neutrinos in the MeV-GeV energy range and have a typical threshold of tens of GeV; this is the price one pays for reaching large detection volume. IceCube[9] is under construction and taking data with a partial array of 1320 ten inch photomultipliers positioned between 1500 and 2500 meter and deployed as beads on 22 strings below the geographic South Pole. Its effective telescope area already exceeds that of its predecessor AMANDA by roughly one order of magnitude. The detector will grow by another 14$\sim$18 strings in the 2007-08 Antarctic summer to be completed in 2011 with 80 strings [10].

The AMANDA experiment has observed neutrinos with energies as high as $\sim$100 TeV, at a rate consistent with the flux of atmospheric neutrinos extrapolated from lower energy measurements; see Fig. 1. The fluxes of cosmic neutrinos shown in the figure at higher energies are, in contrast, a matter of speculation. It is known that non-thermal sources such as supernova remnants, AGN and GRB accelerate electrons to energies close to 1 PeV. Their existence is inferred from the observation of TeV gamma rays whose spectrum extends to $\sim$ 100 TeV in some sources[11]. All observations can be accommodated in models where the origin of the photons, from radio to TeV energy, is synchrotron radiation by the electrons and, at the highest energies, inverse Compton scattering of ambient light, primarily the synchrotron photons themselves. There is no reason why non-thermal sources would not accelerate protons or nuclei along with the electrons, turning them into sources of cosmic rays; unfortunately, at this time, no evidence for such cosmic ray accelerators exists. A conclusive signature for the presence of cosmic rays in the sources is the production of pions on ambient radiation and matter. Pion production is revealed by observing the decay products, photons and neutrinos. While it has been a challenge to disentangle such pionic photons from those produced by purely electromagnetic processes[12], charged pions decaying into neutrinos yield incontrovertible evidence. The anticipated neutrino fluxes are shown in Fig. 1, assuming that AGN or, alternatively, GRB, happen to be the correct guess for the unknown sources of the cosmic rays. If not, the real sources may be revealed by neutrinos that, unlike charged primaries, point back to their site of origin. Neutrino astronomy must eventually succeed because, after all, cosmic rays exist. The critical question is whether our estimate of the level of the neutrino fluxes associated with the observed cosmic rays is robust; it sets the scale of the detector.

Table 1. Built as discovery instruments, neutrinos telescopes nevertheless target a range of particle and astrophysics problems.

Favorite Sources	**Possible Science**
Atmospheric ($\sim 100,000$ per year, up to 1000 TeV, charm?)	Oscillations New interactions Test of relativity and equivalence principle
GRB (successful and failed)	Sources of cosmic rays Test of Lorentz invariance Planck scale physics, quantum decoherence
AGN	Sources of cosmic rays
Starburst galaxies	
Supernova remnants Also microquasars, magetars, PWNe, binaries unidentified Egret sources, plane of the galaxy	Sources of cosmic rays
Cosmic rays ineracting with microwave photons	Identify sources of cosmic rays Neutrino cross section at EeV energy
Dark matter	Annihilation in the sun, mostly spin-dependent
Cosmic rays interacting with the sun	Backrounds to WIMP search
Supernovae explosion	Deleptonization, TeV emission, hierarchy, $\sin\theta_{13}$

Thus the faith of neutrino astronomy is intertwined with cosmic ray physics beyond the traditional subject of GZK neutrinos. While kilometer-scale neutrino detectors are discovery experiments with missions as diverse as particle physics and the search for dark matter — see Table1 — their size as astronomical telescopes is very much anchored to the observed fluxes of galactic and extragalactic cosmic rays. Cosmic accelerators produce particles with energies in excess of 10^8 TeV; we still do not know where or how. The flux of cosmic rays observed at Earth follows a broken power law; see Fig. 2. The two power laws are separated by a feature dubbed the "knee". Circumstantial evidence exists that cosmic rays, up to perhaps EeV energy, originate in galactic supernova remnants. Any association with our Galaxy disappears in the vicinity of a second feature in the spectrum referred to as the "ankle". Above the ankle, the gyroradius of a proton in the galactic magnetic field exceeds the size of the Galaxy and we are witnessing the onset of an extragalactic component in the spectrum that extends to energies beyond 100 EeV. Observation of the GZK feature in the HiRes and Auger spectra near the energy threshold for pion production on microwave photons, confirms the existence of an extragalactic component.

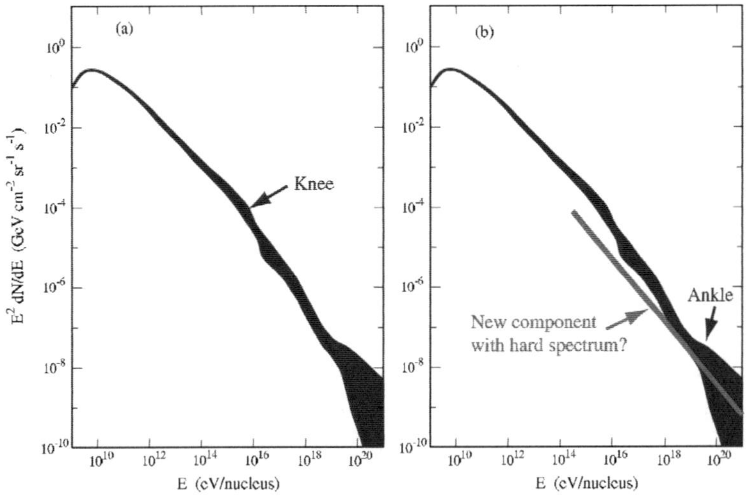

Figure 2. At the energies of interest here, the cosmic ray spectrum consists of a sequence of 3 power laws. The first two are separated by the "knee" (left panel), the second and third by the "ankle". There is evidence that the cosmic rays beyond the ankle are a new population of particles produced in extragalactic sources; see right panel.

2. Neutrinos Associated with the Sources of the Cosmic Rays

2.1. Extragalactic Cosmic Rays

It is routinely emphasized how small cosmic ray particle fluxes are; this may be besides the point. The energetics of the accelerator is likely to be more revealing. A hint in this direction comes from conventional astronomy. While the diffuse universal flux of photons falls by eighteen orders of magnitude between microwave and GeV-energy, the energy carried by the particle flux drops by less than five. Sources are known that emit most of their energy in TeV photons. The energy is the key. The argument has been well advertised for galactic cosmic rays where energetics points at their supernova origin. By integrating the observed flux in Fig. 2 we can obtain the energy density ρ_E of cosmic rays in the galaxy from the relation that flux = velocity × density, or

$$4\pi \int dE \left\{ E \frac{dN}{dE} \right\} = c\rho_E \,. \tag{1}$$

The answer is that $\rho_E \sim 10^{-12}\,\mathrm{erg\,cm^{-3}}$. This is also the value of the corresponding energy density $B^2/8\pi$ of the microgauss magnetic field in the galaxy. The accelerating power needed to maintain this energy density is $10^{-26}\,\mathrm{erg/cm^3s}$ given that the average containment time of the cosmic rays in our galaxy is 3×10^6 years. For a nominal volume of the galactic disk of $10^{67}\,\mathrm{cm}^3$ this requires an accelerator delivering 10^{41} erg/s. This happens to be 10% of the power produced by supernovae releasing 10^{51} erg, or the one percent of a solar mass that is not released in neutrinos, every 30 years. This coincidence is the basis for the idea that shocks produced by supernovae expanding into the interstellar medium are the origin of the galactic cosmic rays[13].

Let's follow the same logic for the extragalactic component in Fig. 2. The flux above the ankle is often summarized as "one 10^{19} eV particle per kilometer square per year per steradian". This can be translated into an energy flux

$$E \left\{ E \frac{dN}{dE} \right\} = \frac{10^{19}\,\mathrm{eV}}{(10^{10}\,\mathrm{cm}^2)(3 \times 10^7\,\mathrm{sec})\,\mathrm{sr}} = 3 \times 10^{-11}\,\mathrm{TeV\ cm^{-2}\,s^{-1}\,sr^{-1}} \,. \tag{2}$$

Following the procedure applied above to the galactic component we obtain an energy density of

$$\rho_E = \frac{4\pi}{c} \int_{E_{\min}}^{E_{\max}} \frac{3 \times 10^{-8}}{E} dE \, \frac{\mathrm{GeV}}{\mathrm{cm}^3} \simeq 10^{-19} \frac{\mathrm{TeV}}{\mathrm{cm}^3} \,, \tag{3}$$

taking the extreme energies of the accelerator(s) to be $E_{\max}/E_{\min} \simeq 10^3$.

The energy content derived "professionally" by integrating the spectrum in Fig. 2 assuming an E^{-2} energy spectrum, typical of shock acceleration, with a GZK cutoff, is $\sim 3 \times 10^{-19}$ erg cm^{-3}. This is within a factor of our back-of-the-envelope estimate recalling that 1 TeV = 1.6 erg. The energy density represents the universe's filling factor in cosmic rays, equivalent to 410 microwave-energy photons per cubic centimeter.

The power required for a population of sources to generate this energy density over the Hubble time of 10^{10} years is $\sim 3 \times 10^{37}$ erg s^{-1} per (Mpc)^3 or, as often quoted in the literature, $\sim 5 \times 10^{44}$ TeV per (Mpc)^3 per year. This works out to[14]

- $\sim 2 \times 10^{44}$ erg s^{-1} per active galaxy, or
- $\sim 2 \times 10^{51}$ erg per cosmological gamma ray burst.

The coincidence of these numbers with the observed energy radiated by these sources in photons, explains why AGN and GRB have emerged as leading candidates for the cosmic accelerators. In either case, it suffices that the shocks associated with acceleration near the black hole dump roughly equal energy in electrons and protons to accommodate the observed coincidence, with the electron energy observed as radiation by synchrotron and inverse Compton scattering.

For GRB the argument is a carbon copy of the one favoring galactic supernova as cosmic ray accelerators. Observations show that, within one gigaparsec cubed, 300 GRB dump about 10^{51} erg of energy into the universe in a single year. They therefore supply roughly 10^{44}erg/yr/Mpc^3 in radiation and, assuming equal energy in protons, we conclude that they represent an environment that can accommodate the observed energetics of the extragalactic cosmic rays. A problem is that the same argument can be made to validate AGN as the sources of the highest energy cosmic rays; see above. In the end, the answer may lay elsewhere.

Where do neutrinos fit into this? The assumption that the energy in neutrinos coincides with the matching energies observed in electromagnetic radiation and cosmic rays yields the level of neutrino fluxes associated with the cosmic rays shown in Fig. 1. In the end the neutrino flux is therefore the flux of Eq. 2. It is often referred to as the Waxman-Bahcall "bound"[15].

The assumption of a matching neutrino flux is rather generic. Cosmic rays are accelerated in regions of high magnetic fields associated with shocked particle flows driven by the gravity of a black hole. They interact with the radiation fields surrounding the black hole. The most important processes are $p+\gamma \rightarrow \Delta^+ \rightarrow \pi^0 + p$ and $p+\gamma \rightarrow \Delta^+ \rightarrow \pi^+ + n$. While the secondary protons may remain trapped in the acceleration region, roughly equal numbers of neutrons and decay products of neutral and charged pions escape. The energy escaping the source is therefore distributed among cosmic rays, gamma rays and neutrinos produced by the decay of neutrons, neutral pions and charged pions, respectively. This generic scenario accommodates the observation of equal energy in cosmic rays and electromagnetic radiation, and extends it to neutrinos. Clearly both GRB and AGN environments can accommodate this scenario although with very dissimilar black holes and radiation targets for pion production. If we take this picture seriously, our previous estimate must be corrected for the fact that the pion takes only 25% of the energy of the secondary neutron thus changing the energy balance between cosmic rays and neutrinos and reducing their flux. In the end we estimate that the muon-neutrino flux associated with the sources of the highest energy cosmic rays is loosely confined to the range

$$E_\nu{}^2 dN/dE_\nu = 1 \sim 5 \times 10^{-8}\,\mathrm{GeV\,cm^{-2}\,s^{-1}\,sr^{-1}} \tag{4}$$

depending on the cosmological evolution of the cosmic ray sources. Model calculations assuming that active galaxies or gamma-ray bursts are the actual sources of cosmic rays yield event rates similar to the generic energetics estimate presented; see Fig. 3.

The anticipated neutrino flux thus obtained has to be compared with the limit of 7.4×10^{-8} GeV cm^{-2} s^{-1} sr^{-1} reached after the first 4 years of operation of the completed AMANDA detector in 2000–2003 [16]. On the other hand, for conservative assumptions for the charm

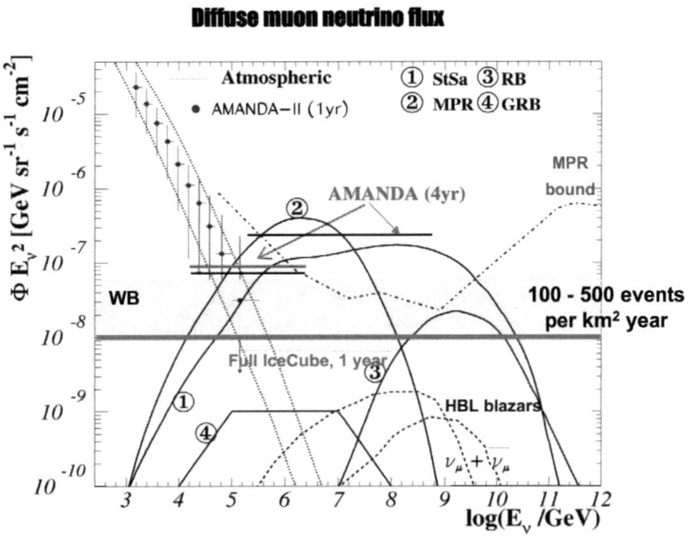

Figure 3. Our estimate of the flux of neutrinos associated with the sources of the highest energy cosmic rays (the shaded range labeled WB) is compared to the limits established by the AMANDA experiment reached with 800 days of data[16]. AMANDA's sensitivity is within a factor of 2 of the most optimistic predictions. Also shown are fluxes predicted by specific models of cosmic ray accelerators: active galaxies labeled StSa[17] and MPR[18], GRB[19] and the diffuse flux produced by cosmic ray producing active galaxies on microwave photons[20] labelled RB. Data for the background atmospheric neutrino flux are from the AMANDA experiment. The IceCube experiment will be sensitive to all predictions after a few years of operation of the full detector. It has sensitivity to the larger fluxes by operating the partially completed detector that already now exceeds AMANDA in instrumented volume.

background and for the detector performance, IceCube has the capability to observe a flux that is an order of magnitude below this limit with 5 σ significance in 3 years[9]. The exact value of the IceCube sensitivity depends on the magnitude of the dominant high energy neutrino background from the prompt decay of atmospheric charmed particles. The level of this background is difficult to anticipate theoretically and little accelerator data is available in the energy and Feynman-x range of interest[21].

The neutrino event rate is obtained by folding the flux predicted with the probability that the neutrino is actually detected in a high energy neutrino telescope; only one in a million neutrinos of TeV energy interacts and produces a muon that reaches the detector. This probability is given by the ratio of the muon and neutrino interaction lengths in the detector medium, λ_μ/λ_ν[4] and therefore grows with energy. For the flux range estimated above we anticipate 100–500 detected muon neutrinos per km^2 per year, with the higher range close to what is already ruled out by AMANDA. In any case, the lower value represents the more realistic estimate as previously argued and the 100 events predicted will be further reduced by the realities of rejecting detector backgrounds, especially for spectra steeper than the E^{-2} assumed throughout. On the other hand, IceCube's effective area for muon neutrinos exceeds 1 km^2 and equal fluxes of electron and tau neutrinos are expected[9]. If IceCube construction remains on schedule, the instrument will accumulate 1 km^2 year of data within the next two years; the confrontation of these arguments with data is imminent, certainly on the 40 year timescale it took to develop the technology for the detectors.

2.2. Cosmic Neutrinos Associated with Galactic Supernova Remnants

Can kilometer-scale neutrino detectors observe neutrinos pointing back at the accelerators of the galactic cosmic rays? It is believed that galactic accelerators are powered by the conversion of 10^{50} erg of energy into particle acceleration by diffusive shocks associated with young (1000–10,000 year old) supernova remnants expanding into the interstellar medium. The cosmic rays will interact with hydrogen atoms in the interstellar medium to produce pions that decay into photons and neutrinos. These provide us with indirect evidence for cosmic ray acceleration. The new twist here is that the eventual observation of pionic gamma rays allows for a straightforward determination of the neutrino flux.

The HESS telescope has opened a new era in astronomy by producing the first resolved images of sources in TeV gamma rays, particularly, in this context, of the supernova remnant RX J1713.7-3946 [22]. While the resolved image of the source reveals TeV gamma ray emission from the whole supernova remnant, it shows a clear increase of the flux in the directions of known molecular clouds. This is suggestive of protons, shock-accelerated in the supernova remnant, interacting with the dense clouds to produce neutral pions that are the origin of the observed increase of the TeV photon signal. The magnitude of the photon flux is consistent with a site where protons are accelerated to energies typical of the main component of the galactic cosmic rays. A similar extended source of TeV gamma rays tracing the density of molecular clouds has been identified near the galactic center. Protons, apparently accelerated by the remnant HESS J1745-290, diffuse through nearby molecular clouds to produce a signal of TeV gamma rays that traces the density of the clouds [23]. Fitting the observed spectrum by purely electromagnetic processes is challenging because the relative height of the inverse Compton and synchrotron peaks requires very low values of the B-field, inconsistent with those required to accelerate the electron beam to energies that can accommodate the observation of 100 TeV photons. Nevertheless, an exclusively electromagnetic explanation of the non-thermal spectrum is not impossible, even favored by some [24]. One can, for instance, partition the remnant in regions of high and low magnetic fields that are the respective sites of acceleration and inverse Compton scattering.

Supernovae associated with molecular clouds are a common feature of associations of OB stars that exist throughout the galactic plane. Although not visible to HESS, possible evidence has been accumulating for the production of cosmic rays in the Cygnus region of the galactic plane from a variety of experiments[25, 26, 27]. Most intriguing is a Milagro report of an excess of events from the Cygnus region with a steady flux of several times that of the Crab.

If the TeV gamma ray signals are indeed of pionic origin, only particle physics establishes the rate of the accompanying neutrinos. Proton-proton collisions yield two charged pions for every neutral pion, with every charged pion decaying into a muon neutrino and antineutrino (one from the pion decay, the other from the decay of the secondary muon) and a neutral pion into two photons. So, the muon neutrino flux would be equal to the photon flux were it not for a factor two reduction from oscillations. The prediction is simple; to first order there is one muon neutrino for every photon produced in the source. Because the protons transfer on average 20% of their energy to secondary pions, and the four leptons in the charged pion decay chain $\pi \to \mu(\to e + \nu_e + \nu_\mu) + \nu_\mu$ take roughly equal energy, neutrinos with 0.05 of the cosmic ray energy are produced. Similarly, photons with 10% of the proton energy originate from the decay of neutral pions. Accelerators producing cosmic rays reaching the "knee" must produce photons with energies up to 100 TeV and neutrinos up to half that energy. This requirement is consistent with observations of RX J1713.7-3946 and MGRO J2019+37 discussed above. They are the targets for neutrino observation of neutrino telescopes located in the Southern and Northern hemispheres, respectively.

Whereas the relation between neutrino and gamma ray fluxes is direct, the information on their spectrum is often limited. This is especially the case for the hotspot MGRO J2019+37 where we have to model the spectrum on the basis of a measurement at a single energy; the spectral slope has not been measured. Uncertainties in the calculation are associated with the propagation of the cosmic rays, with the value of the magnetic fields, and the age of the remnant. After investigating the wide parameter space of models for MGRO J2019+37, it has been shown[28] that the neutrino flux can nevertheless be predicted within a factor of 2 once the model flux is normalized at 12.5 TeV to the Milagro data and a limit at GeV energy is imposed reflecting the fact that EGRET did not observe a GeV counterpart [29]. The neutrino rate from this single source is within the range $2 \leq dN/dt \leq 3.8$ events per year with the IceCube threshold at 50 GeV. The range is bound by the fact that the Milagro observation strongly constrains the flux in the energy range 1–20 TeV, where the neutrino detection probability is highest, resulting in similar predictions for dissimilar SNR characteristics.

The irreducible atmospheric background, due to neutrinos produced in the Northern atmosphere in cosmic ray showers, is calculated using the results of Ref. [30]. In 15 years of operation we predict $4.5\,\sigma \leq N/\sqrt{N_{\rm atmo}} \leq 7.7\,\sigma$ and if the higher end of the predicted event rate range is realized, 5 σ is possible in 4.3 years; see Fig. 4.

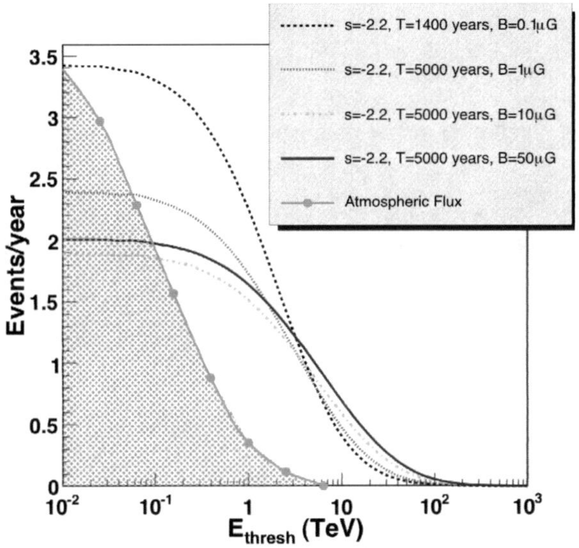

Figure 4. Events due to the gamma-ray spectra with injection $\alpha = -2.2$.

We note that the Milagro collaboration [31] has recently detected multiple additional sources besides MGRO J2019+37, most with fluxes close to 0.5 Crab. The sources with possible counterparts in the GeV range indicate a spectral index of ~ -2.3. If we compute the flux of neutrinos from the Milagro sources (not including the Crab Nebula) detected with post-trial significance of greater than 5 σ assuming a power-law index of -2.3, we get a total event rate in IceCube of 6.9 neutrinos/year. If we also include the more tentative sources, the event rate increases to 11.5 neutrinos/year. In the long run, a correlation analysis of the IceCube and Milagro skymaps should make the detection of these sources likely.

3. Looking Forward

While neutrino "telescopes" are discovery instruments with a variety of missions, the hope is that they may contribute to the resolution of the century old puzzle of the origin of cosmic rays, either by the detection of GZK neutrinos or by directly observing neutrinos from the accelerators. Between now and the next TAUP meeting we can look forward to

- Results from IceCube that, by operating the detector as it grows, will reach a km^2year aperture.
- Enhanced sensitivity of the Auger experiment to neutrinos that initiate horizontal air showers.

- The initial design of a kilometer-scale neutrino detector in the northern hemisphere.
- A new flight of the ANITA experiment as well as a wealth of other ideas and initiatives on detecting the acoustic and radio signatures produced by GZK neutrinos in ice, water, salt, permafrost ...

Acknowledgments

This research was supported in part by the National Science Foundation under Grant No. OPP-0236449, in part by the U.S. Department of Energy under Grant No. DE-FG02-95ER40896, and in part by the University of Wisconsin Research Committee with funds granted by the Wisconsin Alumni Research Foundation.

References

[1] Becker J, *Phys. Rep.* (to be published)
[2] Muenich K 2007 30th ICRC Merida Mexico [325]
[3] Gonzalez-Garcia M.C, Halzen F and Maltoni M 2005 *Phys. Rev. D* **71** 093010 (*Preprint* hep-ph/0502223)
[4] Gaisser T. K., Halzen F and Stanev T 1995 *Phys. Rept.* **258**, 173 [Erratum 1995 **271** 355] (*Preprint* hep-ph/9410384); Learned J.G. and Mannheim K 2000 *Ann. Rev. Nucl. Part. Science* **50** 679; Halzen F and Hooper D 2002 *Rept. Prog. Phys.* **65** 1025 (*Preprint* astro-ph/0204527)
[5] Bouwhuis M [ANTARES Collaboration] 2007 *J. Phys. Conf. Ser.* **60** 239
[6] Babson J *et al* [DUMAND Collaboration] 1990 *Phys. Rev. D* **42** 3613
[7] Balkanov V A *et al* [Baikal Collaboration] 2003 *Nucl. Phys. Proc. Suppl.* **118** 363
[8] Kajita T 2006 *Rept. Prog. Phys.* **69** 1607
[9] Achterberg A *et al* [IceCube Collaboration] 2006 *Astropart. Phys.* **26**, 155 (*Preprint* astro-ph/0604450); Ahrens J *et al* [IceCube Collaboration] 2004 *Astropart. Phys.* **20** 507 (*Preprint* astro-ph/0305196 and http://icecube.wisc.edu)
[10] Karle A 2007 30th ICRC Merida Mexico [1180]
[11] Weekes T 2003 *Bristol UK: IOP* 221 pp
[12] Hoffman W and Hinton J 2007 *Phys. World* **20N3** 30
[13] See for instance Aharonian F, O'C Drury L and and Völk H J 1994 *Astron & Astroph* **285** 645
[14] Gaisser T K, OECD Megascience Forum, Taormina Italy 1997 *Preprint* astro-ph/9707283
[15] Waxman E and Bahcall J N 1997 *Phys. Rev. Lett.* **78** 2292 (*Preprint* astro-ph/9701231)
[16] Hill G 2007 30th ICRC Merida Mexico [1103]
[17] Stecker F W and Salamon M H 1992 *Astrophys. J.* **512** 521 (*Preprint* astro-ph/9808110); Atoyan A and Dermer C D 2001 *Phys. Rev. Lett.* **87** 221102 (*Preprint* astro-ph/0108053 and references therein); Stecker F W 2005 *Phys. Rev. D* **72** 107301 (*Preprint* astro-ph/0510537 for a recent update)
[18] Mannheim K, Protheroe R J and Rachen J P 2001 *Phys. Rev. D* **63** 023003 (*Preprint* astro-ph/9812398)
[19] Guetta D *et al* 2004 *Astropart. Phys.* **20** 429 (*Preprint* astro-ph/0302524)
[20] Rachen J P and Biermann P L 1993 *Astron. & Astrophys.* **272** 161 (*Preprint* astro-ph/9301010)
[21] Gondolo P, Ingelman G and Thunman M 1996 *Nucl. Phys. Proc. Suppl.* **48** 472 (*Preprint* hep-ph/9602402)
[22] Aharonian F A *et al* 2004 *Nature* **432** 75 (*Preprint* astro-ph/0411533); Volk H J, Berezhko E G and Ksenofontov L T *Preprint* astro-ph/0409453
[23] Aharonian F A *et al* 2006 *Nature* **439** 695 (*Preprint* astro-ph/0603021)
[24] For a recent discussion see Katz B and Waxman E *Preprint* arXiv:0706.3485 [astro-ph]
[25] Aharonian F A, Akhperjanian A, Beilicke M, Uchiyama Y and Takahashi T 2002 *Astron. Astrophys.* **393** L37 (*Preprint* astro-ph/0207528)
[26] Knodlseder J *Preprint* astro-ph/0007442; Lang M J *et al* 2004 *Astron. Astrophys.* **423** 415 (*Preprint* astro-ph/0405513); Konopelko A *et al Preprint* astro-ph/0611730; Butt Y M *et al Preprint* astro-ph/0611731
[27] Abdo A A *et al Preprint* astro-ph/0611691
[28] Anchordoqui L, Halzen F, Montaruli T and O'Murchadha A 2007 *Phys. Rev. D* **76** 067301 (*Preprint* astro-ph/0612699); Halzen F and O'Murchadha A *Preprint* arXiv:0705.1723 [astro-ph]
[29] Prodanovic T, Fields B D and Beacom J F 2007 *Astropart. Phys.* **27** 10 (*Preprint* astro-ph/0603618); Kappes A *et al Preprint* astro-ph/0607286; Beacom J F and Kistler M D *Preprint* astro-ph/0607082
[30] Agrawal V, Gaisser T K, Lipari P and Stanev T, 1996 *Phys. Rev. D* **53** 1314 (*Preprint* hep-ph/9509423)
[31] Sinnis G Milagro Collaboration] 2007, *VERITAS First Light Fiesta* Mt. Hopkins Arizona

High-energy gamma ray astronomy

Werner Hofmann

Max-Planck-Institut für Kernphysik, P.O. Box 103980, D 69029 Heidelberg, Germany

E-mail: Werner.Hofmann@mpi-hd.mpg.de

Abstract. High-energy gamma-ray astronomy is a field under very rapid development, addressing issues of astrophysics, of astroparticle physics and of fundamental physics. This contribution summarizes recent results, concentrating mainly on data from ground-based instruments both on Galactic and extragalactic sources.

I will first briefly address the instruments of gamma-ray astronomy, then present the newly emerging VHE gamma-ray sky, followed by a discussion of galactic and extragalactic gamma-ray sources.

1. Instruments

Gamma rays up to some 10 GeV are best observed from space, whereas at very high energies (VHE, $>$ 100 GeV) only ground-based instruments with their larger detection areas provide sufficient detection rates. For space-based gamma-ray detection, 271 sources are known from the 3rd EGRET catalog [1], about 2/3 of which are unidentified. Recent revisions of background models tend to reduce the number of EGRET sources somewhat [2]. New results will soon be provided by the recently launched AGILE instrument, and in 2008 by GLAST, which is expected to detect around 10000 gamma-ray sources. At the time of this meeting, AGILE is still in the commissioning phase, but has already detected first sources, such as the blazar 3C 454.3 [3]. The bulk of the new results in recent years comes from the H.E.S.S. and MAGIC Cherenkov telescopes, which detect the Cherenkov light emitted by particles in gamma-ray induced air-showers. These instruments provide detection areas of about $10^5\,\mathrm{m}^2$, angular resolution for single gamma-rays of 0.1° and better, and flux sensitivities of 1% of the Crab flux and below. Recently, the VERITAS system of four telescopes went into full operation at the Whipple base camp, also reaching a sensitivity of 1% of the Crab nebula in 50 h of observations [4]. Also, the Japanese/Australian CANGAROO III telescope system started to report results [5], clearing up discrepancies between H.E.S.S. and earlier CANGAROO I/II results and essentially in all cases confirming H.E.S.S. data. There is also significant progress in the direct detection of shower particles on the ground: using an improved analysis technique, the MILAGRO collaboration published sky maps (Fig. 1) which which show a number of significant detections at gamma-ray energies in the 10 TeV range [6]; in 6 years of data, the Crab Nebula is seen with $15\,\sigma$ significance. The ARGO-YBJ experiment, a resistive plate chamber carpet of $6000\,\mathrm{m}^2$, started first data taking and reports detection of the Crab Nebula at the $5\,\sigma$ level in 290 h, and of Mrk 421 during a flare [7]. Operating since a couple of years, the Tibet AS-γ instrument also detects the Crab Nebula, Mrk 421 and Mrk 501; a planned upgrade using $10000\,\mathrm{m}^2$ of water-Cherenkov muon detectors is expected to boost sensitivity by a factor 10 above 10 TeV [8].

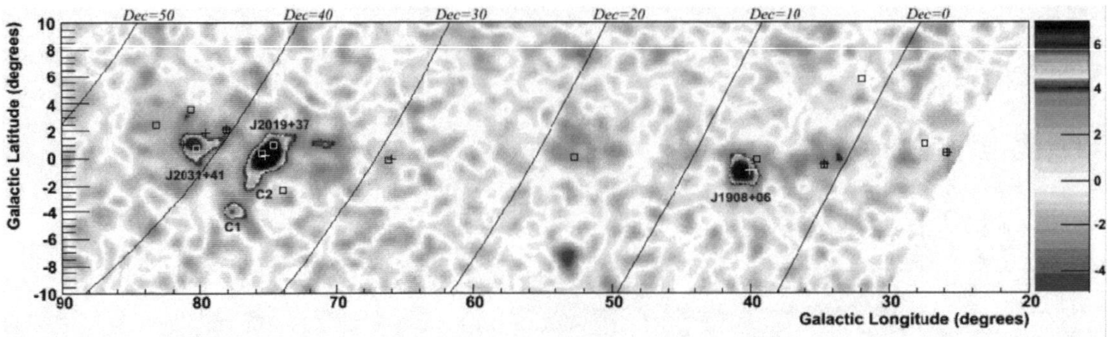

Figure 1. Gamma-ray significance map obtained with MILAGRO, for part of the Galactic plane, showing the three sources MGRO J2031+41, MGRO J2019+37 and MGRO J1908+06 [6].

2. The emerging VHE gamma-ray sky

Over the last years, the number of known VHE gamma-ray sources increased rapidly; the last count gives more than 70 sources (Fig. 2) [9], among them 6 or more supernova remnants, about 20 pulsar wind nebulae and 20 unidentified sources, four binary systems, diffuse emission from clouds and 19 extragalactic sources. The H.E.S.S. Galactic Plane Survey (Fig. 3) shows the Galaxy lined with gamma ray sources, most of them representing extended objects with sizes of some 10 pc.

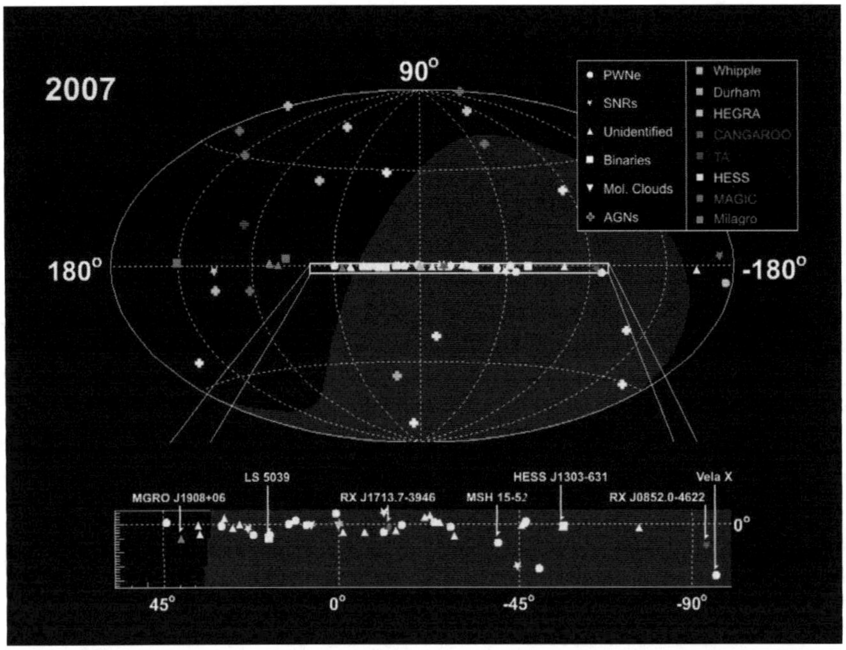

Figure 2. Location and types of VHE gamma-ray sources known by mid-2007, from [9].

The performance of current instruments is illustrated by images like Fig. 4(a), showing resolved emission from a supernova shock wave [10], by the fact that gamma-ray energy spectra

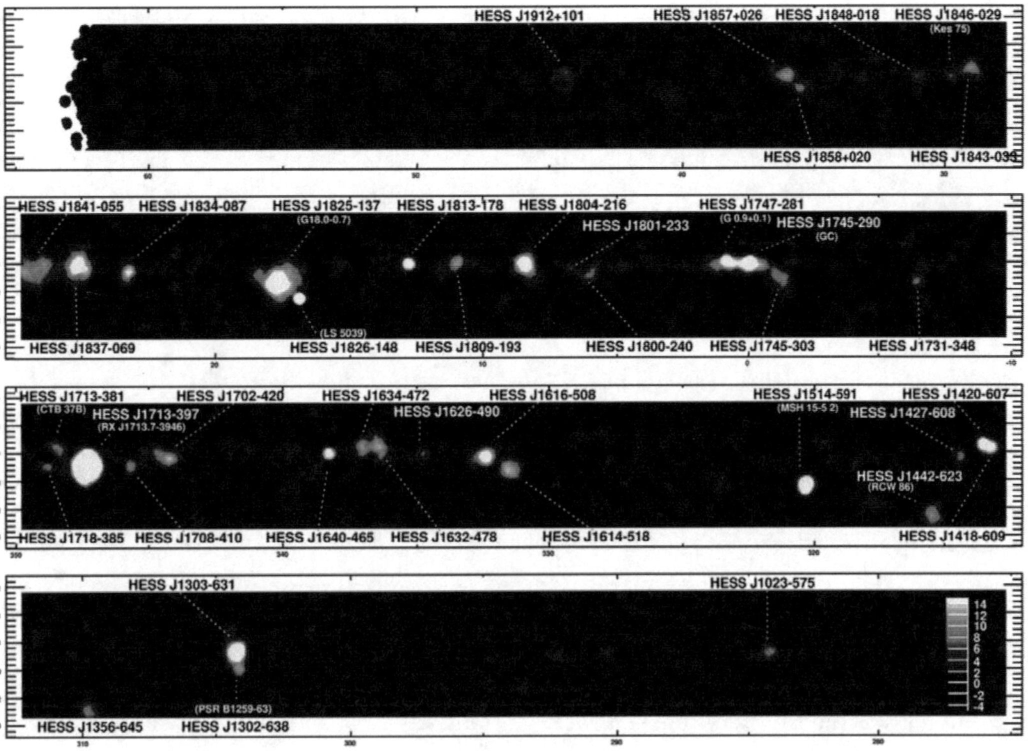

Figure 3. The H.E.S.S. Galactic Plane Survey, imaging the central region of the Milky Way at TeV energies in the region between 60° and −80° Galactic longitude. Color scale indicates the significance of a gamma-ray excess at a certain location, integrated over a region of radius 0.22°.

from this source are measured over more than 2 orders of magnitude in energy and about 6 orders of magnitude in flux [10], or by the observation of periodic emission from the binary system LS 5039 (Fig. 4(b))[11], allowing the determination of the gamma ray period with 3-digit precision to 3.908 ± 0.002 days, well compatible with the orbital period measured in the optical of 3.90603 ± 0.00017 days. Another striking achievement relates to the precision with which source positions can be determined. Using improved techniques to calibrate telescope pointing, H.E.S.S. located the Galactic Center gamma-ray source with a precision of 12", consistent with the black hole Sgr A* but excluding the nearby remnant Sgr A East (Fig 4(c)) [12].

3. Supernova remants

At least six supernova remnants (SNR) are seen in TeV gamma rays: IC 443 [13, 14], RX J0852-4622 [15, 16], RCW 86 [17], RX J1713.7-3946 [10, 18], W28 [19] and Cas A [20, 24], three of those with resolved shell morphology (RX J0852-4622, RCW 86, RX J1713.7-3946). Striking in RX J1713.7-3946 (Fig. 4(a)) but also in RX J0852-4622 [15] is a very close correlation between gamma-ray emission and X-ray emission. This correlation emerges naturally in models where both types of radiation are produced by a single population of electrons, via synchrotron radiation and Inverse Compton scattering. In case of RX J1713.7-3946, a field of about $10\,\mu$G is required to match the relative flux levels [10]. The filaments seen in X-ray images of SNR are often interpreted as evidence for rapid cooling of electrons as they move away from the shock

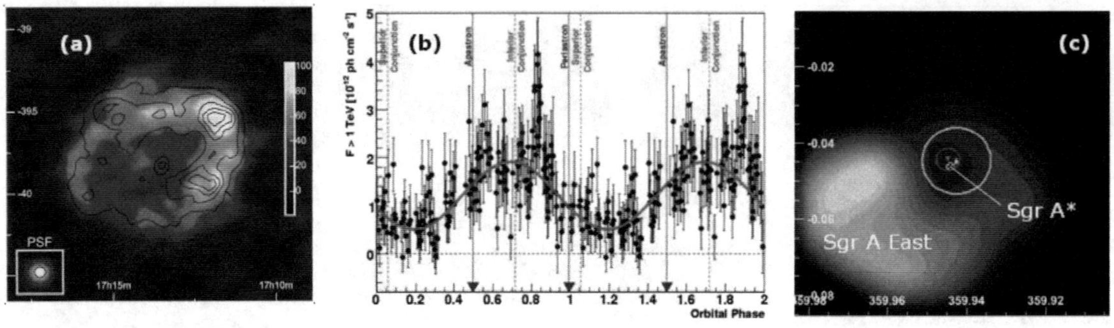

Figure 4. Illustrating the performance achievable with current gamma-ray instruments: (a) Resolved image of the supernova remnant RX J1713.7-3946 [10]. (b) Phase-dependent gamma-ray flux from the binary LS 5039, with the period determined to 3.908 ± 0.002 days [11]. (c) Error box for the location of the Galactic center gamma-ray source: 2004 data - blue circle, 2005/6 data - red circle; the cross shows the location of Sgr A*, the color scale the radio emission from the region, with the remnant Sgr A East as the dominant feature [12].

fronts, which requires much higher fields in the $100\,\mu$G range (e.g. [21]). With such fields, the number of electrons required to sustain the X-ray flux is reduced by a factor O(100), and correspondingly is the Inverse Compton rate. Gamma-ray production by collisions of accelerated protons with gas is then invoked to explain the observed gamma-ray flux, providing a good fit of the energy spectrum [22]. In this model, the correlation between X-ray emission and gamma rays is not quite as straight forward to explain. It has been suggested [23] that turbulent B-fields are dynamically amplified by streaming cosmic rays, resulting in $B^2 \sim \rho v_s^3$, linking magnetic fields (responsible for X-ray production) and gas density (responsible for gamma rays via π^0 production). Two newly discovered supernova remants may help to resolve the question if nucleons are accelerated in significant numbers: both in W28 and in IC 443 (Fig. 5) the gamma-ray emission seems to be concentrated in a region where the supernova shock interacts with dense molecular clouds. Since the latter serve as a target material for protons, this observation is a strong argument for the presence of protons accelerated by the remnant.

4. Pulsar wind nebulae

The first VHE gamma-ray source, the Crab Nebula discovered in 1989, is a pulsar wind nebula (PWN), where are relativistic flow of electrons and positrons is launched near the pulsar and feeds a wind termination shock, from which a power-law distribution of particles convects outward. Recent data from Magic [31] show for the first time the turnover towards low energy of the Crab gamma-ray spectrum, as expected for Inverse Compton radiation. A large number of PWN have recently been discovered; they differ from the Crab Nebula in that they are typically extended sources with a scale of some 10 pc, and surprisingly the gamma-ray PWN is often significantly displaced from its pulsar, with the Vela X PWN [25] or the PWN around PSR J1826+1334 [26] as prominent examples (Fig. 7(a)-(d)). For the latter PWN the morphology depends on gamma-ray energy, with the source shrinking towards the pulsar with increasing energy (Fig. 7(e)) as expected for electrons flowing away from the pulsar and suffering radiative energy losses. This is a clear indication that the gamma-ray source is indeed correlated with the pulsar and not a chance coincidence. A search for gamma-ray emission among 435 pulsars in the range of the H.E.S.S. Galactic Plane Survey shows that pulsars have a high probability to have detectable gamma-ray PWN provided that their spin-down energy flux is high enough,

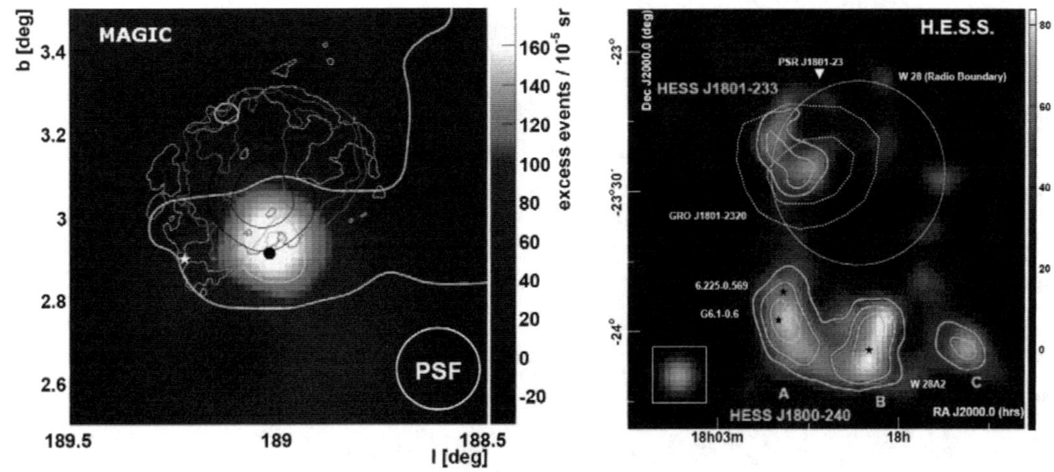

Figure 5. VHE gamma-ray emission associated with the remnants IC 443 [13] and W28 [19]. In both cases, emission is concentrated in regions where the supernova shock wave is believed to interact with dense gas clouds. The gamma-ray sources south of the W28 shell also coincide with gas clouds, possibly illuminated by high-energy nuclei accelerated in the W28 remnant.

$\dot{E}/d^2 > 10^{35}\,\mathrm{ergs\,s^{-1}kpc^{-2}}$ [29]. This implies that typically about 1% of the spin-down energy loss is converted into gamma rays in the VHE energy range. X-ray observations of some of the previously unidentified gamma-ray sources, such as HESS J1640-465, have revealed PWN at their centers (Fig. 8); in these cases, the pulsar itself obviously escaped detection in pulsed radio emission.

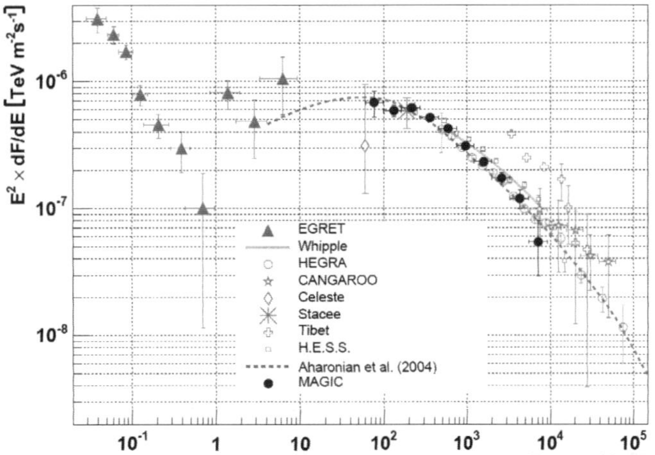

Figure 6. Spectral energy distribution $E^2\mathrm{d}N/\mathrm{d}E$ measured for the Crab nebula, showing the top end of the synchrotron spectrum and the new measurement of the Inverse Compton gamma-ray emission by MAGIC [31].

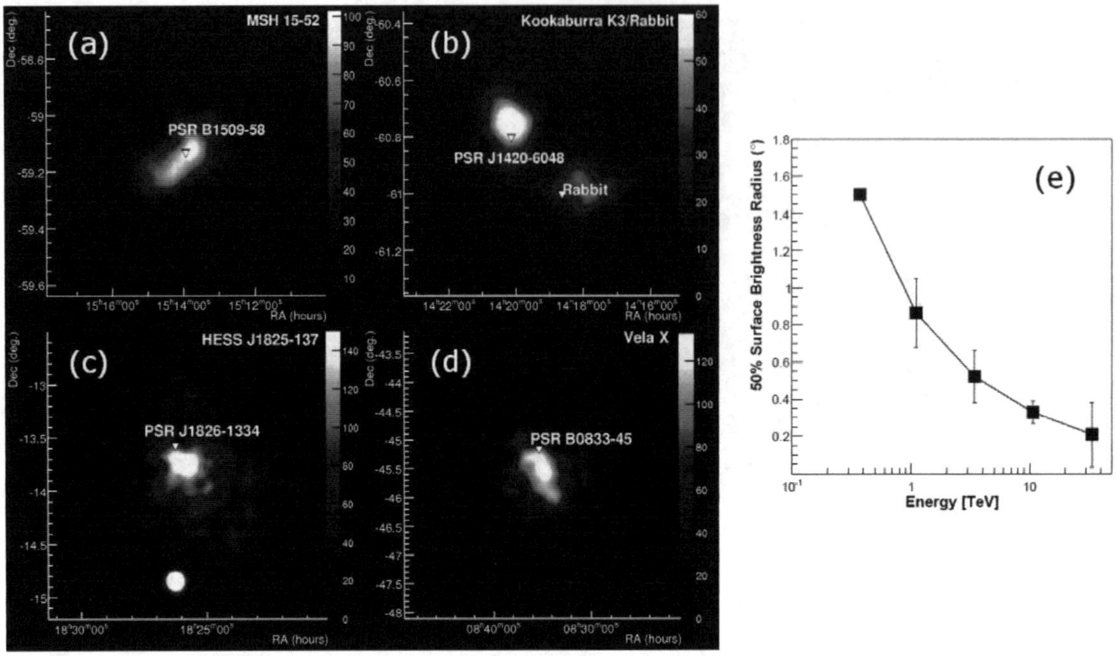

Figure 7. Example of pulsar wind nebulae: (a) the PWN MSH 15-52 associated with the pulsar PSR B1509-58, (b) PWN in the Koobaburra associated with PSR J1420-6047, (c) the PWN HESS J1825-137 associated with the pulsar PSR J1826+1334, (d) the Vela-X PWN emerging from the Vela pulsar [27]. Triangles denote pulsar locations. (e) Distance from the pulsar PSR J1826+1334 at which the surface brightness of the PWN drops to 50% of the flux at the pulsar position, as a function of gamma-ray energy [28].

5. Unidentified sources

A number of TeV gamma-ray sources remain unidentified, not because - like in case of many of the EGRET unidentified sources - there are too many counterpart candidates in the error circle, but rather because there are no plausible counterparts known in radio or X-rays; examples are shown in Fig. 9. Lack of radio and X-ray emission could be caused by absence of high-energy electrons, which would lead one to identify these objects with pure proton accelerators, possibly old supernova remnants, where electrons are suppressed since radiative losses exceed acceleration gains [33]. For at least some of the sources, a consistent multiwavelength description can also be obtained with an electron population with a cutoff in the TeV range, in which case in the KN regime high-energy gamma-rays can still be produced, but the synchrotron radiation peaks below the X-ray range and escapes detection. Also, the recently reported extended MILAGRO sources [6] are unidentified; two of them, MGRO J2031+41 and MGRO J2019+37 (Fig. 10) are in the Cygnus region and could either be associated with enhanced emission from clouds, or represent a superposition of unresolved sources in the Cygnus arm. The latter scenario is supported by the positional coincidence between the source MGRO J2031+41, extended on the 3° scale, and the first unidentified TeV source, TeV J2032+4130, discovered by HEGRA [34] and confirmed by Whipple [35] and MAGIC [36], also extended but only on the 6' scale. Extrapolated to the 20 TeV MILAGRO energy range, TeV J2032+4130 would contribute about 1/3 of the flux measured for MGRO J2031+41. The Tibet AS-γ experiment also reports hotspots in the Cygnus region [37], which are not highly significant after accounting for trials, but two of the

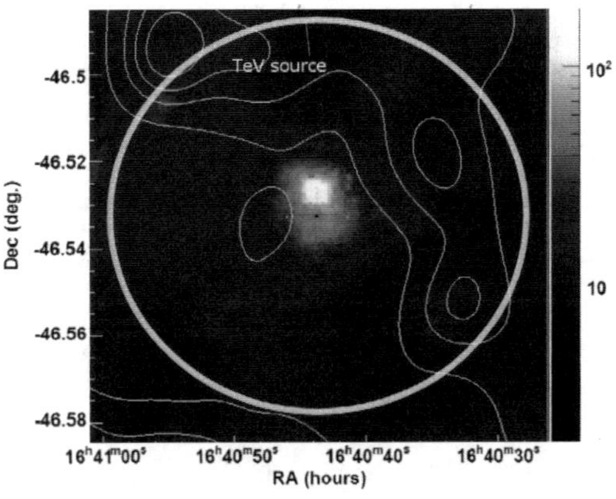

Figure 8. XMM-Newton observations of the VHE gamma-ray source HESS J1640-465 for events above 2 keV (color scale) [30]. The white contours denote the 843 MHz MOST radio data showing the inner edge of the radio shell of G338.3-0.0. The yellow circle indicates the rms extension of HESS J1640-465.

hotspots line up with the MILAGRO sources, the third with a less significant MILAGRO source candidate (Fig. 10).

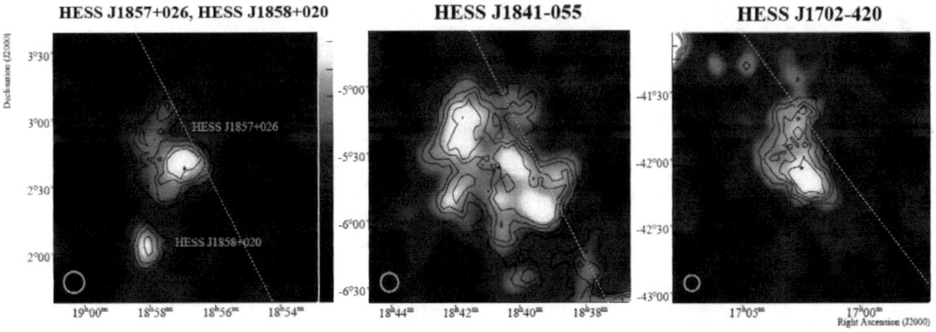

Figure 9. Four examples of VHE gamma-ray sources without known counterparts [32].

6. Binary systems

Four binary systems were so far identified as TeV gamma-ray sources: PSR B1259-63 [38], LS 5039 [11], LS I +61 303 [39, 40] and, most recently, Cygnus X-1 [41]. In LS 5039 and LS I +61 303, the nature of the compact object is still open, it could either be a pulsar, with the pulsar wind driving the gamma-ray emission, or a black hole, in which case the emission would be accretion-powered. PSR B1259-63 is a pulsar in a highly elliptical 3.5 y orbit around a Be-type star, Cygnus X-1 a black hole in a nearly circular 5.6 d orbit. Periodic modulation of gamma-rays from LS 5039 has been established by H.E.S.S., as discussed earlier. New results on LS I +61 303 show periodicity with the orbital period, but with indications of variability from orbit to orbit [39] (Fig. 11). Emission is peaked at phase 0.6-0.7. MAGIC recently reported the detection of a flare from Cygnus X-1 [41] in one half night of 26 nights of observations, at 4σ significance after trials, coincident with an X-ray flare. The importance of binary systems is that along elliptical

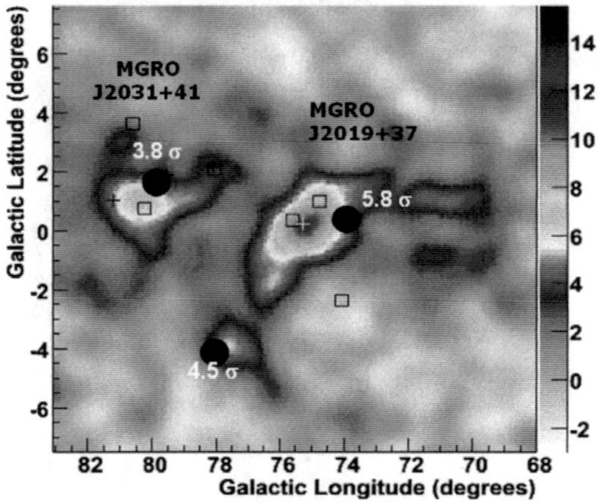

Figure 10. MILAGRO measurements of VHE gamma-ray emission from the Cygnus region, showing significance contours. The two sources MGRO J2031+41 and MGRO J2019+37 represent significant detections even after accounting for number of trials in the search for sources [6]. Black dots: "hotspots" in the Tibet AS-γ array data, with corresponding significance [37].

orbits, conditions such as ambient magnetic and radiation fields as well as particle injection rates vary, allowing the study of how particle acceleration and gamma-ray emission is influenced by these changes, and providing additional constraints for models, as compared to static sources.

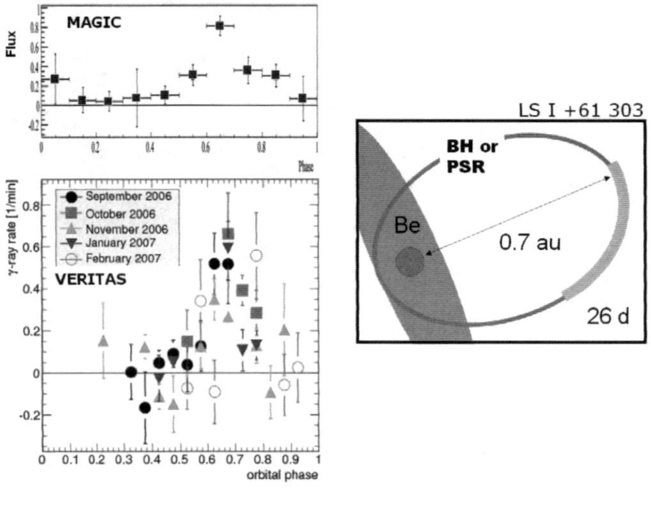

Figure 11. Orbital variation of VHE gamma-ray emission from LS I +61 303, as measured by MAGIC [39] and, over different orbits, by VERITAS [40]. Right: orbit geometry, indicating in blue the region of strong gamma-ray emission.

7. Extragalactic sources of VHE gamma rays

Results discussed in the following concern (i) observations of new sources and new source types, and of very rapid variability of some sources and (ii) results related to the propagation of gamma rays in extragalactic space, where they are subject to absorption by extragalactic background light (EBL) and possibly to a time dispersion due to effects of quantum gravity. A more detailed discussion of acceleration and radiation processes in the sources is left to the review by Dermer [42].

Table 1 lists known extragalactic VHE emitters; the number has grown to 19. Most of them are of the BL Lac type, with a jet pointing at the observer and dominating the emission. Remarkable exceptions are M87, a FR I radio galaxy which surprisingly shows day-scale variability [43], and the recently discovered, by far most distant source, 3C 279, a FSRQ at redshift z=0.536 [44]. In one of 10 nights of observations MAGIC detected a signal from 3C 279, both in the low-energy band from 80-220 GeV and above 220 GeV. The high-energy signal is surprising, since – as discussed later – EBL absorption should dramatically steepen the energy spectrum for such distant objects. Another recent observation which challenges interpretation are the minute-scale variations observed for PKS 2155-304 during a giant outburst in July 2006 [46], with flux levels up to 15 times the Crab flux, and for Mrk 501 in 2005 [47] (Fig. 12). For PKS 2155-305, doubling times around 100 s need to be compared to the size of the supermassive black hole powering the AGN, with r/c of 1...2 $\cdot 10^4$ s, calling for very large Lorentz factors around 100 for the jet motion. It should be pointed out that two of the recent discoveries of sources, Mrk 180 [48] and 1ES 1101-496 [49], were triggered by variation in the optical emission of the objects.

Name	Type	Redshift	Signif. ***: >10	Discovered
M 87	FR I	0.004	***	HEGRA
Mrk 421	BL Lac	0.031	***	Whipple
Mrk 501	BL Lac	0.034	***	Whipple
1ES 2344+514	BL Lac	0.044	***	Whipple
Mrk 180	BL Lac	0.046	5.5	MAGIC
1ES 1959+650	BL Lac	0.047	***	TA
BL Lac	BL Lac	0.069	5.1	MAGIC
PKS 0548-322	BL Lac	0.069	5.8	HESS
PKS 2005-489	BL Lac	0.071	***	HESS
PKS 2155-304	BL Lac	0.116	***	Durham
H 1426+428	BL Lac	0.129	7.5 / 5	Whipple
1ES 0229+200	BL Lac	0.14	6.6	HESS
H 2356-309	BL Lac	0.165	***	HESS
1ES 1218+304	BL Lac	0.182	9 / 6.4	MAGIC
1ES 1101-232	BL Lac	0.186	***	HESS
1ES 0347-121	BL Lac	0.188	***	HESS
1ES 1011+496	BL Lac	0.212	***	MAGIC
PG 1553+113	BL Lac	?	***	HESS/MAGIC
3C 279	FSRQ	0.536	~8 (trials?)	MAGIC

Table 1. Extragalactic VHE gamma-ray sources, with their type, redshift, significance of detection and discovery experiment.

The fast rise times of the bursts of PKS 2155-304 and Mrk 501 allow use for studies of quantum gravity, which predicts energy-dependent propagation speeds, with corrections of order $(E/M_{QG})^q$ where M_{QG} is a characteristic scale of quantum gravity, usually assumed to be related to the Planck scale M_P. Some models allow a linear dependence on energy, $q = 1$, wheras in others the linear term is forbidden by symmetries and only higher orders contribute. In the Mrk 501 flare of July 9, 2005, the MAGIC collaboration reports an energy dependent shift of the peaking time (Fig. 13), with a 4 min lag in arrival times between the lowest and the highest photon energies [47]. If interpreted as an effect of quantum gravity, the shift would correspond

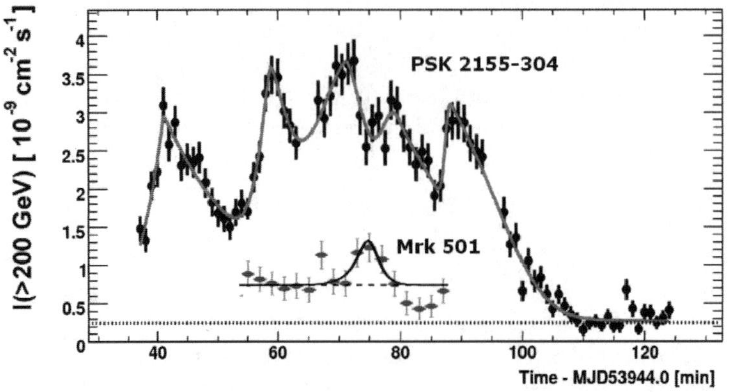

Figure 12. Two AGN flares with minute-scale variability: Gamma-ray flux measured from PKS 2155-305 during the flare on July 28, 2006 [46], and for Mrk 501 on June 30, 2005 [47], shown on approximately the same flux scale (from [9]).

to $M_{QG} = 0.03 M_P$ for $q = 1$ [50]. Of course, acceleration and radiation processes in the source could cause energy dependent peaking times, and the only way to demonstrate that effects are due to quantum gravity is to observe a variation of the lag with distance of the source. Assuming that shifts intrinsic to the source and the QG effect do not conspire and cancel, the result can more conservatively be phrased as a lower limit of $M_{QG} > 0.02 M_P$. Indeed, the analysis of HESS data from the outburst of PKS 2155-304 does not show any energy-dependent peaking, resulting in a lower limit of $M_{QG} = 0.04 M_P$. It should be noted that significantly more stringent limits are obtained from other observations, such as the absence of birefringence in optical AGN spectra as predicted in certain classes of QG models, or on the basis of photon interactions or decays which are influenced by a modified dispersion [51, 52].

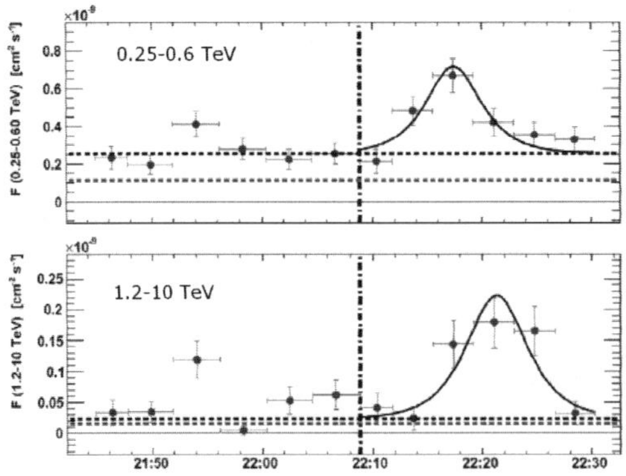

Figure 13. Flare of Mrk 501 on July 9, 2005, as seen in two gamma-ray energy bands [47].

Interactions with EBL quanta cause a modification and in general a steepening of gamma-ray spectra. Such features - or the lack thereof - can be used to measure or put limits on the EBL density, which is of significant cosmological interest and which is - in particlar in the infrared range - difficult to measure directly due to overwhelming foreground light. Since the pair production cross section peaks near the threshold $2E_\gamma E_{EBL} = m_e c^2$, there is an approximate relation between the gamma-ray energy and the wavelength of the EBL photons probed. Particularly constraining are distant gamma-ray sources with hard observed spectra, such as 1ES 0229+200 [53]. Assuming that intrinsic blazar photon indices are not smaller than 1.5 - corresponding to inverse Compton scattering off an E^{-2} electron population - the EBL density must be relatively low in order not to steepen spectra beyond the observed index. Several new analyses and results emerged recently. Raue and Mazin give an analysis which does not assume a specific spectral shape of the EBL [54], resulting in the limits shown in Fig. 14 for the IR EBL, which fall between previous upper limits and lower limits based on galaxy counts. Slightly more stringent limits can be obtained with loose assumptions concerning the shape (e.g. [53]). It has been pointed out that in certain models intrinsic gamma ray spectra with photon indices in the range below 1.5 are possible [55], which would result in less stringent limits; however, it would then be surprising that such hard spectra are not observed for any of the nearby AGN.

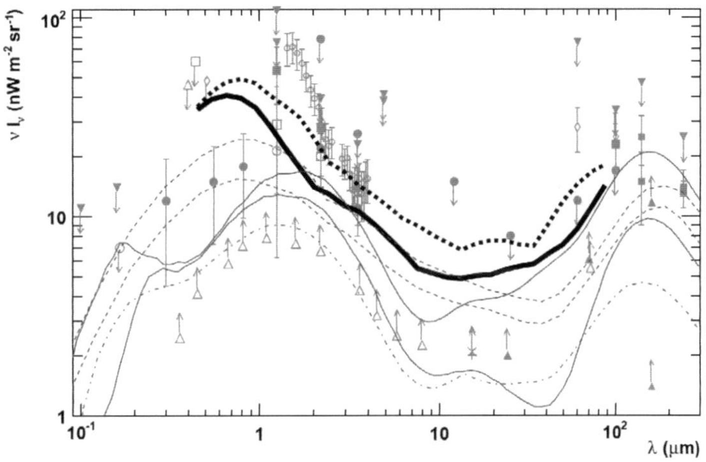

Figure 14. Limits on extragalactic background light (EBL) density in the infrared, derived from gamma-ray observations of AGN, with realistic and extreme assumptions concerning intrinsic AGN spectra (thick full and thick dottet line, resp.). Thin lines show different model predictions for EBL density [54]. Points show direct measurements and upper and lower limits, see [54] for details.

8. Conclusion and outlook

Very high energy gamma-ray astronomy has taken big steps forward in recent years, now providing sky maps, resolved sources, detailed light curves and spectra. Space does not allow to cover limits on gamma-ray emission from other object classes such as GRBs, starburst galaxies, galaxy clusters, or dark matter annihilation in dwarf galaxies. For some of these classes, emission is predicted at levels not too far from current sensitivity. Correspondingly, both HESS and MAGIC are upgrading their instruments, HESS by adding a large 600 m^2 dish, MAGIC by adding a second dish of identical size as the first. The upgrades increase sensitivity by factors 2-3 and help to lower energy thresholds. In the more distant future, arrays of Cherenkov telescopes such as the proposed CTA and AGIS could boost sensitivity at TeV energies by a factor of 10, reaching mCrab sensitivity, and HAWC, based on MILAGRO technology, could provide full-sky coverage at sensitivity better than 0.1 Crab.

The author would like to thank C. van Eldik for his comments on the manuscript.

References

[1] R.C. Hartman et al., ApJS 123 (1999) 79
[2] I. Grenier, J. Casandjian, Bull. AAS 38 (2006) 354
[3] A. Chen et al., Ast. Telegr. 1278 (2007)
[4] G. Maier et al., Proc. 30th ICRC, Merida, Mexico, 2007; arXiv:0709.3654
[5] Y. Yukawa, these proc.
[6] A.A. Abdo et al., ApJ 664 (2007) L91
[7] P. Bernadini, these proc.
[8] K. Kawata, these proc.
[9] J. Hinton, Proc. 30th ICRC, Merida, Mexico, 2007
[10] F. Aharonian et al., A&A 464 (2007) 235
[11] F. Aharonian et al., A&A460 (2006) 743
[12] C. van Eldik et al., Proc. 30th ICRC, Merida, Mexico, 2007; arXiv:0709.3729
[13] J. Albert et al., ApJ 664 (2007) L87
[14] T. B. Humensky, Proc. 30th ICRC, Merida, Mexico, 2007; arXiv:0709.4298
[15] F. Aharonian et al., ApJ 661 (2007) 236
[16] R. Enomoto et al., ApJ 652 (2006) 1268
[17] S. Hoppe et al., Proc. 30th ICRC, Merida, Mexico, 2007; arXiv:0709.4103
[18] H. Muraishi et al., A&A 354 (2000) L57
[19] G. Rowell et al., Proc. 30th ICRC, Merida, Mexico, 2007; arXiv:0710.2017
[20] F. Aharonian et al., A&A 370 (2001) 112
[21] J. Vink, Proc. Symp. 'The X-ray Universe 2005', San Lorenzo de El Escorial, Spain; arXiv:astro-ph/0601131
[22] E.G. Berezhko, H.J. Völk, A&A 451 (2006) 981
[23] S.G.Lucek, A.R. Bell,MNRAS 314 (2000) 65
[24] J. Albert et al., A&A 474 (2007) 937
[25] F. Aharonian et al., A&A 448 (2006) L43
[26] F. Aharonian et al., A&A 460 (2006) 365
[27] S. Funk, Proceedings COSPAR 2006; arXiv:astro-ph/0701471
[28] S. Funk et al., Proc. 30th ICRC, Merida, Mexico, 2007; arXiv:0709.3125
[29] S. Carrigan et al., Proc. 30th ICRC, Merida, Mexico, 2007; arXiv:0709.4094
[30] S. Funk et al., arXiv:astro-ph/0701166
[31] J. Albert et al., arXiv:0705.3244
[32] K. Kosack et al., 30th ICRC, Merida, Mexico, 2007, OG 2.2, 400
[33] R. Yamazaki et al., MNRAS 371 (2006) 1975
[34] F. Aharonian et al., A&A 431 (2005) 197
[35] A. Konopelko et al., arXiv:astro-ph/0611730
[36] E. Oña Wilhelmi, 30th ICRC, Merida, Mexico, 2007, OG 2.2, 532
[37] M. Amenomori, ApJ 633 (2005) 1005
[38] F. Aharonian et al., A&A 442 (2005) 1
[39] J. Albert et al., Science 312 (2006) 1771; T. Jogler, these proc.
[40] G. Maier et al., Proc. 30th ICRC, Merida, Mexico, 2007; arXiv:0709.3661
[41] J. Albert et al., ApJ 665 (2007) L51
[42] C. Dermer, these proc.
[43] F. Aharonian et al., Science 314 (2006) 1424
[44] M. Teshima et al., Proc. 30th ICRC, Merida, Mexico, 2007; arXiv:0709.1475
[45] J. Albert et al., ApJ 666 (2007) L17
[46] F. Aharonian et al., ApJL 664 (2007) L71
[47] J. Albert et al., ApJ 669 (2007) 862
[48] J. Albert et al., ApJ 648 (2006) L105
[49] J. Albert et al., ApJ 667 (2007) L21
[50] J. Albert et al., arXiv:0708.2889
[51] S. Liberati, PoS (P2GC) 018 (2007); arXiv:0706.0142
[52] M. Galaverni, G. Sigl, arXiv:0708.1737
[53] F. Aharonian et al., arXiv:0709.4584
[54] D. Mazin, M. Raue, A&A 471 (2007) 439
[55] F.W. Stecker, S.T. Scully, arXiv:0710.2252

The obscured universe

Charles D. Dermer

Code 7653, Naval Research Laboratory, 4555 Overlook Ave., SW, Washington, DC 20375-5352, USA

E-mail: `dermer@ssd5.nrl.navy.mil`

Abstract. Auger results on clustering of $\gtrsim 60$ EeV UHECR ions and the interpretation of the γ-ray spectra of TeV blazars are connected by effects from the extragalactic background light (EBL). The EBL acts as an obscuring medium for γ rays and a reprocessing medium for ultra-high energy cosmic ray (UHECR) ions and protons, causing the GZK cutoff. The study of the physics underlying the coincidence between the GZK energy and the clustering energy of UHECR ions favors a composition of $\gtrsim 60$ EeV UHECRs in CNO (up to Mg) group nucleons. This has interesting implications for the sources of UHECRs. We also comment on the Auger analysis.

1. Introduction

The recent reports [1, 2] from the Pierre Auger Observatory collaboration of high-significance clustering of UHECRs along the supergalactic plane has opened multi-messenger astronomy. This advance is closely connected to progress in γ-ray and neutrino astronomy. Here we focus on new techniques to unravel the problem of UHECR origin in light of the Auger results. This topic is developed in more detail in my Mérida ICRC contribution [3] and in my book with Govind Menon [4].

2. Charged Particle Astronomy

We begin by reviewing the recent major results from Auger, including the discovery of the clustering of the arrival directions of UHECRs along the supergalactic plane [1]. This discovery was anticipated in the study by Stanev et al. [5], and we return to this paper in the discussion.

2.1. Pierre Auger Observatory

Auger is ~ 3000 km^2 in area, about the size of Rhode Island. It consists of 1600 shower detectors spaced 1.5 km apart, as well as four telescope enclosures each housing 6 telescopes to map Ni fluorescence. The directional point spread function of an UHECR is typically $\lesssim 1°$. The original analysis published in *Science* [1] treated a data set from 1 Jan 2004 to 31 August 2007 consisting of 81 events with total energy $E > 40$ EeV.

2.2. Clustering Result

A probability statistic P corrected for exposure is constructed from the nearest-neighbor angular separation ψ between the arrival direction of an UHECR with energy E and the direction to AGN in the Véron-Cetty & Véron (VCV) catalog [6]. This catalog has 694 active galaxies with

$z \leq 0.024$ or distance $d \lesssim 100$ Mpc (the total catalog has over 10^5 entries) but is incomplete, especially around the plane of our Galaxy, and is also biased by different sample selection techniques.

In the Auger analysis [1], P was minimized for $\psi = 3.1°$, threshold energy $E_{thr} = 56$ EeV, and $z_{max} = 0.018$ ($d_{max} = 75$ Mpc), containing 27 events (the two highest energy events were 90 and 148 EeV). Twelve events correlate within 3.1° of the selected $d < 75$ Mpc AGNs, and another three within the vicinity of one of these nearby AGN. By comparison, 3.2 events would be expected for an isotropic UHECR flux, which it clearly is not. Lack of correlation of $\gtrsim 60$ EeV UHECRs with the Galactic plane rules out Galactic (pulsar, stellar mass black hole) and Galactic halo (including halo dark matter annihilation) models, leaving extragalactic models viable, at least those that follow the matter distribution traced by the supergalactic plane like the nearby AGN in the VCV catalog.

The recent detailed Auger analysis [2] confirms these results, furthermore pointing out that the distance cutoff could range from $\sim 50 - 100$ Mpc and that $\psi \lesssim 6°$. If one accepts the validity of these results, then *UHECRs with $E \gtrsim 60$ EeV originate from AGNs within* 75(±25) *Mpc and are deflected in their travels by at most* $\sim 3° - 6°$ by either the Galactic magnetic field [2] or the intergalactic magnetic field [3].

The lack of a quasi-isotropic background UHECR flux above $E > E_{cl} \cong 60$ EeV is most simply explained if more distant UHECRs fail to arrive here at Earth. Because we know that there are sources of UHECRs with energy up to at least 300 EeV [7], it follows that unless we were in a unusual over-density of UHECR sources within ~ 100 Mpc and an under-density within $\sim 100 - 300$ Mpc or more, then energy losses of UHECRs accelerated by and injected from more sources more distant than $\approx 50 - 100$ Mpc are the cause of the Auger anisotropy.

The clustering energy $E_{cl} = 60$ EeV stands out in this analysis, separating $E \lesssim E_{cl}$ UHECRs formed mainly in the large scale quasi-homogeneous, isotropic universe at distance scales $d \gtrsim 50 - 100$ Mpc from the UHECRs formed in the clumpy structured universe at local, $d \lesssim 100$ Mpc scales.

2.3. GZK Cutoff

A high-significance steepening in the UHECR spectrum at the Greisen-Zatsepin-Kuzmin (GZK) [8, 9] energy $E_{\rm GZK} \cong 10^{19.6}$ eV $\cong 4 \times 10^{19}$ eV was reported at the 2007 Mérida ICRC based on observations taken with the Auger Observatory [10] (see Figure 1), and earlier in 2007 by the HiRes collaboration [11] at $E_{\rm GZK} \cong 10^{19.8}$ eV $\cong 6 \times 10^{19}$ eV.[1] The coincidence $E_{\rm GZK} \cong E_{cl}$ seems likely to originate from underlying physics.

2.4. UHECR Composition

The composition of UHECRs measured from $\approx 4.5 \times 10^{17}$ eV up to 4×10^{19} eV with Auger is neither dominant p nor Fe composition [12], and is not trending toward a lighter composition at the upper energy range, contrary to pre-Auger indications [13]. If the results of Ref. [12] can be extrapolated to $\gtrsim 60$ EeV, then ions must be included in the analysis unless the UHECR composition abruptly shifts to proton-dominated at $E \gtrsim E_{cl}$, as would be the case in neutral beam models of UHECRs [14].

3. GZK Energy

The energy of the GZK cutoff is not precisely defined. Returning to the classic paper by Greisen [8] for inspiration, let us work through the estimates in light of our new knowledge.

[1] The HiRes team also reports a significant dip structure at $E_{dip} \cong 10^{18.6}$ eV.

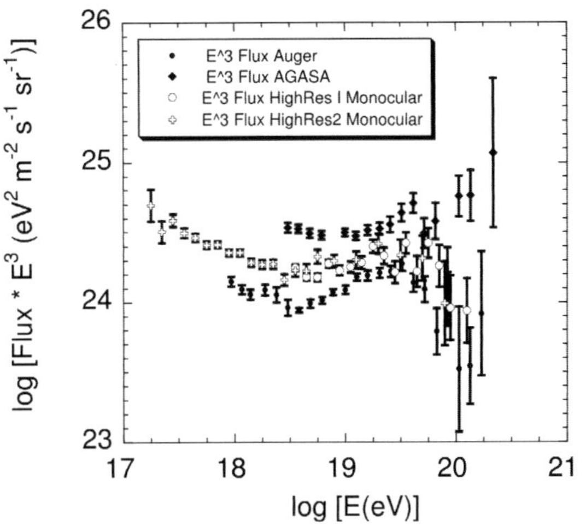

Figure 1. UHECR spectrum, plotted in the form of $E^3 I$, where I is the number intensity, for AGASA (filled diamonds), HiRes (open symbols), and Auger (filled circles).

3.1. GZK Energy: Analytic Analysis

UHECR protons with energy $E_p = \gamma_p m_p c^2 = 10^{20} E_{20}$ eV will undergo strong photopion losses when the threshold condition $\gamma_p \epsilon_{CMBR} \gtrsim m_\pi / m_e \cong 400$ is satisfied, where $\epsilon_{CMBR} \cong 2.70 \times 2.72$ K$/5.9 \times 10^9$ K $\cong 10^{-9}$ is the dimensionless mean photon energy, in units of the electron rest mass, of the CMBR. Strong photopion losses due to interactions with the peak energy photons of the CMB occur when $\gamma_p \gtrsim 4 \times 10^{11}$. The mean free path (MFP) for energy loss of a proton with energy $E_{20} \gtrsim 4$ is $\lambda_{\phi\pi} \cong (1/n_{CMBR}\sigma_{\phi\pi}) \cong 1/(400 \text{ cm}^{-3} \cdot 70\mu\text{b}) \cong 10$ Mpc. From a derivation using a step-function approximation for the product of photopion cross section and inelasticity [14], the mean-free-path for photopion energy losses of UHECR protons with CMBR photons is given by [3]

$$\lambda_{\phi\pi}(E_{20}) \cong \frac{13.7 \exp[4/E_{20}]}{[1 + 4/E_{20}]} \text{ Mpc} . \tag{1}$$

The values of $\lambda_{\phi\pi}(E_{20})$ for $E_{20} = 0.6, 0.8, 1$, and 2 are 1400, 340, 150, and 35 Mpc, respectively. If the UHECR with energies between 60 and 100 EeV were protons, then they could originate over a large distance scale extending to 1 Gpc. This simple estimate suggests that the the UHECRs cannot be protons because the Auger events originate from within 100 Mpc whereas the energy loss MFPs of UHECR protons at these energies are much greater.

Now we estimate the GZK MFPs of ions against photodisintegration on CMBR photons. Note that by contrast with UHECR protons, photopion losses of UHECR nucleons are not effective until $\gamma \langle \epsilon \rangle_{CMB} \gtrsim m_\pi / m_e$ or $E_{20} \gtrsim 3A/(1+z)$, where A is the nucleon mass number. The pion-forming Δ resonance peak energy is $\cong m_e c^2 \epsilon_r \approx 300$ MeV, a much higher energy than the giant dipole resonance (GDR) at ≈ 20 MeV, so photodisintegration breakdown usually dominates photopion losses for ions, especially for large A. The effective photodisintegration energy-loss rate for ions on the CMBR results from both GDR excitation with the emission of one and two nucleons, and from multi-nucleon breakup processes. Because the reaction is a threshold process and UHECRs with $E_{20} \ll 3A/(1+z)$ interact only with the Wien portion of the CMBR radiation field, the GDR is more important than multi-nucleon excitations for $E_{20} \sim 1$ UHECRs.

In this asymptote, we use the δ-function approximation for the photodisintegration cross section of nucleus A, namely

$$\sigma_A(\epsilon_r) \cong \frac{\pi}{2} \sigma_{0,A} \Delta_{GDR} \delta(\epsilon_r - \epsilon_{r,A}) , \tag{2}$$

[19, 20, 15], where

$$\sigma_{0,A} = 1.45A \text{ mb} , \quad \Delta_{GDR} \cong 15.6 , \quad \epsilon_{r,A} \cong 83.5A^{-0.21} ,$$

and $\epsilon_r = \gamma\epsilon(1-\mu)$ is the invariant dimensionless photon energy in the ion's rest frame. In the δ-function approximation for the GDR, eq. (2), the derived inverse of the effective energy loss time scale of an UHECR ion due to photodisintegration processes with photons of the CMBR is

$$t_E^{-1}(E,A) = \frac{2\pi^2 c\Delta_{GDR} k_A \epsilon_{r,A}\Theta}{\gamma^2 A\lambda_C^3} \ln\left([1-\exp(-w_{r,A})]^{-1}\right)$$

$$\cong 3.2\times 10^{-15}\,\frac{k_A A^{1.79}(1+z)}{E_{20}^2}\,\ln\left[\frac{1}{1-\exp(-w_{r,A})}\right] \text{ s} , \tag{3}$$

where

$$w_{r,A} \equiv \frac{\epsilon_{r,A}}{2\gamma\Theta} = \frac{0.83A^{0.79}}{E_{20}(1+z)} .$$

The effective path length for photodisintegration in the δ-function approximation for the GDR is therefore

$$\lambda_E = \lambda_E(E,A) = ct_E(E,A) \to \frac{3.0E_{20}^2 \exp\left(0.83A^{0.79}/[(1+z)E_{20}]\right)}{k_A A^{1.79}(1+z)} \text{ Mpc} \tag{4}$$

in the limit $w_{r,A} \gg 1$, that is, $E_{20} \ll 5(A/10)^{0.79}$.

When a single proton or neutron is ejected, then a fraction A^{-1} of the original energy E is lost to the original nucleon, and for the ejection of two nucleons, a fraction $2/A$ of energy is lost. For multi-nucleon injection, an average fractional energy loss k_A/A is used [17], where $k_A = 1.2/A, 3.6/A$, and $4.349/A$ for $A = 4, 10 \leq A \leq 22$, and $23 \leq A \leq 56$, respectively. The photodisintegration energy-loss MFPs have only a generalized meaning in terms of the mean distance for an UHECR nucleon to be broken up into mostly lower energy protons and neutrons and a nucleon with A about half of the original nucleonic mass. Photodisintegration of Fe, for example, leads to significant fraction of $A \lesssim 15$ nucleon secondaries [17].

Equating eq. (4) with an energy-loss distance $\lambda = 100\lambda_{100}$ for $E = 60E_{60}$ EeV gives $\exp(1.38A^{0.79}/E_{60})/[k_A A^{1.79}] = 333\lambda_{100}E_{60}^2$. It is easily verified that all nuclei with $A \gtrsim 14$ have energy-loss mean free paths at $E_{60} \approx 1$ much longer than 100 Mpc. On the basis of this analytic treatment, one might conclude UHECRs could originate only from nucleons with $4 \lesssim A \lesssim 14$.

3.2. GZK Energy: Numerical Analysis

These results are not conclusive unless additional radiation fields, including IR and optical features from stellar and reprocessed radiation, and energy-loss processes of photopair production and universal expansion are taken into account. Figure 2 shows the energy density of the EBL and cosmic rays, and the predicted upper limit to the cosmogenic neutrino energy density [21].

3.2.1. Extragalactic Background Light The mean intensity of light in intergalactic space is referred to as the EBL. The IR and optical EBL is decomposed at low redshift ($z \lesssim 0.25$) into a dust component and a stellar component as shown in Figure 3. These fits are assumed to span the likely range of the diffuse radiation fields between ≈ 1 and a few hundred μm at low z. The modified blackbody spectral energy density is written in the form

$$\epsilon u(\epsilon) = u_0 \frac{w^k}{\exp(w)-1} = m_e c^2 \epsilon^2 n_{ph}(\epsilon) , \tag{5}$$

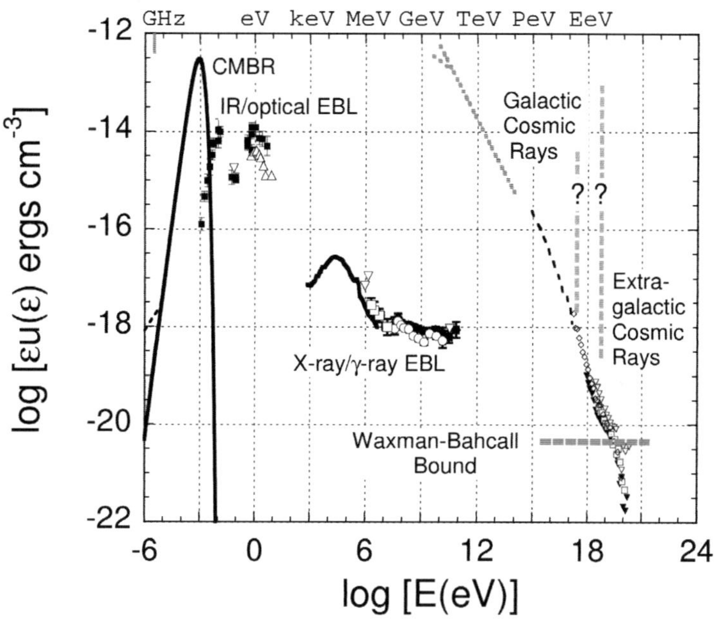

Figure 2. Energy density in intergalactic space of various radiations, including the CMBR, the IR and optical, X-ray, γ ray, extragalactic cosmic ray, and the predicted maximum energy density of cosmogenic neutrinos. Also shown is the near-Earth energy density of cosmic rays; the transition energy between the galactic and extragalactic component is uncertain.

Table 1. Properties of the Dust and Two Stellar Components

Component	T(K)	u_0 (10^{-14} ergs cm^{-3})	k
CMBR	2.72	6.38	4
Dust (Hi)	62	0.819	3.8
Dust (Low)	31	0.273	3.8
Star 1 (Hi)	7100	1.1	2.0
Star 1 (Lo)	7100	0.55	2.0
Star 2	16,600	0.5	3.0

where $w \equiv \epsilon/\Theta$. The stellar component is decomposed into the sum of two modified blackbodies, with the higher-temperature stellar component fixed. The HI EBL has the more intense dust and stellar fields, and the LO EBL has the weaker dust and stellar fields. Two other EBLs are considered here: a high dust/low stellar field and a low dust/high stellar field. Parameters of the CMBR and modified blackbodies are given in Table 1.

The intensity of the diffuse and unresolved γ-ray EBL from EGRET observations and analysis [23, 24] is shown in Figure 4. It is decomposed into contributions from blazars, separated into FSRQs and BL Lacs, and the superposed intensity from numerous faint sources where the γ rays are formed primarily by interactions of cosmic rays accelerated by supernovae in star-forming galaxies and by structure formation shocks in clusters of galaxies (reviewed in [26, 27]). The particle and radiation fields are connected because GZK effects of UHECRs with the EBL form cascade γ rays that contribute a truly diffuse component to the γ-ray EBL [25]. The intensity of this component is sensitive to the transition energy from galactic to extragalactic UHECRs and the formation rate of UHECR sources through time.

The analytic result for the energy-loss MFP of Fe with $A = 56$ on CMB photons is given by eq. (4) and labeled "analytic" in Figure 5. The accuracy is tolerable, but a more detailed calculation, also shown in Figure 5, is needed to draw more reliable conclusions. The numerical calculations for Fe treat photopion and photopair losses with the EBL as well as from universal

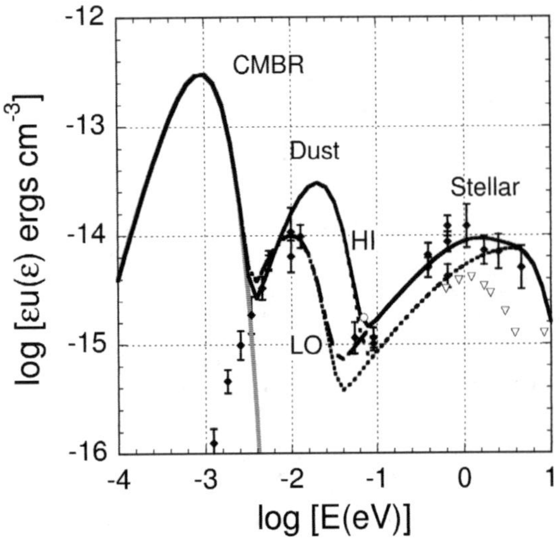

Figure 3. Ranges of EBL used in analysis. Data from Ref. [22]. The HI (LO) EBL refers to a high-(low-) intensity dust component and a high-(low-) intensity stellar component.

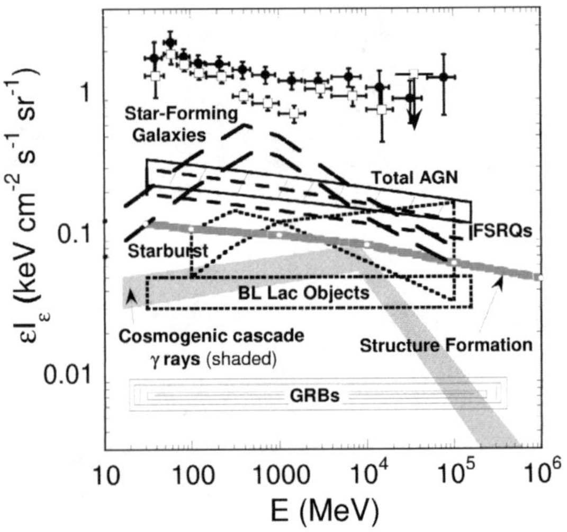

Figure 4. Decomposition of the diffuse/unresolved γ-ray background measured with EGRET [26]. Shaded region shows estimate for cascade γ-rays induced by UHECRs formed by sources evolving as AGNs [25].

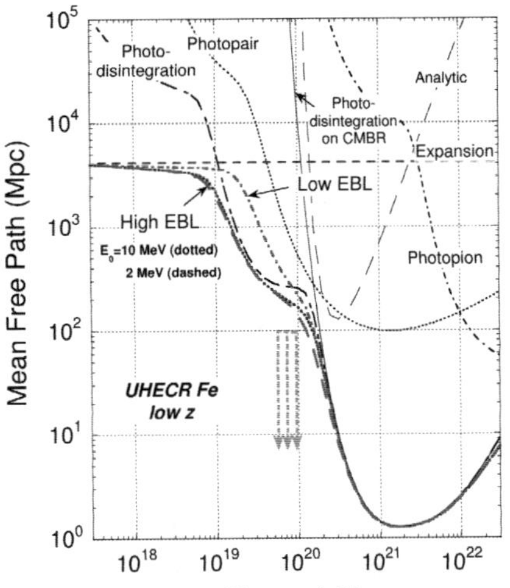

Figure 5. Effective energy-loss mean-free-path of Fe in LO and HI EBLs, with the effective photodisintegration energy-loss MFP calculated according to the prescription of Ref. [17].

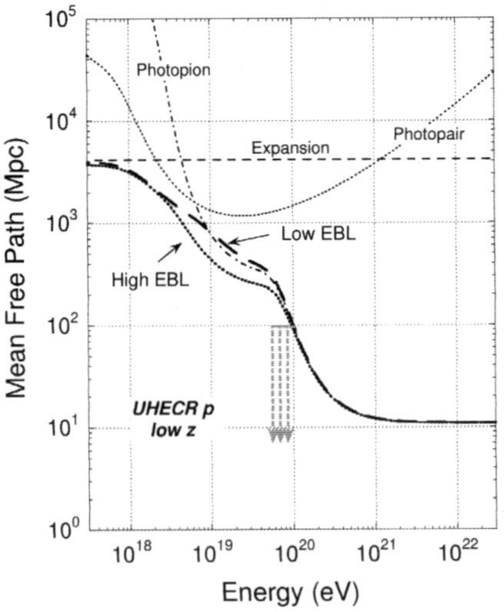

Figure 6. Effective energy-loss mean-free-path of UHECR protons in LO and HI EBLs shown in Fig. 3.

expansion (the Hubble radius is ≈ 4160 Mpc at the present epoch), with the photodisintegration, photopair, and photopion components shown separately for the HI EBL. The total energy loss MFP is also shown for the LO EBL. (See also [28, 29].) The distances from which we should expect to detect 60 – 100 EeV Fe range from ≈ 200 Mpc to 150 Mpc.

Fig. 6 shows the energy-loss MFP of UHECR protons in the LO and HI EBLs. The MFPs, which include photopion, photopair and expansion losses, range for the HI EBL from 200 to 100 Mpc for energies ranging from 60 to 100 EeV, respectively (similar to the values for UHECR Fe). Unless there was a deficit of UHECR sources between ≈ 100 – 200 Mpc, or an enhanced pathlength due to strong, $\gtrsim 10^{-10}$ G intergalactic magnetic fields, it is difficult to account for unless UHECR p and Fe comprise only a small fraction of the $\gtrsim 60 \times 10^{19}$ eV UHECRs detected with Auger.

How else to resolve the quandary that the UHECR proton and Fe MFPs are too long to be commensurate with the Auger data, so that the UHECRs cannot be protons or Fe? Possibly the Auger detector is not properly calibrated, though it is unlikely to be off by more than $\approx 20\%$ compared to the factor of ≈ 2 needed. For instance, Berezinsky [30], using the dip energy seen most clearly in an $E^3 I$ plot (see Fig. 1) as a fiducial, and multiplying by an energy calibration factor 1.2, 0.75, 0.83, and 0.625 for Auger, AGASA, Akeno and Yakutsk data, respectively, puts the various data sets in agreement with each other. Watson [32] argues that there are $\approx 19\%$ and 11% errors in the Auger energy calibration at 10^{17} and 10^{20} eV, respectively, so the correction factor for Auger is $\lesssim 1.2$. If this is true, then only a few percent of the $\gtrsim 6 \times 10^{19}$ eV UHECRs could be protons.

Let us provisionally accept that the Auger clustering results and the calculations presented here require that $\gtrsim 60$ EeV UHECRs are depleted in p and Fe (a full Monte Carlo simulation is required to remove remaining uncertainties). Some models, in particular, the neutral beam model that Armen Atoyan and I developed (for blazars, see [14], and for GRBs, see [31]) were based on the acceleration of protons to $\gtrsim 10^{20}/\Gamma$ eV energies in the shocked fluid frame, where Γ is the flow Lorentz factor. Photopion processes make an escaping neutron and γ-ray beam. If protons do not make up a large fraction of the $\gtrsim 10^{19.6}$ eV UHECRs, then this model no longer has to contend with the difficulty of accelerating protons to such high energies. On the other hand, this or any other model must be able to accelerate ions to ultra-high energies and get them from the accelerator into intergalactic space with the requisite emissivity.

Which ions? Again accepting the approximate integral meaning of an energy-loss MFP, Figures 7 and 8 show that UHECR ions up to Ne ($A = 20$) or at most Mg ($A = 24$) would effectively lose their energy within ≈ 100 Mpc. An abundance of UHECR in light ions gives the most economical explanation of the Auger results. If correct, then composition analyses of 60 – 100 EeV UHECR data will give an average atomic mass number $\langle A \rangle \lesssim 24$ ($\langle \ln A \rangle \lesssim 3.2$), and probably in the range $\langle A \rangle \lesssim 20$. This prediction can be compared with Fig. 10 in Watson [32], giving the measured energy dependence of $\langle \ln A \rangle$ using the the QGSJETII-03 model. According to our calculations, the anisotropy of $\cong 60$ EeV UHECRs with $d_{max} \lesssim 50$ – 100 Mpc could only originate from muclei with $2 \lesssim \langle \ln A \rangle \lesssim 3$ (the lower limit is not firm, and depends on source type).

Let us return to the question of the EBL. The dust EBL primarily attenuates $\gtrsim 4$ TeV photons and the stellar EBL primarily attenuates $\lesssim 4$ TeV. The LO and HI EBLs can be used to calculate $\tau_{\gamma\gamma}$ up to redshifts $z \approx 0.25$ where cosmic evolutionary processes start to change the EBL energy density compared to the present. The exponential attenuation factor $\exp(\tau_{\gamma\gamma})$ as a function of photon energy $\mathtt{E}$ for different redshifts is plotted in Figure 9. We can make a cut defined by $\exp(\tau_{\gamma\gamma}) = 10$ (or any other number, for that matter) to define a relationship between redshift and photon energy (the relation defined by $\tau_{\gamma\gamma}(z, \mathtt{E}) = 1$ is sometimes referred to as the "Fazio-Stecker relation [33]"; see also [34]) Here we make the cut defined by $\exp(\tau_{\gamma\gamma}) = 10$ to give the obscuration energy $\mathtt{E}_{\mathtt{osc}}$ as a function of obscuration redshift $\mathtt{z}_{\mathtt{osc}}$, shown in Fig.

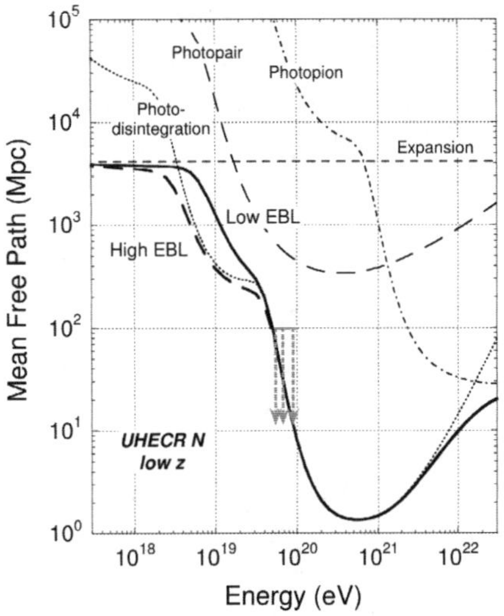

Figure 7. As in Fig. (5), for UHECR N.

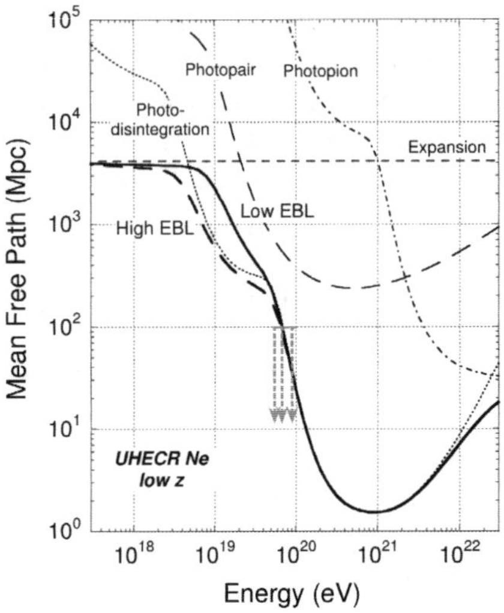

Figure 8. As in Fig. (5), for UHECR Ne.

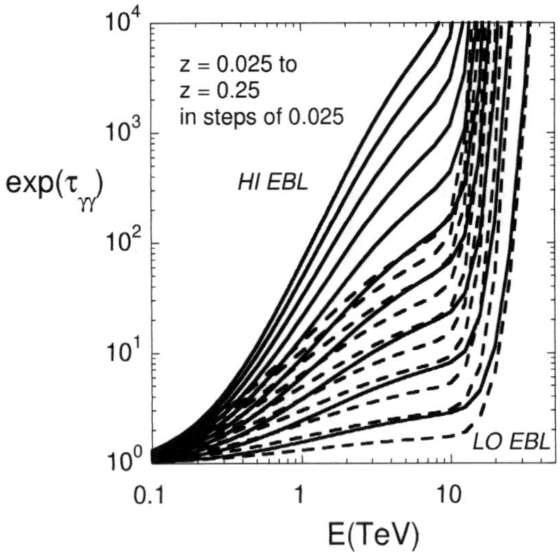

Figure 9. The factor $\exp(\tau_{\gamma\gamma})$ at different redshifts for the LO and HI EBLs shown in Fig. 3, as a function of photon energy $\mathtt{E}$.

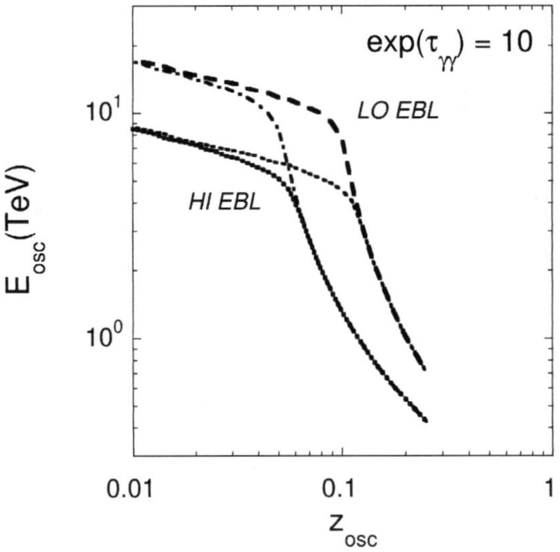

Figure 10. Relation between the obscuration energy $\mathtt{E}_{osc}$ and obscuration redshift z_{osc} defined by $\exp(\tau_{\gamma\gamma}) = 10$ for various high and low dust and stellar EBLs.

10. This figure is not however very useful for comparison with observational data, because the highest measured energies of photons depends on telescope sensitivity, source brightness and spectrum. Nevertheless, the dust component forms an opaque screen to photons with $\mathtt{E} \gtrsim 7$ TeV and $\gtrsim 15$ TeV for the high and low dust EBL at the redshift $z \approx 0.03$ of Mrk 421 and Mrk 501. Significant detection of ≈ 8 TeV from Mrk 421 [35, 36] and Mrk 501 [37], and of $\gtrsim 10$ TeV photons from 1ES 0229+200 [38] ($z = 0.14$) suggests that the EBL intensity at $\sim 10 - 100$ μm is smaller than given by high dust EBL.

We now ask why the sources of UHECRs should be enriched in medium mass (C – Ne) nuclei. Refs. [15, 16] discusses various scenarios for accelerating UHECRs with significant ion content in GRB environments, including capturing supernova shell material by a jetted relativistic wind, external shock capture of circumburst material, and hypernovae [39]. We favor the hypothesis [40] that an external shock is important during the prompt and early afterglow phases of a GRB. GRBs formed by core collapse of C or O Wolf-Rayet stellar progenitors would be surrounded by a wind enriched in these light elements that could be captured and accelerated to ultra-high energies. This could be accomplished by 1st- or 2nd-order Fermi processes at the shock formed by the interaction of a relativistic wind, expelled during the collapse to a black hole, with the surrounding medium.

For blazars and radio galaxies that propel intermittent winds to form internal shocks, the composition of UHECRs would depend on the makeup of the matter accreting onto supermassive black holes. If near-Solar metallicities occur, then proton acceleration would have to be suppressed, perhaps by a gyroresonant mechanism, and the medium mass ions would have to be processed through the black hole environment without breakup due to spallation or photodisintegration.

4. Discussion

The clustering result reported [1] by the Auger Collaboration is a major discovery, anticipated by Stanev, Biermann, Lloyd-Evans, Rachen, and Watson [5] in 1995. In the earlier study, Northern hemispheric data from Haverah Park, AGASA, Volcano Ranch, and Yakutsk with 73 (27), 25 (7), 13 (5), and 32 (3) events above 20 (40) EeV, respectively, and with zenith angles smaller than 45° were analyzed. The average and rms angular distances of the arrival directions of UHECRs from the supergalactic plane were calculated, after correcting for exposure. Compared to an isotropic source flux, a statistical enhancement towards the supergalactic plane at the 2.5 – 2.8 σ level was found for arrival directions of events with $E > 40$ EeV. The supergalactic plane runs through the Virgo Cluster at ≈ 20 Mpc, and contains an assortment of radio glaxies such as M87, Cen A and NGC 315, and the starburst galaxies M82 and NGC 253. Stanev et al. [5] argue that their analysis supports the hypothesis that radio galaxies are the sources of UHECRs accelerted, for exmple, in the inner jets or at the knots or hot spots of radio galaxies. Supporting this contention is the analysis of Shaver and Pierre [41] showing harder integral source count distributions of extragalactic radio sources at large spectral fluxes toward the supergalactic plane in the 408 MHz 1 Jy southern hemisphere Molonglo catalog, and clustering of sources in the 2.7 GHz 2 Jy Wall and Peacock catalog [42] towards the supergalactic plane. This clustering conforms with the association of UHECR arrival directions with radio sources in the supergalactic plane.

By comparison, the Auger team [1, 2] uses the VCV catalog [6] consisting of 85,221 quasars, 1122 BL Lac objects and 21,737 active galaxies (including 9628 Seyfert 1s). Quoting Véron-Cetty and Véron, "This [VCV] catalogue should not be used for any statistical analysis as it is not complete in any sense, except that it is, we hope, a complete survey of the literature." The associations between the arrival directions of UHECRs and directions to nearby AGN in the VCV catalog may only reflect that both the AGNs in the VCV catalog and the sources of UHECRs within $\approx 50 - 100$ Mpc are preferentially found toward the supergalactic plane, and that the large number of sources in the VCV catalog gives an apparent stronger underlying association than would be found if a smaller catalog, such as the Wall and Peacock catalog, were used. Associations made with a mixed catalog are more likely to be spurious, and the removal of statistical biases in such analyses are challenging. Correlating UHECR arrival directions with flux-limited surveys of sources with relatively complete redshift information is the only reliable way of identifying the source population(s) of UHECRs in the absence of high-energy neutrino detection from these sources.

A concern expressed recently [44, 45] is that although some of the UHECRs observed with Auger are associated with the direction toward Cen A, none seem to originate from the Virgo cluster center of our supergalaxy in the direction of M87. The lower exposure of Auger toward M87, by a factor of 3 compared the exposure to Cen A [2], and the Northern Hemisphere detection of UHECRs that could be associated with Virgo [5, 43] could ameliorate this issue. Ref. [45] suggests that if all UHECRs are C, then the short energy-loss MFP of C would mask M87 and explain preferential detection of Cen A. But our calculations show that the effective energy loss MFP of UHECR C with $E \cong 60$ EeV is ≈ 50 Mpc, within which M87, at a distance of ≈ 18 Mpc, falls. The MFP of UHECR He at $60 - 100$ EeV is $\sim 3 - 5$ Mpc, and so could restrict arrival directions to Cen A.

The dip energy, viewed as a consequence of photopair effects on UHECR protons [30] either from GRBs [46] or AGNs [47], could involve photodisintegration and photopair energy losses of ions interacting with photons of the EBL. There is also the possibility, again, that there is a deficit of UHECR sources between ≈ 100 and ≈ 300 Mpc. Whether this is true depends on source type, in particular whether UHECRs originate from specific classes of GRBs or AGNs.

In summary, the Auger team has opened charged particle astronomy by correlating the anisotropic arrival directions of $\gtrsim 60$ EeV UHECRs with AGNs located within ≈ 75 Mpc. Calculations presented here of the effective energy-loss MFP support an origin of UHECR ions in light nuclei with $A \lesssim 24$. The conclusion following from this analysis is that detection of UHECR clustering is most easily understood if most UHECRs are mainly light nucleons. The identification of the actual underlying source population must use complete (at least at high, $|b| \gtrsim 10°$, galactic latitude) flux-limited catalogs.

Acknowledgments

I would like to thank Justin Finke and Soebur Razzaque for many useful conversations and for comments on the paper, Isabelle Grenier for sharing her presentations, and Venya Berezinsky and the organizers of the Topics in Astroparticle and Underground Physics TAUP 2007 Conference in Sendai for the opportunity to attend the conference and visit Japan. This work is supported by the Office of Naval Research, NASA *GLAST* Science Investigation DPR-S-1563-Y, and NASA *Swift* Guest Investigator Grant DPR-NNG05ED411.

References

[1] Auger Collaboration, 2007, Science, vol.318, p.939-943 (9 November 2007)
[2] The Pierre Auger Collaboration 2007, ArXiv e-prints, 712, arXiv:0712.2843
[3] Dermer, C. D. 2007, ArXiv e-prints, 711, arXiv:0711.2804, in 30th ICRC, Mérida, Yucatan, Mexico
[4] Dermer, C. D., & Menon, G., "High Energy Radiation from Black Holes: γ rays, cosmic rays, and neutrinos," to be published in 2008 by Princeton University Press.
[5] Stanev, T., Biermann, P. L., Lloyd-Evans, J., Rachen, J. P., & Watson, A. A. 1995, Physical Review Letters, 75, 3056
[6] Véron-Cetty, M.-P., & Véron, P. 2006, Astronomy and Astrophysics, 455, 773
[7] Bird, D. J., et al. 1995, Astrophysical Journal, 441, 144
[8] Greisen, K. 1966, *Physical Review Letters*, **16**, 748
[9] Zatsepin, G. T. and Kuz'min, V. A. 1966, *JETP Lett.* **4**, 78
[10] Yamamoto, T., & for the Pierre Auger Collaboration 2007, ArXiv e-prints, 707, arXiv:0707.2638
[11] HiRes Collaboration 2007, ArXiv Astrophysics e-prints, submitted to PRL, arXiv:astro-ph/0703099
[12] Unger, M., Engel, R., Schüssler, F., Ulrich, R., & Pierre AUGER Collaboration 2007, Astronomische Nachrichten, 328, 614, arXiv:0706.1495v1
[13] Watson, A. A. 2006, Nuclear Physics B Proceedings Supplements, 151, 83
[14] Atoyan, A. M., & Dermer, C. D. 2003, Astrophysical Journal, 586, 79
[15] Wang, X.-Y., Razzaque, S., & Mészáros, P. 2007, ArXiv e-prints, 711, arXiv:0711.2065, ApJ, in press
[16] Murase, K., Ioka, K., Nagataki, S., & Nakamura, T. 2008, ArXiv e-prints, 801, arXiv:0801.2861
[17] Puget, J. L., Stecker, F. W., & Bredekamp, J. H. 1976, Astrophysical Journal, 205, 638
[18] Stecker, F. W., & Salamon, M. H. 1999, Astrophysical Journal, 512, 521

[19] Karakula, S., & Tkaczyk, W. 1993, Astroparticle Physics, 1, 229
[20] Anchordoqui, L. A., Beacom, J. F., Goldberg, H., Palomares-Ruiz, S., & Weiler, T. J. 2007, Physical Review D, 75, 063001
[21] Waxman, E., & Bahcall, J. 1999, Physical Review D, 59, 023002
[22] Hauser, M. G., & Dwek, E. 2001, Annual Review of Astronomy and Astrophysics, 39, 249
[23] Sreekumar, P., et al. 1998, Astrophysical Journal, 494, 523
[24] Strong, A. W., Moskalenko, I. V., & Reimer, O. 2004, Astrophysical Journal, 613, 962
[25] Kalashev, O. E., Semikoz, D. V., & Sigl, G. 2007, ArXiv e-prints, 704, arXiv:0704.2463
[26] Dermer, C. D. 2007, in The Multi-Messenger Approach to High-Energy Gamma-Ray Sources, eds. J. M. Paredes, O. Reimer, and D. F. Torres (New York: Springer), 127
[27] Kneiske, T. M. 2007, ArXiv e-prints, 707, arXiv:0707.2915
[28] Harari, D., Mollerach, S., & Roulet, E. 2006, Journal of Cosmology and Astro-Particle Physics, 11, 12
[29] Hooper, D., Sarkar, S., & Taylor, A. M. 2007, Astroparticle Physics, 27, 199
[30] Berezinsky, V. 2008, ArXiv e-prints, 801, TAUP 2007 Conference, arXiv:0801.3028
[31] Dermer, C. D., & Atoyan, A. 2003, Physical Review Letters, 91, 071102
[32] Watson, A. A. 2008, ArXiv e-prints, 801, arXiv:0801.2321
[33] Fazio, G. G., & Stecker, F. W. 1970, Nature, 226, 135
[34] Kneiske, T. M., Bretz, T., Mannheim, K., & Hartmann, D. H. 2004, Astronomy and Astrophysics, 413, 807
[35] Fossati, G., et al. 2007, ArXiv e-prints, 710, arXiv:0710.4138
[36] Yadav, K. K., et al. 2007, Astroparticle Physics, 27, 447
[37] Samuelson, F. W., et al. 1998, Astrophysical Journal Letters, 501, L17
[38] Aharonian, F., et al. 2007, Astronomy and Astrophysics, 475, L9
[39] Wang, X.-Y., Razzaque, S., Mészáros, P., & Dai, Z.-G. 2007, Physical Review D, 76, 083009
[40] Dermer, C. D., & Mitman, K. E. 2004, Astronomical Society of the Pacific Conference Series, 312, 301
[41] Shaver, P. A., & Pierre, M. 1989, Astronomy and Astrophysics, 220, 35
[42] Wall, J. V., & Peacock, J. A. 1985, MNRAS, 216, 173
[43] Cunningham, G., Lloyd-Evans, J., Pollock, A. M. T., Reid, R. J. O., & Watson, A. A. 1980, Astrophysical Journal Letters, 236, L71
[44] Gorbunov, D., Tinyakov, P., Tkachev, I., & Troitsky, S. 2007, ArXiv e-prints, 711, arXiv:0711.4060
[45] Fargion, D. 2008, ArXiv e-prints, 801, arXiv:0801.0227
[46] Wick, S. D., Dermer, C. D., & Atoyan, A. 2004, Astroparticle Physics, 21, 125
[47] Aloisio, R., Berezinsky, V., Blasi, P., Gazizov, A., Grigorieva, S., & Hnatyk, B. 2007, Astroparticle Physics, 27, 76

Diffuse high-energy neutrino searches in AMANDA-II and IceCube: Results and future prospects

Kotoyo Hoshina for the IceCube Collaboration

Department of Physics, University of Wisconsin, Madison, U.S.A.

E-mail: hoshina@icecube.wisc.edu

Abstract. The AMANDA-II data collected during the period 2000–2003 have been analysed in a search for a diffuse flux of high-energy extra-terrestrial muon neutrinos from the sum of all sources in the Universe. With no excess events seen, an upper limit of $E_\nu^2 \times dN_\nu/dE_\nu < 7.4\times10^{-8}$ GeV cm^{-2} s^{-1} sr^{-1}was obtained. The sensitivity of the diffuse analysis of IceCube 9 string for 137 days of data is calculated to be $E_\nu^2 \times dN_\nu/dE_\nu < 1.3 \times 10^{-7}$ GeV cm^{-2} s^{-1} sr^{-1}. No excess events are observed, which confirms the AMANDA-II upper limit.

1. Introduction

Extra-terrestrial neutrinos have been regarded as one of the most promising tools to investigate the non-thermal Universe. Due to their low interaction cross section, neutrinos retain their original source information such as direction and energy spectrum. However, the low interaction cross section makes it difficult to detect neutrinos which must interact within or near the telescope. The size of the telescope must thus be large enough to collect a sufficient number of neutrinos to find and probe extra-terrestrial sources.

AMANDA-II is a neutrino telescope consisting of 677 optical modules attached to 19 strings, deployed between 1500 m to 1950 m depth within a 200 m diameter in glacial ice at the South Pole. Following the success of AMANDA-II, the cubic kilometer-sized IceCube experiment started construction in 2005. By the spring of 2007, 22 strings with 1320 digital optical modules (DOMs) out of the 4800 planned had been deployed, at a depth of 1450 m - 2450 m surrounding AMANDA-II.

With these observatories numerous analyses are performed to search for extra-terrestrial neutrinos [1, 2, 3]. The target of diffuse analysis are neutrinos from unresolved sources. If the neutrino flux from an individual source is too small to be detected by point source search techniques [1], it is nevertheless possible to investigate their characteristics by combining events from isotropically distributed sources. The extra-terrestrial neutrinos are predicted to follow a $\Phi \propto E^{-2}$ energy spectrum resulting from shock acceleration processes [4]. Since the atmospheric neutrino flux has a much softer energy spectrum, an excess of events at higher energy over the expected atmospheric neutrino background would be indicative of an extra-terrestrial neutrino flux.

In this paper we present analyses with AMANDA-II data collected during the period 2000–2003 and IceCube data obtained with 9 strings (IC9) in 2006.

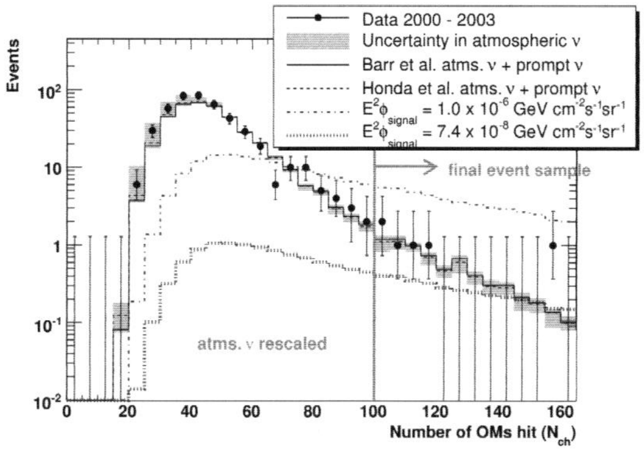

Figure 1. N_{ch} for the AMANDA-II 2000–2003 diffuse muon neutrino analysis compared to atmospheric neutrino expectations [6, 8].

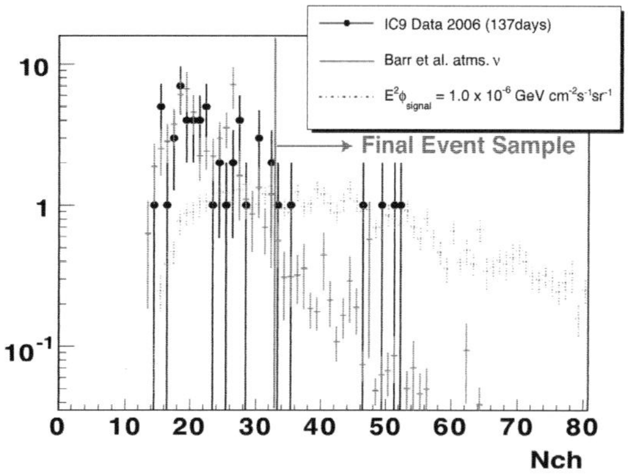

Figure 2. N_{ch} distribution in IC9 for 137 days livetime compared to atmospheric neutrino expectations [7].

2. Search Method

Cosmic ray interactions in the atmosphere create pions, kaons, and charmed hadrons which can later decay into muons and neutrinos. The main background for this analysis consists of atmospheric muons traveling downward through the ice. Diffuse analyses use the Earth as a filter to search for upgoing astrophysical neutrino-induced events. Once the background muons have been rejected, the data set mainly consists of neutrino-induced upward events [5]. To separate atmospheric neutrinos from extra-terrestrial neutrinos, we make use of the number of optical modules that reported at least one Cherenkov photon during an event (N_{ch}) as the energy estimator. The number of events above an N_{ch} cut has been compared with sets of monte-carlo prediction for E^{-2} and atmospheric neutrino models (Bartol [6, 7], Honda [8]). The N_{ch} cut was optimized to produce the best limit setting sensitivity [9]. In order not to bias the analysis, data above the resulting N_{ch} cut were kept hidden from analyzer while the lower N_{ch} events were compared to atmospheric neutrino expectations from Bartol and Honda.

3. AMANDA-II diffuse muon searches

Searches for a diffuse flux have been performed with AMANDA-II data obtained 2000-2003 (807 days livetime) [11]. The observed N_{ch} distribution is compared to the atmospheric neutrino background calculations in Figure 1. For $N_{\text{ch}} > 100$ region, 6 events were seen, while 7.0 were expected. Using the range of atmospheric uncertainty (shaded band in Figure 1) in the limit calculation [10] leads to an upper limit on a $\Phi \propto E^{-2}$ flux of muon neutrinos at Earth of $\text{E}_\nu^2 \times \text{dN}_\nu/\text{dE}_\nu = 7.4 \times 10^{-8}$ GeV cm^{-2} s^{-1} sr^{-1}. This upper limit is valid in the energy range 16–2500 TeV. Figure 3 shows the upper limit on the ν_μ flux from sources with an E^{-2} energy spectrum. The limit from the AMANDA-II 807 days analysis is a factor of four above the Waxman-Bahcall upper bound [4].

4. IceCube 9 String diffuse muon searches

Data from the first nine IceCube strings (IC9), with livetime 137 days, was analyzed to search for a diffuse flux. The atmospheric muon rejection and N_{ch} cut threshold are re-optimized [13] to accommodate the new detector geometry. The finalized N_{ch} cut=33 for the IC9 137 days gives a sensitivity $\text{E}_\nu^2 \times \text{dN}_\nu/\text{dE}_\nu = 1.3 \times 10^{-7}$ GeV cm^{-2} s^{-1} sr^{-1} on the neutrino energy range of 25 TeV to 10 PeV, which is a factor of two above the AMANDA-II 807 day upper limit.

Since the sensitivity of IC9 137 days was well above the AMANDA-II upper limit, we revealed

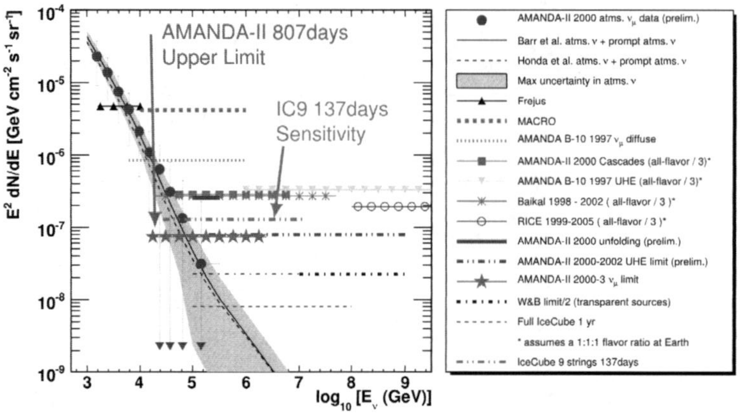

Figure 3. Upper limit on the ν_μ flux from sources with an E^{-2} energy spectrum are shown for single and all-flavor analyses [11, 12].

the high-energy events ($N_{\rm ch}$ >33) for the sake of a verification study of the detector. 6 events were observed while 6.5 were expected from the Bartol atmospheric neutrino model [7], which indicates no signal observation, consistent with the better AMANDA-II upper limit.

5. Outlook for IceCube 22 strings

It is an encouraging sign that the IC9 137 day is within a factor of two of the 807 day AMANDA-II result. With same atmospheric muon rejection and $N_{\rm ch}$ cut as IC9, a sensitivity of IceCube 22 strings (IC22) for 137 days is expected to reach $\mathrm{E}_\nu^2 \times \mathrm{dN}_\nu/\mathrm{dE}_\nu = 4.0 \times 10^{-8}$ GeV cm^{-2} s^{-1} sr^{-1}. This is already a factor of three better than IC9. Improvements of event separation to extract extra-terrestrial neutrinos from atmospheric neutrinos [14] are ongoing together with the optimization of background muon rejection for the IC22 geometry, which should improve the sensitivity further.

6. Acknowledgements

This work is supported by the Office of Polar Programs of the National Science Foundation.

References

[1] T. Montaruli, these proceedings
[2] C. Rott, these proceedings
[3] M. Ono, these proceedings
[4] E. Waxman and J. Bahcall, *Phys. Rev.* D **59**, 023002 (1998).
[5] P. Desiati, these proceedings
[6] G.D. Barr, T.K. Gaisser, P. Lipari, S. Robbins, and T. Stanev, *Phys. Rev.* D **70**, 023006 (2004).
[7] G.D. Barr, T.K. Gaisser, S. Robbins, and T. Stanev, Phys. Rev. D **74**, 094009 (2006).
[8] M. Honda, T. Kajita, K. Kasahara, and S. Midorikawa, *Phys. Rev.* D **70**, 043008 (2004).
[9] G.C. Hill and K. Rawlins, *Astropart. Phys.* **19**, 393 (2003).
[10] Cousins R D and Highland V L 1992 *Nucl. Ins. Meth. Phys. Res.* **A320** 331, Feldman G and Cousins R 1998 *Phys. Rev.* D **57** 3873, Conrad J, Botner O, Hallgren A and de los Heros C 2003 *Phys. Rev.* D **67** 012002, Hill G C 2003 *Phys. Rev.* D 118101
[11] A. Achterberg et al., *Phys. Rev.* D, **76**, 042008 (2007).
[12] K.Münich, ICRC proceedings (2007), arXiv0711.0353
[13] K.Hoshina,J.Hodges, and G.C.Hill, ICRC proceedings (2007), arXiv0711.0353
[14] G.C.Hill, ICRC proceedings (2007), arXiv0711.0353

Status report on the ANTARES Neutrino Telescope

Bertrand Vallage, on behalf of the ANTARES Collaboration
DSM/Dapnia, CEA/Saclay, 91191 Gif sur Yvette Cedex, France
E-mail: bertrand.vallage@cea.fr

Abstract. After a long and fruitful R&D phase to validate its detector concept, the ANTARES collaboration will complete by 2008 the deployment of its 12-string underwater detector at a depth of 2500 m in the Mediterranean Sea, 40 km south-east of the French town of Toulon The 3D array of 900 photomultiplier tubes (PMT) will allow high energy neutrino astronomy with a pointing accuracy better than 1 degree. The detector size grew by steps from 1 line in March 2006, to 5 lines in January 2007, sketching a real 3D detector. Data recorded continuously provide a precise calibration of the charge and arrival time of Cerenkov photons coming from muon tracks. The position and orientation of individual detectors are measured frequently by compasses, tiltmeters and external acoustic triangulation. The data from the 3D detector allow the reconstruction of downward going cosmic ray muons and the search for the first upward going neutrino induced muons. Preliminary results of these analysis are presented.

1. Introduction

One of the main goals of the ANTARES neutrino telescope is the detection of high energy cosmic neutrinos from galactic or extra-galactic sources. For the charged current interaction of a ν_μ, the signature relies on the detection of the Cerenkov light induced by the secondary μ in the water. The detector consists of a large 3D PMT array, where the precise knowledge of the time, amplitude and position of the hits allow the reconstruction of the direction and the energy of the muon. Cosmic neutrino signals can be mimicked by atmospheric neutrinos and muons produced by cosmic rays in the atmosphere. The atmospheric muon component is drastically reduced burying the detector at a large depth under water, 2500 m in the case of ANTARES.

The ANTARES project [1] gathers about 180 physicists, engineers and sea-science experts from 23 institutes of 7 European countries, with the aim of operating a "0.1 km^2" neutrino telescope located 20 km south of the French Porquerolles Island ($42°50'N, 6°10'E$).

The total sky coverage is 3.5π sr, with an instantaneous overlap of 0.5π sr with that of AMANDA, and allows the observation of the Galactic Centre 67% of the day. An intense $R\&D$ programme since 1996 allowed the full characterization of the site, which was found to fulfil the criteria of good water transparency [2], low water current and limited biofouling [3].

2. ANTARES design

The detector (Fig 1) consists of 12 lines arranged in an octogonal shape with a spacing about 70m in between lines. 25 storeys are distributed along 350m, starting 100m above the anchor at the sea-bed, and a buoy on top ensures that the line is vertical. Lines are deployed individually, and therefore require underwater connection to a junction box with the help of either manned or remote control submarines. The junction box itself is connected to the shore station by a 40

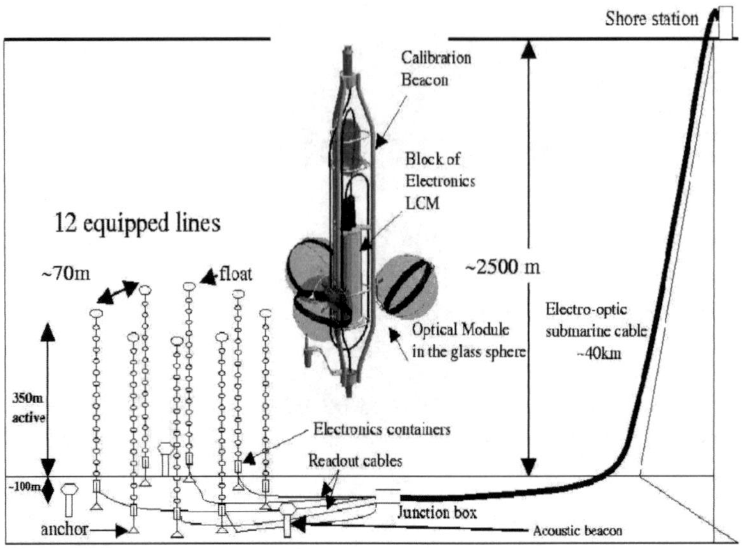

Figure 1. Scheme of the ANTARES detector with a blow-up of a storey.

km long electro-optical cable. The detector size is about equivalent to the one of AMANDA; the main differences between the two experiments being the detection media and the flexibility of the ANTARES lines.

The basic detector element is the storey, comprising 3 Optical Modules (OM) [4], each one made of a 17" glass sphere housing a 10" PMT (R7081-20 from Hamamatsu), looking downwards at 45° from the vertical. Due to the flexibility of the lines, a redundant positioning system is incorporated on each line. It consists of compasses and tiltmeters measuring the roll, pitch and heading of each storey and 5 acoustic hydrophones [5] spread over the line. For time calibration purposes, each line also houses 4 optical beacons [6] equiped with blue LEDs, and 2 green laser beacons are placed at the bottom of 2 central lines All the associated electronics, plus the readout computer and clock system, are contained in a titanium cylinder of a Local Control Module (LCM). The storeys of one line are chained on a mechanical electro optical cable which carries the power, the 20 MHz common clock, and the digitized data on optical ethernet.

3. Detector operation

The first detector line came into operation in March 2006. The second one followed in September 2006, and 3 additional lines were connected on January 29th 2007, forming a real 3D 5-line detector. Since then, 4 detector lines and an instrumented line have been deployed and will be connected before the end of the year.

The DAQ system [7] includes a multi-level trigger. Data are buffered when the hit amplitude exceeds the level-0 threshold (L0), around 1/4 photo-electron. The resulting rate is about 60 kHz per optical module, dominated by the optical background coming about equally from ^{40}K activity in the water and bioluminescence. All L0 data are transmitted to shore through multiplexed optical gigabit links, implying throughputs from 0.2 to 1 Gb/s/line. Onshore, local level-1 coincidence between optical modules of the same storey are used as input to higher level triggers. The average level-2 counting rate is a few Hz. In addition, the detector receives GRB alerts, in which case all the activity of the detector is recorded for a few minutes.

The ^{40}K activity can be derived from the measured salinity, and the estimated background rate in our PMT is 34 ± 7 kHz. The coincidence rate of Cerenkov photons coming from a single decay on 2 PMTs of a storey can be estimated as well to be 13 ± 4 Hz, in very good agreement with the measured value of 14.5 ± 0.4 Hz. Therefore this component of the optical background turns out to be quite useful to monitor the relative PMT efficiencies, as well as to check the timing within a storey.

Contrary to the ^{40}K background, the bioluminescence background suffers from seasonal and annual variations. Since september 2006, the mean rate is 75% of the time below 100 kHz, resulting in efficient DAQ, trigger and event reconstruction. A safeguard against bioluminescence bursts is applied online by means of a "high rate veto", most often set to 500 kHz.

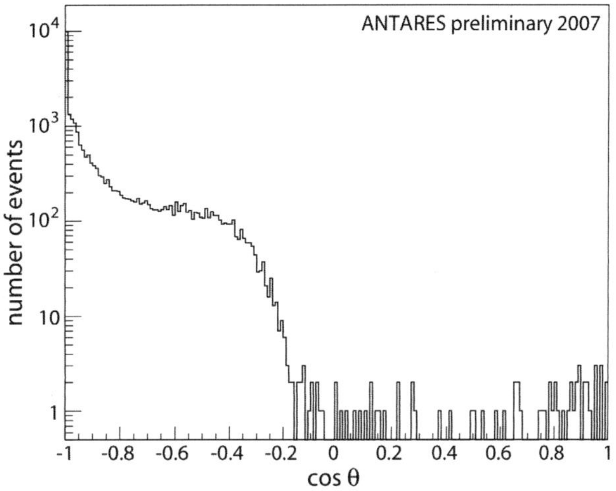

Figure 2. First glimpse analysis Nadir angle distribution ignoring the detector alignment.

A "first glimpse" analysis has been performed on some good quality data sample from February to March, ignoring all storey positioning aspects and assuming straight lines. The 3D reconstruction proceeds in several fitting steps, and provides 5 track parameters after χ^2 minimization [8]. Figure 2 displays a **very preliminary** Nadir angular distribution obtained after reconstruction and quality cuts enhancing the upgoing tracks. The dominance of downgoing tracks from atmospheric muons is still clearly visible, and some upgoing neutrino candidates appear at $cos\theta > 0.1$. The alignment, result of line shape fits using slow control positioning data available every 6 minutes, is currently implemented in the track reconstruction, and will soon allow for a much efficient determination of the neutrino candidates.

4. Summary and outlook

The ANTARES collaboration operates since January 2007 a 5-line 3D detector. The large experimental sample of downgoing muons collected allows detailed studies of the detector response and efficiencies in various conditions. The first neutrino candidates in the Mediterranean sea have been reconstructed. Work is in progress to implement refined position and time calibration to reach the design goals of the apparatus. The deployment of the 12-line ANTARES detector is expected to be completed early 2008.

References

[1] Aslanides E et al 1999 *Preprint* astro-ph/9907432
[2] Aguilar J A et al 2005 *Astropart. Phys.* **23** 131
[3] Amram P et al 2003 *Astropart. Phys.* **19** 253
[4] Amram P et al 2002 *Nucl. Inst. and Meth.* A **484** 369
[5] Niess V 2005 *Proc. of 29th ICRC(Pune)* vol 5 p 155
[6] Ageron M et al 2007 *Nucl. Inst. and Meth* A **578** 498
[7] Aguilar J A et al 2007 *Nucl. Inst. and Meth* A **570** 107
[8] Becherini Y 2007 *Proc. of 30th ICRC (Merida)*

First Results of the IceCube Observatory on High Energy Neutrino Astronomy

Teresa Montaruli* for the IceCube Collaboration [1]

* Department of Physics, University of Wisconsin-Madison, 53706, WI, USA, on leave University of Bari, 70126, Italy

E-mail: `teresa.montaruli@icecube.wisc.edu`

Abstract. The first results of a search for high energy neutrinos in excess of the atmospheric neutrino background, both from any sky direction and from 26 known sources, using 9 strings of IceCube are presented. We summarize also the results of AMANDA-II from 2000 to 2006. The expected performance of future configurations of the growing IceCube detector are discussed.

1. Introduction

The main objective for neutrino telescopes is the detection of astrophysical neutrinos at energies above some tens of GeV, which would prove proton or nuclei acceleration in sources. This discovery would help in identifying the sources of the ultra-high energy cosmic rays observed up to about 10^{21} eV in giant air shower arrays. The analysis techniques illustrated here aim at finding the signal of neutrinos from point-sources on top of the background of atmospheric neutrinos, which is more uniformly distributed in arrival direction. We refer to other contributions at this conference for a general introduction to IceCube and for the detection of diffuse and atmospheric neutrino fluxes [1]. IceCube is now running with 22 strings (IC22) and during this construction season, just started in Nov. 2007, we aim at installing 14 to 18 more strings. When completed in 2011, it will be composed of up to 80 strings (IC80) holding 4800 photomultipliers at depths between about 1.5 and 2.5 km in the Antarctic ice. IceCube can measure all neutrino flavors but muon neutrinos are the only ones that allow a pointing accuracy at the level of a degree. Hence we focus here on this channel. Searches for steady-state emission of neutrinos from point-sources typically require a sample of neutrinos with $< 10\%$ contamination by background atmospheric muons. To achieve this goal we are limited to lower (Northern) hemisphere searches, unless other time or energy-dependent constraints are included in the analysis (eg. for bursting emissions such as gamma-ray bursts (GRBs) or very high energy extra-galactic sources with much harder spectra than atmospheric neutrinos). The Earth acts as a filter and absorbs the background of atmospheric muons. It begins to absorb neutrinos above the PeV range depending on the direction, which determines the column density crossed by neutrinos. The first results of IC9 are summarized in Sec. 2, together with the 2005-6 data of AMANDA-II, which combined with the previous 5 years [2] constitutes the largest sample of high energy neutrinos ever collected for these studies (924+887 in 386 d in 2005-6 + 4282 in 1001 d in 2000-4). These results, together with the anticipated performance for various configurations

[1] IceCube web site at http://icecube.wisc.edu

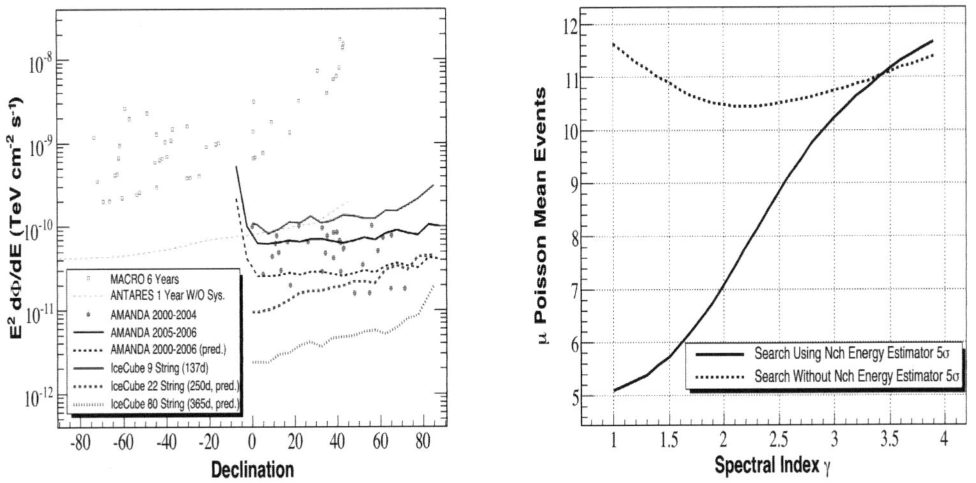

Figure 1. (left) Average (over right ascension) sensitivities vs declination for all sky point-source searches and upper limits for specific source catalogues. Neutrino fluxes are assumed to have E^{-2} differential spectra. Results and predictions for Northern hemisphere experiments: MACRO upper limits [3] and predicted ANTARES sensitivity for 365 d [4]; South Pole experiments (from top to bottom): IC9 (137 d), AMANDA-II 2005-6 (386 d) analyses, dots are upper limits for AMANDA-II 2000-2004 [2] (1001 d). The upper dashed line is the sensitivity estimated for the combined AMANDA-II data from 2000-6. IC22 is expected to take data for 250 d and the predicted sensitivity is better than the cumulative result of AMANDA-II (notice the strong trend with declination of IceCube compared to AMANDA-II due to the higher energy threshold that allows better discrimination of atmospheric neutrinos). The IC80 prediction for 365 d is also shown. (right) Signal events needed for a neutrino source discovery with 50% probability at 5σ significance vs the spectral index of its differential flux using the Nch energy estimator (solid line) and not using it (dashed line) for a period equivalent to the 2005 data.

of IceCube and expected and measured results from other experiments are represented in Fig 1 (left), summarizing the state of the art for neutrino point-like source searches.

2. IC9 and AMANDA-II results and expected performance of future configuration

The data sample for IC9 was obtained in 137.4 d during Jun.-Nov. 2006: 233 neutrino events (227 expected from atmospheric neutrinos and less than 10% from mis-reconstructed atmospheric muons) are extracted by a filtering procedure including two cuts aimed at rejecting coincident events from 2 independent showers (NDir = number of PMTs hit by unscattered light > 8 and σ = event by event accuracy of tracking $< 2.5°$). More details on the analysis are in [5]. We have developed for these events and for the 2005-6 analysis of AMANDA-II a likelihood-based analysis that employs the angular distribution of the background + source hypothesis compared to the background only one. This method can be extended by including further information, as shown in Fig. 1(right) where the two curves allow a comparison of the method using additional energy-based information and not using it. The variable used here that relates to the neutrino energy is Nch, the number of hit photomultipliers along the track. This variable is sensitive to different spectra and allows a better separation of signal from background (atmospheric neutrinos follow an $E^{-3.6}$ power law above about 100 GeV) when the signal spectrum is harder. It can be seen that for an E^{-2} source spectrum the improvement is of the order of 30% compared to the case using only the directional information, and the total improvement compared to more

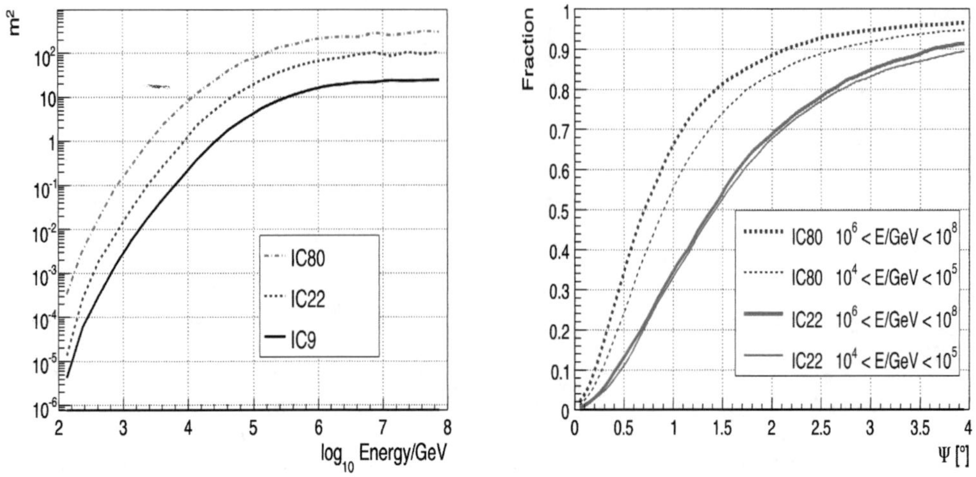

Figure 2. (left) Effective area averaged over the Northern hemisphere for equal contributions of neutrinos and antineutrinos as a function of energy for IC9, IC22 and IC80. (right) Point spread function (cumulative distribution of angular reconstruction errors) for IC22 and IC80 for two different energy ranges.

traditional binned analyses using optimized angular windows [2] is about 45% at declinations of $\sim 40°$. Similar studies will be performed using time dependent information especially in view of multi-wavelength studies involving the enormous amount of information coming from gamma-ray experiments. Other programs ongoing in the experiment are Target of Opportunity programs in which IceCube sends alerts to other experiments [5]. In Fig 2 we show the effective area and point spread function for IC9 and what we expect for IC22 and IC80. These estimates are based on the selection criteria used in the IC9 analysis, while work continues on optimizing the cuts for the latest configuration. With these cuts we expect about 3,700 events in IC22, and 40,000 in IC80 per year from atmospheric neutrinos from the lower hemisphere.

GRBs are extremely promising source candidates for IceCube, which will be at least 10 times more sensitive than AMANDA-II and should detect GRB neutrinos in a few years if fireball models are correct. We refer to [5] for more details.

3. Outlook

Up to now there is no significant detection of astrophysical neutrinos. Nevertheless, we have well-developed methods to achieve a high discovery potential as the detector size approaches the cubic-kilometer scale. At the moment the collaboration is considering extending the detector for high energy studies by deploying some of the outer strings of the final detector at a larger distances than others. There is also interest to deploy an inner core array inside IceCube to enhance the detection capability for atmospheric neutrinos, WIMPs and galactic sources and lower the energy threshold down to about 30 GeV.

References

[1] Hoshina K and Desiati P for the IceCube Collaboration, these proceedings
[2] Achterberg A *et al*, Phys. Rev. **D75** (2007)102001
[3] Ambrosio M *et al.*, Astrophys. J. **546** (2001) 1038
[4] Aguilar JA for the ANTARES Collaboration, proc. of ICRC2007, Merida, Jul. 2007 and arXiv:0710.0252.
[5] The IceCube Collaboration, Proceedings of ICRC2007, Merida, Mexico, July 2007 and arXiv:0711.0353.

The NEMO project: achievements and perspectives

P. Sapienza on behalf of the NEMO collaboration

INFN- Laboratori Nazionali del Sud- Via S. Sofia 62 –95123 Catania Italy

E-mail: sapienza@lns.infn.it

Abstract. The NEMO R&D activities towards an underwater km^3 Cherenkov neutrino telescope are described. These activities include a long-term survey of a 3500 m deep site, located 80 km off the coast of Capo Passero in Sicily, that shows oceanographic and water properties that are optimal for the installation of the detector. A complete feasibility study was carried out and a preliminary design for an underwater km^3 detector was performed on the basis of Monte Carlo simulations aimed at the optimization of the detector performance. The NEMO Phase-1 ended with the successful installation at an underwater test site at 2000 m depth of a technological demonstrator. Moreover, the realization of an infrastructure at the Capo Passero site (3500 m depth) is ongoing.

The opening of the high energy neutrino astronomy will provide new information about the most violent phenomena occurring in the Universe. Motivations for the construction of high energy neutrino telescopes have been strengthened by the observation of several tens of γ-TeV sources detected by the Cherenkov telescopes Hess [1] and Magic [2] in these last years. In particular, several of these sources show a power law spectrum with a spectral index around 2, consistently with a Fermi acceleration mechanism, and characteristics that strongly support the presence of a proton acceleration mechanism. Moreover, several sources don't show any counterpart at other length waves. The detection of neutrinos from γ-TeV sources would provide the smoking gun in favor of hadronic acceleration processes.

Calculations of fluxes for several neutrino candidate sources indicate, for both diffuse and point-like sources, that high energy neutrino astronomy requires detection areas of the order of 1 km^2 and therefore detectors with volume of 1 km^3 or more have to be built.

Two telescopes are needed in the two opposite hemispheres to cover the whole sky. At the South Pole is currently under construction IceCube[3], the cubic kilometer neutrino telescope in the deep Antarctic ice. Also in the Mediterranean Sea the efforts of the collaborations working in the field are eventually merging towards the construction of a km^3 Cherenkov telescope. The neutrino telescope in the Mediterranean Sea will survey the larger part of the Galactic disc, including the Galactic Center which is not visible from the South Pole.

The realization of a km^3-scale high-energy neutrino telescope in the Mediterranean Sea requires the development of very complex technologies to cope with the extreme conditions of the deep sea such as corrosion and very high pressure and to assure the needed reliability for an experiment with a life time of about 10 years. Moreover, the deployment of detector elements, maintenance and remote handling of deep undersea connections are hard technological challenges that require dedicated R&D project.

The high-energy neutrino telescope has to be realized by equipping a volume of water of the order of 1 km^3 with several thousand optical sensors arranged in a 3D lattice.

NEMO [4] is an advanced R&D program aiming at the solution of the technological issues related to the realization of a km^3 neutrino telescope in the deep Mediterranean Sea. The project includes the characterization of an optimal site for detector installation and the construction and test of technological prototypes and their validation at depths up to 3500 m.
With about 30 sea campaigns, the NEMO activity for the characterization of the Capo Passero site (3500 m), at about 80 km off-shore the SE coast of Sicily, provided a long term monitoring of the water properties such as the light transmission properties, the optical background noise, the sedimentation and the biological activity, the water current and the environmental properties (salinity, temperature) in situ. The measurements and site survey demonstrate that both the water optical properties and the oceanographic features are optimal for the installation of the km^3 Cherenkov telescope [5] so that Capo Passero was indicated as a candidate site for the km^3 installation to ApPEC already in January 2003.
One of the main concerns for the feasibility study was to reduce the number of structures that host the optical modules in order to minimize the number of underwater connections thus allowing the operations with Remotely Operated Vehicle (ROV) at depths of about 3500 m. The feasibility study carried out with the support of Monte Carlo simulations aiming at the optimization of detector performance, indicates that a km^3 neutrino telescope can be realized with about 80-100 structures (towers or strings) hosting a total number of 5000-6000 photomultipliers (PMT).

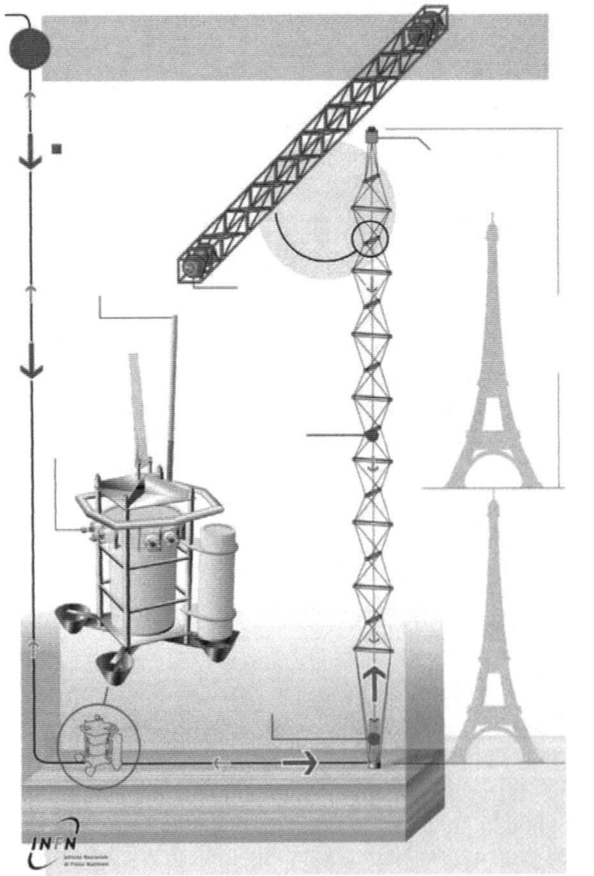

Figure 1. The junction box and the tower of NEMO.

The NEMO-Phase1 project was aimed at the validation of the technological solution proposed for the realization and installation of the km^3 detector through the realization of a technological demonstrator including all the key elements of the km^3. The main elements of the telescope are the electro-optical cable that carries signals and power, the towers and a junction box that distributes power and data from and to shore. A pictorial view of the key elements of the telescope is shown in fig. 1. The design of the junction box is innovative since the pressure-resistant steel vessels containing the electronics are hosted inside a corrosion-resistant fiberglass container filled with mineral oil. This solution decouples the two problems of pressure and corrosion resistance thus allowing to operate in the severe conditions of high-depth salt water, but with a significant cost decrease if compared to standard titanium containers. The tower consists of a three-dimensional semi-rigid structure composed by a sequence of bars, which host the optical modules and the instrumentation, interlinked by a system of ropes and anchored on the seabed. The structure is kept vertical by an appropriate buoyancy on the top. In the layout proposed for the km^3, each tower will consists of a sequence of 12 m long bars spaced vertically by 40 m, with the lowermost one placed at about 150 m above the sea bottom.

NEMO-Phase1 went to completion in December 2006 with the successful deployment and connection of a junction box and a tower made of four 15 m long storeys at the underwater test site of the Laboratori Nazionali del Sud (LNS) of INFN at about 2000 m depth. Fig. 2 shows the preparation phase to the deployment of the junction box.

Figure 2. Preparation to the junction box deployment on the deck of vessel Teliri.

The transport of physics and control data occurs via an electro-optical cable that connects the LNS on-shore station in the Catania harbor to the off-shore station that is about 20 km far.

The data to be recorded are selected with an onshore trigger procedure. Atmospheric muon events have been reconstructed (fig. 3) and the analysis of the muon observed flux with a Monte Carlo comparison is ongoing.

In order to validate the developed technologies at the depth of 3500 m and perform a long term site survey in Capo Passero, the NEMO Phase-2 started. A 100 km electro-optical cable, with characteristics that are compatible with the needs of a km^3 detector, was laid in July 2007 to connect the deep sea infrastructures at 3500 m depth to a shore station. A complete 16 storey NEMO tower is under construction and will be deployed and connected in 2008. The realization of the NEMO project demonstrates that technological solutions exist for the realization of an underwater km^3 detector. The NEMO collaboration participates in the EU Design Study KM3NeT [6] and the following Preparatory Phase KM3NeT-PP. The NEMO experience will contribute to the advancement of the KM3NeT activities. Indeed, both the NEMO program and time schedule are well fitted to the Design Study and Preparatory Phase of KM3NeT.

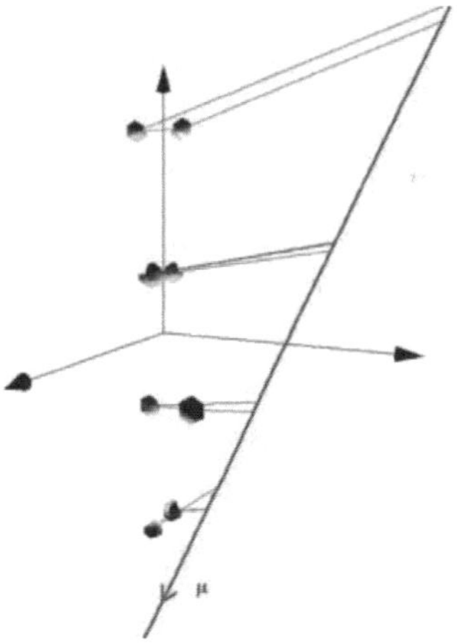

Figure 3. A reconstructed downgoing muon track where 8 PMTs were hit. The size of the optical module fired refers to the charge.

References

[1] F. Aharonian et al., Astrophys. J. 636 (2006) 777.

[2] J. Cortina et al, Astrophys. Space Sci. 297 (2005)245.

[3] The IceCube Collaboration, Astropart. Phys. 26 (2006) 155; see also the web site http://icecube.wisc.edu

[4] E. Migneco et al, Nuclear Instruments and Methods A 567 (2006) 444; see also the web site http://nemoweb.lns.infn.it

[5] G. Riccobene et al, Astropart. Phys. 27 (2007) 1-9

[6] http://www.km3net.org

KM3NeT : a large underwater neutrino telescope in the Mediterranean Sea

Petros A. Rapidis
Institute of Nuclear Physics, NCSR `Demokritos', Agia Paraskevi Attikis, Athens 15310, Greece

For the KM3NeT consortium

E-mail: rapidis@inp.demokritos.gr

Abstract. High energy neutrinos produced in astrophysical processes will allow for a new way of studying the universe. In order to detect the expected flux of high energy neutrinos from specific astrophysical sources, neutrino telescopes of a scale of a km^3 of water will be needed. A Northern Hemisphere detector is being proposed to be sited in a deep area of the Mediterranean Sea. This detector will provide complimentary sky coverage to the IceCube detector being built at the South Pole. The three neutrino telescope projects in the Mediterranean (ANTARES, NEMO and NESTOR) are partners in an effort to design, and build such a km^3 size neutrino telescope, the KM3NeT. The EU is funding a 3-year Design Study; the status of the Design Study is presented and some technical issues are discussed.

1. Introduction

Detecting high-energy neutrinos from astrophysical sources will be a major step towards a more complete understanding of the universe. These neutrinos can be detected by water/ice Cherenkov telescopes. Such detectors can contribute to the study of active galactic nuclei, supernova remnants, micro-quasars, gamma ray bursts, etc. In addition searches for neutrinos from the decay or annihilation of dark matter particles (WIMPs), magnetic monopoles and other exotic particles can be carried out [1]. Studies indicate [2] that such signatures will only be detectable by detectors of sizes of a km^3 of water or larger. Present existing telescopes [3], [4] are too small for this task.

The IceCube [5] telescope being built at the South Pole will have an instrumented volume of ice of a km^3. Neutrino telescopes are mostly downward looking detectors, and as a result the IceCube device will have a limited reach for sources in the southern sky. Thus the Galactic plane with a multitude of possible high energy neutrino sources, such as supernova remnants, microquasars, pulse wind nebulae, as well as several unassociated gamma-ray sources reported by the H.E.S.S. telescope[6] cannot be studied by IceCube.

To address this lack of sky coverage a neutrino telescope in the Mediterranean Sea has been proposed. The three pilot neutrino telescope projects in the Mediterranean (ANTARES [4], NEMO [7], and NESTOR [8]) are partners in an effort to design, and build such a km^3 size neutrino telescope, the KM3NeT [9]. The Mediterranean Sea offers some unique advantages for such a device: deep sites near the shore, clear waters, and periods of good weather needed for sea operations. The KM3NeT consortium consists of 37 Institutes from 10 European Countries [10]. The KM3NeT will be an interdisciplinary research infrastructure, serving as a deep water facility for associated sciences, like marine biology, oceanography, earth and environmental sciences, in addition to being a neutrino telescope.

2. Status of KM3NeT

The EU has initiated a 3-year design phase for KM3NeT which started in February 2006, and will conclude in 2010 with the publication of a Technical Design Report. A Preparatory Phase (PP) will follow, overlapping for one year with the Design Study. The PP will address political, governance, and financial issues of KM3NeT, including the site selection. The PP will also include prototyping work, in view of the start of the telescope construction in 2011. The overall cost of KM3NeT is estimated to be 220 -250 M€. KM3NeT is part of the ESFRI (European Strategic Forum on Research Infrastructures) roadmap [11] for future large scale infrastructures.

3. Detector Layout

The Design Study has as a goal the design of a detector with the highest sensitivity. Minimum requirements are an instrumented volume of at least 1 km^3, with angular resolution of about 0.1^o for neutrino energies above 10 TeV, sensitivity to all neutrino flavors, and a lower energy threshold of a few hundreds of GeV (and around 100 GeV for pointing sources). Neutrino Cherenkov telescopes are lattices of a large number of photo-detection units that detect the Cherenkov light. These units, the optical modules (OM), contain one or more photomultipliers (PMT) and are enclosed in a pressure resistant, waterproof glass sphere. The OMs used to date contain one large PMT (typically 10"). The OMs are arranged so as cover the instrumented volume. The specific choice of PMTs, their arrangement in the OM, as well as the layout of OMs in the instrumented volume are being studied with detailed MC simulations, in order to optimize the sensitivity of the detector.

Various OM configurations and detector layouts have been studied [12]. Examples of possible OMs can be seen in Fig. 1. These include OMs with a single PMT, double OMs, and OMs with many small PMTs inside. A number of detector layouts have been evaluated, with strings of vertically equidistant OMs arranged in cubic, ring, hexagonal, clustered or mixed layouts (Fig. 2). The instrumented volume has been kept constant to 1 km^3 and the total photocathode area was kept constant.

The effective area for neutrinos[13] as a function of energy is shown in Fig. 3 where the KM3NeT configuration-1 detector consists of 127 vertical strings each with 25 OMs arranged in a hexagonal pattern, 100 m horizontal spacing, 15 m vertical spacing between OMs and 3 large PMTs per OM, similar to the ANTARES configuration. The configuration-2 detector is a structure of 225 vertical strings each having 36 OMs, arranged in a tetragonal pattern with inter-line horizontal spacing of 95 m, vertical spacing between OMs 16.5 m, with each OM containing 21 3" PMTs. The effective area calculation includes full simulation of the neutrino interaction, muon propagation, Cherenkov light transmission, track reconstruction and event selection. Optical noise of 40 kHz due to ^{40}K decays in the sea water was added, as measured by NESTOR [14] and NEMO [15] in the respective sites. The higher effective area of the KM3NeT detector configurations compared to Ice-Cube can be explained in terms of the higher photocathode area and the better angular resolution of the water detector.

4. Detector Components and Procedures

The recovery and repair of faulty components from the deep sea is very difficult. Therefore, the reliability of all deep sea components is absolutely essential. Design choices being considered towards improved reliability include simplifying the design and reducing the number of connections that have to be mated in the sea. The possibility of transmitting all data-to-shore via optical fibre and

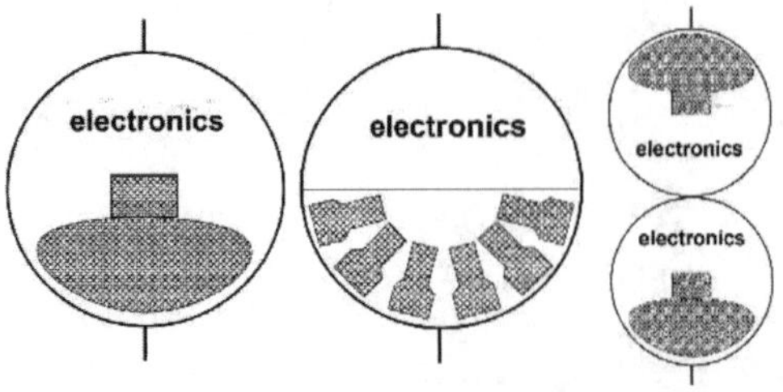

Figure 1: Various types of Optical Modules.

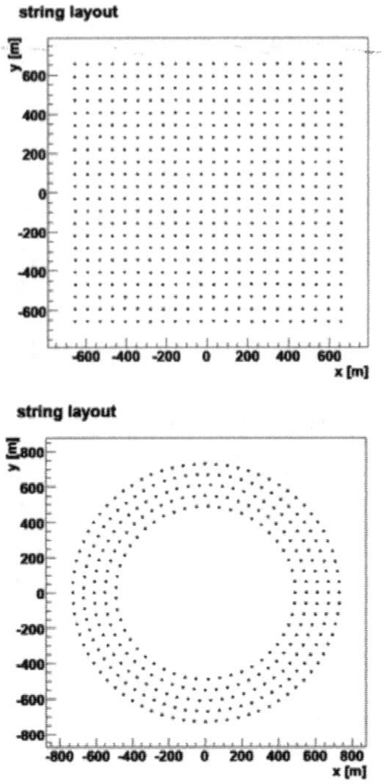

Figure 2: Cubic and ring detector layouts

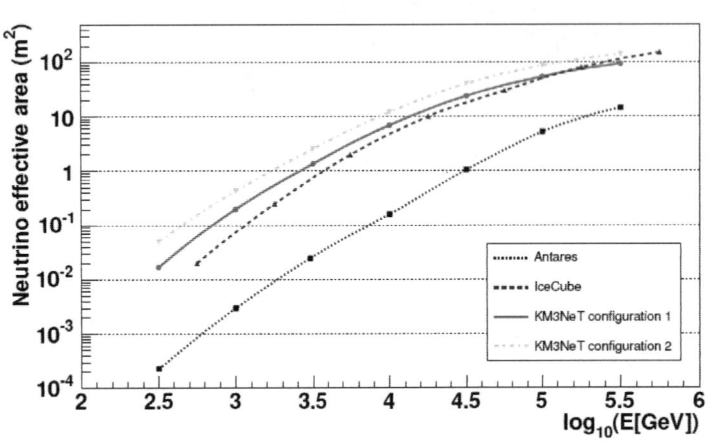

Figure 3: Neutrino effective areas vs. energy for two KM3NeT layouts, IceCube and ANTARES. The details of the KM3NeT configurations are described in the text.

reduction of the in-sea electronics using a photonics-based network [16] is also being studied. In this scenario, the OM signal is impressed with an optical reflector/modulator on an optical signal coming in from the shore. The modulated signal is transmitted to shore and time stamped at the shore end, thus necessitating no lasers or digitization electronics in the deep.

Additional research and development is under way to evaluate the performance of OMs with many smaller PMTs[17]. These have the advantage of higher quantum efficiency, better single photon resolution, smaller transit time spread, and provide directionality which helps in reducing noise from bioluminescense and ^{40}K decays in the sea water. Also under study are 10" PMTs with segmented photocathode area [18], as another way of gaining directionality, and crystal hybrid PMTs[19].

Calibration of angular resolution, absolute position and angular offset, that can be achieved by using a floating sea-top array [20], similar to IceTop is under investigation. The position of the sea-top detector can be determined via GPS, and the signal from extensive air showers can be used to calibrate the underwater detector. Three floating stations separated by distances of 20m equipped with 16m^2 of scintillator each will probably be sufficient for this scheme.

The sheer size of the detector implies that new ways of sea deployment will have to be developed. The ANTARES mode of deployment with separate operations for the line deployment and the cable connection with a submersible remotely operated vehicle may be time consuming and impractical for a detector the size of KM3NeT with around 250 lines. A possible solution could be the method proposed by the NEMO collaboration in which each line is "rolled-up" in a container[21] equipped with a release mechanism which unfolds the line after the container reaches the sea floor. By connecting several of these compact containers together, one sea operation may suffice for the deployment of several lines. Such a deployment model reduces both the time the interconnection of lines and the number of connections that have to be mated in the sea which are a significant source of potential failures. The feasibility of such a deployment model has been succesfully demostrated by the NESTOR collaboration when they deployed a tower module without any underwater operation [22].

5. Associated Marine and Earth Sciences

The KM3NeT research facility will also serve the cause of marine and geophysical sciences. The existence of dedicated and permanent sea to shore connections allows the operation of long term real-time monitoring stations serving these disciplines. There is a separate working group within the Design Study, working on these issues. Besides monitoring instrumentation attached to the strings of the neutrino telescope additional dedicated associated sciences stations are envisaged.

6. References

[1] C. Spiering, 2005 Phys. Scripta, **T121**, 112

[2] A. Kappes *et al.*, 2007 Astrophys.J, **656,** 870; M. D. Kistler and J. F. Beacom, 2006 Phys. Rev. D **74**, 063007

[3] The AMANDA detector at the South Pole, http://amanda.uci.edu

[4] The ANTARES detector in the Mediterranean Sea , http://antares.in2p3.fr

[5] The IceCube home page, http://icecube.wisc.edu/

[6] F. Aharonian *et al.*, H.E.S.S. Coll., 2006 Astron.Astrophys. **449** , 223; 2005 Astron.Astrophys. **437**, L7

[7] The NEMO home page, http://nemoweb.lns.infn.it/

[8] The NESTOR home page, http://www.nestor.org.gr/

[9] The KM3NeT home page, http://www.km3net.org

[10] Cyprus, France, Germany, Greece, Ireland, Italy, The Netherlands, Romania, Spain, UK

[11] The ESFRI home page, http://cordis.europa.eu/esfri/home.html

[12] S. Kuch, Ph.D. Thesis, U of Erlangen, Germany, 2007, available from http://www.opus.ub.uni-erlangen.de/opus/volltexte/2007/692/

[13] The effective neutrino area, A, is defined by $N=\Phi A$, where N is the number of detected neutrino events, and Φ is time integrated flux of incident neutrinos.

[14] G. Aggouras, *et al.*, NESTOR Coll., 2005 Nucl. Inst. Meth. A **552**, 420

[15] T. Chiarusi, *et al.*, NEMO Coll., Environmental parameters and water characteristics from deep-sea surveys at the NEMO sites, in *Proc. International Cosmic Ray Conference* (*ICRC05*), Pune, India, Aug. 2005

[16] P. Kooijman *et al.*, Photonic readout of optical modules in neutrino telescopes, in *Proc. International Cosmic Ray Conference (ICRC07),* Merida, Mexico, July 2007

[17] P. Kooijman *et al.*, Multi-PMT optical module for undersea neutrino telescopes, in *Proc. International Cosmic Ray Conference (ICRC07)*, Merida, Mexico, July 2007

[18] M. Taituti, Optical module for deep-sea neutrino telescopes,in *Proceedings for ICATPP07 Conference*, Como, 2007

[19] A. Braem *et a*l., 2007 Nucl. Inst. Meth. A **570**, 467

[20] A. Leisos, A Sea-Top Infrastructure for Calibrating an Underwater Neutrino Telescope, paper presented at *TeV Particle Astrophysics 2007*, Venice, Italy, 27-31 August 2007

[21] P. Sapienza, Status of the NEMO project, in *Proc. 20th European Cosmic Ray Symposium (ECRS 2006)*, Lisbon, 2006, e-Print: astro-ph/0611105; also contribution to this conference.

[22] E. Anassontzis *et al.*, 2006 Nucl. Inst. Meth. A **567**, 538

Air Shower Radio Detection with LOPES

J. Blümer[a,b], W.D. Apel[a], J.C. Arteaga[a], T. Asch[c], J. Auffenberg[d], F. Badea[a], L. Bähren[e], K. Bekk[a], M. Bertaina[f], P.L. Biermann[g], H. Bozdog[a], I.M. Brancus[h], M. Brüggemann[i], P. Buchholz[i], S. Buitink[j], H. Butcher[e], A. Chiavassa[f], F. Cossavella[b], K. Daumiller[a], V. de Souza[b], F. Di Pierro[f], P. Doll[a], R. Engel[a], H. Falcke[e,j], H. Gemmeke[c], P.L. Ghia[k], R. Glasstetter[d], C. Grupen[i], A. Haungs[a], D. Heck[a], J.R. Hörandel[j], A. Horneffer[j], T. Huege[a], P.G. Isar[a], K.-H. Kampert[d], D. Kickelbick[i], Y. Kolotaev[i], O. Krömer[c], J. Kuijpers[j], S. Lafebre[j], P. Luczak[l], H.J. Mathes[a], H.J. Mayer[a], C. Meurer[a], J. Milke[a], B. Mitrica[h], C. Morello[k], G. Navarra[f], S. Nehls[a], A. Nigl[j], J. Oehlschläger[a], S. Ostapchenko[a], S. Over[i], M. Petcu[h], J. Petrovic[j], T. Pierog[a], S. Plewnia[a], J. Rautenberg[d], H. Rebel[a], M. Roth[a], A. Saftoiu[h], H. Schieler[a], O. Sima[m], K. Singh[j], M. Stümpert[b], G. Toma[h], G.C. Trinchero[k], H. Ulrich[a], J. van Buren[a], W. Walkowiak[i], A. Weindl[a], J. Wochele[a], J. Zabierowski[l], J.A. Zensus[g]

LOPES COLLABORATION

[a] Institut für Kernphysik, Forschungszentrum Karlsruhe, Germany
[b] Institut für Experimentelle Kernphysik, Universität Karlsruhe, Germany
[c] Inst. Prozessdatenverarbeitung und Elektronik, Forschungszentrum Karlsruhe, Germany
[d] Fachbereich Physik, Universität Wuppertal, Germany
[e] ASTRON, Dwingeloo, The Netherlands
[f] Dipartimento di Fisica Generale dell'Università Torino, Italy
[g] Max-Planck-Institut für Radioastronomie Bonn, Germany
[h] National Institute of Physics and Nuclear Engineering, Bucharest, Romania
[i] Fachbereich Physik, Universität Siegen, Germany
[j] Dept. of Astrophysics, Radboud University Nijmegen, The Netherlands
[k] Istituto di Fisica dello Spazio Interplanetario, INAF, Torino, Italy
[l] Soltan Institute for Nuclear Studies Lodz, Poland

E-mail: `Johannes.Bluemer@ik.fzk.de`

Abstract. LOPES is an array of 30 radio antenna co-located with the KASCADE-Grande extensive air shower detector in Karlsruhe, Germany. It is designed as a digital radio interferometer for the detection of radio emission from extensive air showers. LOPES features high bandwidth and fast data processing. A unique asset is the concurrent operation with KASCADE-Grande. We report about the progress in understanding the radio signals measured by LOPES. In addition, the status and further perspectives of LOPES and the large scale application of this novel detection technique are sketched.

Introduction LOPES (= LOFAR prototype station) is an array of relatively simple, quasi-omnidirectional dipole antennas to detect the radio emission of extensive air showers produced

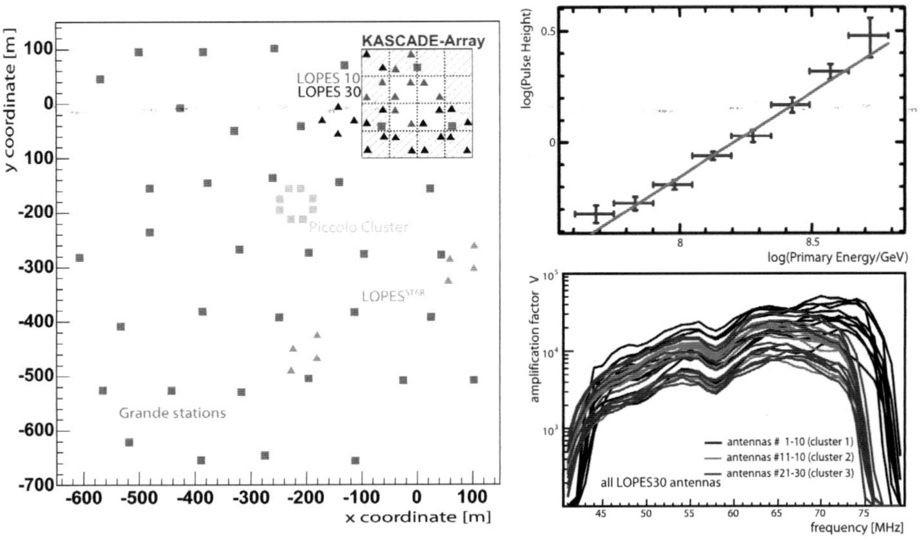

Figure 1. Left: Sketch of the LOPES set-up at the KASCADE-Grande experiment. Upper right: Average radio pulse height of the detected events versus the primary particle energy as reconstructed by KASCADE-Grande. Lower right: Frequency dependent amplification factors for all 30 antennas obtained by the amplitude calibration.

by cosmic ray particles. The radio waves are digitized and sent to a central computer. This combines the advantages of low-gain antennas, such as the large field of view, with that one of high-gain antennas, like the high sensitivity and good background suppression. To demonstrate the capability to measure air showers with these antennas, LOPES has been built as an embedded antenna array within the air shower experiment KASCADE-Grande [1]. In this way we exploit the unique opportunity to cross-calibrate the radio emission of air showers in the energy range from 10^{16} eV to 10^{18} eV.

The theoretical foundations and pioneering works had been conducted from the 1960's. The proof-of-principle was achieved in 2005 using a first deployment stage of 10 LOPES antenna (see ref. [2] and references therein). We have now extended the antenna field to 30 stations [3].

Recent theoretical studies [4] of the radio emission in the atmosphere are embedded in the scheme of coherent geo-synchrotron radiation. Here, electron-positron pairs generated in the shower development gyrate in the Earth's magnetic field and emit radio pulses by synchrotron emission. Such simulations lead to expectations of the relevant radio emission in the frequency range from 10 MHz to 500 MHz, which corresponds roughly to the thickness of the shower front. Newest results suggest that by measuring field strength and lateral behavior of the radio emission simultaneously will allow to determine primary energy and mass of the cosmic rays by this technique [5].

We investigate in detail the correlation of the measured field strengths with the shower parameters, in particular the orientation of the shower axis (geomagnetic angle, azimuth angle, zenith angle), the position of the observer (lateral extension and polarization of the radio signal), and the energy and mass (electron and muon number) of the primary particle. The basic element of the reconstruction is the beam-forming by time shifting, where the data from each pair of antennas is multiplied time-bin by time-bin, the resulting values are averaged, and the square root is taken while preserving the sign. We call this the cross-correlation beam or CC-beam. Although the shape of the resulting pulse (CC-beam) is not really Gaussian, fitting a Gaussian to the smoothed data gives a robust value for the peak strength.

LOPES-10 First measurements during six months of data taking were performed with a set up of 10 antennas. More than 600 showers were detected in the radio domain and basic correlations of the radio signals with shower parameters could be established. In Figure 1 the dependence of the reconstructed radio pulse height with the primary energy of the cosmic particles is depicted. The correlation supports the expectation that the field strength increases linearly with the primary energy of the cosmic rays, i.e. the received energy of the radio signal increases quadratically with the primary energy. The correlation between the reconstructed field strength with the geomagnetic angle suggests a geomagnetic origin of the emission mechanisms. By investigating more distant events an exponential dependence of the signal with the mean distance of the shower axis to the antennas was found [6]. The functional form of this dependence and also the lateral scaling parameter is of high interest for the further development of the radio detection technique. Further interesting features of the radio emission in EAS were investigated by analyzing very inclined showers [7] and events measured during thunderstorms [8].

LOPES-30 The sensitive area of the antenna field has been enlarged using 30 units. Each single antenna has been absolutely calibrated in the amplitude response using a commercial reference antenna [9]. The calibration procedure yields frequency-dependent amplification factors representing the complete system behavior (antenna, cables, electronics) in the environment of the KASCADE-Grande experiment. After one year of measurements using the East-West polarization by all 30 antennas, the LOPES-30 set-up was reconfigured to perform dual-polarization measurements [10]. With the data taken by LOPES-30 it is expected to calibrate the radio emission by air showers in the energy range from 10^{16} eV to 10^{18} eV and to identify the geosynchrotron effect as the dominant emission mechanism in air showers.

LOPES and the Pierre Auger Experiment LOPES is paving the way for an application of this detection technique in large UHECR experiments, like the Pierre Auger Observatory. We are optimizing the antenna design for an application at Auger, including a self-trigger system, named LOPES^{STAR} [11]. In parallel to the activities in Karlsruhe, tests are performed at the Auger Southern site, and first radio signals from showers have been detected [12].

[1] G. Navarra et al. - KASCADE-Grande coll., *Nucl. Instr. Meth. A* **518**, 207 (2004).
[2] H. Falcke et al. - LOPES coll., *Nature* **435**, 313 (2005).
[3] A. Horneffer et al. - LOPES coll., *Int. Journ. Mod. Phys. A* **21 Suppl.**, 168 (2006).
[4] T. Huege, R. Engel, and R. Ulrich, *Astropart. Phys.* **27**, 392 (2007).
[5] T. Huege, R. Engel, and R. Ulrich, *Proc. of 30th ICRC 2007, Merida, Mexico* (2007), in press.
[6] W.D. Apel et al. - LOPES coll., *Astropart. Phys.* **26**, 332 (2006).
[7] J. Petrovic et al. - LOPES coll., *Astronomy & Astrophysics* **462**, 389 (2007).
[8] S. Buitink et al. - LOPES coll., *Astronomy & Astrophysics* **467**, 385 (2007).
[9] S. Nehls et al. - LOPES coll., *Proc. of 30th ICRC 2007, Merida, Mexico* (2007), in press.
[10] P.G. Isar et al. - LOPES coll., *Proc. of 30th ICRC 2007, Merida, Mexico* (2007), in press.
[11] H. Gemmeke et al. - LOPES coll., *Int. Journ. Mod. Phys. A* **21 Suppl.**, 242 (2006).
[12] A. van den Berg et al. - Pierre Auger coll., *Proc. of 30th ICRC 2007, Merida, Mexico* (2007), in press.

JEM EUSO mission on International Space Station

Yoshiyuki Takahashi and the JEM-EUSO Collaboration

Dept. of Physics, The University of Alabama in Huntsville, Huntsville, AL35899, USA;
and RIKEN Computational Astrophysics Laboratory, 2-1 Hirosawa, Wako-shi, Japan

yoshi@cosmic.uah.edu

Abstract. The Extreme-Universe Space Observatory (EUSO) on Japanese Experiment Module (JEM) is a mission on ISS [1], led by Japan Aerospace Exploration Agency (JAXA) to investigate the nature and origin of extremely-high-energy cosmic rays (EHECRs). This is the sole window on the extreme-energy universe. JAXA has selected JEM-EUSO for study as one of two international mission candidates for installation on the JEM Exposure Facility by 2013. JEM-EUSO (hereafter, EUSO) is designated to pioneer measurements of EHECR-induced extensive air showers (EAS) from space. As the detector, it uses the whole earth, observing from the International Space Station (ISS) where a remote sensor is located. The signal is fluorescent and Cherenkov light from the Extensive Air Showers (EAS).

1. Extreme-Energy Astronomy - Astrophysics, Cosmology, and Fundamental Physics Quests at the highest energies

The flux of EHECR protons at energies above ~7 x 10^{19} eV reaching earth from distances exceeding 200 Mpc is so suppressed by the Greisen-Zatsepin-Kuzmin (GZK) effect [1] that we expect to find the dominant sources only nearby (< 90 Mpc at 10^{20} eV). Because these EHECRs will not be much deflected by extra-galactic and galactic magnetic fields (**Fig.1**), charged-particle astronomy is possible above these high energies. The goals of EUSO, having the instantaneous aperture of 0.6 - 3 million km^2 sr, are set as follows:

- Collect a large sample of EHECRs (several x 1000 above 7 x 10^{19} eV) extending to 10^{21} eV and beyond; measure their energies, directions, and particle species; and determine their nature and origins.
- Make astronomical source observations at extreme energies using charged particles (and neutrinos and/or gamma rays, if detected) initiating comprehensive studies of the individual astronomical sources of cosmic ray particles at these extreme energies.

2. Outstanding Problems and Opportunities

2.1 Acceleration mechanism, probing by GZK and super-GZK Particles:

No known models for astrophysical objects accelerate particles beyond 10^{21} eV (ZeV). Even if ~ZeV energies are achievable, it is unclear how the accelerated particles in the shock regions can emerge from the ambient dense radiation fields without significant energy loss. In addition, most of the candidate sources, radio galaxies and AGN, are at distances well beyond the GZK horizon, and so their cosmic ray fluxes suffer photo-pion losses on the CMB. Thus, the mystery of acceleration dynamics that deliver observed EHECRs persists.

2.2 Relatively nearby sources help exploring the acceleration mechanisms

Close sources may deliver multiple events to EUSO's large aperture, as the simulation in **Fig. 1** shows. Dozens of events per single source are possible, sufficient for definitive source identification. These events will also allow an inference of the energy spectrum and its end point, from which we may learn the nature of the hitherto unknown acceleration mechanism, as well as the type of the sources.

A rather different picture emerges if there are no identifiable strong astronomical EHE sources within 50 Mpc. Then, event-clustering about the rare, nearby sources (as suggested in the simulations in **Fig.1**) would not be seen, and the more or less isotropic EHECRs must come from distant sources. Among the known particles, only neutrinos can propagate unimpeded to Earth from distant acceleration sites at super-GZK energies. One possibility for observable fluxes from distant sources is the Z-burst mechanism [2]. In this mechanism, neutrinos from distant sources annihilate on the cosmic neutrino background (CNB) to produce very boosted Z bosons. Those annihilations occurring within the GZK-horizon produce, via Z-decay, cosmic-rays which may arrive at Earth. Since the Z-bursts result from a resonant process, the incident neutrino energy is well-defined, $E_{\nu}^{R} = M_Z^2/2m_n = 4$ (eV/m_n) ZeV. If this flux is nonzero, Z-bursts must occur. Observations of Z-burst events could allow us to make the first direct measurement of the CNB density, liberated only a second after the Big Bang, earlier by 400,000 years than liberation of the CMB in the hydrogen recombination era.

There is a another speculative class of super-GZK sources which do not require acceleration at all: The so-called top-down models, in which EHECRs are produced in the decay or annihilation of supermassive particles (SMPs) or topological defects (TDs), or SUSY or Extra-dimensional dark matter. These dark and heavy relics may have been produced in the hot-phase early-universe (sometimes via a symmetry-breaking phase-transition, sometimes at the termination of the inflationary epoch of the very early universe). The recent Auger limit on the primary photon fraction sets a severe upper-bound on the local SMP contribution, but not on TDs.

2.3 Neutrino Astronomy

Detection of astrophysical neutrinos demands an extraordinarily large volume because of the small cross section. The neutrino-nucleon cross section increases indefinitely with energy in the Standard Model (SM) of electroweak and quantum chromodynamics (QCD) ,$\sigma_{\nu N} \propto E_{LAB}^{0.4\pm0.1}$ above 10 TeV, yielding an ~0.1 microbarn at 10^{20} eV. This extrapolated cross section however is uncertain; we expect it to be large enough for cosmic neutrinos to produce observable numbers of showers in the ~10^{13} tons of atmosphere and earth-crust within the field of view of EUSO.

2.3.1. Horizontal Neutrino Showers (HNS) and Upward neutrino showers (UNS)

HNS can constrain topological defects, mirror matter, and the extra-dimension or low-scale unification theory. Neutrino-hadron cross-section data at the highest accelerator energy was given by the Electron Proton collider HERA experiment, $\sigma_{\nu N}$ ~$2\times10^{-34}\mathrm{cm}^2$ at $\sqrt{s}$ =314 GeV. Extra-dimensional models [3] predict a host of larger cross-sections, of which some are 10-100 times larger than what QCD suggest. According to the standard QCD prediction and cosmogenic neutrino flux calculations, EUSO is predicted to observe 1-10 neutrino events per year, while with extra-dimension, hundreds of events could be observed. If the top-down scenario for super-GZK particles (purple and green lines) is valid, an addition of at least several to hundreds of events per year are also expected. If only a few neutrino events are seen, it would clearly exclude most of the top-down models, as well as the extra-dimensional models.

The idea of UNS for Earth-crust events is that some neutrinos interact near the Earth's surface after skimming through the Earth. While electrons and muons are absorbed by the Earth, sometimes a tau (from ν_τ interactions) will emerge from the Earth and decay to produce a shower which fully develops in earth's atmosphere. It is predicted that such neutrino events through oceans are enhanced x10 relative to land, thereby favoring space-based observations. Such a measurement would establish the value of this cross section at high energies independent of the neutrino flux [4]. The expected number of events should be enhanced by a factor of ~10 depending on energy due to the Landau-Pomeranchuk-Migdal (LPM) effect [5] in the Earth's crust.

From the predicted flux of GZK cosmogenic neutrinos alone, the EUSO mission is expected to observe a few or more events in the Earth's atmospere and crust. If some more speculative predictions are correct, even more neutrino events will be observed.

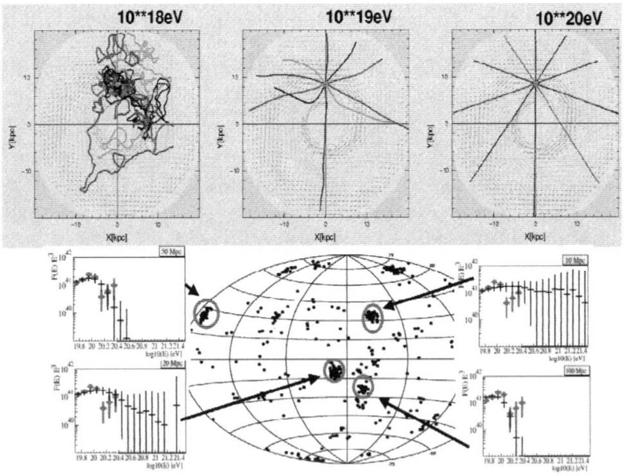

Fig. 1 Magnetic deflections and the source tracing at EHE.

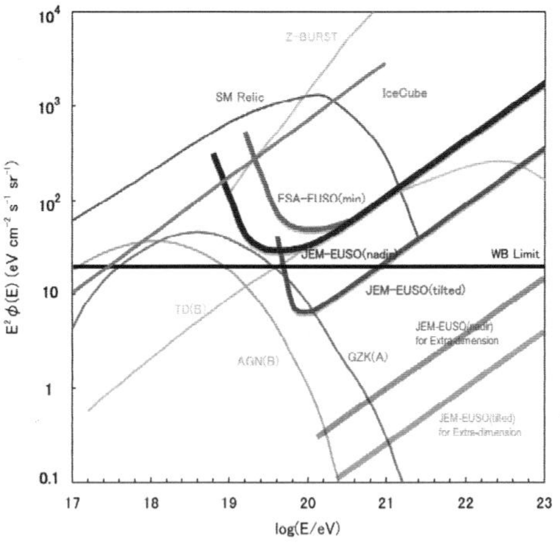

Fig. 2. Observable neutrinos

Fig. 2 shows neutrino fluxes from various models and the sensitivity of EUSO for detecting 1 event/energy-decade/year. An observational efficiency of 25% is assumed. Red Thick Line represents EUSO of previous ESA-Phase-A design; Blue Thick Line: EUSO Nadir; Green Thick Line: EUSO-Tilt. As for Ice cube (Pink line), a few events/energy-decade/10years is assumed. Black Line denotes Waxman-Bahcall limit EUSO sensitivity lines at the lower right part of the graph correspond to the high-cross-section case of the extra-dimension models.

3.4. High-Energy Gamma Rays and Quantum Gravity

The gamma ray window of the known universe is closed in the $10^{13} - 10^{20}$ eV energy interval because they are attenuated by pair production on infrared, microwave, and radio backgrounds, therefore, no gammas from distant sources are expected. However, once one optical length for photons exceeds the GZK length above 10^{20} eV the recovery of the transparency in gamma-ray channel is guaranteed. It has been conjectured further that quantum gravity effects [6] may make the universe far more transparent to such gamma rays. EUSO observations of extremely high energy (EHE) gamma rays originating from distant sources could become an evidence for this effect. EUSO could observe this effect by a time delay between a GRB or AGN flares and the arrival of the high-energy gamma rays from them.

References

[1] Greisen K.: *Phys. Rev Lett.*, Vol. 16, p. 748, 1966; Zatsepin G.T.and Kuzmin V.A.: *Pia'ma Zh. Eksp. Teor. Fiz.*, Vol. 4, p. 78, 1966.
[2] Weiler, T.: *Phys. Rev. Lett.*, 49, 234, 1982; Weiler, T.: *Astropart. Phys.*, 11, 303, 1999.
[3] Anchordoqui L A, Feng L J, Goldberg H, and Shapere A D, *Phys. Rev.* **D65** 124027 (2002).
[4] Palomares-Ruiz S.; Irimia A; and Weiler T. J.: *Phys. Rev. D* **73**, 083003 (2006).
[5] Landau L.D; and Pomeranchuk I.J.: *Dokl. Akad. Nauk., SSSR*, Vol. 92, pp. 535 and 735, 1953..
[6] Kifune T, *ApJ*, Vol. 518, p. L21, 1999; Amelio-Camelia G et al, *Nature*, Vol. 393, p. 763, 1998.

On the Characteristics of Neutrino Events in (Ultra-)High Energy Astrophysics Experiments from the View Point of Energy Estimation

N Takahashi[1], E Konishi[1], N M Budnev[2], I Nakamura[3], A A Kochanov[2], K Mitsui[4], and A Misaki[5]

[1] Graduate School of Science and Technology, Hirosaki University, Hirosaki, 036-8561 Japan
[2] Department of Physics, Irkutsk State University, Irkutsk, Russia
[3] Molecular Analysis and Life Science Center, Saitama University, Saitama, Japan
[4] Faculty of Management and Information, Yamanashi Gakuin University, Kofu, Japan
[5] Advanced Research Institute for Science and Engineering, Waseda University, Tokyo, 169-0092 Japan

E-mail: `taka@cc.hirosaki-u.ac.jp`

Abstract. The energies of muon neutrino events with energies above 10 TeV could not be estimated in principle even in 1 cubic kilometer detectors, whilst those of electron neutrino events could be estimated even at 10 EeV.

1. Introduction

The reliable energy determination of neutrino events in (ultra-) high energy neutrino astrophysics projects, the scale of which exceed over 1 cubic kilometer is indispensable for extracting physically meaningful conclusions from these projects. Only the Fully Contained Events in these experiments provide a reliable energy estimate by their definition. Now, we examine the characteristics of the muon neutrino events and electron neutrino events in (ultra-)high energies from the view point of the energy estimation. These neutrino events are exclusively produced by deep inelastic scattering. In these cases, almost all the neutrino energy is transferred to the produced muons (electrons). Therefore, here, we discuss the characteristics of (ultra-)high energy muons (electrons).

2. Characteristics of (Ultra-) High Energy Muons

(Ultra-)High Energy Muons lose their energy exclusively through processes of direct electron pair production, bremsstrahlung or photo-nuclear interaction in stochastic ways. Due to stochastic characters of these elementary processes, the ranges of the muons are more widely distributed as their energies increase. For example, the range of a 1 EeV muon is distributed from 5 km to 25 km. The Cherenkov photons from the primary muon are proportional to track lengths of charged particles produced whose energies are higher than the threshold energies for the Cherenkov photons. The total track lengths are classified as those from the primary muons and those of the accompanied electron showers due to the muons concerned.

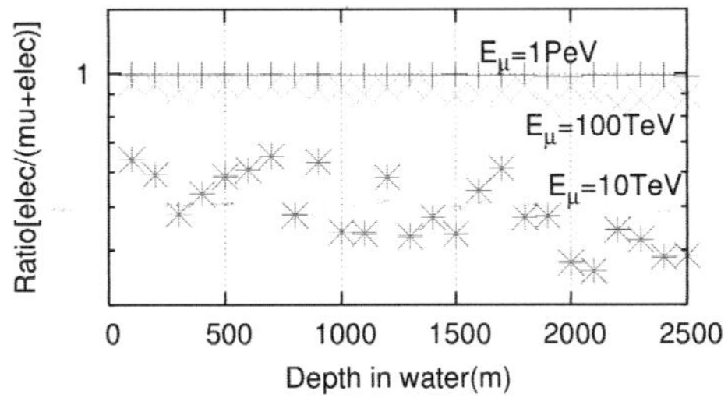

Figure 1. The dependence of the ratio of the electron track length to the total track length on the traversed distance by muons with 10 TeV to 1 PeV.

Namely, accompanied electron showers are generated by direct electron pair production, bremsstrahlung and photo-nuclear interaction. These showers are calculated by the exact Monte Carlo Method in one dimensional way. In Figure 1, we give the ratios of the track lengths due to the electron showers to the sum of those electron track lengths and the track lengths of the muons concerned as functions of the depth traversed. It is clear from the figure that the total track length of a primary muon with energy larger than 10 TeV essentially comes from the accompanied electron showers.

Thus, muons with energy larger than 10 TeV should be the muon dressed with an aggregate of electron showers, not as a single muon. The muons lose their energy through direct electron pair creation, bremsstrahlung and photo-nuclear interactions. These three elementary processes produce electron showers which finally produce the Cherenkov photons in addition to those due to the muons themselves. In Figure 2, how these three elementary processes take place

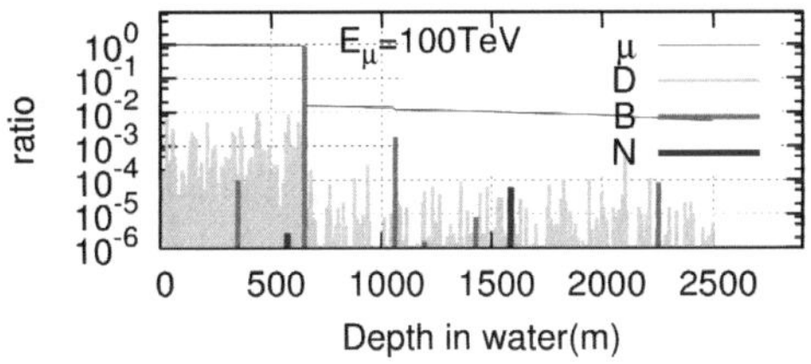

Figure 2. An exactly simulated behavior of the energy loss by the accompanied electron showers for 100 TeV muon. D means due to direct electron pair production, B bremsstrahlung and N photo nuclear interaction.

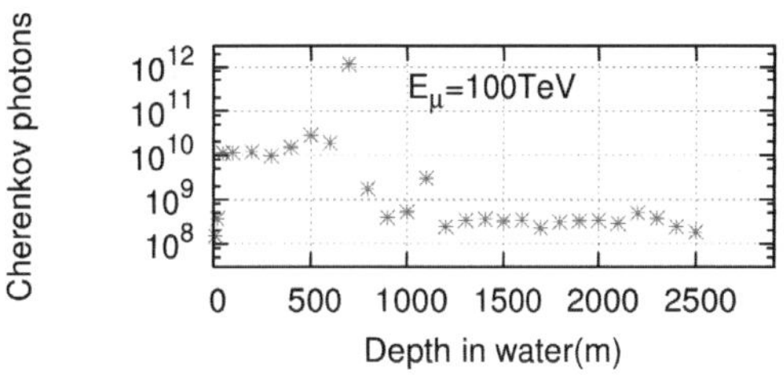

Figure 3. An exactly simulated transition curve for the Cherenkov photon number for a 100 TeV muon which just corresponds to Figure 2.

randomly in a stochastic way is indicated as a function of the depth in water by a muon with primary energy of 100 TeV. In this case, the muon loses 97% of the energy catastrophically at 659 meter due to bremsstrahlung. In Figure 3, we show that the Cherenkov photons produced by electron showers are given as the function of the traversed distance in the corresponding muon in Figure 2. The biggest Cherenkov photons quantity at 705 meter is due to the electron showers initiated by bremsstrahlung. In Figure 4, the Cherenkov photon number distributions are given for muons with single energy at given observation points of 1000 meters. It is clear from the figure that the width of the Cherenkov photon number increases as the primary energy increases, which reaches two orders of magnitude at 1 Pev. It is shown from the figure that we could not determine the primary energy from the measurement of the Cherenkov photons. Furthermore, the real muons concerned may obey some production spectrum. Here, we assume for given $\gamma = 1, 2, 3$, where $E_{min} = 10$ TeV and

$$N(E_\mu)dE_\mu \propto E_\mu^{-(\gamma+1)}dE_\mu \quad (1)$$

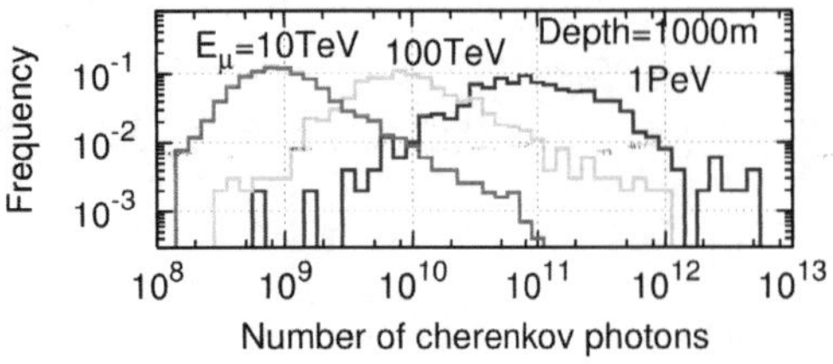

Figure 4. The dependence of the ratio of the electron track length to the total track length on the traversed distance by muons with 10 TeV to 1 PeV.

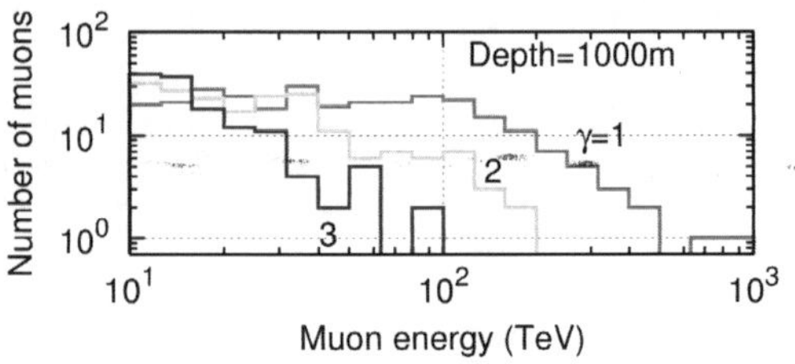

Figure 5. The dependence of the primary muon energy for producing the Chrerenkov photon (1.00×10^{10} to 1.26×10^{10}) for the muon production spectrum.

$E_{max} = 1$ PeV. For given index, γ, we sample the energy of the muon from Eq.(1). We simulate this muon in an exact stochastic way and finally obtain the Cherenkov photons which are produced by the electron showers at various observation points. We repeat this procedure and finally obtain the energy distribution for primary muon energies for different production spectra of the primary muon energy and for the range of the Cherenkov photons. In Figure 5, we give the primary energy distribution for the Cherenkov photons $= 1.0 \times 10^{10}$ to 1.26×10^{10}. It is clear from the figures that the primary energy distributions for muons are strongly dependent on the shapes of their production spectrum and the measured Cherenkov photon number. The energy estimation for the primary energy of the muon from the Cherenkov photons is in principle impossible. This is not because of technical difficulties.

3. Characteristics of (Ultra-)High Energy Electron

Different from the muon neutrino events, electrons produced by electron neutrinos via the deep inelastic scattering produce single electron showers, which could be usually contained inside a 1 cubic kilometer detector, namely, being regarded as Fully Contained Events, whose energies could be determined. However, when the primary energies of the electrons become remarkably high, then, the LPM effect begins to be effective[1]. For example, an electron with energy 10^{21} eV produces electron showers with extremely long tails. In Figure 6, we give electron track lengths for four showers. The track lengths are proportional to the Cherenkov photons produced by these showers. It is clear from the figure that we could not estimate the primary energy of the electrons concerned in a 1 cubic kilometer detector.

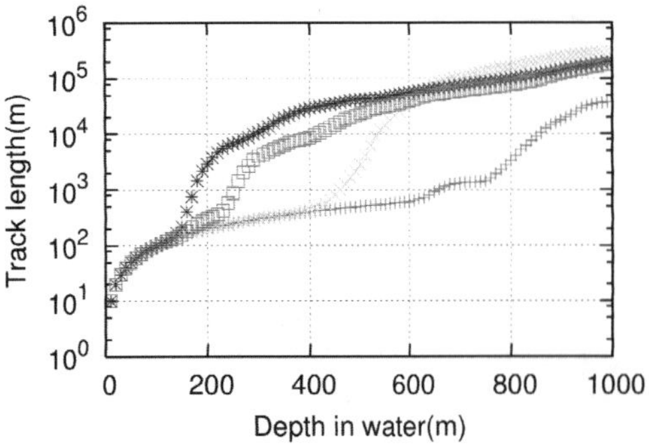

Figure 6. The transition curves for track lengths of electrons in four LPM showers. $E_{prim} = 10^{21}$ eV and $E_{min} = 10^{15}$ eV.

4. Conclusion

The energy estimation for muon neutrino events with energy larger than 10 TeV in principle could be impossible, apart from the technical difficulty, while, the energy determination for electron neutrino events could be possible even in the LPM showers, although technical difficulties may exist.

References

[1] Konishi E, Misaki A and Fujimaki N 1978 *Nuovo Cimento* **48A** 509
Konishi E, Adachi A, Takahashi N and Misaki A 1991 *J.Phys.G:Nucl.Part.Phys* **17** 719
Misaki A 1990 *Fortscr.Phys.* **38** 413

Study of Dynamical Fluctuations in Silicon-Nucleus Collisions at High Energy

Mustafa Abdusalam Nasr and Aisha Khalifa Elwagni
High Energy Physics Group
Department of Physics, Faculty of Science, 7th October University
Misurata-Libya, P.O.BOX 1111- E.Mail: bayio@yahoo.com

Abstract. An attempt is made to study the intermittency in silicon-emulsion collisions at $14.5A\,\text{GeV}\,c^{-1}$. The experimental data used reveals the presence of intermittency and multifractality. An attempt is also made to find Renyi Dimensions also called self-similar dimensions, D_q and multifractal spectrum, f(α). The self-similar multifractal spectrum is observed to be a convex curve with a maximum around q=0.

1. Introduction

The study of fluctuations in multiparticle production in high energy hadronic collisions started since the observation of a relatively high multiplicity cosmic ray event known as JACEE[1] event. In order to explain the observed fluctuations in the rapidity distribution for this event, Bialas and Peschanski proposed [2,3] scaled factorial moments (SFMs) approach to investigate dynamical fluctuations in multiparticle production in high energy nuclear collisions.

2. Details of the Data

A random sample comprising of 595 interactions having $n_s \geq 2$, where n_s represents the number of charged particles produced in an event with relative velocities, $\beta \geq 0.7$, produced in $14.5A\,\text{GeV}\,c^{-1}$ Silicon-Emulsion collisions, is used for carrying out the present analysis. The emission angles of all the relativistic charged particles were measured and their pseudorapidities were determined. All other relevant details about the stacks used, criteria employed for selecting the events and the method of measuring the emission angles may be found elsewhere [4]. Furthermore, for comparing the experimental results with the corresponding values predicted by the Lund model, FRITIOF [5], a sample consisting of 5000 events, identical to the experimental ones were simulated.

3. Results and discussion

The variations of $\ln < F_q >$ with $-\ln\delta\eta$ for the experimental and FRITIOF data are depicted in Fig. 1. It is observed that $\ln < F_q >$ increases linearly with $-\ln\delta\eta$ for both the data samples. In order to study the dependence of factorial moments on the target mass, variation of $\ln < F_q >$ with $-\ln\delta\eta$ is examined separately for CNO, emulsion and AgBr groups of target nuclei; these variations are plotted in Fig, 2. The solid lines in Figs. 1 and 2 are obtained by method of least squares fitting to the data. It is seen that for the three categories of interactions, CNO, emulsion and AgBr a linear rise in the SFMs with decreasing bin width $\delta\eta$, is observed. This behavior supports the presence of intermittency in these collisions. It is also clear from the figures that the values of SFMs are relatively higher for the interactions due to the heavy targets. This might happen due to low multiplicities in the interactions due to CNO in comparison to those for AgBr targets[6].

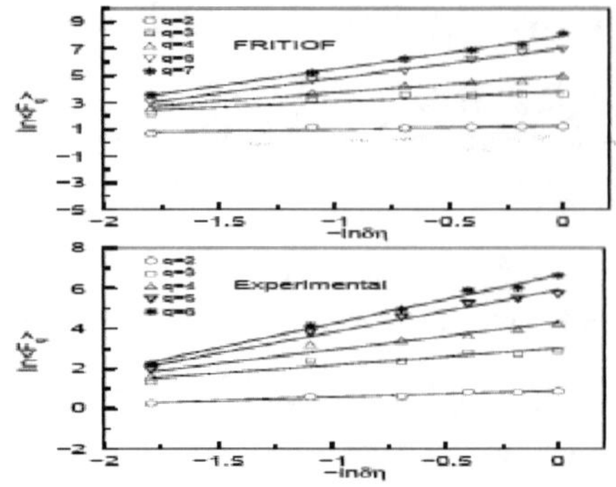

Fig. 1 $\ln < F_q >$ versus $-\ln \delta\eta$.

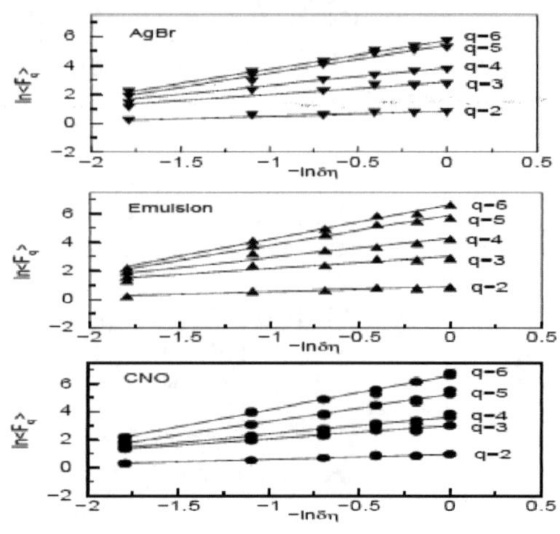

Fig. 2 $\ln < F_q >$ versus $-\ln \delta\eta$ for AgBr, CNO and Emulsion group of nuclei.

The intermittency indices, Φ_q are correlated with strength of the intermittency. The values of Φ_q can be found from the observed linear rise of $\ln < F_q >$ with $-\ln \delta\eta$. It may be mentioned that $\Phi'_q s$ are the slopes of the fits to $\ln < F_q >$ versus $-\ln \delta\eta$ plots. Given in Table 1 are the values of Φ_q for the collisions of Silicon Nuclei with CNO, emulsion and AgBr targets.

Table1: The intermittency indices Φ_q, for the interactions of $14.5A\,\mathrm{GeV}\,c^{-1}$ silicon nuclei with CNO and AgBr groups of targets.

Order of the moment(q)	Φ_q (CNO)	Φ_q (Em.)	Φ_q (AgBr)
2	0.422 ± 0.035	0.388 ± 0.068	0.515 ± 0.141
3	0.975 ± 0.166	0.844 ± 0.168	0.652 ± 0.169
4	1.411 ± 0.155	1.465 ± 0.177	1.298 ± 0.168
5	2.136 ± 0.151	2.115 ± 0.144	1.973 ± 0.082
6	2.685 ± 0.148	2.722 ± 0.227	2.593 ± 0.188

Fig. 3 shows the variations of Φ_q with q for the experimental and simulated data. Statistical errors are also indicated in the plots. In Fig. 3 a linear increase is clearly discernible in the value of Φ_q with the order of the moments, q. It may be stressed that increase in Φ_q with q is quite prominent in the case of the experimental data in comparison to those for the corresponding FRITIOF generated data. However, from Table 1 it is clearly evident that amongst the three groups of targets, the values of $\Phi'_q s$ are relatively higher for the collisions due to CNO targets. Fig. 4 exhibits the variations of anomalous dimensions [6], $d_q (= \frac{\Phi_q}{q-1})$, with the order of the moments, q; the value of d_q is found to increase with q.

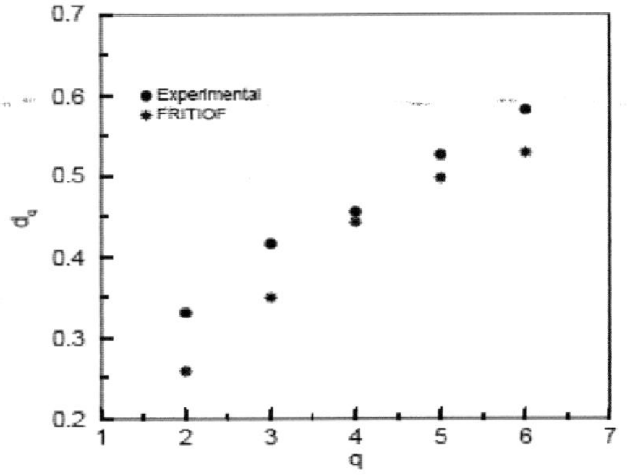

Fig. 4 Variation of d_q with q.

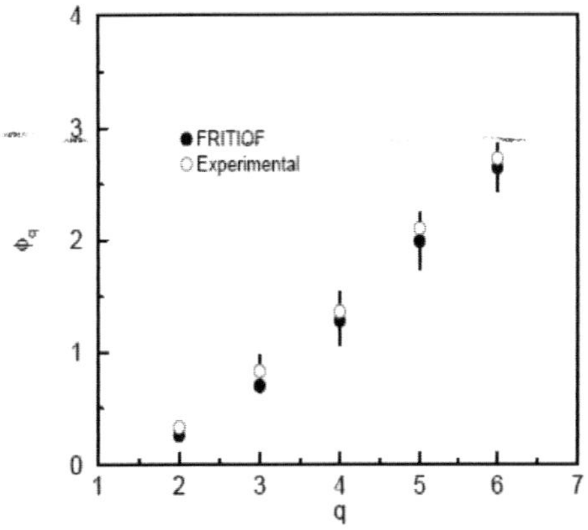

Fig. 3 Variation of Φ_q with q.

According to Levy stability, as a self-similar fractal system, the multifractal final state in high energy nuclear collisions can be described in terms of Levy stability index [7], μ. It is used to investigate the existence of multifractality as a self-similar system in relativistic nuclear collisions. This parameter is envisaged to provide some useful information about elementary fluctuations around the tail [7] of the pseudorapidity distribution. It is also envisaged to be a measure [7] of the degree of multifractality.

4. Conclusions

It is clearly seen from the present investigation that the data used reveals the presence of intermittency and multifractality in silicon-nucleus collisions. Moreover, analyses for the experimental and simulated data show good compatibility.

Reference:

[1]. T.H.Burnett et al.(JACEE Collaboration), Phys. Lett.50(1983)2062.
[2]. A.Bialas and R. Peschanski, Nucl. Phys., B273(1986)703.
[3]. A.Bialas and R. Peschanski, Nucl. Phys., b308(1988)857.
[4]. Shakeel Ahmad, M. M.Khan, Ahmad, Zafar and M.Irfan. Act. Phys. Pol., 35,(2004)809.
[5]. B. Andersson, G. Gustafson, Ingelman and T. Sjostrand, Phys. Rep., 97,(1983)31.
[6]. D. Ghosh et al, Phys. Rev., D49(1994)3113.
[7]. Liu Feng, Liu Fuming and Liu lianshou, Phys. Rev., D59,(1999)1140

The High-Energy Neutrino Background from Gamma-Ray Bursts

Kohta Murase

Yukawa Institute for Theoretical Physics, Kyoto University, Sakyo-ku, Kyoto, 606-8502, Japan

E-mail: `kmurase@yukawa.kyoto-u.ac.jp`

Abstract. High-energy neutrinos from gamma-ray bursts (GRBs) have been expected in various scenarios. Many predictions for the prompt emission and afterglow emission were done in the pre-*Swift* era. In this paper, we see predictions of the high-energy neutrino emission in the *Swift* era. Recently, *Swift* has shown several novel phenomena such as flares, which allows new possibilities for the high-energy neutrino emission from GRBs. High-energy neutrino signals may be detected by future-coming large neutrino detectors such as IceCube and KM3Net. If detected, they are useful for revealing the nature of GRBs and connection between ultra-high-energy cosmic rays (UHECRs) and GRBs.

1. Introduction

Gamma-ray bursts (GRBs) are the most energetic phenomena in the universe. Their emission is usually explained by dissipation of kinetic energy of relativistic shells. In both of the internal shock model for the prompt emission and external shock model for the afterglow (AG), accelerated electrons in the (mildly) relativistic shocks are considered to radiate high-energy gamma rays. If relativistic shells contain baryons, not only electrons but also protons (and possibly heavy nuclei) will be accelerated in the both models. The cosmic rays accelerated up to very high energies exceed the photomeson production threshold, and produce pions and kaons via the photomeson production. Pions and kaons will decay to neutrinos as they cool down due to various cooling processes. We take into account these cooling processes in order to calculate neutrino spectra in detail. We also use Geant4 combined with experimental data to treat the photomeson production. The method is described in Murase [1]. Expected muon events from one burst are too few to be detected. Therefore, we will need many GRB events as the neutrino background, which can be calculated by assuming that the GRB rate traces the star formation rate for the given local GRB rate. By observing neutrinos simulataneously with gamma rays, various models of the neutrino emission from GRBs will be tested by future-coming neutrino observations.

2. The Expected High-Energy Neutrino Background

Waxman & Bahcall [2] predicted neutrino bursts under the internal shock model, assuming that ultra-high-energy cosmic rays (UHECRs) come from GRBs. Since the origin of UHECRs (above the ankle below the GZK cutoff) is still unknown, observations of neutrinos will be important in the sense that they enable us to test the GRB-UHECR hypothesis. Although many authors studied the neutrino emission under this model, most of the calculations were based on the Δ-resonance approximation. We studied the neutrino emission in detail without this approximation, by exploiting the numeical method [1, 3]. We also introduced the nonthermal baryon loading factor $\xi \equiv U_p/U_\gamma$. This is because the recently estimated local GRB rate requires the large nonthermal baryon loading factor for GRBs to explain observed UHECRs [3]. Even if GRBs are not main sources of UHECRs, high-energy neutrinos from GRBs may be detected by future-coming neutrino detectors such as IceCube. For example, $\xi = 10$ leads to the detectable neutrino background flux, as long as collision radii are sufficiently small. Observations have been giving us useful insights into the amount of accelerated baryons [4, 5]. The high-energy neutrino emission is also expected in the afterglow phase. Waxman & Bahcall [6] predicted neutrino afterglows under the reverse shock model, while Dermer [7] did under the forward shock model. In the former model, strong infrared/optical flashes from short-lived reverse shocks are expected. However, recent observations have suggested the lack of such infrared/optical flashes. Several mechanisms for supression of infrared/optical flashes are proposed. In terms of recent observational results, the high-energy neutrino emission from the reverse shock is discussed in detail by Murase [1]. Even if we can expect very-high-energy neutrinos in the afterglow phase, the expected background flux is usualy low. Because cosmogenic neutrinos will mainly contribute to the very-high-energy neutrino background, the simulataneous observations with gamma rays are still required. In Fig.1, we show the pre- era predictions of high-energy neutrinos.

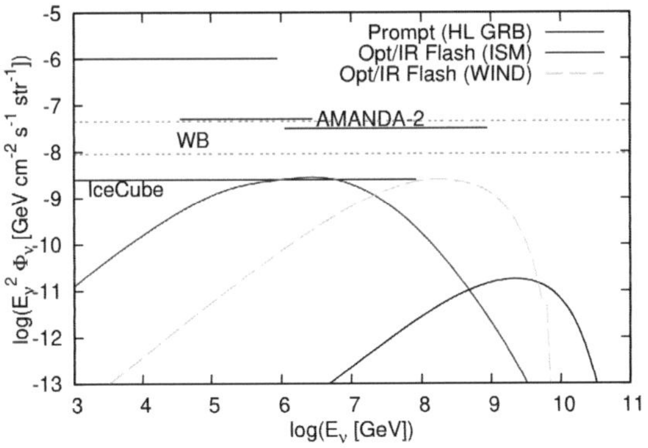

Figure 1. The predictions of the neutrino background in the pre- era (where high-energy cosmic rays have the comparable energy to that of the prompt emission).

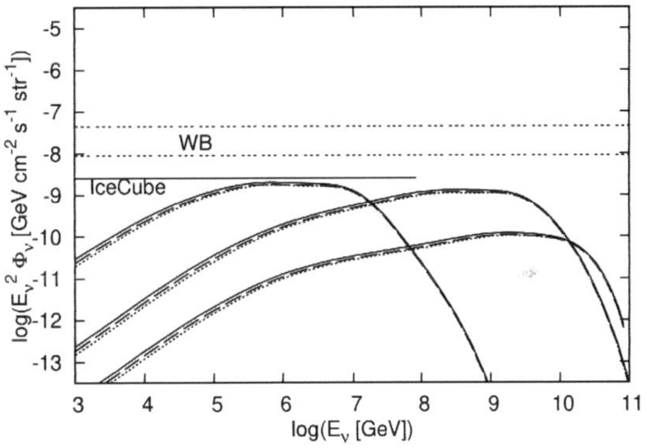

Figure 2. The optimistic predictions of the netrino background from early afterglows in the era. Curves are Late Prompt, AG (WIND) and AG (ISM) from top to bottom.

The satellite was launched to solve several outstanding questions in the pre- era. has presented indeed very fruitful results until now. First, one of the novel results is the discovery of XRF 060218. Motivated by the discovery of this nearby event associated a supernova, which had the low luminosity (LL), some recent studies have suggested the existence of a different population from conventional high luminosity (HL) GRBs. If true, such a population may give the significant contribution to the observed ultra-high-energy cosmic ray flux and neutrino background flux. Murase et al. [8] first suggested that LL GRBs could be the origin of UHECRs. Even though it may be difficult for protons to be accelerated above ~ 10 eV, heavy nuclei could achieve such energies. The LLGRB-UHECR hypothesis can be tested by future-coming neutrino observations [8]. Second, discovered the existence of the mysterious shallow decay emission in the early afterglow phase. Although various interpretations have been proposed, its origin is not conclusive yet. Implications to the neutrino background, based on the late prompt emission model and reverse shock model, are described in Murase [1] (see Fig.

2). Third, many bursts have flares in the early afterglow phase. Flares are also energetic, and they imply the late activities of the central engine. They have lower peak energies and are expected to have lower Lorentz factors, which leads to the more efficient neutrino production. We showed the high-energy neutrino emission from flares can be important [9]. In Figs. 3 and 4, we compare results from several scenarios described above. HL GRBs have luminosity $L \sim 10^5$ ergs s^{-} and duration $T \sim 10$ s. On the other hand, we assume $L \sim 10$ ergs s^{-} and $T \sim 100 - 1000$ s for LL GRBs, and $L \sim 10$ $^{-}$ ergs s^{-} and $T \sim 100$ s for the late prompt emission such as flares (Late Prompt). The typical pion production efficiency is $\sim 0.1 - 1$ for HL GRBs, $\sim 0.01 - 0.1$ for LL GRBs, and ~ 1 for Late Prompt. In the case of Late Prompt, the high efficiency is expected so that the highest-energy cosmic rays will be depleted and neutrino production is efficient. The local long GRB rate is typically ~ 1 Gpc^{-} yr^{-} for HL GRBs. On the other hand, it might be higher $\sim 100 - 1000$ Gpc^{-} yr^{-} for LL GRBs. The high rate of LL GRBs could compensate their lower luminosity. In HL GRBs and Late Prompt, neutrino signals will be correlated with the prompt emission and the early afterglow phase, respectively. But we cannot expect such correlations for most LL GRBs because they are dim. Note that the contribution to the very high energy background from Late Prompt and LL GRBs could dominate that from HL GRBs due to their lower magnetic field strength. Although neutrino afterglows from the external shock can be important in the very high energies, fiducial parameter sets (as shown in Fig. 3) predict the insufficient background flux for detection by IceCube. High-energy neutrinos will be the very evidence of the acceleration of baryons in GRBs. The described models will be tested by IceCube and KM3Net. Once neutrino signals are detected, we can obtain useful information on physics of GRBs.

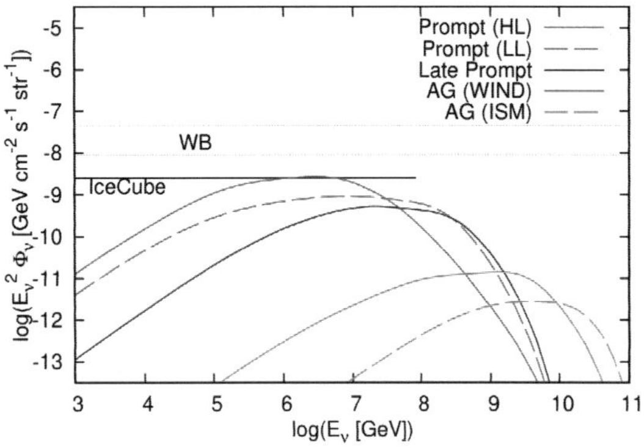

Figure 3. The fiducial predictions of the neutrino background from GRBs in the era [1].

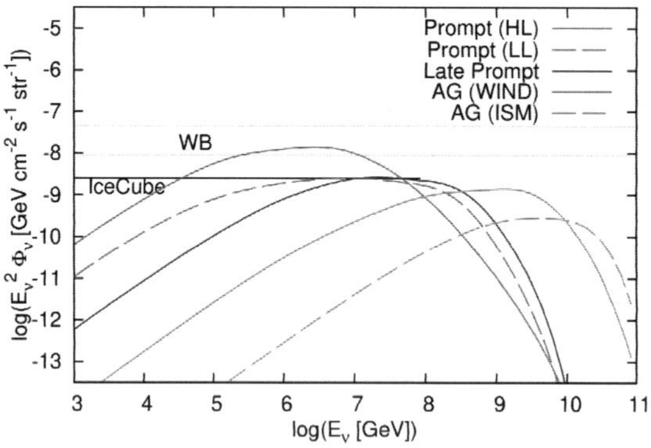

Figure 4. The optimistic predictions of the neutrino background from GRBs in the era [1].

References

[1] Murase K 2007 *Phys. Rev. D* in press
[2] Waxman E and Bahcall J 1997 *Phys. Rev. Lett.* **78** 2292.
[3] Murase K and Nagataki S 2006a *Phys. Rev. D* **73** 063002
[4] Achterberg et al. 2007a *Astrophys. J.* **664** 397
[5] Achterberg et al. 2007b *arXiv*:0705.1186
[6] Waxman E and Bahcall J 2000 *Astrophys. J.* **541** 707
[7] Dermer 2002 *Astrophys. J.* **574** 65
[8] Murase K Ioka K Nagataki S and Nakamura T 2006 *Astrophys. J.* **651** L5
[9] Murase K and Nagataki S 2006b *Phys. Rev. Lett.* **97** 051101

GLAST The Gamma Ray Large Area Space Telescope

C. Cecchi on behalf of the GLAST LAT Collaboration
Univeristy of Perugia and INFN Sezione di Perugia
Via A.Pascoli 06124 Perugia

E-mail: Claudia.cecchi@pg.infn.it

Abstract. The Large Area Telescope (LAT) has been integrated and is currently under test to be ready for launch by NASA in 2008, as the main instrument on the Gamma-ray Large-Area Space Telescope (GLAST) mission. It will survey the sky in the energy range from 20 MeV to 300 GeV. The detector subsystem of the LAT will provide much better angular and energy resolution, effective area and field of view for gamma rays with respect to previous instruments. We present the LAT integration status, tests data taking and its scientific objectives together with the description of science tools preparation.

1. Introduction

The EGRET instrument [1] on the Compton Gamma Ray Observatory performed the first all-sky survey above 50 MeV (Third Egret Catalogue) obtaining information on high energy gamma ray blazars, pulsars, unidentified sources and delayed emission from gamma ray bursts. Furthermore the instrument studied diffuse radiation, galactic and extragalactic components. Despite all this, many questions remain still open. The GLAST observatory is composed of two instruments: the Large Area Telescope (LAT), sensitive to energies from 20 MeV to 300 GeV, and the Gamma ray Burst Monitor (GBM), to cover the region from 8 KeV to 30 MeV. The Large Area Telescope (LAT) [2] will enable studies of a broad range of scientific topics: transient phenomena and variable sources, Gamma Ray Bursts (GRB's) and Active Galactic Nuclei (AGN's), pulsars, unidentified sources, galactic and extragalactic diffuse emission and even more exotic subject like dark matter annihilation. The LAT (figure 1) is a conversion pair telescope measuring direction, energy and arrival time of photons through the reconstruction of the electron-positron pair, emerging from the photon conversion, in a silicon tracker/converter and a CsI calorimeter. The detector shows a very good angular resolution for source localization and multiwavelength studies, a huge field of view which allows a complete sky survey every two orbits (~ 3 hours), and high sensitivity to transient phenomena, as well as good energy resolution, large effective area (>8000 cm^2) and a Point Spread Function (PSF) a factor of 3 better than EGRET at high energy.

2. The detector

The LAT is an array structure of 4x4 towers each one composed of a precision gamma ray tracker-converter and a calorimeter. The whole detector is covered by a segmented anticoincidence detector shield (ACD).

LAT tracker-converter

Each tracker module has 18 x,y planes, consisting of two layers of single-sided silicon strip detector (AC-coupled, 400μm thick, 228μm pitch) and high-Z converter material (tungsten). Support material for silicon and converter is made of carbon-composite panels, called trays. Alternate trays are mounted rotated by 90° to allow the reconstruction in both x and y directions. Incident photons convert in the converter foils and the two e^+e^- tracks are reconstructed through the silicon planes.

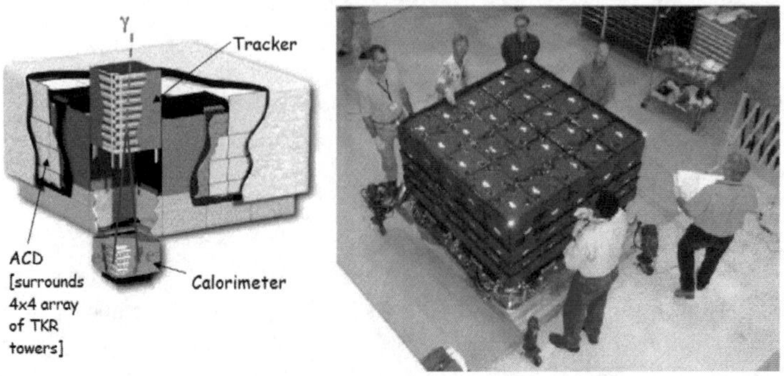

Figure 1: The LAT: schematic layout (left) and assembled detector (right).

LAT calorimeter

Each calorimeter module has 96 CsI(Tl) crystals (2.7cm x 2.0cm x 32.6cm) arranged in 8 horizontal layers of 12 crystals each. The total amount of material is of 8.6 radiation lengths. Each layer is rotated by 90° with respect to its neighbours forming an hodoscopic array. It measures the energy of the incoming electrons and positrons and reconstructs also the direction of the shower using information from the two coordinates plus the measurement of the light yield asymmetry at both ends of the crystals.

LAT performance

The performance of the LAT are related mostly to three factors: design of the hardware, event reconstruction and background rejection. The detector is already completed and; the optimization of reconstruction algorithms and background rejection are being pursued. As an example, figure 2 shows the expected point spread function and effective area (see https://www-glast.slac.stanford.edu/software/IS/glast_lat_perfomance.htm for more details).

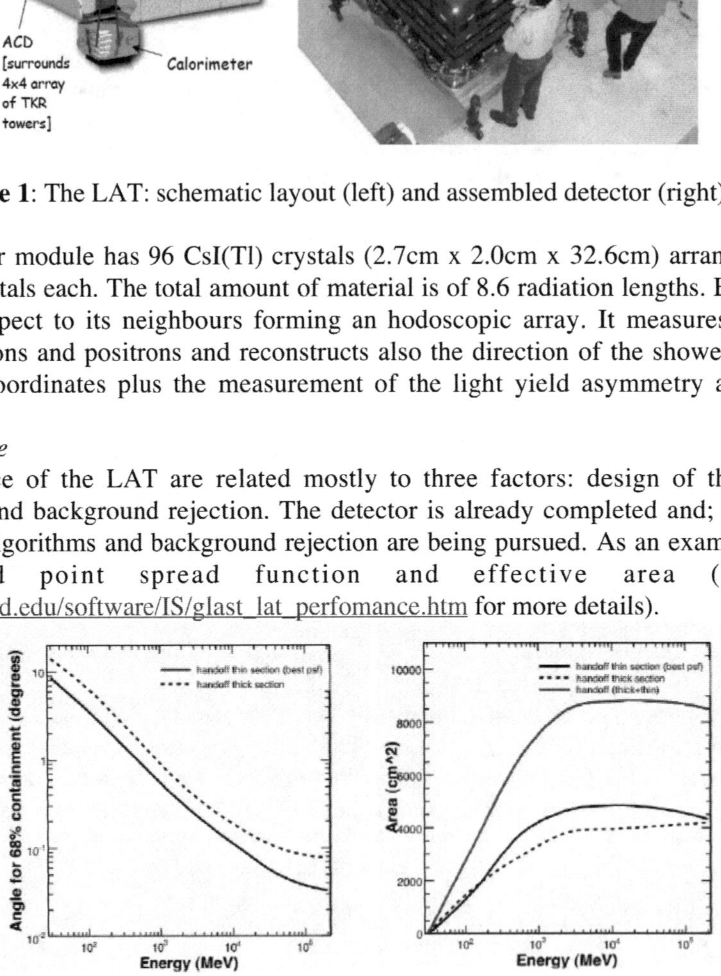

Figure 2: example of the performance of the LAT.

3. GLAST LAT Science

Among the several science topics of the LAT we will discuss in more details GRB, AGN, and the multiwavelength campaign.

GRB's are the outcome of the most powerful explosions in the Universe and about 1500 of them have been discovered by the Compton Gamma Ray Observatory. High energy emission, up to GeV energies has also been observed by EGRET. Data shows that GRB's are isotropic in the sky, and come in two distinct classes of short and long bursts with distributions crossing at about 2 *s* duration, finally X-ray afterglows of long bursts have also been observed [3]. Concerning GLAST, a dedicated Gamma Ray

Burst Monitor (GBM) with energy covering the spectrum energy from 8 KeV to 30 MeV will promptly localize a GRB and perform its spectral analysis. At the same time, complemented by other space- and ground- based high energy photon detectors, GLAST will analyse the last spectral window of the GRB. The estimated LAT detection rate is of 50/year. The overlapping between GLAST and Swift will also be very important, in order to cover the full spectral and temporal emission of the GRB. GRB's discovered by the LAT can be localized with accuracy of 20 *arcmin* by Swift, and with precision ranging in 1-10 *arcmin* when also the afterglow is detected. On the other hand, GRB's discovered by Swift can be followed by the LAT which, covering a wider energy range, will follow the long lived GeV component.

Active Galactic Nuclei (AGN) are extragalactic sources in some cases associated with nuclei of galaxies, characterized by emission processes highly anisotropic and with large time variability [4]. The spectral energy distribution (SED) shows two components: the first one peaking in the UV-X ray band or in the optical band depending on the AGN class, the second peak can be in the GeV or in the TeV band. Despite intensive studies, blazar emission is not very clear. Competing models are based on synchrotron self-Compton (SSC), external Compton (EC) or hadronic processes. Owing to its large field of view, GLAST will observe a large number of blazars and study their spectra and variability on timescales from day to year; the time scales of a few hours can be studied for the brightest sources. Combining the LAT data with those at other wavelengths will be crucial to answer many of the open questions about the structure and the origin of blazars.

Multiwavelength observations are key to many science topics of GLAST (http://glast.gsfc.nasa.gov/science/multi). For gamma ray observations multiwavelength are essential for the identification of the sources, since emission is always expected across most of the electromagnetic spectrum. Infact, except for very bright sources, identification is impossible with gamma ray observations only. Counterparts at other wavelengths are necessary to provide information on source location, mass, distance and energy spectrum.

The primary observing mode of GLAST will be sky survey, covering the sky after two orbits (about 3 hours). In addition to scheduled pointed observations, target of opportunity (TOO) observations will also be supported. After 60 days of initial orbit checkout, verification and calibration, the first year will be devoted to sky survey mode (Phase1). During Phase1, burst alerts and repoints will be enabled as well as extraordinary TOOs. In subsequent years the default mode of observation will still be survey, but some time may be devoted to peer review selected pointed observations. During Phase1 all GBM bursts will be released and LAT data for about 20 sources will be provided. Information on flaring sources will be also released. The data will be released through the GLAST Science Support Center (GSSC)

4. Summary

The final detector has been assembled and environmental tests are being performed. The preparation for science and operation is also well advanced. GLAST will address many important questions about unidentified sources, the origin of the gamma ray emission, the origin of the extragalactic gamma-ray background, the energy behaviour of GRB's, the emission mechanism of AGN's and more exotic phenomena like annihilation of dark matter, or further surprises in the unexplored region between 30 and 300 GeV. An essential component of the GLAST science will come from multiwavelength observations.

Acknowledgments

I wish to thank P. Michelson, S.Ritz, J. M.C. Enery and C. Dermer for the useful discussions.

References

[1] Thompson D J et al. 1993 *Ap. J. S.* **86** 629-656
[2] Atwood B et al. 1994 *Nucl. Instr. Meth.* **A342** 302
[3] Meszaros P, Rees M J 1997 *Ap. J.* **476** 232
[4] Mattox J R et al. 1997 *Ap. J.* **476**

Recent Results from CANGAROO-III

Yohei Yukawa for the CANGAROO Collaboration

Institute for Cosmic Ray Research, University of Tokyo, 5-1-5 Kashiwa-no-Ha
Kashiwa City, Chiba 277-8582, Japan

E-mail: yukawa@icrr.u-tokyo.ac.jp

Abstract. The CANGAROO-III telescope system for very-high-energy γ-ray astrophysics consists of four 10-m atmospheric Cherenkov telescopes located near Woomera, South Australia. We have been observing southern-sky objects since March 2004. Here we report on the status of the system and some recent results from CANGAROO-III observations.

1. Introduction: CANGAROO-III telescopes

CANGAROO, Collaboration of Australia and Nippon for a GAmma Ray Observatory in the Outback, has studied southern sky gamma-ray sources for 10 years by Imaging Atmospheric Cherenkov Telescopes located near Woomera, South Australia (31°S, 137°E). Our latest telescope array, CANGAROO-III, has been operational since March 2004 with 10-m mirror and stereoscopic coincidence to achieve higher sensitivity compared to the previous (10-m later) single telescope (CANGAROO-II).

The CANGAROO-III stereoscopic system consists of four optical telescopes. Each telescope has a prime-focus segmented parabolic reflector with 57 m^2 light collecting area and nanosecond-order time accuracy. The prime mirror consists of 114 spherical mirrors mounted on a parabolic frame with a focal length of 8m [1].

The first telescope, T1, which was originally the CANGAROO-II telescope [2] and upgraded later, is not used for recent observations due to its smaller field of view and higher energy threshold. The second, third, and fourth telescopes (T2, T3, and T4) have identical cameras and data-acquisition systems [3]. Inter-telescope stereoscopic trigger system, which was introduced in December 2004, enables n-fold stereoscopic coincidence (usually we use 2-fold). The telescopes are located at the east (T1), west (T2), south (T3) and north (T4) corners of a diamond with sides of 100m [4].

2. Analysis method

Here, we briefly describe CANGAROO-III analysis procedures, recent observations and results. More details about analysis procedures can be found in [5].

First, every PMT images for each telescope are cleaned by removing isolated hit-pixels and reduced into image moments, which indicate the incident gamma-ray direction [6]. Then the direction is determined event by event as the point from which three images can be seen narrowest.

Second, we use Fisher discriminant (hereafter FD) [7] to Discriminate gamma-ray evens from hadronic events. We calculate a linear combination of image moments:

$$FD = \vec{\alpha} \cdot \vec{P}$$

where $\vec{P} = (W2, W3, W4, L2, L3, L4)$ are energy-corrected widths and lengths for the T2, T3 and T4 and α is a constant vector. FD should show particular one-dimensional distribution for given P distribution. P_{signal} and $P_{background}$ will be best separated when

$$\vec{\alpha} = \frac{\vec{\mu}_{sig} - \vec{\mu}_{BG}}{\vec{E}_{sig} - \vec{E}_{BG}}$$

Finally we calculate the number of gamma-ray like events by one-parameter fitting assuming $FD_{signalregion} = x\, FD_{gamma} + (1-x)\, FD_{background}$, where x is a fraction of gamma-ray signals.

3. Recent observations and results

Recent CANGAROO-III observations and results are summarized in table 1, which contains four detections and nine upper limits.

3.1. Extragalactic sources

3.1.1. *PKS2155-304.* [8] A nearby blazer PKS 2155-304, which was first detected in TeV gamma-rays by Durham group in 1999, is a time-variable TeV source. HESS group reported historically high TeV flux in 2006 July. Just after HESS alert we began follow-up observations and continued to August 2006. We detected gamma-ray excess. The 8 hour time difference between CANGAROO-III site and H.E.S.S. site provide better time coverage of flux monitoring.

3.1.2. *Abell 3667 and Abell 4038.* [9] Clusters of Galaxies are candidate sites of the origin of ultra high energy cosmic rays. Some theories predict TeV gamma-ray emission from Clusters of Galaxies by proton-photon and proton-proton interaction. We picked up two targets from the Abell Catalogue; Abell 3667 and Abell 4038. We obtained 2-sigma upper limits on their integral fluxes, which limit model parameters.

3.1.3. *Centaurus A.* [10] We observed a nearby Active Galactic Nuclei, Centaurus A for 10 hours in 2004. We found no gamma-ray excess above 420 GeV over one-degree radius from the pointing centre. This is consistent with the HESS result in 2004. Obtained flux upper-limit implies that Centaurus A is darker in TeV gamma-ray than expected from optical luminosity, assuming gamma-ray/optical ratio is same to our galaxy.

3.1.4. *SN1987A.* [11] There is a prediction of TeV gamma-ray from SN1987A Increasing with time after explosion. We observed this target by CANGAROO-II in 2001 and by CANGAROO-III in 2004 and 2006. We had no excess in 2001, 2004 and 2006, including some other candidate sources in this region of Large Magellanic Cloud. Thought the predicted flux is under our sensitivity now, next generation Cherenkov telescope array such as CTA is expected to detect gamma-ray signals from SN 1987A near future.

3.2. Galactic sources

3.2.1. *MSH15-52.* [12] TeV gamma-ray emission from Pulsar Wind Nebula MSH15-52 was detected by HESS group in 2004. We had follow-up observations by CANGAROO-III from April 2006 to June and took 48.6 hours data and 582 excess events. Morphology and flux were consistent with H.E.S.S.

3.2.2. *HESS J1804-216.* [13] This source was found by HESS galactic plane survey in 2004. We had follow up observation by CANGAROO-III from 2006 June to August. We had 62.5 hours data and 920 excess events. Morphology and flux ware consistent with H.E.S.S.

Table 1. Summary of recent results from CANGAROO-III.

Detection	Target	Category	Comment	Obs hours [hours]	
Detection	MSH15-52	PWN?	PSR 1509-58	48.6	[12]
Detection	PKS2155-304	Blazer	z=0.117	10.5	[8]
Detection	HESS J1804-216	UnID	HESS source	62.5	[13]
Detection	HESS J1303-631	UnID	HESS source	35	[14]
Upper limit	Abell 3667	Clusters of Galaxies	z=0.055	29.7(On) / 23.7(Off)	[9]
Upper limit	Abell 4038	Clusters of Galaxies	z=0.028	23.8(On) / 17.7(Off)	[9]
Upper limit	Centaurus A	AGN	z=0.0018	10	[10]
Upper limit	ω Centaurus	Globular Cluster		10	[10]
Upper limit	SN1987A	SNR	LMC	10.5 (2004) / 21.8 (2006)	[11]
Upper limit	SNR N157B	SNR	LMC	21.8	[11]
Upper limit	30 Dor C	SNR	LMC	21.8	[11]
Upper limit	LMC X-1	X-ray binary	LMC	21.8	[11]
Upper limit	PSR B0540-69	Pulsar	LMC	21.8	[11]

4. Conclusion

We summarized recent CANGAROO-III results, which were obtained by the CANGAROO-III standard analysis. There were four detections and nine upper limits, including confirmations of H.E.S.S. results.

References

[1] Kawachi, A., et al. 2001, Astropart. Phys., 14, 261

[2] Enomoto, R., et al. 2002a, Nature, 416, 823

[3] Kabuki, S., et al. 2003, Nucl. Instrum. Meth., A500, 318

[4] Enomoto, R., et al. 2002b, Astropart. Phys., 16, 235

[5] "A Search for sub-TeV Gamma-rays from the Vela Pulsar Region with CANGAROO-III", Enomoto, R. et al., Astrophys. J. 638, 397-408 (2006)

[6] Hillas, A. M. Proc. 19th Int. Cosmic Ray Conf. (La Jolla) 3, 445

[7] "The use of multiple measurements in taxonomic problems", Fisher, R. A., 1936, Annals of Eugenics, 7, 179

[8] "Detection of 2006 TeV-outburst of PKS2155-304 with the CANGAROO-III telescope", Sakamoto, Y. et al., Proc. 30th ICRC(Merida, Mexico), 2007

[9] "Observation of clusters of galaxies with the CANGAROO-III telescope system", Kiuchi, R. et al., Proc. 30th ICRC(Merida, Mexico), 2007

[10] "CANGAROO-III Search for Gamma Rays from Centaurus A and the ω Centauri Region", Kabuki, S. et al., Astrophys. J., 668, 968-973 (2007)

[11] "CANGAROO-III Search for Gamma Rays from SN 1987A and the Surrounding Field", Enomoto, R. et al, Astrophys. J., in press, Dec 10 issue 2007

[12] "Observations of extended VHE gamma-ray emission from MSH 15-52 with CANGAROO-III", Nakamori, T. et al., Proc. 30th ICRC(Merida, Mexico), 2007

[13] "Observation of VHE gamma rays from HESS J1804-216 with CANGAROO-III Telescopes", Higashi, Y. et al., Proc. 30th ICRC(Merida, Mexico), 2007

[14] "Observation of HESS J1303-631 with the CANGAROO-III telescopes", Kushida, J. et al., Proc. 30th ICRC(Merida, Mexico), 2007

Observations of the γ-ray binary LS I +61 303 with MAGIC

T. Jogler[1], V. Bosch-Ramon[2], J. Cortina[3], J. M. Paredes[2], M. A. Perez-Torres[4], M. Ribó[2], J. Rico[3], N. Sidro[3] and D. F. Torres[3,5] for the MAGIC collaboration

[1] Max-Planck-Institut für Physik, D-80805 München, Germany
[2] Universitat de Barcelona, E-08028 Barcelona, Spain
[3] IFAE, Edifici Cn., E-08193 Bellaterra (Barcelona), Spain
[4] Instituto de Astrofisica de Andalucia IAA-CSIC
[5] ICREA & Institut de Cienciès de l'Espai (IEEC-CSIC), E-08193 Bellaterra, Spain

E-mail: `jogler@mppmu.mpg.de`

Abstract. The binary system LS I +61 303 is one of four γ-ray binaries known. The system was studied intensively with the MAGIC telescope between September 2005 and December 2006. The total amount of 166 hours of data were taken during two observation campaigns of 6 and 4 month duration. We produced the most detailed light curve of the system in γ-rays with energy above 100 GeV. We found a clear variation of the flux level in different orbital phase bins and on timescales of about one day. The spectral energy distribution can be described well by a power law with spectral photon index of 2.6 ± 0.2. We found no significant change in the spectral index between different orbital phases or during the different observation campaigns.

1. Introduction

The binary system LS I +61 303 consists of a compact object (either a neutron star or a stellar-mass black hole) and a B0Ve star with a circumstellar disc. The compact object exhibits a highly eccentric orbit ($e = 0.73$) [1] with a period of 26.5 days where the periastron passage occurs at phase 0.23. The ephemerides are taken from Gregory [2].

LS I +61 303 was first detected by COS-B in γ-rays and later in radio [3] and X-rays. The MAGIC collaboration discovered the system as an emitter of variable Very High Energy (VHE,E > 50 GeV) γ-rays[4]. Recently the VERITAS collaboration confirmed this detection [5].

In the radio and optical waveband LS I +61 303 shows a periodic emission with a period of $P = 26.496$ d. Evidence for periodical emission was found at X-ray energies [6], with a period compatible with the optical and radio emission, but showing a maximum in a different phase. The source was detected by the EGRET detector on board of the CGRO satellite. The emission in the GeV energy γ-rays might be variable [7].

Despite many observations of LS I +61 303 the nature of the system is still unknown. It is classified as a High Mass X-ray Binary (HMXB) and was considered a microquasar for some time. Recent high resolution VLBA measurements show a complex and changing morphology that departs from a clear jet like radio emitting feature [8],[9]. This new measurements favor the explanation that LS I +61 303 hosts a neutron star with a pulsar wind nebula. Up to now, no final statement can be made concerning its classification. In this paper we report mainly on the followup observation performed with the MAGIC telescope in autumn 2006.

2. Observations and analysis

MAGIC is an Imaging Atmospheric Cherenkov Telescope with a 17m diameter reflector. The telescope is located at the Canary Island of La Palma at 2200m asl, from where LS I +61 303 can be observed at zenith angles larger than 32°. The observations took place in two observation campaigns. The data analysis of the first campaign (OC1) is described in [4] and due to lack of space we will describe only the analysis of the 2nd observational campaign (OC2) here. The OC1 took place between September 2005 and March 2006 and the OC2 was performed between September 2006 and December 2006. The OC2 resulted in a total amount of 112 hours of good quality data, taken on 54 days. The zenith angle ranges from $32° - 44°$, which yields an analysis energy threshold of about 200 GeV. The observation was performed in the so called wobble mode, in which the source is tracked off axis from the camera center and three symmetrical placed off-source positions are used for background estimation.
The data was analyzed using the Magic Analyzing and Reconstruction Software (MARS). The recorded shower images were characterized using the Hillas parameters [10] which were fed into a Random Forest [11]. The Random Forest estimates the energy for each shower image and calculates a parameter called hadronness which is linked to the probability that a shower image is of hadronic origin.
The selection of the γ-rays was done by demanding an 80% γ-ray efficiency for the cutvalue in hadronness and 70% γ-ray efficiency for the cutvalues in θ^2, where θ^2 is the squared angular distance between the source position and the reconstructed origin position of the shower image. The cutvalues were determined from simulated MC γ-rays. After selection of the γ-ray candidates a daily LC and differential energy spectra for the orbital phase bins 0.5–06 and 0.6–0.7 were produced. The differential energy spectra were unfolded taking into account the full detector dependent energy response [12].

3. Results

After averaging the fluxes of the different orbital cycles the LC of OC1 shows a rather broad peak with the maximum emission level between the phase 0.5–0.7 [4]. In the OC2 we saw a sharper peak of the maximum emission level in the phase bin 0.6–0.7 (see Fig.1). The sampling rate of the different orbital phases in one orbital cycle is much higher in the OC2. This leads to a more stable average in most phase bins because each average value is calculated from more individual measurements. Therefore the maximum flux level is located in most orbital cycles in phase bin 0.6–0.7. There is one exception in the last observed orbital cycle of OC2, where we see the highest emission in the phase bin 0.8–0.9, while in the other orbital cycles no significant signal was detected in this phase. It is quite interesting that in the same phase bin and the same orbital cycle the X-ray satellite Swift detected a very high flux level [13].
The produced differential energy spectrum from the OC1 is shown together with the averaged spectra of the phase bins 0.5–0.6 and 0.6–0.7 in Fig. 2. All spectra can be described with pure power laws and the obtained spectral photon indices are compatible with each other within their errors. The spectral index of the spectrum obtained from OC2 between phase 0.6–0.7 exhibits the highest significant detection and can be described by the power law $\frac{dF}{dE} = (2.6 \pm 0.3) \cdot 10^{-12} \frac{E}{1\,\mathrm{TeV}}^{-2.6\pm0.2} [\mathrm{TeV}^{-1}\,\mathrm{cm}^{-2}\,\mathrm{s}^{-1}]$. We found no evidence for significant changes in the spectral index between different phase bins. The flux level changes between the different orbital phase bins up to a factor of ~ 3 and we found changes within timescales of one day. However we found no significant evidence for variability on timescales of 30 to 75 minutes. All intranight LC from the different observing nights can be described by constant flux levels as determined from χ^2 fits of constants to the individual LC's.

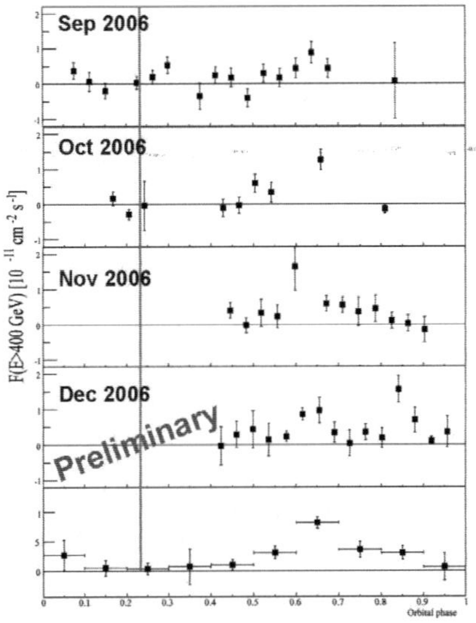

Figure 1. Shown are the LCs of the OC2 for the different orbital cycles and the averaged flux level per phase bin in the last row. The vertical line marks the periastron passage of the compact object.

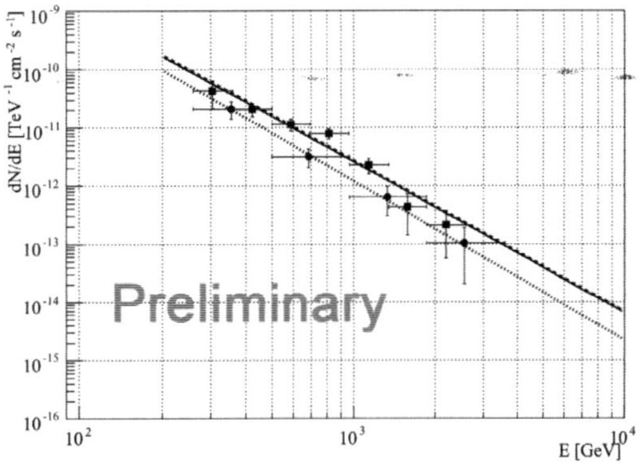

Figure 2. Shown are the differential energy spectra together with their fitted power laws. The spectrum with dashed line is the averaged spectrum from OC1. The dotted line is the OC2 averaged spectrum of phase 0.5–0.6 and the solid line represents the spectrum derived for phase 0.6–0.7 in OC2.

4. Conclusions

LS I +61 303 is a variable γ-ray binary with its highest flux in the phase bin 0.6–0.7. Its spectral energy distribution can be described by a pure power law and its spectral photon index does not change between the highest flux phase bins and is stable on timescales of one year. While we found variability in the flux level on timescales down to one day there is no evidence of intranight variability.

Acknowledgments

We thank the IAC for the excellent working conditions at the ORM. The support of the German BMBF and MPG, the Italian INFN, these Spanish CICYT, the ETH Research Grant TH 34/04 3 and the Polish MNiI Grant 1P03D01028 is gratefully acknowledged.

References

[1] Casares J, Ribas I, Paredes J M, Martí J and Allende Prieto C 2005 *MNRAS* **360** 1105–1109
[2] Gregory P C 2002 *ApJ* **575** 427–434
[3] Gregory P C and Taylor A R 1978 *Nature* **272** 704–706
[4] Albert J *et al.* 2006 *Science* **312** 1771–1773
[5] Maier G 2007 *astro-ph/07093661*
[6] Paredes J M, Marti J, Peracaula M and Ribo M 1997 *A&A* **320** L25–L28
[7] Tavani M, Kniffen D, Mattox J R, Paredes J M and Foster R 1998 *ApJL* **497** L89+
[8] Dhawan V, Mioduszewski A and Rupen M 2006 *VI Microquasar Workshop: Microquasars and Beyond*
[9] Albert J *et al.* 2007 *in preperation*
[10] Hillas A M 1985 *ICRC* (*ICRC* vol 3) ed Jones F C pp 445–448
[11] Albert J *et al.* 2007 *astro-ph/07093719*
[12] Albert J *et al.* 2007 *astro-ph/07072453*
[13] Esposito P, Caraveo P A, Pellizzoni A, de Luca A, Gehrels N and Marelli M A 2007 *A&A* **474** 575–578

Diffusive gamma-rays and antiprotons

Toru Shibata

Department of Physics and Mathmatics, Aoyama-Gakuin University, Kanagawa 229-8558, Japan

E-mail: `shibata@phys.aoyama.ac.jp`

Abstract.

Based on our three-dimensional cosmic-ray (CR) propagation model in the Galaxy, we study the diffusive gamma-rays (Dγ's) and antiprotons ($\bar{p}$'s), and compare our numerical results with recent experimental data, particularly those obtained by EGRET and BESS, each for Dγ data and $\bar{p}$ data respectively. Galactic parameters such as the gas density and diffusion constant are determined from the most recent data for the CR energy spectra, particularly those for protons and helium nuclei from AMS and BESS with excellent accuracy over the 1–100 GeV per nucleon range, and from JACEE and RUNJOB covering the very high energy region up to 500 TeV per particle. Using these parameters, we find that the EGRET data are well reproduced in the energy region $E_\gamma \lesssim 2$ GeV, in both the spectral shape and the absolute value, and also that the BESS data are compatible with our propagation model in the energy region $\gtrsim 1$ GeV. However, there are still two open questions in these components, 1) the intensity of the Dγ's gives a significant enhancement, by approximately the factor 2–3, at $\gtrsim 2$ GeV as compared to those expected from the CR spectrum, and 2) we could not confirm whether or not BESS data with the energies less than 1 GeV are consistent or not with those expected from other CR components.

1. Introduction

We have studied the CR propagation for stable and the unstable elements (typically ^{10}Be) based on our three-dimensional propagation model [1-3], and it reproduces satisfactorily all CR observables nowadays available, assuming appropriate Galactic parameters. So it is quite interesting and important to see how such model with these parameters are in harmony with Dγ's and $\bar{p}$ data. This is because both components have cosmological interests as well as the astronomical ones, for instance the pair annihilations of massive particles in dark matter, so called WIMPs, (for instance, see Edsjö & Gondolo [4]; Bergström et al. [5]; Boer [6]), might produce γ's and antiparticles such as $\bar{p}$ and/or e^+, resulting in the enhancement feature in the energy spectra. In fact Egret data [7] present a clear enhancement of Dγ's intensity around 2–10 GeV, much higher than expected from CR data[1].

Therefore this problem is quite challenging to both astrophysics and particle physics. Because of the limited space, we don't touch the details of our propagation model, but present the essence of the model shortly in the next section, particularly the differences from those currently used.

[1] Most recently Stecker, Hunter and Kniffen [8] reports that the most likely explanation of the EGRET "GeV anomaly" is an error in the estimation of the EGRET sensitivity at the energies above ~ 1 GeV.

2. CR propagation model

Three Galactic parameters, the diffusion coefficient, D, the gas density of the ISM, n, and the source density of CRs, Q, are the most essential for the study of the CR propagation in the Galaxy, and we assume they have following forms depending on space position $\boldsymbol{r}(r,z)$:

$$D(\boldsymbol{r}) \equiv D(r,z) = D_0 \exp[+(r/r_D + |z|/z_D)], \tag{1a}$$

$$n(\boldsymbol{r}) \equiv n(r,z) = n_0 \exp[-(r/r_n + |z|/z_n)], \tag{1b}$$

$$Q(\boldsymbol{r}) \equiv Q(r,z) = Q_0 \exp[-(r/r_Q + |z|/z_Q)], \tag{1c}$$

where D_0, n_0 and Q_0 correspond to the diffusion coefficient, the gas density and the source density of CR at the Galactic center (GC), $\boldsymbol{r}(0,0)$, respectively. In the present work, we often deal with the values of these three parameters in the Galactic plane, which are simply expressed as $D_r = D(r,0)$, $n_r = n(r,0)$, and $Q_r = Q(r,0)$, and for those at the SS, we express as $D_\odot = D(r_\odot,0)$, $n_\odot = n(r_\odot,0)$, and $Q_\odot = Q(r_\odot,0)$, with $r_\odot = 8.5$ kpc.

In order to study the spatial gradient of these parameters, we introduce two *average* scale heights, $\bar{r}$ and $\bar{z}$, appearing not in the separate forms of (r_D, r_n) and (z_D, z_n), but always in the coupled forms defined by

$$\frac{1}{\bar{r}} = \frac{1}{2}\left(\frac{1}{r_D} + \frac{1}{r_n}\right); \quad \frac{1}{\bar{z}} = \frac{1}{2}\left(\frac{1}{z_D} + \frac{1}{z_n}\right). \tag{2}$$

We further assume the source CR density Q and the diffusion coefficient D have the following rigidity-dependences,

$$Q(\boldsymbol{r};R) = Q(\boldsymbol{r})R^{-\gamma}; \quad D(\boldsymbol{r};R) = D(\boldsymbol{r})vR^{\alpha}, \tag{3}$$

where R is the rigidity of the CR particle in GV, and v its velocity in units of the velocity of light c.

We further take into account the stochastic reacceleration by hydromagnetic turbulence, introducing a parameter ζ_0, which corresponds to the the *effective* cross-section for the occurrence of collision of CR's with the magnetic turbulence. The magnitude of ζ_0 is as large as 50–60 millibarn (mb), indicating that the *effective* latitudinal scale height, z_a, of the reacceleration occurrence is as large as (2-3)$\times z_n \approx$ 500-800 pc [2,3].

3. Results and discussions

First we present the CR spectra for protons and helium nuclei [9] in figure 1(a), which are the most effective ones for the study of Dγ and $\bar{p}$ components, and the secondary-to-primary ratio in figure 1(b). Here and in what follows, we assume $\beta = \gamma + \alpha = 2.75$, $\alpha = 1/3$ (Kormogorov-type spectrum of hydromagnetic turbulence), $\zeta_0 = 50$ mb, $z_n/z_D = 0.1$, and $\bar{\sigma}_\odot \simeq D_\odot/[n_\odot c z_n z_D] = 180$ mb with $n_\odot = 0.832$ H atoms cm^{-3}. See references [1–4] for the parameter determination. One finds our curves reproduce very well the experimental data. We further confirmed that four kinds of [unstable-nucleus]/[stable-nucleus], ^{10}Be/^{9}Be, ^{26}Al/^{27}Al, ^{54}Mn/Mn, and ^{36}Cl/Cl in the energy range, 0.05-0.5 GeV/n, are in good agreement with our parameterization [3], while we don't present the result here because of limitted space.

Now based on these results obtained by the CR data nowdays available, we extend the numerical results to the $\bar{p}$'s and Dγ's data which are shown in figures 2(a) and 2(b) respectively. In order to perform numerical calculations for these components, we need the production cross-sections of $\bar{p}$ and Dγ in nuclear interactions, particularly in proton-proton interaction. See references [10] and [11] respectively for the detail of them.

It is remarkable that our curves for $\bar{p}$ reproduce nicely in the energy region, $\gtrsim 1$ GeV, and those for Dγ in the region, $\lesssim 2$ GeV, where the numerical curves for EB (electron bremsstrahlung) and IC (inverse Compton) appearing in figure 2(b) are based on the simulational results by Hunter et

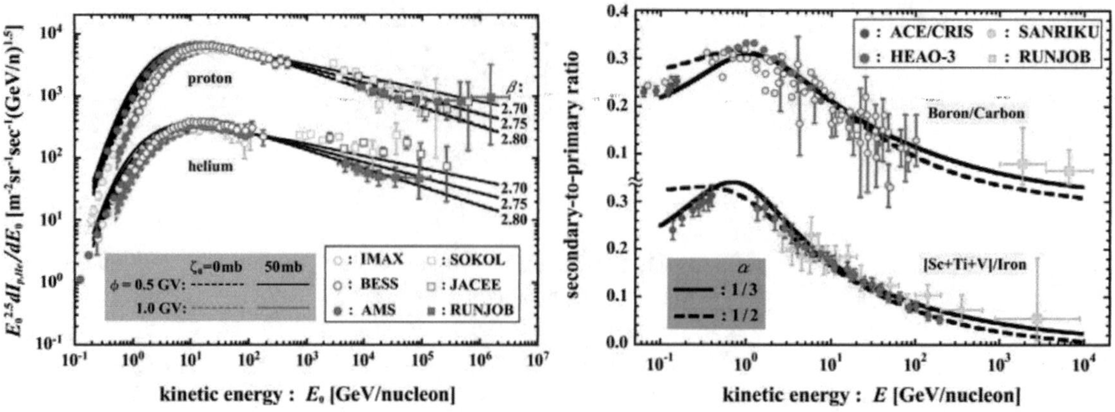

Figure 1. CR intensity for individual elements summarized by RUNJOB group (a) and the secondary-to-primary ratio (b). See reference [9] and the references therein for the data.

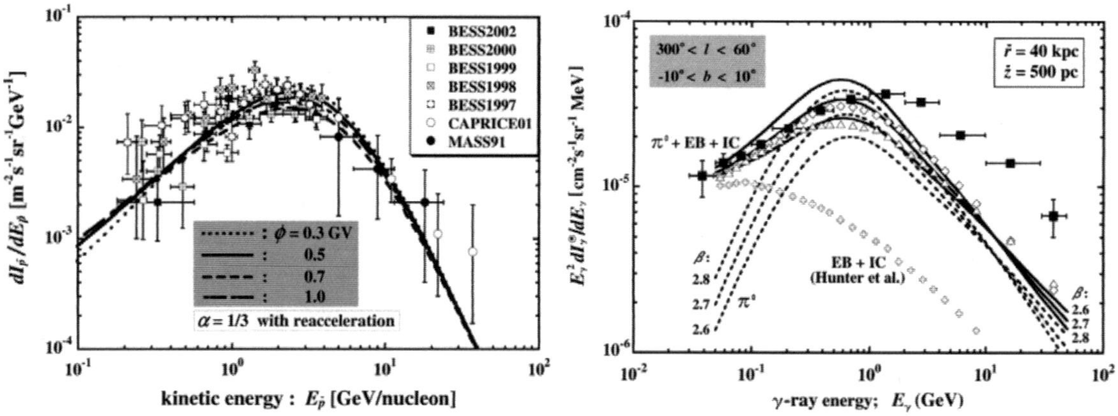

Figure 2. (a) $\bar{p}$-flux obtained by BESS [15], MASS [16], and CAPRICE [17], and (b) Dγ-flux obtained by EGRET [7].

al. [7]. In figure 2(a), we present several curves corresponding to different modulation parameter ϕ, based on the force field approximation [12].

We checked further two *spatial* distributions of Dγ's, longitudinal and latitudinal ones, in the energy interval, $E_\gamma = 1$–$2\,$GeV, as shown in figures 3(a) and 3(b) respectively. We find that EGRET data are again in good agreement with those expected from our model with $(\bar{r}, \bar{z}) = (40\text{kpc}, 0.6\text{kpc})$ (see equation [2]), both in shape and absolute intensity, physical meaning of which is discussed in reference [13].

On the other hand, it is somewhat difficult in figure 1(a) to conclude if the numerical results in the low energy region, $\lesssim 1\,$GeV, are compatible or not with the BESS data. We need more statistics to discuss a possibility of the novel origin such as the evaporation of primordial black holes and/or the annihilation of supersymmetric dark matter particle. However we may say safely that our model with the reacceleration effect matchs well the data, contrary to the previous understanding (for instance Ptuskin [14]) that the reacceleration process doesn't favour the $\bar{p}$ spectrum.

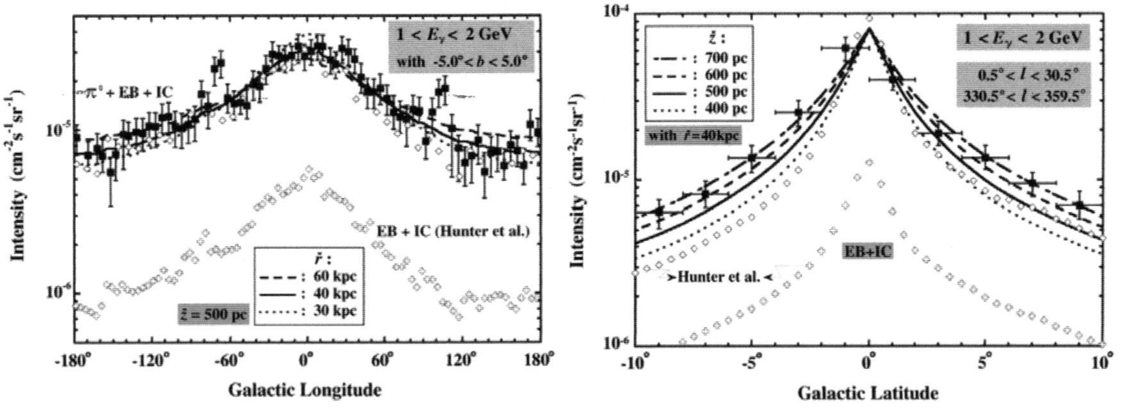

Figure 3. Spatial distributions of Dγ's for (a) the longitudinal and (b) the latitudinal directions in the energy interval $E_\gamma = 1$–2 GeV near the Galactic plane.

About the Dγ's as presented in figure 2(b), we find again a significant discrepancy between the EGRET data for $\gtrsim 2$ GeV and the curves expected from CRs intensities, which has been pointed out by many authors. However, this "GeV anomaly" is recently reexamined by EGRET team [8] as mentioned in the footnote in the introduction, and it might be due to the systematic error in the sensitivity determination of the EGRET detector. Anyway, we need better statistics to answer the above questions, and await for the future experiment mentioned above, particularly BESS-PolarII [19], AMS [20], and PAMELA [21] for $\bar{p}$, and GLAST [18] for Dγ data. With reliable Galactic parameters as well as progress in the quality of the experimental data, we might detect a signal of novel sources for $\bar{p}$'s and Dγ's other than the secondary products in the near future.

4. References

[1] Shibata, T., Hareyama, M., Nakazawa, M., & Saito, C. 2004, ApJ, 612, 238
[2] Shibata, T., Hareyama, M., Nakazawa, M., & Saito, C. 2006, ApJ, 642, 882
[3] Shibata, T., & Ito, T. 2007, ApJ, 655, 892
[4] Edsjö, J., & Gondolo, P. 1997, Phys. Rev. D, 56, 1879
[5] Bergström, L., Ullio, P., & Buckley, J. H. 1998, Astropart. Phys., 9, 137
[6] Boer, W. 2005, CERN Courier, 45, 17
[7] Hunter, S. D., et al., 1997, ApJ, 481, 205
[8] Stecker, F. W., Hunter, S. D., & Kniffen, D. A., arXiv:0705.4311v1 [astro-ph] 29 May 2007
[9] Derbina, V. A., et al. 2005, ApJ, 628, L41
[10] Shibata, T. & Futo, Y., 2008, Astropart. Phys., in press
[11] Suzuki, R., Watanabe, J., & Shibata, T. 2005, Astropart. Phys., 23, 510
[12] Gleeson, L. J., & Axford, W. I. 1968, ApJ, 154, 1011
[13] Shibata, T., Honda, N., & Watanabe, J. 2007, Astrop. Phys., 27, 411
[14] Ptuskin, V. 2005, Proc. 29th Int. Cosmic Ray Conf. (Pune), 10, 317
[15] Haino, S., et al. 2004, Phys. Lett. B, 545, 1135
[16] Hof, M., et al. 1996, ApJ, 467, L33
[17] Bergström, D., et al. 2000, ApJ, 534, L177
[18] Kamae 2004, Proc. 2nd Int. Symp. on H. E. Gamma-Ray Astronomy, edited by Aharonian, F. A., Völk, H. J., & Horn, D., (Heidelberg), 210
[19] Yamamoto, A., Invited talk at the workshop on "Cosmic-rays and High Energy Universe" held at Aoyama-Gakuin University, March, 2007 (Proceedings will be published soon)
[20] Battiston, R., 2005, Proc. 29th Int. Cosmic Ray Conf. (Pune), 10, 151
[21] Boezio, M., et al., 2001, ApJ, 561, 787

ARGO-YBJ experiment in Tibet

P. Bernardini, on behalf of the ARGO-YBJ Collaboration

Dipartimento di Fisica, Università del Salento, and INFN, Sezione di Lecce, 73100 Lecce, Italy

E-mail: paolo.bernardini@le.infn.it

Abstract. The ARGO-YBJ detector is an extensive air shower array consisting of a carpet of Resistive Plate Chambers (RPCs). The main scientific goal of this Chinese-Italian experiment is the observation of very high energy (VHE) gamma-rays from astrophysical sources. High altitude and full coverage ensure a low energy threshold (few hundreds of GeV for gamma-initiated showers). High duty-cycle and large field of view allow a continuous sky survey. The data collected with the central carpet during the detector debugging have been analysed. First results about detector performance, gamma astronomy and cosmic ray physics are presented.

1. Introduction

The ARGO-YBJ (Astrophysical Radiation Ground-based Observatory at YangBaJing) experiment is located in Tibet at an altitude of 4300 m [1]. The detector consists of a single layer of RPCs operated in streamer mode, on a total area of about 110×100 m^2 (Fig. 1). The central part (78×74 m^2) is fully active and surrounded by a sampling ring with other 1000 m^2 (20% of the outer ring) equipped with RPCs. The experiment is in data-taking with the central carpet since June 2006. The sampling ring has been completely mounted and will be put in data acquisition quite soon. In the next months the RPCs will be covered with a layer of lead as photon-converter, in order to enhance the experimental sensitivity.

The detector is logically divided in 154 clusters, each made by 12 RPCs, with a dedicated Local Station for the DAQ. The digital read-out of the RPCs is performed by means of inductive strips (6×62 cm^2) well suited to detect small air showers. The fast-OR of 8 strips is called "pad" and defines the space-time pixel of the detector, with a time resolution of ~ 1 ns. In order to extend the measurable energy range, each RPC is equipped also with two large electrodes, called "big pads", which provide an analog signal proportional to the collected charge.

ARGO-YBJ collects data in shower and scaler mode. The first one works when the number of pads fired in a time-window of 420 ns exceeds the multiplicity required by the trigger condition. The event (arrival direction, core position and so on) is fully reconstructed looking at the space-time pattern (Fig. 2). The scaler mode does not require any trigger, it records the rate for four multiplicities (≥ 1, ≥ 2, ≥ 3 and ≥ 4) on each cluster in a time window of 0.5 s. The scaler mode allows the detection of low energy transient phenomena (e.g. GRBs and solar flares) observed as non-statistical fluctuations of the background [2].

The high altitude location, the full coverage with high space-time granularity, and the RPC time resolution allow an excellent angular resolution and reduce the energy threshold in shower mode to few hundreds of GeV. The continuous sky survey in the declination band from $-10°$ to $+70°$ is possible thanks to the duty-cycle close to 100% and the field of view greater than 2 sr.

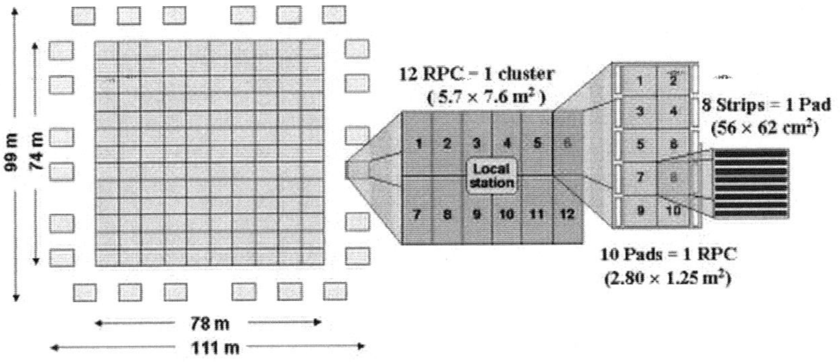

Figure 1. Detector setup. The present results are based on data collected with the 130 clusters (ARGO-130) of the central carpet.

2. VHE gamma-astronomy

The data collected with ARGO-130 have been analysed with the chess-board method after the timing-calibration of the pads [3]. The results confirm the simulations and the detector capability as a gamma-telescope with an angular resolution lower than 0.5° for internal events with more than 200 fired pads [4]. The Moon shadow in the cosmic ray flux has been observed (Fig. 3) with a statistical significance $n_\sigma \sim 10$ in 560 hours (T) with the Moon zenith angle (θ) lower than 45° [5]. The angular spread of the Moon shadow results compatible with the estimated angular resolution. Also the Sun shadow has been observed ($\sim 6\ \sigma$ in 208 hours of observation). As expected the ratio $n_\sigma/\sqrt{T}$ is the same for both the shadow sources.

Gamma Sources - The Crab Nebula has been observed with an evidence of $\sim 5\ \sigma$ in 50 days, requiring $\theta < 40°$ and a number of fired pads (N_{pads}) higher than 200 [1]. On the basis of this result the detector sensitivity is of the order of 0.3 Crab/year. The search has been performed also for Markarian 421 in the period July-August 2006, in coincidence with a flare in the X-band. In 80 hours the signal was detected with $n_\sigma = 5.5$ (Fig. 4). The analysis was performed requiring $N_{pads} \geq 60$, corresponding to a median gamma energy of 500 GeV.

Gamma Ray Bursts - The GRB spectra, well studied by satellite detectors in the keV-MeV range, are still quite unknown at higher energies. Only two observations indicate some emission in the GeV range [6]. By means of the scaler mode the ARGO-YBJ detector covers the range from 1 to 100 GeV. A first search for signals in coincidence with GRB satellite observations has been performed [7]. No significant emission has been detected and fluence upper limits ($\sim 10^{-3} \div 10^{-5}$ erg/cm^2) were set for GRBs in the ARGO-YBJ field of view.

3. Cosmic ray studies

The features of the ARGO-YBJ detector offer a unique chance to study deeply the hadronic component of cosmic rays in the energy range (1-1000 TeV) connecting direct and EAS measurements. Many analyses are in progress [8, 9]. The high space-time granularity allows unprecedented studies on EAS phenomenology (different topologies and time structures) [10].

The decrease of the shower frequency with the zenith angle, when primary energy and shower age are determined, gives a measurement of the flux attenuation at that energy. The flux attenuation is controlled by the absorption length, related to the interaction length. Therefore it is possible to estimate the proton-air cross section. Two sub-samples of events with mean energy of 3.9 and 12.7 TeV have been selected. The results of this analysis are shown in Fig. 5 [11]. Extension of this measurement to unexplored regions is possible, selecting higher energies by means of the charge analog read-out.

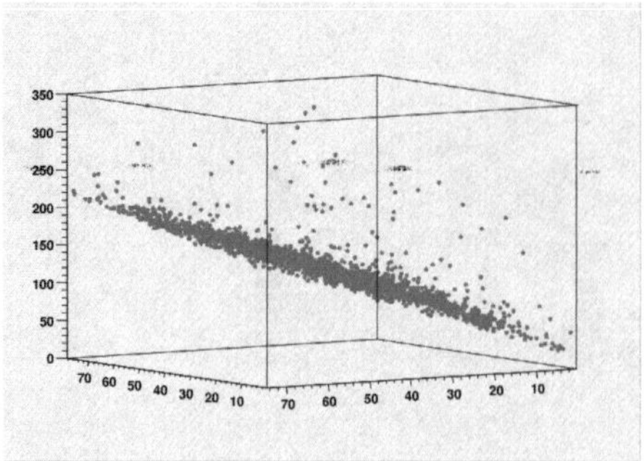

Figure 2. Space-time view of an extensive air shower.

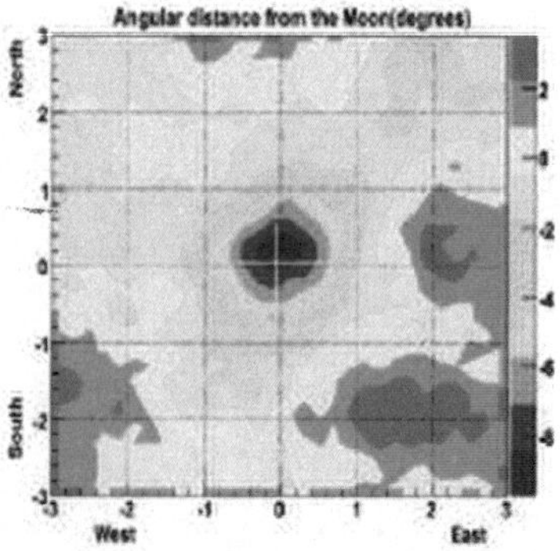

Figure 3. Moon shadow in the cosmic ray flux.

4. Conclusions

The ARGO-YBJ detector is almost completed (130/154 clusters are currently taking data). The data collected during the detector setting-up/debugging phase have been analysed. The detector performance is in agreement with the design requirements (angular resolution, energy threshold, sensitivity). Some preliminary results have been presented: Moon and Sun shadows, detection of the Crab Nebula, Markarian 421 flare observed in summer 2006, upper limits on the GRB fluence, proton-air cross section. Future implementations: guard ring in DAQ, Pb γ-converter, γ-hadrons discrimination [12].

[1] Martello D 2007 *Proc. of the 30th Intern. Cosmic Ray Confer. (Merida)* 1029
[2] Cappa A et al 2007 *Proc. of the 30th Intern. Cosmic Ray Confer. (Merida)* 0857
[3] He HH et al, 2007 *Astroparticle Physics* vol 27 p 528 (also *physics*/0701291)
[4] Di Sciascio G et al 2007 *Proc. of the 30th Intern. Cosmic Ray Confer. (Merida)* 0573
[5] Wang B et al 2007 *Proc. of the 30th Intern. Cosmic Ray Confer. (Merida)* 0458
[6] Hurley K et al 1994 *Nature* vol 371 p 652; Atkins R et al 2000 *Ap. J. Lett.* vol 533 p L119
[7] Di Girolamo T et al 2007 *Proc. of the 30th Intern. Cosmic Ray Confer. (Merida)* 1034
[8] Chen SZ et al 2007 *Proc. of the 30th Intern. Cosmic Ray Confer. (Merida)* 0826
[9] Cheng N et al 2007 *Proc. of the 30th Intern. Cosmic Ray Confer. (Merida)* 0904
[10] Calabrese Melcarne AK et al 2007 *Proc. of the 30th Intern. Cosmic Ray Confer. (Merida)* 1100
[11] De Mitri I et al 2007 *Proc. of the 30th Intern. Cosmic Ray Confer. (Merida)* 0950
[12] Dattoli M 2007 *Proc. of the 30th Intern. Cosmic Ray Confer. (Merida)* 1033

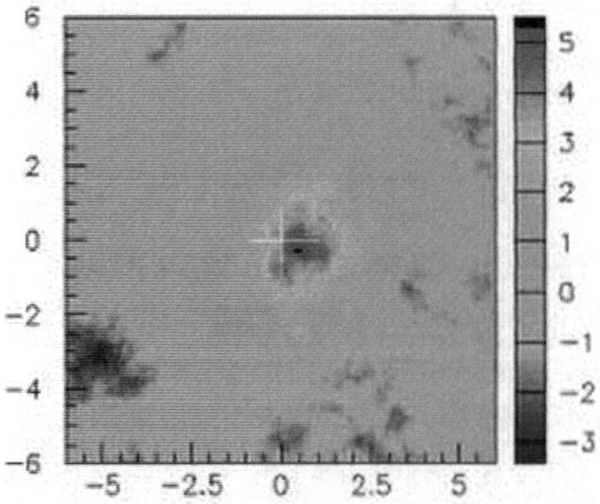

Figure 4. Markarian 421 signal during the summer 2006 flare.

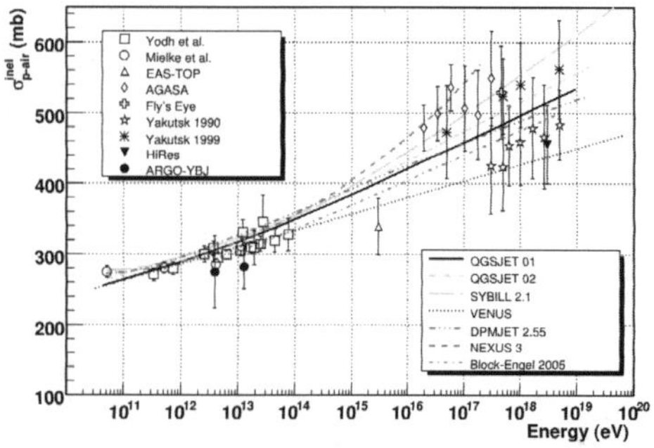

Figure 5. Proton-air inelastic cross section, the ARGO-YBJ measurements are the black dots.

The cosmic ray primary spectrum in the transition region between direct and indirect measurements (10 TeV - 10 PeV)

M. Bertaina[1], **G. Battistoni**[2], **S. Muraro**[3], **G. Navarra**[1], **A. Stamerra**[4]

[1] Dipartimento di Fisica Generale dell' Università di Torino and INFN, Via P. Giuria 1, 10125 Torino, Italy

[2] INFN Milano and Laboratori Nazionali di Frascati dell'INFN, Via Celoria 2, 20133 Milano, Italy

[3] Dipartimento di Fisica dell' Università and INFN Milano, Via Celoria 2, 20133 Milano, Italy

[4] Dipartimento di Fisica dell' Università di Pisa and INFN, Via Buonarroti 2, 56010 Pisa, Italy

E-mail: bertaina@to.infn.it

Abstract. The cosmic ray spectrum between 10 TeV and 10 PeV is a key region to understand the origin and the acceleration processes of the galactic radiation. Around 100 TeV measurements performed by balloons and air shower arrays partially overlap. The experimental results recently obtained by both techniques are summarized. The picture that comes out from a common interpretation of the data is a steeper proton spectrum compared to the helium and CNO groups with a dominance of the helium component in the knee region. The agreeement on the proton spectrum by balloons and air shower arrays is remarkable.

1. Introduction

The knowledge of the energy spectrum of the different elements of the primary cosmic rays is a main tool for astrophysical purposes and for the calculation of the atmospheric neutrino flux. In the TeV region, measurements are performed by means of balloons and satellite experiments, however, at higher energies the statistics becomes poor and the energy determination, being non calorimetric, depends on the interaction parameters and their fluctuations. At energies higher than 1 PeV only Extensive Air Showers (EAS) arrays operate. By means of multi-component and multi-parameter studies, the energy of the primary particle can be inferred in a quite reliable way, however, the composition study can be retrieved only through the comparison with complicate simulation studies. A lot of effort has been put in the past decades to increase the experimental data in the 10 TeV - 10 PeV region, on one side by means of more refined balloon experiments with longer duration flights (i.e. ATIC, JACEE, RUNJOB, CREAM), and, on the other side, multi-component arrays (i.e. GRAPES, TIBET, KASCADE, EAS-TOP) have been installed with detectors (Cherenkov, hadrons, emulsions) sensitive to specific components of the primary spectrum in the TeV region to overlap the energy region covered by direct experiments. In parallel, more refined versions of the simulation codes have been developed in the past years [1] benchmarked by the most recent experimental data obtained at RHIC and by the results of cosmic ray experiments. Despite the intrinsic uncertainties of each techinque, it is possible to extract a global picture in the 10 TeV - 10 PeV region.

2. Techniques and results

Tab. 1 shows a summary of the key characteristics and techniques exploited by the experiments whose results are presented in Fig. 1. The interpretation of the data should take into account the intrinsic uncertainties of the different techniques. The typical exposure so far reached by balloon experiments is $A \approx 1000$ m$^2 sr$. The thickness of the employed calorimeters is $X \approx 1$ λ_{int}. Despite the clear charge separation (typical charge resolution is $\Delta Z = 0.2$ charge unit) there are differences among the fluxes measured by the experiments on specific components (the most striking one is the He flux detected by JACEE and RUNJOB), which, however, are less significant if the quoted uncertainties are properly taken into account.

The main advantage of the present ground based detectors is the possibility of accumulating much higher statistics and to perform multi-component measurements. Despite the fact that the interpretation of their results relies on simulations which are based on extrapolations of data obtained at lower energies, the all particle spectrum obtained by several experiments ([13] - [18]) employing different methods (N_e or Cherenkov) and at various altitudes is in good agreement ($\approx$ 10 % energy uncertainty). The most challenging task is the extraction of the abundances of the single components from the all particle spectrum. New refined techniques are being employed in the new generation of detectors which are sensitive to specific components of the cosmic ray radiation to overcome such difficulties (see Tab. 1). As an example, the EAS-TOP and MACRO technique [12], with an acceptance 20 times higher than typical balloon experiments, was able to select an almost pure light beam (p+He at 80 TeV), eliminating the intrinsic difficulties in the deconvolution of the primary spectrum based on simulations.

In parallel, simulation codes are considerably improving. As an example, the reliability of the FLUKA code [19], which is probably the most benchmarked code at low energies, has been recently checked by comparing the simulated μ^+ and μ^- spectra at different energies, atmospheric depth and geomagnetic rigidities with the experimental data of the BESS and L3+C experiments, showing a general agreement inside 10-15% uncertainty [19]. In conclusion, balloons and ground based experiments are complementary techniques, and the most recent simulation codes such as FLUKA offer a more reliable base for the interpretation of the experimental results.

Fig. 1 shows a compilation of results for the proton, helium and CNO groups. The agreement of all techniques on the proton flux in the entire energy range is remarkable and gives some confidence on the reliability of both direct and indirect measurements. A fit to the data provides a flux $\frac{dN}{dE} = (8.8 \pm 0.5) \cdot 10^{-2} \cdot E^{-2.75 \pm 0.01}\ m^{-2} s^{-1} sr^{-1} TeV^{-1}$. The discrepancy between JACEE and RUNJOB on the helium spectrum seems to be solved in favour of JACEE. The new data of ATIC-2 and CREAM seem to be very well in agreement with the indirect measurements obtained by L3+C experiment interpreted through the FLUKA simulation, EAS-TOP/MACRO, GRAPES and KASCADE. A fit to these data provides a flux $\frac{dN}{dE} = (6.2 \pm 0.5) \cdot 10^{-2} \cdot E^{-2.61 \pm 0.01}$ $m^{-2} s^{-1} sr^{-1} TeV^{-1}$. It is also worthwhile observing that the uncertainties on the flux of the single components obtained by the N_e-N_μ technique at the knee are of the same level of those observed on the data of the balloon experiments. The spectrum of the CNO group shows more uncertainties partly due to the slightly different definition of CNO between direct and indirect measurements. Moreover, the limitations that ground based detectors have in dividing the all particle spectrum into mass groups turns into a natural tendence of overestimating the flux of the single components. However, also the CNO group seems to have a flatter spectrum compared to the proton one: $\frac{dN}{dE} = (2.3 \pm 0.3) \cdot 10^{-2} \cdot E^{-2.61 \pm 0.01}\ m^{-2} s^{-1} sr^{-1} TeV^{-1}$.

What is interesting to observe from Fig. 1 is that the extrapolation of the spectra of the single components into the knee region indicates that around 3-5 $\times$ 10^{15} eV the flux is dominated by the helium component which seems to be, therefore, the main responsible of the knee in the primary spectrum, in accordance with several experimental results at the knee [15, 17].

Table 1. Experiments whose data are used in the present analysis (ref. [2] - [16]).

Type	Technique	Energy Range	Experiment and Sensitive Components
dir.	Spectrometer	1-200 GeV	AMS(p-He), BESS(p,He), HEAO(CNO-Fe)
dir.	Calorimeter	30GeV - 500 TeV	ATIC(all), CREAM(all)
dir.	Emulsion chambers	10-500 TeV	JACEE, RUNJOB (all)
ind.	Hadron calorimeter	500 GeV -1 PeV	KASCADE, EAS-TOP (p)
ind.	Muon spectrometer	100 GeV - 10 TeV	L3+C(mostly p and He)
ind.	Cherenkov + TeV μ	50-300 TeV	EAS-TOP/MACRO(p,He,CNO)
ind.	N_e-N_μ	100 TeV - 10 PeV	GRAPES, KASCADE, EAS-TOP (all)
ind.	Emulsion chambers	5-300 TeV	Tibet ASγ (p,He)

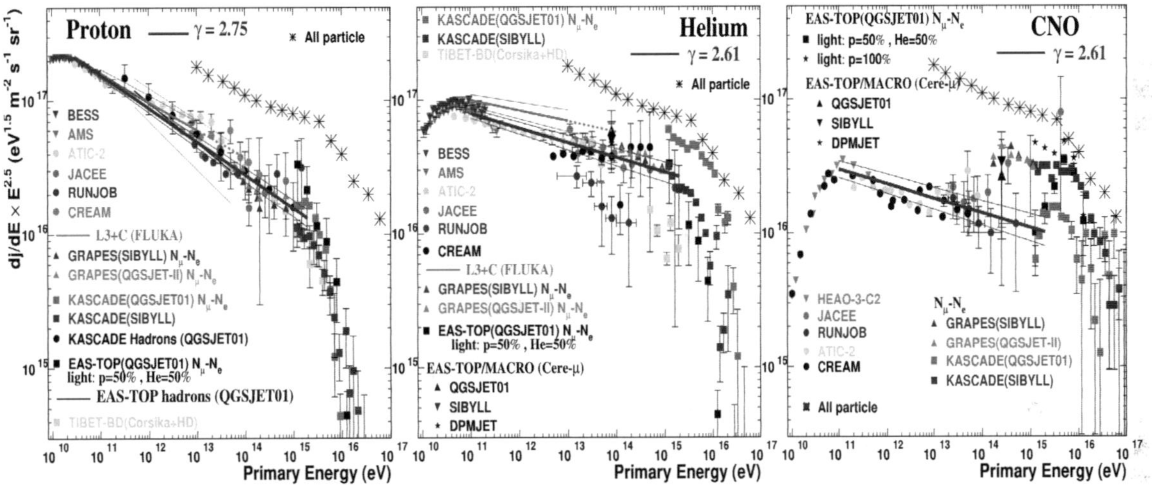

Figure 1. The spectra of H, He and CNO group; the blue line indicates a fit to the data.

References

[1] Ostapchenko S. et al. 2006 *astro-ph/0610788v1*
[2] Alcarez J. et al. 2000 *Phys. Lett.* B **472** 215
[3] Sanuki T. et al. 2000 *Astrophys. J.* **545** 1135
[4] Binns W. et al. 1981 *Astrophys. J.* **247** L115
[5] Wefel J.P et al. 2005 *Proc. 29th ICRC* **3** 105
[6] Asakimori K. et al. 1998 *Astrophys. J.* **502** 278
[7] Derbina V.A. et al. 2005 *Astrophys. J.* **628** L41
[8] Yoon Y.S. et al. 2007 *Proc. 30th ICRC* **OG 1.1**
[9] Antoni T. et al. 2004 *Astrophys. J.* **602** 914
[10] Aglietta M. et al. 2003 *Astrop. Phys.* **19** 329
[11] L3 Collaboration 2004 *hep-ex/0408114*
[12] Aglietta M. et al. 2004 *Astrop. Phys.* **21** 223
[13] Hayashi Y. et al. 2005 *Proc. 29th ICRC* **10** 243
[14] Amenomori M. et al. 2006 *Phys. Lett.* B **632** 58
[15] Aglietta M. et al. 2004 *Astrop. Phys.* **583** 21
[16] Antoni T. et al. 2005 *Astrop. Phys.* **24** 1
[17] Chernov D. et al. 2006 *Int. J. Mod. Phys.* A **20** 6799
[18] Arqueros F. et al. 2000 *Astron. Astroph.* **359** 682
[19] A. Ferrari et al. 2005 *CERN-2005-010*; Muraro S. 2006 *PhD Thesis, University of Milano*

Tibet Air Shower Array: Results and Future Plan

The Tibet ASγ Collaboration
M. Amenomori[1], X. J. Bi[2], D. Chen[3], S. W. Cui[4], Danzengluobu[5], L. K. Ding[2], X. H. Ding[5], C. Fan[6], C. F. Feng[6], Zhaoyang Feng[2], Z. Y. Feng[7], X. Y. Gao[8], Q. X. Geng[8], H. W. Guo[5], H. H. He[2], M. He[6], K. Hibino[9], N. Hotta[10], Haibing Hu[5], H. B. Hu[2], J. Huang[11], Q. Huang[7], H. Y. Jia[7], F. Kajino[12], K. Kasahara[13], Y. Katayose[3], C. Kato[14], K. Kawata[11,*], Labaciren[5], G. M. Le[15], A. F. Li[6], J. Y. Li[6], Y.-Q. Lou[16], H. Lu[2], S. L. Lu[2], X. R. Meng[5], K. Mizutani[13,17], J. Mu[8], K. Munakata[14], A. Nagai[18], H. Nanjo[1], M. Nishizawa[19], M. Ohnishi[11], I. Ohta[20], H. Onuma[17], T. Ouchi[9], S. Ozawa[11], J. R. Ren[2], T. Saito[21], T. Y. Saito[22], M. Sakata[12], T. K. Sako[11], M. Shibata[3], A. Shiomi[9,11], T. Shirai[9], H. Sugimoto[23], M. Takita[11], Y. H. Tan[2], N. Tateyama[9], S. Torii[13], H. Tsuchiya[24], S. Udo[11], B. Wang[8], H. Wang[2], X. Wang[11], Y. Wang[2], Y. G. Wang[6], H. R. Wu[2], L. Xue[6], Y. Yamamoto[12], C. T. Yan[11], X. C. Yang[8], S. Yasue[25], Z. H. Ye[15], G. C. Yu[7], A. F. Yuan[5], T. Yuda[9], H. M. Zhang[2], J. L. Zhang[2], N. J. Zhang[6], X. Y. Zhang[6], Y. Zhang[2], Yi Zhang[2], Zhaxisangzhu[5] and X. X. Zhou[7]

[1]Department of Physics, Hirosaki University, Hirosaki 036-8561, Japan.
[2]Key Laboratory of Particle Astrophysics, Institute of High Energy Physics, Chinese Academy of Sciences, Beijing 100049, China.
[3]Faculty of Engineering, Yokohama National University, Yokohama 240-8501, Japan.
[4]Department of Physics, Hebei Normal University, Shijiazhuang 050016, China.
[5]Department of Mathematics and Physics, Tibet University, Lhasa 850000, China.
[6]Department of Physics, Shandong University, Jinan 250100, China.
[7]Institute of Modern Physics, SouthWest Jiaotong University, Chengdu 610031, China.
[8]Department of Physics, Yunnan University, Kunming 650091, China.
[9]Faculty of Engineering, Kanagawa University, Yokohama 221-8686, Japan.
[10]Faculty of Education, Utsunomiya University, Utsunomiya 321-8505, Japan.
[11]Institute for Cosmic Ray Research, University of Tokyo, Kashiwa 277-8582, Japan.
[12]Department of Physics, Konan University, Kobe 658-8501, Japan.
[13]Research Institute for Science and Engineering, Waseda University, Tokyo 169-8555, Japan.
[14]Department of Physics, Shinshu University, Matsumoto 390-8621, Japan.
[15]Center of Space Science and Application Research, Chinese Academy of Sciences, Beijing 100080, China.
[16]Physics Department and Tsinghua Center for Astrophysics, Tsinghua University, Beijing 100084, China.
[17]Department of Physics, Saitama University, Saitama 338-8570, Japan.
[18]Advanced Media Network Center, Utsunomiya University, Utsunomiya 321-8585, Japan.
[19]National Institute of Informatics, Tokyo 101-8430, Japan.
[20]Tochigi Study Center, University of the Air, Utsunomiya 321-0943, Japan.
[21]Tokyo Metropolitan College of Industrial Technology, Tokyo 116-8523, Japan.
[22]Max-Planck-Institut für Physik, München D-80805, Deutschland.
[23]Shonan Institute of Technology, Fujisawa 251-8511, Japan.
[24]RIKEN, Wako 351-0198, Japan.

[25]School of General Education, Shinshu University, Matsumoto 390-8621, Japan.
* Paper presented at the conference by K. Kawata

E-mail: kawata@icrr.u-tokyo.ac.jp

Abstract. The Tibet air shower array, which has an effective area of 36,900 m^2, has been in operation at Yangbajing in Tibet, China at an altitude of 4,300 m above sea level. In this paper, we will briefly introduce the recent gamma-ray observation with the present Tibet air shower array and our future plan which is called the Tibet muon detector (MD) project.

1. Introduction

The Tibet air shower (AS) experiment has been successfully operated at Yangbajing (90.522° E, 30.102° N, 4,300 m above sea level) in Tibet, China since 1990. The array was constructed first in 1990 and gradually upgraded by increasing the number of counters, then the Tibet AS array was consists of 533 plastic scintillation counters of 0.5 m^2 placed on a 7.5 m square grid to detect high-energy (> a few TeV) cosmic-ray showers. In the late fall of 2003, the area of the Tibet AS array was further enlarged up to 36,900 m^2 by adding 256 counters. This final array configuration has been successfully in operation, triggering AS events at a rate of 1,700 Hz. Using these arrays, we already successfully detected gamma rays from Crab, Mrk 501 and Mrk 421 [1, 2, 3]. Also, we set stringent upper limits to gamma rays from Galactic plane at 3 TeV and 10 TeV [4, 5].

2. Recent results

Recently, we updated the gamma-ray energy spectrum of the Crab Nebula [6]. We found no evidence for time variability of flux intensity from the Crab Nebula at multi-TeV energies. Also the gamma-ray energy spectrum of the Crab Nebula is consistent with our previous results and other observations by the atmospheric imaging Cherenkov telescopes. We also have successfully observed gamma-ray sources and precise large-scale cosmic-ray anisotropy in the northern sky [8, 9, 10]. In these results, we first pointed out new small anisotropies in Cygnus region at multi-TeV energies [10]. One of them is coincident with MGRO J2019+37 which the Milagro experiment recently established to gamma-ray source [11]. After that, we first present the energy spectrum of MGRO J2019+37 in multi TeV region [12].

The absolute gamma-ray energies in multi-TeV region observed by the Tibet AS array are verified by the Moon's shadow observation [7]. As primary cosmic rays are shielded by the Moon, we observed a deficit in cosmic rays called the Moon's shadow, and the center of the Moon's shadow shifts westward depending on primary cosmic-ray energies due to the geomagnetic field. Using this effect, the systematic error of absolute energy scale is estimated to be less than approximately $\pm 10\%$ level.

3. Future plan

We are planning to build a water-Cherenkov-type muon detector array (Tibet MD array) around the Tibet AS array [13, 14, 15]. Each muon detector is a waterproof concrete pool, 7.2 m wide $\times$ 7.2 m long $\times$ 1.5 m deep in size, equipped with two 20 inch-in-diameter PMT (HAMAMATSU R3600). The Tibet MD array consists of 192 muon detectors set up 2.5 m underground as shown by gray areas in Figure 1. Its total effective area amounts to be 10,000 m^2 approximately for muon detection with an energy threshold of 1 GeV.

The Monte Carlo simulation including the AS generation and responses of the Tibet AS array and MD array was done to estimate the discrimination between gamma rays and hadrons based on counting the number of muons accompanying an air shower [14]. As a result, hadron-induced air showers are suppressed by 99.99% around 100 TeV, while gamma-ray-induced air showers remain by more than 95%. Finally, we calculate the integral flux sensitivity of the Tibet AS+MD array to point-like gamma rays as shown by the thick solid curve in Figure 2. Note that our sensitivity above 200 TeV is defined as a flux corresponding to 10 gamma-ray events, since the background events are fully suppressed to less than one event.

The Tibet AS+MD array will have the sensitivity to gamma rays in the 100 TeV region by an order of magnitude e better than any other previous existing detectors in the world. We can expect to discover a dozen new point-like sources and diffuse gamma rays from Galactic plane with extremely low background level in the northern sky. In the near future, the MAGIC and VERITAS experiments, together with the Tibet AS+MD array will contribute to a deeper understanding of the origin and acceleration mechanism of cosmic rays.

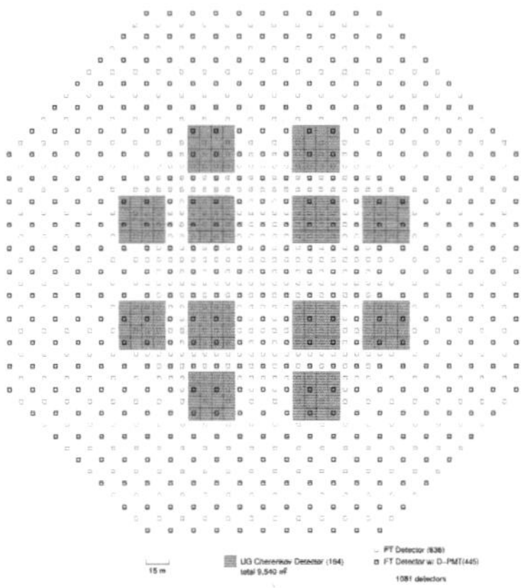

Figure 1. Schematic view of the Tibet AS+MD array.

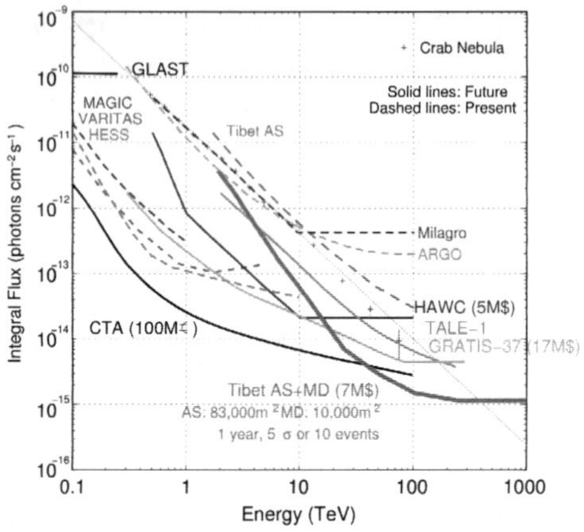

Figure 2. Integral flux sensitivities to point-like gamma-ray sources.

Acknowledgments

The collaborative experiment of the Tibet Air Shower Arrays has been performed under the auspices of the Ministry of Science and Technology of China and the Ministry of Foreign Affairs of Japan. This work was supported in part by Grants-in-Aid for Scientific Research on Priority Areas (712) (MEXT), by the Japan Society for the Promotion of Science, by the National Natural Science Foundation of China, and by the Chinese Academy of Sciences.

References

[1] Amenomori M et al. 1999 *Astrophys. J.* **525** L93
[2] Amenomori M et al. 2000 *Astrophys. J.* **532** 302
[3] Amenomori M et al. 2003 *Astrophys. J.* **598** 242
[4] Amenomori M et al. 2002 *Astrophys. J.* **580** 887
[5] Amenomori M et al. 2002 *Advances in Space Research* **37** 1932
[6] Amenomori M et al. 2007 *Proc. 30th Int. Cosmic Ray Conf.* **OG.2.2**
[7] Amenomori M et al. 2005 *Proc. 29th Int. Cosmic Ray Conf.* **6**, 53
[8] Amenomori M et al. 2005 *Astrophys. J.* **626** L29
[9] Amenomori M et al. 2005 *Astrophys. J.* **633** 1005
[10] Amenomori M et al. 2006 *Science* **314** 439
[11] Abdo A A et al. 2007 *Astrophys. J.* **658** L33
[12] Amenomori M et al. 2007 *Proc. 30th Int. Cosmic Ray Conf.* **OG.2.2**
[13] Amenomori M et al. 2007 *Astrophys. and Space Science* **309** 435
[14] Amenomori M et al. 2007 *Proc. 30th Int. Cosmic Ray Conf.* **OG.1.5** (*Preprint* astro-ph/0710.2757)
[15] Amenomori M et al. 2007 *Proc. 30th Int. Cosmic Ray Conf.* **OG.1.5** (*Preprint* astro-ph/0710.2752)

Propagation of UHE protons through a magnetized large scale structure

Santabrata Das[1], Hyesung Kang[2], Dongsu Ryu[3], Jungyeon Cho[3]

[1]ARCSEC, Sejong University, Seoul 143-747, Korea.

[2]Department of Earth Sciences, Pusan National University, Pusan 609-735, Korea.

[3]Department of Astronomy & Space Science, Chungnam National University, Korea.

E-mail: `sbdas@canopus.cnu.ac.kr`, `kang@uju.es.pusan.ac.kr`, `ryu@canopus.cnu.ac.kr`, `cho@canopus.cnu.ac.kr`

Abstract. The propagation of ultra high energy cosmic rays is affected by the extragalactic magnetic fields (EGMFs) that were produced during the course of large scale structure formation of the universe. We adopt a novel model where the large scale EGMFs are estimated from local dynamic properties of the gas flows in hydrodynamic simulations of a concordance ΛCDM universe. Using the model EGMFs, we calculate the deflection angle, time delay and energy spectrum of protons with $E > 10^{19}$ eV that are injected at cosmological sources and then travel through the large scale structure of the universe, losing the energy due to interactions with the cosmic background radiation. Implications of this study on the origin of UHECRs are discussed.

1. Introduction

In recent years, the enigma of the nature and origin of ultrahigh energy cosmic rays (UHECRs) with energy E $>$ 10 EeV (1EeV $= 10^{18}$eV) has received considerable attention, both observationally and theoretically [1-2]. Yet the acceleration mechanism and the corresponding astrophysical 'accelerators' of those energetic particles are still largely unknown. At these extremely high energies, UHECRs cannot be confined within the Galactic plane, which points more towards their extragalactic origin. In addition, the isotropic distribution of arrival directions suggests that large number of UHECR sources might be distributed at sufficiently large distances [2-3]. The composition analysis of UHECRs indicates that most of them are lighter nuclei at higher energies [4]. However, according to the recent report of Pierre Auger Observatory (Auger), UHECRs are dominated by hadrons instead of photons and the mass composition is mixed, possibly getting heavier above 30 EeV [5]. UHECRs from cosmological sources loose energy during their propagation through the intergalactic space due to the interactions with the cosmic background radiation, resulting in the Greisen-Zatsepin-Kuzmin (GZK) cutoff in the energy spectrum [6]. Observations of Yakutsk EAS array and HiRes confirmed the presence of GZK suppression, although AGASA experiment failed to identify this cutoff [2, 7-8]. A similar suppression is also observed in the energy spectrum obtained by recent Auger experiment [9], which seems to have ended the controversy of the presence of GZK cutoff.

Trajectories of charged UHECRs are deflected by the large scale EGMFs, leading to the deviation of arrival directions from their sources and consequently time delays with respect to photon travel time [10-11]. Therefore, finding source locations from arrival directions of UHECRs is not straightforward. Indeed, the main challenge is to estimate the accurate EGMFs, since it is very difficult to measure them through observations. From this standpoint, we employ a novel model of EGMFs in which the strength of magnetic fields is estimated from local vorticity and turbulence, using the results of cosmological hydrodynamic simulation [12]. We calculate the trajectories of UHE protons from active galactic nuclei (AGN) propagating through the model EGMFs, considering all relevant energy loss processes. We then study the angular deflection, time delay and energy spectrum of UHE protons that arrive in the regions similar to the Local Group.

2. Numerical Models and Calculation

We performed a set of six cosmological hydrodynamic simulations of a concordance ΛCDM universe each with different initial realizations. A cubic box of comoving size $100h^{-1}$ Mpc is simulated using 512^3 grid zones for gas and gravity and 256^3 particles for dark matter. Using the simulation data at the present epoch ($z = 0$), the magnetic field energy is calculated from local vorticity and the turbulent kinetic energy of flow motion, based on a turbulence dynamo model [12]. We identify X-ray clusters of galaxies with $kT > 1.0$ keV in the simulation as 'host' clusters that contain an AGN-like source of UHECRs. With this selection criterion, there are 20-30 such clusters in the simulated volume of $(100h^{-1}\text{Mpc})^3$, corresponding to the source density of $n_s = 2 - 3 \times 10^{-5}h^3\text{Mpc}^{-3}$ and the mean separation of $l_s \sim 40h^{-1}$Mpc. The field strength at the source locations ranges mostly $0.1 < B_s(\mu\text{G}) < 2$ with $\langle B_s \rangle \sim 1\mu$G. UHE protons with a power-law energy spectrum of $E^{-\gamma}$ for 10 EeV $\leq E \leq 10^3$EeV are distributed among the sources and launched into random directions. We select the groups of galaxies with 0.05keV $< kT <$ 0.5keV as observer locations, assuming they are similar to the Local Group. These groups are mostly located along the filaments, following the matter distribution of the large scale structure. We identify 1000-1400 such groups in the simulation volume. The field strength at the observer locations is typically $10^{-4} < B_{obs}(\mu\text{G}) < 0.1$ with $\langle B_{obs} \rangle \sim 10$nG. Once a particle visits a mock observer modeled as a sphere of $R_{obs} = 0.5h^{-1}$ Mpc, the arrival direction, time delay and energy of the particle are registered as an observed event. We let the particle continue its journey, visiting several groups during its full flight, until its energy is reduced to 10 EeV. With 3×10^4 protons injected for each model at sources, about 2.6×10^5 events are recorded in total for all six models.

3. Result

Fig. 1a shows the volume fraction in the ρ_{gas} - B plane. On average the EGMFs are correlated with the gas density: the strongest around clusters of galaxies and the weakest in voids. According to Fig. 1b, the volume filling factor with $B > 10^{-8}$G is $f(> 10^{-8}\text{G}) \approx 0.02$. Since the gyroradius of a proton is $r_g = 10\text{kpc}\ (E/10\text{EeV})(B/\mu\text{G})^{-1}$, the trajectories of ~10 EeV protons could be severely deflected when they leave the sources with $\langle B_s \rangle \sim 1\mu$G and when they fly by the clusters/groups with $B \gtrsim 10^{-8}$G. On the other hand, they can fly almost rectilinearly in the low density void regions with $B < 10^{-12}$G.

Fig. 1c shows the distribution of deflection angle, θ, with observed energy. Here, the results of the cases with $\gamma = 2.7$ and the minimum source-observer distance $D_{\text{min}} = 3$ Mpc are presented. On average, the deflection angle decreases with energy, showing a transition from diffusive to rectilinear propagation regime at $E \sim 60$ EeV. For $E \geq 60$ EeV, 35 % of observed events arrive with $\theta \leq 5°$ from source positions in the sky. Since the deflected path length is longer than the rectilinear distance, the particle travel time is delayed compared to the light travel time. Fig. 1d shows the distribution of delay time (t_d) with observed energy. The time delay decreases with increasing energy with a wide spread at a given energy. On average, time delay $\langle t_d \rangle \sim 10^{9.5}$yrs around 10 EeV and $\langle t_d \rangle \sim 10^7$yrs above 100 EeV.

Fig. 2 compares the predicted energy spectra of UHE protons in several models with AGASA, HiRes I and HiResII data [2, 6, 8]. The spectra for $\gamma = 2.0, 2.4, 2.7$ and $D_{\text{min}} = 3$ and 10 Mpc are

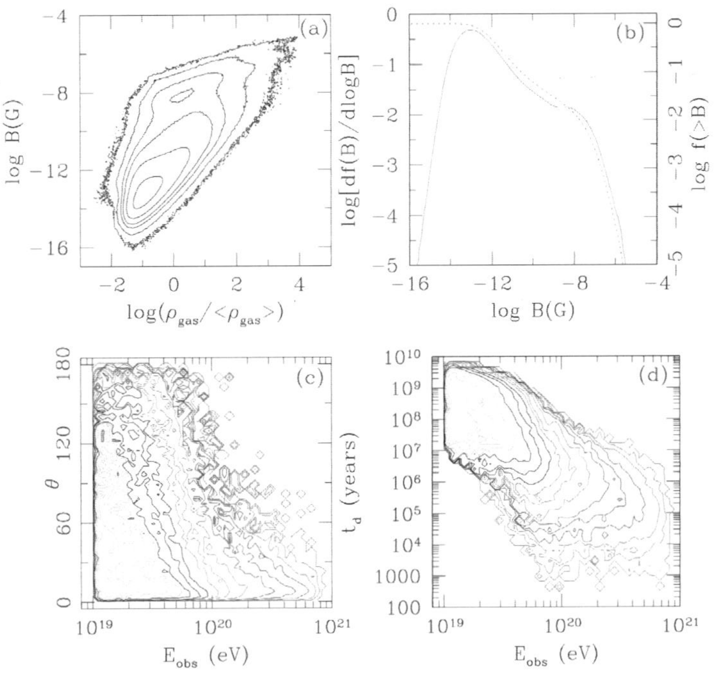

Figure 1. (a) Volume fraction in the gas density-EGMF strength plane. (b) Volume fraction (solid line) and its cumulative distribution (dotted) as a function of EGMF strength. (c) Distribution of events in the plane of observed energy and deflection angle (d) Distribution of events in the plane of observed energy and time delay. Protons are injected with a power-law spectrum of $\gamma = 2.7$ at sources.

presented. Here, $J(E)$ is given in an arbitrary unit, as the amplitude of the injected spectrum at sources is arbitrary. The presence of GZK suppression above 60 EeV is obvious, and therefore, AGASA data is not consistent with our results. Above 100 EeV, the flux is much higher for the model with $D_{\rm min} = 3$ Mpc than that with $D_{\rm min} = 10$ Mpc, indicating the importance of contributions from nearby sources. The injection spectra with γ =2.4-2.7 are consistent with observed data in the considered energy range.

4. Conclusion

We study the propagation of UHE protons through the EGMFs correlated with the large scale structure of the universe. We estimate the EGMFs from cosmological hydrodynamic simulations of a ΛCDM universe, adopting a turbulence dynamo model. Sub-GZK protons have interaction lengths up to 1 Gpc and they are significantly scattered by the EGMFs during their long journey to observers, resulting in a wide range of deflection angle almost independent of source to observer distance. Super-GZK protons, on the other hand, come from sources within 100 Mpc and so could avoid visiting strong field regions. Above 60 EeV, about 35 % of UHE protons arrive within 5° of their source positions in the sky. The average time delay ranges $10^7 - 10^{9.5}$ years, although it could be less than 10^7 years for highest energy protons. Thus, in the present scenario UHE proton astronomy may be possible only above the GZK cutoff energy. This is consistent with the recent results of Auger collaboration that showed a correlation of UHECRs above 60 EeV with neighboring AGN [13]. The predicted energy spectrum of UHE protons exhibits the GZK suppression, and fits the HiRes observations for the injection spectrum index

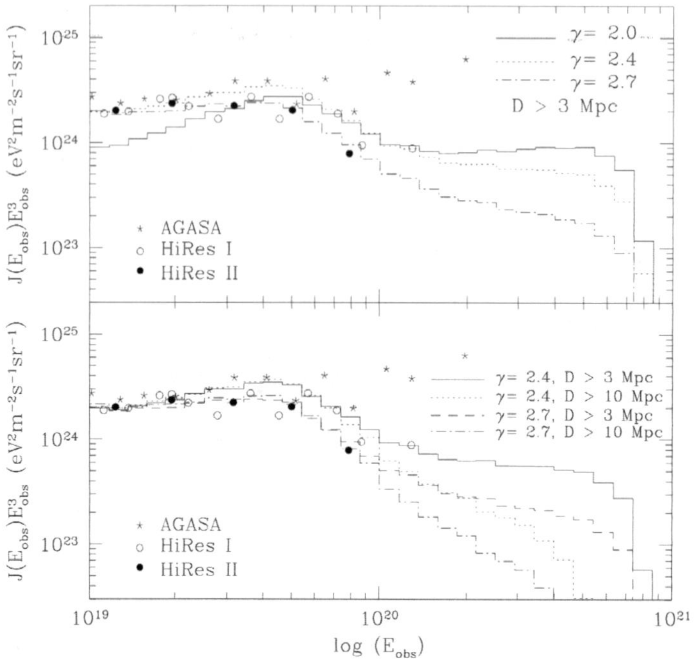

Figure 2. Predicted energy spectra of UHE protons. *Top panel:* the cases with $(\gamma, D_{\min}) =$ (2.0, 3 Mpc) (solid), (2.4, 3 Mpc) (dotted), (2.7, 3 Mpc) (dashed) are shown. *Bottom panel:* the cases with $(\gamma, D_{\min}) =$ (2.4, 3 Mpc) (solid), (2.4, 10 Mpc) (dotted), (2.7, 3 Mpc) (dashed), (2.7, 10 Mpc) (dot-dashed) are shown. Filled circles are the data from HiRes-2 [8], open circles from HiRes-1, and stars from AGASA [2].

of $\gamma = 2.4 - 2.7$ [14].

Acknowledgments: SD and HK were supported by KOSEF through Astrophysical Research Center for the Structure and Evolution of the Cosmos. DR was supported by the Korea Research Foundation Grant funded by the Korean Government (MOEHRD) (KRF-2004-015-C00213).

References

1 Cronin J W 1999 *Rev. Mod. Phys.* **71** 165
2 Nagano A and Watson A A 2000 *Rev. Mod. Phys.* **72** 689
3 Burgett W S and O'Malley M R 2003 *Phys. Rev. D* **67** 092002
4 Abbasi R U et al 2005 *ApJ* **622** 910
5 Unger M et al (Auger Collaboration) 2007 *proceedings for the 30th ICRC (Merida, Mexico)* arXiv0706.1495
6 Berezinsky V S, Gazizov A and Grigor'eva S I 2006 *Phys. Rev. D* **74** 043005
7 Pravdin M I et al Yakutsk collaboration 1999 *proceedings for the 26th ICRC (Salt Lake City, USA)* **3** 292
8 Zech A 2004 *eprint arXiv:astro-ph/0409140*
9 Facal P et al (Auger Collaboration) 2007 *proceedings for the 30th ICRC (Merida, Mexico)* arXiv0706.4322
10 Armengaud E, Sigl G and Miniati F 2005 *Phys. Rev. D* **72** 043009
11 Dolag K, Grasso D, Springel D and Tkachev J 2005 *Journal of Cosmology and Astroparticle Physics* **1** 009
12 Ryu D, Cho J, Kang H, and Das S 2007 *in preparation*
13 Pierre Auger Collaboration 2007 *Science* **318** 939
14 Kang H, Das S, Ryu D & Cho J 2007 *proceedings for the 30th ICRC (Merida, Mexico)* arXiv:0706.2597

The KASCADE-Grande Experiment

J. Blümer[a,b], W.D. Apel[a], J.C. Arteaga[b], F. Badea[a], K. Bekk[a], M. Bertaina[c], H. Bozdog[a], I.M. Brancus[d], M. Brüggemann[e], P. Buchholz[e], A. Chiavassa[c], E. Cantoni[c], F. Cossavella[b], K. Daumiller[a], V. de Souza[b], F. Di Pierro[c], P. Doll[a], R. Engel[a], J. Engler[a], M. Finger[b], D. Fuhrmann[f], P.L. Ghia[g], H.J. Gils[a], R. Glasstetter[f], C. Grupen[e], A. Haungs[a], D. Heck[a], J.R. Hörandel[b], T. Huege[a], P.G. Isar[a], K.-H. Kampert[f], D. Kickelbick[e], H.O. Klages[a], Y. Kolotaev[e], P. Luczak[h], H.J. Mathes[a], H.J. Mayer[a], C. Meurer[a], J. Milke[a], B. Mitrica[d], A. Morales[a], C. Morello[g], G. Navarra[c], S. Nehls[a], J. Oehlschläger[a], S. Ostapchenko[a], S. Over[e], M. Petcu[d], T. Pierog[a], S. Plewnia[a], H. Rebel[a], M. Roth[a], H. Schieler[a], O. Sima[i], M. Stümpert[b], G. Toma[d], G.C. Trinchero[g], H. Ulrich[a], J. van Buren[a], W. Walkowiak[e], A. Weindl[a], J. Wochele[a], J. Zabierowski[h]

KASCADE-Grande Collaboration

[a] Institut für Kernphysik, Forschungszentrum Karlsruhe, Germany
[b] Institut für Experimentelle Kernphysik, Universität Karlsruhe, Germany
[c] Dipartimento di Fisica Generale dell'Università Torino, Italy
[d] National Institute of Physics and Nuclear Engineering Bucharest, Romania
[e] Fachbereich Physik, Universität Siegen, Germany
[f] Fachbereich Physik, Universität Wuppertal, Germany
[g] Istituto di Fisica dello Spazio Interplanetario, INAF, Torino, Italy
[h] Soltan Institute for Nuclear Studies, Lodz, Poland
[i] Department of Physics, University of Bucharest, Romania

E-mail: `Johannes.Bluemer@ik.fzk.de`

Abstract. KASCADE-Grande is an extensive air shower array co-located with the original KASCADE air shower experiment at Forschungszentrum Karlsruhe, Germany. The multi-detector system allows to investigate the energy spectrum, composition, and anisotropies of cosmic rays in the energy range extended up to 10^{18} eV. An overview on the performance of the apparatus and first results are presented.

Introduction. The major goal of KASCADE-Grande (covering a primary energy range of $10^{14} - 10^{18}$ eV) is the observation of the so-called 'iron-knee' in the cosmic-ray spectrum, which is expected at around 10^{17} eV following previous KASCADE results, whereby the energy positions of the knees of individual mass or charge groups suggest a rigidity dependence. KASCADE-Grande will allow to reconstruct the energy spectra of various mass groups similar to KASCADE, which will give the possibility to distinguish between astrophysical models for the transition region from cosmic rays of galactic to extragalactic origin. Examples are models of the type claimed by Berezinsky [1] predicting a pure extragalactic proton composition already at energies around 10^{18} eV and models which have an extension of the galactic component up to the ankle [2]

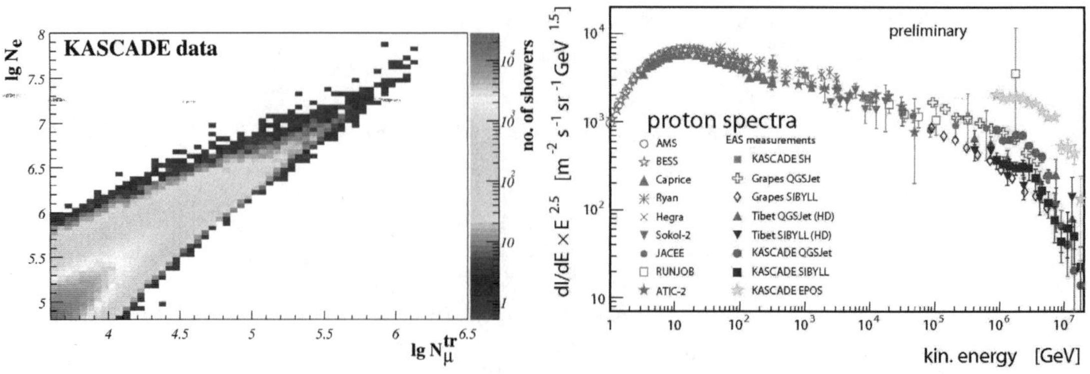

Figure 1. Left: correlation of electron and muon numbers measured with KASCADE. Right: proton spectra unfolded from the data presented in the left panel, based on different hadronic interaction models. The results are compared to direct measurements.

or assuming a mixed composition of the extragalatic flux [3]. In the two latter cases one expects a mixed composition in the energy range of KASCADE-Grande.

Unfolding procedures have been applied to the shower size data correlating the electron and muon numbers observed with KASCADE (Figure 1, left panel). The resulting energy spectra for different primary mass groups exhibit knee-like features, most clearly in the all-particle spectrum as well as for proton and helium primaries [4]. As an example, proton spectra are shown using the hadronic interaction models SIBYLL [6], QGSJet [7] and EPOS [8]. EPOS is not in agreement with results from direct measurements [9]. Improvements to the hadronic interaction models underlying the analysis are required as well as measurements extending to higher energies, in particular to energies where the iron-knee is expected.

KASCADE-Grande. KASCADE is located at 49.1°n, 8.4°e, 110 m a.s.l.. It is taking data since 1996, [10] with an effective area of 0.04 km^2. It was extended to KASCADE-Grande [11] in 2003 by installing a large array of 37 scintillation detectors at an average spacing of 137 m. Each station has a scintillation detector of 10 m^2 area read out by 16 photomultipliers. The dynamic range is $1 : 10^5$ from 1/3 to 30000 charged particles per station for the measurement of particle densities and arrival times. KASCADE-Grande provides an area of 0.5 km^2 and operates jointly with the KASCADE detectors. While the Grande detectors are sensitive to all charged particles, the KASCADE detectors measure the electromagnetic component and the muonic components separately. Muon tracking detectors allow to investigate the muon production height in the atmosphere at three different threshold energies.

The core position can be reconstructed within ≈ 15 m and a direction resolution of $\approx 0.5°$ is reached. KASCADE-Grande is fully efficient at energies above $3 \cdot 10^{16}$ eV, thus providing a large overlap with the KASCADE energy range. The total muon number in showers is obtained from KASCADE detectors, which is a particular asset of KASCADE-Grande compared to other experiments in the energy range of the expected iron-knee. Figure 2 shows the data quality of KASCADE-Grande for a single event and a similar correlation matrix as Figure 1. It was recently demonstrated that the muon number can be reconstructed for showers with inclination angles up to 70° with sufficient accuracy [12]. Hence, additional sensitivity to composition and hadronic interactions is derived from muon density measurements, muon tracking at different muon energy thresholds, and from an extended range in zenith angle. In addition, a large data set has been analysed in terms of searches for anisotropies. The bottom panel of figure 2 shows the Rayleigh amplitude using two different methods [13].

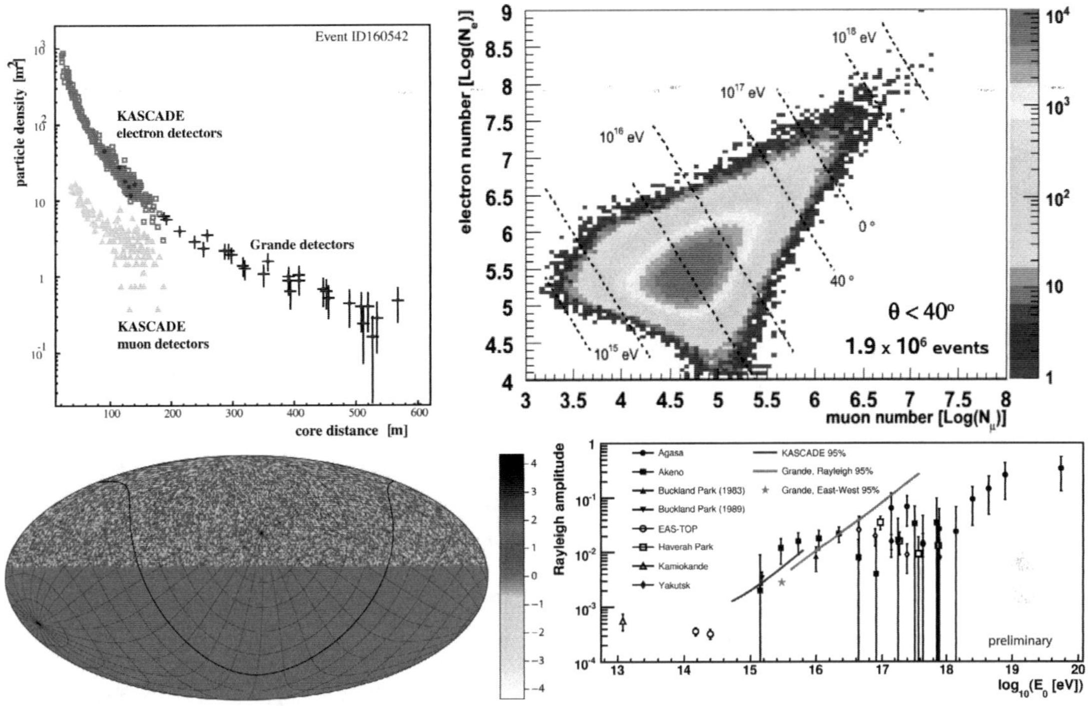

Figure 2. Top-left: lateral distribution of an event measured with KASCADE-Grande. Top-right: correlation between inferred electron and muon numbers in showers measured with KASCADE-Grande, similarly to the left panel of Figure 1. Bottom-left: Significance map of local excesses of particle arrival directions; the distribution is Gaussian and exhibits no significant excess. Bottom-right: Rayleigh amplitude of the harmonic analyses of KASCADE-Grande data compared with results of other experiments.

Conclusion. We have illustrated the capabilities of KASCADE-Grande to perform studies of air showers up to energies of 10^{18} eV, including inferences on the nature of the primary particles, energy spectra for mass groups, tests of hadronic interaction models and anisotropy analyses.

[1] R. Aloisio et al., *Astrop. Phys.* **27**, 76 (2007).
[2] M. Hillas, *J. Phys. Conf. Ser.* **47**, 168 (2006).
[3] D. Allard et al., *Astron. Astrophys.* **473**, 59 (2007).
[4] T. Antoni et al. KASCADE Coll., *Astrop. Phys.* **24**, 1 (2005).
[5] D. Heck et al., *FZKA-Report 6019* Forschungszentrum Karlsruhe, (1998).
[6] J. Engel et al., *Phys. Rev. D* **50**, 5013 (1994).
[7] N.N. Kalmykov and S.S. Ostapchenko, *Phys. Atom. Nucl.* **56** 346 (1993).
[8] K. Werner, F.M. Liu and T. Pierog, *Phys. Rev. C* **74** 044902 (2006).
[9] H. Ulrich et al. - KASCADE-Grande coll., *Proc. of 30th ICRC 2007, Merida, Mexico* (2007), in press.
[10] T. Antoni et al. - KASCADE coll., *Nucl. Instr. and Meth. A* **513**, 490 (2003).
[11] G. Navarra et al. - KASCADE-Grande coll., *Nucl. Instr. Meth. A* **518**, 207 (2004).
[12] A. Haungs et al. - KASCADE-Grande coll., *Proc. of 30th ICRC 2007, Merida, Mexico* (2007), in press.
[13] S. Over et al. - KASCADE-Grande coll., *Proc. of 30th ICRC 2007, Merida, Mexico* (2007), in press.

The Telescope Array Experiment: Status and Prospects

H Tokuno , R U Abbasi , T Abu-Zayyad , R Azuma , J W Belz , T Benno , D R Bergman[5], S A Blake , O Brusova , R Cady , Z Cao , B G Cheon , J Chiba , M Chikawa , I S Cho , T Chung , F Cohen , K Doura , T Doyle , A Endo , H Fujii , T Fukuda , M Fukushima , Y Hayashi , K Hayashi , N Hayashida , K Hibino , K Hiyama , K Honda [5], P Huentemeyer , G A Hughes[5], T Iguchi , D Ikeda , K Ikuta [5], N Inoue , T Ishii [5], C C H Jui , K Kadota , F Kakimoto , T Kanbe [5], K Kasahara , H Kawai , S Kawakami , S Kawana , E Kido , Y Kondo , Y J Kwon , S Machida , K Martens , T Matsuda , T Matsuyama , J A J Matthews , J N Matthews , M Minamino , K Miyata , M Mostafa , Y Murano , T Nakamura , S Nam , T Nonaka , S Ogio , S Oh , M Ohnishi , H Ohoka , A Ohshima , T Okuda , J Ormes , S Ozawa , I H Park , J H Park , D Rodriguez , H Sagawa , N Sakurai , L M Scott[5], T Shibata , N Shimizu , H Shimodaira , G Sinnis , J D Smith , P Sokolsky , R W Springer , S R Stratton[5], S Suzuki , M Takeda , A Taketa , M Takita , Y Tameda , K Tanaka , M Tanaka , H Tanaka , M J Taylor , M Teshima , J R Thomas , S B Thomas , G B Thomson[5], T Tomida [5], R Torii , Y Tsunesada , Y Uchihori [5], S Udo , Y Unno , Y Wada , V B Wickwar , L R Wiencke , T D Wilkerson , Y Yamakawa , H Yamaoka , J Yang , S Yoshida , H Yoshii , and Y H Yun

[1] Institute for Cosmic Ray Research, University of Tokyo, Kashiwa, Japan
[2] University of Utah, Salt Lake City, USA
[3] Tokyo Institute of Technology, Tokyo, Japan
[4] Kinki University, Osaka, Japan
[5] Rutgers University, Piscataway, USA
[6] Hanyang University, Seoul, Korea
[7] Tokyo University of Science, Noda, Japan
[8] Yonsei University, Seoul, Korea
[9] Ewha Womans University, Seoul, Korea
[10] Utah State University, Logan, USA
[11] Saitama University, Saitama, Japan
[12] Institute of Particle and Nuclear Studies, KEK, Tsukuba, Japan
[13] Osaka City University, Osaka, Japan
[14] Kanagawa University, Yokohama, Japan
[15] Yamanashi University, Kofu, Japan
[16] Los Alamos National Laboratory, Los Alamos, USA
[17] Musashi Institute of Technology, Tokyo, Japan
[18] Pusan National University/ARCSEC, Pusan, Korea
[19] Advanced Research Institue for Science and Engineering, Waseda University, Tokyo, Japan

[20] Chiba University, Chiba, Japan
[21] University of New Mexico, Albuquerque, USA
[22] Kochi University, Kochi, Japan
[23] University of Denver, Denver, USA
[24] Chungnam National University, Daejeon, Korea
[25] Hiroshinma City University, Hiroshima, Japan
[26] Max-Planck-Institute for Physics, Muenchen, Germany
[27] National Institute of Radiological Sciences, Chiba, Japan
[28] Ehime University, Matsuyama, Japan

E-mail: htokuno@icrr.u-tokyo.ac.jp

Abstract. Telescope Array (TA) is a hybrid detector of a surface detector array and fluorescence telescopes. This hybrid detector will measure the energy spectrum, anisotropy and composition of ultra-high energy cosmic rays (UHECRs) to identify their origin. The almost construction of the detector has been completed in May 2007, and the detector is running under test and adjustments. The first hybrid observation with the full configuration is planned in beginning of 2008. In this paper the status and prospects of TA detector is described.

1. Introduction

Ultra high energy cosmic rays (UHECRs) will interact with the cosmic microwave background. Therefore a cutoff in their energy spectrum (GZK cutoff) is expected at the energy 10 $^{.5}$ eV by Greisen, Zatsepin and Kuzmin [1]. However, the AGASA result shows that there are 11 events beyond the GZK cutoff [2]. In contrast, HiRes group reported that there is a GZK cut-off in their observed energy spectrum [3]. AGASA and HiRes claims their uncertainty in energy determination is 18% and 17% respectively. The rescaling spectra of AGASA and HiRes within their uncertainties in agree well below 10 eV. It seems that the inconsistency is due to the systematic uncertainties of both experiments in energy determination. In order to make clear the difference, and to get definite answer on the origin of UHECRs, we will observe extensive air showers (EASs) with a hybrid detector which consists of an AGASA type surface detector (SD) array and with HiRes type air fluorescence detectors (FD) simultaneously. Using the hybrid detector the difference of the energy spectrum observed by AGASA and HiRes can be studied on event by event above 10 eV. The detection area of SD array is of about 700 km , which is 8 times larger than AGASA detection area. A sufficient statistics gives us a fine structure of the energy spectrum at GZK cutoff region, and it allows to study the distribution and property of UHECRs sources.

2. Status and Prospects

The configuration of TA detector is shown in Fig.1. It consists of a Surface detector (SD) array and three stations of air Fluorescence Detectors (FDs) surrounding SD array. This hybrid detector is located in Utah, USA (39.3° N, 112.9° W) with an average altitude of 1400 m.

The SD array consists of 512 SDs in a grid with 1.2 km spacing covering the ground area of $\sim$ 700 km . From our simulation studies, the trigger efficiency is expected to be $\sim$ 100 % for cosmic rays with the primary energy more than 10 eV. Each SD has 2 layers plastic scintillator (thickness: 1.2 cm, area: 3 m). These layers are separated optically, and PMT at each layer collects signals with wavelength shifting fibers. In order to reduce trigger frequency, local electronics find a coincidence signal between upper and lower scintillator. Waveforms of PMT output recorded by local electronics are transmitted to a central DAQ system via wireless LAN. The 95 % of installation have been completed. We are adjusting equipments, and checking performances of the array. Remaining surface detectors ($\sim$20) will be installed soon, and full operation of SD array will be planed from December 2007. For more detail see[4].

Three air Fluorescence Detector (FD) stations are surrounding the SD array, and the separation of each station is $\sim$ 30 km (Fig.1). The FD consists of spherical mirror optics, PMT camera and recording electronics system. The spherical mirror with a diameter of 3.3m consists of 18 segment mirrors whose curvature radius is 6m. The PMT camera has 256 PMTs of 16×16 matrix, and FOV (field of view) of each PMT is $\sim 1^\circ$. The FOV of one telescope is $18^\circ \times 15.6^\circ$, and the total FOV for one station is $3^\circ - 33^\circ$ in elevation and 108° in azimuth. The waveforms of PMT are digitized by Electronics with 10MHz, 12 bit resolution and 14 bit dynamic range, and stored on HDDs. Dead time of sampling sequence is 1% at the normal trigger rate $\sim$ 1 Hz.

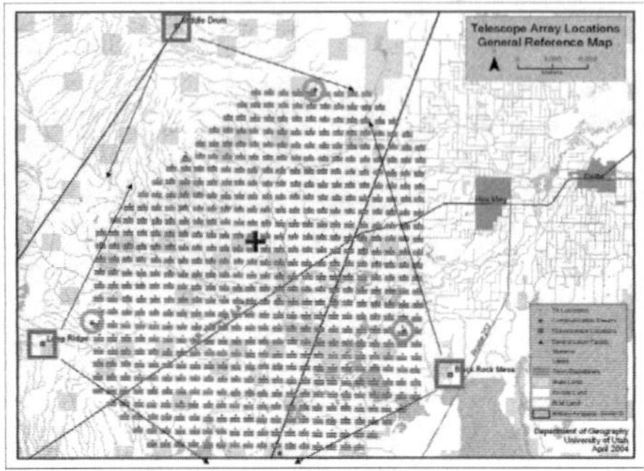

Figure 1. Detector configuration of TA. The locations of SDs are indicated by small numbers. The locations of 3 FD stations are marked by square.

From our simulation studies, trigger efficiencies of stereo events are expected to be $\sim$ 100 % for cosmic rays with the primary energy more than 10 eV. HiRes-I have been moved to the third station located on North in Fig.1. The configuration of FOV is different from HiRes-I, however, optics and electronics is same as HiRes-I. To calibrate FDs optics and monitor for transparency of the atmosphere various calibration methods will be applied[5]. For End to End calibration which includes air fluorescence yield to electronics performances, 40 MeV electron beam system will be installed on the front of the Southeast station[5]. The installation of telescopes for the first and second station have been completed and we are starting observation with stereo mode in June 2007. More details are found in[5].

SD with full operation will be started on the beginning 2008 as originally scheduled, and we should obtain the AGASA equivalent exposure by SD before TAUP 2009. At that time we will have more than 800 events with the energy > 10 eV, and more than 10 events with the energy > 10 eV with assuming the AGASA flux. The systematic uncertainties and differences of SD, FD, and Hybrid method will be understood from event by event analysis at 10 eV. We will conclude the energy spectrum at the GZK cutoff region based on the results. Also we have a plan to extend to low Energy region below 10 eV. It consists of infill SD array, FDs with higher altitude of FOV, and additional FDs covered the region of infill SD array (TALE experiment)[6].

Acknowledgments

The Telescope Array experiment is supported by a grant-in-aid for Scientific Research from the Ministry of Education, Culture, Sports, Science and Technology of Japan, and by the U.S. National Science Foundation through awards PHY-0307098 and PHY-0601915 (University of Utah) and PHY-0305516 (Rutgers University). The Dr. Ezekiel R. and Edona Wattis Dumke Foundation, The Willard L. Eccles Foundation and The George S. and Dolores Dore Eccles Foundation all helped with generous donations. HT is supported the 21st Century COE Program at University of Tokyo "Quantum Extreme Systems and Their Symmetries". We would like to thank all the technical staffs of our home institutions.

References

[1] Greisen K 1966 *Phys. Rev. Lett.* **16** 748, Zatsepin T and Kuzmin V A 1966 *JETP Lett.* **4** 178
[2] Takeda M, *et al.* 1998 *Phys.Rev.Lett.* **81** 1163, 2003 *Astropart. Phys.* **19** 447
[3] Abu-Zayyad T *et al.* 2004 *Phys. Rev. Lett.* **92** 151101, Abu-Zayyad T *et al.* 2005 *Astropart. Phys.* **23** 157
[4] H.Sagawa *et al.* 2007 *Proc. of 30th ICRC* in press
[5] S.Ogio *et al.* 2007 *Proc. of 30th ICRC* in press
[6] Thomson G B *et al.* 2007 *Proc. of 30th ICRC* in press

The LHCf experiment ∼ Verification of interaction models of cosmic rays at 10^{17} eV

Yoshitaka Itow for the LHCf collaboration

Solar-Terrestrial Environment Laboratory, Nagoya University, Furo-cho, Chikusa-ku, Nagoya, Aichi, 464-8601, Japan

E-mail: `itow@stelab.nagoya-u.ac.jp`

Abstract. Observations of ultra high energy cosmic rays (UHECR) depend on particle production models at very forward region. However yet no experimental verification has not been done at such high energy. It brings uncertainty in energy scale or composition measurements. The LHCf experiment will provide unique data to calibrate existing hadron interaction models at 10^{17} eV foreseen at LHC. Experimental overview and expected results are presented.

1. Introduction

Ultra high energy cosmic rays (UHECRs) is one of attractive enigma in astrophysics and particle physics. Their energy spectrum, compositions etc. have been studied with increased statistics by many recent air shower experiments [1, 2, 3]. For further improvement of analysis, it is indispensable to improve air shower simulations in which hadron interaction models at primary vertexes play a essential role. It is known that secondary particles produced at a very forward region largely contribute to development of air showers. However there are not enough knowledge on hadron interaction at forward region where only phenomenological models verified at lower energy are extrapolated by several orders.

The LHCf experiment is a measurement of neutral particles at zero degree of 7 TeV proton collisions in the LHC corresponding 10^{17}eV at a laboratory frame. The experiment is scheduled to take data in the commissioning phase of the LHC machine in 2008. These data will verify various hadron interaction models such as QGSJET, DPMJET, SYBILL and EPOS at interesting energy for cosmic ray physics above the knee to GZK region. Here we briefly introduce the LHCf experiment. More detail information can be found in the LHCf TDR [4].

2. The LHCf experiment

The two small EM calorimeters (Detector-1 and Detector-2) will be installed in the massive iron shields, TAN, located ±140m apart from IP1 where the ATLAS detector exists. Inside the TAN, the beam pipes makes a transition from a single to double forming "Y-shape" vacuum chambers. The two LHCf detector (Detector-1 and Detector-2) will be inserted in each TAN at their crotch of the Y-shape. The particles produced at zero degree in IP1 travel through the beam pipe then be detected by the detectors.

The Detector-1 has two towers of diamond shaped tungsten sampling calorimeters (2cm×2cm and 4cm×4cm) stacked vertically. Each tower consists of 22 tungsten plates ($2X_0$) interleaved with plastic scintillators. The Detector-2 is similar but has rectangle shapes (2.5cm×2.5cm and

3.2cm×3.2cm) in cross section. Each tower has 44 X_0 and 1.3 interaction length. In addition Detector-1 and Detector-2 have 4 layers of scintillating fibers and Si strip detectors providing incident position of showers with accuracy of 200 μm and 75 μ, respectively.

The range of shower energy measured by LHCf is above 100 GeV up to several TeV. Energy resolution is $3\%/\sqrt{E(TeV)} + 1.2\%$ for EM showers by correcting shower leakage deduced from their incident positions. The π^0 mass can be reconstructed in the invariant mass distribution of two γ's, one each hitting the two tower calorimeters. Hadronic showers can be discriminated by measuring the longitudinal shape of shower development. Energy resolution of hadron showers is expected to be about 30% at 6TeV due to longitudinal shower leakage out the back of the calorimeters. These basic performance has been well confirmed by test beam experiment at CERN-SPS H4 beam line[5].

Because acceptance of each calorimeter tower is small, multiplicity of hits will be $\sim$ 1 event at 0 degree. The X_F-distributions of γ's, π^0's will be measured cleanly without contamination of multi particles. Figure 1 shows energy distribution of single γ events for the 2cm×2cm calorimeter of Detector-1 at the center of the neutral particle flux, i.e. at 0 degree. Here 4 different distributions expected from various interaction models, QGSJET, QGSJET II, DPMJET3 and SIBYLL are shown. One can see the differences in the shape of energy distributions among these models. We tested discrimination based only on the differences in the shape of energy spectra. Here we defined the χ^2 as,

$$\chi_1^2(\alpha, \beta) = \sum_i \frac{(\alpha N_i^{data}(\beta) - N_i^{model})^2}{(\sigma_i^{data})^2 + (\sigma_i^{model})^2} \tag{1}$$

where $N_i^{data}(\beta) = N(\beta E_\gamma)$ is i-th bin of the energy spectrum of γ's scaled by factor of β for a reference model (QGSJET-II), $N_i^{model} = N(E_\gamma)$ is i-th bin of energy spectrum of γ's for tested models (QGSJET,DPMJET3 and SIBYILL), α is a normalization factor, β is a scale factor for the energy scale, σ_i^{data} and σ_i^{model} are statistical errors of the i-th bin for the reference model and the tested model, respectively. Here we conservatively set α and β as free parameters. We use 4×10^5 Monte Carlo events of the QGSJET-II model for the reference model and 9×10^5 Monte Carlo events for each of the tested models. The minimum-χ^2 values obtained for the QGSJET, DPMJET3 and SIBYLL models are χ_1^2=58, χ_1^2=73 and χ_1^2=135, respectively with DOF=67. Therefore the SIBYLL model can be discriminated against the other two models by such a shape analysis, since the SIBYLL model gives a rather softer spectrum for the high energy γ's than that from the QGSJET-II model at zero degree.

Similar analysis will be done for π^0 energy spectrum by reconstructing two γ's, one each hitting the two tower calorimeters. The reconstructed mass of π^0 will be used also for good calibration of absolute energy scale and its time variation during beam exposure.

Measurement of energy spectrum of very forward neutrons is a compartmental to the γ measurement and provide another important information for inelasticity in interaction models. Due to limited interaction length, energy resolution for hadronic shower is no excellent. But 30% energy resolution is expected by selecting showers which starts their shower development near the front of the detector. Figure 2 shows the energy spectra measured by LHCf taking into account 30% energy resolution. Even with 30% energy resolution there are significant differences between the models. Below 3 TeV contamination by neutral kaons becomes significant therefore we will restrict the neutron analysis to events $>$ 3 TeV which still covers a significant part of the spectrum.

3. Status and running scenario

Construction of detector has been done and calibrated by 100-220 GeV electron, proton and muon beams at the CERN SPS H4 beam-line in the summer 2007. The detectors will be finally installed in the TANs during Jan 2008. The first operation of LHC in colliding beam mode is envisioned with low luminosity ($\sim 10^{30}$cm^{-2}s^{-1}) with 43 equally spaced bunches per beam ($\sim 2\mu$sec between bunches). The LHCf is being scheduled to take data shortly during this early mode of LHC operation. As we will see below operating with such low luminosity still assures an adequate event rate for LHCf. Event rate of γ's is about 500 Hz even at very low luminosity $\sim 10^{29}$cm^{-2}s^{-1}. Those for π^0s and neutrons are about 5 Hz at the same luminosity. Thus quite quick measurements can provide enough statistics for the analysis. The LHCf experiment will give an indispensable calibration data for UHECR physics just after the first 7 TeV collision which would be foreseen at some time in 2008.

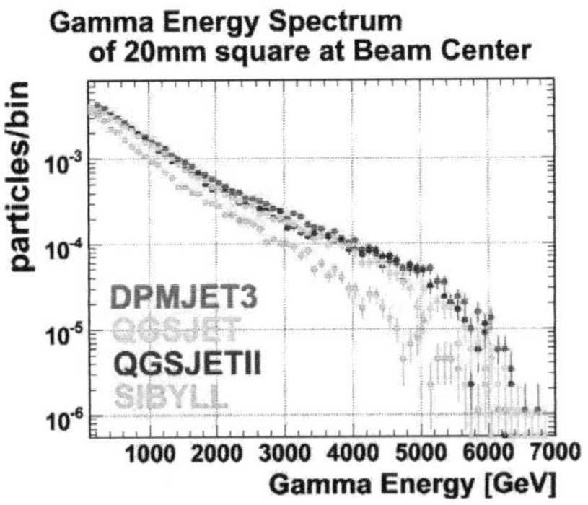

Figure 1. Energy distribution of γ's measured by the 2cm×2cm calorimeter of Detector-1 at 0 degree. Here 4 different distributions expected from DPMJET3,QGSJET II, QGSJET and SIBYLL (from top to bottom) are shown.

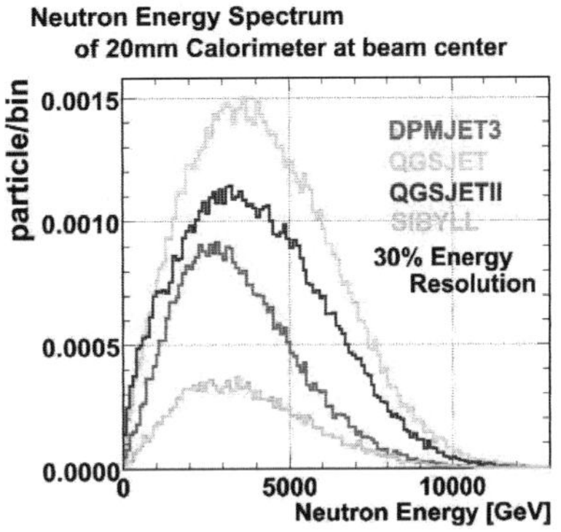

Figure 2. Energy distribution of neutrons measured by the 2cm×2cm calorimeter of detector-1 at 0 degree. Here 30 % energy resolution is taken into account by simple Gaussian convolution. Here 4 different distributions expected from QGSJET, QGSJET II, DPMJET3 and SIBYLL (from top to bottom) are shown.

References

[1] M. Takeda et al., Phys. Rev. Lett., 81 (1998) 1163.

[2] P. Sommers et al., Proc. of 29th Int. Cosmic Ray Conf., Pune (ICRC 2005), astro-ph/0507150.

[3] T.Yamamoto et al., Proc. of 30th Int. Cosmic Ray Conf., Merida (ICRC 2007), astro-ph/0707.2638

[4] O.Adriani et al, Technical design report of the LHCf experiment: Measurement of photons and neutral pions in the very forward region of LHC,CERN-LHCC-2006-004

[5] T.Sako et al, Nucl.Instrum.Meth A578:146-159,2007

Identification of Extremely-high energy starting neutrino events with the IceCube observatory

M. Ono, A. Ishihara, and S. Yoshida for the IceCube collaboration

Department of Physics, Chiba University, Chiba 263-8522, Japan

E-mail: `mio@hepburn.s.chiba-u.ac.jp`

Abstract. The IceCube neutrino observatory is capable of detecting Extremely High Energy (EHE) neutrinos with energies beyond 10^8 GeV originated in the highest energy cosmic rays. The high energy muon bundles associated with cosmic ray air showers constitute a major background and a reliable rejection mechanism must be developed. A possible approach is to identify events induced by neutrinos inside the detector instrumentation volume. The initial simulation study is briefly reported.

1. Introduction

Extremely high energy (EHE) neutrinos are expected to play a key role in understanding the origin of EHE cosmic rays (EHECRs), because energetic neutrinos can be produced by collisions of EHECRs and the cosmic microwave background photons, a process known as GZK mechanism [1]. In the initial stage, IceCube neutrino observatory [2] has performed the search for these EHE neutrinos by looking for extremely luminous events as the signal [3, 4]. Rejection and estimation of atmospheric muon background events is central to the analysis. However, the simulation of atmospheric muon events is associated with rather incomplete knowledges of the hadronic interactions in the relevant energy range, which leads to a hard-to-reduce uncertainty in muon background identification and rejection. One way to reduce this uncertainty is by selecting the neutrino events which are induced inside the detector instrumented volume, which in principle minimizes the dependence on our knowledge about the atmospheric muon bundles.

We report here the initial study of starting EHE neutrino event identification with full IceCube using Monte Carlo simulation. The MC sample studied consists of isotropic muons and starting ν_μ events with energies from 10^5 GeV to 10 GeV, following E^- spectrum.

2. The methods to identify neutrino-induced events

We can expect that the location of ν_μ-N interaction is close to the center of the IceCube detector, so are the positions of the IceCube digital optical modules (DOMs) which first records the Cherenkov photons from the neutrino induced charged lepton or hadron secondaries in the event, while the earliest hits in atmospheric muon events are always found in the outer layer in-ice detectors. Fig. 1 shows the distribution of the first hit DOM locations of the simulated events in cylindrical coordinates where z is depth and ρ the distance from the central axis of the array. Only the DOMs recording more than 5 photo-electrons contribute in the plot to reduce

[1] Starting neutrino events are defined as the ones with the neutrino interaction vertex point within sphere of 400 m radius from the array center.

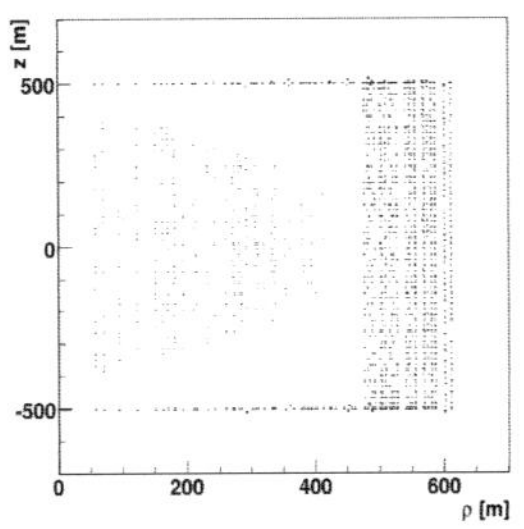

Figure 1. Distribution of the first hit DOM from cylindrical coordinates. Blue dots are muon events and red dots are starting muon-neutrino events.

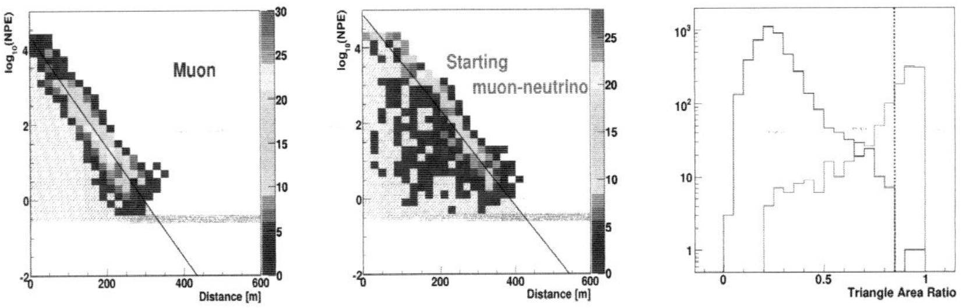

Figure 2. Examples of the lateral distribution for a muon event (left) and a starting muon-neutrino event (middle). Green dots are DOMs bellow a threshold (0.3 photo-electron). Color histogram shows distribution of DOMs with signal. Black solid line is derived by fitting the distribution. The right triangles are shaded in black and red. Distribution of the triangle area ratio obtained by the lateral distribution is plotted in the right panel for though going muon events (blue) and the starting ν_μ events (red). The triangle ratio cut is fixed at 0.85 (black dotted line).

the chance of mistaking the earliest hit from noise as signal. This threshold was determined by the real data generated from nitrogen laser deployed in the deep ice together with the DOMs [5]. One can see that μ and starting ν_μ events are clearly separated.

Another approach is based on the fact that, unlike muon which emits Cherenkov photons throughout its track, trajectories of neutrinos are invisible to the optical sensors and thus a starting neutrino induced event leaves sizable number of DOMs without photon signal near the neutrino trajectories. This can be observed in the characteristic lateral distributions. Shown on the left and middle in Fig. 2 is the number of photo electrons recorded in each DOM as a function of lateral distance (LD) of the DOMs from the MC-truth trajectories of the through-going μ and ν_μ respectively. The green band at the bottom of each distribution represents those DOMs recording less than 0.3 photo-electron. To describe the difference, we define the two right triangles shaded in black and red shown in the left panel of Fig. 2. One has the right apex at zero LD, and the other at the LD of the closest LD DOMs with no photon signal recorded. Taking the ratio of the areas of these triangles event-by-event, it is observed that neutrino induced events are more likely to give this ratio close to unity because DOMs without photon hit exist at smaller lateral distances. The area ratio distributions are shown in the right panel of Fig. 2. The plot indicates that this method identifies approximately 70 % of the starting neutrino events with the area ration cut of 0.85, while rejecting 99.95 % of muon events.

3. Discussion

The first hit DOM method shows high efficiency but it relies on a single DOM information and requires state-of-the-art level of understanding of the detector response. The lateral distribution method can be more reliable, but a high accuracy of the track trajectory reconstruction is necessary. Note that the present study uses the simulation truth trajectories. We also found that irreducible background to this approach is a muon decay event which very rarely occurs in EHE range. A combination of the two methods described here might provide a more reliable way to identify neutrino induced events with less dependence on detailed detector response.

References

[1] V. S. Berezinsky and G. T. Zatsepin, Phys. Lett. **28B**, (1969) 423.
[2] J. Ahrens *et al.*, Astropart. Phys. **20** 507 (2004); http://icecube.wisc.edu/.
[3] A. Ishihara for the IceCube collaboration, Nucl. Phys. **B** (Proc. Suppl.) (2007) pp. 200-206; astro-ph/0611794.
[4] A. Ishihara for the IceCube collaboration, astro-ph/07110353 (2007) pp. 87-90.
[5] J. Kiryluk for the IceCube collaboration, astro-ph/07110353 (2007) pp. 95-98.

The 20th Anniversary of SN1987A

ATSUTO SUZUKI
KEK, High Energy Accelerator Research Organization,
Oho 1-1, Tsukuba, Ibaragi, 305-0801, Japan
atsuto.suzuki@kek.jp

ABSTRACT

Observation of a neutrino burst from the supernova, SN1987A opened a new window of observational astronomy by neutrinos. And the history showed that the SN1987A neutrino burst observation was the vanguard of successive discoveries of neutrino properties by Super-Kamiokande, SNO, K2K, KamLAND and so on. On the occasion of the SN1987A 20th anniversary, the backstage story up to the discovery of the SN1987A neutrino bursts is summarized, tracing the Kamiokande log-note and including the IMB, LSD and Baksan data.

1. Before February 1987

1.1. Kamiokande-II

Kamiokande is an imaging water Cherenkov detector with 3000 tons of pure water in the main detector tank located at a depth of 2700 m.w.e. in the Kamioka mine. The unique feature of this experiment is the use of 1000 20-inch diameter (the world-largest photosensitive aperture) photomultiplier tubes (PMT's), which provide a large coverage of the photosensitive area over the inner detector surface. The original motivation which stimulated the construction of the Kamiokande detector was searches for nucleon decay. The experiment started on July 6, 1983. After running the detector for several months, the careful analysis on cosmic-ray muons which are stopped in the detector and are accompanied by $\mu \rightarrow$ e decay electrons, gave us hints to detect the ^{8}B solar neutrinos. From Fig.1 one can see that the lower end of the energy spectrum reaches down to $\sim$ 15 MeV. Below this energy, the number of background events rises up enormously. This figure implies the possibility to detect the ^{8}B solar neutrinos with a maximum energy of 14 MeV, with the aid of reasonable reduction of background events. This was the motivation for upgrading the Kamiokande detector (Kamiokande-II, Fig.2).

The Kamiokande-II construction began in September 1984. In April 1985 the renewed water purification system turned on, although the detector water was insufficient in quality of radio impurity. The trigger rate under a 8.5 MeV detection threshold energy decreased rapidly with the same decay time as that of ^{222}Rn as shown in Fig.3. However we often suffered from water leak troubles in the purification system. This obligated us to supply fresh water into the detector tank. Consequently a big jump-up of the trigger rate occurred due to much ^{222}Rn resolving in the fresh mine-water, shown also in Fig.3. Fortunately such troubles calmed down until the end of December 1986. Fig.4 is the trigger rate change in 1986 and 1987.

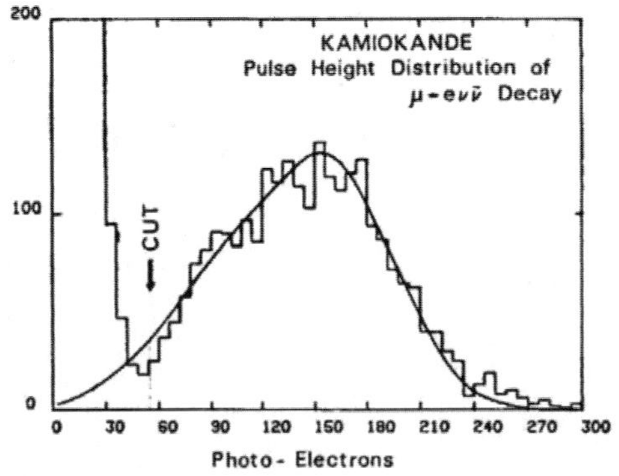

Figure 1: Energy spectrum of observed electrons from cosmic-ray $\mu \to e\nu\bar{\nu}$ decays. 110 photoelectrons correspond to $\sim$ 30 MeV.

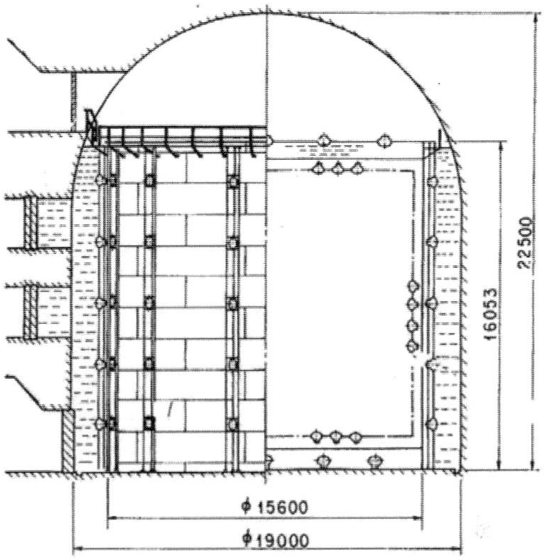

Figure 2: Kamiokande-II detector.

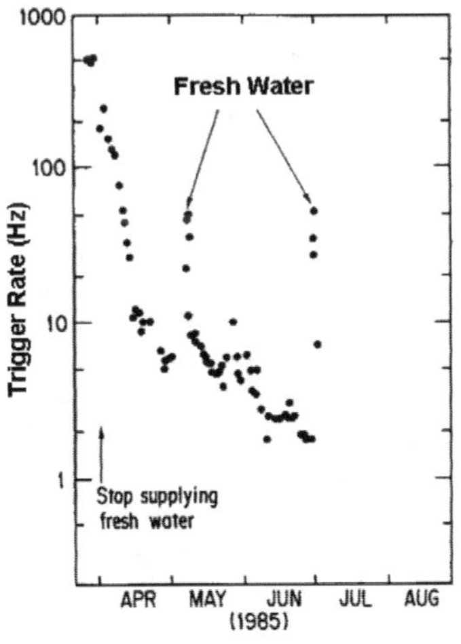

Figure 3: Trigger rate in 1985.

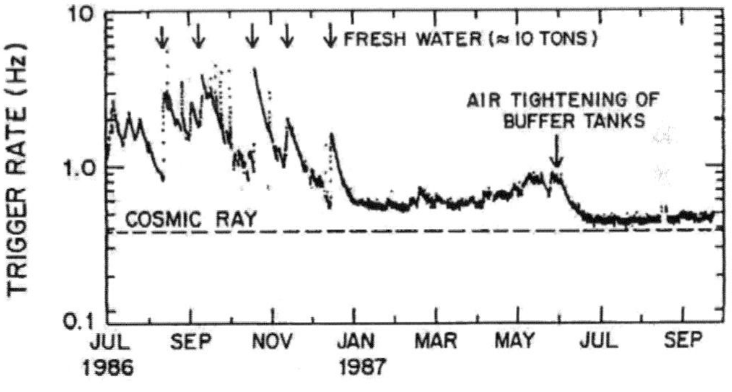

Figure 4: Trigger rate in 1986 and 1987.

1.2. IMB-3

The IMB detector, designed to search for proton decay, is located in the Morton-Thiokol salt mine near Fairport, Ohio at a depth of 1570 m.w.e. It consists of a rectangular tank ($22.5 \times 17 \times 18$ m^3) filled with 8000 ton purified water. The six sides are instrumented with 2048 5-in. PMT's arranged on an approximate 1-m grid. The

IMB detector started operation in July 1982. Since the early phases of the experiment (IMB-1 and IMB-2) did not have a sufficient performance to allow an efficient particle identification, the detector-upgrade by replacing 5-in. PMT's by 8-in. PMT's with a wave shifter plate (60cm × 60cm) shown in Fig.5 was proposed in September 1983. In September 1986, IMB-3 was ready, having 4 times the light collection of IMB-1. The inner view of the IMB-3 detector is shown in Fig.6.

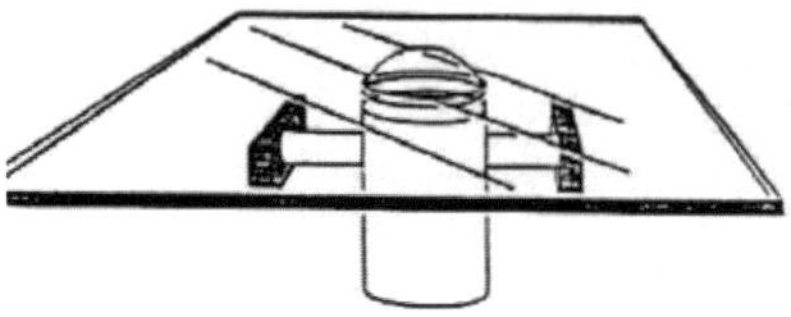

Figure 5: Photosensor unit : 8-in. PMT with a wave length shifter plate.

Figure 6: IMB-3 detector (from J. van der Velde, Hawaii 2007).

1.3. LSD and Baksan

The Mont Blanc Neutrino Observatory is located in the Mt. Blanc tunnel at a vertical depth of 1930 m.w.e., is a 90 ton Liquid Scintillation Detector (LSD) (Fig.7) consisting of 72 counters ($1.0 \times 1.5 \times 1.0$ m^3) in 3 layers, arranged in a parallelepiped shape with (6×7 m^2) area and 4.5 m height. The low-energy local radioactivity background from the surrounding rock is reduced by shielding each counter and the whole detector with more than 200 tons of Fe slabs. The liquid scintillator is watched from the top of each counter by 3 PMT's (15 cm diameter). LSD has been running since January 1985.

The Baksan scintillation telescope shown in Fig.8 is located at a depth of 850 m.w.e. at the North Caucasus. It consists of 3156 standard detectors arranged in eight planes forming a parallelepiped shape with two internal horizontal layers. Each of 3156 standard detectors is an aluminum box ($0.7 \times 0.7 \times 0.3$ m^3) containing oil-based liquid scintillator that is viewed by one 15 cm PMT. The total target mass is 330 tons. The data taking has been continuing since June 1980.

1.4. Summary of this section

The detection efficiencies of the above 4 detectors at the beginning of 1987 are

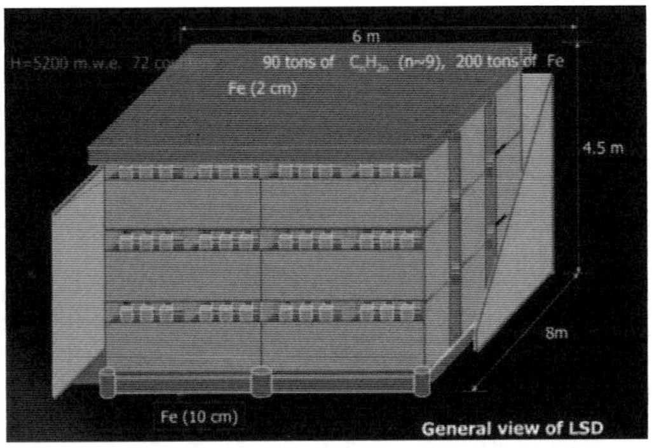

Figure 7: Liquid scintillator detector :LSD (from O. Saavedra, Hawaii 2007).

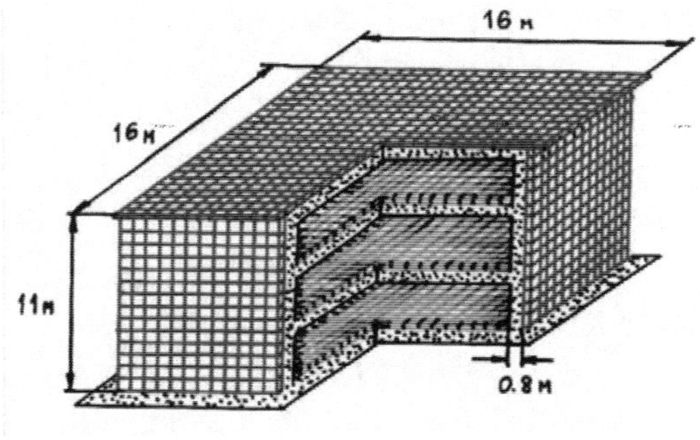

Figure 8: Baksan underground scintillation telescope (from E. N. Alexeyev, Hawaii 2007).

summarized in Fig.9 as a function of event energy. The energies of 50 % detection efficiency were 5.5 MeV for LSD, 7.5 MeV for Kamiokande-II, 10 MeV for Baksan and 36.5 MeV for IMB-3. Thus 4 detectors were ready for detecting supernova neutrinos.

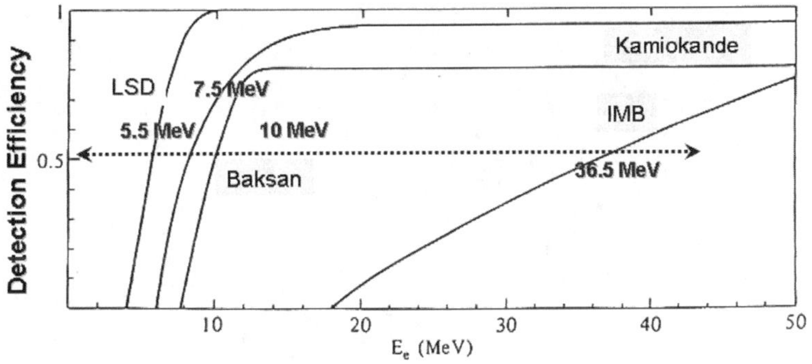

Figure 9: Kamiokande-II trigger efficiency as a function of electron energy, which is compared with the IMB, LSD and Baksan trigger efficiencies.

2. On 23-25 February 1987

On February 23, 1987 bursts of light and neutrinos reached the earth from the brightest supernova (named SN1987A, Fig.10) in 383 years since the Kepler's supernova. Light from the explosion, ~170,000 light-years away in the Large Magellanic Cloud (LMC), a satellite galaxy of our own, was visible only in the Southern Hemisphere. The light was discovered by I. Shelton at the Las Campanas Observatory in northern Chile[1)] on a photograph of LMC taken in February 24.23 UT, 1987. R. McNaught, however, later reported[2)] that SN1987A had already been there at magnitude 6.5 in a photograph taken on 23.443 UT. On February 25, the presence of hydrogen was observed in the spectra of SN1987A. This feature indicates that SN1987A is a type-II supernova caused by the gravitational collapse of a massive star. By comparing the supernova's position with older photographs of LMC, the progenitor star was identified to be a hot blue supergiant star, called Sanduleak -69 202. According to the supernova theory, this means the birth of a neutron star which is accompanied by the emission of ~99 % of the released gravitational energy in the form of neutrinos within a few seconds. Thus the neutrino burst signals the time of core collapse, which is earlier than the time of brightening by 1~ 10 hours, depending on the size of the progenitor.

Figure 10: Supernova SN1987A in LMC. The left shows the immediate neighborhood of the supernova before (1977) and the right after the explosion. (photographs from European Southern observatory's 1-m Schmidt Telescope at La Silla in Chile)

3. On 25 February 1987

The curtain of SN1987A neutrino drama was raised by the fax of February 25 sent by Sidny Bludman (Univ. of Pennsylvania) to E. Beier (Univ. of Pennsylvania), one of the Kamiokande-II US collaborators. This fax says *"SENSATIONAL NEWS ! SUPERNOVA WENT OFF 4-7 DAYS AGO IN LARGE MAGELLENIC CLOUD, 50 KPC AWAY. NOW VISIBLE MAGNITUDE 4~5, WILL REACH MAXIMUM MAGNITUDE IN A WEEK. CAN YOU SEE IT ? THIS IS WHAT WE HAVE BEEN WAITING 350 YEARS FOR ! "* (see Fig.11) E. Beier and I were working inside the mine to construct the air-tight structure on the detector roof.

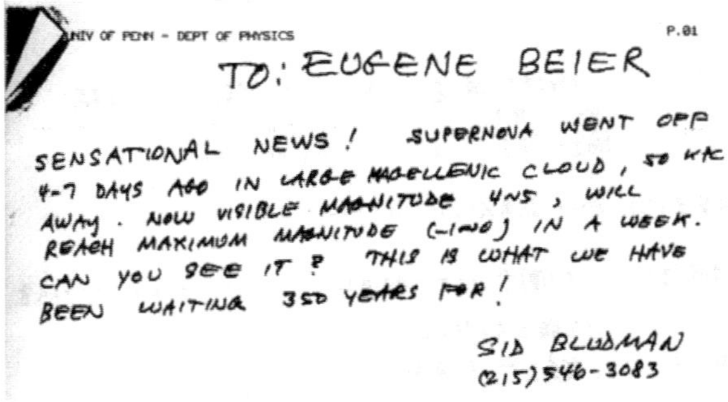
UNIV OF PENN - DEPT OF PHYSICS P.01

TO: EUGENE BEIER

SENSATIONAL NEWS! SUPERNOVA WENT OFF 4-7 DAYS AGO IN LARGE MAGELLENIC CLOUD, 50 KPC AWAY. NOW VISIBLE MAGNITUDE 4~5, WILL REACH MAXIMUM MAGNITUDE (-1~0) IN A WEEK. CAN YOU SEE IT? THIS IS WHAT WE HAVE BEEN WAITING 350 YEARS FOR!

SID BLUDMAN
(215)546-3083

Figure 11: Fax from Sid Bludman to E. Beier

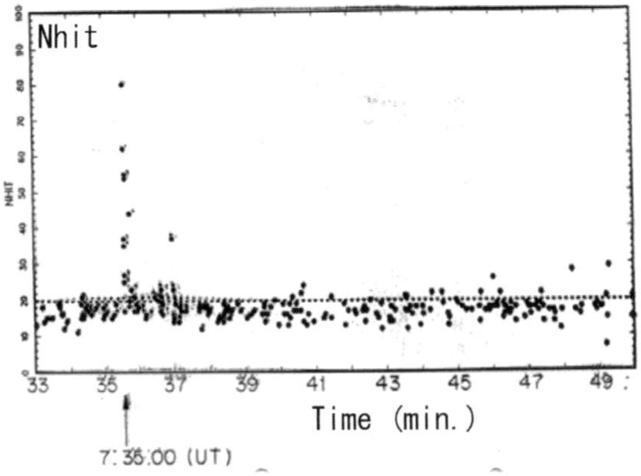

Figure 12: Scatter plot of N_{hit} against time relative to 7:35:35 UT, February 23, 1987.

Figure 13: Printer output around SN1987A.

4. On 26 February 1987

The data-tapes taken in February 20-25 were sent to Tokyo by a door-to-door delivery service. The analysis team in Tokyo rushed into developing utility- software-programs for finding neutrino burst events, using previous data samples. The team found that $N_{hit}-Time$ plot for space-reconstructed events was useful for this purpose, where N_{hit} is proportional to an event energy and $Time$ stands for event-production-time.

5. On 27-28 February 1987

The data-tapes arrived at Univ. of Tokyo. Analysis of the data started from the evening of February 27. Normal event reduction procedures and a subsequent

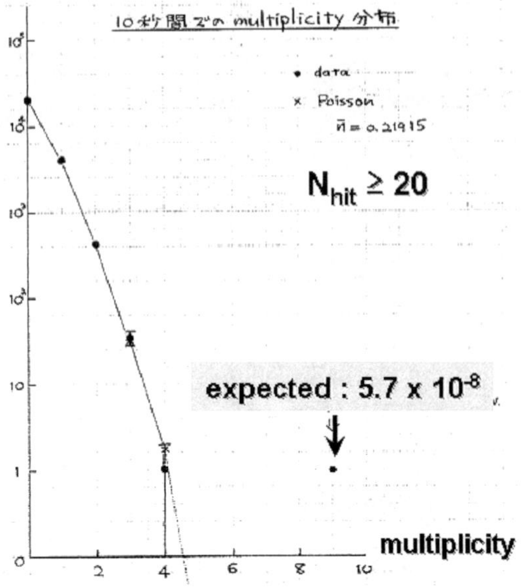

Figure 14: Poisson distribution of the number of events with $N_{hit} \geq 20$ per 10 sec. in a 2.7-day period.

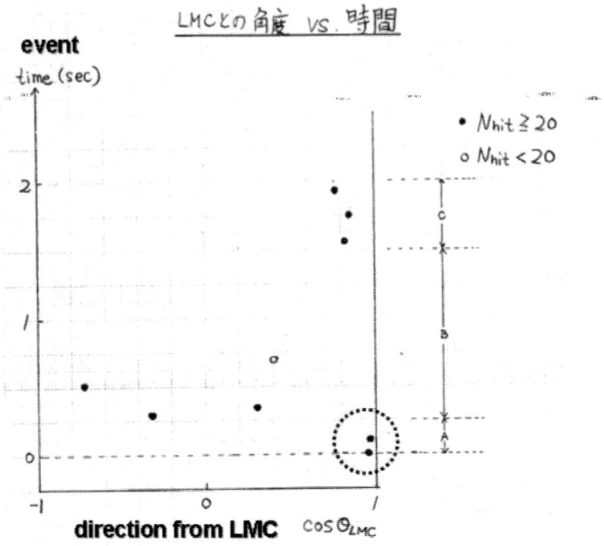

Figure 15: Scatter plot of the detected event time and the cosine of the angle between the measured electron direction of the Large Magellanic Cloud.

supernova-burst-search were carried out all day long. The team could easily found out the event-burst around 7:35 UT in the $N_{hit} - Time$ plots. The team said it was an easy work to pick up a cluster of 11 events which lasted for about 10 sec. from 500 pages of print-outs. Fig.12 and Fig.13 show the scatter plot of N_{hit} against event time. The neutrino burst at 7:35:35 UT is evident in this figure. The data over a period of 2.7 days from 21 February to 24 February was searched to determine the statistical significance of the burst at 7:35:35. This is indicated by the Poisson distribution shown in Fig.14 for events with $N_{hit} \geq 20$. It is seen that the probability of occurrence of a burst with 9 events per 10 sec., based on the observed distribution in Fig.14 is less than 5.7×10^{-8}. For events with $N_{hit} \geq 30$ the probability is less than 5.2×10^{-11}. The burst was so significant. Fig.15 is the scatter plot of the burst event time and the angular correlation between the burst event and LMC. It is seen that the first 2 events point back to LMC

6. On 1-2 March 1987

The whole story of the supernova data analysis was informed to Prof. Koshiba by M. Nakahata. Nakahata expected Prof. Koshiba's smile for the discovery of historical event. On the contrary Prof. Koshiba did not smile and ordered Nakahata to analyze more data back to January 1 in order to make the burst sure for more significant confirmation. We rushed again into data analysis. Meanwhile on 2 March,

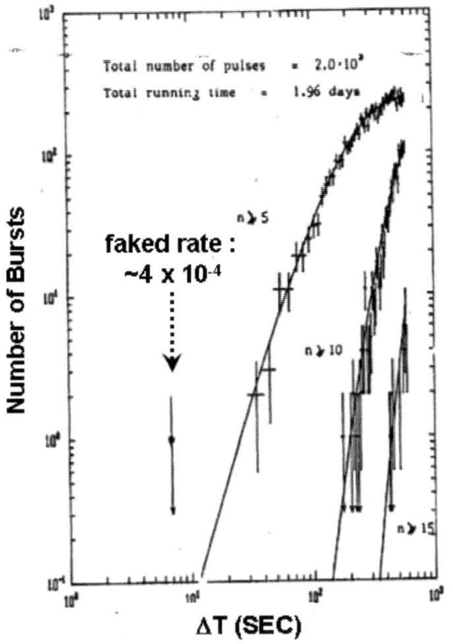

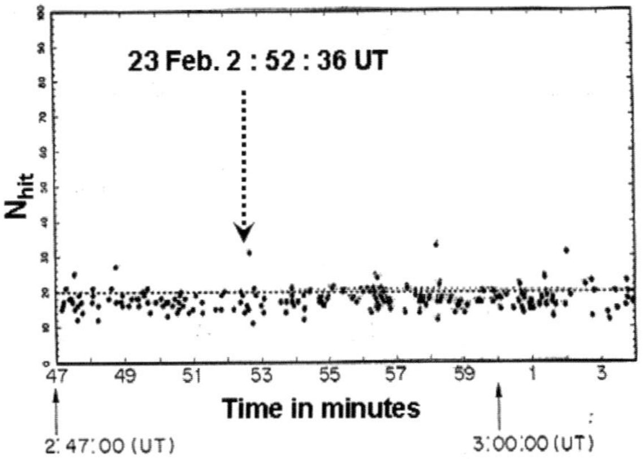

Figure 16: (Left) Distributions of bursts of pulses with multiplicity $n \geq 5, n \geq 10$ and $n \geq 15$, as a function of burst duration time Δt in seconds. The full lines are the corresponding expected distributions, assuming randomness of the background. (from the LSD fax of March 5, 1987)

Figure 17: (Top) The same as for Fig.14, but including the LSD event-time.

the news of SN neutrino detection arrived from LSD. LSD reported the burst of 5 pulses at 2:52:36 UT on 23 February 1987. This burst time is 4.5 hours before that of Kamiokande. The probability of occurrence of a burst with 5 events per 7 seconds is found to be $\sim 4 \times 10^{-4}$ (see in Fig.16).

We searched the event rate around the LSD event-time, but did not find any significant signals as shown in Fig.17. The expected event number in Kamiokande is 27 ± 12, using the Kamiokande detection efficiencies for the LSD event energies, the fiducial mass difference and the difference of the number of free protons (see also Fig.18). More detailed analysis was carried out with an additional trigger system in Kamiokande called a low-level trigger. This trigger system has a 50 % efficiency for 6.5 MeV energy deposit and is used to monitor a trigger rate precisely, particularly for a noise-event rate. Kamiokande observed 20 events at the LSD burst time during 7-second period. On the other, hand 92 events are expected, based on the LSD data. Kamiokande did not confirm the neutrino-burst at the LSD time.

7. On 2-7 March 1987

Until 6 March the data analysis was completed and the preprint was prepared. We were ready for submitting the paper. The abstract in this paper[3)] entitled "Observation of a Neutrino Burst from the Supernova SN1987A" says that *"A neutrino burst was observed in the Kamiokande-II detector on 23 February 1987, 7:35:35 UT*

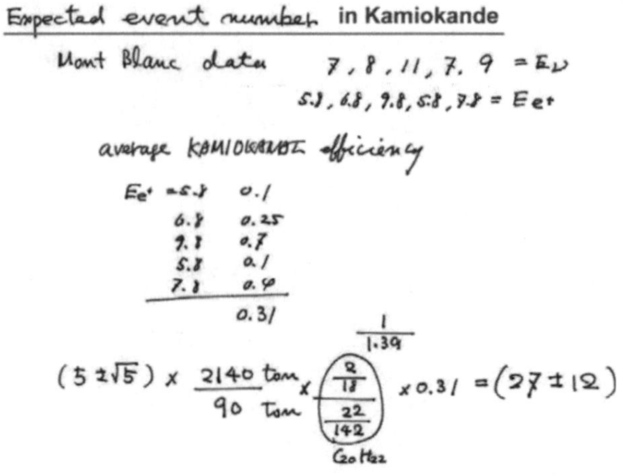

Figure 18: Quick calculation of the event rate at the LSD time, based on the Kamiokande data (from my note).

(±1min.) during a time interval of 13 seconds. The signal consisted of electron events of energy 7 to 36 MeV, of which the first two point back to the Large Magellanic Cloud with angles 18°±18° and 15°±27°." Typical burst event-patterns are shown in Fig.19. On 7 March 1987, the paper was posted for a submission to Phys. Rev. Lett. We thought all the analysis works were finished. We had a small party in a seminar room. Our satisfied time was blown off, once one of graduate-students came into this room. He told us that our definition of the coordinate system was wrong. Soon after this, we found he was correct. We were not familiar with the right ascension-declination coordinate. Although the analysis went back to the beginning, fortunately our mistake gave only a sign-change, and what's more we took this mistake two times. Thus (-1)×(-1) = (+1) led a minimum modification on the results. In this workshop, I heard the similar trouble had happened in IMB.

8. On 8 March 1987

The burst-analysis report of IMB was faxed to us on 8 March 1987. The report starts from "*As everyone by now, we have good evidence for having seen the supernova in the Large Magellanic Clouds from the salty depths of Cleveland ...*" Then "*First of all we began by knowing the time that Kamiokande was reporting, namely that they had a neutrino burst on 2/23 at 7:35:35 UT. Fig.20 shows the results of a scan of our data from Tape 2601 with a sliding time window of 10 seconds over a period of 20 minutes around the Kamiokande time. Note that the only data cut was to take events with* $N_{\rm hit} < 100$. *One sees a nice peak of 6 events in 2 seconds at just about the trial time. ......, so that the probability of coincidence is something less than* 4×10^{-3}*......*"

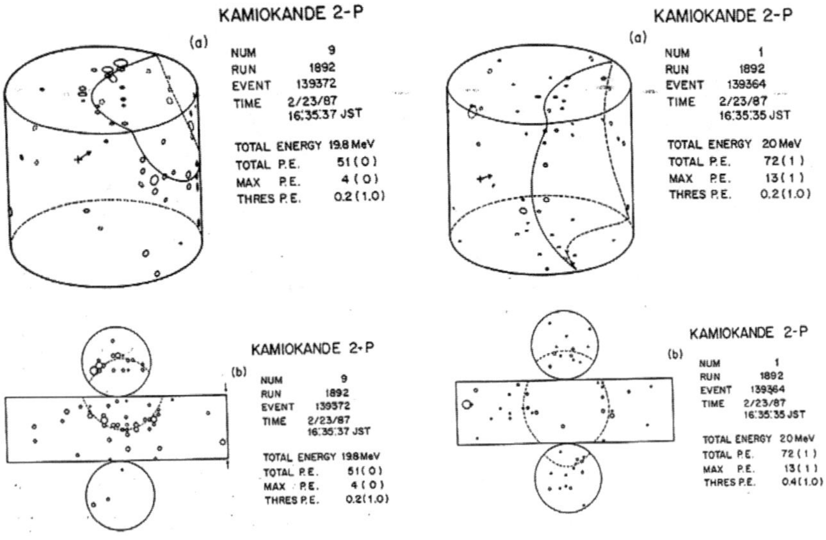

Figure 19: Tree-dimensional reproduction and exploded view of the Kamiokande burst events.

The final result of IMB was summarized in the paper entitled "Observation of a Neutrino Burst in Coincidence with Supernova SN1987A in the Large Magellanic Cloud"[4]. The observed signals of IMB consist of 8 neutrino events with energies in the range of 20-40 MeV, spread over an interval of 6 seconds. The burst time is 23 February, 7:35:41.37 (±50 msec.) UT. The patterns of PMT-hits for the burst events are shown in Fig.21.

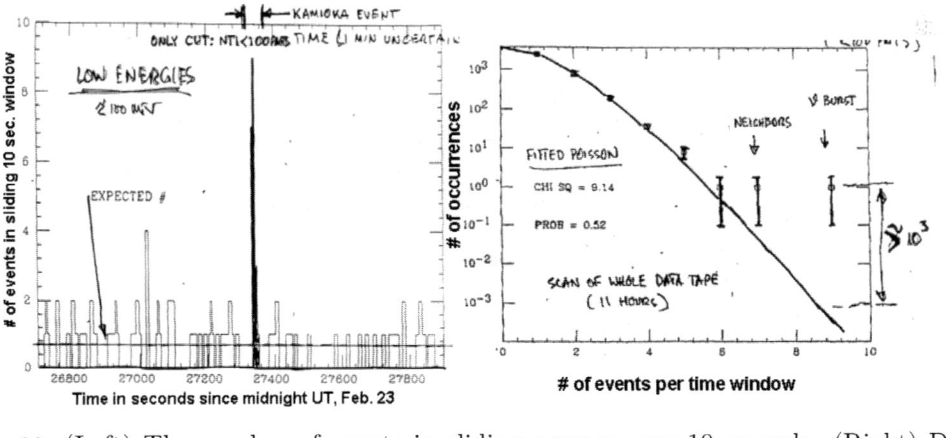

Figure 20: (Left) The number of events in sliding average over 10 seconds. (Right) Distribution of the number of occurrence for 10-second window. (from the IMB report of March 8 1987)

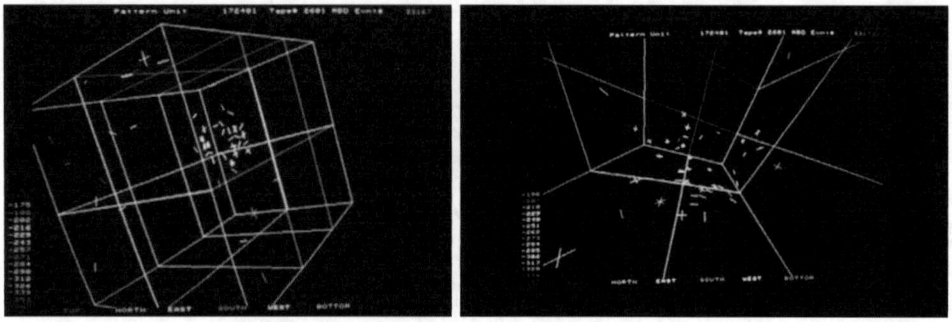

Figure 21: PMT hits patterns of the IMB burst events. (from the IMB web-site)

9. On 9 March and On 6 April 1987

Kamiokande and IMB submitted their papers to Phys. Rev. Lett. on 9 March 1987. Kamiokande had a press release at the Ministry of Education (see Fig.22). Both papers of Kamiokande and IMB were published in the April 6 1987 issue of Phys. Rev. Lett.

Figure 22: The Kamiokande results were opened to the public on 9 March 1987.

10. Baksan Data and Data Summary

Afterwards we received the report on "Detection of the Neutrino Signal from SN1987A Using the INR Baksan Underground Scintillation Telescope". A signal of 5 events within 9.1 seconds was found in the fiducial mass of 200 tons at February 23, 7:36:11.8 (+2-54)sec. The random background probability was 1.5×10^{-5}. The time sequence of events detected by the Kamiokande-II, Baksan[6], IMB-3 and LSD[5]

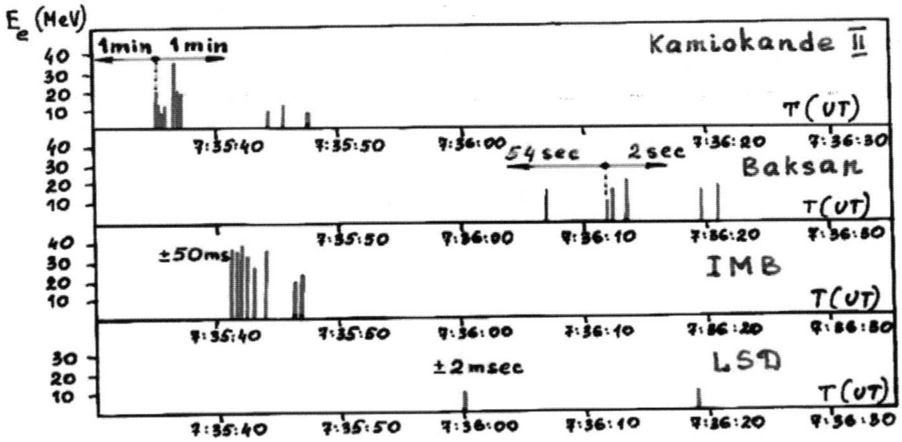

Figure 23: The time-sequence structures of burst events observed in Kamiokande-II, Baksan, IMB-3 and LSD. (from E.N. Alexeyev, Hawaii 2007)

at 7:35 UT is listed is shown in Figs.23 and 24[7] where the earliest events of three groups (Kamiokiande-II, Baksan and IMB-3) are assigned at t=0.

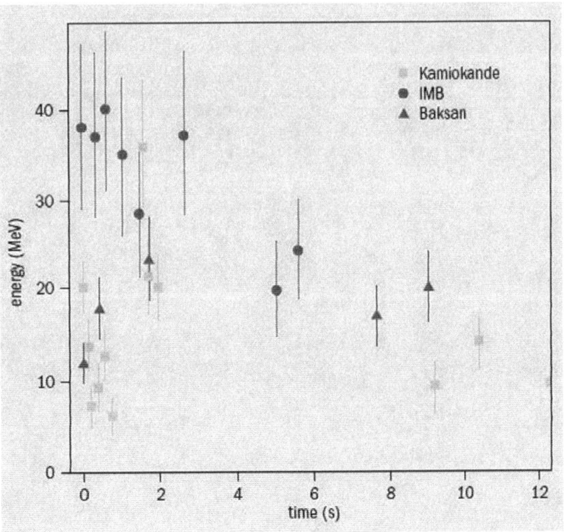

Figure 24: SN1987A neutrino events observed by Kamiokande, IMB and Baksan showed that the neutrino burst lasted about 13 seconds. (see ref.7)

11. Lucky !?

Kamiokande observed a neutrino burst from the supernova SN1987A, escaping through several accidents and happenings. We learned much about the difficulty to continue data-taking under a good detector condition which is requisite for observatories. Fig.25 shows the most serious accident in Kamiokande. The supernova burst arrived ~2 minutes after a data-taking blind period due to an electronics gain calibration. Gain calibration of the Kamiokande electronics, generating test pulses into all of the PMT channels, was repeated every 2 hours. It took about 3 minutes each. Based on this experience, the procedure of gain checks was modified by carrying out channel by channel instead of all channels simultaneously. Thus there has been no dead time for the gain check with a victim of one rest channel.

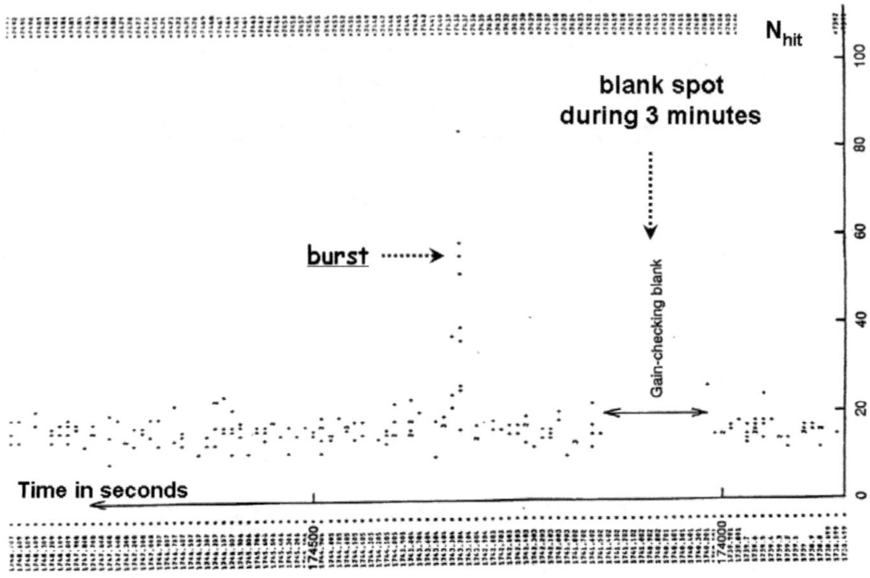

Figure 25: Computer print-out of the scatter plot; Nhit vs. the event detection time. (see ref. 8)

IMB also got out of one of attempts to thwart the supernova detection. On 23 February, 5:00:00.001 UT, ~2 hours before the neutrino burst, high-voltage power supplies tripped off, shutting down one-forth of the detector's 2048 PMT's. This also caused the on-line data-filtering routines to stop functioning. Fortunately, the raw data could be recovered from the partially malfunctioning detector and suffered no serious biases due to the missing PMT's.

With the above mentioned stories, people might say that Kamiokande and IMB would be so lucky. However according to the Japanese proverb, catching good luck is a kind of ability.

12. Conclusion; On 11 June 1988

On 11 June 1988, Prof. L. Okun had the summary talk of Neutrino'88 in Boston. He made comments on " *To predict the year of explosion of a supernova is not harder than to predict the year of funding a big accelerator or a big detector. I expect that the date of the next supernova is 2003 ± 15 years.*" We have to be more serious to detect next supernovae. We say "Another SN soon - nearer, but not too near".

13. References

1) I. Shelton, *IAU Circular* **4316** (1987).
2) R. McNaught, *IAU Circular* **4316** (1987).
3) K. Hirata et al., *Phys. Rev. Lett.* **58** (1987) 1490.
4) R. M. Bionta et al., *Phys. Rev. Lett.* **58** (1987) 1494.
5) M. Aglietta et al., *Europhys. Lett.* **3(12)** (1987) 1315.
6) E. N. Alexeyev et al., *Proc. of the Leptonic Session of the 22nd Recontre De Moriond* (1987) 739.
7) M. Nakahata, *CERN COURIER* **47** (2007) 23.
8) M. Koshiba, *Phys. Rep.* **220** (1992) 229.

Underground laboratories

A. Bettini

Dipartimento di Fisca G. Galilei. Via Marzolo 8. 35131 Padova Italy
(Padua University and INFN Section and Laboratorio Subterráneo de Canfranc Spain)

Bettini@pd.infn.it

Abstract. Underground laboratories provide the low radioactive background environment necessary to explore the highest energy scales that cannot be reached with accelerators, by searching for extremely rare phenomena. To advance the effective-high-energy frontier we need to struggle for background control and reduction. After having discussed the most important characteristics of an underground laboratory, all of the world facilities that are presently operational or in an advanced status of approval are briefly described. None of the projects for very large facilities are described.

1. Introduction. Characteristics of the underground laboratories

The only known phenomena beyond the Standard Model are in neutrino physics: neutrinos are massive and flavour lepton numbers are not conserved. This was discovered in underground laboratories with natural sources (Sun and cosmic rays). In general, underground experiments provide an indirect reach to the highest energy scales. For example, a 50 meV neutrino mass corresponds, via seesaw mechanism, to a 10^{16} GeV energy scale. The higher is the energy the more rare are the corresponding phenomena. Consequently, pushing forward the high-energy frontier requires every backgrounds to be reduced. There are important physical and practical differences between the existing facilities. These range from fully developed laboratories to simple underground sites.

The muon flux decreases with the thickness of the rock overburden, roughly, but not exactly, exponentially. Under a flat surface the flux is 10^{-3} m^{-2}s^{-1} at a depth of 2.3 km w.e. (1 km water equivalent is about 300 m of rock), 10^{-4} m^{-2}s^{-1} at 3.7 km w.e. and 10^{-5} m^{-2}s^{-1} at 5.3 km w.e. Under a mountain, like at Kamioka or Gran Sasso, the angular dependence of the muon flux is complicated, due to the shape of the surface. It must be measured to provide the input to the background simulations needed by the most delicate experiments. The flux is time dependent with seasonal variations of several percent.

Neutrons originate mainly from (α,n) and fission processes (U and Th) in the rocks. The energies range from thermal to several MeV, and consequently are not difficult to shield. The neutron flux is substantially independent of the depth (if >100 m or so). It depends on the local geology. However, in practice the n flux is around a few 10^{-2} m^{-2}s^{-1} in all laboratories.

Muons interactions in the rocks produce dangerous neutrons at a depth-dependent rate. The fluxes are typically 3-4 orders of magnitude smaller than the main n flux. The energies are large, up to several GeV, demanding thicker shields. Even more dangerous are the neutrons produced in the shields, in the detector and in the materials around it. In the case of fast reactions, namely if a neutron immediately follows the muon, the background can be reduced by anticoincidence. Neutrons from metastable nuclides can be eliminated by working at a larger depth. Their effect depends on the details of the experiment, and is generally more severe for high-Z materials.

Radon (^{222}Rn) is a radioactive, volatile gas that is always present in the atmosphere, being continuously produced by the decay of ^{226}Ra present in the rocks. An important source of Rn is ground water. The Rn activity, which is typically 10 – 20 Bq/m^3 in the open air, is larger by two orders of magnitude or more in closed underground cavities. It is reduced by ventilation. The equilibrium activity depends on the emanation rate and on the ventilation speed. The input air duct, which may be a few km long, should be made of SS or similar materials to avoid Rn collection input.

All experiments need to be shielded. Indeed, the shield thickness determines the physical size of the most sensitive experiments. For example, the frontier double beta decay experiments presently under construction, CUORE and GERDA, have diameters of about 15 m. The next generation may need halls 20-30 m in diameter and height. Please note that the maximum safe diameter decreases while the costs increase with depth.

Laboratories differ in many other important aspects: horizontal or vertical access, interference with nearby activities (mine work, road traffic, etc.), quality of the support infrastructures (laboratories, office space, assembly halls, etc.) and personnel on the surface, degree of internationality of the user community and programme of the advisory committee, policy of space and time allocation.

Scientific sectors different from astro-particle physics, such as biology, geology and engineering can profit of the very special underground environment provided by the laboratories and their facilities. Space does not allow these important issues to be discussed here.

The facilities are now reviewed in geographical order, from W to E. For details see "The world underground science facilities. A compendium", available at arXiv:0712.1051. The, more relevant, historical order of the starting dates of the oldest projects is the following. 1966 BNO, 1979 LNGS and LSM, 1983 KAMIOKA, 1984-85 Slovotvina and LSC (the elder).

2. **BUL. Boulby Palmer Laboratory (UK).** http://astro.ic.ac.uk/Research/ZEPLIN-III/boulby.htm
The site was developed starting in 1988 by N. Spooner and collaborators initially from RAL. An active potash mine at 1000 m depth under a flat surface hosts it. The access is through a shaft. The salt environment limits the cavities width to about 5 m. A clean area of approximately 1500 m^2 is available to experiments. The neutron flux with E>0.5 MeV is 1.7×10^{-2} s^{-1}m^{-2} and the muon flux is 4.5×10^{-4} m^{-2} s^{-1}. A building on the surface (200 m^2) hosts laboratories for computing, electronics and chemistry, offices, a conference room, changing rooms, mess rooms, a mechanical workshop, storage and construction rooms. The scientific programme is focused on dark matter search with ZEPLIN II, to be completed by 2008, ZEPLIN III, to be completed by 2010 and DRIFT II, in the R&D phase. There are low radioactivity measurements and geophysics research. About 30 scientists work at the laboratory.

3. **LSC. Laboratorio Subterráneo de Canfranc (Spain).** http://ezpc00.unizar.es/lsc/index2.html
The first underground facility located under the Pyrenees, close to a dismissed railway tunnel, was created in the 1980s by A. Morales and the Nuclear and High-Energy Physics Department of the Saragossa University. Taking advantage of the excavation of a parallel road tunnel, the new laboratory was built later. The underground structures have been completed in 2005. However, more recently several design and construction defects have emerged and necessary reparation work is under way. LSC is managed by a Consortium between the Spanish Ministry for Education and Science, the Government of Aragon and the University of Saragossa. The surface building has been funded and is presently being designed. It will contain headquarters, administration, a library, a meeting room, offices, laboratories, storage areas and a mechanical workshop, safety structures and management, for a total of approximately 1500 m^2. A dozen of employees are being hired.

The access is horizontal, via one of the tunnels. The entrance must be communicated to the freeway tunnel control. The spaces available underground are: Hall A, measuring 40×15×12(h) m^3, Hall B of 15×10×8(h) m^3, Clean room of 45 m^2 and Services for 215 m^2. The old lab area is 100 m^2. The maximum rock coverage is 850 m. The muon flux is between 2×10^{-3} and 4×10^{-3} m^{-2} s^{-1}, depending on the location; the *n* flux is 2×10^{-2} m^{-2} s^{-1}. The Rn activity in the air is 50-80 Bq/m^3 with a ventilation of 11 000 m^3/h, i.e. one lab volume in 40'.

The scientific programme is presently being defined now. A provisional Scientific Policy Committee has collected expressions of Interest. After the appointment of the first director, a new International Scientific Committee has been appointed. A call for proposals will be launched soon.

4. LSM. Laboratoire Subterrain de Modane (France). http://www-lsm.in2p3.fr/

The Laboratory is operated jointly by CNRS/IN2P3 and CEA/DSM. Excavation of the Laboratory started in 1979, and was completed by 1982, to host a 900 t iron tracking calorimeter to search for proton decay. The "Frejus" experiment was finished in 1988.

The access is horizontal through the Frejus roadway tunnel. Intervention of the tunnel control is needed to stop traffic at the entrance or exit of a vehicle from the lab. The Main Hall is 30×10×11(h) m^3, the Gamma Hall has an area of 70 m^2, two smaller halls have 18 m^2 and 21 m^2 areas, for a total of 400 m^2. The surface building includes offices (100 m^2), a warehouse and workshop (150 m^2) and a flat. The personnel include 8 technicians and engineers and one post doc.

The rock overburden is 1700 m. The muon flux is 4.7×10^{-5} $m^{-2}s^{-1}$. The *n* flux is 5.6×10^{-2} $m^{-2}s^{-1}$. A low Radon activity in the air, 15 Bq/m^3, is obtained by in taking fresh air at the rate of 1.5 lab volumes/hour. An "antiradon factory" produces 150 m^3/hr of air with 10 mBq/m^3. The laboratory is almost full with NEMO 3 (double beta decay), EDELWEISS (dark matter), which should run at least up to 2010, and a low-radioactivity counting facility. About 100 scientists work at the lab.

The Ulisse project profits by a unique opportunity given by the construction of a new tunnel approved by the French and Italian Governments to increase the safety conditions of the traffic. Two large halls are foreseen: A of 100×24 m^2 and B of 18×50 m^2. An extremely low-background environment will be obtained in Hall B by surrounding its central volume with a water shield and by artificially producing a very low Rn content atmosphere (0.1 mBq/m^3).

5. LNGS. Laboratorio Nazionale del Gran Sasso (Italy). http://www.lngs.infn.it/

LNGS is a national laboratory of the INFN. It is the largest in the world, serving the largest and most international scientific community. In 1979 the President of the INFN A. Zichichi proposed to the Parliament to build a large underground laboratory close to the Gran Sasso freeway tunnel then under construction (an opportunity that substantially reduced the cost). In 1982 the Parliament approved the construction, which was completed by 1987.

Access is horizontal, through the freeway. The underground laboratories consist of three main halls (called A, B and C), about 100×20×18(h) m^3 plus ancillary tunnels, providing space for services and small-scale experiments. Two 90 m long tunnels were built for two Michelson interferometers to be used for geology studies. The total area is 17 300 m^2, the total volume 180 000 m^3.

Services hosted on the surface campus include offices, a mechanical workshop, storage facilities, a chemical lab, an electronic workshop, an assembly hall, computers and networking, a library, a canteen, sleeping rooms, conference rooms, headquarters, administration. Special care has been given to the development of structures, procedures and training activities in matter of safety, of the users and of citizens, as well as environmental impact. A number of outreach activities and visits to the lab are systematically organised by a dedicated Service. Personnel (physicists, engineers, technicians, administration) include a permanent staff of 64 and 23 non-permanent positions.

The rock overburden is 1400 m. The muon flux is 3×10^{-4} m^{-2} s^{-1}. *n* flux is 3.78×10^{-2} $s^{-1}m^{-2}$. Radon in the air is 50-120 Bq/m^3 with a ventilation system providing one lab volume of fresh air in 3.5 hr. Major civil engineering work has been performed in 2004-7 to upgrade the safety conditions of the interacting structures of the free-way, the water-collection systems and the laboratory.

LNGS is operated as an international laboratory. In the process of approval of the proposals, an international Scientific Committee, appointed by INFN, advises the Director. Underground space and other resources are allocated to experiments for a definite amount of time, in order to guarantee turnover. The rich experimental programme includes dark matter searches, with LIBRA, CRESST2, XENON10, WARP; Double Beta Decay with COBRA, CUORICINO, GERDA; Solar neutrinos (geo-neutrinos) with BOREXINO; Supernova neutrinos with LVD; Nuclear astrophysics with LUNA2; CNGS with OPERA and ICARUS; a test of the Equivalence principle, VIP. A special facility is dedicated to low-radioactivity measurements. The laboratory also supports several experiments on geology, biology and environmental issues. Almost all of the experiments are second-generation ones and have been approved for several years of data taking. The scientific user community is 752 strong, involving scientists from 26 countries.

6. CUPP. Centre for Underground Physics in Pyhäsalmi (Finland). http://cupp.oulu.fi/
The Centre is hosted in a working mine. Several cavities, dismissed by the mine, are available at different depths down to 980 m, for a total area of more than 1000 m^2. Presently, the mine works at between 1000 m and 1400 m depth. Access is both via a shaft and an inclined tunnel. The EMMA experiment is being installed. Small lab and office space is available in a surface building. A guest-house is also available. The personnel consist of about 3 people on site and 3 at Oulu University.

7. SUL. Solotvina Underground Laboratory (Ukraine). http://lpd.kinr.kiev.ua/LPD_SUL.htm
The Laboratory was constructed in 1984 under the leadership of Yuri Georgievich Zdesenko by the Lepton Physics Department of the Institute for Nuclear Research (Ukrainian National Academy of Sciences) in a salt mine. The laboratory space is divided into a Main Hall: 25×18×8(h) m^3 and four chambers 6×6×3(h) m^3. The total is area about 1000 m^2. On the surface, three living rooms are available. Staff members consist of 14 technicians and engineers. Access is vertical by the mine cage, taking into account the time-table of the mine.

The lab is 430 m deep in salt (≈1 km w.e.). The muon flux is 1.7×10^{-2} m^{-2} s^{-1}. The n flux is 2.7×10^{-2} m^{-2} s^{-1}. The radon concentration in air is 33 Bq m^{-3}. Eleven researchers and PhD students of the LPD work at SUL mainly on double beta decay, preparing a new ^{116}Cd experiment using 1-2 kg $^{116}CdWO_4$ higher quality crystal scintillators and developing R&D projects on scintillators and for SuperNEMO.

8. BNO. Baksan Neutrino Observatory (Russia). http://www.inr.ac.ru/INR/
The Laboratory is operated by the INR of the Russian Academy of Sciences. It is managed as an observatory, with very long-duration experiments. It is the oldest facility in the world built specifically for scientific research. M. Markov, the Head of Nuclear Physics Division of the Academy of Sciences of the USSR, obtained in 1966 a special Decree of the Soviet Government, and construction of the Baksan Neutrino Observatory started under Mount Andyrchi in the Caucasus. A new village, called 'Neutrino', was built as a part of the original project in a previously empty space with personnel providing the necessary services (heating station, water-supply system, first medical help, transportation, safety, etc.). The staff directly related to science is 50-60. The scientific activity started under the leadership of Alexander Chudakov and George Zatsepin. The access is horizontal via two dedicated tunnels, with train transportation

A large hall, 24×24×16 m^3 in volume, 300 m deep, hosts the Baksan Underground Scintillation Telescope. BUST has been ready to observe neutrinos from galactic supernovae since 1978. The Laboratory of the Gallium Germanium Neutrino Telescope hosting SAGE, has a volume of 60×10×12 m^3 at a vertical depth of 2 100 m. In this lab the muon flux is $3.03\pm0.19\times10^{-5}$ $m^{-2}s^{-1}$. The neutron flux (E>1 MeV) is 1.4×10^{-3} $m^{-2}s^{-1}$. The Rn activity is 40 Bq/m^3 with a fresh air input of 60 000 m^3/h.

The construction of a larger and deeper hall, about 40 000 m^3 in volume, was started in 1990, and stopped in 1992, when the Soviet Union collapsed. Further fate of this not finished construction is presently under discussion. Low-background Chambers with a volume from 100 to 300 m^3 are used for the R&D of dark matter and double beta decay search as well as for gravitational wave search and for some geophysics measurements. The number of users is 30-35.

9. Y2L. YangYang Laboratory (Korea). http://dmrc.snu.ac.kr/

The lab is operated by the Dark Matter Research Centre (DMRC) of Seoul National University. The presently available area is 100 m^2. Expansion to 800 m^2 is planned, but is not yet funded. 100 m^2 space for office space, computing and detector test facility is available on surface. Access is horizontal by car. The lab utilizes the space in the tunnel of the host YangYang Pumped Storage Power Plant. Researchers are requested to obey the safety and security regulation of the Plant.

The rock overburden is 700 m with a μ flux of 2.7×10^{-3} $m^{-2}s^{-1}$. The neutron flux is 8×10^{-3} $m^{-2}s^{-1}$ for 1.5 MeV < E_n < 6.0 MeV. The radon activity is 40-80 Bq/m^3. The underground space is mostly occupied by the Korea Invisible Mass Search (KIMS) experiment, currently taking data for a WIMP search with 100 kg CsI(Tl) crystal detectors. Other activities include R&D for double beta decay and background measurements with a HPGe counter. The scientific users are about 30.

10. **Oto Cosmo Observatory (Japan).** http://wwwkm.phys.sci.osaka-u.ac.jp/info/syoukai/oto-e.html

The laboratory has been developed by H. Ejri of the Osaka University and collaborators. Its area of consists of Lab. 2 (50 m^2) hosting ELEGANT V and MOON-1 on dark matter search with NaI and 0ν2β decay of ^{100}Mo. Lab. 1 (33 m^2) hosts ELEGANT VI on 0ν2β decay of ^{48}Ca and dark matter search with CaF_2. The rock coverage is 470 m, with a muon flux of 4×10^{-3} $m^{-2}s^{-1}$. The neutron flux is 4×10^{-2} $m^{-2}s^{-1}$. The radon inside "radon free" containers is 10 Bq m^{-3}. The access is horizontal through a non-used railway tunnel, which also provides the non-forced ventilation.

11. Kamioka Observatory (Japan). http://www-sk.icrr.u-tokyo.ac.jp/index_e.html

The Kamioka Observatory is operated by the Institute for Cosmic Ray Research, University of Tokyo. It was established in 1983 by M. Koshiba as Kamioka Underground Observatory. The original purpose of this observatory was to conduct the KamiokaNDE experiment, to which Super-Kamionkande followed. The present facilities have been designed for Super-K, the largest existing underground experiment. Recently, an enlargement has started, in order to accommodate more experiments. The KamLAND experiment is operated by the Neutrino Centre, Tohoku University. Buildings for offices and computer facilities are available on the surface. The staff members are 13 scientists, 2 technical support units, one for administration.

The coverage is 1000 m and the muon flux is 3×10^{-3} $m^{-2}s^{-1}$. The thermal neutron flux is $8.25\pm0.58 \times 10^{-2}$ $m^{-2}s^{-1}$, the non-thermal is $11.5\pm1.2 \times 10^{-2}$ $m^{-2}s^{-1}$. The ventilation is 3000 m^3/h. The access is horizontal by car, with no interference with the mining activity. The underground structures are as follows: Hall SK (50 m diameter) hosting Super-Kamiokande, to be continued for 15 years, at least. Clean room (10×5 m^2) with XMASS prototype. Hall 40 (L-shape, 40 m×4 m arm) hosting the purification tower for XMASS and the NEWAGE experiment on dark matter. Hall 100 (L-shape, 100 m×4 m arm) with CLIO, a prototype of Gravitational Antenna (to be terminated in 2013) and a Laser displacement detector. The new Hall A (15×21 m^2) hosting XMASS 800 kg (untill 2012) with space available for another experiment. The new Hall B (6×11 m^2) hosting CANDLE on double beta decay, to be occupied untill 2012. Small areas are available in the dismissed mine. The scientific users are more than 200 in number. A budget request for the underground large cryogenic gravitational antenna LCGT has been submitted.

12. INO. India based Neutrino Observatory (India). http://www.imsc.res.in/~ino/

One of the two experiments that first observed atmospheric neutrinos in 1964 was located at 2700 m depth in the Kolar Gold Mine in India. The India based Neutrino Observatory is the project to create an underground laboratory in southern India. It will be located near the PUSHEP hydroelectric pumping station, under a 1300 m rock overburden.

The lab will be organized with international laboratory standards with a Scientific Advisory Committee, services to the users, environmental, safety, security and outreach activities. Two main underground cavities are foreseen: Lab1 with a volume of 26× 135× 25(h) m^3 and Lab 2 with 53.4× 12.5× 8.6(h) m^3 plus connection tunnels and services. Access will be horizontal through a dedicated 2 km tunnel. On the surface it is planned to have: a 1400 m^2 building for administration, offices, shops, etc., a 2750 m^2 building with a lecture hall and a guest house and a residential complex with 20 quarters. Personnel will be 50 to 100.

The main foreseen experiment is ICAL, a 50 kt magnetized Fe tracking calorimeter for atmospheric and very long base-line accelerator neutrinos. It will occupy only a fraction of Lab1.

13. SNO-Lab (Canada). http://www.snolab.ca/ also http://www.sno.phy.queensu.ca/

The SNO experiment has completed its glorious life and its cavity, 200 m^2 area, is now being freed for further experimental activity. New structures are under construction: a Main hall of volume 18× 15× (15 to 19.5 height) m^3, a service hall of about 180 m^2 and a number of narrow (6-7 m) volumes, called "ladder labs". The construction of another structure, called cryopit, has been approved recently. This hall is designed to cope with the safety issues surrounding large volumes of cryogenic fluids. The total area will be 7 215 m^2, of which 3 055 m^2 is available for the experiments; the total volume will be 46 648 m^3, of which 29 555 m^3 will be available for the experiments. The access will be vertical, through the shaft of the working mine, available daily. All the laboratory will be clean, class 1500.

On the surface a 3159 m^2 building will host a clean room, laboratories, staging and assembly areas, office space (60 users), meeting rooms, control rooms, an IT server room, an emergency generator, high-speed network link off site, high-speed network link surface/underground, safety structures and management. Staff will be of 30 full-time people.

The rock coverage is 2000 m under a flat surface. The μ flux is 3×10^{-6} $m^{-2}s^{-1}$, the thermal neutron flux is 4.7×10^{-2} $m^{-2}s^{-1}$, the fast neutron flux is 4.6×10^{-2} $m^{-2}s^{-1}$. The radon in the air is high, 120 Bq/m^3. The ventilation in the smaller lab spaces provides 10 air changes per hour, in the larger ones, 5 air changers per hour.

PICASSO, searching for dark matter (2 kg) with the super-heated bubbles technique, is already running. SNO+ will be hosted in the former SNO cavity; it will be based on liquid scintillator for low energy solar neutrinos, geoneutrinos and double beta decay, by dissolving ^{150}Nd in the liquid. Dark matter search includes DEAP/CLEAN with noble liquids, which is getting ready to install prototype, and a LoI from superCDMS with bolometers. More LoIs are expected to be reviewed by the Experimental Advisory Committee.

14. SUL. Soudan Underground Laboratory (USA). http://www.soudan.umn.edu/

The underground structures include: the Soudan lab (20×7×10(h) m^3) that hosts: a) CDMSII, expected to run until 2009; b) a low-background counting facility that currently occupies 5x5x3 m^3 and will expand to 25×14×14(h) m^3, if funded. The MINOS lab that hosts: a) MINOS that occupies 35×16×14(h) m^3 and is expected to run until 2009 or longer with a 2-year decommissioning period at the conclusion; b) the high-purity copper fabrication facility that occupies 4x6x3(h) m^3 and expects to run for at least another two years. The users are 265 in number.

The access is vertical via a two-compartment slightly angled shaft. Diameters in excess of 1m and lengths in excess of 10m pose a problem. Access outside normal operating hours is possible. There is an access charge paid to our host institution, Soudan Underground Mine State Park. Normal laboratory safety requirements are in place. The laboratory coexists with an historic State Park, which offers mine tours during the summer months to the public, and winter tours to school groups. Some tours utilize a visitor's gallery available in the MINOS laboratory. There is no active mining activity.

The overburden is 700 m of rock. The muon flux is 2×10^{-3} m^{-2} s^{-1}. The neutron interaction rates are approximately 10 kg^{-1} d^{-1} (from U/TH, low energy) or 0.01 kg^{-1} d^{-1} (muon generated in the rock). The radon concentration is seasonal, varying from 300 Bq/m^3 in the winter to 700 Bq/m^3 in the summer. The mine has natural ventilation, about 550 m^3/h for the level of the laboratories. Half of this is diverted to ventilate the MINOS and Soudan spaces. This results in a complete air change every 110'.

The major facility on the surface is a building of approximately 650 m^2 with offices, a kitchen and sanitary facilities.

The laboratory has a staff of 9, including secretarial and accounting assistance and network and computer maintenance personnel. It is staffed 10 hours/day, 5 days per week, but the staff is on-call during the balance of the time and responds to requests for emergency access.

15. DUSEL. Deep Underground Science and Engineering Laboratory (USA)

Solar neutrino physics started in the Homestake mine in South Dakota with the experimental work of R. Davis and the theoretical sun model developed by J. Bahcall. After a long and complex process, in spring 2007, NSF selected amongst several proposals the Homestake mine in South Dakota as the site in which the Deep Underground Science (physics, biology, and geology) and Engineering Laboratory (DUSEL) should be designed. The project foresees a funding by NSF of about 250 M$ for the facility plus a contribution to the initial set of experiments, costing about 250 M$. In addition, to prepare the site, SD provides 46 M$ on its own and 70 M$ from a donation by T. D. Sanford. NSF is expected to fund with 15 M$ the design of the facility, which will take three years. For the time being, the facility is a SD State Laboratory, funded primarily from SD-controlled money. Water constantly flows into the mine at a rate of 1.2 Mt/yr. Pumps had stopped in spring 2003. Rehabilitation work started in January 2007, securing the level at 1450 m depth with pumps at 1600 m.

On the surface several existing buildings will be rehabilitated for about 10 000 m^2 to host offices, support structures and laboratories. A major science education centre has been funded by Sanford. The initial staff is estimated to be 30-50 strong, to increase to 100-150 when the State Lab will become DUSEL. The Scientific Committee has been established with an USA composition in 2006.

Laboratory spaces will be built separately for biology, geology and physics. Service and R&D structures (e.g. electroforming) will be available for physics at a 100 m deep level. Two main campuses are foreseen for physics at about 1450 m and 2200 m deep. Each will contain a number (4 and 3 respectively) of standard modules of $50\times20\times15$ m^3 plus service areas. A staged construction of the upper campus using SD and private funding is foreseen. For the lower campus, NSF and Congress approval will be necessary.

A site-independent study group on the science of the DUSEL has been appointed by NSF with B. Sadoulet as chair. The work has been completed at the time of writing and presented at NSF on November 2, 2007.

16. Outlook and acknowledgements

Physics beyond the Standard Model was discovery of Underground Laboratories (solar and atmospheric neutrinos). The physics programme for the next decades appears to be extremely rich and challenging. The interest of the scientific community is increasing. We have several facilities, but only a few are managed as international laboratories, with a really international Scientific Committee, full-scale services, turnover of the experiments. New laboratories and upgrades of existing facilities are being created along these lines. They will provide very good working conditions in the next decades. Existing ‘single-Institution’ facilities may still contribute with R&D activity, which however should be well focussed on frontier experiments, training of the students, and low-background measurements, which are badly needed in material screening for the experiments. However, new sub-critical facilities should not be created.

Co-ordination actions are occurring inside the European Union. In particular EU, has funded with 7.5 M€ over 5 years the ILAS project on gravitational waves, dark matter and double beta decay. The actions include: a) “networking”, in practice exchange of experience and common developments in matters of safety and public awareness; b) Joint R&D projects (low-background techniques, double beta decay, noise in gravitational wave detectors); c) international access to underground laboratories, funding per diems to foreign scientists in their work at the laboratory. These actions are useful and are having positive impacts. However, while directing LNGS I learnt that only the merging of contributions of different (scientific) cultures on a global, not only Western European, scale can create a fully developed scientific environment. Indeed, this is the case of the accelerator laboratories.

It is a pleasure to thank the organisers for the very interesting conference and for giving me an opportunity to contribute. The preparation of the present report would have been impossible without the help of the Directors or their delegates. I am indebted in particular to N. Spooner, S. Paling, J. Morales, F. Piquemal, E. Coccia, R. Antolini, J. Peltoniemi, F. Danevich, V. Gavrin, N. Mondal, S-K. Kim, H. Ejiri, Y. Suzuki, K. Lesko, E. Peterson, T. Nobele and G. Gerbier.

AUTHOR INDEX

AUTHOR INDEX

AUTHOR INDEX

AUTHOR INDEX

AUTHOR INDEX

AUTHOR INDEX

AUTHOR INDEX

AUTHOR INDEX

AUTHOR INDEX

AUTHOR INDEX